WITHDRAWN

R 895.16 .P37 1986

Parker, Herbert M.

Herbert M. Parker

BRADNER LIBRARY
SCHOOLCRAFT COLLEGE
LIVONIA, MICHIGAN 48152

HERBERT M. PARKER

Publications and Other Contributions to Radiological and Health Physics

Edited and annotated by:

Ronald L. Kathren
Raymond W. Baalman
William J. Bair

Battelle, Pacific Northwest Laboratories
Richland, Washington

BATTELLE PRESS
Columbus • Richland

COVER

Portrait of Herbert M. Parker by Fabian Bachrach overlies a sketch of an electrometer from his earliest laboratory notebook.

PART DIVIDERS

Reproductions of pages from a notebook kept by Herbert Parker during his years at Manchester University are used as illustrations on the Part dividers; the cover of this notebook is pictured in Part VI, Pictorial Miscellany. The pages selected have no particular reference to the subject matter of the respective Parts.

DESIGN

Fishergate Publishing Company, Inc., of Annapolis, Maryland, designed the book and prepared it for printing.

Printed in the United States of America.

Library of Congress Cataloging-in-Publication Data

Parker, Herbert M.
   Herbert M. Parker: publications and other contributions to radiological and health physics.

   Includes index.
   1. Radiology, Medical—Collected works. 2. Radiation—Safety measures—Collected works. 3. Parker, Herbert M. I. Kathren, Ronald L., 1937–
II. Baalman, Ray W. III. Bair, William J., 1924–
IV. Title.
R895.16.P37   1986   616.07'57   86-22351
ISBN 0-935470-36-0

Copyright ©1986 Battelle Memorial Institute

All rights reserved. No part of this book may be reproduced or transmitted in any form or by any means, electronic or mechanical, including photocopying, recording or by any information storage and retrieval system, without permission from the publisher.

FOREWORD

During Herbert M. Parker's last years, several of us suggested to him on more than one occasion that he compile his published and unpublished papers and reports in a bound volume for distribution not only to his friends and colleagues but also to those interested in the historical development of radiological physics and radiation protection. I believe Herb was flattered by these suggestions. Although he surely knew the significance of his scientific contributions, I sensed he was not convinced that there would be sufficient interest to warrant the effort such a production would require.

For whatever reason, Herb never compiled his papers for publication in a bound volume. In his absence, we decided to do this for him, recognizing that he would probably have been embarrassed by this publication but, yet, also pleased and proud.

Three people, in addition to Herb, are largely responsible for this book. First, we are indebted to his wife, Margaret, for granting Battelle, Pacific Northwest Laboratories permission to reprint Herb's papers. Second, Ronald L. Kathren, long an ardent admirer of Herb, devoted many hours to searching out the papers, writing the introductions and biosketch, and reviewing Herb's illustrious career. Finally, Raymond W. Baalman provided the editorial expertise required to produce this volume.

William J. Bair

Richland, Washington
July 1986

CONTENTS

Foreword iii

Preface ix

A Brief Scientific Biography xiii

PART I: The Manchester Years 1

    Introduction 3
I-1   A Dosage System for Gamma Ray Therapy 5
I-2   The Dependence of Back-Scattering 47
I-3   The Distribution of Radiation in Deep X-Ray Therapy . 59
I-4   A System of Dosage for Cylindrical Distributions
      of Radium 73
I-5   Dosage Measurements by Simple Computations 97
I-6   Radium Implant Reconstructor 105
I-7   Radium Dosage: The Manchester System
      (Covers and Prefaces to the first and second editions) 111
I-8   The Physical Basis of the Planar Mould 117

PART II: Medical Physics After Manchester 135

    Introduction 137
II-1 Radium Element Versus Radon 139
II-2 A Physical Evaluation of Supervoltage Therapy 145
II-3 Limitations of Physics in Radium Therapy 153
II-4 The Manchester System: 161
      Comments Prepared for the Afterloading Conference .. 162
      Notes Prepared for a Lecture 170
II-5 Supervoltage Roentgentherapy (title page) 175

**PART III: Contributions to the Evolution of
      Radiological Units** 179

Introduction 181
III-1 Notes on Radiation Dosimetry 183
III-2 Tentative Dose Units for Mixed Radiations 193
III-3 Principles Underlying the Measurement of Radiation ... 199
III-4 Radiological Units 205
III-5 Radiation Units (HW-SA-2536) 213
III-6 Units, Radiation: Historical Development 243

PART IV: Radiation Protection: The Manhattan District and Beyond 251

Introduction 253
IV-1 Possible Hazards of Repeated Fluoroscopies in Infants .. 255
IV-2 The Tolerance Dose 267
IV-3 Tolerance Dose (Teaching notes—1945) 291
IV-4 Tolerance Dose Memo dated Feb. 22, 1946 301
IV-5 Status of Health and Protection at the Hanford Engineer Works 305
IV-6 Action Taken on Spot Contamination 315
IV-7 Radiation Exposure Data 323
IV-8 Health Physics, Instrumentation, and Radiation Protection 327
IV-9 Absorption of Plutonium Fed Chronically to Rats: Title Page 391
IV-10 Radiation Protection in the Atomic Energy Industry ... 393
IV-11 Effects of Reactor Accidents on Plant and Community .. 403
IV-12 Health Problems Associated with the Radiochemical Processing Industry 407
IV-13 Selected Materials on Radiation Protection Criteria and Standards: Their Basis and Use 413
IV-14 Radiation Standards, Including Fallout: Oral and Prepared Statements 427
IV-15 Radiation Protection Standards: Theory and Application 449
IV-16 Radiation Exposure Experience in a Major Atomic Energy Facility 487
IV-17 Radiation Exposure from Environmental Hazards 493
IV-18 Environmental Effects of a Major Reactor Disaster 501
IV-19 Personnel Dosimetry for Radiation Accidents 507
IV-20 Radiation Exposure of Uranium Miners: Critiques and Correspondence Related to Dr. Evans' "CORD" Document 525
IV-21 The Dilemma of Lung Dosimetry 541
IV-22 Festina Lente 551
IV-23 An Intrauterine Dosimeter 567
IV-24 Plutonium, Industrial Hygiene, Health Physics, and Related Aspects 571
IV-25 Unwanted Radiation Exposures—The Meaning for Man 627
IV-26 Principles of Standard Setting: The Radiation Experience 631

IV-27 Plutonium — Health Implications for Man 637
IV-28 Statement on Proposed Rule-Making on Environmental
       Radiation Standards for the Uranium Fuel Cycle ... 645
IV-29 Protection Programs of the Plutonium Project 663
IV-30 The Squares of the Natural Numbers in
       Radiation Protection 671
IV-31 X-Ray Measurement and Protection 1913-64
       (Book review) 701

PART V: Radioactive Waste Management 705

      Introduction 707
V-1   Speculations on Long-Range Waste Disposal Hazards ... 709
V-2   Hazards of Waste Storage at Hanford Works 721
V-3   North American Experience in the Release of
      Low-Level Waste to Rivers and Lakes 779
V-4   Radioactive Waste Management in Selected
      Foreign Countries 789
V-5   Nuclear Waste Disposal 795

PART VI: Pictorial Miscellany 813

PREFACE

The modern-day practice of medical and health physics was shaped to a large extent by Herbert M. Parker, whose career extended from the early 1930s to the mid 1980s. His earliest scientific contributions helped to systematize and mature clinical radiation therapy; these alone would have been sufficient to ensure him a place in the history of radiological sciences. Beyond those early contributions, he was instrumental in the establishment of the profession that has come to be called health physics. He emphasized and reemphasized the importance of scientifically based radiation protection measures for the environment as well as for workers. As a scientist, he was precise and demanding of himself, striving for uncompromising quality. As a manager, he demanded the same uncompromising quality from his subordinates and encouraged sound productive and pragmatic research.

In compiling this work, it has been our goal to create more than just another memorial collection of limited utility destined to gather dust on a hidden shelf in a seldom visited library. We have tried, in a real sense, to put together a textbook, a history, and a potentially important reference source for practitioners of health and medical physics. For the scientist—particularly for the younger scientist newly entering the field—it can serve to instruct, providing valuable lessons in the fundamentals of radiological physics. For the historian, it can provide a clear picture of the origins and evolution of the science and art of health physics and of the maturation of clinical radiation therapy. The seminal papers brought together here from diverse and sometimes obscure sources should make this volume valuable as a reference to those in the practice of both medical and health physics. We hope, therefore, that it will serve not only as a suitable memorial to Herb Parker but also as a useful and valuable addition to the literature of the radiological sciences.

To Herb's long-time friend and colleague William J. Bair at Battelle, Pacific Northwest Laboratories goes much of the credit for making this work a reality. Shortly after Herb's death in the spring of 1984, Bill began to develop plans for a suitable memorial, soliciting ideas from Herb's friends and associates. Largely through Bill's efforts and the generosity of Battelle Memorial Institute and the U.S. Department of Energy, the Herbert M. Parker Lecture was established as an annual presentation in conjunction with the Hanford Life Sciences Symposium, which Herb had long supported. In addition to sponsoring these annual lectures, Battelle agreed to assist with publication of this memorial volume, to be distributed in conjunction with the First Annual Parker Lecture.

Gathering the contributions of Herb Parker was both an honor and privilege for the editors, but it was not a simple and straightforward task; Herb left behind not even a complete bibliography, let alone a compendium of his collected works. However, through the kindness of his wife, Margaret, his personal files were made available to the editors, and these yielded prime copies of his earlier and least easily obtained scientific papers.

Although we would have liked to have included all of Herb's contributions in these pages, to do so simply was not possible because of space limitations and our desire to have the work available for distribution at the 1986 Hanford Life Sciences Symposium. Thus, the editors offer their apologies for excluding, reluctantly, a few of Herb's many publications but note that all his major published works and several heretofore unpublished historically significant memoranda and notes have been included. We hope the selection will prove satisfactory.

The papers have been divided into five groups, based primarily on subject matter. Although Herb left behind a legacy of seminal ideas and contributions to medical and health physics, compared to many scientists, particularly those who are forced to write by the "publish or perish" rubric, he was not a prolific publisher of scientific papers and reports in the open literature. However, those papers prepared for formal publication were uniformly of high quality and are indicative of his masterful command of both science and the English language.

Many of his ideas, however, were put forth anonymously or semi-anonymously in publications of committees on which he had served as a member and to which he contributed a lion's share of the writing. These include such important documents as Report No. 39 of the National Council on Radiation Protection and Measurements, *Basic Radiation Protection Criteria*, originally published in 1971. Others were put forth orally, perhaps in informal meetings, and thus exist only in the form of unpublished notes or memoranda, if they exist at all. Some take the form of testimony before Congress or of oral remarks transcribed for publication in conference proceedings.

Where and what Parker chose to publish is a rather revealing insight into his personality. His early publications appeared largely in the medical literature, which was the logical vehicle for his contributions to medical physics. Ironically, perhaps, although an acknowledged founder of the profession of health physics, he did not publish in the journal *Health Physics* until 1969, nearly fifteen years after its inception, and then only sporadically. Doubtless this was due in part to his conviction that health physics was properly a part of medical physics. Thus, his later publications, which largely dealt with radiation protection, appeared in a variety of places, but relatively few were in peer-reviewed scientific journals. This should not be construed as indicating that these writings were not worthy of publication in peer-reviewed journals; it is probably more indicative of his impatience with the often ponderously slow scientific publication process and his desire to disseminate his ideas to the most appropriate audience in a timely manner, even if this meant more limited distribution of the work. Clearly, he was not writing with an eye toward carving out a niche in the history of science; he did not keep even a simple up-to-date list of his publications! He was, after all, basically a shy man, but his shyness was overwhelmed by his extraordinary contributions to the radiological sciences.

Ronald L. Kathren
Raymond W. Baalman

Richland, Washington
July 1986

Photo by Fabian Bachrach

A BRIEF SCIENTIFIC BIOGRAPHY

Herbert M. Parker was one of those rare individuals who was blessed with not only a keen mind, but the ability to apply his intellectual gifts to the solution of complex and oft-times vexing problems that would lead toward improved and safe applications of ionizing radiation to the betterment of mankind. A clear and logical thinker, Parker made contributions to radiological science that spanned two continents and a half century, and in the process helped to develop and mature two related but distinct applied-physics specialties: medical or radiological physics and health physics. These parallel fields of applied physics owe their origins to the discovery of x rays and radioactivity, events that occurred in the closing years of the nineteenth century. Of the two, medical physics—the science and art of the application of ionizing radiations for medical purposes—is clearly the older, tracing its beginnings to the earliest days of x-ray usage. The early medical physicists were also the first health physicists, performing radiation protection functions as a sideline to their medical application of ionizing radiation. Health physics—the science and art of protecting people and the environment from the hazards of ionizing radiations—grew into a profession in its own right during World War II, evolving out of the need for occupational radiation protection in the wartime projects associated with the production of nuclear weapons.

Herbert Myers Parker was born in Accrington, England, on April 13, 1910. At an early age, he evidenced an intellectual ability that brought him recognition in the form of several academic scholarships and awards. In 1927, at the age of seventeen, he was named Manchester Ship Canal Scholar and, while a student at the University of Manchester, received no less than four scholarships—The Dalton Scholarship, Bridge Mechanical Scholarship, State Scholarship, and the Lancashire County Scholarship. As a student, he was also the recipient of several awards that presaged his enormously productive scientific career. In 1928, he was awarded the Higginbotham Prize at the University of Manchester, and the following year he received the H. G. Mosely Prize.

In 1930, he was granted the Bachelor of Science in Physics from the University of Manchester and received the M.Sc. the following year. His M.Sc. thesis research, an x-ray analysis of iron pyrites using Fourier methods, was of high quality and was published in the November 1932 issue of *Philosophical Magazine*, Series 7, Volume XIV, coauthored with W. J. Whitehouse. Although he wanted to continue his studies and pursue an academic career, his fellowship from the Department of Scientific and Industrial Research ran out in 1932, and, with the depression of the 1930s in full force, he was obliged to give up formal study. Thus, at the invitation of Ralston Paterson, a physician specializing in radiotherapy, he joined the staff of the Holt Radium Institute in 1932 to work on the development of a unified system for radium dosimetry.

Early Contributions to Medical Physics

James Ralston Kennedy Paterson, some thirteen years Parker's senior, was already a well-established radiotherapist and researcher at Christie Hospital and Holt Radium Institute, serving as Director of the latter. In collaboration with Byrl R. Kirklin, a noted American radiologic educator and practitioner, he had published a paper on the roentgenologic manifestations of lung carcinoma[1] and, in conjunction with his physician colleague Margaret C. Tod and secretarial assistant Marion Russell, was in the process of compiling statistical data on patients that would, in 1939, be published as the first of an important series of statistical reports on cancer therapy results at Manchester.

In 1932, Paterson, the experienced radiotherapist, and the youthful physicist Parker began collaborative research in the scientific aspects and application of radiotherapy for malignancies. Their attention was directed at radium, then the only practical available radionuclide source of penetrating photon radiation, and one which had gained wide radiotherapeutic use, particularly for gynecological applications. There was, however, no uniformity or standard technique for the application of radium to the treatment of malignancy; each radiotherapist had his own method based on experience, and often confusion rather than order was the rule.

Paterson and Parker attacked the problems of radium therapy from both the medical and physics viewpoints. At Paterson's insistence, Parker was fully involved in the medical aspects, even to the extent of inserting radium needles into patients. In this way, the youthful medical physicist became cognizant of the clinical realities faced by the therapist and gained an understanding and appreciation of the medical aspects firsthand. Together, they developed a series of tables showing the relative dosage delivered at specified distances from applicators of specific geometry, filtration, and source strength. They published their results in a series of three papers, which are reproduced in this work.

The Paterson-Parker system, as it soon came to be called, was made possible by Parker's magnificent mathematical derivations and lengthy calculations. To fully appreciate the difficulties he faced, it must be recalled that this work was done long before the age of computers or other electronic calculational aids; mathematical calculations were performed by hand. Parker spent a full year performing various numerical integrations that today could be done in a matter of minutes with the aid of a desktop microcomputer. These laborious theoretical calculations were confirmed with experimental measurements made with specially designed cavity ionization chambers. One of his laboratory notebooks from that era has survived and shows his careful attention to detail and precision in his measurements. Pages from this notebook serve to illustrate the cover and section dividers of this book.

The Paterson-Parker system revolutionized radium therapy, making possible the uniform irradiation of malignant tumors while minimizing the dose to healthy surrounding tissue. An important advance offered by the system was the development and application of what came to be a fundamental and universally accepted radiological dose constant for a point source of radium in equilibrium with daughters: 8.4 roentgens per milligram-hour at a distance of 1 cm. Thus, what had once been an art whose success was dependent on the skill of the artist was transformed into a scientifically based therapy regimen, with tables expressing tissue dose in roentgens (then the only unit of dose) for a given exposure in terms of milligram-hours for linear sources, planar applicators, and volume distributions.

The Paterson-Parker techniques were recognized and applied worldwide; indeed, they are in use today, more than a half century after publication of the initial and now classic paper.[2] Inclusion of the Paterson-Parker system in contemporary radiological physics textbooks, such as the widely used *The Physics of Radiology*,[3] is a testimonial to its fundamental importance. The noted American radiologist Ira Kaplan[4] referred to them as "pioneering." Otto Glasser, the doyen of American medical physicists, and his coauthors[5] included the Paterson-Parker system as one of the "Milestones of Radiology," in so doing, ranking it with the discovery of artificial radioactivity, and Parker with the Joliot-Curies and Enrico Fermi. Parker was indeed in good company.

The original Paterson-Parker techniques and tables were expanded into what has become known worldwide as the Manchester System of radium dosage. The original publication in book form appeared in 1947 and was reprinted in 1949 and 1958. A second edition was published in 1967.[6] A major portion of these books is devoted to the physical aspects and includes several chapters authored by Parker which provide the physical bases of dosimetry for line, cylindrical, and mould sources, and of interstitial treatment. These clearly written chapters not only are a testimonial to Parker's writing skill but also to his mathematical abilities.

While at Manchester, Parker also made other contributions to radiological physics, including studies of x-ray backscatter, deep therapy, and dosimetry. In collaboration with W. J. Meredith, who also became a radiological physicist of note, he developed an apparatus for the exact three-dimensional reconstruction of the arrangement of needles used for a radium implant, thereby lending even greater accuracy to the Paterson-Parker method. This work well illustrates Parker's quest for perfection and also gives a measure of the man as a practical engineer as well as a physicist. He presented an important lecture on simplified methods of dose calculation and measurement of tissue doses at the Fifth International Congress of Radiology held in Chicago in 1937, which was published in full.[7] This presentation marked Parker's first visit to the United States, which was in a few years to become his adopted homeland, and of whose citizenship he was quite proud.

Supervoltage Therapy and the Years at Swedish

In 1938, after six highly productive years at Manchester, Parker accepted an invitation to pursue his interests in deep therapy under the tutelage of American radiologist Simeon T. Cantril at the Swedish Hospital Tumor Institute in Seattle. At thirty, Cantril was only two years older than Parker, but he had recently assumed responsibility for the Tumor Institute and sought a physicist to replace John E. Rose, who had left for other duties. Although research in supervoltage therapy had been ongoing since the 1920s, and the Turner Institute itself had had an operational 800-k&p General Electric x-ray generator since 1932, there was still skepticism among radiotherapists regarding the merits of what was then considered to be a high-voltage form of therapy.

At Swedish Hospital, Parker was placed in charge of radiological physics at the Tumor Institute, where he embarked on an energetic program of research in collaboration with Cantril and Franz Buschke, a German-born radiologist. One of his first tasks was the physical characterization of supervoltage therapy, which was the subject of a supplement to the *Staff Journal of Swedish Hospital Tumor Institute* published in May 1941. He also published, in conjunction with Cantril and Buschke, an article on the applications of high-voltage therapy in the more broadly available periodical *Radiology*.[8]

The work at Swedish culminated in the book *Supervoltage Roentgentherapy*, jointly authored with Buschke and Cantril and published in 1950, some eight years after Parker had departed from the hospital to join the war effort as a part of the Manhattan District. Although delayed by the war, the book was the first of its kind and has been identified as one of the "Memorable Books" in the comprehensive history of radiology by the late radiologist-historian E. R. N. Grigg.[9] Oddly, other than a brief quote by Cantril made in 1955 and a single reference, no mention is made of this pioneering book and the work on which it was based in the "official" history of radiology prepared for the American College of Radiology by Edward and Ruth Brecher.[10]

Largely clinical in its orientation, the book includes an introductory chapter on physical considerations, presumably written by Parker, which included a section on protection of both the patient and others. The authors noted the absence of radiation sickness in patients treated by supervoltage techniques and concluded that this form of treatment had an important, albeit limited, place in the treatment of cancer. They also optimistically noted the possible curative benefit of even higher energy radiations, a prediction borne out by the widespread use of the 1.25-MeV photons from ^{60}Co for therapy a decade later.

As always, Parker's interests were broad and at Swedish Hospital were not confined to supervoltage therapy. While there, he became concerned about the hazards of excessive x-ray exposure to children, and, in an article coauthored with his colleague Franz Buschke on this subject,[11] warned of the dangers of repeated fluoroscopies in infants. This article presaged his later great contributions to the science and art of radiation protection. Another important contribution was his discussion of the limitations of physics in radium therapy, published in *Radiology* in October 1943. This critique of his own Paterson-Parker method is revealing of Parker's scientific character, showing him both as the perfectionist and the pragmatist.

The War Years

Although he had been at Swedish Hospital for only four years, in 1942 Parker responded to an urgent request from Simeon T. Cantril, who had left the Tumor Institute a few months earlier to take a position as Health Director for Clinton Laboratories near Oak Ridge, Tennessee, then operated by the University of Chicago Metallurgical Laboratory as part of the super secret Manhattan District project that would lead to the development of the atomic bomb. Although his work at Swedish Hospital was far from complete, Parker, then still a British subject, joined the Metallurgical Laboratory as Chief of the Physical Measurements Section of the Health Division under Cantril. This was a turning point in his career, for in this role he left behind him the clinical aspects of physics to become heavily involved in the development of a radiological safety program of unprecedented proportion and scope — one that would have to protect large numbers of people against the as yet largely unknown hazards from hitherto unknown radioelements.

Parker was thus to become the principal architect of the health physics program at Clinton Laboratories, providing for development of suitable instruments and standards for the measurement and control of radiation. He served as head of the fledgling Clinton Laboratories Health Physics organization in 1943, and, along with Ernest O. Wollan, a cosmic-ray physicist, and Carl C. Gamertsfelder, he was one of the original three to bear the title of "health physicist," a designation which he personally deplored.*

---

* The etymology of the term "health physics" can be traced to the need for secrecy during the development of the atomic bomb. A name was needed for the type of work done by the radiation protection group that would not provide a clue to the work of the Manhattan District project. As most of the original members of Parker's group were physicists working on problems related to health, the name "health physics" was coined. Even today, the name is rather obscure. Parker's objections were in part based on his feeling that not only was the name inappropriate but that health physics, as a profession, should be closely tied to or even a part of the classical medical or radiological physics community rather than a separate discipline.

His group later included several who would make their mark in the new profession of health physics, in no small measure due to his influence — Karl Z. Morgan, long-time head of the Health Physics Division at Oak Ridge National Laboratory and sometimes erroneously referred to as the father of health physics; the late James C. Hart, who died in office as President of the Health Physics Society and who made significant behind-the-scenes contributions to the administrative aspects of health physics and to the formation of the International Radiation Protection Association; Carl C. Gamertsfelder, who, as a young physicist under Parker, was present at the start-up of the first reactor under the stands at the University of Chicago, making ionization chamber measurements of radiation levels; and Jack W. Healy, who followed Parker to Hanford, and who became known for his significant contributions to internal dosimetry and biokinetic modeling and is fittingly honored as the first Herbert M. Parker Lecturer. Ironically, one of the original eight members of his group was physicist John E. Rose, whom he had replaced at Swedish Hospital, and who ultimately concluded his career as a health physicist at Argonne National Laboratory.

Having organized the program at Oak Ridge, Parker was not to stay long; it was a case of having done his job too well. The reactors for the production of plutonium and the associated chemical separation plants and research facilities were under construction at a large site centered around what had been the town of Hanford in sparsely populated south-central Washington State. Based on discussions with Robert Stone, Medical Director of the Manhattan Project, Arthur Holly Compton communicated his concern about radiation protection on the plutonium production project in a letter to Roger Williams of E. I. duPont de Nemours and Company, the responsible contractor for the Hanford site operations. In this letter, Compton mentioned that Herbert M. Parker was the best man available to take charge of the radiation exposure measurements at Hanford and that Stone was amenable to releasing him for this purpose.[12] Parker thus left the Clinton Laboratories, already indelibly stamped with his mark in the form of a well-organized operational health physics program built around instrument development and measurement, to take over at Hanford in the summer of 1944.

The Hanford task was an enormous challenge. The thousand or so curies of radium that had up to that time been separated and used for radioluminous compounds had resulted in many deaths in perhaps a few thousand occupationally exposed dial painters. At Hanford, there was potential for exposure of tens of thousands of workers to the millions of curies of radium-like radionuclides that would be formed in the plutonium production reactors. And there was plutonium itself, an element of largely unknown but predictably high hazard potential, which would be present in large quantities.

At Hanford, Health Physics was one of four sections within the Medical Department. Under Parker's direction, it evolved quickly from a group concerned with shielding estimates to one concerned primarily with detection and measurement of radiation, and the translation of these data into necessary protective measures. Details of the operation of the program have been briefly described by Cantril and Parker.[13]

The health physics program Parker developed at Hanford presciently considered the environmental effects of the Hanford operations, including early investigations of diffusion of radioactivity into the biosphere. Parker was among the first to undertake quantitative assessment of the effects of reactor operations, presenting a landmark paper on this subject at the first United Nations Conference on the Peaceful Use of Atomic Energy in 1955. In the late 1940s, he emphasized environmental studies, and as a result of his influence, the Hanford site has the longest continuous operational environmental radioactivity surveillance program of any nuclear facility in the world. He was also concerned early on with the management and disposal of radioactive wastes and made significant contributions in this area as well, including a clear exposition in a yearbook of *The Encyclopedia Britannica*.[14] Most importantly, he organized an extraordinarily effective program for the operational control of radiation hazards, which even today serves as a model for contemporary health physics programs.

Some of Parker's specific contributions bear a closer look. Despite the pressures of organizing and administering the Health Physics Section at the Hanford Works, as it was then called, during the latter months of World War II, Parker was never very far removed from the scientific work of the group and indeed played a leading role in several important developments. Given the general lack of radiological monitoring instrumentation, a major function of the section was the development of suitable instruments, an area in which Parker played a leading role. His understanding of the basic physics underlying radiation dose led to his insistence on ionization chambers as the primary means of measuring dose.

Recognizing even then the limitations of the roentgen for quantification of tissue dose, particularly from different types of radiations, Parker was instrumental in the conceptual development of a new unit with a purely physical basis, the *rep*. Its name derived from the acronym for *roentgen equivalent physical*, which was originally defined as the absorption of 83 ergs/g of tissue or the production of 1.61×10^{12} ion pairs per gram, regardless of the type of exposing radiation. This unit was later redefined as 93 ergs/g, and came also to be called the *parker*.[15] For many, the rep really stood for "roentgen equivalent *parker*," and this definition was often given by old-timers in the health physics field who were cognizant of Parker's contributions. The rep or parker was subsequently replaced by the *rad*, which was numerically more satisfying in that it was equal to 100 ergs/g, although it was conceptually identical to the rep. Thus, the rep is the direct ancestor of the modern dose unit, the *gray*.

Parker was also instrumental in the development of a unit of biological dose that took into account the differing biological effectiveness of equal physical doses of different radiations. This unit was originally called the *reb* (roentgen equivalent biological), but during one of his early presentations of the new unit, Parker was suffering from a cold, which led to difficulty in differentiating it from the rep. Accordingly, the name of the unit was changed to *rem*, an acronym that might imply "roentgen equivalent man, mouse, mammal, or medium" depending on the particular effect under consideration. This unit was conceptually identical to later versions of the rem and hence is the direct forerunner of the modern unit of dose equivalent, the *sievert*. Parker was an early advocate of the use of the rem for specification of dose from mixed radiation fields.

One of Parker's major scientific contributions to health physics, made while at Hanford before the end of World War II, was the introduction of the first maximum permissible concentration (MPC) for any radionuclide. The radionuclide that Parker chose was one of the chief products of the Hanford Works, ^{239}Pu. In April 1944, based on his calculations of the dose that would be received by lung tissue, Parker proposed a maximum permissible level of 3.1×10^{-11} μCi/cm^3 for ^{239}Pu in air. This value is strikingly close to the current occupational MPC for insoluble ^{239}Pu of 4×10^{-11} μCi/cm^3 recommended by ICRP Committee II in 1959.

The Postwar Years

After the war, Parker chose to remain at Hanford, serving as Assistant Superintendent of the Medical Department with responsibility for health physics for two years before assuming the management of an independent department with responsibility for health physics. He ultimately rose to the position of Manager of the entire Hanford Laboratories in 1956, a position he held until the operation of the Laboratories was assumed by Battelle Memorial Institute in 1965. Parker remained on with Battelle as a full-time consultant to the Director, retiring in 1971 from full-time activities at Battelle. He retained an office at Battelle but divided his efforts between there and his consulting company, H.M.P. Associates.

Throughout his career in senior-level mangement positions, Parker maintained interest and currency in both health and medical physics and continued to contribute seminal ideas and publications to both fields. Ever mindful of the need for well-qualifed health physicists, he organized a fifteen-part lecture series at Hanford, which served as the basic theoretical introduction for new health physicists there; he presented two of the 2½-hour lectures himself. Those fortunate enough to have attended these will recall the clarity and informative nature of his presentations, for he was an excellent lecturer.

Examination of his publications reveals the breadth of his interests in the two parallel professions and to some extent the significance of his contributions. One of his primary interests, and an area in which he made major contributions to both medical and health physics, lay in the area of dosimetry. His conception and application of both physical and biological dose units suitable for mixed radiations was originally put forth in the mid-1940s in a broader-scope internal technical report of the Manhattan District dealing with health protection coauthored with Cantril (MDDC-1100). Subsequently, Parker refined his ideas and presented them in a paper at the thirty-fourth annual meeting of the Radiological Society of North America in December 1948; the written version of that work is one of the classics of radiation dosimetry and appeared in the February 1950 issue of the journal *Radiology*.[16] It was one of only twenty-two papers—two of which were by Parker—chosen for reprinting because of its importance to radiation protection in the Twenty-Fifth Anniversary Issue of the journal *Health Physics*.[17]

Another important contribution to dosimetry was his report to the American College of Radiology Commission on Radiologic Units, Standards and Protection (CRUSP). This lengthy document, the so-called CRUSP report entitled "Some Background Information on the Development of Dose Units," was submitted to the College in November 1955. It not only traced the historical development but also provided a clear exposition of the definition and application of radiation quantities and units. In itself, it is a monumental contribution, all the more so when one considers that Parker prepared this comprehensive document virtually unaided and while he was serving in a senior management role at the Hanford Laboratories. Although this definitive and exceptionally clear discussion was never published per se, it was reprinted in 1981 in National Bureau of Standards Publication 625 and is still fresh today.

Parker's contributions to the radiological sciences were not only in the area of science per se. Under his leadership as Manager of the Hanford Laboratories—then an approximately $100,000,000 facility with an annual budget of $35,000,000—the Laboratories achieved world renown for research in the radiological sciences and, in particular, radiation biology. Their outstanding operational health physics program was also widely recognized. During Parker's tenure, he strongly supported the Hanford Radiobiology Symposia, which have evolved into the annual Hanford Life Sciences Symposium. He was also instrumental in the establishment of an educational program by which Hanford scientists and engineers could work part-time toward advanced degrees while regularly employed; this grew into the Tri-Cities University Center in Richland, a facility that now offers both undergraduate and graduate degrees through its five participating universities.

Professional Activities, Honors, and Awards

Parker was an active member of more than a dozen scientific and professional societies covering a broad spectrum consistent with his eclectic interests. He was elected a Fellow in no less than six: the American Association for the Advancement of Science, American College of Radiology, American Nuclear Society, American Physical Society, American Public Health Association, and the Institute of Physics (Great Britain). He was a charter member of both the Radiation Research Society and the American Association of Physicists in Medicine, serving a term as a member of the Board of Directors of the latter. He was also a member of the Society for Nuclear Medicine and the Society of Sigma Xi.

Given his strong view that radiation protection was a specialty area of medical physics and his aversion to the term "health physics," it is not surprising that he was not among the charter members of the Health Physics Society when that body was formed in 1956. However, in his mellower later years, recognizing that the health physics profession was indeed viable on its own, he quietly joined the Health Physics Society and was quickly elected a member of its Board of Directors. He also served on the Board of Directors of the American Nuclear Society and was named a Diplomate in Radiological Physics of the American Board of Radiology in 1947 and of the American Board of Health Physics when that body was formed in 1960.

Interestingly, he is doubtless the only member of the health or medical physics professions to be featured on the cover of a major national magazine. The May 6, 1961, issue of *Business Week* showed a full-length picture of a youthful Parker standing on the Hanford desert soil with an experimental reactor in the background. The magazine carried a feature story on nuclear energy and the Hanford site, including Parker's yet to be realized belief that plutonium fuel reactors would one day be a commercial reality. A major portion of that article was devoted to the safety aspects of reactors, and the article concluded on a strong note of optimism for the future of nuclear power.

Many of Parker's contributions were made through service on numerous national and international committees. In this milieu, he gave freely of his ideas, and much of his original work was directly incorporated into the publications of these committees. He also chaired the NCRP Scientific Committee 1 (SC-1) from 1959 to 1972. Under his leadership, the committee prepared the comprehensive and scientifically based NCRP Report No. 39 entitled "Basic Radiation Protection Criteria," which was published in 1971. Parker's fine hand is apparent throughout—in the clarity and succinctness of the prose, the organization of the document, and in the completeness of the dose-limiting recommendations, which were tabulated on a single page with references to paragraphs in the text where they were discussed.

He also served on several important government committees and boards, including the Atomic Energy Commission Safety and Industrial Health Advisory Board and the Advisory Panel on Radioisotopes. He chaired the National Academy of Sciences Panel on Foreign Activities and served on several NAS committees, including the Committee on Waste Disposal of the Biological Effects of Ionizing Radiation (BEIR) study. More recently, he was a consultant to the U.S. Advisory Committee on Reactor Safeguards and to the Director of the Bureau of Radiological Health. In the international arena, he served as technical advisor to the U.S. Delegation on the Peaceful Uses of Atomic Energy in Geneva in 1955 and as Chairman of the International Commission on Radiological Protection Subcommittee on Isotopes and Waste Disposal. By any standards, the quality and quantity of his service on scientific committees was outstanding.

His career was punctuated by numerous professional honors and many kinds of recognition, including appointments as consulting physicist to the Staffordshire Royal Infirmary (1936–1938) and the Ashton under Lyme Royal Infirmary. He also held academic appointments at Manchester College of Technology (Lecturer in Radiological Physics, 1937 to 1938) and the University of Washington (Assistant Clinical Professor of Radiology, 1952 to 1960). Those responsible for the latter appointment were somewhat embarrassed in later years, believing that the rank of assistant professor did not befit his contributions.

Finally, there were the major awards. Election to membership in the recently formed National Academy of Engineering took place in 1978. Other accolades included honorary membership on the Faculty of Radiologists in Great Britain, the Janeway Medal of the American Radium Society in 1955, the Distinguished Scientific Achievement Award (and life membership) in the Health Physics Society in 1971, and the William D. Coolidge Award of the American Association of Physicists in Medicine in 1979. The Columbia Chapter of the Health Physics Society created the Herbert M. Parker Award in his honor, and at the time of his death, he was a candidate for the National Technology Medal.

A Personal Glimpse

In a professional sense, Herb Parker was in many ways an extraordinary man. Colleagues and subordinates recall him with great respect for his knowledge and professional abilities, tinged with a mixture of fondness, awe, and even an occasional sprinkling of fear and mortification, for he did not suffer fools gladly and was quick to decimate a half-baked idea or ill-prepared presentation with his pungent British prose. He was also a man of uncompromisingly high ethics.

Herb Parker presented a commanding appearance; his voice was rich and resonant, and he was noted for his speaking ability and his dry English wit. Classically educated in the British manner, he was intellectually comfortable with the humanities as well as the sciences.

In 1936, while still in Britain, he married Margaret Fawphrop whom he had met while he was a graduate student and she a freshman physics student at the University of Manchester. Together, they emigrated to the United States in 1938 and became the parents of two sets of twins: John and Elizabeth, and Henry and Linda. In 1946, he became a naturalized U.S. citizen. He was fond of some ordinary pursuits, such as ballroom dancing, contract bridge, and stamp collecting, and he enjoyed a good dry martini. But he was in many respects basically a shy man who, with his wife, developed prize-winning irises in the large garden with its variety of trees that surrounded the old farmhouse on the Columbia River that was his home. He died a month before his seventy-fourth birthday on March 5, 1984, in Richland, Washington, the small city that serves as the residential community for most of the Hanford work force, and where he had made his home for more than 40 years.

References

1. Paterson, R., and B. R. Kirklin, 1928. "The Roentgenologic Manifestations of Primary Carcinoma of the Lung. I. Parenchymal Type." Am. J. Roentgenol. 19:20–27.

2. Paterson, R, and H. M. Parker, 1934. "A Dosage System for Gamma-Ray Therapy." *British Journal of Radiology*.7:592–632.
3. Johns, H. E., and J. R. Cunningham. 1983. *The Physics of Radiology*. Fourth Edition. Springfield: Charles C. Thomas.
4. Kaplan, I. I. 1949. *Clinical Radiation Therapy*. 2nd Ed. New York: Paul B. Hoeber, Inc., p. 74.
5. Glasser, O., E. H. Quimby, L. S. Taylor, and J. L. Weatherwax. 1952. *Physical Foundations of Radiology*. New York: Paul B. Hoeber, Inc., p. 14.
6. Meredith, W. J., Ed. 1967. Radium Dosage: The Manchester System; E & S Livingston, Ltd., Edinburgh and London.
7. Parker, H. M. 1939. "Dosage Measurements by Simple Computations." Presented before the Fifth International Congress of Radiology. *Radiology* 32:591–597.
8. Cantril, S. T., F. Buschke, and H. M. Parker. 1941. "Irradiation in Cancer of the Cervix Uteri." *Radiology* 36:534–542.
9. Grigg, E. R. N. 1965. *The Trail of the Invisible Light*. Springfield: Charles C. Thomas, p. 839.
10. Brecher, R., and E. Brecher. 1969. *The Rays. A History of Radiology in the United States and Canada*. Baltimore: Williams and Wilkins.
11. Buschke, F., and H. M. Parker. 1942. "Possible Hazards of Repeated Fluoroscopies in Infants." *The Journal of Pediatrics* 21:524–533.
12. Seaborg, G. T. 1978. "History of Met Lab Section C-1." Lawrence Berkeley Laboratory, Publication 112, Vol. II, p. 407.
13. Cantril, S. and H. M. Parker. 1951. "Status of Health and Protection at the Hanford Engineer Works." In *Industrial Medicine on the Plutonium Project*, R. S. Stone, Ed., New York: McGraw-Hill, p. 476–484.
14. Parker, H. M. 1978. "Nuclear Waste Disposal." *Science and the Future. The 1978 Encyclopedia Britannica Yearbook.*, pp. 201–217.
15. Lapedes, D. N., ed. 1976. *McGraw-Hill Dictionary of Scientific and Technical Terms*. New York: McGraw-Hill Book Co.
16. Parker, H. M. 1950. "Tentative Dose Units for Mixed Radiations." *Radiology* 54:257–262.
17. Parker, H. M. 1980. "Tentative Dose Units for Mixed Radiations." *Health Physics* 38:1021–1024.

tive E

F.E

A B

F.E + differential

Electrometer Connections

F.E

E or H switch

B
A
safety resistances

+30

Voltage selector switch.

−30

10v

30ω

400ω

F.E.

Plate connections.

PART I

THE MANCHESTER YEARS

Apparatus.

Lindemann electrometer connected in special way. Case is raised to a 'floating earth' of about 230 volts, and additional twelve volts is applied to needle. In this way instrument is used in a relatively insensitive condition (5 divs/volt) while full 240 v. is applied across chamber. Electrometer plates can be connected at will to high or low sensitivity ranges.

Chambers.

Elektron metal case & centre rod. Amber insulation. Screw cap permits access to centre rod, space between constitutes the chamber proper. Vol. about 20 c mms. Advantage over Sievert spherical chamber - total dimensions 3 mms × 3 mms approx as against 10×10 mms. This prevents difficulties in high space gradients and approximates more to a __point__ measurement. Cannot be used nearer than 3 mms to a needle & prob. not accurate at < 5 mms.

INTRODUCTION

Herbert Parker's first major contributions to radiological physics were made in the 1930s during his years at Manchester. In that decade he published three papers, coauthored with British radiologist Ralston Paterson, on radium dosimetry and applications in therapy. These papers serve as the basis for the widely used Paterson-Parker system of radium dosage and are as fresh today as when they were written more than a half century ago.

In 1947, more than a decade after publication of the first paper, a book based on these and related works entitled *Radium Dosage: The Manchester System* was published because of the enormous demand for reprints. A second edition of this book was published years later and still remains in print. A chapter from the first edition is included and shows the clarity and style of his mathematical derivations.

This section also includes a treatise on dosimetry for radiologists, which was presented at the Fifth International Congress on Radiology, held in Chicago in May 1937. This was the first of many papers Parker presented in the United States. Another paper worthy of special note is the one published in 1939 describing a device of Parker's own design for reconstruction of the spatial distribution of implanted radium needles in vitro, a work which illustrates not only his quest for perfection but gives a measure of the man as a practical engineer as well as a physicist.

A Dosage System for Gamma Ray Therapy

The British Journal of Radiology, Volume VII

October 1934

Reprinted with permission of the British Journal of Radiology.

A DOSAGE SYSTEM FOR GAMMA RAY THERAPY

PART I
By RALSTON PATERSON, M.D., F.R.C.S., D.M.R.E.

PART II
By H. M. PARKER, M.Sc.

(Received June 7, 1934)

592   *Ralston Paterson and H. M. Parker*

A DOSAGE SYSTEM FOR GAMMA RAY THERAPY

PART I

By RALSTON PATERSON, M.D., F.R.C.S., D.M.R.E.

(Received June 7, 1934)

THE evolution of scientific dosimetry in relation to radium therapy must occur in two stages. The first stage is the physical one, concerned with methods of measurement of dose. The second, dependent on the first, is the therapeutic stage concerned with the determination of the optimum therapeutic dose. With the exception of brief comment at the end on certain biological findings, this paper deals with the first of these stages and is an attempt to reduce to simple terms the answers which the physicist provides to various problems, and to present them in such a form that they can be applied in everyday clinical use—not just in single cases, but over the whole therapeutic field. Many papers have already been published dealing with this subject, but physical studies have, for the most part, been definitely technical, and clinical studies to a great extent purely empirical. No physically accurate system of dose measurement has, as yet, been described in a complete form, which is yet sufficiently simple to allow of routine clinical use, especially by those not mathematically minded.

In general, the study of dosage in practical radium therapy raises two main questions. How much radium will be required, and how must it be arranged?

The first question ultimately involves the need for measurement of the amount of energy of radiation delivered at a point in or throughout a treated region. Dosage in terms of milligramme-hours, or millicuries destroyed, gives no real knowledge of quantity of radiation delivered, and has too long hindered progress. It is essential that one or other of the many units of quantity of radiation be brought into use in such a way that the clinician can think and talk in terms of true dosage.

The second aspect of the problem is equally vital, and concerns the need for arrangement of the radioactive sources in tissue, or on applicators, in such a way as to produce uniform radiation throughout the treated volume or over the treated area. This might be called the problem of distribution.

The correlation of a method of *dosage* measurement with a scheme of distribution giving homogeneity constitutes what amounts to a physical *system of dosage*. The clinician having decided from biological considerations

A Dosage System for Gamma Ray Therapy

what quantity of radiation he needs to give is, by the application of this system, enabled to apply homogeneous radiation of known physical intensity in the majority of ordinary anatomical situations. This article is essentially a description of such a system of dosage as it is employed at the Radium Institute in Manchester. The system has been gradually evolved over the last three years, and is applicable to all forms of radium therapy other than certain types of interstitial implantation. It is sufficiently simple for ordinary clinical use and is believed to be physically sound.

Dosage

The unit of radiation chosen was the international "r" unit of quantity of radiation, the value of which, in gamma radiation, can be derived mathematically from known physical data.

A unit of *dosage* rather than of *intensity* was preferred, firstly because of the belief that total dose is the most important single factor in therapy (provided due consideration is given to the duration of radiation). It was also chosen because the clinician finds it very much easier to think in terms of total dose delivered rather than in terms of intensity and time, separately. Any of the numerous units of dosage described in the literature would have served equally well, but the 1mc.-hr. and the "r" unit were considered as most likely to command ultimate international recognition. The "r" unit was selected in the belief that gamma rays and X rays must ultimately be measured in the same unit. The evaluation of the "r" unit is discussed fully in Part II, Section A, a value of 8·4 "r" per hour being accepted as the intensity of radiation at a distance of one centimetre from a one milligramme point source of radium filtered by 0·5 millimetre of platinum.

Dosage Charts

Actually 1,000 "r" was selected as the working unit and a series of graphs was constructed—Part II, Section C, p. 627—showing the amount of radium required to produce 1,000 "r" in various situations.

The *area graphs* (Charts 1 and 2) show in milligramme-hours the amount of radium required on flat surface applicators to give 1,000 "r" over any area at various treatment "distances." It is assumed that the radium is spread over an area on the applicator parallel to, and of the same size as, the treated area, in such a manner as to produce homogeneous radiation at the stated distance. These graphs are similar to those published by Mayneord[1] for intensities of radiation.

The *tube graphs* (Charts 3 and 4) give, similarly, for tubes of various radii the radium (expressed in milligramme-hours) required along the central axis to give 1,000 "r" at the surface of such tubes. These graphs must be read for the over-all active length.

The graph readings for zero area or zero length may be used to ascertain the radium needed to give this dose at various distances from a *point source.*

The graphs are used as follows: the dose in 1,000 "r" having been decided on, the graph reading for the area or length under consideration is multiplied by that number (of thousand "r"); this figure is then corrected, if necessary, for filtration; this gives the number of milligramme-hours required. The actual amount of radium element or radon (emanation) is found by dividing this figure, either by the number of hours intended, if element is to be used, or by the number of millicurie-hours per initial millicurie for the time in question, if radon is to be employed.

It is shown in the rules for distribution, which are described later, how these graphs either are already suitable, or can be made suitable, for all normal clinical situations. Even where the radium is not distributed satisfactorily the graphs indicate the "mean" dose received over the whole area treated. In this way they can be used to assess treatments already given, to determine dose at other levels than that for which homogeneity is planned, to estimate dosage from single layer implants, to measure treatments fully described in publications by other therapists, and for various other problems.

Comparison with Other Units

It is of interest to compare the "r" unit used in this way with other units of dosage.

1,000 "r" = 119 intensity-millicurie-hours—Sievert, Stockholm.

= 225,000 ergs/cm.³—Murdoch and Stahel, Brussels.

= 11 "D" units—Mallet, Paris.

= approx. 1·1 "erythema dose"—Memorial Hospital, New York.

5,000—7,000 "r" = range of *maximal* dosages given on "mcd. per square centimetre" basis at Fondation Curie, Paris.

Distribution

The problem of distribution of radium is in many respects not only a more difficult, but a more acute problem. A study of Fig. 1 will illustrate its importance. Various aspects of this subject have been studied by other observers, *viz.*, Sievert,[2] Murdoch and Stahel,[3] Mayneord,[4] and others. Most of the accurate studies are almost entirely physical in nature, and either deal with single examples or are difficult to carry out expeditiously in ordinary clinical routine.

The author has also to acknowledge indebtedness to various sources,

A Dosage System for Gamma Ray Therapy 595

other than published papers, for this portion of the study, but particularly to Dr. Murdoch and Dr. Suzanne Simon, in Brussels, for many of the basic ideas.

The problem was met for surface therapy by the evolution of a series of *Rules*, outlining definite methods of arrangement of radium on applicators in lines or circles. The distance between the lines and the size and number

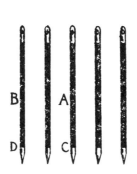

INTENSITY

	r PER HOUR
A =	42·5
B =	30·5
C =	30·5
D =	23·0

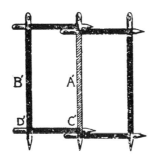

	r PER HOUR
A' =	20·0
B' =	18·5
C' =	20·0
D' =	19·2

Fig. 1.
Diagram illustrating the extreme importance of correct arrangement. The diagrams show the intensity produced at 1 cm. by two different radium distributions over the same area.

of the circles bear fixed relationships to the distance between radium and treated area. These *Rules* are as follows:—

RULES GOVERNING DISTRIBUTION OF RADIUM ON APPLICATORS AND MOULDS

Definition.—The thickness of the applicator or the distance separating the plane on which the radium is mounted, and the plane at which the dose is assessed (*i.e.*, the treated area) will be referred to as the "distance."

The amount of radium to be used is first determined from the "Dosage Graphs," and that amount is then arranged upon the applicator in one of the following ways. All of these ensure homogeneous radiation over the treated surface.

Circles

Use circles wherever possible. Arrange the radium uniformly round the circumference, employing as many radioactive foci as possible; a circular arrangement may, however, be considered as obtaining if, with a minimum of *six* containers, a space not exceeding the distance exists between the active ends of each tube or needle.

A single circle alone is sufficient where the diameter is less than *three* times the distance. The circle whose diameter is 2·83 times the distance is "ideal."

Where the diameter is from three to six times the distance, 5 per cent of the radium should be placed at the centre.

For large areas use two concentric circles and a centre spot as follows:—
A. Put 3 per cent of the total radium at the centre.
B. Use percentages of radium for the outer circle as indicated in this table.

	6	7½	10
Diameter divided by "distance"			
Per cent Radium outer circle	80%	75%	70%

C. Distribute the remainder round a circle of half diameter.

For circles at small "distances" the last arrangement is not practical and the following should be substituted, but is less accurate.

Diameter 6—7 times "distance" = 10 per cent total radium at centre.
Diameter 7—9 times "distance" = 20 per cent total radium at centre.

Squares

The radium should be distributed in a line round the periphery with uniform linear density (*i.e.*, the number of milligrammes per centimetre). An attempt should be made to have active radium along the whole length of any line, but a linear arrangement is considered as obtaining if a space not exceeding the "distance" exists between the active ends of each needle or tube in line.

If the length of the side of the square does not exceed twice the "distance" no further radium is required.

A Dosage System for Gamma Ray Therapy 597

Additional lines should be added, if required, parallel to the side to divide the area into strips of width twice the "distance."

For one added line, linear density to be half that of periphery.

For two or more lines, linear density to be two-thirds that of periphery.

Rectangles

The charts apply strictly for circles and squares. For rectangles proceed as for squares, adding the additional lines parallel to the longer side and make a correction in the direction of increased milligramme-hours, as follows:—

Ratio of sides of rectangle	2 : 1	3 : 1	4 : 1
Percentage to be added	5%	9%	12%

Irregular Areas

Distribute the radium uniformly round the periphery and either, if the irregular area be roughly rectangular, add lines parallel to the longer length to divide the area into strips of width twice the mould distance as for rectangles, or, if the area be roughly elliptical, add a centre spot appropriate to the mean of the two diameters as for circles.

Convex Areas

These rules apply strictly to flat areas, applicator area and treated area being equal. For curvatures up to a degree corresponding with a hemisphere or a semi-cylinder, all the above rules may be applied with the following proviso. The amount of radium to be used should be ascertained from the graphs for the *area treated*, but should be spread over the larger, but corresponding area on the applicator.

Concave Areas

Again, the rules as applicable to flat areas may be used, but the area measured for dosage purposes is the *area of the applicator* regardless of the area treated. Care should be taken to include the correction for the elongation of rectangles. If the applicator area, however, becomes small relative to the treated area, the arrangement should be discarded and a greater "distance" employed so as to permit use of a linear source (tube graphs) or of a point source.

Linear or Tubal Applicators

Radium should be mounted along the central axis and the tube graphs employed. If possible, no interval should be left between the active ends of each container or needle, but a space between active ends not exceeding the radius of the tube may be allowed. For tubes which are long in relation to their

598   Ralston Paterson and H. M. Parker

radius, the applicator should be carried somewhat beyond the treated zone, or, if this is not possible, an addition of 4 per cent of the total radium must be put at each end of the tube.

Single Foci

A radium container or group of containers should be considered as a single focus (a point source) if of small size relative to the "distance."

Dose at Other Levels

It is frequently necessary to determine the dose at a plane or planes other than the "treated area," but parallel to it, either to assess the dose being delivered at the base of a tumour (depth dose) or in calculation for "sandwich" applicators.

This is obtained in 1,000 "r" by dividing the number of milligramme-hours actually used by the graph reading at the *new* distance for the area in question.

In doing this the fact that at such other distances radiation may not be so homogeneous must be kept in mind.

Filtration

The *area graphs* are given for 0·5 millimetre of platinum. If other filtrations be used the following corrections require to be made.

Platinum—Thickness	0·8	1	1·5	2
Per cent correction—*Add*	5%	10%	20%	35%

Gold	As platinum.
Lead and Silver	As half their thickness in platinum.
Monel, Brass, etc.	As one-third their thickness in platinum.

The *tube graphs* are given for filtrations of 0·5 platinum and of 1·5 platinum, respectively. The appropriate graph should be employed. For filtrations of 1 millimetre of platinum the mean of the readings from both these graphs should be used. For other filtrations correction factors as for the area graphs may be employed, but are not quite accurate.

Useful Tables
Radon (Radium Emanation)

Days	2	3	4	5	6	7	8	9	10	11	12	Permanent
Mc.-Hrs. per initial mc.	40·3	55·5	68·5	79	88	95·5	102	107	111	115	118	133

A Dosage System for Gamma Ray Therapy

CIRCLES

Diameter (cm.)	1½	2	2½	3	3½	4	4½	5	5½	6	6½
Area (sq. cm.)	2	3	5	7	9.5	12.5	16	19.5	24.0	28.5	33
Circumference	4.7	6.3	7.9	9.4	11.0	12.6	14.1	15.7	17.3	18.9	20.4
Diameter (cm.)	7	7½	8	8½	9	9½	10	10½	11	11½	12
Area (sq. cm.)	38.5	44	50.5	57	63.5	71	78.5	86.5	95	104	113
Circumference	22.0	23.6	25.1	26.7	28.2	29.8	31.4	33.0	34.6	36.1	37.7

Rules

These "Rules" are to a large extent self-explanatory, but some further discussion is advisable. As a starting point it was accepted that to demand

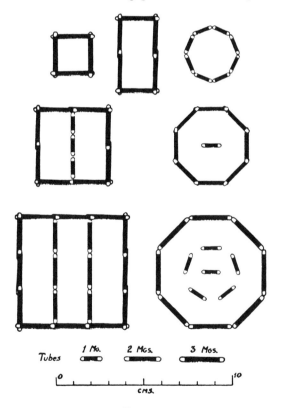

Fig. 2.

Diagram illustrating the "Rules for Distribution," depicting typical arrangements to produce homogeneity at 1 cm.

absolute homogeneity was impractical, and that some variation of intensity from point to point over the treated field had to be allowed It was decided that the limit of this variation should not be more than ± 10 per cent, and no generalisation was allowed unless this condition was met. Actually, the variation resulting from an absolute application of the rules very seldom exceeds ± 5 per cent—a striking degree of accuracy for clinical work. The

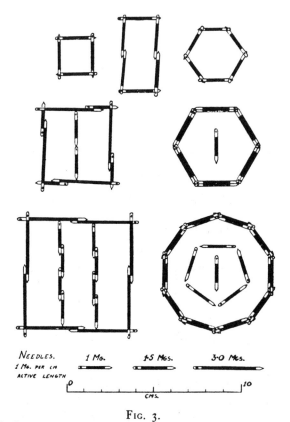

FIG. 3.
As Fig. 2. To illustrate how the system can be adapted to whatever types of radium container are available. This figure should be studied in relation to the discussion in Part II (D) on distribution.

details of the mathematical and physical data on which each Rule is based is given in Part II, Section B, and therefore will not be further discussed here.

The generalisations made in the "Rules" for curved surfaces (convex and concave areas) have not been mathematically substantiated for irregular surfaces, but have been proved to hold good for numerous examples of symmetrical form, *e.g.*, sectors of cylinders and of spheres of varied curvatures at varied distances. It is probable, therefore, that they hold good even for a considerable degree of curvature, and it was decided to accept them subject to the limitations stated.

Figs. 2 and 3 show diagrams of a number of distributions illustrative of the application of the "Rules." While these are exact plans of distributions to produce homogeneous intensity at 1 cm. they may be equally considered as plans to scale of distributions to produce homogeneous intensity at 2 cm., if all dimensions are doubled, and similarly they may be used as a basis for 3 cm. distributions with dimensions tripled; and so on.

PRACTICAL APPLICATION OF THE SYSTEM

In planning skin applicators, or dental moulds, an attempt is made to treat an area at least one to two centimetres wide of apparent tumour, and to use a distance which will give adequate dose at sufficient depth. In practice it is found that the "circle" distributions are very much easier to employ, and these are used as frequently as possible (Fig. 5, A, B, C). It is quite often practical, particularly with small lesions, to choose as a suitable distance that distance which allows for the use of a single ideal circle, namely, where the diameter of the treating circle is 2·83 times the applicator "distance" (Fig. 5, B). Where this is possible, an unexpected clinical advantage arises, *viz.*, the ability to expose the lesion for accurate placing and for dressings, etc., through a hole in the middle of the applicator.

The linear distributions are used rather less frequently, but are found useful for large areas at a short distance, such as applicators for post-operative skin recurrences of breast carcinoma (Fig. 5, D). It would appear, on physical grounds, that for areas which are large relative to the "distance," the rectangular distributions may be preferable to the circular.

In the treatment of skin lesions such regular arrangements can usually be employed, but within the mouth limitations of space are much greater, and here the rules for irregular areas find frequent use for intra-buccal moulds (Fig. 5, F).

The rules for curvature allow these general principles to be applied fairly extensively (Fig. 5, G), but they are not applicable with safety to very marked degrees of curvature, such as is found, for instance, with an applicator enclosing the lip. (This is calculated as a type of the sandwich applicator discussed later.)

The tube graphs indicate the dosage at the surface of radium-bearing tubes, as used for œsophagus, body of uterus, rectum (Fig. 5, H), etc., and for radium "corks," such as vaginal "colpostats." They are equally of value in assessing dosage resulting from any type of linear applicator.

It is frequently necessary to ascertain the dosage at any distance from a point source of radiation. This, for instance, occurs where a sorbo ball containing radium is sewn into the bladder or other cavity. It will be obvious how the reading for zero area or zero length of each set of graphs gives the dosages at various distances from a point source.

Dose in Tissue

So far we have been considering surface applicators, and it will be apparent that the dose is assessed at a single plane. It is, however, easy to obtain the dose delivered at any depth below such "treated surface" by reversing the process of calculation and working back from the milligramme-hours actually used, to find the dose delivered at the new distance from the applicator.

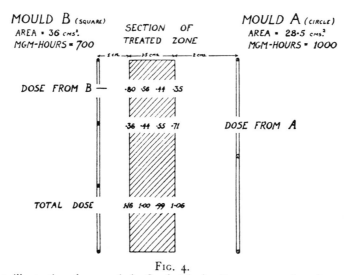

Fig. 4.
Diagram illustrating the use of the Sandwich Applicator to produce homogeneous dosage through a thickness of tissue. Dosage is stated in terms of 1,000 "r."

Sandwich Applicators

It is frequently possible, and where possible infinitely more desirable, to "sandwich" a growth between two moulds in such a way that, to all intents and purposes, homogeneous dosage is given through and through a whole thickness of tissue, the fall of intensity of each mould being balanced by the rise of intensity of the other. To do this, an applicator is planned consisting of two moulds parallel to each other and enclosing between them the tumour-bearing tissue. For either mould a probable figure for milligramme-hours is now chosen, and the dose produced is calculated for the adjacent surface and for each successive half centimetre level of the underlying tissue. The same process is then carried out for the other mould, the doses being calculated for the same half-centimetre levels, except that, obviously, they are approached from the opposite side. The doses calculated for each level are then summated and, using the figures obtained as a guide, the milligramme-hours, or the distances can be adjusted until on re-calculation a uniform dosage is arrived at. The radium is usually mounted so as to produce homogeneity on the surface nearest to it, to which it contributes the bulk of the radiation. A detailed study of Fig. 4 will explain this process more clearly than further description. This type of calculation is the only one described which is in any way complicated; it is, however, only used for extensive treatments, for which it is invaluable and which would otherwise be impractical. Moreover, once the principle is understood, it is not nearly so complex as it appears on first reading. (For example, see calculation for Fig. 5, J.) The "sandwich" mould principle, suitably modified, is used in making applicators for lip, cheek, alveolus (Fig. 5, G, H), and similar sites.

Interstitial Radiation

While it is not intended to deal fully here with the application of measured dose to interstitial work, it will be seen that for any single-layer implantation the rules for surface dosage can be employed to assess the intensity of radiation on a plane at any particular distance from the plane of implantation; and the distribution rules can be used to plan the arrangement of seeds or needles in such a way that at this "distance" the radiation is homogeneous. In practice this is usually assessed for a "distance" of $\frac{1}{2}$ cm., and the assumption is made that, apart from the local zones of intense radiation immediately adjacent to the needles or seeds, the same intensity may be considered as present at the actual plane of the implant, and therefore through and through the one centimetre thickness of tissue containing the implant (Fig. 5, E). The further extension of this method of calculation to two or more parallel planes of needles, one to two centimetres apart, is also obvious, the calculation being on similar lines to that employed for a sandwich applicator.

In practice many lesions of appreciable depth lend themselves to "sandwiching" between an external applicator and a light interstitial implantation deep to the tumour. The applicator supplies the greater portion of the radiation, the implant being used to build up depth dose. The calculation for this arrangement is also similar to that required for a sandwich applicator. The whole question of volume radiation is, however, going to be a subject for separate consideration later.

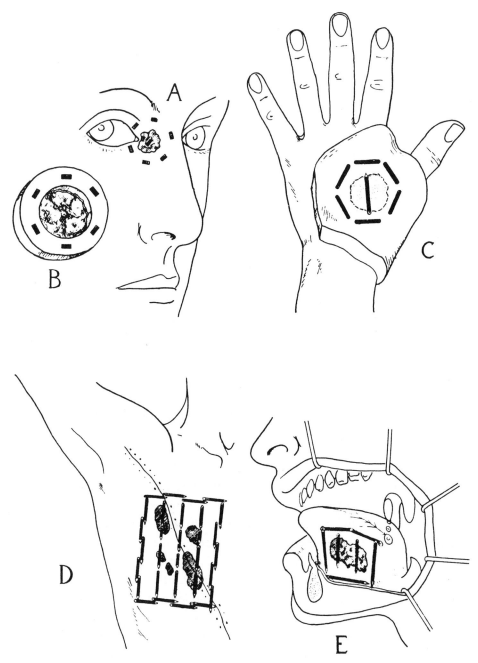

Fig. 5.—Examples of the System—for details see text.

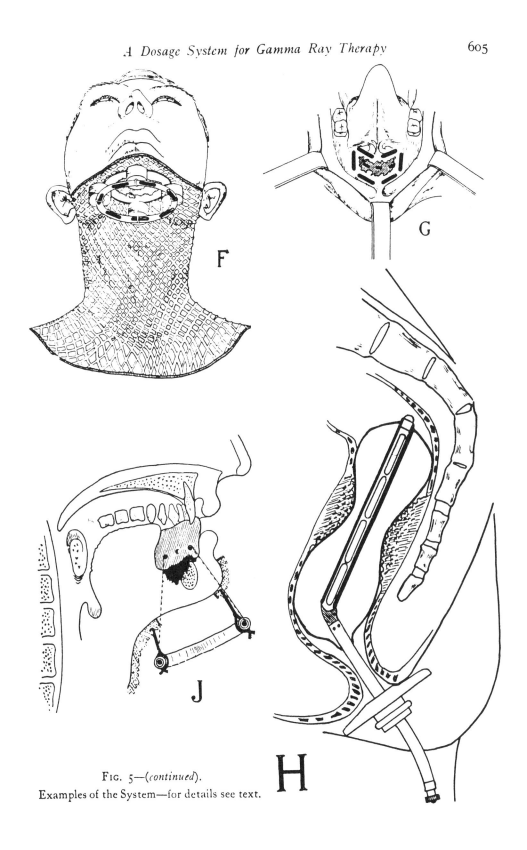

Fig. 5—(continued). Examples of the System—for details see text.

606   *Ralston Paterson and H. M. Parker*

Examples

Fig. 5 shows diagrams illustrating a variety of treatments, the calculations for which are given in detail as concrete examples of the application of the method.

A. *Rodent Ulcer Face.*—Lesion inner canthus, size 1·2 diameter.
 Treatment by gold seed implantation (non removable) at ½ cm. deep to skin—to give 6,500 "r."
Calculation
Treat circular area 2·5 diameter = 5 sq. cm.
1,000 "r" (at ½ cm.) = 155 mg. hrs. 6,500 "r" = 1,010 mg. hrs.
Mc. hrs. / initial mc. = 133 ∴ Radon = $\frac{1010}{133}$ = 7·6 *millicuries.*
Gold seeds ·5 mm. wall, therefore no correction for filtration.
Distribution
Diameter = 5 times "distance."
Use single circle with 5% centre spot.
Use 6 × 1·1 mc. seeds plus 1 weak seed central.

B. *Nævus of Face*—On cheek, 4 cm. diameter.
 Treatment by elasto-plast Sorbo applicator, using silver radon seeds—to give 1,000 "r" in 5 days.
Calculation
"Ideal" circle 4 cm. diameter is at 1½ cm. "distance," area 12·5 sq. cm.
1,000 "r" (at 1½ "distance") = 710 mg. hrs.—No filtration correction.
Mc. hrs. per initial mc. in 5 days = 79·1 ∴ $\frac{710}{79 \cdot 1}$ = 9 mcs.
Use 6-1·5 mc. silver seeds (1 mm. silver filtration).

C. *Epithelioma Dorsum of Hand*—Size of lesion 3·5 diameter.
 Treatment by 1 cm. mould. To give 6,000 "r" in 20 days (the mould being worn 16 hours per day).
Calculation
Treat area 5 cm. diameter = 19·5 sq. cm.
Mould distance = 1 cm.
1,000 "r" (at 1 cm.) = 630 mg. hrs. 6,000 "r" = 3,780 mg. hrs.
Correction for filtration 1 mm. platinum, add 10% ∴ 4,158 mg. hrs.
4,158 mg. hrs. = 13 mg. in 320 hours.
Distribution.
Diameter 5 times distance.
Therefore single circle plus 5% centre spot.
Use 6-2 mg. tubes in a circle, plus 1 mg. tube as centre spot (all 1 mm. Pt.).
Dose at ½ cm. below skin.
At new distance of 1½ cm. 1,000 "r" = 880 + 10% (for filtration) = 968 mg. hrs.
Mg. hrs. actually used = 4,158.
∴ depth dose = $\frac{4158}{968} \times \frac{1000 \text{ "r"}}{1}$ = 4,300 "r."

D. *Breast Recurrences in Post-operative Scar*—over fairly wide area.—Desired to treat an area 8 × 10 with a one-centimetre mould to 5,500 "r" in 6 days.

A Dosage System for Gamma Ray Therapy 607

Calculation
Area 8 × 10 = 80 sq. cm.
1,000 "r" (at 1 cm.) = 1,450 mg. hrs.   5,500 "r" = 7,975 mg. hrs.
7,975 mg. hrs. = 56 mgs. for 143 hours. Needles ·5 Pt. ∴ no filtration correction.

Distribution
Rectangle at 1 cm. ∴ additional lines of ⅔ intensity 2 cm. apart.
1 mg. per cm. for 36 cm. of periphery and ⅔ mgs. per cm. for 30 cm. of lines = 56 mgs.
∴ sides = 5-2 mgs. and 4-2 mgs. respectively.
∴ lines = 3-2 mgs. + 1-·66 mgs. each.

Dose at 1 cm. Below Skin.
At new distance of ·2 cm. 1,000 "r" = 2,580 mg. hrs.
Mg. hrs. actually used = 7,975.
∴ depth dose = 3,100 "r."

E. *Tongue Epithelioma*—Superficial lesion side of tongue, size 2 × 1½ cm.
Single plane implantation to 7,000 "r" in about 8 days decided upon.

Calculation
Treat area 3 × 3 = 9 sq. cm.
1,000 "r" (at ½ cm.) = 215 mg. hrs.   7,000 "r" = 1,505 mg. hrs.
Needle filtration = ·8 mm. platinum ∴ add 5% ∴ 1,580 mg. hrs.
1,580 mg. hrs. = 8·5 mgs. in 186 hrs.

Distribution
Square area at ½ cm.
∴ addition lines 1 cm. apart of ⅔ intensity periphery.
Use 3-1·5 mg. needles 3 cm. active, and 4-1 mg. needles 2 cm. active, as illustrated.

F. *Lupus Epithelioma Chin*—Size 5 × 3½. Treatment by Nidrose applicator.—To give 6,000 "r" in about 10 days (continuous radiation).

Distribution
Area to be treated is a *convex surface*, therefore for dosage purposes measure *area treated*.
Treat an ellipse 7½ × 6, *i.e.* mean diameter 6¾.

∴ "ideal" distance to treat with single ring $= \dfrac{6\cdot 75}{2\cdot 83} = 2\frac{1}{2}$ cm. approximately.

Calculation
Area = ellipse 7½ × 6 = 35·4 sq. cm.
Dose per 1,000 "r" (at 2½ cm.) = 2,000 mg. hrs. ∴ 6,000 "r" = 12,000 mg. hrs.
Filtration 2 mm. platinum ∴ add 35% makes 16,200 mg. hrs.
16,200 mg. hrs. = 70 mgs. for 231 hrs.
∴ use 7-10 mg. tubes (2 mm. Pt.) as single ring over the larger area corresponding to the skin ellipse.

G. *Epithelioma Alveolus*—Superficial lesion, irregular in shape, about 3 × 1½.
Treatment by single dental mould to 12,000 "r" surface dose in from 10 to 12 hours per day for 12 days.

Calculation
Area to treat approximately 4 × 2 = 8 sq. cm.
Dose per 1,000 "r" (½ cm. graph) = 200 mg. hrs. ∴ 12,000 "r" = 2,400 mg. hrs.
Filtration 1 mm. platinum ∴ add 10% = 2,640 mg. hrs.
Area roughly rectangle sides 2 : 1 ∴ add 5% = 2,772 mg. hrs.
2,772 mg. hrs. = 21 mgs. in 132 hrs.

Distribution
Rectangular area 4 × 2 at ½ cm. distance.
∴ one additional line down centre of ½ intensity periphery.
Use 6-3 mg. tubes, plus 3-1 mg. tubes, as sketched.

608   *Ralston Paterson and H. M. Parker*

H. *Carcinoma Rectum*—Annular lesion about 3 centimetres long.
Treatment by Todd's applicator (*i.e.* an inflatable dumbell-shaped applicator, the diameter of which in the middle is two centimetres, but at the ends is four centimetres. Desired to give 10,000 "r" to the surface of tumour in about 5 days with maximum limit of 5,000 "r" on normal mucosa, as further treatments are intended.

Calculation
(1) *Tumour Surface*
Active length = 8 centimetres.
1,000 "r" (1·5 tube graph at 1 cm.) = 500 mg. hrs. ∴ 10,000 "r" = 5,000 mg. hrs.
5,000 mg. hrs. = 40 mgs. for 125 hours ∴ use 4 × 10 mg. tubes (1·5 Pt.).
(2) *Mucosal Surface*
1,000 "r" (tube graphs at 2 cm.) = 1,130 mg. hrs.
Mg. hrs. actually used = 5,000.
Hence dose on mucosa = $\frac{5000}{1130} \times \frac{1000 \text{ "r"}}{1} = 4{,}420$ "r."

J. *Epithelioma Floor of Mouth and Alveolus*—Lesion as Fig. F, but invading bone.
Treatment by "Sandwich Mould"—Desired to give *total* dose 11,000 "r" on mucosal surface, and 6,000 "r" on skin surface, with a minimum of 5,500 "r" in the tissue.

Moulds F. and G. have been chosen so that in 8 days they achieve this result. Calculations for the individual moulds, therefore, need not be repeated.

Calculations for the sandwich is as follows: Distance between skin and alveolus = 2½ cm.
Mould G.—100 hrs., *i.e.* 12 hrs./day for 8 days:

1,000 "r" at area 4 × 2 corrected for filter.   *Mc. hrs. actually used* = 2,100 ∴ *dose.*

    ½ cm. =   230 mg. hrs.   9,130   mucosa.
    1 cm. =   449   ,, ,, ...   4,680
    1½ cm. =   679   ,, ,, ...   3,100
    2 cm. =   965   ,, ,, ...   2,180
    2½ cm. = 1,320   ,, ,, ...   1,590
    3 cm. = 1,710   ,, ,, ...   1,230   skin.

Mould F. 200 hrs. continuous:
1,000 "r" at area 35·4 *corrected for filter.*   *Mg. hrs. actually used* = 14,000 ∴ *dose.*

    2½ cm. = 2,720 mg. hrs.   .. ..   5,150 "r"   skin.
    3 cm. = 3,400   ,, ,, ...   4,110 "r"
    3½ cm. = 4,040   ,, ,, ...   3,470 "r"
    4 cm. = 4,860   ,, ,, ...   2,880 "r"
    4½ cm. = 5,400   ,, ,, ...   2,590 "r"
    5 cm. = 6,290   ,, ,, ...   2,200 "r"   mucosa.

Combined Dose

	½ cm. spaces					
	Skin					Mucosa
Internal mould ..	1230	1590	2180	3100	4680	9130
External mould ..	5150	4110	3470	2880	2590	2220
Total dose ..	6380	5700	5650	5980	7270	11350

Telecurietherapy

The opportunity has not been available for applying the "r" unit method of measurement clinically to "bomb" therapy—this Institute has no bomb. It would obviously be easy, however, to assess both surface and depth dose obtainable by beam therapy on this basis from a knowledge of the construction of any particular bomb.

Experimental Verification

The outline of this whole dosage system was first evolved on a biological and clinical basis, the details of this evolution being now unimportant. As must be apparent, it has since been mathematically substantiated, and in the process rendered much more accurate. Some form of experimental proof was also indicated. Attempts at photometric methods of studying the problem were first made, but were discontinued as the photographic approach was shown conclusively to be unsatisfactory (H.M.P., in work not yet published). Within the last few months, however, small ionisation chambers of the Sievert condenser type have been constructed, and with these it has been possible to verify experimentally a large number of examples of the above distributions. Agreement with theory has been found to be excellent.

It is fully realised by the authors that no cognizance is taken of the fact of scatter from, or absorption by, normal tissues, and that the whole system has been calculated on an "in air" basis, it would appear, however, from such information as has been published, that this is not likely to prove an important factor. It is proposed, moreover, to investigate this problem shortly, using the ionisation chambers.

It is equally appreciated that the "r" unit may not be accepted ultimately as the international unit of gamma radiation. It will be obvious, however, that conversion to any other unit which may be established will be merely a piece of simple arithmetic. The value of the system as a whole will not, therefore, be invalidated.

Biological Aspects

A dosage system along these lines has been in clinical use in this Institute for well over two years. Because it provides an accurate measure of actual radiation, it has been possible to establish much more definitely than has previously been done the tolerance of normal skin and other normal tissues. Moreover, the determination to a reasonable degree of accuracy of the *in vivo* lethal dose for certain types of tumour has also been achieved.

Experience shows that changes in both malignant and normal tissues as a result of radiation depend not only on total dose, but upon the period

610   *Ralston Paterson and H. M. Parker*

over which that dose is spread (or with intermittent treatment, upon the duration of each session and the interval between sessions). The dose required to produce lethal change rises as the total period of radiation increases. Thus, any statements regarding the reactions of normal tissue, or relative to the lethal dose for tumours must be qualified by a statement of the total length of treatment and of the duration of sessions. The following statements are made in reference to a total period of irradiation in the region of eight days, the radiation being either *continuous*, or, if intermittent, lasting for at least *twelve hours in each twenty-four*. The statements must be read solely as in reference to that period. Discussion of the extent to which they have to be modified for other periods is not relevant to this paper.

Skin

The reaction of skin depends to some extent upon the site, condition, and vascularity. For normal, healthy skin on face, trunk, etc., the following may be taken as an average of the reaction to be expected.

- 3,000 "r"   Faint erythema.
- 4,500 "r"   Definite erythema.
- 6,000 "r"   Moist desquamation—"radio-epidermitis" or "Schute." The reaction period lasts for about six weeks, but return to normal is complete.
- 7,500 "r"   A very marked reaction of similar type, lasting for a longer period and borderline for skin safety.
- 9,000 "r"   Extremely severe reaction, resulting in a percentage of permanent radium necroses, either immediate or late.

Certain sites—for instance, fingers, hands, feet and shins, post-operative scar-tissue (*e.g.* after breast amputation), lupus scar-tissue, etc.—tolerate rather lower dosages (or conversely respond vigorously to these dose levels). There is also some relationship between the size of the area radiated and the severity of the reaction. Larger areas respond somewhat more vigorously to the same dose than small areas. As far as present evidence goes this, however, does not appear to be a very important factor.

Buccal Mucosa

The response of buccal mucosa to radiation is one which is difficult to grade. The typical response is well-known, and consists of a temporary erythema followed by the formation of a yellowish, white membrane surrounded by a well defined red margin. This reaction zone indicates the area radiated. Between dosages of about 4,000 "r" and about 12,000 "r" the appearance does not vary greatly, but the reaction period lengthens with

A Dosage System for Gamma Ray Therapy 611

increase in dose, and the duration of the reaction may be taken as an index of dose. Most areas in the mouth will tolerate doses of as much as 12,000 "r" (applied as a surface dose only), and return ultimately to normal. Over 12,000 "r" there is a very real liability to the formation of permanent necroses.

Other Tissues

It is scarcely as easy to be definite about other tissues. It would appear, however, that vaginal mucosa tolerates fully as much as buccal mucosa, possibly rather more. The tolerance of the interior of the uterine cavity is phenomenal, and certainly doses of at least 20,000 "r"—30,000 "r" can be successfully given at the surface of an ordinary intra-uterine tube. Data *re* rectal mucosa are less easily obtained, but the tolerance is lower than that of vaginal mucosa. The limit of applicable dose is estimated to be under 10,000 "r." The tolerance of the mucous membrane of the œsophagus, as estimated from its response to intra-œsophageal radium bougies, is about the same level.

Epithelioma

In assessing the response of malignant tissue to radiation it is no longer possible to consider merely surface dose, but it becomes necessary to attempt to assess the minimum dose received at any part of the tumour, as obviously this can be the only measure of the lethal or sublethal dose. This is possible either with superficial lesions, in which the dose level as assessed at the skin is also that at the base of the tumour, or with deeper lesions where double moulds are used, allowing through and through dosage, or by assessing "depth" dose. It has been possible to accumulate considerable information about the response of epitheliomata in various sites of the body to treatment by measured dose. It has been found that, *subject to the proviso regarding duration of exposure* already discussed, a dose of 6,000 "r," if delivered to the whole of a tumour and tumour-bearing zone, causes permanent resolution of the great majority of epitheliomata. This figure may, therefore, be taken as a serviceable *in vivo* "tumour-lethal" dose. The absolute minimum lethal dose is probably slightly under that figure. Experience with skin and normal tissue tolerance appears to show that this dose can be given in some way or other in the majority of normal situations to the whole of an accessible tumour of moderate size without destruction of normal tissue.

Other Tumours

Data regarding other tumours are not so extensive. Rodent ulcer (basal-cell carcinoma) if previously untreated is, of course, more sensitive than epithelioma, but at the eight day period is not as strikingly sensitive as is

612   *Ralston Paterson and H. M. Parker*

usually believed. Disappearance occurs after 3,000—4,000 "r," but for permanent response a dose in excess of 5,000 "r" appears necessary. Breast malignancy is a problem by itself. Sensitivity may be assessed from the treatment of recurrent nodes in skin; a striking response is frequently obtained with doses as low as 4,000 "r," and a high percentage of responses occurs if doses between 5,000—6,000 "r" are employed. The already accepted beliefs regarding the resistance of certain growths, particularly adeno-carcinomata, and the sensitivity of other groups of tumours, have been maintained, but it is not possible, as yet, to obtain sufficient data regarding exact lethal dose levels.

PART II

By H. M. Parker, M.Sc.

(Received June 7, 1934)

Section A

Calculation of the Intensity of the γ-radiation from Radium, in terms of the "r" unit

The derivation is based on the figure for the total number of ions produced in air by the γ-radiation from 1 gram of radium in equilibrium with its decay products. This value has been the subject of careful experimental work by Rutherford[5] and his collaborators, and is given as $2 \cdot 13.10^{15}$ pairs of ions per second (L. H. Gray). This leads to a value of 8·4 "r" per hour for the intensity at 1 cm. from a point source of 1 milligramme of radium, filtered by 0·5 mm. platinum. More recent determinations, quoted in a paper by Mayneord and Roberts,[6] to which reference should be made, agree closely with the value that we had accepted.

Section B

Distribution Rules

The rules are based on the assumption that the γ-radiation from a point source strictly obeys an inverse square law. They therefore ignore such factors as change in quality of radiation and scattering and absorption in air, which almost certainly have no serious effect under practical conditions. The radioactive sources discussed are, therefore, the mathematical abstractions of linear sources with line density of ρ mgs. per cm. The special case of a ring source of this nature has been previously investigated by Sievert,[2] Souttar,[7] and Mayneord,[4] and that of a rectilinear source by Sievert,[2] and Mayneord.[4]

A Dosage System for Gamma Ray Therapy 613

The general problem is to state rules for the distribution of sources over a particular surface (the applicator) such that the intensity of radiation over a second surface (the treated area) at a constant distance from it is uniform and calculable. For the purposes of this paper, "uniform" or "homogeneous" radiation is to be understood in the restricted sense of suffering variations of not more than ± 10 per cent. The problem appears to be soluble with the following definition of the relation between applicator and treated area:—

Let A (Fig. 6) represent any surface (*e.g.* the skin). Let C be any selected region (the treated area) on the surface A. Through any point P on A draw the normal PP^1 of constant length h. As P moves over the surface A, P^1 will define a second surface A^1, and in particular, the second region C^1 as P moves over C. Then C^1 represents the surface applicator and h is the constant distance between applicator and treated area.

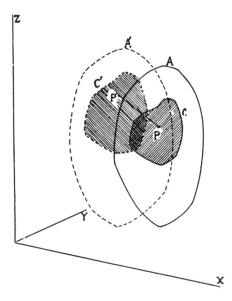

Fig. 6.
The relation between a treated area C and its surface applicator C^1 at a distance h.

Let the figure be referred to a set of rectangular axes X, Y, Z, so that the coordinates of P and P^1 are (Px, Py, Pz) and (Px^1, Py^1, Pz^1). Let there be $M_1, M_2, M_3, \ldots$ mg. of radium at the points (x_1, y_1, z_1) (x_2, y_2, z_2) etc. Then the intensity at any point (x, y, z) on C is:

$$I = \Sigma \frac{M_1}{(x - x_1)^2 + (y - y_1)^2 + (z - z_1)^2} \text{ over all the active points.}$$

Also $h = [(Px - P^1x)^2 + (Py - P^1y)^2 + (Pz - P^1z)^2]^{1/2}$

Now suppose we choose a new applicator and treated area referred to axes X^1, Y^1, Z^1, and fulfilling the conditions

$$\left.\begin{array}{l} x^1 = sx \\ y^1 = sy \\ z^1 = sz \end{array}\right\} \text{ and } \left\{\begin{array}{l} M_1^1 = s^2 M_1 \\ M_2^1 = s^2 M_2 \\ \ldots = \ldots \end{array}\right.$$

for every point of the figure.

Then the intensity at the corresponding point of the new treated area is:

$$I^1 = \Sigma \frac{M_1^1}{(x^1 - x_1^1)^2 + (y^1 - y_1^1)^2 + (z^1 - z_1^1)^2} \text{ over all points.}$$
$$= I \qquad \qquad \text{from the given relations.}$$

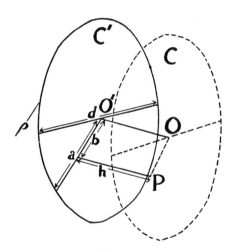

Fig. 7.
Ring applicator and treated area.

Also $h^1 = sh$
for all pairs of corresponding points, and therefore the distributions of intensity are identical. It follows that we need only calculate the distribution for one particular value of h, and the actual intensity for any other value may be found by taking into account the scale-factor s.

Stage 1.

Plane Circular Applicators

If the surface A is plane and C is a circle, the applicator C^1 becomes a plane circle of the same size. The scale-factor relation reduces to the simple form that the distribution of intensity over C is determined solely by the ratio d/h where d is the diameter of the circle.

A Dosage System for Gamma Ray Therapy   615

(*A*) If the radium is arranged as a ring of line density ρ around the periphery of C^1 the calculation of the intensity at any point P is made from the equation:

$$I_p = \frac{2\pi a \rho}{[(a^2 + b^2 + h^2)^2 - 4a^2b^2]^{1/2}} \text{ (Sievert}^2\text{)}.$$

Where a = radius of ring (Fig. 7).
       h = normal distance of P from plane of ring.
       b = distance from foot of normal from P on the plane of the ring.
       = distance of P from centre of C.

$$I_p = \frac{I_0}{[(a^2 + b^2 + h^2)^2 - 4a^2b^2]^{1/2}}$$

Where I_0 = intensity at 1 cm. from a point source of $2\pi a \rho$ mgm., the total amount of radium in the ring.

For particular values of a and h, the variation of intensity across any diameter of C is a function of b, and by differentiation we have

$$\frac{dI}{db} = \frac{2b[(a^2 - h^2) - b^2]}{[(a^2 + b^2 + h^2)^2 - 4a^2b^2]^{3/2}}$$

The denominator is essentially positive. Therefore, to find maximum and minimum values of I we need only consider the sign of the numerator.

Case 1. $a < h$

     $[a^2 - h^2 - b^2]$ is negative and never zero for real values of b.

$$\frac{dI}{db} = 0 \text{ if } b = 0, \text{ is positive if } b < 0, \text{ and negative if } b > 0$$

(regarding b as positive in one sense from O and negative in the other.)

I, therefore, has a maximum value at $b = 0$, and no other turning point. Hence the minimum intensity over the treated area C occurs at $b = a$.

$$\text{At } b = 0, \; I = \frac{I_0}{a^2 + h^2}$$

$$\text{at } b = a, \; I = \frac{I_0}{h(4a^2 + h^2)^{1/2}}$$

The variation of intensity is greatest when the ratio R of these two limits is greatest.

$$R = \frac{h(4a^2 + h^2)^{1/2}}{a^2 + h^2} = \frac{(1 + 4p^2)^{1/2}}{1 + p^2} \quad \text{where } p = \frac{a}{h}$$

$$\frac{dR}{dp} = \frac{2p(1 - 2p^2)}{(1 + p^2)^2 (1 + 4p^2)^{1/2}}$$

whence R has maximum value if $p = \dfrac{1}{\sqrt{2}}$

The variation is then $\pm 7\cdot2\%$.

Hence, subject to our definition of "uniform," all rings with $d/h \leqslant 2$ give "uniform" intensity over the treated area.

The variation for other values is:

d/h .. =	0	·5	1	1·5	2
Variation.. =	0	±3%	±5%	±6·75%	±5%

Case 2. $a > h$

$\dfrac{dI}{db} = 0$ if $b = 0$ and this is now a minimum turning point.

Also $\dfrac{dI}{db} = 0$ if $b = \pm \sqrt{a^2 - h^2}$

and these are maximum turning points, symmetrical about the origin.

The maximum intensity over the treated area occurs at $b = \sqrt{a^2 - h^2}$ and is $\dfrac{1}{2ah}$.

The minimum occurs either at $b = 0$ or $b = a$, according as

$$\dfrac{1}{h^2(1+p^2)} \quad < \text{ or } > \quad \dfrac{1}{h^2\sqrt{1+4p^2}}$$

Hence the minimum value is at $b = a$ if $p < \sqrt{2}$ and at $b = 0$ if $p > \sqrt{2}$.

If $p < \sqrt{2}$ the variation between the limits $\dfrac{1}{2ph^2}$ and $\dfrac{1}{h^2\sqrt{1+4p^2}}$ is readily shown to decrease as p increases. It has its greatest value ± 5 per cent for $p = 1$, and its least of ± 3 per cent. at $p = \sqrt{2}$. All such circles are therefore satisfactory. If $p = \sqrt{2}$ the circle is the one which Mayneord has already emphasised as a particularly good arrangement. If $p > \sqrt{2}$ the divergence between the limits $\dfrac{1}{2ph^2}$ and $\dfrac{1}{h^2(1+p^2)}$ steadily increases as p increases.

(B) If $p > \sqrt{2}$ the intensity has a minimum value at the centre of the treated area, and it becomes possible to improve the distribution by adding a certain amount of radium to the centre of the ring. Suppose we place $k.2\pi a\rho$ mgm. at the centre. Then the combined intensity at a point of C is:

$$\frac{kI_0}{h^2 + b^2} + \frac{I_0}{[(a^2 + b^2 + h^2)^2 - 4a^2b^2]^{1/2}}$$

There exists a variety of methods for determining that value of k which makes this expression most nearly constant over the range $b = 0$ to $b = a$, but the following is perhaps the simplest.

To make equality at $b = 0$ and $b = a$,

$$\frac{1}{1+p^2} + k = \frac{1}{\sqrt{1+4p^2}} + \frac{k}{1+p^2} \text{ or } k = \frac{1+p^2-\sqrt{1+4p^2}}{p^2\sqrt{1+4p^2}}$$

To make equality at $b = 0$ and $b = \sqrt{a^2 - h^2}$ (max. point).

$$\frac{1}{1+p^2} + k^1 = \frac{1}{2p} + \frac{k^1}{p^2} \text{ or } k^1 = \frac{p(p-1)}{2(p+1)(p^2+1)}$$

e.g. if $p = 2$, $k = 0.053$ and $k^1 = 0.067$

This case is illustrated in Fig. 8, in which curve A shows the intensity due to the ring alone,

Curve B — ring and centre spot with $k = 0.053$.

Curve C — ring and centre spot with $k^1 = 0.067$.

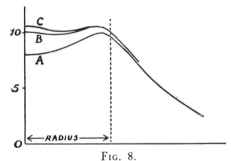

Fig. 8.
Distribution of intensity along a radius of a circular treated area.
A. Due to ring applicator with $d/h = 4$.
B. Ring + centre spot with $k = 0.053$.
C. Ring + centre spot with $k = 0.067$.

B shows a variation of ± 4 per cent, C of ± 2 per cent.

Curves of this type show at a glance how the intensity varies across the treated area, and the best curve can easily be obtained by trying values over the range k to k^1. For $p = 2$, the value $k^1 = 0.067$ is also the best value of k.

The percentage of the total radium to be placed at the centre is

$$x = \frac{100\,k}{1+k}$$

618 Ralston Paterson and H. M. Parker

The following table is drawn up in this manner:

d/h =	3	4	5	6
x =	4%	6%	6·5%	6·5%
Percentage variation =	±3·5%	±2%	±5·5%	±8%

Over the range $d/h = 3$ to < 6 we may set $x = 5$ per cent and remain within our limit of ± 10 per cent.

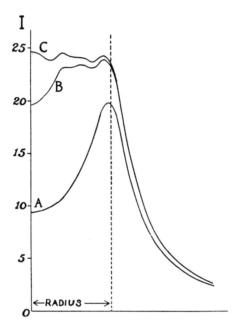

Fig. 9.
Distribution of intensity along a radius of a circular treated area.
    A. Ring applicator with $d/h = 8$.
    B. Outer ring + inner ring.
    C. Outer ring + inner ring + centre spot.

(C) If $d/h = 6$ the best value of centre spot produces a distribution with intensity variation ± 8 per cent, and for any larger value of d/h the variation is greater. These cases can be improved by the addition of a further ring of radium of half the previous diameter. The amounts to be placed at the centre and in the inner ring are obtained approximately by calculating the points $b = 0$, $b = a/2$, $b = \sqrt{a^2 - h^2}$ for equality. Then by inspection of the resultant graph it is possible to decide whether small variations from the provisional values are likely to improve the distribution. Fig. 9 refers to the case $d/h = 8$. Curve A represents the intensity along a radius for the outer ring alone. Curve B for the inner and outer rings, and Curve C the final intensity. The variation is only ± 2 per cent.

The following distributions are obtained in this way:

d/h .. .. =	6	8	10	12	14
% of radium in centre =	$3\frac{1}{2}$%	$2\frac{1}{2}$%	3%	3%	3%
% of radium at inner ring .. .. =	16%	$23\frac{1}{2}$%	$26\frac{1}{2}$%	28%	28%
% variation .. =	±$2\frac{1}{2}$%	±2%	±4%	±$7\frac{1}{2}$%	±10%

This permits the simplification made in the mould rules (Part. I).

Circles with $d/h \geqslant 14$ are not satisfactory under this method, and they are not required in practice. For the completeness of the dosage charts discussed later the range has been extended to $d/h = 20$ by the use of three rings and a centre spot.

Note.—An alternative solution of the plane circular applicator has been given by Mayneord[1] on the basis of a combination of a radioactive disc and peripheral ring. It is interesting that for equal areas both methods require the same radium content to produce the same intensity. This is probably a proof that both methods give satisfactory homogeneity. In a sense, the rings and centre spot may be regarded as a practical approximation to an ideal radioactive disc and peripheral ring.

Stage 2
Plane Rectangular Applicators

Isodose curves of the intensity in a plane 1 cm. from a rectilinear source of line density ρ and of lengths from 1 to 30 cm. have been constructed from Sievert's tables. The intensity from four lines arranged as a rectangle is then obtained by summation of effects over the treated area. A radioactive square of side $a = 2h$ gives uniform distribution analogous with the ring of $d/h = 2\sqrt{2}$. Rectangles of sides a and b ($b > a$) give sufficient homogeneity with $a = 2h$ and $b \leqslant 4a$. In rectangular treated areas the intensity in the extreme corners only falls below the general level and is excluded from the permissible ± 10 per cent variation.

620 *Ralston Paterson and H. M. Parker*

The simple rule for larger rectangles is to add radioactive lines parallel to the longer side with density $\frac{1}{2}\rho$ for one added line, and $\frac{2}{3}\rho$ for more than one line. The rule is based on the analysis of forty different rectangles at 1 cm. distance. Three typical cases are shown in Tables I, II and III.

TABLE I

Square 2 cm. × 2 cm. at 1 cm. ∴ *no added lines. Variation* = ± 5 *per cent excluding corners.*

	0·0	·25	·5	·75	1·0	1·25	1·5	1·75	2·0
0	*(291)*	*(300)*	325	330	335	330	325	*(300)*	*(291)*
·25	*(300)*	320	334	342	353	342	334	320	*(300)*
·5	*325*	334	350	355	349	355	350	334	*325*
·75	*330*	342	355	344	344	344	355	342	*330*
1·0	*335*	353	349	344	344	344	349	353	*335*
1·25	*330*	342	355	344	344	344	355	342	*330*
1·5	*325*	334	350	355	349	355	350	334	*325*
1·75	*(300)*	320	334	342	353	342	334	320	*(300)*
2·0	*(291)*	*(300)*	325	330	335	330	325	*(300)*	*(291)*

(horizontal axis: cm. →; vertical axis: cm. ↓)

TABLE II

Rectangle 16 cm. × 4 cm. at 1 cm. distance ∴ *1 added line of density $\frac{1}{2}\rho$.*
Variation = ± 6 *per cent excluding corners.*

b \ a	0	·5	1	1·5	2	2·5	3	3·5	4
0	*(343)*	385	401	416	422	416	401	385	*(343)*
1	*377*	393	395	402	411	402	395	393	*377*
2	*403*	391	379	377	*389*	377	379	391	*403*
3	*401*	397	375	375	*379*	375	375	397	*401*
4	*408*	391	377	371	*384*	371	377	391	*408*
5	*408*	397	371	373	*383*	373	371	397	*408*
6	*410*	398	373	373	*386*	373	373	398	*410*
7	*412*	409	386	382	*386*	382	386	409	*412*
8	*416*	408	387	382	*393*	382	387	408	*416*
9	*412*	409	386	382	*386*	382	386	409	*412*
10	*410*	398	373	373	*386*	373	373	398	*410*
11	*408*	397	371	373	*383*	373	371	397	*408*
12	*408*	391	377	371	*384*	371	377	391	*408*
13	*401*	397	375	375	*379*	375	375	397	*401*
14	*403*	391	379	377	*389*	377	379	391	*403*
15	*377*	393	395	402	411	402	395	393	*377*
16	*(343)*	385	401	416	422	416	401	385	*(343)*

Figures in italic fall directly below active lines.

A Dosage System for Gamma Ray Therapy

TABLE III

Square 16 cm. × 16 cm. at 1 cm. ∴ 7 added lines each of line density ⅔ρ.
Variation = ± 10 per cent excluding corners:

⟶ cm.

	0	1	2	3	4	5	6	7	8
0	(*405*)	*473*	*532*	*546*	*561*	*550*	*569*	*559*	*563*
1	*464*	*476*	*524*	517	*543*	530	*553*	541	*567*
2	*471*	*474*	*512*	493	*531*	508	*540*	519	*546*
3	*471*	*468*	*497*	484	*517*	500	*529*	507	*535*
4	*480*	*465*	*505*	478	*516*	494	*528*	500	*533*
5	*477*	*465*	*493*	481	*518*	492	*524*	497	*526*
6	*480*	*471*	*509*	486	*526*	495	*533*	499	*533*
7	*485*	*469*	*508*	483	*523*	494	*534*	504	*542*
8	*490*	*480*	*512*	500	*527*	504	*539*	516	*545*

This table shows the intensity over ¼ of the square.
The remainder is obtained by mirror-reflection about the row and column for 8 cm.

Indefinitely elongated rectangles cannot conform to the general rule since the limiting case of a single line of radium exhibits a marked fall of intensity opposite the ends, unless we exclude the ends as the logical extension of the excluded corners. The rule is quite satisfactory up to an elongation of 4 : 1 (*vide* example 16 cm. × 4 cm.), and is still within the limits ± 10 per cent at a ratio 6 : 1. Such strips, however, are not required in practice.

It is worthy of note that the rectangular applicators have a certain amount of latitude in respect of the amount of radium to be placed in the added lines. Thus in many cases the strength may be reduced from ⅔ρ to ½ρ without serious loss of accuracy. For example, the 16 cm. × 16 cm. square at 1 cm. has a variation of ± 12 per cent instead of ± 10 per cent if the added lines have strength ½ρ, and for the range of 4—6 added lines it is practically immaterial which strength is used. This feature is of some value with a limited stock of radium.

Stage 3

The General Plane Applicator

An important fact arises from a comparison of the actual intensity in "r" per milligramme-hour from circular and rectangular applicators of the same area. Table IV shows such a comparison for circles, squares, and rectangles, having ratio of sides 2 : 1. Circles and squares agree exactly and the rectangles diverge only slightly. It follows that, provided the radium has been distributed correctly, the intensity depends on the *area* of the applicator rather than on the shape. This fact, coupled with the established flexibility of the rectangle rules, suggests a tentative method of distribution of radium for any irregular plane area, *viz.*, outline the periphery with a linear source of constant density ρ, and add lines parallel to the longer direction to divide the area into strips of width $2h$, following the rectangle rule.

TABLE IV

Comparison of Square and Circular Applicators at 1 cm. skin distance.

AREA	Square r/mg.hr.	Circle r/mg.hr.
1	4·95	5·00
4	2·92	2·86
16	1·46	1·43
36	0·926	0·952
64	0·654	0·666
144	0·385	0·385

Comparison of Rectangular (2:1) and Circular Applicators.

AREA	Rect. r/mg.hr.	Circle r/mg.hr.	% diff.
2	4·00	4·16	4%
8	1·89	2·00	6%
32	0·971	1·01	4%
72	0·581	0·613	6%
128	0·403	0·406	1%

It is scarcely to be expected that the generalisation will in all cases lead to homogeneity as good as that achieved by the regular distributions. Nevertheless, the risk of serious error must be quite small, and a number of cases measured by condenser chambers of the Sievert type has given reasonably good results. Figures which are unduly elongated or which have sharp re-entrant angles are not included in the generalisation.

Stage 4
Extension to Curved Surfaces
A. Convex Surfaces

Suppose we have a treated area C, originally plane, and an applicator C^1 of equal area at a constant distance h. If the treated area is bent to present a convex surface to the applicator, in such a way that the treated area and h remain constant, two effects will be noticed:—(1) The folding of the applicator about the treated area tends to increase the intensity by a "cross fire" effect;

A Dosage System for Gamma Ray Therapy 623

(2) The area of the applicator increases by the definition of the relation between C and C^1. Hence the same amount of radium is extended over a larger surface and the intensity over the treated area will be reduced. These two effects act in opposite senses, and the possibility arises that the distribution and actual intensity will remain sufficiently constant whether the treated area be plane or convex. It is only possible to test the hypothesis by the examination of a number of examples. The appropriate analysis for the bending of a rectangular applicator to cylindrical curvature is given below:—

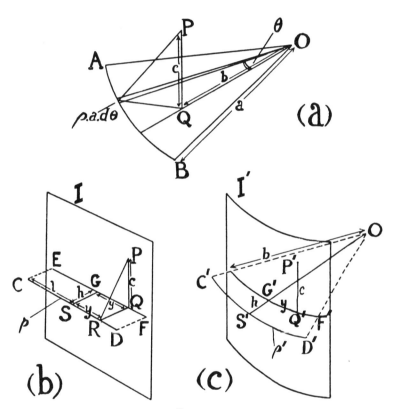

Fig. 10.
(a) Diagram for the calculation of intensity from radium distributed uniformly along the arc AB.
(b) and (c) Curvature of a linear source about a cylinder.

Consider first the intensity of gamma radiation from an unfiltered line source, line density ρ in the arc of a circle AB. (Fig. 10a).

Let O be the centre of curvature of AB.

P is any point in space and PQ of length c is normal to the plane ABO

The intensity at P is: $\displaystyle Ip = \int_A^B \frac{\rho.a.d\theta}{a^2 + b^2 + c^2 - 2ab\cos\theta}$

$$= \frac{2\rho.a}{\{[(a+b)^2 + c^2][(a-b)^2 + c^2]\}^{1/2}} \left[tan^{-1}\left(x_A \sqrt{\frac{(a+b)^2 + c^2}{(a-b)^2 + c^2}}\right) \right.$$
$$\left. + tan^{-1}\left(x_B \sqrt{\frac{(a+b)^2 + c^2}{(a-b)^2 + c^2}}\right) \right] \quad \ldots\ldots\ldots\ldots\text{I}$$

where $x_A = tan\dfrac{A\hat{O}Q}{2}$, $x_B = tan\dfrac{B\hat{O}Q}{2}$.

by standard integration.

Next consider a straight line source CD (dens: ρ) of length $2l$ parallel to the plane I and at distance h from it. (Fig. 10b). P is any point on the plane, PQ is $\perp'$ CD and RQ is normal to the plane. S is the medial point of CD. $SR = y$, $RQ = h$ and $PQ = c$. EGF is the projection of CD on the plane. The intensity at P is:

$$Ip = \frac{\rho}{h}\left[tan^{-1}\frac{l+y}{h} + tan^{-1}\frac{l-y}{h} \right] \quad \ldots\ldots\ldots\ldots\text{II}$$

In Fig. 10c the line has been bent to the arc of a circle so that plane I has a radius of curvature of b and centre O.

$E^1F^1 = 2l$, $C^1E^1 = S^1C^1 = D^1F^1 = h$, from the relation between applicator and treated surface. The new line density $\rho^1 = \dfrac{b}{b+h}\cdot\rho$ keeping total amount of radium constant.

Writing $a = b + h$ and $\rho^1 = \dfrac{b}{b+h}\cdot\rho$ in equation I.

$$Ip^1 = \frac{2b\rho}{\{[(2b+h)^2 + c^2][h^2 + c^2]\}^{1/2}} \left[tan^{-1}\left(x_A \sqrt{\frac{(2b+h)^2 + c^2}{h^2 + c^2}}\right) \right.$$
$$\left. + tan^{-1}\left(x_B \sqrt{\frac{(2b+h)^2 + c^2}{h^2 + c^2}}\right) \right] \quad \ldots\ldots\ldots\text{III}$$

where $x_A = tan\left(\dfrac{l+y}{2b}\right)$, $x_B = tan\left(\dfrac{l-y}{2b}\right)$

Now any rectangle can be progressively curved about a cylindrical surface, and the intensity of each component part of the applicator may be calculated from Equation II or III according to whether the part is $\|'$ or $\perp'$ to the

A Dosage System for Gamma Ray Therapy 625

axis of the cylinder. The equations are of such form that ρ, l, h, y, and c are constant for a particular point P. The parameter b determines the degree of curvature. In addition, y and c vary for different points of the treated area. Apparently, rectangles may be bent in this way almost to a semi-cylindrical form before the alteration in distribution and dosage becomes severe. Some typical results are set out in Table V. The calculations for circular applicators applied to surfaces of spherical curvature are easily made, and show that the curvature may be extended almost to the hemispherical form. More irregular curvatures are not amenable to mathematical discussion, but it is reasonable to assume that the compensation will be satisfactory over a considerable range.

TABLE V

Folding of a treated area 6×6 cm. about a cylinder. $h = 1$ cm.

Radius of curvature, b, in cms. =	∞	10	5	3	2
Angle subtended at centre =	0°	34·4°	68·8°	114·6°	171·9°
Average intensity referred to plane as 1·00 .. .. =	1·00	·98	·97	·97	·97
Ratio of central intensity to peripheral intensity .. =	1·00	1·00	1·01	1·02	1·05

B. Concave Areas

If the treated area is curved in the concave sense, the applicator diminishes in size. It may be shown as above that if the area of applicator be taken as a constant factor, the curvature will then produce a negligible effect on distribution and dosage over the treated area. In this case the rule holds good up to the limiting cases of a point source for spherical curvature and a rectilinear source for cylindrical curvature, when the distance h becomes the radius of curvature.

Section C

Dosage Charts

The distribution rules which are derived in the previous section are valid irrespective of the system of units in which the γ-radiation is measured. This follows at once since the distribution is concerned only with the relative variation of intensity over an irradiated region, and not with absolute values.

The actual intensity at any point of a treated area may be computed in terms of the "r" unit by introducing the result of Section A for the intensity at 1 cm. from a point source. In general it is more convenient to know the amount of radium and the required time (*i.e.* the number of milligramme-hours) to produce a given dose (intensity × time), and the results are therefore expressed in this form.

1. *Area Charts* (Charts 1 and 2).

When radium is arranged on an applicator to give uniform intensity over the treated area, it becomes possible to assign a meaning to the intensity and hence to the dose delivered to the treated area. For clinical convenience the dosage is incorporated in a series of graphs of area (ordinate) against "milligramme-hours to produce 1,000 r" (abscissa) with the parameter h for the family of curves. The basic points are obtained from a consideration of the series of circular applicators at $h = 1$. The intensity used is the mean value (by integration of annuli) over the treated circle. It has already been shown that square areas will fall exactly on the same graph. Also rectangles and irregular areas may be assessed from the same chart provided that a small correction is added for elongation as follows:—

Elongation	2 : 1	3 : 1	4 : 1
Percentage mgm.-hours to be added..	5%	9%	12%

From the previous section it is, moreover, apparent that the dosage over curved surfaces is given by the same graph, with the proviso that treated area is read for a convex surface, and area of applicator for a concave surface. Hence the dose for any treated area may be obtained from the single dosage chart provided that due attention is paid to the correct arrangement of the active sources. From the graph for $h = 1$, the family of curves for convenient values of h is directly obtained by means of the "scale-factor" previously described.

Filtration

The mathematical basis is restricted for simplicity to unfiltered sources. To make the charts of clinical value provision is made for the calculation appropriate to standard filtrations from 0·5 mm. to 2·0 mm. of platinum. This is effected by allowing for the general reduction of intensity due to the absorption in platinum which has been assumed to have a constant absorption coefficient $\mu = 2·0$. No account is taken of angled filtration. This is an effect which cannot easily be estimated for the compound applicators, and it is believed to be small, as the intensity at any point of the treated area may

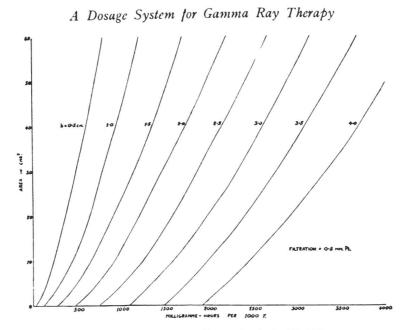

CHART 1. DOSAGE GRAPHS—SMALL AREAS
Giving dosage for areas up to 60 sq. cm. at distances from ½—4 cm.

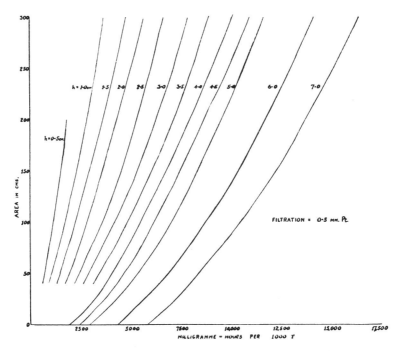

CHART 2. DOSAGE GRAPHS—LARGE AREAS
Giving dosage for areas up to 300 sq. cm. at distances from ½—7 cm.

628 *Ralston Paterson and H. M. Parker*

be regarded as the summation of a large number of contributions of which only a few can be simultaneously reduced to any extent by oblique filtration.

2. *Charts for Rectilinear Sources.* (Charts 3 and 4.)

The graphs in this case are developed from a direct conversion of Sievert's tables to the "r" unit. They indicate the dose opposite the mid-point of the line at the stated distance h. Unfortunately the effect of oblique filtration for a single line source is not a negligible factor. There is some difficulty in publishing these charts without loss of accuracy on the one hand, or introducing too many complications for the clinician on the other. It is hoped that the compromise adopted by the authors will be of general value.

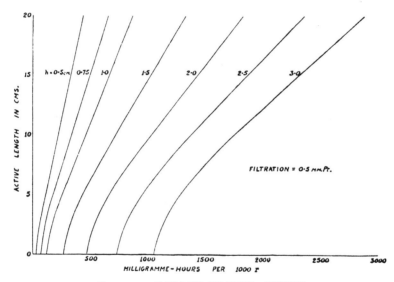

CHART 3. DOSAGE GRAPHS—TUBES
(·5 platinum filter) giving dosage at the surface of tubal applicators of lengths up to 20 cms. and for radii from ½ to 3 cm.

The charts are given separately for filtrations of 0·5 mm. Pt. and 1·5 mm. Pt., the two values which are probably in most common use. Then it so happens that the values for 1·0 mm. Pt. are exactly the means of the corresponding values for 0·5 mm. and 1·5 mm. as in the appended table.

Active length in cm.	0	2	4	8	12	16	24	32
Actual mg.-hrs. to produce 1,000 "r"	131	172	250	445	645	850	1265	1680
Mean of values for 0·5 mm. and 1·5 mm.	132	172	251	443	643	852	1272	1687

A Dosage System for Gamma Ray Therapy 629

The dosage appropriate to 0·6 mm. and 0·8 mm. Pt. may be obtained with reasonable accuracy from the ordinary filtration correction. A department habitually using the charts at filtrations other than 0·5 mm. and 1·5 mm. would be advised to reconstruct the charts accordingly.

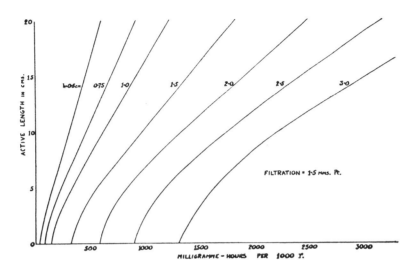

CHART 4. DOSAGE GRAPHS—TUBES
(1·5 platinum filter)—as Chart 3, but for the heavier filtration. For filtration 1 mm. of platinum the mean of the readings from Charts 3 and 4 should be taken.

Section D
Some Notes on the Practical Application of the Distribution Rules

The rules postulate the use of continuous rings and lines of radium. In the practical construction of applicators some approximation to ideal conditions has to be permitted, and it is clearly important to the constructor to know precisely which arrangements provide satisfactory approximations. General guiding principles are given below, but in the interests of brevity the proofs are omitted.

1. *A Ring of Radium of Diameter d at Distance h.*

(a) The most satisfactory arrangement is to outline the circumference with radium tubes or needles in such a way that the active lengths form a continuous polygonal figure. Six tubes forming a hexagon constitute the minimum permissible number. Tubes of different lengths may be used provided that no length exceeds $d/2$.

(b) The tubes may be arranged as an open polygon provided that the spacing between adjacent active ends of tubes does not exceed the distance h. Six tubes constitute the minimum requirement.

(c) Two types of tubes of equal length with line densities ρ and 2ρ may be used alternately to constitute a closed polygon, provided that the length of each tube does not exceed $2h$, or to form an open polygon with intervals of length h, provided that the length of a tube does not exceed h. Minimum number of tubes is four of each type.

(d) With point sources (*e.g.* radon seeds) the distance between adjacent points should not exceed h. Minimum number of points = 6.

d/h =	2	4	8	12
Number of point sources	6	12	24	36
Maximum variation from ideal	1%	1%	1%	1%

A spacing of $1 \cdot 5h$ may be used if necessary, introducing a variation of ± 5 per cent approximately.

The alteration in distribution of intensity in any of the arrangements (a), (b), (c), and (d) is only apparent in the immediate neighbourhood of the periphery of the treated area.

2. *Inner Ring and Centre Spot of the Compound Circular Applicators*

The inner ring contributes about half the total intensity to the regions in which its "ripple," due to departure from the ideal ring, might be serious. Consequently the variations tend to be smoothed out, and the spacing may be extended to $2h$ with a ripple of ± 5 per cent. Nevertheless, no fewer than six active foci should be used if possible. The "centre point" behaves as a point source if its active length does not exceed h.

3. *Rectangular Applicators*

(a) The straight lines which are the components of the rectangular applicators should preferably be made active throughout their length by overlapping the inactive ends of tubes in the plane of the applicator.

(b) Where necessary, an active line may be constructed by a number of tubes provided that the intervals between adjacent active ends do not exceed h. For two tubes constituting a line the inactive gap should not exceed $h/2$.

(c) Different strengths of tubes may be used alternately if the length of each does not exceed $2h$ and the number of tubes is not less than five.

(d) A long row of point sources is valid if the spacing is not more than h.

A Dosage System for Gamma Ray Therapy   631

3. *Single Line Sources*

The conditions in this case are more critical because there is no smoothing out of ripples by contributions from other parts of an applicator. A distance between adjacent active ends not exceeding $h/2$ is recommended. With an odd number of tubes (five or more) a spacing of h is just permissible.

Two tubes with an inactive gap form a bad approximation, and should be avoided.

In conclusion, Mr. H. M. Parker wishes to express his sincere thanks to the Manchester Committee on Cancer for the Research Grant which has enabled him to carry out this work.

SUMMARY

A practical system of dosage measurement is described, applicable to all forms of radium therapy other than certain types of interstitial implantation. The international "r" unit is accepted as a satisfactory unit of quantity of gamma radiation, and the Imc.Hr. of Sievert assessed as equal to 8·4 "r." Graphs are submitted showing for various distances the amount of radium required on any applicator to produce a dose of 1,000 "r" over any desired area. A system of "Rules" is also given, defining how, in the situations met with in actual clinical practice, radium must be distributed upon such applicators to produce a reasonably homogeneous radiation over the whole of a treated surface. The effects of certain dosages of gamma radiation on normal and malignant tissue are described. The lethal dose for squamous epithelioma appears to be 6,000 "r" delivered as continuous radiation for a period of eight days; normal healthy skin safely tolerates the same dose delivered over a like period, and normal mucosa considerably more than that. The physical and mathematical data on which the system is based are outlined.

ZUSAMMENFASSUNG

Die Verff. beschreiben ein praktisches System der Dosismessung, welches für alle Arten der Radiumtherapie ausser gewissen Typen der lückenhaften Implantation verwendbar ist. Die internationale Einheit "r" wird als geeignete Quantitätseinheit der gammastrahlung und das Imc. Hr. von Sievert gleich 8.4 "r" angenommen. Graphische Darstellungen zeigen für verschiedene Entfernungen die notwendige Menge Radium auf jedem Applikator, um eine Dosis von 1,000 "r" über jede gewünschte Reichweite hervorzubringen. Ein System von "Regeln" wird auch gegeben, welche definieren wie in Fällen, die in der klinischen Praxis vorkommen, der Radium auf solchen Applikators verteilt werden muss, um eine verhaltnissmässig gleichartige Bestrahlung über die

632   *Ralston Paterson and H. M. Parker*

ganze behandelte Fläche zu befördern. Die Verff. beschreiben die Wirkungen gewisser Dosis der Gammabestrahlung auf normalen und bösartigen Geweben. Die tötliche Dosis fur schuppige bösartige Geschwülste scheint 6,000 "r" zu sein, gegeben als ununterbrochene Bestrahlung für 8 Tage. Die normale gesunde Haut erträgt dieselbe Dosis während der gleichen Periode und normaler Schleim eine beträchtlich grössere Dosis. Dis physichalischen und mathematischen Grundlagen, worauf das System begründet ist, werden angezeigt.

Résumé

Les auteurs décrivent un système pratique de mensuration de dose, applicable à tout genre de Curiethérapie excepté certains types d'implantation interstitielle. Ils acceptent l'unité internationale "r" comme unité satisfaisante de radiation gamma, et estiment 1mc. Hr. de Sievert comme égal a 8.4 "r." Des graphiques montrent pour diverses distances la quantité de radium nécessaire à produire sur tout applicateur une dose de 1,000 "r" pour toute étendue désirée. Un systeme de "règles" détermine comment, dans les situations recontrées en pratique clinique, le radium doit être distribué sur des applicateurs pareils, pour produire une radiation plus ou moins homogène sur toute la surface traitée. On décrive les effets de certaines doses de radiation gamma sur les tissus normales et malins. La dose mortelle pour l'épitheliome squameux parait être 6,000 "r," donnée comme dose continuelle pendant huit jours; la peau normale tolère sans danger la même dose pour la même période, et le mucus normal tolère une dose plus grande. Les fondations physiques et mathématiques du système sont ebauchées.

REFERENCES

1. W. V. Mayneord, *Acta Radiologica*, xiv, 1933, p. 95.
2. R. Sievert, *Acta Radiologica*, i, 1921, p. 89, and xi, 1930, p. 249.
3. J. Murdoch, S. Simon, E. Stahel, *Acta Radiologica*, xi, 1930, p. 350.
4. W. V. Mayneord, *British Journal of Radiology*, v, 57, p. 677.
5. Rutherford, Chadwick and Ellis, *Radiations from Radioactive Substances*.
6. Mayneord and Roberts, *British Journal of Radiology*, vii, 1934, p. 158.
7. H. S. Souttar, *British Journal of Radiology*, iv, 1930, p. 681.

The Dependence of Back-Scattering

Acta Radiologica, Volume XVI

March 1935

EXCERPTUM

ACTA RADIOLOGICA

Redactores

T. DALE   L. EDLING   P. FLEMMING MØLLER   G. FORSSELL
Oslo   Lund   København   Stockholm

R. GILBERT   L. G. HEILBRON   J. W. S. HEUKENSFELDT JANSEN
Genève   Amsterdam   Amsterdam

S. A. HEYERDAHL   C. G. JANSSON   J. JUUL
Oslo   Helsingfors   København

H. R. SCHINZ   G. A. WETTERSTRAND
Zürich   Helsingfors

Editor

GOSTA FORSSELL
Stockholm

H. M. Parker

THE DEPENDENCE OF THE
BACK-SCATTERING OF ROENTGEN RAYS
IN A PHANTOM ON FOCAL DISTANCE,
QUALITY OF RADIATION AND
FIELD SIZE

Vol. XVI. Fasc. 6   N:o 94

Stockholm: *P. A. Norstedt & Söner*

THE DEPENDENCE OF THE BACK-SCATTERING OF ROENTGEN RAYS IN A PHANTOM ON FOCAL DISTANCE, QUALITY OF RADIATION AND FIELD SIZE[1]

by

H. M. Parker, M.Sc.

Holt Radium Institute, Manchester

Introduction

Although considerable attention has been devoted to the question of the percentage back-scatter of roentgen rays at the surface of a scattering medium, the more general problem concerning the amount of scatter at various points within the scattering body has been relatively neglected. As this problem is by no means without interest, both from the physical point of view, and in its practical applications to roentgen therapy, a continuation of the work of THORÆUS (1) has been considered justifiable. In accordance with recent tendencies in therapy the investigations have been extended to much harder radiations. The convenient method of presentation developed by THORÆUS has been followed in toto. The true primary intensity at the focal distance F is taken to be unity with the scattering phantom removed. Then the true primary intensity at a depth d is written $I_P \times \delta m$, where I_P is obtained by applying the inverse square law to the relation $I_P = 1$ at $d = 0$, and δm is the transparency of the absorbing depth d for the particular radiation used. Denoting the total measured intensity by I_T, the scattered radiation is $I_S = I_T - I_P \cdot \delta m$. The ratio I_S/I_T constitutes a convenient measure of the amount of scattered radiation in terms of the total radiation.

With these definitions, the values of I_S/I_T for the central axis of the irradiated zone have been tabulated for different focal distances, qualities of radiation, and field sizes. The completely general proposition involving

---

[1] Submitted for publication March 15th, 1935.

measurements off the central axis is too complex for treatment in a single paper.

Experimental Arrangements

A continuously evacuated roentgen tube of the type described by ALLIBONE and BANCROFT (2) was used as the source of radiation. The tube is provided with a gold anticathode and operates at constant potentials up to 250 K. V. at 10 m.a. The output is steady and reproducible within the limits of accuracy of measurement. The filtration effect of the steel wall of the tube sets a lower limit to the available quality at H.V.L. = 0.32 mm. Cu, consistent with the delivery of reasonable intensities. The upper limit, H.V.L. = 3.3 mm. Cu, is attained by a filter of (Tube wall + 1.05 mm. Sn + 0.4 mm. Cu + 1 mm. Al) at 250 K.V. As no radical differences were detected in the measurements at the same H.V.L. of radiations obtained by different combinations of voltage and filter, H.V.L. in Cu has been accepted provisionally as a sufficiently exact designation of quality.

The depth dose measurements involved were made with four standard commercial dosimeters. Three of the ionisation chambers were cylindrical and the fourth spherical, the diameters varying from 0.8 cm. to 1.8 cm. Parallelism of the chambers with the standard air-chamber over a very wide range of quality is claimed by the makers. No systematic difference was found in the responce of the four chambers in air between H.V.L. = 0.32 mm. Cu and 3.3 mm. Cu. Nevertheless, the measured depth doses were found to vary in a manner which could apparently be related to the chamber size. The figures finally chosen represent an attempted extrapolation to zero size. These results are to be published in detail elsewhere. The measuring phantom consisted of slabs of wax of unit density each 40 cm. × 40 cm. and of graded thicknesses totalling 35 cm. A preliminary comparison with measurements in a large water phantom ensured that no error was introduced in substituting the special wax for water. Wax was preferred mainly because two of the chambers were not suitable for extensive measurements in a fluid, but other advantages included more exact positioning at the zero level (no meniscus effect) and the possibility of measuring with the beam not in a vertical direction. The zero position was read with the chamber axis in the scattering surface (i. e. »half immersed»).

The measurements of the primary intensity in air were repeated with the anticathode, aperture, and chamber rotated as a whole through 90° on either side of the vertical position to direct the beam on to distant walls. The invariability of the intensity under these conditions established that no appreciable amount of scatter from surrounding objects was being included. Moreover, rapid alternations of the surface readings with and without back-scatter could be made by rotating the beam and resetting the chamber without moving the heavy phantom. Estimations of surface back-scatter with the four chambers were reasonably consistent. Circular fields of diameters 5—20 cm. and rectangular fields from 6×4 cm. to 15×20 cm. were used. In order to simulate therapeutic conditions as closely as possible applicators lined with lead, copper, and aluminium (to absorb the lead characteristic radiation) and closed at the proximal end with a thin layer of wood were used to define the beam. The percentage back-scatter at the surface and the depth doses proved to be smooth functions of the area of the field provided that rectangles of elongation greater than 3:2 were not included.

The half value layers in copper were measured with the smallest field, the distance between the chamber and the filters being at least 35 cm. The values of the transparency δm were taken from Thoræus and the tables of Grebe and Nitzge (3) for the harder radiations, the two sources agreeing well over the common range.

Experimental Results

1. The dependence of I_S/I_T at any depth on focal distance

From the depth dose and surface back-scatter at $F = 40, 60, 80$ and 100 cm. for different radiations, I_S/I_T at depths down to 12 cm. has been calculated by the Thoræus method. The results for H.V.L. $= 1.5$ mm. Cu and four fields are collected in Table 1. It is immediately apparent that for each setting of A and d, I_S/I_T is unaltered by the change in F. Similar results are shown in Table II for the extreme qualities to demonstrate that the independence of I_S/I_T on F is a general property. The physical interpretation of this observation is evidently that the scattering block acts on the beam which would exist without scatter in a constant manner. This operational effect may be turned to good account in the construction of complete isodose curves for the distribution of radiation throughout the scattering medium. The accurate determination of these curves is a laborious process, particularly if it is undertaken for a number of fields at various focal distances. It now becomes possible to deduce the complete set of curves from measurements at one focal distance, combined with a knowledge of the surface back-scatter for the fields used, by proceeding as follows:

From the set of measurements at focal distance F, I_T and I_P may be evaluated for any point in the phantom. For the corresponding point at the new focal distance F', I'_P may be written down at once, and I'_T

Table I

H.V.L. = 1.5 mm. Cu.

Depth cm.		F = 40 cm.			F = 60 cm.			F = 80 cm.			F = 100 cm.		
		I_T	I_S	I_S/I_T	I_T	I_S	I_S/I_T	I_T	I_S	I_S/I_T	I_T	I_S	I_S/I_T
A = 19.6 cm²	0	1.135	.135	.12	1.135	.135	.12	1.135	.135	.12	1.135	.135	.12
	1	1.07	.27	.25	1.08	.27	.25	1.09	.27	.25	1.10	.275	.25
	2	.93	.30	.32	.955	.295	.31	.965	.295	.31	.97	.30	.31
	4	.68	.26	.38	.715	.275	.38	.745	.285	.38	.75	.29	.38
	6	.49	.22	.45	.53	.24	.45	.55	.25	.45	.56	.25	.45
	8	.35	.17	.48	.39	.185	.49	.41	.20	.49	.42	.205	.49
	10	.26	.14	.54	.29	.15	.52	.305	.155	.51	.315	.155	.51
	12	.19	.11	.58	.205	.115	.56	.22	.12	.55	.225	.12	.55
A = 80 cm²	0	1.37	.37	.27	1.37	.37	.27	1.37	.37	.27	1.37	.37	.27
	1	1.32	.52	.39	1.34	.535	.39	1.36	.54	.39	1.365	.54	.395
	2	1.17	.54	.46	1.22	.56	.46	1.24	.57	.46	1.24	.57	.46
	4	.93	.51	.55	.99	.55	.555	1.01	.55	.55	1.02	.56	.55
	6	.73	.46	.63	.80	.51	.64	.805	.505	.63	.81	.50	.62
	8	.57	.39	.69	.64	.44	.69	.645	.435	.68	.65	.445	.68
	10	.42	.30	.71	.49	.35	.71	.50	.35	.70	.51	.35	.69
	12	.32	.24	.75	.37	.28	.75	.38	.28	.74	.39	.29	.74
A = 150 cm²	0	1.41	.41	.29	1.41	.41	.29	1.41	.41	.29	1.41	.41	.29
	1	1.41	.61	.43	1.405	.59	.42	1.41	.59	.42	1.41	.585	.42
	2	1.29	.66	.51	1.29	.63	.49	1.30	.63	.49	1.30	.63	.49
	4	1.05	.63	.60	1.08	.64	.59	1.10	.64	.58	1.12	.66	.59
	6	.835	.565	.68	.89	.60	.68	.92	.62	.68	.94	.63	.68
	8	.65	.47	.725	.71	.515	.73	.735	.525	.72	.74	.525	.72
	10	.51	.39	.76	.56	.42	.75	.57	.42	.74	.585	.425	.74
	12	.395	.315	.795	.43	.34	.795	.445	.345	.78	.45	.345	.77
A = 300 cm²	0	1.52	.52	.34	1.52	.52	.34	1.52	.52	.34	1.52	.52	.34
	1	1.52	.72	.47	1.52	.70	.46	1.52	.70	.46	1.52	.70	.46
	2	1.41	.78	.55	1.42	.76	.54	1.42	.75	.53	1.43	.76	.54
	4	1.19	.77	.65	1.23	.79	.64	1.23	.77	.63	1.26	.80	.64
	6	.97	.70	.72	1.03	.74	.72	1.03	.73	.71	1.05	.74	.71
	8	.79	.61	.77	.85	.65	.77	.85	.64	.76	.875	.66	.76
	10	.64	.52	.81	.68	.54	.80	.70	.55	.79	.72	.56	.79
	12	.51	.43	.84	.56	.47	.84	.57	.47	.83	.57	.47	.83

Table I. The independence of I_S/I_T on focal distances at H.V.L. = 1.5 mm. Cu. Values of δm and the depth doses can be calculated from the columns given.

may be calculated from the relation $\dfrac{I'_T}{I'_P} = \dfrac{I_T}{I_P} = \dfrac{\delta m}{(1 - I_S/I_T)}$ which is constant for corresponding points. The values of I'_T can then be translated into the new set of isodose curves. A complication is introduced, especially at large depths, by the change in the geometric width of the beam. Ex-

THE DEPENDENCE OF THE BACK-SCATTERING OF ROENTGEN RAYS

Table II

a) H.V.L. = 0.32 mm. Cu. $A = 19.6$ cm².

Depth cm.	F = 40 cm.			F = 60 cm.			F = 80 cm.			F = 100 cm.		
	I_T	I_S	I_S/I_T	I_T	I_S	I_S/I_T	I_T	I_S	I_S/I_T	I_T	I_S	I_S/I_T
0	1.235	.235	.19	1.235	.235	.19	1.235	.235	.19	1.235	.235	.19
1	1.21	.49	.40	1.23	.49	.40	1.23	.49	.40	1.235	.48	.39
2	1.06	.52	.49	1.08	.525	.49	1.10	.535	.49	1.12	.55	.49
4	.77	.46	.60	.79	.465	.59	.82	.48	.59	.84	.50	.60
6	.545	.365	.67	.58	.38	.66	.61	.40	.66	.625	.415	.665
8	.37	.26	.705	.415	.29	.70	.43	.30	.70	.45	.31	.69
10	.26	.19	.73	.30	.22	.73	.32	.23	.72	.335	.245	.73
12	.19	.15	.79	.22	.17	.775	.23	.18	.78	.24	.19	.78

b) H.V.L. = 3.3 mm. Cu. $A = 19.6$ cm².

Depth cm.	F = 40 cm.			F = 60 cm.			F = 80 cm.			F = 100 cm.		
	I_T	I_S	I_S/I_T	I_T	I_S	I_S/I_T	I_T	I_S	I_S/I_T	I_T	I_S	I_S/I_T
0	1.075	.057	.07	1.075	.075	.07	1.075	.075	.07	1.075	.075	.07
1	1.03	.215	.21	1.05	.215	.205	1.05	.215	.205	1.06	.22	.21
2	.925	.24	.26	.955	.25	.26	.97	.255	.26	.97	.25	.26
4	.68	.215	.32	.73	.235	.32	.75	.24	.32	.755	.235	.31
6	.50	.19	.38	.55	.21	.38	.575	.22	.38	.585	.22	.38
8	.375	.155	.415	.41	.17	.415	.44	.185	.42	.455	.185	.405
10	.28	.13	.465	.32	.15	.47	.34	.16	.47	.35	.16	.46
12	.21	.11	.52	.245	.125	.51	.26	.13	.50	.27	.135	.50

Table II. The independence of I_S/I_T on focal distance at the extreme qualities.

periment shows, however, that suitable interpretation of »corresponding points» obviates the difficulty. If the central axis and two intersecting axes at right angles in the surface be chosen as axes of co-ordinates, the point in the second system corresponding to (d, x, y) in the first becomes simply (d, kx, ky) where $k = \dfrac{F\ (F' + d)}{F'\ (F + d)}$. This represents, in general, quite a small correction. Outside the geometric beam I_P and I'_P are actually zero but the method may be applied with success by using symbolically the values that would exist with the limiting aperture removed.

The Tables I and II exhibit a small reduction in I_S/I_T for the greater depths as F is increased. This is presumably due to the reduction in the total irradiated volume. There is, moreover, a slight increase in I_S/I_T

Table III

Quality	I	II	III	IV	V	VI
H.V.L. mm. Cu.	0.32	0.8	1.5	1.8	2.54	3.3

$\dfrac{I_S}{I_T}$ at $A = 19.6$ cm^2

Depth cm.	I	II	III	IV	V	VI
0	.19	.195	.12	.105	.10	.07
1	.40	.35	.25	.24	.23	.21
2	.49	.41	.32	.285	.27	.26
4	.60	.48	.38	.375	.345	.32
6	.67	.54	.45	.44	.42	.38
8	.70	.58	.48	.48	.47	.415
10	.73	.63	.54	.53	.50	.465
12	.79	.66	.58	.57	.55	.52

$\dfrac{I_S}{I_T}$ at $A = 80$ cm^2

Depth cm.	I	II	III	IV	V	VI
0	.31	.305	.27	.21	.20	.18
1	.50	.40	.39	.355	.33	.315
2	.60	.53	.46	.41	.40	.38
4	.70	.61	.55	.53	.50	.46
6	.77	.68	.63	.60	.57	.54
8	.81	.73	.69	.66	.62	.58
10	.83	.775	.71	.705	.655	.63
12	.87	.805	.75	.74	.695	.675

$\dfrac{I_S}{I_T}$ at $A = 150$ cm^2

Depth cm.	I	II	III	IV	V	VI
0	.345	.335	.29	.26	.245	.23
1	.52	.49	.43	.41	.38	.36
2	.62	.57	.51	.475	.46	.43
4	.73	.67	.60	.58	.56	.52
6	.80	.73	.68	.66	.625	.60
8	.84	.77	.725	.71	.68	.645
10	.86	.81	.76	.76	.72	.70
12	.89	.84	.79	.79	.755	.745

$\dfrac{I_S}{I_T}$ at $A = 300$ cm^2

Depth cm.	I	II	III	IV	V	VI
0	.41	.39	.34	.32	.29	.275
1	.58	.535	.47	.46	.42	.41
2	.68	.62	.55	.52	.51	.48
4	.78	.715	.65	.63	.61	.58
6	.83	.77	.72	.71	.68	.66
8	.87	.82	.77	.765	.73	.71
10	.89	.85	.81	.805	.76	.755
12	.92	.88	.84	.84	.805	.80

Table III. The variation of I_S/I_T with quality of radiation.

at $d = 0$ as F increases, produced by the slightly increased I_P values at neighbouring points, but this only affects the third figure, which has not been included in the tables.

2. The variation of I_S/I_T at any depth with quality

The values of I_S/I_T at depths down to 12 cm. for four field sizes and six qualities of incident radiation are set out in Table III. It follows from the previous section that the results will apply equally to any focal distance. The table shows at once that there is a diminution in I_S/I_T at

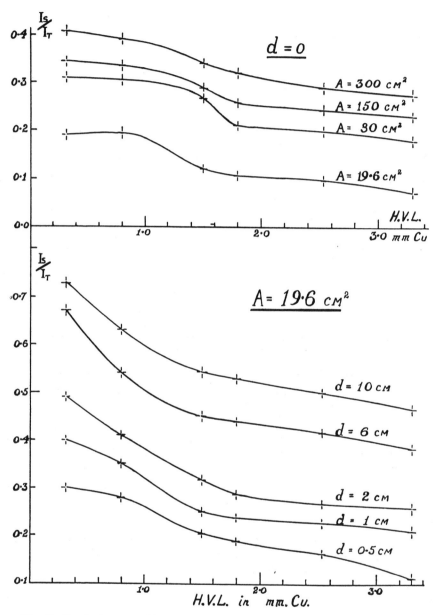

Fig. 1. Typical curves of the variation of I_S/I_T with quality of radiation.

any depth with increasing hardness of incident radiation, over the range considered. The mode of variation with H.V.L. is more readily seen from a graphical representation and a few typical curves are given in

Fig. I. All the curves for depths greater than 1 cm are of the same general shape, viz. a rapid fall of I_s/I_T between the qualities I and III with a subsequent slower diminution. This type of curve is to be expected qualitatively from the nature of the scattering processes involved. As the effective wave length of the beam decreases the scattering coefficients and the relative amounts scattered at different angles change in the direction of decreasing I_s/I_T much more rapidly at first than the gain due to increasing transparency of the phantom with respect to the scattered rays. The curves for $d=0$ seem to be of a different type and indicate rather a sharp fall of I_s/I_T between H.V.L. = 1.1 mm. Cu and 1.8 mm. Cu, with only slow changes in the other regions investigated. This special variation persists to a depth of about 0.5 cm. and is presumably due to the fact that the superficial layers can only collect scattered rays which have been sent back at angles greater than 90°. The dependence of this radiation on quality is, of course, different from that of the more general scatter at deeper points of the phantom.

3. The dependence of I_s/I_T on the field size

It is self evident that for each value of d, the back-scatter at a point on the central axis must increase as the field size A increases. The precise mode of increase may perhaps merit closer study and for this purpose it is convenient to consider I_s, rather than I_s/I_T. Some values for $d=0$ and $d=10$ are set out in Table IV. Equal increments of A have been chosen, representing approximately equal increments in the irradiated volume. Under these conditions the contributions to the scatter at the central axis should be given by a summation of terms of the type $\frac{\delta m}{r^2} f(\theta)$, neglecting multiple scattering. r is the distance of a scattering point from the axial point considered, δm the transparency of the scattered radiation along this path, and $f(\theta)$ takes account of the relative scattering power in this direction. As the field size is increased $\frac{1}{r^2}$ and δm will obviously decrease, and on the whole θ will tend more nearly to 90° giving a reduction in $f(\theta)$ over the range $\lambda e = 0.08$ Å to 0.24 Å. Consequently for equal increments in the scattering volume one would expect rapidly falling contributions to the total scatter. This is qualitatively the case at $d=0$ for Qual. VI, but for Qual. III the increments above $A=100$ are approximately constant, and for Qual. I, there is even some indication of a rise in ΔI_s above $A=150$. It is unwise to base too many deductions on the small experimental values of ΔI_s but it may be pointed out that the discrepancy between theory and experiment could be accounted for

THE DEPENDENCE OF THE BACK-SCATTERING OF ROENTGEN RAYS

Table IV

$d = 0$ cm.

Field size cm²	Qual. I.		Qual. III.		Qual. VI.	
	I_s	Increment ΔI_s	I_s	Increment ΔI_s	I_s	Increment ΔI_s
0	0	—	0	—	0	—
50	.37	.37	.30	.30	.155	.155
100	.47	.10	.38	.08	.25	.095
150	.515	.045	.41	.03	.31	.06
200	.575	.06	.45	.04	.325	.015
250	.625	.05	.485	.035	.335	.01
300	.695	.07	.515	.03	.34	.005

$d = 10$ cm.

Field size cm²	Qual. I.		Qual. III.		Qual. VI.	
	I_s	Increment ΔI_s	I_s	Increment ΔI_s	I_s	Increment ΔI_s
0	0	—	0	—	0	—
50	.29	.29	.24	.24	.20	.20
100	.375	.085	.33	.09	.285	.085
150	.43	.055	.39	.06	.345	.06
200	.48	.05	.435	.045	.39	.045
250	.53	.05	.48	.045	.425	.035
300	.58	.05	.52	.04	.46	.035

Table IV. The increments of I_s for equal increments of field size.

by the existence of multiple scattering. Experiments to determine the amount of multiple scatter are at present in progress.

Discussion

1. Physical

The values of I_s/I_T differ from those of some other writers, being for the most part higher and showing a steady decrease as the radiation hardens. These differences must be ascribed solely to the different values of the percentage back-scatter at $d = 0$, since the depth dose figures agree tolerably well. As regards the surface values the serious discrepancies between the results of various observers presumably arise from the different experimental conditions and types of ionisation chambers used.

An analysis of these effects is beyond the scope of the present paper. It is interesting to note, however, that Quimby (4) and collaborators, employing a widely different technique have obtained results in surprisingly close agreement with those of the present writer. Preliminary calculations based on scattering formulae also seem to support, qualitatively, the dependence of I_s/I_T on type of radiation derived in this series of experiments.

2. Applications to therapy

A number of points of interest in deep therapy arises from the measurements of I_s/I_T. In the first place the dose at a depth in the phantom receives a far greater contribution from back-scatter than might be supposed. Even with a field as small as $A = 19.6$ cm. and at a normal H.V.L. $= 1.5$ mm. Cu, scatter at 10 cm. deep constitutes 54 % of the total radiation. At $A = 300$ the contribution has risen to 81 %. It follows that in the application of the rays to the human body, the presence of bone masses or air cavities may have an appreciable influence on the dose delivered to a tumour as compared with that calculated from isodose charts. Hence extensive measurement of the radiation at various points of a cadaver or fairly complete models of the body might profitably be undertaken as an aid to more exact dosage.

The rapid diminution of I_s/I_T in the superficial layers between H.V.L. $= 1.1$ mm. Cu and 1.8 mm. Cu offers suitable conditions for an examination of the possible variation of erythema with quality of radiation, a problem which does not yet appear to have been satisfactorily solved. In going to the harder of these two radiations not only is the primary hardness increased but advantage is taken of the notable decrease in back-scatter, which must necessarily be of a softer character. The average composition of the radiations in the first 5 mm. should therefore be quite different. Furthermore, with a roentgen ray set operating at 250 K.V. it is easily possible to obtain the two radiations at equal intensity and the range is so limited that errors in dosimetry are not likely to be important.

Increase of field size produces a less rapid increase of I_s/I_T for the hard radiation, in accordance with the increased forward concentration of the scattered radiation. The cut-off at the edge of the beam is therefore better defined for the hard radiation. This may prove to be a valuable asset of high voltage radiation, especially where narrow fields have to be directed just to miss vital structures. Reduced scattering out of the main beam is automatically combined with increased homogeneity in it.

The application of the results to the construction of isodose curves has already been described.

SUMMARY

Values of the back-scattering coefficient I_S/I_T of roentgen rays from H.V.L. = 0.32 — 3.3 mm. Cu are given for different focal distances, field sizes, and depths in a scattering block. I_S/I_T at any depth is independent of the focal distance and decreases steadily with increasing hardness of radiation. Therapeutic applications are briefly mentioned.

ZUSAMMENFASSUNG

Die Werte des Rückstreuungskoeffizients I_S/I_T der Röntgenstrahlen von H.W.S. = 0.32 — 3.3 mm. Cu werden für verschiedene Brennweiten, Feldgrössen und Tiefen in einem zerstreuenden Phantom gegeben. I_S/I_T in irgend einer Tiefe ist von der Brennweite unabhängig und vermindert sich beständig mit der zunehmenden Härte der Strahlung. Heilkräftige Anwendungen werden kurz erwähnt.

RÉSUMÉ

Les valeurs de la proportion, I_S/I_T de rayonnement diffusé au rayonnement total des rayons de roentgen de l'épaisseur de demi-valeur 0.32— 3.3 mm. Cu sont données pour des distances focales différentes, des dimensions du champ et des profondeurs dans un bloc de cire. I_S/I_T à n'importe quelle profondeur est indépendante de la distance focale et diminue régulièrement à mesure que la dureté du rayonnement s'augmente. On a parlé en peu de mots des applications thérapeutiques.

REFERENCES

1. THORÆUS, R. Acta Rad. Supplement XV. 1932.
2. ALLIBONE, T. E. and F. E. BANCROFT. Brit. Journ. Radiol. VII. 1934. p. 65.
3. GREBE, L. and K. NITZGE. Strahlenther. Sonderband XIV, 1930.
4. QUIMBY, E. H., C. DE F. LUCAS, A. N. ARNESON and W. S. MACCOMB. Radiology 23. 1934. p. 743.

The Distribution of Radiation in Deep X-Ray Therapy

The British Journal of Radiology, Volume VIII

November 1935

Reprinted with permission of the *British Journal of Radiology*.

THE
DISTRIBUTION OF RADIATION
IN DEEP X-RAY THERAPY

By H. M. Parker, M.Sc. and Joan Honeyburne, B.Sc.
Holt Radium Institute, Manchester

(*Received June* 13, 1935)

November 1935

THE DISTRIBUTION OF RADIATION IN DEEP X-RAY THERAPY

By H. M. PARKER, M.Sc. and JOAN HONEYBURNE, B.Sc.

Holt Radium Institute, Manchester

(Received June 13, 1935).

INTRODUCTION

THE distribution of radiation in a body exposed to a beam of X-rays is a problem of major importance in therapy, and the results of many observations in phantoms of water or other scattering media have already been published. Recent tendencies in deep X-ray therapy have introduced changes in technique which render many of the previous investigations incomplete. The use of voltages up to 250 K.V. (constant potential) with filters of about 1 mm. of copper has become common practice, representing qualities somewhat harder than those most fully investigated. In addition, the choice of low intensity treatments has encouraged the more frequent use of greater focal distances since modern X-ray sets are capable of delivering a satisfactory intensity at distances up to 100 cm. and advantage can thus be taken of the improvement in depth dose.

Improvements in the stability of X-ray equipment, and particularly in the design of reliable dosimeters have led to a demand for increased accuracy in the charting of the dose delivered to any part of an irradiated body. In this connection, also, the general use of isodose charts in conjunction with a series of anatomical cross sections seems to require the extension of such charts to greater depths than has hitherto been customary, in order to assess the various contributions to remote surfaces in multiple field techniques. These several factors have combined to encourage the authors to produce a set of isodose charts and depth dose tables which may be of routine use to the majority of X-ray therapists.

Limitations of space prohibit the inclusion of isodose charts at various focal distances, but one of us (H.M.P.) has given elsewhere a method (to be published in *Acta Radiologica*) by which the distribution at any focal distance may be approximately calculated from the charts given in this paper, together with a knowledge of the percentage back-scatter at the surface of the phantom. The measurements have been confined to cases of normal incidence on a complete scattering body with a plane surface. The conditions under which these measurements are valid in practical therapy may be inferred from papers by Quimby[1], May[2], and others.

Experimental Method

A continuously evacuated Metropolitan-Vickers X-ray tube of the type

Distribution of Radiation in Deep X-Ray Therapy

described by Allibone and Bancroft[3], and operating up to 250 K.V. (constant potential) at 10 ma. was used throughout the measurements. The tube was provided with a gold anticathode and the rays were emitted through a steel window 1·0 mm. thick in the continuous steel wall of the tube. The normal filter consisted of this window with sufficient copper to make a total equivalent

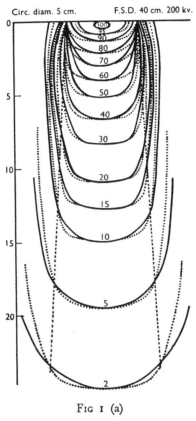

Fig 1 (a)

Isodose Charts of the distribution of radiation in a large phantom. Charts drawn for planes containing the target and any diameter of the circular applicators. Full line shows distribution at 200 K.V., dotted line at 250 K.V. F.S.D.=40 cm. Filter=1 mm. Cu (equiv.) + 1 mm. Al.

of approximately 1 mm. Cu, together with 1 mm. of aluminium. The half value layers were 1·5 mm. Cu and 1·8 mm. Cu respectively at 200 K.V. and 250 K.V. measured by inserting copper foils in the beam defined by the smallest applicator, the ionisation chamber being 40 cm. from the foils and 250 cm. from the nearest scattering wall. The kilovoltage was in each case checked against the spark gap between large spheres complying with British standard specifications. The tube could be moved vertically through a range of 155 cm. and the beam rotated through nearly 360° in a vertical plane. As the tube had no adjustment at right angles to this plane it was considered advisable to use a solid scattering medium to obtain correct alignment by tilting the phantom if necessary. The phantom was accordingly constructed of unit-density wax assembled in sheets 40 cm. × 40 cm. of various thicknesses. The lower block, 15 cm. thick, had its upper surface accurately plane. On this rested a sheet uniformly 2 cm. thick with a cavity in the upper surface exactly fitting the ionisation chamber in the "half-sunk" position. This sheet also carried two pointers registering against a scale attached to the lower block. The chamber could be traversed across the field at right angles to its long axis by reference to the scale and pointers. For large displacements of the chamber the overhang of the surface sheet was supported by an additional block of wax and the corresponding gap on the other side filled in by strips 2 cm. thick. Measurements at a depth in the phantom were obtained by raising the tube by a certain distance and inserting the corresponding thickness of wax above the chamber, the first sheet having a cavity to fit the upper half of the chamber. In the traverses, to obtain the distribution of intensity along any line parallel to the surface of the phantom, only the two sheets holding the chamber were moved. With wide incident beams it was necessary to add temporary blocks to the phantom so that the distance of its edge from the geometric edge of the beam was not less than 15 cm. in order to ensure complete scattering. Sufficient unit density wax was available to carry the measurements to a depth of 20 cm., readings to 35 cm. being achieved somewhat less accurately by the addition of sheets of a cheaper, less dense wax. The measuring phantom has been described at some length on account of its unconventional design. The special method of obtaining positions below the surface was preferred to the more usual practice of inserting the chamber in successive holes at different depths, for three reasons:—(1) it is difficult to obtain and maintain an accurate fit in the holes below the surface, (2) holes not in use have to be plugged with wax or boxwood dummies of the chamber, (3) four different chambers were used, necessitating four sets of holes. With the method chosen it was only necessary to prepare four sets of the traversing sheets. The choice of a solid phantom was forced by the problem of tilting and by the fact that two of the chambers (detachable condenser type) were unsuitable for immersion in water. The absence of a meniscus effect and the exact positioning of the chamber in the wax blocks seemed to offer incidental advantages. These desiderata outweighed the obvious disadvantages of relative slowness of adjustment, tackiness of the wax, which could be overcome by coating the

Vol. VIII, No. 95

Distribution of Radiation in Deep X-Ray Therapy   687

sheets with collodion, and deformation of the sheets, which could be reduced to a minimum by careful handling.

Four different ionisation chambers were used to eliminate the effect of any particular chamber construction. The relative response of the chambers to radiations from 0·33—3·33 mm. H.V.L. in Cu was approximately constant,

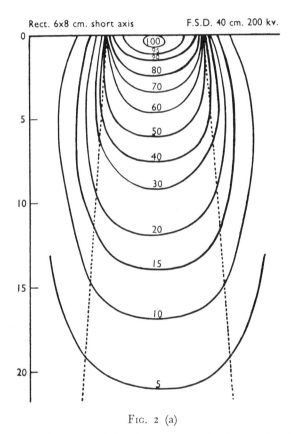

Fig. 2 (a)

Isodose Charts for planes containing the target and short axis of the rectangular applicators. 200 K.V., F.S.D.=40 cm.

so that it was reasonable to suppose that all the chambers were essentially "air-walled." The instruments used were:—

    1. Victoreen r-meter, detachable condenser type, range 0-100 r. This had the smallest ionisation chamber (diameter 8 mm., length 10 mm.).

    2. Victoreen r-meter, detachable type, range 0-25 r.

    3. Siemen's dosimeter, spherical carbon-walled chamber.

    4. Mekapion integrating dosimeter.

November 1935

688   *H. M. Parker and J. Honeyburne*

Technique of Measurement

The determinations were made as far as possible under normal treatment conditions with the beam defined by applicators lined with lead, copper, and aluminium, and provided with a thin wooden base. It was to be expected, therefore, that the results might differ from those obtained by the more usual method employing lead apertures to define the field size. A complication

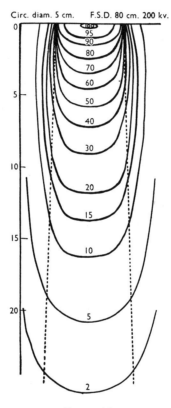

Fig. 3 (a)
Circular applicators at 200 K.V., F.S.D. = 80 cm.

was introduced by the use of the standard applicators in that it became impossible to measure directly the surface intensity at the exact distances 40, 60, 80 and 100 cm. on account of the finite size of the ionisation chamber. The measurements were carried out entirely at focal distances of (40+R), (60+R) etc. (R=chamber radius) and then interpolated with respect to focal distance and field size to obtain the true values at 40, 60 cm. etc. The quicker but less reliable method of obtaining the true surface intensity at the proper distance by the extrapolation of readings at slightly greater distances, and combining

Distribution of Radiation in Deep X-Ray Therapy

these with direct measurements of the depth intensities was also used. The alternative methods gave results concordant within the experimental error.

Depth Doses

The depth dose measurements were made for a large number of fields at 200 K.V. [filter=1 mm. Cu (equiv.) +1 mm. Al] and at 250 K.V. with the same filter.

Five readings were taken at each depth and constant reference was made to the surface value as it affected all the derived depth doses. Apparent changes in the surface intensity were generally traced to an aberration of the measuring instrument, as compared with a radium standard, rather than to a change in the output of radiation.

The depth doses as measured by the different chambers were not identical, the larger ones giving higher readings at all depths. In particular, values in excess of 100 per cent which have frequently been reported for a depth of 1 cm. were almost entirely absent in the measurements with the two smallest chambers. The nature of the variation of apparent depth dose with chamber diameter was such that the measurements with the smallest chamber available appeared to be practically identical with the attempted extrapolation to zero chamber size, except possibly for depths less than 2 cm. All the readings were accordingly reduced to correspond with those of the smallest chamber, and hence effectively to a point measurement.

The depth doses at 200 K.V. and 250 K.V. were found to be identical within the experimental error when measured with the small chamber. With the largest chamber a definite but unimportant difference was observed, the values being slightly less at depths less than 5 cm. and slightly greater for depths greater than 5 cm. for the harder radiation. The measurements were extended to still harder radiations (H.V.L.=3·3 mm. Cu) by the use of tin filters. Only a small increase in the depth doses was obtained, in agreement with the results of Mayneord[4] and others over the same range. The depth doses at 40, 60, 80 and 100 cm. for 200-250 K.V. are set out in the Tables I and II. The curves of depth dose against depth are very nearly exponential at depths greater than 5 or 6 cm.

The horizontal traverses

In order to complete the isodose curves for the principal planes it was necessary to obtain the distribution of intensity along principal axes parallel to the phantom surface at different depths. In this case it was to be expected that the readings near the edge of the irradiated zone would be influenced by the finite size of the chamber, particularly at small depths. The measurements were therefore made exclusively with the smallest chamber. The fall of intensity at the edge of the field was then found to be exceedingly rapid, and was still further augmented by applying a correction for the size of the chamber. With the chamber axis exactly at the edge of the field only half of the chamber was exposed to the direct beam. At 1·6 mm. outside or

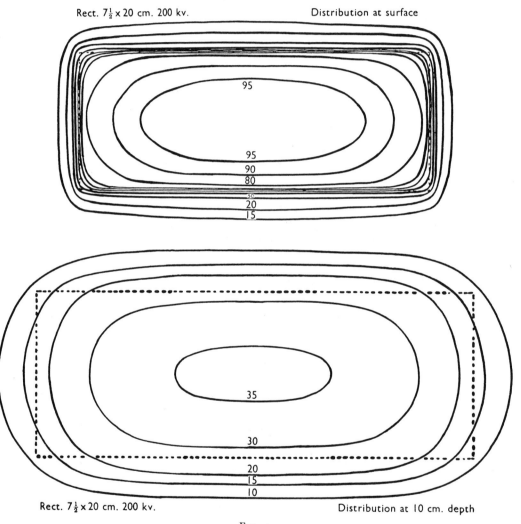

FIG. 4

Typical example of the distribution of radiation in planes parallel to the phantom surface. 200 K.V., F.S.D.=40 cm.

inside the beam respectively one-quarter and three-quarters of the chamber were directly irradiated, and at 4 mm. displacement or more, the chamber was either entirely outside or within the beam. From these considerations it was

Distribution of Radiation in Deep X-Ray Therapy

possible to determine approximately what the readings would be if the chamber had been entirely outside the beam when its axis was outside, and within when the axis was within. Below the surface no simple correction was applicable on account of the rapidly increasing effect of the scattering processes. At depths greater than 5 cm. the edge of the beam was so badly defined that the larger chambers could be used to take advantage of the rapidity of measurement without loss of accuracy. In all cases the traverses were continued right across the field, the true centre being deduced from the symmetrical curves thus obtained. It was usually found that the chamber had been adjusted by eye to within 1 mm. of the true centre. It was also found that the traverses led to identical results whether they were carried out in a line parallel to the cathode—target direction—or at right angles to it. In other words, the radiation from the target was spherically symmetrical within the cone transmitted by the largest applicator, a feature which is not valid for all types of X-ray tube, and which leads to a considerable simplification of the measurements.

The isodose charts reproduced in Figs. 1 and 2 refer to the measurements at 200 K.V. and 40 cm. F.S.D. In Fig. 3 a few typical examples of the distribution with 80 cm. F.S.D. are given. The remaining isodose charts for 80 cm. and those at 60 cm. and 100 cm. may be interpolated with tolerable accuracy, except near the edge of the beam, from the charts given and from the depth dose tables. The dotted curves of Fig. 1 show the modification introduced by increasing the voltage to 250 K.V. Although the depth doses on and near to the central axis are unaltered owing to the balancing of reduced scatter against increased penetration, the shape of the isodose curves near the geometric edge of the beam is definitely changed. The radiation at the higher voltage is much more sharply confined to the directly irradiated zone, and it may be that this improved definition alone would merit the use of still harder radiations in therapy.

A full solution of the distribution of intensity would involve measurements off the principal planes in addition to the customary isodose charts. For the circular applicators the problem is automatically solved as the isodose curves in planes parallel to the scattering surface are obviously concentric circles whose positions may be determined from the distribution in the vertical plane. The use of circular applicators, therefore, offers an important practical advantage in those cases in which multiple fields with non-coplanar central axes have to be used. An example of the isodose curves for the rectangular applicators in the scattering surface and at 10 cm. deep is given in Fig. 4. The curves near the corner of the irradiated zone in the surface measurements

NOVEMBER 1935

692   *H. M. Parker and J. Honeyburne*

are unreliable on account of the uncertainty in the correction for the size of the chamber at such points. The general nature of the curves for other field sizes and depths may be approximately estimated, where necessary, from the examples given.

In conclusion, the authors wish to express their thanks to Dr. Ralston Paterson, F.R.C.S., D.M.R.E., Director of the Holt Radium Institute, for his continued interest throughout the work, and for his advice concerning the selection of measurements of most practical importance.

Summary

Depth dose tables for X-rays at 200 K.V. and 250 K.V. [filter 1 mm. Cu (equiv.) +1 mm. Al] for F.S.D. =40, 60, 80 and 100 cm. are given for the circular and rectangular fields normally employed. Isodose curves in the principal planes are given for F.S.D. =40 cm. at 200 K.V., with some examples at 200 K.V. (80 cm. F.S.D.) and 250 K.V. (40 cm. F.S.D.). The distribution at right angles to the principal planes is also indicated. The experimental method is described in detail.

Zusammenfassung

Die Arbeit gibt Tiefendosentabellen für Rontgenstrahlungen von 200 K.V. und 250 K.V. (1 mm. Cu +1 mm. Al) bei Hautabständen von 40, 60, 80 und 100 cm. Die üblichen Feldgrossen, rund und rechteckig, wurden gewahlt. Isodosencurven wurden bestimmt für die Hauptebenen für 40 cm. Hautabstand und 200 K.V. sowie in ein paar Fällen von 200 K.V. bei 80 cm. Fokushautabstand und von 250 K.V. und 40 cm. Hautabstand. Ebenfalls gegeben wird die Dosenverteilung rechtwinkelig zür Haupteinfallsebene. Die Versuchsanordnung wird genau beschrieben.

Résumé

Les doses profondes pour les rayons X à 200 K.V. et 250 K.V. (filtre 1 mm. Cu (equiv.) +1 mm. Al) pour les distances focales de 40, 60, 80 et 100 cm. sont données pour les champs circulaires et rectangulaires employées en pratique.

Courbes isodoses dans les plans principaux sont données pour D.F. de 40 cm. à 200 K.V. avec quelques exemples à 200 K.V. (80 cm. D.F. et 250 K.V. (40 cm. D.F.). La distribution perpendiculaire aux plans principaux est aussi indiquée. La méthode expérimentale est décrite en détail.

Distribution of Radiation in Deep X-Ray Therapy

TABLE I

PERCENTAGE DEPTH DOSES WITH CIRCULAR APPLICATORS

200–250 K.V. (constant potential).   Filter=1 mm. Cu (equiv.)+1 mm. Al.

F.S.D.	Depth cm.	Diameter in cm.					
		5	7·5	10	12·5	15	20
40 cm.	0	100·0	100·0	100·0	100·0	100·0	100·0
	1	94·0	96·0	97·8	99·0	100·0	100·0
	2	81·5	84·5	87·0	90·1	92·0	92·6
	5	51·0	56·8	61·5	65·0	68·1	71·2
	10	23·0	27·1	31·2	35·0	37·7	42·4
	15	10·2	12·3	16·0	18·0	20·0	24·9
	20	4·7	5·7	8·1	9·3	10·8	13·8
	25	2·0	2·7	4·3	4·9	5·9	7·9
	30	0·9	1·2	2·2	2·5	3·1	4·6
60 cm.	0	100·0	100·0	100·0	100·0	100·0	100·0
	1	95·2	97·0	98·5	98·7	99·8	100·0
	2	83·7	87·1	89·0	90·7	92·1	92·6
	5	54·3	60·4	64·3	67·7	70·6	72·7
	10	24·8	29·2	33·8	36·4	39·9	44·5
	15	11·2	13·4	17·2	19·3	21·6	26·1
	20	4·9	6·2	9·2	10·1	11·9	15·5
	25	2·2	2·9	4·7	5·5	6·5	9·2
	30	1·0	1·3	2·4	2·9	3·5	5·5
80 cm.	0	100·0	100·0	100·0	100·0	100·0	100·0
	1	96·0	97·0	99·0	99·2	99·7	100·0
	2	84·9	88·0	90·7	91·6	92·4	93·6
	5	56·0	60·8	65·5	68·9	71·5	74·7
	10	26·3	30·4	36·0	38·7	41·1	46·3
	15	12·1	14·6	18·4	20·3	22·1	27·2
	20	5·6	7·1	9·7	10·7	12·5	16·2
	25	2·6	3·5	5·0	5·6	6·9	9·7
	30	1·2	1·7	2·6	3·0	3·8	5·8
100 cm.	0	100·0	100·0	100·0	100·0	100·0	100·0
	1	96·6	97·0	99·4	99·6	99·8	100·0
	2	85·6	88·6	92·0	92·3	92·7	94·4
	5	56·4	61·2	66·3	69·9	72·6	76·3
	10	26·9	30·9	36·6	39·1	42·1	47·2
	15	12·5	14·8	19·3	20·8	23·0	28·0
	20	5·8	7·3	10·3	11·2	12·9	16·8
	25	2·7	3·6	5·5	6·0	7·1	10·1
	30	1·3	1·8	2·9	3·2	3·9	6·0

November 1935

H. M. Parker and J. Honeyburne

TABLE II

PERCENTAGE DEPTH DOSES WITH RECTANGULAR APPLICATORS

200–250 K.V. (constant potential) Filter = 1 mm. Cu (equiv.) + 1 mm. Al.

F.S.D.	Depth.	6×4	6×8	10×8	10×12	7.5×20	10×15	15×20
40 cm.	0	100.0	100.0	100.0	100.0	100.0	100.0	100.0
	1	94.5	96.4	98.0	99.0	99.2	100.0	100.0
	2	81.8	84.6	87.0	90.0	90.4	91.5	92.5
	5	51.7	57.2	61.5	64.9	65.5	66.2	71.0
	10	23.3	26.9	31.2	34.9	35.5	36.5	42.1
	15	10.7	12.7	16.0	17.9	19.1	19.3	24.1
	20	4.8	5.9	8.1	9.3	10.0	10.1	13.6
	25	2.1	2.8	4.3	4.9	5.3	5.4	7.7
	30	1.0	1.3	2.2	2.5	2.8	2.8	4.4
60 cm.	0	100.0	100.0	100.0	100.0	100.0	100.0	100.0
	1	96.0	97.3	98.5	98.7	98.9	99.0	100.0
	2	84.5	87.3	89.0	90.6	90.9	91.2	92.5
	5	55.1	60.7	64.4	67.6	68.7	69.9	72.5
	10	26.0	29.5	33.9	36.3	38.4	39.7	44.3
	15	11.5	13.7	17.5	19.2	21.1	21.4	25.9
	20	5.2	6.5	9.2	10.0	10.9	11.1	15.2
	25	2.3	3.1	4.7	5.4	5.8	5.9	8.9
	30	1.1	1.5	2.4	2.9	3.2	3.2	5.2
80 cm.	0	100.0	100.0	100.0	100.0	100.0	100.0	100.0
	1	96.5	97.8	99.0	99.1	99.3	99.5	100.0
	2	85.0	88.0	90.7	91.5	91.7	92.0	93.5
	5	56.0	61.7	65.5	68.8	70.3	71.3	74.5
	10	26.8	31.0	36.1	38.6	39.9	40.5	46.0
	15	12.5	15.0	18.5	20.2	21.4	21.8	27.0
	20	6.0	7.3	9.8	10.6	11.3	11.6	16.0
	25	2.8	3.6	5.0	5.6	6.0	6.2	9.3
	30	1.3	1.8	2.6	3.0	3.4	3.4	5.5
100 cm.	0	100.0	100.0	100.0	100.0	100.0	100.0	100.0
	1	96.9	98.2	99.4	99.5	99.7	99.9	100.0
	2	85.6	88.9	92.0	92.2	92.4	92.6	94.2
	5	56.5	62.5	66.3	69.8	71.7	72.5	76.1
	10	27.4	32.1	36.7	39.0	40.5	41.1	46.9
	15	13.2	15.7	19.4	20.7	22.1	22.5	27.8
	20	6.3	7.7	10.4	11.1	11.8	12.1	16.4
	25	3.0	3.8	5.5	5.9	6.4	6.6	9.7
	30	1.4	1.9	2.9	3.2	3.6	3.6	5.8

REFERENCES

[1] Quimby, E. H., M. M. Copeland and R. C. Woods, *Am. Journ. Roentgen.*, xxxii, 1934, p. 534.
[2] May, E. A., *Radiology*, xxii, 1934, p. 559.
[3] Allibone, T. E., and F. E. Bancroft, *Brit. Journ. of Radiol.*, vii, 1934, p. 65.
[4] Mayneord, W. V., and J. E. Roberts, *Brit. Journ. of Radiol.*, vi, 1933, p. 321.

Vol. VIII, No. 95

Distribution of Radiation in Deep X-Ray Therapy

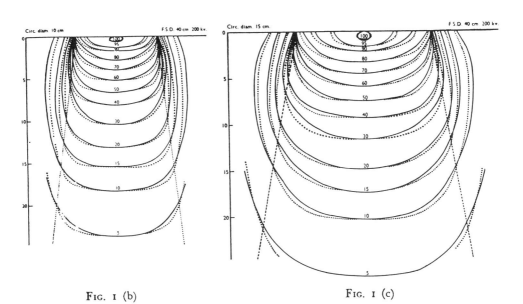

Fig. 1 (b) Fig. 1 (c)

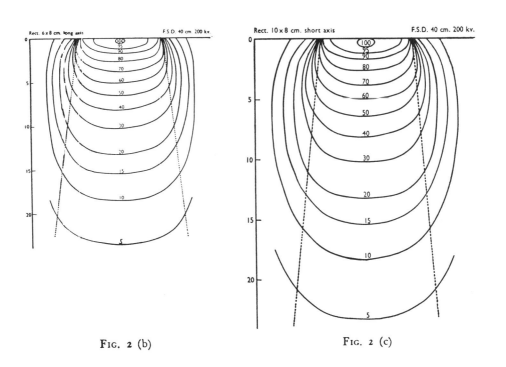

Fig. 2 (b) Fig. 2 (c)

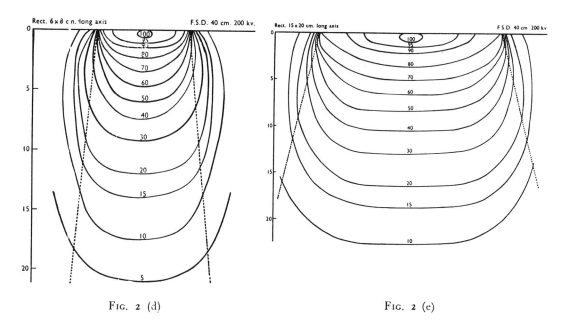

Fig. 2 (d)

Fig. 2 (e)

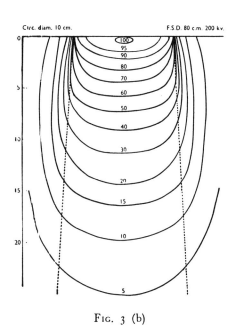

Fig. 3 (b)

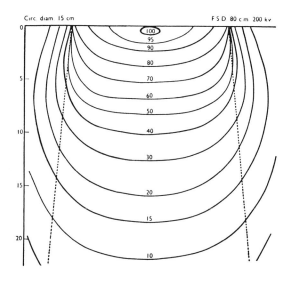

Fig. 3 (c)

A System of Dosage for Cylindrical Distributions of Radium

The British Journal of Radiology, Volume IX

August 1936

Reprinted with permission of the *British Journal of Radiology*.

A SYSTEM OF DOSAGE FOR CYLINDRICAL DISTRIBUTIONS OF RADIUM

By Ralston Paterson, M.D., F.R.C.S., D.M.R.E., H. M. Parker, M.Sc.,
Holt Radium Institute, Manchester, and
F. W. Spiers, Ph.D., The General Infirmary, Leeds.

(Received March 7, 1936)

A SYSTEM OF DOSAGE FOR CYLINDRICAL DISTRIBUTIONS OF RADIUM

By RALSTON PATERSON, M.D., F.R.C.S., D.M.R.E., H. M. PARKER, M.Sc.,
Holt Radium Institute, Manchester, and
F. W. SPIERS, Ph.D., The General Infirmary, Leeds.

(*Received March* 7, 1936)

INTRODUCTION

In a previous paper by two of the present authors, a system of dosage dealing with most of the problems of the superficial application of radium was put forward. At that time one section, *viz*., that relating to cylindrical distributions, was deliberately omitted in order to avoid undue emphasis on a section of the work which is slightly more complicated and, in practice, constitutes only a small fraction of the whole. The function of the present article is to complete the dosage system for superficial therapy. In the interests of brevity, a knowledge of the previous contribution will be assumed. Dr. F. W. Spiers has recently applied more satisfactory mathematical analysis to some of the problems to be discussed, and it is believed that a joint publication will be the most useful method of presenting the results.

As in the previous article on dosage, two distinct problems require consideration:—

(*a*) The distribution of radium on the applicator to give uniform irradiation over the treated surface.

(*b*) The assessment of dosage at the treated surface or at any tissue depth.

These problems are again approached by the statement of "Distribution Rules" and by the presentation of "Dosage Tables."

PART I

By RALSTON PATERSON

CLINICAL ASPECTS

The use of cylindrical applicators arises when a part has to be completely surrounded by an applicator carrying radium in the form of a single ring or of a cylinder. It will be apparent that radium may be mounted on such an applicator in two essentially different ways, either as a series of coaxial rings or as a series of equidistant lines parallel to the axis of the cylinder and at a constant distance from it. A single ring may be regarded as the limiting case for a cylinder of zero length. In general, the former arrangement of radium gives a more uniform distribution of intensity and is the method of choice where applicable.

Distribution Rules

1. Place the radium in a series of coaxial rings at a distance apart of twice the distance between the radium and the treated surface (*i.e.*, at twice the "thickness" of the applicator). If two rings are required the total radium is divided equally between them (1 : 1). If three rings are required the centre ring has half the radium content of an end ring (1 : $\frac{1}{2}$: 1). If four or more rings are required, each inner ring has two-thirds the radium content of an end ring (1 : $\frac{2}{3}$: $\frac{2}{3}$: 1) (1 : $\frac{2}{3}$: $\frac{2}{3}$: $\frac{2}{3}$: 1) *ad. inf.*

A ring is preferably composed of a continuous chain of radium needles or tubes. If necessary the needles or tubes, provided they are short, may be placed equidistantly at right angles to the plane of the proposed ring, to form a short belt.

2. *Aliter.*—Mount the radium in a series of straight lines to the required length, parallel to the cylinder axis and at intervals round the circumference, equal to twice the "thickness" of the applicator. Add a ring or short belt of radium at each end. The amount of radium per cm. of the straight lines has to be two-thirds of the amount per cm. in the end rings.

N.B.—Cylinders of the second type are intended to be used only when a sufficient stock of the short needles or tubes required in the first method is not available. The alternative method demands short sources for the end rings only. A useful expression in the operation of the Type 2 cylinder is:—

Radium content of each straight line =

$$\frac{\text{Length of cylinder} \times \text{radium content of each end ring}}{5 \times \text{diameter of cylinder.}}$$

Vol. IX, No. 104

A System of Dosage for Cylindrical Distributions of Radium

Dosage

As in the previous article the unit of dosage is 1,000 r (gamma), and the dosage tables are designed to give the number of milligramme-hours required to produce that dose under prescribed circumstances. Table I states the

TABLE I

Milligramme-hours to produce 1,000 r at the centre of rings of different diameters at standard filtrations.

Diameter D. in cm.	Filtration				
	0·5 mm. Pt.	0·8 mm. Pt.	1·0 mm. Pt.	1·5 mm. Pt.	2·0 mm. Pt.
2	119	126	132	145	161
3	268	284	296	327	362
4	476	506	526	581	643
4·5	603	640	666	736	814
5	744	790	822	908	1,000
5·5	900	956	995	1,100	1,220
6	1,070	1,140	1,180	1,310	1,450
6·5	1,260	1,340	1,390	1,540	1,700
7	1,460	1,550	1,610	1,780	1,970
7·5	1,670	1,780	1,850	2,040	2,260
8	1,900	2,020	2,100	2,320	2,570
8·5	2,150	2,280	2,380	2,620	2,900
9	2,410	2,560	2,660	2,940	3,250
9·5	2,690	2,850	2,970	3,280	3,620
10	2,980	3,160	3,290	3,630	4,020
10·5	3,280	3,480	3,630	4,010	4,430
11	3,600	3,820	3,980	4,400	4,860
11·5	3,940	4,180	4,350	4,800	5,310
12	4,290	4,550	4,740	5,240	5,780
13	5,030	5,340	5,550	6,140	6,790
14	5,830	6,200	6,440	7,120	7,870
15	6,700	7,120	7,410	8,180	9,050
16	7,620	8,080	8,420	9,300	10,300
17	8,600	9,140	9,500	10,500	11,600
18	9,640	10,200	10,700	11,800	13,000
19	10,700	11,400	11,900	13,100	14,500
20	11,900	12,600	13,100	14,500	16,100
21	13,100	13,900	14,500	16,000	17,700
22	14,400	15,300	15,900	17,600	19,400
23	15,700	16,700	17,400	19,200	21,200
24	17,100	18,200	18,900	20,900	23,100
25	18,600	19,700	20,500	22,700	25,100

The values are given correct to three significant figures on the basis of 1 mc. = 8·4 r per hour.

number of milligramme-hours to produce 1,000 r at the centre of a ring of any chosen diameter for the more common filtrations. These figures are obtained directly from the inverse square law and the known absorption in platinum. The amount of radium required to produce 1,000 r at any given surface inside a radium cylinder (*i.e.*, generally the skin surface of the treated part) may be derived from the basic figure for a ring of the same diameter, obtained from Table I, by multiplying it by the appropriate factor from Table II.

TABLE II

Factors for Cylinders of Types 1 and 2.

$\dfrac{L}{D}$	$\dfrac{d}{D}$						
	0·0	0·2	0·4	0·5	0·6	0·7	0·75
0·0	1·00	0·96	0·84	0·75	0·64	0·51	0·44
0·1	1·01	0·98	0·86	0·78	0·68	0·56	0·49
0·2	1·06	1·02	0·92	0·84	0·75	0·64	0·57
0·3	1·12	1·09	1·00	0·93	0·83	0·73	0·64
0·4	1·18	1·15	1·07	1·00	0·90	0·80	0·74
0·5	1·24	1·21	1·13	1·07	0·97	0·87	0·82
0·6	1·31	1·28	1·20	1·12	1·04	0·94	0·88
0·8	1·45	1·41	1·34	1·28	1·18	1·08	1·01
1·0	1·60	1·56	1·49	1·43	1·33	1·21	1·14
1·2	1·73	1·69	1·62	1·55	1·45	1·33	1·26
1·4	1·86	1·83	1·74	1·68	1·58	1·46	1·39
1·6	2·00	1·96	1·87	1·80	1·69	1·57	1·49
1·8	2·15	2·11	2·02	1·94	1·83	1·69	1·61
2·0	2·31	2·27	2·18	2·09	1·97	1·82	1·74

Two items have to be considered in the use of Table II.

(*a*) The ratio of the length L of the radium cylinder to its diameter D.

(*b*) The ratio of the diameter d of the treated part to the diameter D of the radium cylinder. The usual values of this ratio $\dfrac{d}{D}$ appear in the table as column headings.

Incomplete cylinders

In certain instances incomplete cylinders or rings, particularly semi-cylinders, are useful. These are, however, with the exception of the three-quarter ring quoted in the following discussion of neck packs, so uncommon that codified statements of the method are not required. Some information on the manipulation of these special cases is given in the physical sections of the article.

Applications and examples

In actual clinical practice, a ring or cylindrical type of applicator is of

Vol. IX, No. 104

A System of Dosage for Cylindrical Distributions of Radium

value in a relatively limited number of cases, *viz.*, carcinoma in the neck—either malignant secondary glands or primary pharyngeal or laryngeal carcinomata, carcinoma of the penis, and tumours (either bone or skin tumours) of the arms, fingers, or legs. A few practical examples of the operation of the dosage system as applied to these parts will be the simplest method of clarifying the foregoing general statement of the method.

1. *Neck Packs.*—A former type of neck pack consisted of an applicator, usually of Columbia paste, moulded to fit the neck exactly, with the radium evenly distributed over two surfaces corresponding to the main gland-bearing areas. With a mould thickness of one or two centimetres the dose below the surface soon became inadequate and the neck packs were valueless, except for highly sensitive growths such as lympho-sarcoma. It is doubtful whether this method of treatment does ever achieve more than temporary resolution.

Certainly distances of from 3 to 5 cm. are required to obtain anything approaching a sufficient depth dose. It may be shown (Part II) that this distribution can be effectively replaced by a single ring of radium surrounding, or almost surrounding, the neck, without any real diminution of adequately irradiated surface and with definite improvement in depth dose. If sufficient radium is available, a distance of 4 cm. is definitely to be preferred to 3 cm. distance. Except for patients with extremely thin necks, a distance of 5 cm., although essential to give a really satisfactory depth dose, necessitates the use of such a large amount of radium that the method becomes dangerous because of the general systemic effects produced.[2]

Ex. A. Ca. Larynx.—Neck of average diameter, 9 cm., to be treated by a single ring at $4\frac{1}{2}$ cm. Required dose is 6,500 r in 21 days, and the available radium is filtered by 1.5 mm. Pt.

Dose calculations:—$d = 9$ cm. $D = 9 + 4\frac{1}{2} + 4\frac{1}{2} = 18$ cm. $\frac{L}{D} = 0.0$ $\frac{d}{D} = 0.5$ at surface of neck.

The reading from Table I for $D = 18$ at 1.5 mm. Pt is 11,800 mg. hrs. to produce 1,000 r.
From Table II, for $\frac{L}{D} = 0.0$, and $\frac{d}{D} = 0.5$, factor = 0.75.

∴ at surface of neck, 1,000 r is given by 0.75 × 11,800 mg. hrs. = 8,850 mg. hrs.
∴ 6,500 r is given by 57,500 mg. hrs.
Use 17 × 6.66 mg. tubes for 21 days (continuous).

Dose at centre of neck ($d = 0$) is $\frac{0.75}{1.00} \times 6,500$ r = 4,880 r, and dose at larynx ($d = 2.5$ cm. approximately) is $\frac{0.75}{0.98} = 4,970$ r.

Distribution:—Applicator shaped to fit the neck. Single rubber tube carrying the sources at the $4\frac{1}{2}$ cm. distance. Each tube is 22 mm. long and the total length of the seventeen tubes is 374 mm., whereas the circumference of the ring is 564 mm. Hence the tubes need to be maintained at equal intervals round the rubber ring by seventeen distance pieces, each about 11 mm. long (Fig. 1A).

August 1936

Ex. B.—Neck of diameter, 12 cm., to be treated at 4 cm. by a three-quarter ring (see Part II) to avoid full radiation of the posterior part of neck. Dose required is 6,500 r in 21 days.

Dose calculation:—$d = 12$ cm. $D = 12 + 4 + 4 = 20$ cm. $\frac{L}{D} = 0$. $\frac{d}{D} = 0.6$.

The reading from Table I for $D = 20$ at 1.5 mm. Pt is 14,500 mg. hrs. per 1,000 r.

From Table II, at $\frac{L}{D} = 0$ and $\frac{d}{D} = 0.6$. Factor = 0.64.

∴ at surface 1,000 r is given by 9,280 mg. hrs., and 6,500 r requires 60,400 mg. hrs. Use 18×6.66 mg. tubes for 21 days.

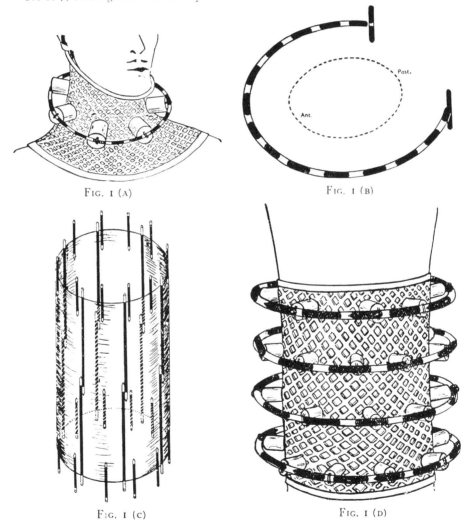

Fig. 1 (A) Fig. 1 (B)

Fig. 1 (C) Fig. 1 (D)

Distribution:—Shaped applicator as before. One-ninth of the total radium is at each end, *i.e.*, two 6.66 mg. tubes at right angles to each end of the three-quarter ring, with the remaining fourteen placed evenly round it.

This requires 13 distance pieces, each 9 mm. long, alternating with the radium tubes (Fig. 1B).

2. *Penis.*—Epithelioma of the glans penis lends itself admirably to treatment by means of a cylindrical mould. This method of treatment, as used in Manchester, has already been described in detail by Dr. R. G. Hutchison[3] as a standard applicator suitable for most normal cases. The following illustration of a mould at a larger distance will serve to demonstrate the general principle and the procedure under the second rule.

Ex. C.—Penile shaft, average diameter 3 cm., to be treated over a length of 9 cm. at a distance of 1·5 cm. Dose required is 7,000 r over 21 days.

Dose calculation:—$d = 3$ cm. $D = 6$ cm. $L = 9$ cm. $\frac{L}{D} = 1\cdot5$. $\frac{d}{D} = 0\cdot5$.

Reading from Table I for $D = 6$ at 0·5 mm. Pt is:—1,070 mg. hrs. per 1,000 r.

From Table II, for $\frac{L}{D} = 1\cdot5$ and $\frac{d}{D} = 0\cdot5$. Factor = 1·74.

∴ at treated surface, 1,000 r is given by $1\cdot74 \times 1{,}070 = 1{,}860$ mg. hrs.
∴ 7,000 r is given by 13,000 mg. hrs.
Time = 16 hours per day for 21 days. ∴ Ra. = *38·7* mg.
Along the cylinder axis (*i.e.*, at urethra) $d = 0$ and the factor is 1·93.

∴ dose = $\frac{1\cdot74}{1\cdot93} \times 7{,}000$ r = 6,320 r (90 per cent).

Distribution:—Using the second rule, the circumference of the cylinder is 19 cm., and twice the mould thickness is 3 cm. Use six lines parallel to the axis, with a ring at each end. Assume first 1 mg. per cm. in the end rings. Then the lines have $\frac{2}{3}$ mg. per cm., and the total radium is $2 \times 19 + 6 \times 9 \times \frac{2}{3} = 74$ mg. Hence the linear density should be 0·52 mg. per cm., *i.e.*, approximately 10 mg. in each end ring and 3 mg. in each line. Total = 38 mg.

Aliter:—Put unit radium in each ring. Then amount in each line = $\frac{L}{5D} \times 1 = 0\cdot3$. Total Ra. = $2 + 6 \times 0\cdot3 = 3\cdot8$ units, *i.e.*, each unit is 10 mg. approx. Use 10 mg. in each ring and 3 mg. in each line as before. Use two 1·5 mg. needles, active length 4·5 cm. end to end for each line, and ten 1 mg. needles round the circumference at each end. (Fig. 1C).

3. *Extensive Epithelioma of leg.*—This will illustrate the general application of cylindrical moulds to the treatment of arms, legs or fingers—sites in which treatable lesions are not commonly met, but which present, by any other method, distinctly difficult problems of adequate dosage and measurement.

Ex. D.—Extensive superficial lesion covering three-quarters of the circumference of the leg and extending some five or six centimetres along it. Treat by means of a cylindrical applicator to 6,000 r in 8 days. A high depth dose is not desirable, so that a small thickness is chosen.

Diameter of leg = 7·5 cm. Treated length = 7·5 cm. Distance = 1·25 cm., *i.e.*, $d = 7\cdot5$ cm. $D = 10$ cm. $L = 7\cdot5$ cm. $\frac{L}{D} = 0\cdot75$, $\frac{d}{D} = 0\cdot75$.

The reading from Table I for $D = 10$ cm. at 1·0 mm. Pt is 3,290 mg. hrs. per 1,000 r.

From Table II, at $\frac{l}{D} = 0\cdot75$ and $\frac{d}{D} = 0\cdot75$, factor = 0·98.

∴ at treated surface, 1,000 r is given by 3,220 mg. hrs. and 6,000 r by 19,300 mg. hrs.
Time = 8 days. ∴ Ra = 100 mg.

August 1936

R. Paterson, H. M. Parker and F. W. Spiers

At 1 cm. below the surface $d' = 5.5$ cm. and $\dfrac{d'}{D} = 0.55$, so that the new factor is 1·19

∴ depth dose at 1 cm. $= \dfrac{0.98}{1.19} \times 6{,}000 = 4{,}940$ r. The dose at 2 cm. deep is 4,520 r.

Distribution:—Nidrose applicator completely surrounding the part, with four coaxial rings of radium contained in rubber tubes 2·5 cm. apart. Or the mould may conveniently be made of two semi-cylinders of Stent's composition held together by bolted flanges. The relative contents of the rings are respectively $1 : \tfrac{2}{3} : \tfrac{2}{3} : 1$.

The distribution is therefore 30 : 20 : 20 : 30, giving the required 100 mg. About ten to twelve sources are required in each ring to avoid "ripple" round the circumference. The end rings may be composed of six 3·33 mg. tubes with five 2 mg. tubes alternating as far as possible. Ten 2 mg. tubes form a satisfactory inner ring, but many variations of the sub-division may be devised.

Summary

A practical dosage system for the use of cylindrical applicators is described. Rules are given for the distribution of radium on such applicators to obtain approximately uniform irradiation over the treated part. The dose at a point within the cylinder is derived from the dose at the centre of a single ring of the same diameter by means of a table of factors, depending on the shape of the cylinder and the distance of the point considered from the axis, relative to the total diameter. The system is demonstrated by practical examples.

Part II

By H. M. Parker

AN OUTLINE OF THE PHYSICAL BASIS OF THE DOSAGE SYSTEM

Distribution

The proposed arrangements of radium for cylindrical applicators were derived by considering the bending of rectangular frameworks, previously described,[1] until a pair of opposite sides coincided. The distribution of radiation from the resulting series of coaxial rings was readily obtained by the repeated application of the ring formula due to Sievert.[4] It was found that the intensity over a coaxial surface within the radium cylinder, up to a maximum diameter determined by the condition that the distance between adjacent rings should not exceed twice their distance from the surface, was constant within limits of ± 10 per cent. An exception had to be made in the case of the extreme ends of long cylinders. It was thus possible to evaluate a mean intensity over the cylindrical surface at each value of $\dfrac{d}{D}$ where d is the diameter of the surface and D is the total diameter of the radium cylinder.

The percentage rise of mean intensity radially from the axis was evaluated for cylinders of different shapes and constructions. Curves showing the radial

A System of Dosage for Cylindrical Distributions of Radium 495

distribution for six cylinders each of length L equal to the diameter D are given in Fig. 2. Three of the cylinders are of the ring type, set respectively for skin distances of $\frac{D}{8}$, $\frac{D}{6}$ and $\frac{D}{4}$, and the remaining three are of the alternative Type 2 with straight lines of radium parallel to the axis, and end rings, for the

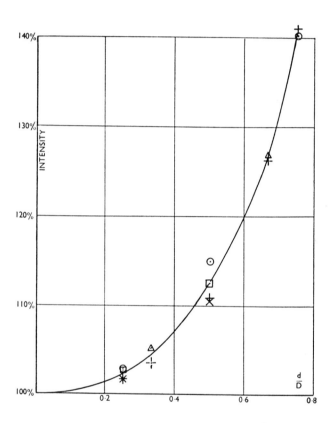

FIG. 2. The radial variation of intensity for cylinders of $\frac{L}{D} = 1\cdot 0$. The circle, square and triangle refer to Type 1 cylinders set homogeneous for skin distances of $\frac{1}{8} D$, $\frac{1}{6} D$ and $\frac{1}{4} D$ respectively. The crosses refer to the corresponding Type 2 cylinders.

same skin distances. The points relating to the various cylinders fall on a single smooth curve to within 3 per cent. Fig. 3, which includes analogous curves for $\frac{L}{D} = 0\cdot 5$ and $2\cdot 0$, demonstrates that the existence of a unique curve of radial distribution is a property extending over the range of cylinder shape required. It follows that subsequent calculations involving radial

August 1936

distribution may be made in general terms, depending only on $\dfrac{L}{D}$ and not on the particular construction of cylinder to be used.

NOTE.—The calculations for Type 2 were obtained by combining the figures for two end rings with those for a uniformly radioactive cylinder in lieu of the series of lines employed in practice. The relative amounts of radium in the separate parts were determined by the number of lines required to

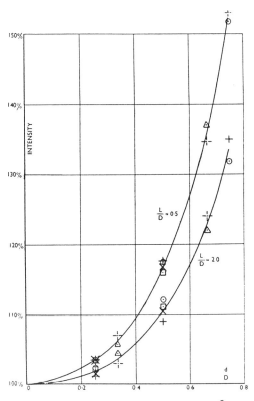

FIG. 3. The radial variation of intensity for cylinders of $\dfrac{L}{D} = 0\cdot 5$ and $\dfrac{L}{D} = 2\cdot 0$.

maintain the interval of twice the mould thickness round the circumference, the linear density in each being two-thirds of that in the end rings.

Dosage

The mean axial intensities per milligramme of radium on cylinders of a constant diameter D, plotted against the elongation $\dfrac{L}{D}$ gave rise to a smooth curve. The deviation from this curve of points relating to cylinders of Types

Vol. IX, No. 104

A System of Dosage for Cylindrical Distributions of Radium 497

1 and 2 did not exceed 3 per cent and 5 per cent respectively over the range $\frac{L}{D} = 0$ to 2. This fact, in conjunction with the results of the preceding paragraph, made it possible to relate the intensity per milligramme at any point in the volume bounded by the ends and the surface of maximum diameter for which the cylinder was designed to give uniform irradiation to the intensity at the centre of a single ring of diameter D, by means of a single conversion factor. For practical convenience, the same relation has been expressed inversely in Table II by multiplication factors operating on the number of milligramme hours read from Table I in order to obtain a dose of 1,000 r at the required point within the cylinder.

TABLE III

Factors for Cylinders of Type 3.

$\frac{L}{D}$	$\frac{d}{D}$						
	0·0	0·2	0·4	0·5	0·6	0·7	0·75
0·0	1·00	0·96	0·84	0·75	0·64	0·51	0·44
0·1	1·00	0·96	0·84	0·76	0·65	0·52	0·45
0·2	1·01	0·98	0·86	0·77	0·67	0·54	0·48
0·3	1·03	0·99	0·87	0·79	0·69	0·58	0·52
0·4	1·05	1·02	0·90	0·83	0·74	0·63	0·58
0·5	1·08	1·05	0·94	0·87	0·78	0·67	0·63
0·6	1·13	1·11	1·01	0·94	0·84	0·76	0·71
0·8	1·20	1·17	1·08	1·01	0·93	0·85	0·81
1·0	1·28	1·26	1·18	1·12	1·05	0·96	0·92
1·2	1·39	1·36	1·28	1·23	1·14	1·06	1·01
1·4	1·50	1·48	1·39	1·34	1·26	1·16	1·12
1·6	1·61	1·58	1·52	1·46	1·39	1·30	1·24
1·8	1·74	1·72	1·66	1·60	1·52	1·42	1·37
2·0	1·88	1·86	1·81	1·73	1·65	1·55	1·48

In some cases, especially in interstitial applications, it is impracticable to place a ring or short belt at each end of the cylinder. The distribution then reduces to a series of active lines all parallel to the axis, an arrangement which may be calculated as a uniformly active cylindrical surface, as fully described in Part III. The practical aspects of dosage in this case may be covered by a set of factors (Table III) as before, but these will now apply only to the maximum dose delivered in the central plane of cross-section.

Limitations of the method

In the preparation of a dosage system of the type described here, some

August 1936

498   R. Paterson, H. M. Parker and F. W. Spiers

inaccuracy has to be tolerated in order to achieve a measure of generality. It is important, however, to emphasise the conditions under which the approximations remain valid. In the first place, the factors of Table II were evaluated without allowance for the effects of oblique filtration. This appears to be justified for cylinders composed of coaxial rings. When short belts or the Type 2 cylinders are used, an error increasing with $\dfrac{L}{D}$ and $\dfrac{d}{D}$ is introduced. For a filtration of 0·5 mm. pt the error is small except for long cylinders, where it may reach 8 per cent. The diminution in dose by oblique filtration may become quite large in the case of more heavily filtered sources, which should, therefore, not be used except for short cylinders. The factors in Table III are corrected for obliquity at 0·5 mm. pt. Separate tables would be required for accuracy at other filtrations (*cf.* tube graphs in a previous publication).

Modifications introduced by the absorption of radiation in tissue have been entirely neglected. It is believed that this does not lead to gross error in dosage calculation unless the radiation has to traverse considerable thicknesses of tissue. Accurate dosage in neck packs or long cylinders applied to limbs would be achieved most conveniently by direct ionisation measurements. It should be pointed out that absorption in tissue and oblique filtration both tend to reduce the delivered dose.

The number of active sources required in the rings and straight lines to provide a sufficiently close approximation to the complete rings and lines of the theoretical discussion may be inferred from the guiding principles given before.[1]

Special cases

In the treatment of laryngeal or pharyngeal lesions it is sometimes satisfactory to substitute a single ring of radium for a cylinder of the required treated length. The effectively treated length (*i.e.*, that over which the intensity variation does not exceed ± 10 per cent) is readily calculable. For an average case it proves to be some 6 or 7 cm., assuming the neck to be cylindrical, and consequently somewhat greater for an actual neck. The effectively treated length along the axis is 9 or 10 cm.

If it should be undesirable to irradiate the whole circumference of the neck to the full dose, a three-quarter ring with one-ninth of the total radium at each end, may be used. There is a reduction of approximately 25 per cent in the intensity opposite the missing segment. The dose per milligramme at the centre of such an applicator is necessarily identical with that at the centre of a complete ring of the same diameter, and the radial distribution is nearly the same. The dosage calculation may therefore be made from Tables I and II. A similar arrangement with a semi-circle having one-sixth of the total radium at each end gives a uniform surface dose. The calculation, however, no longer conforms to Table II, the factor for $\frac{d}{D} = 0.5$ being 0.60 instead of 0.75 for a complete ring.

TABLE IV

Factors for semi-cylinders.

$\frac{L}{D}$	$\frac{d}{D}$						
	0.0	0.2	0.4	0.5	0.6	0.7	0.8
0.0	1.00	0.77	0.57	0.47	0.38	0.29	0.24
0.2	1.01	0.80	0.60	0.50	0.40	0.32	0.27
0.4	1.05	0.85	0.63	0.54	0.45	0.37	0.33
0.6	1.13	0.92	0.68	0.61	0.52	0.44	0.40
0.8	1.20	1.01	0.76	0.68	0.59	0.51	0.47
1.0	1.28	1.09	0.84	0.75	0.67	0.58	0.54
1.2	1.39	1.18	0.94	0.85	0.74	0.65	0.61
1.6	1.61	1.38	1.11	1.03	0.93	0.83	0.79
2.0	1.88	1.63	1.33	1.21	1.11	1.00	0.95

An abridged set of factors relating to semi-cylinders of the interstitial type is given in Table IV. The factors refer only to the maximum dose at any particular $\frac{d}{D}$ value, and the distribution is by no means uniform over the coaxial cylindrical surfaces. Such arrangements are specially useful in certain cases when supplemented by an axial tube and additional lines corresponding to the one-sixth parts at the ends of a semi-circle. A proper discussion of these cases would be irrelevant to the present article.

In conclusion, Mr. H. M. Parker wishes to express his appreciation of the support given by the Manchester Committee on Cancer during the period in which the above calculations were made.

Summary

The physical evidence for the validity of the distribution rules and of the tables of factors for the dosage calculation is given. Some special cases relating to incomplete rings and cylinders are briefly discussed.

August 1936

500   R. Paterson, H. M. Parker and F. W. Spiers

PART III

By F. W. SPIERS

THE RADIATION FIELD OF A CONTINUOUS CYLINDRICAL APPLICATOR

The present section adds to the foregoing treatment of cylindrical applicators a method of calculation applicable to any distribution of radium which can be regarded as an approximately continuous cylinder. The intensity distribution in and around such a cylinder can be obtained by integration of the simple ring formula, and the expressions derived admit of fairly rapid numerical computation in terms of standard mathematical tables. The calculations bear out the principles put forward in Parts I and II, and where numerical comparison has been made very satisfactory agreement has been obtained between the two methods of treatment. The results are represented by graphs suitable for dosage determination, and are particularly applicable to interstitial cases, where it becomes important to assess the dose at general points inside the cylinder.

Method of calculation

(a) *A point in the end plane of the cylinder.*—The calculation is based upon the equation given by Sievert[4] for the intensity due to a ring source of radium. In the first case a point P in the end plane of the cylinder will be considered (Fig. 4a). The whole distribution is symmetrical about the cylinder axis, and hence the intensity at any point on a co-axial circle through P will be the same as that at P.

The cylindrical surface is regarded as made up of a large number of rings, each of radius a and thickness dh. If the surface density of the radium on the cylinder is σ mgrs. per sq. cm., then the line density of a ring such as X is $\sigma \cdot dh$ mgrs. per cm. Putting this in the equation for a ring source (4), the intensity at P in r per hour due to the ring X is:—

$$dI_P = \frac{2\pi a (\sigma \cdot dh) I_o}{\{(a^2 + y^2 + h^2)^2 - 4a^2 y^2\}^{1/2}} \quad \ldots \ldots \ldots (1)$$

where a = the radius of the cylinder;
dh = the thickness of the ring;
h = the distance of the ring from the end plane;
y = the distance of P from the cylinder axis;
I_o = the intensity in r per hour at 1 cm. from 1 mg. of radium for the filtration in question.

Vol. IX, No. 104

A System of Dosage for Cylindrical Distributions of Radium   501

The intensity at P due to the whole cylinder is obtained by integrating the expression (1) over the length of the cylinder:—

$$I_P = 2\pi a\sigma I_o \int_0^{h_1} \frac{dh}{\{(a^2 + y^2 + h^2)^2 - 4a^2y^2\}^{1/2}}$$

$$= 2\pi a\sigma I_o \int_0^{h_1} \frac{dh}{[\{(a+y)^2 + h^2\}\{(a-y)^2 + h^2\}]^{1/2}} \quad \ldots \quad (2)$$

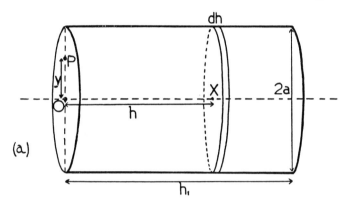

(a)

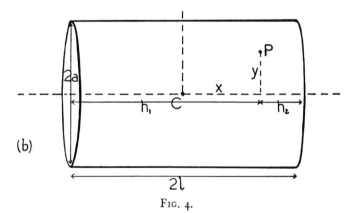

(b)

Fig. 4.

By substituting $z = h/(a - y)$ in this expression and reducing, the integral becomes:—

$$I_P = \frac{2\pi a\sigma I_o}{a + y} \int_0^{z_1} \frac{dz}{\sqrt{(1 + z^2)(1 + c^2z^2)}} \quad \ldots \ldots \ldots (3)$$

where $c = (a - y)/(a + y)$*.

---

*The integration is only possible if $\sqrt{1-c^2}$ lies between 0 and 1; but since $\sqrt{1-c^2} = \dfrac{2\sqrt{ay}}{a+y}$, this condition is satisfied for all values of a and y.

AUGUST 1936

502   R. Paterson, H. M. Parker and F. W. Spiers

The part to be integrated in equation (3) is an "elliptic integral" which can be evaluated from standard tables compiled for the purpose (5). Values of the integrals are tabulated in terms of two variables, ϕ_1 and a, related to the cylinder dimensions in the following manner:—

$$\phi_1 = tan^{-1} z_1 = tan^{-1} \frac{h_1}{a-y}, \quad a = sin^{-1} \sqrt{1-c^2} = sin^{-1} \frac{2\sqrt{ay}}{a+y} \quad \ldots (4)$$

The intensity at P can therefore be calculated from the equation:—

$$I_P = \frac{2\pi a \sigma I_o}{a+y} \cdot F(\phi_1 a) \quad \ldots \ldots (5)$$

where $F(\phi_1 a)$ is the elliptic integral obtained from the tables using the appropriate values of ϕ_1 and a determined by the expressions (4).

(b) *A general point inside the cylinder.*—The method can now be extended to the general case of any point P inside the cylinder which has coordinates (x,y) referred to the centre of the cylinder C (Fig. 4b).

For the intensity at P the effects of the two cylinders h_1 and h_2 must be added, giving:—

$$I_P = \frac{2\pi a \sigma I_o}{a+y} \left\{ F(\phi_1 a) + F(\phi_2 a) \right\} \quad \ldots \ldots (6)$$

where a is the same as in (4) and

$$\phi_1 = tan^{-1}\left(\frac{h_1}{a-y}\right) = tan^{-1}\left(\frac{l+x}{a-y}\right), \quad \phi_2 = tan^{-1}\left(\frac{h_2}{a-y}\right) = tan^{-1}\left(\frac{l-x}{a-y}\right).$$

These expressions are much more convenient in form if the coordinates of the point P are stated in terms of the half-length l and the radius a of the cylinder. The coordinates of P are now taken to be:—

$$m = x/l \text{ and } b = y/a, \text{ and further let } l/a = k.$$

This ratio k is the ratio length/diameter, and expresses the "elongation" of the cylinder; it is identical with the ratio $\frac{L}{D}$ used in Parts I and II. Similarly, $b = y/a$ is equivalent to the ratio $\frac{d}{D}$ used earlier. The final equations for determining the intensity at any point P now are:—

$$I_P = \frac{2\pi \sigma I_o}{1+b} \left\{ F(\phi_1 a) + F(\phi_2 a) \right\} \quad \ldots \ldots (7)$$

where $a = sin^{-1}\left(\frac{2\sqrt{b}}{1+b}\right)$, $\phi_1 = tan^{-1}\left(k \frac{1+m}{1-b}\right)$ and $\phi_2 = tan^{-1}\left(k \frac{1-m}{1-b}\right)$ (8).

Equations (7) and (8) apply equally well to points beyond the end of the cylinder; here $x/l > 1$ and the second integral must be subtracted. For points altogether outside the cylinder, for which $y/a > 1$ the equations still hold with $\phi = tan^{-1} k(1 \pm m)/(b-1)$. For a point on the axis, the expression (7) reduces to that for a line source, and the intensity is simply proportional to the angle subtended at the point by the cylinder. It will be seen that in equations (7) and (8) the variables k, m and b are relative figures, expressing the elongation of the cylinder and the relative position of the point P, and that the actual magnitude of the cylinder is only involved in the quantity σ, the amount of radium per sq. cm. of the surface. It thus becomes possible to construct a set of curves showing the relative variation of intensity in the cylinder with reference to some point for which the absolute intensity can readily be calculated. Accordingly, the equations are used to relate the intensity at any point in a cylinder to the intensity at the centre of the corresponding ring, which has the same radius as the cylinder and the same amount of radium distributed uniformly on it.

No account of oblique filtration is included in the above calculation, but consideration of this question in Part II has shown that, except for heavily filtered needles, the effect is not a serious one

(c) *The radiation outside the cylinder.*—The radiation falling on a point in the central plane of the cylinder, but outside the cylindrical shell, will now be considered. For such a point the distance y from the axis is greater than the radius a, and the intensity, according to equation (6), will be:—

$$I_P = \frac{2\pi a \sigma I_o}{a+y} \{2 F(\phi_c a)\} = \frac{2\pi a \cdot 2l \cdot \sigma I_o}{2l(a+y)} \{2 F(\phi_c a)\}$$

$$= \frac{Q I_o}{2l(a+y)} \{2 F(\phi_c a)\} \quad \ldots \quad (9)$$

where $Q = 2\pi a \cdot 2l \cdot \sigma$ = total amount of radium on the cylinder,

$$\phi_c = tan^{-1}\left(\frac{l}{y-a}\right) = tan^{-1}\left(\frac{k}{b-1}\right) \text{ and } a = sin^{-1}\left(\frac{2\sqrt{ay}}{a+y}\right) = sin^{-1}\left(\frac{2\sqrt{b}}{1+b}\right).$$

Since the expression (9) does not include the effect of oblique filtration, it will be most useful to compare it with the expression for the intensity at the same point had all the radium Q been placed on the cylinder axis of length $2l$. Using the formula for a simple line source with no oblique filtration, the intensity would be (6):—

$$I'_P = \frac{Q I_o}{2 l y} \{2 \phi'\} \quad \ldots \ldots \ldots \ldots (10)$$

where $\phi' = tan^{-1}\left(\dfrac{l}{y}\right) = tan^{-1}\left(\dfrac{k}{b}\right)$.

Hence, from (9) and (10) the ratio I_p/I'_p is:—

$$\frac{\text{Intensity for cylinder}}{\text{Intensity for line}} = \frac{y}{a+y} \cdot \frac{F(\phi_c a)}{\phi'} = \frac{b}{1+b} \cdot \frac{F(\phi_c a)}{\phi'} \quad \ldots (11)$$

This equation can be used to test the approximation involved in considering the cylinder as a line source for purposes of intensity calculation.

Results.

(a) *Radiation inside a cylinder.*—Using equations (7) and (8), graphs have been constructed to present the variation of intensity inside a cylinder in a manner suitable for application to dosage problems.

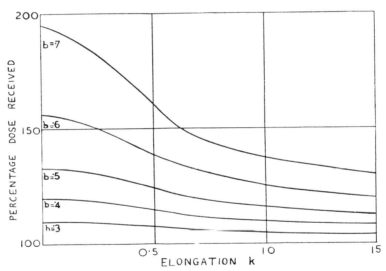

FIG. 5. Variation of dose in the central plane of the cylinder as a function of the elongation k. Dose at the centre of the cylinder taken as 100.

Fig. 5 gives the variation along any radius in the central plane of the cylinder as a function of the elongation k. The intensity at the centre of the cylinder is taken to be 100 per cent in order to indicate the effect of the factor k on the radial distribution. Continuous curves are drawn for points on the radius corresponding to $b = y/a = 0.3$ up to $b = 0.7$. It will be seen that short cylinders (k small) show a rapid variation along the radius, the centre point being undesirably low, but that the distribution of intensity is greatly improved by using cylinders with k equal to 0.7 or more.

The sets of curves given in Figs. 6 and 7 are drawn to give the actual dose distribution inside a cylinder as the factor k is varied. The intensity

A System of Dosage for Cylindrical Distributions of Radium

figures are arranged to give the actual dose in r on the basis that the dose at the centre of the corresponding ring of the same diameter is 1,000 r. Hence, Table I, which gives the number of milligramme-hours to produce 1,000 r at the centre of the corresponding ring, will give the number of milligramme-hours required to produce the actual doses at points inside a given cylinder, as read off from Figs. 6 and 7. These sets of curves show the dose variation along the axis and along parallel lines at distances $y/a = 0.3$ and $y/a = 0.6$ from it, and give the dose at a sufficient number of points to indicate the distribution for most cylinders occurring in practice.

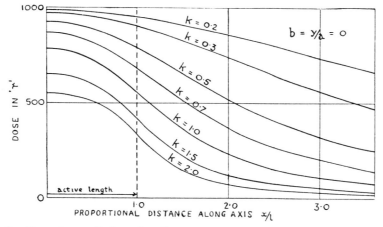

FIG. 6. Dose along cylinder axis. Dose at centre of corresponding circle = 1,000 r.

A gradual fall of the dose along the axis is shown by the short cylinders, while a steep fall in the dose occurs immediately outside the active length in the case of long cylinders. The long cylinders, therefore, confine their radiation field more strictly to the geometrical limits of the cylinder. In practice the active length of the cylinder should cover the tumour adequately; the fall of dose beyond the active length and outside the cylinder itself then ensures a reasonably well-defined field of radiation with the minimum damage to the surrounding tissues.

(b) *Radiation outside a cylinder.*—It is sometimes necessary to assess the dose outside a cylindrical distribution of radium, and this can be done, at any rate approximately, by considering all the radium to be concentrated uniformly along the cylinder axis. Sievert[7] has studied the distribution from line sources in great detail, and has evaluated tables which take into account both the distribution of the radium throughout a cylindrical volume and the oblique filtration of the platinum container. Two of the present authors[1] have used

506   R. Paterson, H. M. Parker and F. W. Spiers

Sievert's tables to construct a series of "Tube Graphs," from which the magnitude of the dose can be determined for clinical purposes. These graphs give the required relation between milligramme-hours and the dose in r for points at various distances from line sources of radium. The dose from a

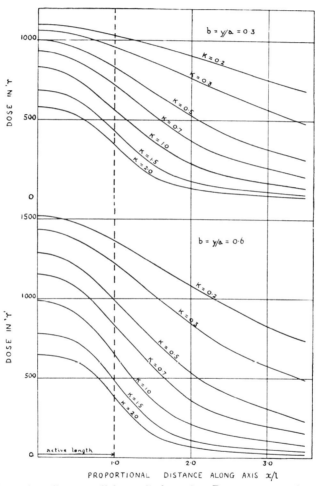

FIG. 7. Dose along lines parallel to cylinder axis. Dose at centre of corresponding circle = 1,000 r.

cylindrical distribution is, however, greater than that calculated by using the tube graphs, and the magnitude of the discrepancy involved depends upon the elongation of the cylinder and the distance of the point in question from the axis. With the aid of equation (11) the necessary corrections have been determined and are tabulated in Table V. The figures given are the percentages to be *added* to the dose obtained by the tube graph calculation.

A System of Dosage for Cylindrical Distributions of Radium

(c) *The equivalence of a narrow belt of radium and a ring source.*—In Parts I and II reference has been made to the occasional necessity of employing a belt of short radium needles rather than the usual ring source which was considered for the purpose of calculation. Table VI compares the radiation field of a typical short cylinder, 1·5 cm. long and 5 cm. diameter, with the

TABLE V

Radiation Field outside a Cylinder: Percentage Corrections to be *added* to the Tube Graph Dose.

Value of $b = \dfrac{y}{a} = \dfrac{d}{D}$	Value of the Cylinder Elongation $k = \dfrac{L}{D}$				
	1	2	3	5	10
1·3	62	39	33	29	27
1·5	44	28	22	19	17
2·0	25	17	13	10	8
2·5	16	11	8	7	5
3·0	11	8	7	5	4
3·5	9	7	5	4	3
4·0	7	5	4	3	2

TABLE VI

Comparison of Ring, 5 cm. diameter, and Cylinder, 5 cm. diameter, 1·5 cm. long.

Distance from the axis	Distance along the axis in cm.						
	0	0·5	1	2	3	4	5
1·5 cm.	147·8 / *160·8*	137·3 / *142·9*	112·1 / *110·4*	66·3 / *64·3*	41·7 / *40·7*	28·0 / *27·6*	20·0 / *19·7*
0·75 cm.	109·0 / *113·1*	104·5 / *107·5*	93·0 / *93·7*	64·4 / *63·4*	42·7 / *41·9*	29·1 / *28·6*	20·7 / *20·4*
0 cm.	100·0 / *102·9*	96·7 / *99·0*	87·7 / *88·7*	63·3 / *62·8*	42·9 / *42·2*	29·4 / *28·9*	20·9 / *20·6*

Intensity at centre of cylinder taken as 100·0 and figures for ring are printed in italics.

field of a ring source of the same diameter. The intensity at the centre of the cylinder is in this case taken as 100·0 and the figures for the ring source are printed in italics. The comparison is extended over the volume usually considered in constructing a compound applicator and shows satisfactory agreement between the two sets of figures. At distances greater than 1·5 cm.

from the axis there are considerable differences between the two radiation fields.

In conclusion, the author has much pleasure in thanking Mr. A. A. D. La Touche, F.R.C.S., and Mr. H. M. Parker, M.Sc., for their interest in this problem and for much helpful discussion.

Summary

The distribution of radiation in and around a continuous radium cylinder is calculated by a method of integration based on the formula for a radioactive ring, and the results are represented by a series of graphs which enable the dose to be determined at various points inside the cylinder.

REFERENCES

[1] Paterson, R., and Parker, H. M., *British Journal of Radiology*, vii, 1934, p. 592.
[2] Goodfellow, D. R., *British Journal of Radiology*, viii, 1935, p. 669.
[3] Hutchison, R. G., *British Journal of Radiology*, viii, 1935, p. 306.
[4] Sievert, R., *Acta Radiologica*, i, 1921, p. 89.
[5] Suitable tables are to be found in the following publications:—
    (a) *Synopsis of Applicable Mathematics*, by L. Silberstein, published by Bell and Sons, Ltd., London, 1923.
    (b) *Tables of Functions*, by E. Janke and F. Emde, published by B. G. Teubner, Leipzig and Berlin, 1933.
[6] Mayneord, W. V., *British Journal of Radiology*, v, 1932, p. 677.
[7] Sievert, R., *Acta Radiologica*, 11, 1930, p. 249.

Dosage Measurements by Simple Computations

Radiology, Volume 32

May 1939

Reprinted with permission of the Radiological Society of North America.

DOSAGE MEASUREMENTS BY SIMPLE COMPUTATIONS[1]

By HERBERT M. PARKER, M.Sc., Christie Hospital and Holt Radium Institute, Manchester, England[2]

THE logical basis of tissue dosage, to the physicist, would be the actual increase of energy per unit volume of tissue brought about by irradiation. Direct methods of measuring dosage in these terms are not practicable, but the same object is served if an indirect means giving equivalent results can be set up. The conditions under which ionization measurements can be used for this purpose have been the subject of exhaustive investigations, notably by L. H. Gray. Briefly, Gray's conclusions may be summarized as follows:

The energy absorption per unit volume of a medium in which quantum radiation is being absorbed is proportional to the ionization per unit volume arising in a small air-filled cavity in the medium. The cavity must be situated at a depth in the medium not less than the maximum range of secondary electrons in the medium, and the linear dimensions of the cavity must be small compared with the range of the corpuscular radiation. Transcribing these conditions to conform with measurements in small ionization chambers it is seen that the wall thickness must be equal to the range of secondary electrons and that the proper linear dimensions vary from a few millimeters for gamma rays to one-tenth of a millimeter for x-rays generated at 200 kv. In principle then, it is possible to measure tissue doses by observations on the ionization in small chambers having tissue-equivalent walls. The limiting dimensions of the chambers are so small for x-rays at normal voltages that grave, perhaps unsurmountable, experimental difficulties are introduced. For this reason it has been proposed that all dosage measurements should be related to observations in chambers having air-equivalent walls, and a revised definition of the international roentgen has been put forward on this basis. With air-equivalent walls the total ionization is strictly proportional to the volume of the chamber which can, therefore, be chosen of more practicable size.

Now under most ordinary conditions the tissue to be treated consists entirely of elements of low atomic number and the specific gravity is reasonably constant, so that the observed ionization per cubic centimeter in a minute tissue-wall chamber would be exactly proportional to the energy absorption in tissue, and further, this ionization per cubic centimeter would differ by a constant amount (about 4 per cent) from that in an air-wall chamber of any size. Hence the proposal to measure tissue dose in roentgens by means of air-wall chambers is acceptable as a general principle. Nevertheless circumstances do arise in which the energy absorption in tissue is not directly correlated with the ionization in an air-wall chamber, as, for example, in the neighborhood of skin or bone where elements giving an appreciable photo-electric effect are to be found. It is felt that any new phrasing of the definition of the roentgen should not preclude the possibility of allowing for these special effects in the statement of tissue dose by a too rigorous insistence on the merits of measurements with air-wall chambers.

The problem of tissue-dose measurement is by no means solved when we have laid down satisfactory experimental conditions. It is not feasible to insert ionization chambers in tissue to measure the dose in roentgens directly, and if it were the labor involved in exploring the complete radiation field for each individual case would be prohibitive. The question is not so much how to measure tissue dose, but how to avoid its measurement. I want,

---

[1] Presented before the Fifth International Congress of Radiology, at Chicago, Sept. 13–17, 1937.
[2] Present address, Swedish Hospital, Seattle, Washington.

then, briefly to review the methods by which this can be accomplished in the light of the principles of measurement that I have outlined.

Although there is no sharp delineation in the theory of ionization measurements between x-rays and gamma rays, there is in practice a rather marked differentiation in the approach to dosage simplification for the two agents. For x-ray treatments the customary method is to use a series of isodose charts appropriate to one particular type of tube, together with tables of dosage rate for all the variations of tube voltage, current, filter, focus-skin distance, and field size to be employed. Many of these variables can be treated collectively if we postulate a change in the method of construction of isodose charts. Instead of writing 100 per cent as the surface dosage rate for each chart, let 100 per cent represent the dosage rate at a given point in air under standard conditions. The figures on isodose charts then give at once the tissue dose for the given incident quality of radiation for the focus-skin distance and field size in question. This assumes that the contribution of scattered radiation at any point depends only on quality, as measured by half value layer, and not on the spectral distribution. This appears to be valid over the range of variation that occurs in therapy. It assumes also that independent observations of the dosage rate in a measuring phantom under identical conditions of irradiation will agree. Figure 1 shows

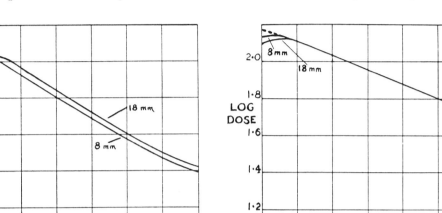

Fig. 1. Influence of chamber diameter on depth dose curves.

Fig. 2. Logarithmic depth dose curves set coincident at a depth of 5 cm.

curves of depth dose for 200 kv. radiation incident on a large block of unit density wax, as recorded by two chambers of diameter 8 mm. and 18 mm., respectively. Both chambers had air-equivalent walls and their response to radiations of H.V.L. 0.3 mm. Cu to 3.0 mm. Cu ran exactly parallel. Yet the two curves are quite different in shape. The initial reading in each case was made with the chamber half immersed so that it is not surprising that there should be a marked difference in behavior over the range in which each chamber proceeds to the fully immersed position. It is more important that the increased depth dose persists at all depths. Suppose the curves are redrawn (Fig. 2) to make them coincide at an arbitrary depth, which here is 5 cm. The curves now coincide exactly except near the surface and one is tempted to derive a true depth dose curve by extrapolating the common part to zero depth. The method would acquire physical significance if the actual dosage rates at this depth were equal but this is not so. The value was 3 per cent more in the large chamber at 5 cm. depth, and 3 per cent less at the surface. The difference at 5 cm. depth is simply explained by the reduced thickness of wax to be traversed in the case of the larger chamber.

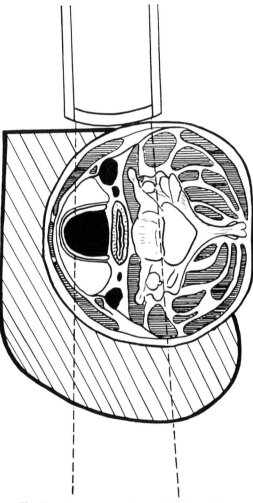

Fig. 3. Arrangement of wax block to bring a neck irradiation into approximate agreement with large phantom measurements.

We should expect an indefinitely small chamber to give readings some 3 per cent less again and this was verified by measurements in a chamber of 2 mm. diameter. At the same time, the apparent surface dosage rate rose by 3 per cent. The estimated shape of depth dose curve for an indefinitely small chamber approximated closely to the straight line in Figure 2, although the readings within 2 mm. of the surface were unreliable. The dependence of dosage rate or chamber size is given in Table I for a few special cases. To make isodose charts of general utility it will be seen that all the observations must either

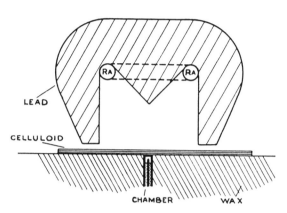

Fig. 4. Measurement of skin dose under conditions which generate copious secondary radiation.

be made with chambers much smaller than those commonly employed or corrected from measurements with two chambers. The use of small chambers is specially advisable when measuring the effects near the geometric edge of a beam the gradient of whose dosage rate is high.

The therapist equipped with a set of isodose charts derived from readings in a large phantom with beams at normal incidence desires to know to what extent they will be valid under the varying treatment conditions arising in practice. Better still, the physicist should define conditions under which the same charts may be used directly, so that the estimation of dosage can be kept in its simplest terms. The use of auxiliary scattering bodies to insure the restoration of equivalence between experi-

TABLE I.—DEPENDENCE OF DOSE ON CHAMBER DIAMETER

Chamber Diameter	Relative Dose	
	Near surface	5 cm. deep
V. small	143	100
8 mm.	138	103
18 mm.	135	106

ment and practice in irradiations at oblique incidence, or of limited masses of tissue, such as limbs and necks, is a familiar procedure. General conditions, which should be regarded as an integral part of a dosage system, can be laid down. Thus for 200 kv. radiation full ionization is received at any point in the irradiated field when the scattering medium extends 8 cm. laterally beyond the geometrical limits of the beam and to 8 cm. beyond the point considered. For most purposes a thickness of 5 cm. is sufficient. The same thickness can be used over a wide range of radiations on account of the compensation between the penetration of the beam and the abundance of scatter. A typical application of the principle is shown in Figure 3.

Compensation for anomalous scattering regions within tissue is not feasible, but it appears, as, for example, in the measurements by Mrs. Quimby and her collaborators on cadavers, that the presence of bone or of small cavities does not seriously disturb the estimation of tissue doses from isodose charts. In some special cases, notably for lung cavities, a discrepancy arose which could be accommodated by a simple correction curve in such a way that the dosage for individual cases could be derived with a probable error not exceeding a small percentage. It may be claimed then that the tissue dose, which would be

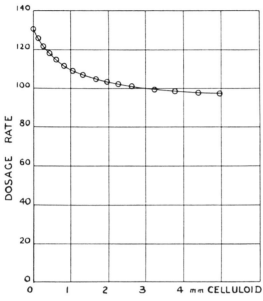

Fig. 5. Variation of skin dose with thickness of celluloid screen in Figure 4.

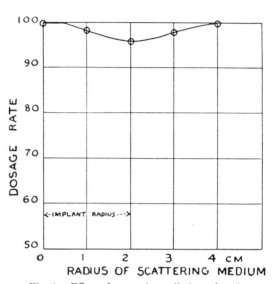

Fig. 6. Effect of scattering cylinders of various radii on the dose within a cylindrical radium implant. (Chamber wall = 1 mm. electron metal.)

recorded by a small air-walled chamber, can be written down within perhaps 5 per cent under all treatment conditions by simple computations from the basic charts and a knowledge of tube output.

It remains to be seen how far these measurements would be affected by taking into account the presence of skin, bone, and cartilage associated with tissue. It has been assumed so far that the energy absorption in tissue will be directly proportional to the measured ionization equivalent in air-walled chambers. It is evident that the true absorption in these particular substances may be quite different, mainly on account of the photo-electric absorption in the sulphur of skin, and the calcium and phosphorus of bone. L. H. Gray has estimated, for example, that the ionization in a minute chamber with skin-equivalent walls may well be 50 per cent in excess of that in a tissue-wall chamber for x-rays generated at 200 kv. Moreover, the range of this additional effect extends to about

80μ from the actual sulphur-bearing layers. Blood capillaries are found within from 40μ to 140μ of such layers. Consequently it may be expected that the additional effects will play a part in the production of visible erythema reactions. It will follow that tissue doses, regarded as energy absorption, but determined by convention in terms of the ionization in air-walled chambers, may be seriously in error for soft radiation in just the cases which have been subject to the most careful clinical investigation. For the present, the best one can do is to regard an erythema dose as not yet having graduated to the status of a physical dose. Ultimately it may be possible to make a sufficiently close approximation to the photo-electric contribution, which could readily be applied to isodose charts of the type that I have described. In the meantime, it would be unfortunate if the anomalies of skin dose were allowed to discourage the statement of tumor doses in roentgens.

The analogous effects in bone and cartilage do not give rise to problems of physical interest. It is obvious that zones of high dose will occur within and in the immediate vicinity of these bodies. These zones constitute a potential danger in treatment and their appreciation can safely be left to the clinician.

In comparison with the x-ray problems the measurement of absolute tissue dose with radium is straightforward. The difficulties diminish rapidly as the wave length of the incident radiation diminishes. The permissible dimensions of the chamber increase inversely as λ^2, and the disturbing photo-electric effect diminishes as λ^3. The choice of material for chamber walls is also much less critical, and well-established corrections can be employed to convert readings in such convenient materials as elektron metal to the conventional equivalents for air walls. Against this one has to set the increased wall thickness of the chambers required to exclude secondary corpuscles generated at distant points from the chamber. It is well known that a thickness equivalent to some four millimeters of graphite is necessary and sufficient for this purpose. This condition modifies the conception of tissue doses under two circumstances, namely, near a skin surface or close to a radium needle used interstitially. Figure 4 illustrates a typical experiment on the magnitude of the first effect. An apparatus similar to a radium bomb was used to irradiate a block of wax in which was inserted a small cylindrical chamber with its axis at right-angles to the surface. The walls of the chamber consisted of elektron metal and the upper end was machined to a thickness equivalent to 0.7 mm. of tissue. Successive layers of celluloid were applied to the phantom surface and the ionization was noted at each stage. The ionization was reduced by approximately 25 per cent by the addition of 3 mm. of celluloid (Fig. 5). The conditions in this experiment were exceptionally favorable for the generation of extraneous secondary particles, but in some radium treatments the state of affairs is not dissimilar. For the sake of consistency, an adequate layer of celluloid, wax, or similar material should be placed on a skin surface during irradiation by external sources, so that the dose at a superficial point is derived entirely from the ionization produced by corpuscular radiation generated in the immediate vicinity.

The analogous effect round a radium needle in tissue is of little importance. It occurs in a region where the gamma-ray dose is already in excess of the general value and the extra ionization arising from secondary β particles produced in the needle wall merely shortens the total time for which a given needle can be safely left *in situ*.

Apart from these two effects the estimation of tissue dose for gamma radiation presents the fullest scope for simple calculation. Once the dosage rate in roentgens at one centimeter from a point source of one milligram of radium with standard filtration has been determined, the tissue dose at any point can be written down in general without further reference to ionization measurements. This basic value has been the subject of exhaustive investigation by radiophysicists, and the more recent measurements have converged to a figure which can probably be written as 8.3 ± 0.2 roentgens per hour with 0.5 mm. Pt filtration.

Three types of radium treatment, which are amenable to different methods of dosage simplification, can be distinguished. These are the so-called bomb therapy, irradiation by radium moulds, and interstitial therapy. Dosage estimation in bomb therapy proceeds by means of isodose charts constructed in a manner identical with that for x-ray therapy. With radium moulds the determination of dose is particularly easy. One is normally concerned with conditions in the first few centimeters of tissue and within this range the net effect of absorption and scattering on the dose received at any point is so small that it can, in general, be neglected. The determination of dose thus becomes a purely mathematical exercise. Simple solutions depend upon the possibility of stating rules for the distribution of radium sources so that the dosage rate shall be approximately uniform and calculable over sheets parallel to the radium applicator. The complexity of any such system of rules is fixed primarily by the degree of accuracy to which the dose is made uniform over a given surface. Considering the inevitable error in the measurement of a treating distance of the order of one or two centimeters, a margin of error of ± 10 per cent is not unreasonable. This gives sufficient scope for generalizations which lead to a system of dosage calculation whereby the tissue dose at any relevant point can be written down at once when the area of the radium applicator, the treatment distance, and the effective radium content are known.

The extent to which the absorption of the primary radiation in tissue will influence a tissue dose calculated in this way depends very much on the geometrical arrangement of each particular case. It is probably for this reason that the reported values of the effective absorption are not consistent. It seems to be agreed that the net absorption is not more than 4 per cent per centimeter; measurements that I have made suggested that a value of 2 per cent per centimeter would be a fair average. In view of the incidental errors of this type of therapy, it seems unnecessary to apply an absorption correction to depth doses up to 3 cm. deep.

Conditions in interstitial radium therapy are rather different. On paper, at least, it is possible to give rules for the distribution of radium sources in space in such a way that the dose will be approximately uniform throughout the volume contained, except within a few millimeters of each source, and can be calculated from the total volume and the effective radium content. These proposed arrangements have been set up in the laboratory and the radiation field explored by small ionization chambers. The dose in air has invariably been found to be consistent with the calculated value and the change in dose by immersion of the whole arrangement in a scattering medium of unit-density has been shown to be negligible. Actually, the dose reaches a minimum when the scattering medium just occupies the space contained by the needles (Fig. 6). With a phantom approximating to real conditions the tissue dose is never far from the air dose. At first sight, then, the calculation of dose in interstitial therapy would appear to be the simplest possible case. This would be true if radium needles could be inserted in predetermined positions, which is clearly impossible. The problem of dosage thus becomes the purely geometrical one of determining the actual configuration of sources in a radium implant. In some cases a study of two radiographs of the implant is sufficient for this purpose, but more generally it is essential to reconstruct the implant in space. To take an extreme case, the series of implants shown in plan in Figure 7 can be arranged to give identical radiographs in both anteroposterior and lateral films. The dose at a point A changes in this series by a factor of five. In real practice one can expect errors up to perhaps 50 per cent, if the dose is estimated from radiographs alone. For this reason we have to regard the spatial reconstruction, which can be readily performed by optical means, as an integral part, and in fact the principal part, of the mechanism of dosage estimation.

the only source of error of any importance being connected with special effects in skin. One essential feature stands out—that

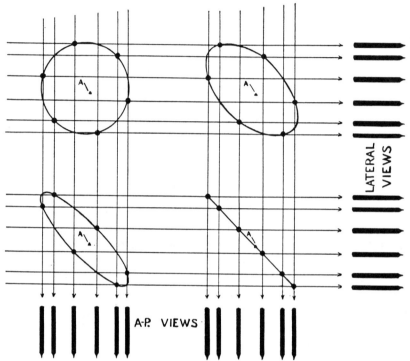

Fig. 7. Identity of the radiographs of four different implants, viewed by parallel beams of x-rays. With the normal divergent beams small differences arise which are inappreciable to the eye but can be analyzed geometrically or, more conveniently, by a special "implant reconstructor."

SUMMARY

To summarize this somewhat disconnected survey of the possibilities of reducing the determination of tissue doses to a system of elementary computations, it may be claimed that for radium treatments a simple system, sufficiently accurate for clinical requirements, can be established. For x-ray treatments, with which may be grouped radium bomb therapy, satisfactory results can be achieved by the compilation of isodose charts in a special way, much closer collaboration between the physicist and the radiotherapist is called for. Simplicity in dosage cannot be achieved if the therapist uses his physical agents in any arbitrary manner, nor can the therapist tolerate restrictions on his technic so prescribed as to limit its utility to a few selected cases. Only by a careful consideration of the requirements of both specialists can a general scheme of treatment and dosage be built up, and not until such a system is available equally well to the smallest clinic as to the largest can we be said to have solved the problem of tissue dosage.

Radium Implant Reconstructor

The British Journal of Radiology, Volume XII

August 1939

Reprinted with permission of the *British Journal of Radiology*.

A RADIUM IMPLANT RECONSTRUCTOR

By H. M. Parker, M.Sc., F. Inst. P., and W. J. Meredith, M.Sc.
Holt Radium Institute, Manchester

(Received April 14, 1939)

A RADIUM IMPLANT RECONSTRUCTOR

By H. M. PARKER, M.Sc., F.Inst.P., and W. J. MEREDITH, M.Sc.

Holt Radium Institute, Manchester

(Received April 14, 1939)

1. FUNCTION

THE application of γ rays in radium therapy is normally performed by one of three methods, *viz.*:—

(1) Teleradium therapy.
(2) Use of external applicators or moulds.
(3) Interstitial, including intra-cavitary applications.

In all cases a precise knowledge of the position of the radium sources with respect to the tissue to be treated is required. In the first and second methods this can be obtained by direct calculations. The third method introduces a complication, due to the inaccessibility of the radium sources, which are inserted in the tissue. Attempts have been made to estimate the dosage under these conditions by measurements with microionisation chambers, inserted in the tissue, but it is seldom found feasible to obtain a reading at more than one random point in view of the trauma caused by the insertions. This paper describes an apparatus for the accurate reconstruction in three dimensions of the arrangement of needles forming a "Radium Implant." It was developed in connection with the methods of distribution for interstitial radium therapy, devised by R. Paterson and one of us,[1] and has become a routine agent for the determination of the proper length of treatment. The models produced by the instrument have emphasised certain trends in implantation, and have led to improvements which will be described elsewhere.

2. IDEAL TECHNIQUE

If two radiographs of the needles *in situ* are obtained, taken at the same instant by two X-ray tubes fixed at right-angles to one another, and each at a known, fixed distance from films in fixed holders, the production of a model is quite simple. In this case, the reconstructor would simply consist of two units mounted at right-angles, each consisting of a lamp, which corresponds to the X-ray tube, and a carrier for the radiograph, or a tracing of it, at a distance from the lamp equal to the distance between the X-ray target and the film. Dummy needles, identical in size to the radium needles used in the implantation, and mounted on thin wires, would then be manipulated in front of the tracings until the shadows cast by the needles exactly

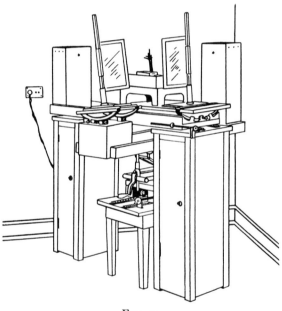

FIG. 1.
General view of Reconstructor.

fitted the tracing. The success of the operation depends on a knowledge of:—

(a) The distance from target to film.
(b) The point where the normal (central) ray strikes the film.
(c) The true lengths of the radium needles.

The Reconstructor

The ideal of two simultaneous exposures is hardly ever attainable, however, the normal procedure being to take one exposure and then turn the patient through a nominal right-angle for the second exposure. This movement of the patient and accompanying necessary movement

Vol. XII, No. 140

H. M. Parker and W. J. Meredith

of films, introduces six "degrees of freedom" in the taking of the radiographs, all of which have been allowed for in the reconstructor, which is illustrated in Figs. 1 and 2.

The movements provided in the reconstructor are:—

This movement is necessary because the central ray to one film may be at varying distances from the second film.

4 and 5. It must be possible to tilt each film carrier (C) about its centre to allow for any tilt of the patient with reference to the film. The

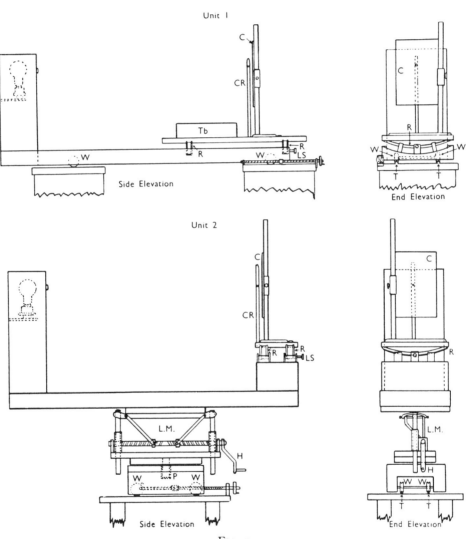

FIG. 2.
Details of construction of the Reconstructor.

1. Unit 2 may be rotated with respect to Unit 1, thus allowing reconstruction from radiographs taken at any angle. The unit pivots about the rod P.

2 and 3. Each unit is mounted on wheels (W), on rails (T), and can be moved parallel to its longest axis, under the control of a screwed rod.

carriers are mounted on curved rails (R), and may be locked in any position by the lock screws (LS).

6. The height of Unit 2, relative to Unit 1, may be varied by means of the lift mechanism (LM), which is controlled by means of the handle (H). This movement allows for any

AUGUST 1939

A Radium Implant Reconstructor

slipping of the patient on the couch, or alteration of posture if standing, whilst being turned between the exposures; it also allows for the fact that the needles may not be in the same position relative to the central ray in each picture.

3. Procedure

The procedure for the production of a model from two radiographs, of a radium needle implantation, is as follows:—

Tracings of the needle shadows, with the point of contact of the central ray on each film marked on them, are made, the corresponding shadows in each are identified, and the tracings mounted in the holders (C). These are adjusted until the central ray to each (defined by a hole in the rod (CR) mounted in front of each holder) falls on to the traced central point. Then a dummy needle, mounted on a wire, placed in a sandwich of plasticine between sheets of perforated zinc placed on the table (Tb), is manipulated until its shadow exactly fits the tracing of a needle in holder C1. Unit 2 is then adjusted until a fit is obtained between the shadow on its holder and the corresponding needle tracing—the usual adjustments are the cross-tracking of Unit 1, the movement of Unit 2 to obtain the correct magnification, and then some small tilt of C2, with a possible adjustment of the angle between the two units. A perfect fit having been obtained for the tracings on both films, a second needle is introduced and manipulated to give shadow fits for both tracings. Should it be found impossible to do this, the settings of the various movements for the first fit must have been slightly at fault, and adjustments are made in these and in the position of the first needle until register is obtained for both needles on both tracings. The setting of the apparatus is now almost certainly correct (especially if the two needles chosen are definitely non-parallel), and the placing of the other needles in their correct positions is a comparatively simple process and only consists in their manipulation in front of the tracings.

The process outlined above may be greatly simplified, however, if some marker of known shape is placed on the patient's skin during radiography and included in the radiographs. A tracing of its shadow is included on each tracing, and the actual marker may be brought on to the reconstructor, and the various movements and its position adjusted until a fit in both tracings is obtained. If the marker is of suitable shape—the one usually used is a metal ring with one diameter—no further adjustments of the instrument are necessary, and the correct positioning of the needles may be done quite quickly. All the dummies being in position, plaster of Paris is massed round the bases of the supporting wires, and when this is dry the plasticine may be removed and the model stored. The plasticine supports the needles adequately during the reconstruction, but a noticeable sagging would occur after a few hours if the model were not supported by a more rigid medium.

A model produced as above can now be measured for the purposes of dosage calculation, but it is of little value for any dosage measurements, owing to the excessive exposure to the worker that the placing of radium needles in the position of the dummies would entail. However, by the use of a special device this can be overcome. Instead of mounting the dummy needles directly on to the supporting wires, they are placed in small celluloid tubes, themselves mounted on wires. The model is made as before, and when completed, the dummies are removed from their sheaths and radium needles substituted. The dosage rate may then be measured by means of micro-ionisation chambers. For refined measurements the actual needles from the implant, in their correct relative positions, may be used to allow for the non-uniform distribution of radium in the needles.

The advantages of taking radiographs with two fixed tubes and a fixed patient have already been pointed out, and many of these still remain if the patient may remain fixed for the two exposures, even though the tube has to move. This may be done in the taking of stereoscopic radiographs, when the tube is moved through the requisite stereoscopic "shift" and the patient remains stationary. A reconstructor for this type of radiography has been developed and is shown in Fig. 3. It simply consists of two lamps at a distance apart equal to the "shift" of the X-ray tube, and at a distance from a tracing holder equal to the target-film distance. Tracings are made as before, but in this case both on the same piece of paper, in such a way that the point of contact of the central ray for each film is on the same point of the tracing for each drawing, and the shift axes coincident. This tracing is then mounted on the holder with its central axis and central spot over those of the holder, and in front of it dummy needles are manipulated as before. Alternatively, the needles, or tubes, may conveniently be suspended from their centres by loosely coiled soft copper wires mounted in a cage-like device. Such tubes take up any desired new position at a touch. This modification is particularly valuable in the reconstruction of certain intra-cavitary applications where three or four tubes in a small bundle are commonly employed. Whilst the stereo-method is less

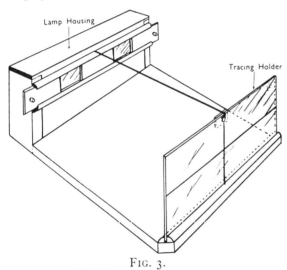

Fig. 3.
The Stereo-reconstructor.

accurate than the first one, it is easier to use, and is adequate for the reconstruction of insertions into soft parts where some movement of the radium itself is liable to occur.

4. Technical Details

The X-ray target may be simulated by either a clear projector lamp with a small filament, or by a small diaphragm illuminated by an extended source. In the latter method the diaphragm, not the lamp, should be set up at the target-film distance from the film carrier. For a distance of two feet a 6-volt, 3-amp. clear bulb is adequate. For distances up to six feet a 100-watt pearl lamp with a circular aperture of 3 mm. radius is suitable. In either case the shadows will be sufficiently defined in a dimly-illuminated room.

The choice of target-film distance is a compromise. The models are more rapidly made if the distance is short, but the films become too distorted for ordinary inspection. A distance of

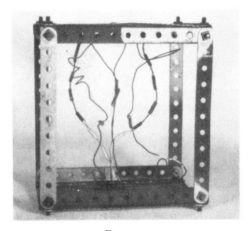

Fig. 4a.
Model from Stereo-reconstructor showing supporting wires.

Fig. 4b.
(*Left*) Incomplete model, showing plasticine and zinc base, also celluloid sheaths; (*Right*) Completed model with Plaster of Paris base.

three feet has been standardised here, and under these conditions the reconstructor is rather bulky. Economy of space could be effected by the permanent use of convex lenses, the effective distance being adjusted from measurements of the magnification of the needles. A further advantage of the lens method is that by suitable adjustment of the lens position, the reconstructions from films taken at any distance can be

August 1939

A Radium Implant Reconstructor

made. The stereo constructor uses a two-feet distance with an eight-inch tube shift, this being the largest available without re-design of the tube stand.

The limit of accuracy seems to depend on the factors (a), (b), and (c) of paragraph two, and is probably of the order of 0·2 mm. in the location of the centre of each needle. Direct tests on models of known geometry indicate that the error is less than 0·5 mm. Radiography with Lysholm grids is unsuitable, as the grid lines may obliterate the true end points of the needles.

The relation of the needles to the anatomical structures may be determined whenever the structures are sufficiently prominent to be fixed precisely in the radiographs. Sometimes it is possible to insert small radio-opaque objects (*e.g.*, very short lengths of silver or gold rod) at definite anatomical positions, whilst opaque catheters may be introduced into canals such as the ureters, and these may be accurately indicated in the model. In the same way the distance of the needles from specific points on the skin can be fixed by markers. It should be noted that it is *essential*, and not merely convenient, to use an external marker (paragraph three) when the implant to be reconstructed consists of small, radon sources (gold capillary tubing) of indeterminate lengths, in order that the primary adjustments of the apparatus may be made.

Accounts of various devices for the localisation of foreign bodies in tissue have appeared in the literature. The authors have thought it worth while to describe this special form of localisation, designed to serve a specific function in radium therapy problems, but which might find more general use in problems involving the spatial relationship of inaccessible objects.

Acknowledgment

The authors wish to acknowledge the contributions of their former colleague L. W. Ball, M.Sc., to the development of the apparatus, which, together with the accessories, was built by Mr. J. Lister of the Holt Radium Institute.

Summary

Two devices for the reconstruction of the spatial distribution of a number of radium needles implanted into tissue are described, and the methods of their manipulation outlined. The question of optimum conditions is discussed, and the possible use of the apparatus outside the field for which it was designed is stressed.

Zusammenfassung

Beschreibung zweier Vorrichtungen zur Rekonstruktion der räumlichen Verteilung einer Anzahl von im Gewebe implantierten Radiumnadeln und deren Handhabung. Die Frage der optimalen Bedingungen wird besprochen und auf die anderweilige Verwendungsmöglichkeit der Apparate hingewiesen.

Résumé

Deux procédés pour la répartition d'un certain nombre d'aiguilles de radium implantées dans le tissu sont décrits et la technique de leur manipulation est expliquée. La question des conditions optima est discutée et l'usage possible de l'appareil en dehors du champ pour lequel il était désigné est invisagée.

REFERENCE

[1] Paterson, R., and Parker, H. M., *British Journal of Radiology*, vii, p. 592, 1934.

Radium Dosage:
The Manchester System

First and Second Editions

1947 and 1967

Reprinted with permission of the Williams and Wilkins Company.

RADIUM DOSAGE
The Manchester System

COMPILED FROM ARTICLES BY

RALSTON PATERSON,
M.D., F.R.C.S., F.F.R.

H. M. PARKER,
M.SC., F.INST.P.

F. W. SPIERS,
PH.D.

M. C. TOD,
F.R.C.S., F.F.R.

S. K. STEPHENSON,
B.SC.TECH.

W. J. MEREDITH,
M.SC., F.INST.P.

EDITED BY

W. J. MEREDITH
M.SC., F.INST.P.
*Christie Hospital and Holt Radium Institute
Manchester*

BALTIMORE
THE WILLIAMS AND WILKINS COMPANY
1947

RADIUM DOSAGE

The Manchester System

COMPILED FROM ARTICLES BY

RALSTON PATERSON,
M.D., F.R.C.S., F.F.R.

F. W. SPIERS,
PH.D.

S. K. STEPHENSON,
B.SC.TECH.

H. M. PARKER,
M.SC., F.INST.P.

M. C. TOD,
F.R.C.S., F.F.R.

W. J. MEREDITH,
M.SC., F.INST.P.

EDITED BY

W. J. MEREDITH
M.SC., F.INST.P.

Christie Hospital and Holt Radium Institute
Manchester

BALTIMORE
THE WILLIAMS AND WILKINS COMPANY
1947

PREFACE TO THE FIRST EDITION

THE papers which form the basis of this volume are still in great demand, although the first of them was published as long ago as 1934. It was, therefore, felt desirable to present them in one volume as a complete dosage system covering all phases of mould, intra-cavitary and interstitial gamma-ray therapy.

The volume is divided into two main parts. Part I consists of the clinical aspects of the constituent papers re-written to form a continuous whole. Nothing that is essential to the use of the system has been omitted, but unification has allowed the omission of repetitions, whilst some small additions have been made in the light of clinical experience. More attention has been paid to examples of the application of the various rules and, in view of the aim of the National Radium Commission to standardize their containers, the examples are almost always worked with tubes and needles appearing on the Commission's list. It must be emphasized that the examples are meant to illustrate the use of the rules and tables and do not necessarily represent the most suitable treatment for the part considered. Every effort has been made, however, to keep close to current clinical practice. In Part I no attempt has been made to prove or justify any of the rules or data set forth since this is unnecessary for their clinical application. The purely physical aspects of the work can be found in Part II which consists of the physical sections of the papers presented almost completely in their original form.

Acknowledgments and thanks are due to the Editorial Committee of the British Institute of Radiology and to the respective authors for their permission to reprint much of their material from the *British Journal of Radiology*. Thanks are also due to Dr. F. W. Spiers of Leeds for preparing additional graphs for his chapter on Cylindrical Distributions, and to Mr. J. B. Bradbury for preparing many of the new diagrams.

W.J.M.

March 1947.

RADIUM DOSAGE

The Manchester System

COMPILED FROM ARTICLES BY

RALSTON PATERSON,
C.B.E., M.D., F.R.C.S., F.F.R.

F. W. SPIERS,
C.B.E., D.Sc.

S. K. STEPHENSON,
B.Sc.TECH., F.INST.P.

H. M. PARKER,
M.Sc., F.INST.P.

M. C. TOD,
F.R.C.S., F.F.R.

W. J. MEREDITH,
D.Sc., F.INST.P.

EDITED BY

W. J. MEREDITH
D.Sc., F.INST.P.
Christie Hospital and Holt Radium Institute
Manchester

SECOND EDITION

E. & S. LIVINGSTONE LTD.
EDINBURGH AND LONDON
1967

PREFACE TO THE SECOND EDITION

ALTHOUGH, in the twenty years that have passed since this book first appeared, megavoltage beam therapy has produced a major change in the overall pattern of radiotherapy, the types of treatment here described are still widely used. Nor have the applications of computer techniques shaken the basic tenets of the system. In fact just the reverse has happened : the great number of calculations that can be performed quickly with the computer have confirmed the essential accuracy of the method. The book continues to be in demand, and since stocks of the original edition were exhausted, it seemed appropriate to produce a new edition.

In this there have been some extensive alterations in text in order, it is hoped, to improve its clarity. New examples, in closer keeping with modern methods, have been introduced to illustrate the basic principles, which are absolutely unchanged. A new chapter has been added explaining how the now available artificially produced gamma-ray emitting isotopes may be used in the system, and how roentgens can be converted into rads.

It might have been expected that the preparation of a new edition would be the occasion for going over from the roentgen (now the unit of " exposure ") to the rad (the unit of " absorbed dose "), and there is something to be said for such a course. However, it must be stressed that dosage consistency rather than absolute accuracy is most important in radiotherapy. There is nothing to be gained clinically by such a change and confusion would certainly be caused. The magnitudes of the roentgen and the rad are almost identical, whilst this type of treatment not only forms a compact and consistent whole, but is so different in its time pattern that dosage comparisons with other methods are unlikely to be of great value. On balance, it was decided to retain the old pattern but to add in Chapter XIII, information on how the changes can be made by those who wish to do so.

W.J.M.

Manchester 1967.

The Physical Basis of the Planar Mould

Radium Dosage: The Manchester System

1947

RADIUM DOSAGE
The Manchester System

COMPILED FROM ARTICLES BY

RALSTON PATERSON,
M.D., F.R.C.S., F.F.R.

F. W. SPIERS,
PH.D.

S. K. STEPHENSON,
B.SC.TECH.

H. M. PARKER,
M.SC., F.INST.P.

M. C. TOD,
F.R.C.S., F.F.R.

W. J. MEREDITH,
M.SC., F.INST.P.

EDITED BY

W. J. MEREDITH
M.SC., F.INST.P.

*Christie Hospital and Holt Radium Institute
Manchester*

BALTIMORE
THE WILLIAMS AND WILKINS COMPANY
1947

CHAPTER VII

THE PHYSICAL BASIS OF THE PLANAR MOULD AND LINE SOURCE SYSTEMS

(By H. M. Parker, M.Sc., F.Inst.P.)

CALCULATION OF THE INTENSITY OF THE γ-RADIATION FROM RADIUM, IN TERMS OF THE r UNIT

THE derivation is based on the figure for the total number of ions produced in air by the γ-radiation from 1 gram of radium in equilibrium with its decay products. This value has been the subject of careful experimental work by Rutherford and his collaborators (1930) and is given as $2 \cdot 13.10^{15}$ pairs of ions per second (L. H. Gray). This leads to a value of $8 \cdot 4\ r$ per hour for the intensity at 1 cm. from a point source of 1 milligramme of radium, filtered by 0·5 mm. platinum. More recent determinations, quoted in a paper by Mayneord and Roberts (1934), to which reference should be made, agree closely with the value that we had accepted.

DISTRIBUTION RULES

The rules are based on the assumption that the γ-radiation from a point source strictly obeys an inverse square law. They therefore ignore such factors as change in quality of radiation and scattering and absorption in air or tissue, which almost certainly have no serious effect under practical conditions. The radio-active sources discussed are, therefore, the mathematical abstractions of linear sources with line density of ρ mg. per cm. The special case of a ring source of this nature has been previously investigated by Sievert (1921 and 1930), Souttar (1931), and Mayneord (1933), and that of a rectilinear source by Sievert (1921 and 1930) and Mayneord (1933).

The general problem is to state rules for the distribution of sources over a particular surface (the applicator) such that the intensity of irradiation over a second surface (the treated area) at a constant distance from it is uniform and calculable. For the purposes of this paper, " uniform " or " homogeneous " irradiation is to be understood in the restricted sense of suffering variations of not more than ± 10 per cent. The problem appears to be soluble with the following definition of the relation between applicator and treated area :—

Let A (Fig. 12) represent any surface (e.g. the skin). Let C be any selected region (the treated area) on the surface A. Through any point P on A draw the normal PP' of constant length h. As P moves over the surface A, P' will define a second surface A', and in particular, the second region C' as P moves over C. Then C' represents the surface applicator and h is the constant distance between applicator and treated area.

48   RADIUM DOSAGE: THE MANCHESTER SYSTEM

Let the figure be referred to a set of rectangular axes X, Y, Z, so that the co-ordinates of P and P' are (Px, Py, Pz) and $(P'x, P'y, P'z)$. Let there be

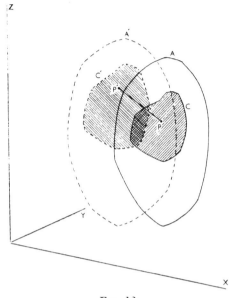

Fig. 12

The relation between a treated area C and its surface applicator C' at a distance h.

$M_1, M_2, M_3, \ldots$ mg. of radium at the points (x_1, y_1, z_1) (x_2, y_2, z_2), etc. Then the intensity at any point (x, y, z) on C is:

$$I = \Sigma \frac{M_1}{(x-x_1)^2+(y-y_1)^2+(z-z_1)^2} \text{over all the active points.}$$

Also $h = [(Px-P'x)^2 + (Py-P'y)^2 + (Pz-P'z)^2]^{\frac{1}{2}}$.

Now suppose we choose a new applicator and treated area referred to axes X', Y', Z', and fulfilling the conditions

$$\left. \begin{array}{l} x' = sx \\ y' = sy \\ z' = sz \end{array} \right\} \quad \text{and} \quad \left\{ \begin{array}{l} M_1' = s^2 M_1 \\ M_2' = s^2 M_2 \\ \ldots = \ldots \end{array} \right.$$

for every point of the figure.

Then the intensity at the corresponding point of the new treated area is:

$$I' = \Sigma \frac{M_1'}{(x'-x_1')^2+(y'-y_1')^2+(z'-z_1')^2} \text{over all points.}$$

$$= I \qquad \text{from the given relations.}$$

Also $h' = sh$

for all pairs of corresponding points, and therefore the distributions of intensity are identical. It follows that we need only calculate the distribution for one particular value of h, and the actual intensity for any other value may be found by taking into account the scale-factor s.

Stage 1—Plane Circular Applicators

If the surface A is plane and C is a circle, the applicator C' becomes a plane circle of the same size. The scale-factor relation reduces to the simple

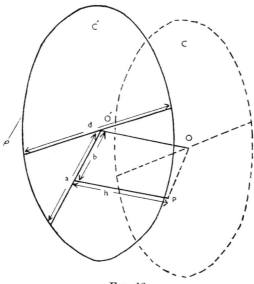

Fig. 13
Ring applicator and treated area.

form that the distribution of intensity over C is determined solely by the ratio d/h where d is the diameter of the circle.

(A) If the radium is arranged as a ring of line density ρ around the periphery of C' the calculation of the intensity at any point P is made from the equation:

$$I_p = \frac{2\pi a \rho}{[(a^2+b^2+h^2)^2 - 4a^2b^2]^{\frac{1}{4}}} \ldots \ldots (\text{Sievert, 1921 and 1930}).$$

Where a = radius of ring (Fig. 13).
    h = normal distance of P from plane of ring.
    b = distance from foot of normal from P on the plane of the ring.
      = distance of P from centre of C.

$$I_p = \frac{I_0}{[(a^2+b^2+h^2)^2 - 4a^2b^2]^{\frac{1}{4}}}.$$

Where I_0 = intensity at 1 cm. from a point source of $2\pi a \rho$ mg., the total amount of radium in the ring.

For particular values of a and h, the variation of intensity across any diameter of C is a function of b, and by differentiation we have

$$\frac{dI}{db} = \frac{2b[(a^2-h^2)-b^2] \cdot I_0}{[(a^2+b^2+h^2)^2 - 4a^2b^2]^{\frac{3}{2}}}.$$

The denominator is essentially positive. Therefore, to find maximum and minimum values of I we need only consider the sign of the numerator.

D

50 RADIUM DOSAGE: THE MANCHESTER SYSTEM

Case 1. $a<h$

$[a^2-h^2-b^2]$ is negative and never zero for real values of b.

$\frac{dI}{db}=0$ if $b=0$, is positive if $b<0$, and negative if $b>0$

(regarding b as positive in one sense from O and negative in the other).

I, therefore, has a maximum value at $b=0$, and no other turning point. Hence the minimum intensity over the treated area C occurs at $b=a$.

$$\text{At } b=0, \, I=\frac{I_0}{a^2+h^2}$$

$$\text{at } b=a, \, I=\frac{I_0}{h\,(4a^2+h^2)^{\frac{1}{2}}}.$$

The variation of intensity is greatest when the ratio R of these two limits is greatest.

$$R=h\,\frac{(4a^2+h^2)^{\frac{1}{2}}}{a^2+h^2}=\frac{(1+4p^2)^{\frac{1}{2}}}{1+p^2} \text{ where } p=\frac{a}{h}$$

$$\frac{dR}{dp}=\frac{2p\,(1-2p^2)}{(1+p^2)^2\,(1+4p^2)^{\frac{1}{2}}}$$

whence R has maximum value if $p=\frac{1}{\sqrt{2}}$.

The variation is then $\pm 7\cdot 2\%$.

Hence, subject to our definition of "uniform", all rings with $d/h \leqslant 2$ give "uniform" intensity over the treated area.

The variation for other values is:

d/h . . =	0	0·5	1	1·5	2
Variation . =	0	$\pm 3\%$	$\pm 5\%$	$\pm 6\cdot 75\%$	$\pm 5\%$

Case 2. $a>h$

$\frac{dI}{db}=0$ if $b=0$ and this is now a minimum turning point.

Also $\frac{dI}{db}=0$ if $b=\pm\sqrt{a^2-h^2}$

and these are maximum turning points, symmetrical about the origin.

The maximum intensity over the treated area occurs at $b=\sqrt{a^2-h^2}$ and is $\frac{I_0}{2ah}$.

The minimum occurs either at $b=0$ or $b=a$, according as

$$\frac{1}{h^2\,(1+p^2)}< \text{ or } >\frac{1}{h^2\,\sqrt{1+4p^2}}.$$

Hence the minimum value is at $b=a$ if $p<\sqrt{2}$ and at $b=0$ if $p>\sqrt{2}$.

PHYSICAL BASIS OF THE PLANAR MOULD

If $p < \sqrt{2}$ the variation between the limits $\dfrac{I_0}{2ph^2}$ and $\dfrac{I_0}{h^2\sqrt{1+4p^2}}$ is readily shown to decrease as p increases. It has its greatest value ± 5 per cent. for $p=1$, and its least of ± 3 per cent. at $p=\sqrt{2}$. All such circles are therefore satisfactory. If $p=\sqrt{2}$ the circle is the one which Mayneord has already emphasized as a particularly good arrangement. If $p > \sqrt{2}$ the divergence between the limits $\dfrac{I_0}{2ph^2}$ and $\dfrac{I_0}{h^2(1+p^2)}$ steadily increases as p increases.

(B) If $p > \sqrt{2}$ the intensity has a minimum value at the centre of the treated area, and it becomes possible to improve the distribution by adding a certain amount of radium to the centre of the ring. Suppose we place $k.2\pi a\rho$ mg. at the centre. Then the combined intensity at a point of C is:

$$\frac{kI_0}{h^2+b^2} + \frac{I_0}{[(a^2+b^2+h^2)^2-4a^2b^2]^{\frac{1}{2}}}.$$

There exists a variety of methods for determining that value of k which makes this expression most nearly constant over the range $b=0$ to $b=a$, but the following is perhaps the simplest.

To make equality at $b=0$ and $b=a$,

$$\frac{1}{1+p^2} + k = \frac{1}{\sqrt{1+4p^2}} + \frac{k}{1+p^2} \text{ or } k = \frac{1+p^2-\sqrt{1+4p^2}}{p^2\sqrt{1+4p^2}}.$$

To make equality at $b=0$ and $b=\sqrt{a^2-h^2}$ (max. point),

$$\frac{1}{1+p^2} + k' = \frac{1}{2p} + \frac{k'}{p^2} \text{ or } k' = \frac{p(p-1)}{2(p+1)(p^2+1)}$$

e.g. if $p=2$, $k=0.053$ and $k'=0.067$.

This case is illustrated in Fig. 14, in which curve A shows the intensity due to the ring alone.

Curve B—ring and centre spot with $k=0.053$.

Curve C—ring and centre spot with $k'=0.067$.

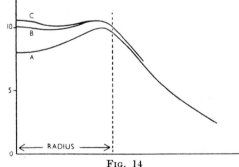

Fig. 14

Distribution of intensity along a radius of a circular treated area.
    A. Due to ring applicator with $d/h=4$.
    B. Ring+centre spot with $k=0.053$.
    C. Ring+centre spot with $k'=0.067$.

52   RADIUM DOSAGE : THE MANCHESTER SYSTEM

B shows a variation of ± 4 per cent., C of ± 2 per cent.

Curves of this type show at a glance how the intensity varies across the treated area, and the best curve can easily be obtained by trying values over the range k to k'. For $p=2$, the value $k'=0\cdot067$ is also the best value of k.

The percentage of the total radium to be placed at the centre is

$$x = \frac{100\ k}{1+k}.$$

The following table is drawn up in this manner :

d/h . . . =	3	4	5	6
x . . . =	4%	6%	6·5%	6·5%
Percentage variation =	$\pm 3\cdot5\%$	$\pm 2\%$	$\pm 5\cdot5\%$	$\pm 8\%$

Over the range $d/h=3$ to <6 we may set $x=5$ per cent. and remain within our limit of ± 10 per cent.

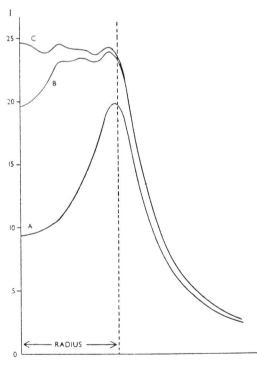

Fig. 15

Distribution of intensity along a radius of a circular treated area.
    A. Ring applicator with $d/h=8$.
    B. Outer ring + inner ring.
    C. Outer ring + inner ring + centre spot.

PHYSICAL BASIS OF THE PLANAR MOULD

(*C*) If $d/h=6$ the best value of centre spot produces a distribution with intensity variation ± 8 per cent., and for any larger value of d/h the variation is greater. These cases can be improved by the addition of a further ring of radium of half the previous diameter. The amounts to be placed at the centre and in the inner ring are obtained approximately by calculating the points $b=0$, $b=a/2$, $b=\sqrt{a^2-h^2}$ for equality. Then by inspection of the resultant graph it is possible to decide whether small variations from the provisional values are likely to improve the distribution. Fig. 15 refers to the case $d/h=8$. Curve A represents the intensity along a radius for the outer ring alone. Curve B for the inner and outer rings, and Curve C the final intensity. The variation is only ± 2 per cent.

The following distributions are obtained in this way:

d/h . . . . =	6	8	10	12	14
% of radium in centre . =	$3\frac{1}{2}\%$	$2\frac{1}{2}\%$	3%	3%	3%
% of radium at inner ring =	16%	$23\frac{1}{2}\%$	$26\frac{1}{2}\%$	28%	28%
% variation . . . =	$\pm 2\frac{1}{2}\%$	$\pm 2\%$	$\pm 4\%$	$\pm 7\frac{1}{2}\%$	$\pm 10\%$

This permits the simplification made in the mould rules (Chap. II).

Circles with $d/h \geqslant 14$ are not satisfactory under this method, and they are not required in practice. For the completeness of the dosage tables discussed later the range has been extended to $d/h=20$ by the use of three rings and a centre spot.

Note.—An alternative solution of the plane circular applicator has been given by Mayneord (1933) on the basis of a combination of a radio-active disc and peripheral ring. It is interesting that for equal areas both methods require the same radium content to produce the same intensity. This is probably a proof that both methods give satisfactory homogeneity. In a sense, the rings and centre spot may be regarded as a practical approximation to an ideal radio-active disc and peripheral ring.

Stage 2—Plane Rectangular Applicators

Isodose curves of the intensity in a plane 1 cm. from a rectilinear source of line density ρ and of lengths from 1 to 30 cm. have been constructed from Sievert's tables. The intensity from four lines arranged as a rectangle is then obtained by summation of effects over the treated area. A radio-active square of side $a=2h$ gives uniform distribution analogous with the ring of $d/h=2\sqrt{2}$. Rectangles of sides a and b ($b>a$) give sufficient homogeneity with $a=2h$ and $b \leqslant 4a$. In rectangular treated areas the intensity in the extreme corners only falls below the general level and is excluded from the permissible ± 10 per cent. variation.

54   RADIUM DOSAGE: THE MANCHESTER SYSTEM

The simple rule for larger rectangles is to add radio-active lines at separation=$2h$ parallel to the longer side with density $\frac{1}{2}\rho$ for one added line, and $\frac{2}{3}\rho$ for more than one line. The rule is based on the analysis of forty different rectangles at 1 cm. distance. Three typical cases are shown in Tables 5, 6, and 7.

TABLE 5

Square 2 cm. × 2 cm. at 1 cm. ∴ no added lines. Variation = ±5 per cent. excluding corners

	0·0	·25	·5	·75	1·0	1·25	1·5	1·75	2·0
0	*(291)*	*(300)*	*325*	*330*	*335*	*330*	*325*	*(300)*	*(291)*
·25	*(300)*	320	334	342	353	342	334	320	*(300)*
·5	*325*	334	350	355	349	355	350	334	*325*
·75	*330*	342	355	344	344	344	355	342	*330*
1·0	*335*	353	349	344	344	344	349	353	*335*
1·25	*330*	342	355	344	344	344	355	342	*330*
1·5	*325*	334	350	355	349	355	350	334	*325*
1·75	*(300)*	320	334	342	353	342	334	320	*(300)*
2·0	*(291)*	*(300)*	*325*	*330*	*335*	*330*	*325*	*(300)*	*(291)*

TABLE 6

Rectangle 16 cm. × 4 cm. at 1 cm. distance ∴ 1 added line of density $\frac{1}{2}\rho$
Variation = ±6 per cent. excluding corners

	0	·5	1	1·5	2	2·5	3	3·5	4
0	*(343)*	385	401	416	422	416	401	385	*(343)*
1	*377*	393	395	402	411	402	395	393	*377*
2	*403*	391	379	377	*389*	377	379	391	*403*
3	*401*	397	375	375	*379*	375	375	397	*401*
4	*408*	391	377	371	*384*	371	377	391	*408*
5	*408*	397	371	373	*383*	373	371	397	*408*
6	*410*	398	373	373	*386*	373	373	398	*410*
7	*412*	409	386	382	*386*	382	386	409	*412*
8	*416*	408	387	382	*393*	382	387	408	*416*
9	*412*	409	386	382	*386*	382	386	409	*412*
10	*410*	398	373	373	*386*	373	373	398	*410*
11	*408*	397	371	373	*383*	373	371	397	*408*
12	*408*	391	377	371	*384*	371	377	391	*408*
13	*401*	397	375	375	*379*	375	375	397	*401*
14	*403*	391	379	377	*389*	377	379	391	*403*
15	*377*	393	395	402	411	402	395	393	*377*
16	*(343)*	385	401	416	422	416	401	385	*(343)*

Figures in italics fall directly below active lines.

PHYSICAL BASIS OF THE PLANAR MOULD 55

TABLE 7

Square 16 cm. × 16 cm. at 1 cm. ∴ 7 added lines each of line density $\frac{2}{3}\rho$
Variation = ± 10 per cent. excluding corners

cm.

	0	1	2	3	4	5	6	7	8
0	(405)	473	532	546	561	550	569	559	563
1	464	476	524	517	543	530	553	541	567
2	471	474	512	493	531	508	540	519	546
3	471	468	497	484	517	500	529	507	535
4	480	465	505	478	516	494	528	500	533
5	477	465	493	481	518	492	524	497	526
6	480	471	509	486	526	495	533	499	533
7	485	469	508	483	523	494	534	504	542
8	490	480	512	500	527	504	539	516	545

This table shows the intensity over ¼ of the square.
The remainder is obtained by mirror-reflection about the row and column for 8 cm.

Indefinitely elongated rectangles cannot conform to the general rule since the limiting case of a single line of radium exhibits a marked fall of intensity opposite the ends, unless we exclude the ends as the logical extension of the excluded corners. The rule is quite satisfactory up to an elongation of 4 : 1 (*vide* example 16 cm. × 4 cm.), and is still within the limits ±10 per cent. at a ratio 6 : 1. Such strips, however, are not required in practice.

It is worthy of note that the rectangular applicators have a certain amount of latitude in respect of the amount of radium to be placed in the added lines. Thus in many cases the strength may be reduced from $\frac{2}{3}\rho$ to $\frac{1}{2}\rho$ without serious loss of accuracy. For example, the 16 cm. × 16 cm. square at 1 cm. has a variation of ±12 per cent. instead of ±10 per cent. if the added lines have strength $\frac{1}{2}\rho$, and for the range of 4—6 added lines it is practically immaterial which strength is used. This feature is of some value with a limited stock of radium.

STAGE 3—THE GENERAL PLANE APPLICATOR

An important fact arises from a comparison of the actual intensity in r per milligramme-hour from circular and rectangular applicators of the same area. Table 8 shows such a comparison for circles, squares, and rectangles having ratio of sides 2 : 1. Circles and squares agree almost exactly and the rectangles diverge only slightly. It follows that, provided the radium has been distributed correctly, the intensity depends on the *area* of the applicator rather than on the shape. This fact, coupled with the established flexibility of the rectangle

56   RADIUM DOSAGE : THE MANCHESTER SYSTEM

rules, suggests a tentative method of distribution of radium for any irregular plane area, *viz.* outline the periphery with a linear source of constant density, and add lines parallel to the longer direction to divide the area into strips of width $2h$, following the rectangle rule.

TABLE 8

Comparison of Square and Circular Applicators at 1 cm. Skin Distance

Area	Square ; r/mg.-hr.	Circle ; r/mg.-hr.
1	4·95	5·00
4	2·92	2·86
16	1·46	1·43
36	0·926	0·952
64	0·654	0·666
144	0·385	0·385

Comparison of Rectangular (2 : 1) and Circular Applicators

Area	Rect. ; r/mg.-hr.	Circle ; r/mg.-hr.	% diff.
2	4·00	4·16	4%
8	1·89	2·00	6%
32	0·971	1·01	4%
72	0·581	0·613	6%
128	0·403	0·406	1%

It is scarcely to be expected that the generalization will in all cases lead to homogeneity as good as that achieved by the regular distributions. Nevertheless, the risk of serious error must be quite small, and a number of cases measured by condenser chambers of the Sievert type has given reasonably good results. Figures which are unduly elongated or which have sharp re-entrant angles are not included in the generalization.

STAGE 4—EXTENSION TO CURVED SURFACES

A. Convex Surfaces.

Suppose we have a treated area C, originally plane, and an applicator C' of equal area at a constant distance h. If the treated area is bent to present a convex surface to the applicator, in such a way that the treated area and h remain constant, two effects will be noticed :—(1) The folding of the applicator about the treated area tends to increase the intensity by a " cross fire " effect ; (2) The area of the applicator increases by the definition of the relation between

PHYSICAL BASIS OF THE PLANAR MOULD 57

C and C'. Hence the same amount of radium is extended over a larger surface and the intensity over the treated area will be reduced. These two effects act in opposite senses, and the possibility arises that the distribution and actual intensity will remain sufficiently constant whether the treated area be plane or convex. It is only possible to test the hypothesis by the examination of a number of examples. The appropriate analysis for the bending of a rectangular applicator to cylindrical curvature is given below :—

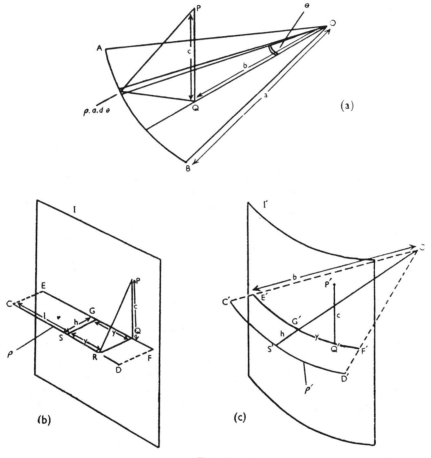

Fig. 16

(a) Diagram for the calculation of intensity from radium distributed uniformly along the arc AB.
(b) and (c) Curvature of a linear source about a cylinder.

Consider first the intensity of gamma radiation from an unfiltered line source, line density ρ, in the arc of a circle AB. (Fig. 16a).

Let O be the centre of curvature of AB.

P is any point in space and PQ of length c is normal to the plane ABO.

58   RADIUM DOSAGE: THE MANCHESTER SYSTEM

The intensity at P is : $I_p = \int_A^B \dfrac{\rho.a.d\theta}{a^2+b^2+c^2-2\,ab\,\cos\theta}$

$$= \dfrac{2\,\rho.a.}{\{[(a+b)^2+c^2][(a-b)^2+c^2]\}^{\frac{1}{2}}} \left[tan^{-1} \left(x_A \sqrt{\dfrac{(a+b)^2+c^2}{(a-b)^2+c^2}} \right) \right.$$
$$\left. + tan^{-1} \left(x_B \sqrt{\dfrac{(a+b)^2+c^2}{(a-b)^2+c^2}} \right) \right] \quad\dotsb\text{I}$$

where $x_A = tan\,\dfrac{A\hat{O}Q}{2}$, $x_B = tan\,\dfrac{B\hat{O}Q}{2}$

by standard integration.

Next consider a straight line source CD (dens $:\rho$) of length $2l$ parallel to the plane I and at distance h from it (Fig. 16b). P is any point on the plane, PR is $\perp^r CD$ and RQ is normal to the plane. S is the medial point of CD. $SR = y$, $RQ = h$ and $PQ = c$. EGF is the projection of CD on the plane. The intensity at P is :

$$I_p = \dfrac{\rho}{\sqrt{h^2+c^2}} \left[tan^{-1} \dfrac{l+y}{\sqrt{h^2+c^2}} + tan^{-1} \dfrac{l-y}{\sqrt{h^2+c^2}} \right] \dotsb\text{II}$$

In Fig. 16b the line has been bent to the arc of a circle so that plane I has a radius of curvature of b and centre O.

$E'F' = 2l$, $C'E' = S'G' = D'F' = h$, from the relation between applicator and treated surface. The new line density $\rho' = \dfrac{b}{b+h} \cdot \rho$ keeping total amount of radium constant.

Writing $a = b+h$ and $\rho' = \dfrac{b}{b+h} \cdot \rho$ in equation I.

$$I_{p'} = \dfrac{2b\rho}{\{[(2b+h)^2+c^2][h^2+c^2]\}^{\frac{1}{2}}} \left[tan^{-1} \left(x_A \sqrt{\dfrac{(2b+h)^2+c^2}{h^2+c^2}} \right) \right.$$
$$\left. + tan^{-1} \left(x_B \sqrt{\dfrac{(2b+h)^2+c^2}{h^2+c^2}} \right) \right] \dotsb\text{III}$$

where $x_A = tan\left(\dfrac{l+y}{2b}\right)$, $x_B = tan\left(\dfrac{l-y}{2b}\right)$

Now any rectangle can be progressively curved about a cylindrical surface, and the intensity of each component part of the applicator may be calculated from Equation II or III according to whether the part is $\|^l$ or $\perp^r$ to the axis of the cylinder. Equation III is of such form that $\rho, l, h, y,$ and c are constant for a particular point P. The parameter b determines the degree of curvature. In addition, y and c vary for different points of the treated area. Apparently, rectangles may be bent in this way almost to a semi-cylindrical form before the alteration in distribution and dosage becomes severe. Some typical results are set out in Table 9. The calculations for circular applicators applied to

PHYSICAL BASIS OF THE PLANAR MOULD

surfaces of spherical curvature are easily made, and show that the curvature may be extended almost to the hemispherical form. More irregular curvatures are not amenable to mathematical discussion, but it is reasonable to assume that the compensation will be satisfactory over a considerable range.

TABLE 9

Folding of a treated area 6×6 cm. about a cylinder. $h = 1$ cm.

Radius of curvature, b, in cms. =	∞	10	5	3	2
Angle subtended at centre . . =	0°	34·4°	68·8°	114·6°	171·9°
Average intensity referred to plane as 1·00 . . . . =	1·00	·98	·97	·97	·97
Ratio of central intensity to peripheral intensity . . . =	1·00	1·00	1·01	1·02	1·05

B. Concave Areas.

If the treated area is curved in the concave sense, the applicator diminishes in size. It may be shown, as above, that if the area of applicator be taken as a constant factor, the curvature will then produce a negligible effect on distribution and dosage over the treated area. In this case the rule holds good up to the limiting cases of a point source for spherical curvature and a rectilinear source for cylindrical curvature, when the distance h becomes the radius of curvature.

DOSAGE TABLES

The distribution rules which are derived in the previous section are valid irrespective of the system of units in which the γ-radiation is measured. This follows at once since the distribution is concerned only with the relative variation of intensity over an irradiated region, and not with absolute values.

The actual intensity at any point of a treated area may be computed in terms of the r unit by introducing the result of page 47 for the intensity at 1 cm. from a point source. In general it is more convenient to know the amount of radium and the required time (i.e. the number of milligramme-hours) to produce a given dose (intensity $\times$ time), and the results are therefore expressed in this form.

(1) AREA DOSAGE

When radium is arranged on an applicator to give uniform intensity over the treated area, it becomes possible to assign a meaning to the intensity and hence to the dose delivered to the treated area. For clinical convenience the data for dosage calculations are best set out in a series of graphs of " area " (ordinate) against " milligramme-hours to produce 1000 r " (abscissa) with the parameter h for a family of curves. Figures from which such curves can be drawn are given in Table A of the Appendix. The basic points are obtained from a consideration of the series of circular applicators at $h=1$. The intensity used is the mean value (by integration of annuli) over the treated circle. It has already been shown that square areas will fall exactly on the same graph. Also rectangles and irregular areas may be assessed from the same chart provided that a small correction is added for elongation as follows:—

Elongation	2 : 1	3 : 1	4 : 1
Percentage mg.-hrs. to be added . .	5%	9%	12%

From the previous section it is, moreover, apparent that the dosage over curved surfaces is given by the same graph, with the proviso that treated area is read for a convex surface, and area of applicator for a concave surface. Hence the dose for any treated area may be obtained from the single dosage chart provided that due attention is paid to the correct arrangement of the active sources. From the graph for $h=1$, the family of curves for convenient values of h is directly obtained by means of the " scale-factor " previously described.

Filtration.—The mathematical basis is restricted for simplicity to unfiltered sources. To make the charts of clinical value provision is made for the calculation appropriate to standard filtrations from 0·5 mm. to 2·0 mm. of platinum. This is effected by allowing for the general reduction of intensity due to the absorption in platinum which has been assumed to have a constant absorption co-efficient $\mu=2\cdot0$. No account is taken of oblique filtration. This is an effect which cannot easily be estimated for the compound applicators, and it is believed to be small, as the intensity at any point of the treated area may be regarded as the summation of a large number of contributions of which only a few can be simultaneously reduced to any extent by oblique filtration.

(2) DOSAGE FROM RECTILINEAR SOURCES

The values (given in Table B of the Appendix) of the number of milligramme-hours required to give 1000 r for any tube source are developed from a direct conversion of Sievert's Tables to the r unit. They indicate the dose opposite the mid point of the line source at a stated distance h. Unfortunately the effect of oblique filtration for a single line source is not a negligible factor. There is some difficulty in publishing values without loss of accuracy on the one hand or introducing too many complications for the clinician on the other. It is hoped that the compromise adopted by the authors will be found of general value.

The values are given separately for filtrations of 0·5 mm. Pt. and 1·0 mm. Pt., the two values which are likely to be in most common use. The dosage appropriate to 0·6 mm. and 0·8 mm. Pt. may be obtained with reasonable accuracy

PHYSICAL BASIS OF THE PLANAR MOULD

from the 2% per 0·1 mm. rule. A department habitually using radium at other than 0·5 mm. or 1·0 mm. Pt. would be wise to reconstruct the charts accordingly.

SOME NOTES ON THE PRACTICAL APPLICATION OF THE DISTRIBUTION RULES

The rules postulate the use of continuous rings and lines of radium. In the practical construction of applicators some approximation to ideal conditions has to be permitted, and it is clearly important to the constructor to know precisely which arrangements provide satisfactory approximations. General guiding principles are given below, but in the interests of brevity the proofs are omitted.

1. *A Ring of Radium of Diameter d at Distance h.*

(*a*) The most satisfactory arrangement is to outline the circumference with radium tubes or needles in such a way that the active lengths form a continuous polygonal figure. Six tubes forming a hexagon constitute the minimum permissible number. Tubes of different lengths may be used provided that no length exceeds $d/2$.

(*b*) The tubes may be arranged as an open polygon provided that the spacing between adjacent active ends of tubes does not exceed the distance h. Six tubes constitute the minimum requirement.

(*c*) Two types of tube of equal length with line densities ρ and 2ρ may be used alternately to constitute a closed polygon, provided that the length of each tube does not exceed $2h$, or to form an open polygon with intervals of length h, provided that the length of a tube does not exceed h. Minimum number of tubes is four of each type.

(*d*) With point sources (e.g. radon seeds) the distance between adjacent points should not exceed h. Minimum number of points $=6$.

d/h . . . . =	2	4	8	12
Number of point sources .	6	12	24	36
Maximum variation from ideal	1%	1%	1%	1%

A spacing of $1\cdot5h$ may be used if necessary, introducing a variation of ±5 per cent. approximately.

The alteration in distribution of intensity in any of the arrangements (*a*), (*b*), (*c*), and (*d*) is only apparent in the immediate neighbourhood of the periphery of the treated area.

2. *Inner Ring and Centre Spot of the Compound Circular Applicators.*

The inner ring contributes about half the total intensity to the regions in which its " ripple ", due to departure from the ideal ring, might be serious. Consequently the variations tend to be smoothed out; and the spacing may be extended to $2h$ with a ripple of ± 5 per cent. Nevertheless, no fewer than six active foci should be used if possible. The " centre point " behaves as a point source if its active length does not exceed h.

3. *Rectangular Applicators.*

(*a*) The straight lines which are the components of the rectangular applicators should preferably be made active throughout their length by overlapping the inactive ends of tubes in the plane of the applicator.

(*b*) Where necessary, an active line may be constructed by a number of tubes provided that the intervals between adjacent active ends do not exceed h. For two tubes constituting a line the inactive gap should not exceed $h/2$.

(*c*) Different strengths of tubes may be used alternately if the length of each does not exceed $2h$ and the number of tubes is not less than five.

(*d*) A long row of point sources is valid if the spacing is not more than h.

4. *Single Line Sources.*

The conditions in this case are more critical because there is no smoothing out of ripples by contributions from other parts of an applicator. A distance between adjacent active ends not exceeding $h/2$ is recommended. With an odd number of tubes (five or more) a spacing of h is just permissible.

Two tubes with an inactive gap form a bad approximation, and should be avoided.

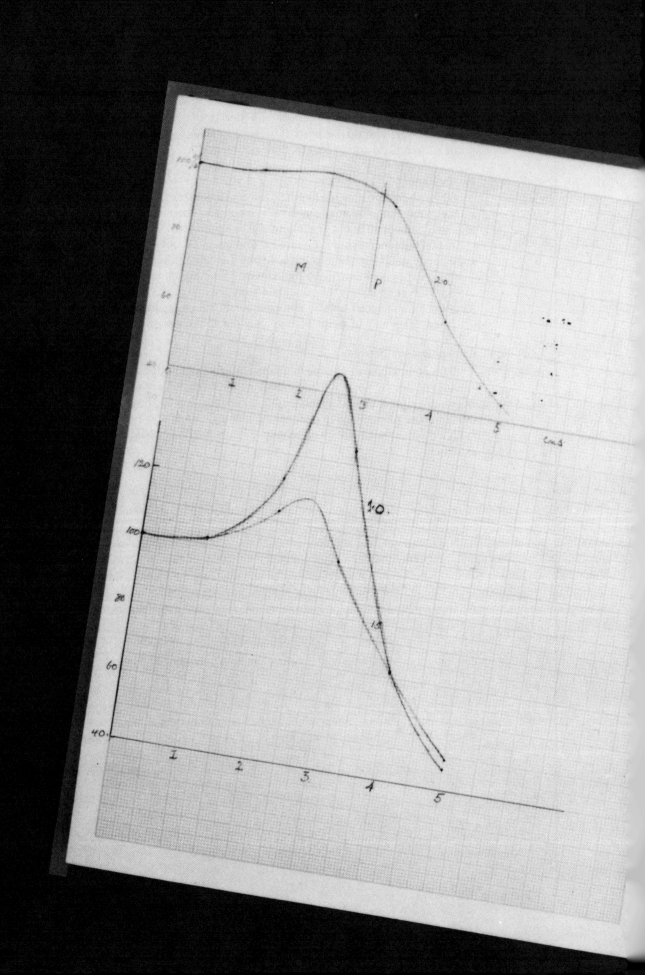

PART II
MEDICAL PHYSICS AFTER MANCHESTER

INTRODUCTION

Although Parker continued in full-time clinical medical physics work for only a short time after leaving Manchester, he never lost his interest in therapy. While at the Swedish Hospital and Tumor Institute in Seattle, he was heavily involved in the application of supervoltage x rays to the treatment of malignant neoplasms. This work culminated in a book, *Supervoltage Therapy*, coauthored with his two physician colleagues, Franz Buschke and Simeon Cantril in 1950, several years after he left Swedish and while he was busily involved in management at the Atomic Energy Commission's Hanford Laboratories, operated by the General Electric Company.

Another significant post-Manchester contribution to medical physics was presented on the eve of Parker's entry into work in the Manhattan District in December 1942. Published in 1943, this paper entitled "Limitations of Physics Radium Therapy" is of timeless value.

Radium Element Versus Radon

Radiation Therapy A Supplement to the
Staff Journal of the Swedish Hospital
Seattle, Washington from the Tumor Institute

February 1940

RADIATION THERAPY

A SUPPLEMENT
to the
STAFF JOURNAL
of the
SWEDISH HOSPITAL
SEATTLE, WASHINGTON

From

THE TUMOR INSTITUTE

No. 1   February, 1940

RADIUM ELEMENT VERSUS RADON

H. M. Parker

RADIOACTIVE sources for radium therapy are commonly prepared in two ways, viz.: (1) as *Radon* sources in which radon, the gaseous emanation from radium is collected either in glass or metal capillaries. Glass capillaries are enclosed in platinum sheaths to screen off undesired beta radiations. Thick walled sheaths with robust screw end suffice for external applications. Needles with detachable trocar point or detachable eye are required for interstitial work. Metal capillaries (i. e.: gold or silver 'seeds') may be inserted directly into tissue.

(2) as *Radium Element* sources in which standard quantities of radium are permanently sealed in platinum cells mounted in platinum tubes or needles. Occasionally the inner cells are made available for temporary assembly in platinum capsules, but the technique is to be deprecated except in small departments where the added flexibility is essential. The sources are normally available in amounts of 0.5 to 3 milligrams in needles, and 1 to 25 milligrams in tubes.

The radon method is characterized by the following features, of which the first two are favorable and the remainder unfavorable:

(a) The sources have no intrinsic value and the risk of loss is confined to the radon plant. Also radon sources may be applied to outpatients, an important economy in hospital beds.

(b) Theoretically the sources can be prepared in advance in any strength in containers of any size or shape.

(c) The activity diminishes comparatively rapidly. A treatment cannot be postponed for a few days if a patient's condition warrants such a procedure, unless new sources can be obtained. Moreover, the dose is not delivered at a steady rate. In an 8-day treatment two-thirds of the dose is delivered in the first four days. A 12-day treatment gives more than half the total dose in the first four days. Some therapists believe that the diminishing dosage rate is an advantage, but their evidence seems inconclusive.

(d) Each individual source has to be measured and its subsequent decay noted. Small numbers of tubes must be differentiated by a color code; large numbers, such as gold seeds, may be adequately used in groups rated by average strength. The process as usually performed involves considerable exposure to the operator, especially as the measurements are made with the tubes at maximum activity. Some attempts have been made to obtain the measurements before the sources have acquired full strength, but their reliability is not generally accepted. The purchaser of radon from an outside institution has to depend on the accuracy of the stated strengths and, although the figures are usually sound, there are cases on record where the errors might have been large enough to vitiate a treatment.[1]

---

[1] Edith H. Quimby and A. U. Desjardins (Journal of the American Medical Association, Vol. 112, p. 1822, 1939), have reported an impartial examination in conjunction with the National Bureau of Standards of 130 radon seeds supplied by radon companies through the regular channels. Both companies claimed an accuracy of plus or minus 1 per cent in the radon content of the seeds. It was found that seeds from one company were commonly 4 or 5 per cent wrong (probably clinically inappreciable). Another set showed errors up to 30 per cent!

These seeds were presumably obtained from reputable companies. It seems reasonable to suppose that sources from other companies or private institutions might well be subject to greater errors. Some institutions still measure seeds in groups of twenty or more and assume that the radon is equally divided between them. The writer has known one instance in which such a set, nominally 2.0 millicuries each, in fact ranged from zero to 5.0 millicuries. In a discussion of radon technique it is proper to emphasize that these errors are not inherent in the method. It would even be feasible for a physician to check the radon content of the sources he purchases. In practice, this precaution is seldom, if ever, attempted.

(e) The radon plant is a potential source of danger to the operator and to people in adjacent rooms. As the normal output of radon is only 90% of the theoretical yield, it is cogent to inquire what happens to the remainder. Some is safely left in the radon flask by incomplete pumping, some is removed by ventilating fans, a portion is absorbed by the oil pumps constituting a separate radiation hazard, and some escapes into the building. In assessing the danger of the technique it should be remembered that the dispersal of half a millicurie of radon throughout a room 15x12x10 feet would produce a "tolerance atmosphere" (J. Read and J. C. Mottram: Brit. J. of Radiol. XII p. 54. 1939), and such a dispersal would be only 1 or 2% of the total known loss in a normal pumping. It is not difficult to demonstrate the presence of radon at a considerable distance from a radon plant despite the use of exhaust fans. Clearly a radon plant should preferably be operated in an isolated out-building, a distinction which it seldom receives. The risk of damage to health by exposure to radon is believed to be high, as continued inhalation of the gas results in the formation of a cumulative radioactive deposit in the lungs. There the tissue may be assailed by the combined alpha, beta and gamma radiations, a local activity perhaps a millionfold greater than that of the gamma rays alone, the only hazard when radium element is used.

(f) Protection of personnel against harmful radiation effects is complex. Primarily the most insidious danger is the release of radon into the atmosphere from the plant, from a broken glass capillary or from a defective gold seed. Precautions must also be taken against beta rays emerging from glass capillaries or through gold seeds less than 0.5 mm. thick. The protection against the penetrating gamma rays is virtually identical with radon or element sources. In interstitial work the radium therapist is at a disadvantage with radon because the initial activity to which he is exposed has to be greater on account of the subsequent decay.

(g) In practice an adequate supply of sources of the proper strength is seldom available at the right time. One of the major treatment problems is to distribute the sources to obtain uniform dose over a stated region. This becomes more difficult with the random strengths usually available. It can always be solved, at the expense of additional manipulations, for external applications. High grade interstitial work, necessitating adjustments after the needles are in situ is virtually impossible.

(h) It is difficult to insert gold seeds in the proper position because the operator loses control of them as soon as they emerge from the seed introducer. Very often they will execute one of two possible turning movements about the tip of the introducer bringing them out of position. The seeds may also wander in tissue during the course of treatment.

(i) Reliable authorities quote that approximately 3% of radon seeds are incompletely sealed thus permitting the escape of radon. The writer believes the figure is an overstatement, but perhaps 1% may be faulty.[2] A leaking seed is an ideal focus for the formation of a necrotic region. Consequently if 10 seeds per case is a representative number, one in every ten cases may ac-

---

[2]The report referred to in footnote ([1]) indicates that 12 of the 130 seeds leaked, mostly by small amounts. The authors conclude as follows: "If, as shown by the measurements of the National Bureau of Standards, the measurements of radon content made by the companies supplying radon to physicians are so inaccurate (sometimes as much as 30 per cent), it is idle to quibble about occasional leakage of less than 1 per cent." In the opinion of the present writer, this last statement is a misconception of the danger attached to leaking seeds. Although it is true that the reduction of radon content per se is unimportant, it does not follow that the effect of small leakages of radon directly into tissue is not damaging. It would be presumptuous to maintain that minor leaks have demonstrably produced ill effects. Nevertheless, until the contrary proposition can be proven, a conservative policy would require the exclusion of defective sources.

RADIUM ELEMENT VERSUS RADON

quire a gratuitous necrosis in addition to those arising from the usual errors in distribution and dosage.

(j) The thin platinum walls of interstitial needles have a fine screw thread engaging the detachable point or eye. There is definite risk of accidentally detaching the end while the needle is in tissue. At the worst the glass cell inside may be left in tissue. This is possibly an unfair criticism of the radon method, but it is a practical danger worthy of note.

(k) Long sources of radon in general have quite uneven activity per unit length. This further complicates the dosage estimation which has to be based on the assumption of uniform strength.

The use of radium element sources contrasts favorably with the above, except for items (m) and (n) as follows:

(l) The sources have considerable value. Hence the issue of and manipulations with the tubes or needles have to be carefully controlled against loss. All patients must be hospitalized, except those who make daily visits for the wearing of radium moulds.

(m) A poor initial choice of permanent sources can be a serious handicap. Especially if the tubes and needles have a high activity per unit length it will be difficult to construct some types of moulds and to make implants of long duration. With a small stock of radium (250 mgm. or less) the principles of treatment have to be decided in advance. It would not be possible to change, say, from an eight-day technique to a 24-hour technique with the same needles.

(n) Sources are available in standard strengths ranging from 0.5 milligrams to 25 milligrams. Once certified, the activity remains constant unless a needle is brutally ill-treated.

(o) Well constructed sources have uniform activity per unit length. The packing is easily checked initially and defective needles can be replaced.

(p) Standard sources can be distributed to give approximately uniform irradiation of known amount in all cases. This is a simple routine procedure for external applications. Interstitial work requires access to radiographic equipment and the services of a physicist.

(q) Protection problems for personnel are clearly defined, and nowadays excessive exposure can only arise from gross negligence.

(r) The dosage-rate is constant. For moulds the daily dose can obviously be controlled at will. For implants the total dose only can be regulated by means of the total time. As it is not self-evident why non-decaying sources help, a numerical example may be cited.

Ex.: To irradiate an extensive lesion of tongue in the form of an elliptic cylinder 4x3x3 cm. a dose of 5500 r (given by 1700 milligram-hours) is required in about 7 days. Approximating to whole numbers one might use (a) 10 mgms. radium element for 7 days, or (b) 18 mcs. radon in needles for 7 days, or (c) 13 mcs. permanent gold seed implant. The completed implant is found to measure in fact 3½x2½x2½ cm., a reduction which is very frequent in practice. To keep the same dose the milligram-hours must be reduced to 1200. The element should then be removed after 5 days, whereas the radon needles come out after 4 days. Furthermore as the total time has been reduced from 7 days to 4 days it may be advisable to reduce the dose. So the time is further reduced to 3½ days. Now the time is even shorter and possibly the dose should be further reduced, and so the final treatment may be a grotesque distortion of the original plan. Whether the short treatment is as effective is a clinician's problem, but in a busy department it may well be that the treatment time has elapsed before it is demonstrated that the needles should have been removed. Note that the gold seed implant is not removable and in this example the dose would go to 7800 r, almost surely a necrotic dose.

Conversely if the implant had proved larger than the one planned, as it usually does, for example, in rectal implants there would be a pro rata increase in the radium time, or a very large extension in the radon time (on account of the diminishing daily contributions). It might be that the radon dose would never reach the required value, necessitating further traumatic interference by the addition of more sources.

When the relative merits of the two systems are compared it is evident that the two main problems, namely the delivery of a successful treatment to a patient, and the reasonable protection of the health of those using the radioactive material, are more favorably approached with element sources. Good work has been done with radon, and poor work has been done with radium, but the chances of consistent success are so much greater with the element that the choice of a department beginning radiation work seems to be clearly indicated. The Tumor Institute, believing that the emphasis is stronger than that, has recently taken the bolder step of converting its radium solution, hitherto used for the preparation of radon, into the solid form for permanent assembly in tubes and needles.

A brief account of the new arrangements may be of some interest. The recovery of the radium salt was in principle the simple matter of driving off the water by gentle heat. Elaborate precautions had to be taken against the emission of radiations and the escape of radon. The final product was a pile of dirty white powder, which could just have been heaped on to a nickel. The value of radium is emphasized when it is noted that this same pile was worth half a million nickels!

The new sources have been obtained as needles containing 0.5 mgm., 1 mgm. (short), 1 mgm. (long) and 2 mgms., and as tubes containing 5, 10 or 20 mgms. This set will suffice to treat all the types of lesion contemplated. It is not a universal set, lacking in particular 3 mgm. needles and short 2 mgm. tubes. Just the lesions for which these would be used are more amenable to treatment by the special x-ray facilities available.

The new supply is housed in a radium safe with some original features. Twelve drawers are used to keep the amount in each compartment low. Each drawer has a small pocket containing a lead block drilled to take the different needles. The main length of the drawer is filled with lead, six inches thick before and behind the radium. The drawer cabinet is surrounded by lead, the minimum protection in any direction being $5\frac{1}{2}$ inches of lead, more than adequate for the present radium supply. Each drawer bears an index plate to register the number of needles inside. The radium officer thus knows how many needles are available without taking radiation by opening drawers. A new device automatically detects radon leaks so that damaged needles can be located. The needles are normally removed from their separate holes one at a time. Provision has been made to empty the entire safe in less than a minute in case of emergency.

To reduce radiation exposure, protective measures based on rapid handling, maintenance of distance from the radium, and the use of lead screens have been freely employed. Although some of the details appear trivial, collectively there is an appreciable gain in protection. Thus radium moulds are assembled and fitted with dummy tubes until the dimensions are exactly correct before the real tubes are removed from the safe. Moulds temporarily not in use are stored in a subsidiary lead-lined safe. Implant needles are held in special lead blocks to be threaded, and are then boiled in a lead-lined sterilizer. Many of these protective devices are modelled after the designs of Dr. J. R. Nuttall (Am. J. Roentg. and Rad. Ther. XLI p. 98. 1939). From the same author an unpublished system of bookkeeping, whereby the utilization of the radium is conveniently checked and counter-checked, has been adopted.

RADIUM ELEMENT VERSUS RADON

The necessity for the precautions outlined above is admittedly a handicap, but with proper care the control can be reduced to a routine, which is neither time-consuming nor involved enough to create errors. In one sense the added caution used with the intrinsically valuable sources is even an advantage, since the sources, be they element or radon, have adequate potency for undesired damage. In all other respects it is anticipated that the conversion will bring the best available weapons to bear on the problems of consistently effective radium treatment and the safeguarding of the personnel connected with it.

A Physical Evaluation of Supervoltage Therapy

Radiation Therapy A Supplement to the Staff Journal of the Swedish Hospital Seattle, Washington from the Tumor Institute

May 1941

RADIATION THERAPY

A SUPPLEMENT
to the
STAFF JOURNAL
of the
SWEDISH HOSPITAL
SEATTLE, WASHINGTON

From

THE TUMOR INSTITUTE

A PHYSICAL EVALUATION OF SUPERVOLTAGE THERAPY

H. M. Parker

The original choice of voltages in excess of 400 K.V. for roentgentherapy was encouraged by two reasons: the hope of obtaining improved depth dose to render deep-seated lesions accessible to radiation, or an appreciation of the publicity value of a powerful new weapon. At the present time the popular press has transferred its activities to the cyclotron and to artificial radioactive sources, undiscovered when supervoltage roentgen tubes were first built. We are, therefore, free to analyze the merits of supervoltage therapy in true perspective.

In a recent survey on "Possibilities of Improved Therapy for Cancer Patients" Voegtlin[1] writes as follows on the subject of supervoltage roentgen therapy: "It appears now, however, that the results of supervoltage have been more or less disappointing. The recent careful studies of Packard have shown that relatively little is gained, as far as the ratio of depth to surface dose is concerned by increasing the voltage from about 400 to 900 kilovolts. This is in agreement with the results obtained by Stone. In fact, Stone found the biologic effect the same, if the same dose is delivered to the patient from opposite sides, the one side receiving 200 kilovolt roentgen rays and the other 1,200 kilovolt rays. A few other investigators have made somewhat more favorable reports. Neverthless, it appears at present that, while supervoltage roentgen rays may perhaps prove of value in the treatment of some deep-seated tumors, no great advance can reasonably be expected from this therapy."

The publication of this report by so eminent an authority on cancer problems is regrettable. We find ourselves in agreement only with the final statement above. No *great* advance can reasonably be expected, if by this we mean that supervoltage therapy will not suddenly provide the solution to radiation therapy problems. It seems in the nature of things that, with any type of radiation, the margin between success and failure is small. If the chance of success is a little greater with supervoltage therapy this might well be classed as a great advance. It is the purpose of this article to point out that there are sound physical reasons for expecting an advance of this nature.

The work of Packard quoted above purports to show that percentage depth dose does not increase appreciably as the voltage increases from 400 K.V. to 900 K.V. These measurements were based on measurements on Drosophila eggs on the premise that physicists are not yet able to measure doses for radiations at voltage much in excess of 200 K.V. This misconception arose from the wording of the original definition of the roentgen, which was fortunately revised in 1937. It is possible to state doses for radiations up to and including gamma-radiation (equivalent to 2000 K.V.) at all points excluding boundaries (near the skin, adjacent to bone or body cavities). It is in fact easier to perform the measurements for high voltage radiation than for low. The reasons for these exceptions apply to measurements with Drosophila eggs, which might well have been abandoned years ago as a basis for dosage estimation. Nevertheless, it is true that the increase in depth dose with increasing hardness of radiation is not nearly so striking as the early proponents of supervoltage therapy had hoped. The expected increase could be pre-calculated by physical methods, and it agrees well with current observations. Increase in depth dose should, however, not be the only criterion of the value of hard radiation. The composition of the roentgen ray beam traversing a large tissue mass is quite different at high voltages. The beam is composed mainly of the primary radia-

A PHYSICAL EVALUATION OF SUPERVOLTAGE THERAPY

tion at the original wavelength with some contribution from scattered radiation at wavelengths not much degraded. At 200 K.V. with average conditions of field size and focus-skin distance the radiation at a point 10 cm. deep is 75% scattered radiation, much of it at appreciably longer wavelengths. The high voltage beam is consequently much "purer", a point to which we shall return later. It is this rapid reduction in scattered radiation which accounts for the slow increase of depth dose as the voltage is increased, for it partially compensates the rise due to the increased penetration of the primary radiation.

The reduction of scatter has an important corrolary, the full value of which has not previously been stated. It follows that the radiation in tissue is confined more closely within the geometric confines of the beam. This has long been recognized and is familiar, for example, in isodose charts (Fig. I).* The sharper limitation of the beam is of real advantage in firing past structures which it is desirable not to irradiate, and in multiple field techniques, where the dangers from overlapping contributions outside the primary beam is quite real. It has also been pointed out that a smaller volume of tissue is irradiated by the sharper beam. The difference seems almost negligible on the cross section of the beams shown in isodose charts. It is of the first importance in the actual patient where this difference persists throughout a large annulus surrounding the beam.

Consider a tumor mass as in Fig. I, located more or less centrally in a body 25cm. thick and calculate the energy absorbed in this mass and in the whole patient at the different voltages, by the ingenious method recently given by Mayneord.³ Of the total energy absorbed by the patient we find the following amounts absorbed by the tumor mass: A (at 200 K.V.) 5%, B (at 800 K.V.) 10%, C (at 2000 K.V.) 11%, D (at ultra-high voltages) 12%. The total radiation absorbed by the patient must be closely connected with undesirable systemic effects. For every unit of energy absorbed by the tumor in Fig. I, the total energy contributing to the systemic effect will be: A, 20 units; B, 10 units; C, 9 units; D, 8 units. The general impression that supervoltage radiation is more readily tolerated is thus well substantiated by physical investigations. If this were the only factor involved there would be no upper limit to the optimum voltage for this particular tumor mass although for practical expediency most of the gain has been accomplished at 800 K.V. For other tumors the relative figures can be very different. In particular for a superficial tumor 2½ cm. thick the total energy per unit energy to the tumor becomes A, 13.5; B, 13.5; C, 18; D, 24; respectively. Clearly it would be absurd to utilize supervoltage radiation at 100 cm. focus-skin distance for such a lesion. (Note that for other reasons it would be interesting to study the effect of supervoltage radiation on superficial tumors at very short focal length, a suggestion already considered by Dresser.⁴ Modern construction of

---

*The charts show the distribution of radiation through a plane containing the central axis of the beam. The incident field is a circle 7.5 cm. diam. Focus-skin distance is 100 cm. As usual the curves show the percentage depth dose referred to 100% at the center of the field on the skin surface.

A. 200 K. V. Constant potential. Filter: 1 mm. Cu + 1 mm Al.
B. 800 K. V. p. Grid-biassed pulsating potential. Filter: 4.5 mm Pb + Sn, Cu, Al.
C. 2000 K. V. nominal; i.e.: well filtered gamma radiation.
D. Hypothetical ultra-penetrating radiation.

The isodose curves have been split along the central axis and reassembled in pairs to make the successive changes in configuration more apparent.

In order to express A, B, C and D in terms of the same field size and focus-skin distance the labor of redetermining some of the curves has been avoided by a method given before.² The extrapolations in this case are rather wide, and the curves should be regarded as "representative" rather than actual isodose charts.

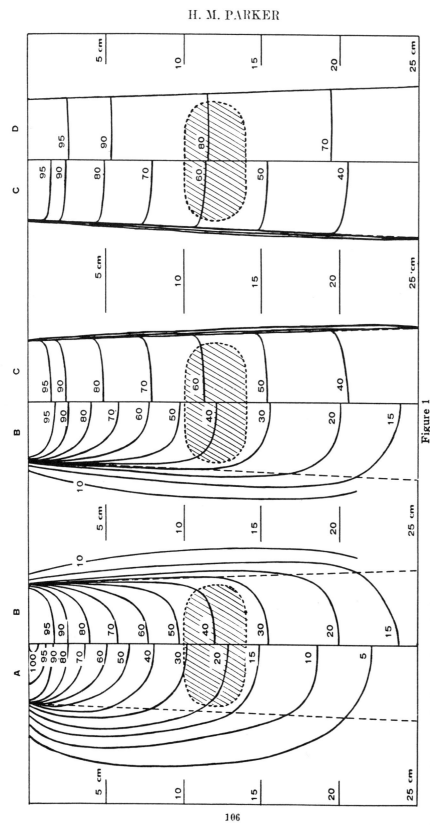

Figure 1

Representative Isodose Curves in a large tissue mass. In comparing A and B (200 K.V. and 800 K.V. radiation) note especially how the radiation in B is confined much more closely to the geometric beam (shown by the broken lines). This change is as important as the increase in depth dose. See footnote on page 105.

A PHYSICAL EVALUATION OF SUPERVOLTAGE THERAPY

high voltage apparatus would make the experiment possible especially if an adequate mass of gold replacing the customary lead protective shield could be obtained.)

The argument of the preceding paragraph is strictly valid only if the biological effect of radiation on tissue is independent of the wavelength. For otherwise the systemic effects produced might bear no proportionality to the energy absorbed. The composition of the beams in terms of wavelength varies from point to point in the tissue; in particular the radiation in the penumbral regions outside the geometric beam is entirely composed of scattered rays of considerably increased wavelength. The net effect is a complex function which the physicist can only evaluate by making simplifying assumptions. When this is done it turns out that the ratios given above are probably not seriously disturbed, although under some conditions the margin of difference could be appreciably reduced.

The whole question of the dependence of biological effect on wavelength is pertinent to the relative merit of supervoltage therapy. At the present time no final answer can be given. In the first place it can be stated definitely that real "specificity" in the same sense that some wavelengths in the ultraviolet region exert specific action on bacteria cannot exist. Until the mechanism of the interaction of radiation on living cells is more fully understood it is difficult for the physicist to draw reliable conclusions about the action of different radiations. At the present time it seems almost certain that the effect depends on the ionization produced in the tissue. In this case it would be improbable that major differences could arise in soft tissue. For although at higher voltage faster ionizing particles are first produced, the final ionization picture is quite similar in all cases. If the effect depends on a direct hit of one or more ionizing particles on a sensitive cell element, or if it is a chemical action or colloidal-chemical action, there is room for specificity of a type, arising from the different ion density along the ionized tracks in the various beams. Present evidence suggests that over a wide range of biological actions, the effectiveness of roentgen-rays generated at voltages of the order of 200 K.V. ranges from one to two times that of the physically equivalent dose of gamma-rays. In the final analysis it is the differential action on malignant and normal tissue which is important in the cancer problem. This is so far removed from the field of physics that the present writer would be foolhardy to do more than mention it. It does seem astonishing that sufficient clinical evidence from cases in which known doses of radiation have been delivered has not yet accumulated to provide an approximate answer. One is tempted to deduce that the variation of clinical effect with wavelength is at most not greater than the changes produced by variation of total volume irradiated, the dosage-rate at any particular wavelength, and the fractionation of the dose. If this were the case it becomes self-evident that for deep tumors, at least, supervoltage radiation, with its minimal systemic effect, would be the treatment of choice.

Certain exceptions at physical boundaries namely the air-skin-soft tissue, soft tissue-bone, soft tissue-mucosa-air interfaces have already been mentioned. Of these three, the first has always commanded a special position. It is on skin that the effects of radiation are most apparent and in the early days of radiation therapy it was the effect on skin which formed the basis for comparison of dose. Even today the use of erythema dose has not quite disappeared. In fairness to the roentgentherapist the physicist must admit that he is partly to blame for this, because it is just here that he is most hesitant to ascribe a real meaning to his own methods of dosage measurement. The problem is complicated by two factors: The skin contains elements, notably sulphur, which are not found in soft tissue. This sulphur, when subjected to radiation gives rise to photoelectric ionization of considerable magnitude, which is not included in the normal methods of estimating dosage. Since this additional ionization changes rapidly with the wavelength of the primary beam, becoming negligible for short wavelengths (gamma-rays) it has been felt for some time that this might be a full physical explanation of the observed dependence of erythema dose on wavelength. Numerically, we can quote figures from recent experiments by L. H. Gray.[5] Under conditions in which the corpuscular radiation in soft tissue arising from irradiation by gamma-rays, 190 K.V. roentgen rays, and 119 K.V. roentgen-rays would be the same, the corpuscular radiation in pure sulphur would be respectively 1.00, 2.49 and 6.00. The amount of sulphur in skin may be about 14% by weight, which would reduce the above figures to 1.00, 1.21, 1.70 in skin. These ratios would reasonably account for observed differences in the skin effect. However, Gray's measurements in dried mouse skin lead to ratios of 1.00, 1.09, 1.36. It is unfortunate that the actual sulphur content of this particular skin was not determined. J. C. Mottram and L. H. Gray[6] exposed the skin of mouse tails to the same three radiations and obtained the following measures of biological effectiveness: for erythema and desquamation 1.0, 1.3, 1.3; for exudation and epilation 1.0, 1.6, 1.6. It is evident that the agreement between physical expectation and biological findings is poor. Although the experiments of Mottram and Gray are open to some criticisms it is unlikely that the two sets of figures can be reconciled. Nevertheless, except for the dissenting opinion of Stone, quoted by Voegtlin, it is a matter of universal observation that the skin reaction for equal dosage is progressively less for shorter wavelengths in the deep therapy range. The sulphur content of skin undoubtedly contributes to this effect but it has to be conceded that it is not the only factor. Another physical factor may be that the layer of skin above the cells that give rise to skin erythema is too thin for the full complement of corpuscular radiation to be built up for the harder radiations. It would be a necessary corrolary that fast secondary electrons could enter from outside the body, ie. the erythema dose would be a function of the external scattering conditions. That this does occur is well known to anyone who has operated telecurietherapy apparatus. It has also been reported for million-volt X-rays by Failla and Edith H. Quimby.[7] Other factors outside the realm of physics may contribute to the skin reaction, as for example the irradiation of the blood supply over an extensive region. The reduced skin reactions at higher voltage, for which two contributing physical causes have been given, remain an indisputable fact which give supervoltage therapy a pronounced advantage over other forms. The radiotherapist should be primarily concerned with the desired effect on the tumor, but where this can be obtained equally well by a choice of radiation type, that which gives the minimum damage to skin with its attendant diminution of discomfort to the patient and risk of infection should clearly be used.

To summarize briefly the merits of supervoltage therapy we have:

1. An increased depth dose, not sufficient in itself to justify the use of supervoltage.

2. A beam of radiation more closely restricted to the geometric beam. This helps in the irradiation of tumors lying close to structures which should not be irradiated. It also simplifies multiple field techniques.

3. For deep tumors it concentrates the energy absorbed in the tumor compared with the total energy absorbed in the body (a measure of deleterious systemic effects) to a marked degree.

4. Skin damage for a given dose to the tumor is physically expected and clinically verified to be less. If the required tumor dose for supervoltage radiation were greater, this advantage might be diminished or even reversed. The answer to this point rests entirely on clinical evidence which

A PHYSICAL EVALUATION OF SUPERVOLTAGE THERAPY

has not yet been adequately presented. The general impression is that somewhat higher dosage is required with hard radiation, but not enough to compensate for the difference in skin effect. The physicist has hoped that equal dosage in roentgens would produce the same effect in soft tissue regardless of the generating voltage. This simplification should not yet be abandoned although mechanisms of action of radiation on tissue can be postulated which would render it invalid.

No discussion of supervoltage therapy is complete without brief mention of its practicability in the tumor clinic. During the past four years, at least, the 800 K.V. equipment at the Tumor Institute has had a better performance record than the standard type of 200 K.V. machine! More modern types of equipment operating at one-million volts should at least equal this record. They have in addition the merits of compactness and mechanical flexibility of adjustment, features lacking in the earlier machines.

The day seems quite close when every physician can switch on his office million-volt tube as easily as he now uses his 100 K.V. machine. When this time comes the physicist who has written favorably about supervoltage may well regret it. For clear as these advantages may seem to him, they are not so outstanding, but that it requires radiotherapeutic skill of the first degree to utilise them for real advance in cancer therapy.

[1] Voegtlin, C: Possibilities of Improved Therapy for Cancer Patients. J. A. M. A. 116: 1491 (April 5) 1941.
[2] Parker, H. M.: The Dependence of the Back-scattering of Roentgen Rays in a Phantom on Focal Distance, Quality of Radiation and Field Size. Acta Radiol. XVI. Fasc. 6. 705. 1935.
[3] Mayneord, W. V.: Energy Absorption. B. J. R. XIII, 235. 1940.
[4] Dresser, R.: Supervoltage Roentgen Therapy. Conference on Applied Nuclear Physics. Cambridge, Mass. 1940.
[5] Gray, L. H.: Physical Investigation of the Photo-Electrons from Sulphur to X-ray Ionisation. B. J. R. XIII. 25. 1940.
[6] Mottram, J. C. and Gray, L. H.: The Relative Response of the Skin of Mice to X-radiation and Gamma Radiation. B. J. R. XIII, 31, 1940.
[7] Failla, G. and Edith H. Quimby: Decrease of Skin Damage by Deflecting Secondary Electrons from a Beam of One Million-Volt X-rays. Amer. Phys. Soc. Washington Meeting, April 1940.

Limitations of Physics in Radium Therapy

Radiology, Volume 41

October, 1943

Reprinted with permission of the Radiological Society of North America.

[Reprinted from RADIOLOGY, Vol. 41, No. 4, Pages 330–336, October, 1943.]
Copyrighted 1943 by the Radiological Society of North America, Incorporated

Limitations of Physics in Radium Therapy[1]

H. M. PARKER, M.Sc., F. Inst. P.

Tumor Institute of the Swedish Hospital, Seattle, Wash.

THE PROBLEMS of greatest interest to the hospital physicist are not necessarily those of the greatest practical importance. It is therefore well to review the field occasionally to decide whether one's activities are properly directed. A favorite problem has been the evaluation of gamma radiation in terms of roentgens. Although this evaluation was logically desirable, it has to be conceded that radium therapy *per se* has gained little from it. Another class of problems is concerned with the methods of distributing radium to obtain uniform and calculable irradiation. This paper is to discuss this class with special reference to the solutions offered by Paterson and Parker (1, 2, 3).

In the first place, absorption and scattering of gamma radiation in tissue are small enough to justify their deferment as a later correction. It is then possible to set up a mathematical theory. The problem can be defined as the uniform calculable irradiation of a prescribed region of tissue without the over-irradiation of other parts. There is a sharp differentiation between interstitial and superficial arrangements. In the former, the prescribed region is clearly a certain volume of tissue. In the latter, although there is still a volume of tissue to be adequately irradiated, one particular layer, the skin, requires more critical treatment than the rest. The region of uniformity is restricted to a single surface. Approximate uniformity is acceptable in the depth.

Under these conditions it appears that the operation of fluid distributions of radium for the interstitial case can be made to give irradiation as uniform as desired. Inasmuch as all distributions except the most symmetrical have to be evaluated by numerical integration, the actual labor of determining the solutions would be prohibitive.

For superficial arrangements manipulation of two-dimensional sheets of radium is sufficient. The problem is always soluble for plane surfaces. Exact solutions can be obtained for many classes of regular curved surfaces, but irregular curves are not amenable to treatment. It should be observed that even for the simplest case of a plane circle it is impossible to achieve uniformity under the premises originally implied in the Paterson-Parker system, namely, that the periphery of the radium arrangement coincides with the normal projection of the treated area. This is self-evident because, for small rings, complete peripheralization of the radium still leaves the dosage rate high in the center. The present definition overcomes this by the artifice of having the "treated area" larger than the "prescribed region."

The Paterson-Parker system falls short of the mathematical ideal by permitted tolerances. It has to be investigated whether these tolerances are acceptable in view of other errors in radium therapy, or whether a more rigid solution should be sought.

The interstitial case is easily dismissed. Instead of operating with a generalized fluid, two density regions were chosen and general distribution propositions set up to a tolerance of ±10 per cent. The fluid was then condensed to discrete sources. The distribution remained good except around the individual foci. These regions were confidently ignored on clinical grounds, because it was possible with weak sources to keep the dosage below the necrotic level, while the balance of the tumor area was adequately irradiated. It is evident that in this field physics is well ahead of clinical practice. The limitations of good implantation become:

1. The number of sources that can be

---

[1] Presented before the Radiological Society of North America, at the Twenty-eighth Annual Meeting, Chicago, Ill., Nov. 30–Dec. 4, 1942.

introduced without excessive trauma.
2. The skill with which they are inserted in postulated positions.
3. The reconstruction of the implant for accurate dosage calculation.

The development of the dosage system for external applicators was quite different. A direct approach was made with rings and straight lines of radium, rather than the mathematically more satisfying fluid sheets. In effect, the selected tolerance of ±10 per cent represented one solution of the condensation of exact sheet distributions to line arrangements. The choice of ±10 per cent tolerance was entirely arbitrary, and as it would be theoretically possible to reduce the tolerance almost to zero, it has to be determined whether other considerations warrant such a variation. As soon as the mathematical system is modified by practical details it becomes impossible to generalize. Examples have to be selected which illustrate either average or the most unfavorable circumstances as required.

A perfect distribution would be modified by absorption and scatter in tissue, and it would be unwise to elaborate a system to closer limits than the changes due to these causes. Roberts and Miss Honeyburne (4) have studied the case of a radium ring. If the ring was fully surrounded by scattering medium the dosage rate was equal to that in air. When the ring was at or near the surface of the medium there was a net absorption of 4 per cent per centimeter depth from the ring. The author has found comparable though smaller effects in a series of clinical applicators. Scatter is more complete along the axis, so that central dose in any parallel plane tends to increase relative to the peripheral dose. Roberts and Honeyburne suggested that this would increase the ratio of diameter to skin distance for the ideal circle from $2\sqrt{2}$ to 3. It seems more probable from their own measurements in air that most of the difference was due to oblique filtration. The author has used a ring of gold seeds, and later a single rotating source, and has obtained close agreement between calculated and air-measured values. Changes with back-scatter were of the order of 2 per cent. The actual dose over a skin surface will depend on the construction of the applicator—whether the radium tubes are essentially in air or at the surface of a wax block. With the present methods of using radium there might be an error of perhaps 5 per cent in the quoted dose.

The effect of substituting discrete radium tubes for the long lines or rings used in the original system has been discussed (1) and preliminary rules given. The basis for these rules was that the variation thus introduced should not exceed the original variation. For example, the replacement of a ring by a number of gold seeds was designed to keep the circumferential variation less than the radial variation. Table I shows this for some rings. The rule was simplified to the requirement that the distance between sources should not exceed the skin distance, a somewhat more stringent rule.

TABLE I: NUMBER OF GOLD SEEDS REQUIRED IN A RING OF DIAMETER d AT RADIUM-SKIN DISTANCE h

d/h	2	3	4	5	6
Radial variation	±5%	±3½%	±2%	±5½%	±8%
Number of seeds by Rule 1	6	8	10	10	10
by Rule 2	6	10	13	16	19

Rule 1. Circumferential variation shall not exceed radial variation.
Rule 2. Distance between seeds shall not exceed h.

The combined variation in a practical case can exceed the ±10 per cent tolerance. In fact, the method of regulating the number of sources is illogical to the extent that arrangements initially poor are allowed greater laxity. The rules for large areas, especially rectangles, could well be improved. In general, with applicators commonly used, the use of discrete sources *per se* need add little or nothing to the existing error. Oblique filtration, a necessary corollary of discrete sources, merits separate discussion.

The original system took cognizance of

oblique filtration only in special treatments with long single sources. It was believed that the effect in most cases would be less than that of other inherent errors. This belief was doubtless encouraged by the complexity of oblique filtration calculations, and in review it effect in many cases. Figure 2 shows an extreme case of a square 10 × 10 cm. at 1 cm. distance. The reduction by oblique filtration was calculated for different filters at two points (A and B). Allowance was made for the normal filtration reduction. At 1.5 mm. Pt the general dose level fell

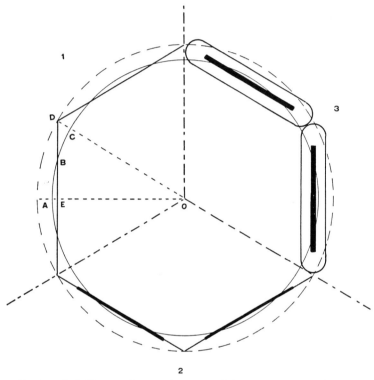

Fig. 1. Six 6.66-mg. tubes as a hexagon. Distance 14 mm. (1) Location of calculated points. (2) Tubes without platinum wall. Filtration correction for 1.5 mm. Pt. (3) Real tubes giving rise to oblique filtration.

DOSAGE RATE (r/hr.)

	No Oblique Filtration	Oblique Filtration
A	47.7	45.4
B	46.6	44.6
C	45.7	44.0
D	43.6	40.6
E	48.4	46.4
O	47.5	47.5

cannot be completely substantiated. In all cases the dosage rate in some parts of the irradiated area will be reduced by oblique filtration. Figure 1 illustrates a familiar example of 6 radium tubes as a hexagon. Near the center of the treated area the reduction in dosage rate is negligible, but near the periphery it is of the order of 4 per cent. This is typical of the some 10 per cent below the direct value. Such a change should be corrected.

Calculations with radium tubes are based on the assumption that the tubes are properly filled. The total radium content should be certified to approximately 1 per cent. The actual strength of modern tubes should be within 2 per cent of the nominal value. These values remain correct unless the tubes are grossly mishandled to produce a major radon leak, under which conditions it would be simpler to seek a new technician rather than a new dosage system. Nevertheless, H. D. Griffith (5) has demonstrated that needles may be so badly loaded that the radiation

Curved surfaces are not amenable to good mathematical treatment, and they are also less satisfactory in clinical practice. The system rules are entirely empirical. For convex areas, for example, they depend on the spread of the radium over a larger area to compensate for the increased

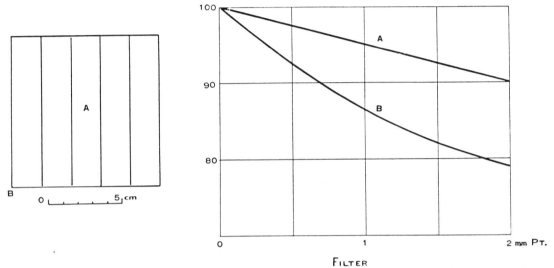

Fig. 2. The graph shows the relative dosage rates at points A and B in the treated area of a 10-cm. × 10-cm. rectangle, for different radium filtration, after the routine filtration correction has been made.

pattern is seriously disturbed. He chose four bad needles to make an applicator, compared this experimentally with four good needles, and shows the striking results in an illustration.[2] The effects of bad packing could clearly outweigh other errors discussed.

Radium tubes should be examined for packing initially, and again after one or two years, by autoradiograms. The eye can detect packing faults of the order of 20 to 30 per cent. Photometry would be preferable. Griffith's applicator has been reconsidered with relative strengths in the two halves of 1.2 to 1 and 1.4 to 1. Figure 3 shows the dose along diagonals at 1.5 cm. (to conform to the 2h rule). At 1.2 to 1 the error is not important. Nor is it excessive at 1.4 to 1 when the improbability of so unfavorable an arrangement is considered.

dosage rate due to cross-fire. The compensation is surprisingly good over the regular curves that have been investigated. Nevertheless, certain special cases give results not within accepted standards. The most familiar case is the lip mold, as in Figure 4. Comparison of the dose with that over the equivalent plane indicates an increase in some regions of 15 per cent. Here a curve is followed by two planes not expanded by curvature. The difficulty can be met, after the manner of Murdoch and Stahel (6), by an individual calculation, or more readily by substituting two parallel planes for the curved mold. Concave areas are not well treated when the applicator area is much less than the treated area. Treatments with small skin distance or treatments at the radius of curvature of a part are sound.

Another difficulty, present in all cases, is exaggerated for curved surfaces. That is the determination of the treated area, which could frequently be in error by 10

---

[2] In the interest of wartime economy, the reader is referred to Griffith's article to determine the details of the packing error.

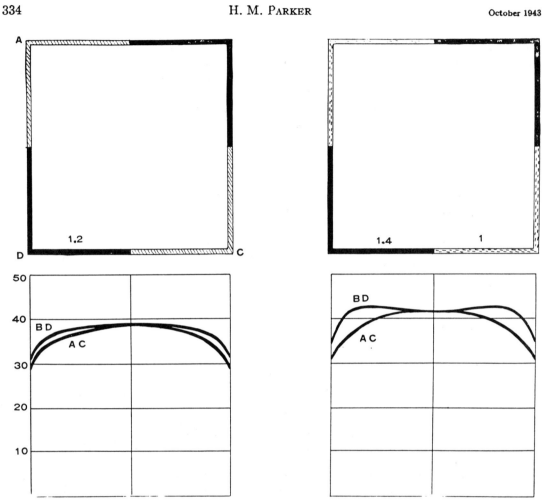

Fig. 3. Radium applicator 3 cm. × 3 cm. made with 8 radium tubes. On the left side the 4 dark tubes have 1.2 mg./cm. On the right the dark tubes have 1.4 mg./cm. In each case the lightly shaded tubes have 1.0 mg./cm. The variation of dose along the diagonals AC and BD in the treated area is shown below each square.

per cent. This would give a dosage error of 5 per cent. Referring to Figure 1, there is some indecision even in this simple case. The effective treated area lies between that of the hexagon (area 12.6 cm.²) and the inscribed circle (11.4 cm.²). The corresponding dosage readings from the area charts differ by 5 per cent. In this particular case, one can consider an equivalent circle, which proves to have a radius of 1.98 cm. and an area equal to the hexagon (by coincidence only). If gold seeds were used as the sources, one would invariably operate in terms of the circle through them.

The last error to be discussed is one of the most serious. It relates to the mechanical difficulties in placing the radium tubes at the prescribed distance from the skin. It is not easy to mount radium tubes on wax or similar applicators so that they be within 0.5 mm. of the correct distance. Customary methods are not accurate to much better than 1 mm. The error has two parts, a systematic deviation from the true thickness plus a random fluctuation. The systematic error is numerically equal to the distance error (approximately). Applicators at 10 mm. give 5 per cent dosage error. Applicators at 10 mm. or less are open to serious objections on these grounds. Where the general thickness is correct and a single tube is misplaced, the error in its vicinity is approximately half that quoted.

When the several errors are marshalled together, it seems probable that there is more influence on the dosage rate than on the dosage distribution. The latter is influenced mainly by oblique filtration (which can sometimes improve the distribution), faults of loading or placing individual tubes, and the errors of curvature. The former, on the whole, should fall below the theoretical value, because absorption and oblique filtration both tend to reduce it.

Applicators can be divided into two arbitrary classes.

net error, but a variation of ±15 per cent with a dosage rate error of 5 to 10 per cent seems probable. Under these conditions it is debatable whether improved arrangements of radium should be developed. The variation could be reduced appreciably with little change in the dosage rate error.

It is always advisable to supplement dosage calculation by direct ionization measurement. It is laborious to check the full distribution, but the readings at

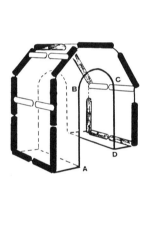
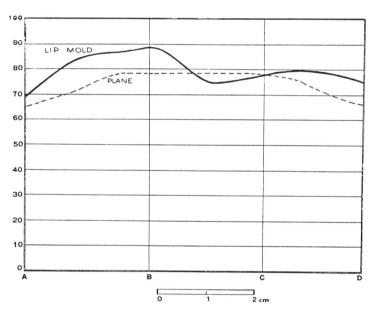

Fig. 4. A conventional lip mold. Shaded tubes have 3 units of radium per cm. Clear tubes have 2 units of radium per cm. The relative dose along a line similar to ABCD, but in the mid-section, is shown in the full curve. The broken curve gives the dose for the equivalent plane, *i.e.*, a plane of length ABCD and width equal to the original. This would be the calculated dose by the Paterson-Parker system. Note that the true dose is high principally over the outer surface of the lower lip A to B, rather than over the actual curved region B to C.

1. Distance 1.5 cm. or more: The distribution rules can be closely obeyed and the geometry of the mold kept accurate. Practically all cases can be treated with a theoretical variation of ±5 per cent. The subsequent errors are estimated to be of this same order.

2. Distance less than 1.5 cm.: The original system may use up the full ±10 per cent tolerance. It is more difficult to reproduce a theoretical arrangement with discrete tubes. At the same time oblique filtration and the faults of loading and position have their greatest effect. It is difficult to ascribe numerical limits to the

selected points can remove the principal dosage rate errors. The author has found closer agreement between measured and calculated doses than would be expected on the basis of the foregoing discussion. A deviation of more than ±3 per cent has been exceptional. It is believed that this is due to the fact that applicators required in the clinic happen to fall in those parts of the system that are inherently accurate. Applicators at distances less than 7.5 mm. are liable to be faulty both in calculation and measurement. In these cases the finite size of the measuring device may become important (7). Here again, clinical practice favors the system. Such applicators are needed either for lesions like hemangioma, where the dose is low, or in the buccal cavity, where reaction is not very sensitive to change of dose. In neither case is there danger in the treatment or loss of information of scientific interest.

Tumor Institute, Swedish Hospital
Seattle, Wash.

REFERENCES

1. PATERSON, R., AND PARKER, H. M.: Brit. J. Radiol. **7**: 592–632, October 1934.
2. PATERSON, R., PARKER, H. M., AND SPIERS, F. W.: Brit. J. Radiol. **9**: 487–508, August 1936.
3. PATERSON, R., AND PARKER, H. M.: Brit. J. Radiol. **11**: 252–266, 313–340, 1938.
4. ROBERTS, J. E., AND HONEYBURNE, J. M.: Brit. J. Radiol. **10**: 515–526, July 1937.
5. GRIFFITH, H. D.: Brit. J. Radiol. **9**: 404–411, June 1936.
6. MURDOCH, J., SIMON, S., AND STAHEL, E.: Acta radiol. **11**: 350–397, 1930.
7. SPIERS, F. W.: Brit. J. Radiol. **14**: 147–156, May 1941.

The Manchester System

Comments Prepared for the Afterloading Conference

Notes Prepared for a Lecture

May 1971

THE MANCHESTER SYSTEM

Comments Prepared for the Afterloading Conference

The hyphen got into the relationship Paterson-Parker in the fall of 1932, so this pre-history will date from that time. It would be grossly unfair to Dr. Paterson and his clinical colleagues, or for that matter to equally distinguished groups elsewhere, to imply that this hyphen caused a quantum jump in the success of radiotherapy.

At the time in question there were several effective centers for radium treatment. Both the distribution of radium sources and the "dosage" of radium had been empirically developed so that experienced therapists had a good feeling for a sensible approach to treatment.

For example, for surface applicators, the Brussells system of Murdoch, Simon and Stahel was a highly successful combination of medicine and physics because this team was willing to spend literally days in empircially adjusting sources and making hand calculations to get even irradiation. I have often speculated that if computers had been available at that time to do the calculations in minutes or seconds rather than hours, we might have stayed with their methods. In fact, I sometimes suspect that now that we have such computers, some clinics are almost going back to that approach.

2

The basic purpose of the Paterson-Parker system was to permit derivation of a sound distribution of radium sources to get effective treatment over the prescribed tissue region, and to know what the dose was, without each clinician having to repeat the learning curve of experience.

The 'distribution' part of the work is the more interesting. So let us get the primary dose method out of the way first. The time was 1932, four years after the introduction of the roentgen for x-ray doses. Contemporary attempts to correlate radium gamma ray measurements with the more conventional x-ray measurements were fascinating, but not always convincing. Dr. Quimby will tell us about some of these.

We took a method which, in the physics of its time, would have been called a derivation of Eve's Number. Dr. Eve himself was so sure that this would become a standard that he cleverly named his daughter "Millie Eve". It is one of the personal tragedies of our field that we do not recognize his contributions by name. The experiments we used were those of L. H. Gray in which he measured the energy absorption as measured by ionization in a large graphite block surrounding a radium source. We derived a so-called K value of 9.3 R per hour. As more generally applied, we used 8.4 R per hour through 0.5 mm Pt filter. Probably the best measurements

3

today would give 8.25 R per hour.

You may have seen a paper by E. H. Porter in BJR September 1970 entitled "How many rads per Paterson-Parker roentgen?" Like too many histories, it is exactly backwards and the question should be "How many roentgens per rad?" Our derivation was a Bragg-Gray Principle application before L. H. Gray became interested in hospital physics (eventually becoming one of the world's best, if not the best) and also, of course, before the prior statement of the Principle by Bragg (ca. 1910) had been resuscitated.

Enough about dose. From the one figure for the dose-rate at 1 cm from 1 mg of radium, it is simply an exercise in the inverse square law to determine the dose from any configuration. This is the essence of the intensity-millicurie hour (Imc-hr) system of Sievert.

Let us now return to the distribution problem. The work started with surface applicators. Distribution patterns were best known for uniform circular rings of radium. Since many surface lesions are more or less circular, it was natural to develop simple arrangements of multiple concentric rings to achieve surface homogeneity of dose, accepting a <u>purely arbitrary</u> variation of ± 10% over the treated area.

4

Obtaining a flat distribution was somewhat of an obsession, perhaps because the skin, in 1932, was indeed the principal barrier to getting enough radiation to all needed parts without seriously damaging skin. Of course, one never really treats a "surface". There is some sort of volume to be considered, depending on the thickness of the tumor.

In retrospect, it is a curious omission of the Manchester System that no simple rules for assuring adequate uniformity in depth, especially as that would have been much easier than getting the surface right. I was quite shaken, when putting these notes together, to discover how cavalierly we had dismissed this point. To quote:

"In planning skin applicators, or dental molds, an attempt is made to treat an area at least one or two centimeters wide of apparent tumor, and to use a distance which will give adequate dose at sufficient depth." Translation: "When I say 'adequate' and 'sufficient' I know what I mean, and you should, too." However, if you rework the original examples given, you will find stated depth doses in the range of 56% to 72%. If we had specifically addressed ourselves to the problem, I suspect we would have said that the depth dose at the greatest depth needing treatment should not be less than 80% of the calculated surface dose.

In other words, the worked examples give depth doses with a margin of depth, just as we had taken a margin in the radius.

Once we had worked out the flat system for circles, the next step was to do it for squares, not because squares are the next 'logical' shape—allipses might perhaps have been that—but because we had a field distribution data for lines of any length, courtesy of Dr. Mayneord for unfiltered sources and Dr. Sievert for sources requiring oblique filtration corrections.

There is no <u>logical</u> relationship between the patterns for compound circular applicators and for the squares with added "lines". We simply took existing practice and found a reasonable value for the maximum spacing of lines (2h) and for their linear activity relative to the periphery.

The arrangements got interesting when we found that the same dose came from equally loaded squares or circles of the same area. It followed that a common dose curve would do for <u>any</u> area, with a rather small correction for elongation. Extending the argument to curved surfaces was the next obvious step. The device of calculating for the "area treated" for convex surfaces and the "area of the applicator" for concave surfaces is about as arbitrary and empirical as one could get. It just happens to work in most cases. If we had had a computer to run out many more cases in detail, we might have been discouraged and missed the practical simplification!

In the interest of time, I will skip what the Manchester System did with the cylindrical applicators and with the cervix treatments.

The systems for interstitial therapy were decidedly more interesting. We tried innumerable ways to come to some volume dose system by combinations of the existing plane figures--none showed us the generalizations we needed.

We had to go back to the elegant solution of Souttar for the case of a sphere. In nearly all branches of science and technology there are some practitioners whose contributions do not make the impact on the literature that they should have. Souttar was a bioengineer decades ahead of his time. While practising as a highly respected clinician, his mind was that of a first-class mathematician.

It was Souttar who finally persuaded me that there was no alternative but to extend the simple mathematics of the sphere to all required shapes. Only the sphere could be done as a perfect integral and all the rest required laborious numerical integration. I spent the greater part of a year on this. A few years ago I had my computer associates estimate what it would take on a computer--about two minutes was the answer--so we have made some progress.

However, the work was fun for its time, as it led to the one simple formula to treat a volume, V, of any reasonable size and shape:

$$M = 34.1 \, V^{2/3} \, e^{0.07(E-1)}$$

As a minor historical note, I think I derived a completely disproportionate satisfaction from the purely empirical elongation correction. Perhaps it was some compensation for dreary hours of calculation to contemplate that a cigar body whose length is ten times its diameter gives the same dose when treated on the two-shell principle as a discus whose diamter is ten times its thickness, when both are adjusted to the same volume and radium content.

All parts of the Manchester System were made to apply generally regardless of <u>scale</u> by ignoring scattering and absorption of the gamma radiation in tissue. A number of people have worked on this phase with results that are not in notably good agreement. Our own measurements, which were quite extensive, have never been published. We made measurements with chambers that were quite small for their time--22 cubic millimeters. We also made some of the best "mathematical" sources, such as rings, by mounting a single gold seed on a minimally scattering lathe device and rotating the source. The chamber was arranged for measurements on any traverse of the configuration. Until I looked at my 1934 notes I didn't realize that I had checked some of the convex surface applicators by having the chamber free to turn on an arc to describe the curved surface. My tissue absorption measurements were second class in that it was inconvenient in my horizontal axis device to use water as the tissue substitute. Plastics of that period were not up to present day standards. Some measurements of that time implied about a 4% absorption and scattering in tissue per cm. I could never approach so high a value, which would have required corrections for large applicators and large volume implants. In those implants--bigger than any clinically used--I found, with tissue added, a slight <u>increase</u> in dose in the central region and a slight <u>decrease</u> near the periphery. This seemed to be consistent with scattering theory. The real reason is often overlooked. In any implant, the dose received at any point comes very largely from the nearest sources. Those sources that are far enough away to suffer appreciable tissue absorption are already done in by the inverse square law.

However, these are antique measurements, and after 35 years we have better instruments, better materials, and better physicists, which together can give superior answers on these points.

HM Parker
May 1971

MANCHESTER SYSTEM

Two important features:

1. Actual dose delivered
2. Distribution of radium sources to make dose uniform enough for various circumstances.

PRIOR STATE OF THE ART

Dose

1. Empirically established - usually as amount of radium to give so many ESDs.
2. X-ray treatments meanwhile had been put on a fairly reasonable physics basis in terms of roentgens.
3. A few renegades recognized that γ-rays aren't anything but X-rays under another trade name and sought ways to put Ra doses into roentgens. Large chamber results 7.5 — 10 R/hr/mg @ 1 cm.
4. The M/r system did this from L. H. Gray's measurements on cavity ionization. Essentially this was the Bragg-Gray Principle before it was specifically formulated.

Gave K value of 9.3 R/hr or 8.4 R/hr through 0.5 mm Pt. without high-faluting individual calculations.

Surface Applicators

Simplest arrangements of rings

and squares

accepting a variation of dose of $\pm$ 10% over the treated surface.

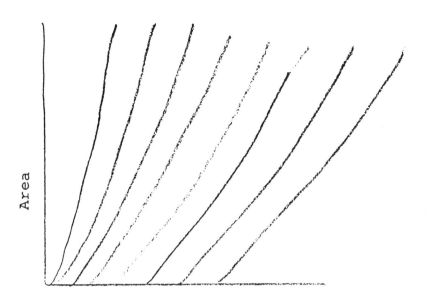

Coman curve at 1 cm.

If solved for 1 cm, can calculate for any treating distance provided we ignore absorption and scatter and oblique filtration.

Rectangles - small correction factor for elongation.
Ellipses - the same and hence generalize to any area.

Then start bending the surfaces

Modern value is close to 8.25 R/hr.

Distribution

Beginning in 1920s but mostly in 1930s, pattern of radiation dose around simple geometrical forms - lines or rings of radium was worked out by Sievert and Mayneord.

Sievert was the first, I believe, to solve the equations for filtered sources.

Applications were mostly developed on purely empirical grounds and the simplest idea was to simulate a uniform disc or square of radium. This happens not to be the right way, as it turns out that the Ra should be quite strongly peripheralized.

In this period, the best applications were prepared by the Brussels group, Murdoch, Simon and Stahel, who took literally days to compute trial distributions. If computers had existed then, this method would probably still be in use because the answers could have been run off in seconds.

Manchester people are notoriously simple-minded so they sought a system that would suit all cases.

Empirically find that if you take the "treated area" for convex surfaces and the applicator area for concave surfaces you get nearly enough the right answer. This had been found empirically true by Paterson before we proved it mathematically. So we are in business with surface applicators of any needed shape and size.

Cylinders - special case of rectangular applicator folded until it joins edges

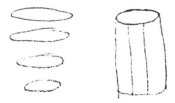

but needs special calculation for surface and depth doses.

Line Sources - only case that needs large correction for oblique filtration. Ovoids for cervix - never much interested in vaginas.

Implants - First developed empirically by the clinicians from multiple planes. Simple solution for all volumes was not obtained until we went back to basic principles for fluid distributions of Ra.

First done by Souttar for a sphere

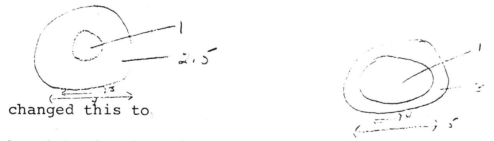

changed this to

Then laboriously calculated equivalent for ellipsoids, cylinders, truncated cones and rectangular blocks. It took me one solid year of calculating. For fun, I set it up in computer form about 10 years ago. Solutions took 2 minutes - probably less now.

-4-

The formula for dose as a simple function of volume (V 2/3) with an elongation factor is really quite an exciting one.

$$M = 21.57 \, V^{2/3} \, e^{0.07 (E-1)} \longrightarrow 34.1 \, V^{2/3}$$

NOT THE SAME V

Cigar versus discus

10 x 10 x 1

agreement is most striking

Limitations - Handbook p. 203
Other nuclides - ^{60}Co etc. better than Ra

Supervoltage Roentgentherapy (Title Page)

Book published by
Charles C. Thomas, Publisher

1950

From Franz Buschke, M.D., Simeon T. Cantril, M.D., Herbert M. Parker, M.Sc., *Supervoltage Roentgentherapy*, 1950. Courtesy of Charles C. Thomas, Publisher, Springfield, Illinois.

SUPERVOLTAGE ROENTGENTHERAPY

By

FRANZ BUSCHKE, M.D.
SIMEON T. CANTRIL, M.D.
HERBERT M. PARKER, M.Sc.

From
The Tumor Institute of
The Swedish Hospital
Seattle, Washington

CHARLES C THOMAS · PUBLISHER
Springfield · Illinois · U.S.A.

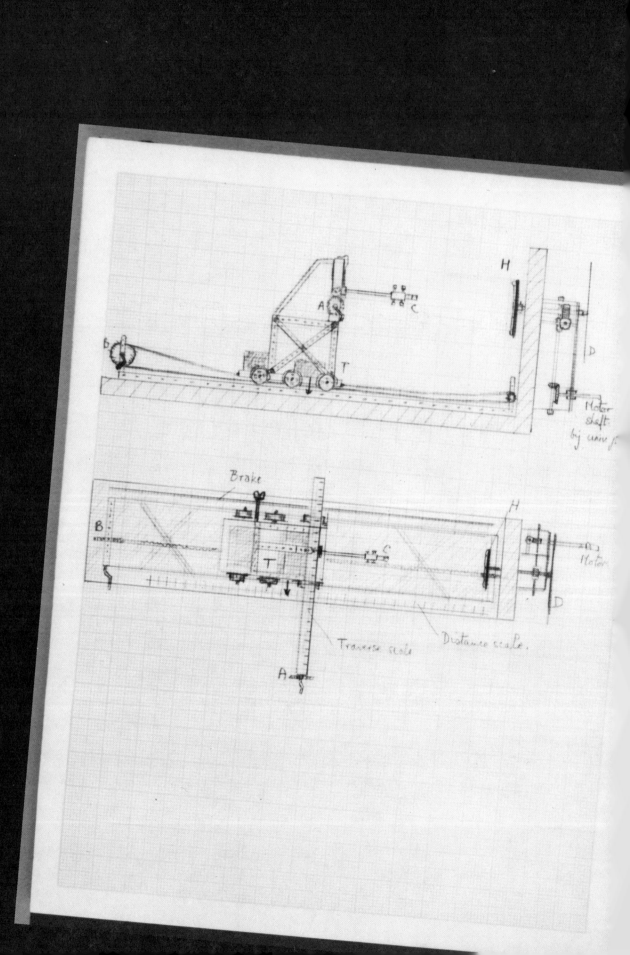

PART III

CONTRIBUTIONS TO THE EVOLUTION OF RADIOLOGICAL UNITS

33

<u>Technique of measurement of intensity from plane moulds & hollow cylinders</u>

Measuring bench constructed of Meccano parts. A trolley runs on a track, which defines the direction of the central axis. The trolley can be moved by sprocket & chain to take up quickly any position, which may be read on the distance scale. Mounted rigidly at rt ∠s to the trolley is a traverse movement controlled by A. The traverse plate carries a horizontal rod to which the measuring chamber may be attached. Amt of traverse indicated on a horizontal scale. Radium mould mounted on headstock H which can be rotated if required by motor connected up by 400:1 reduction gear via universal joint to take strain off bench. Amt of rot^n can be measured on disc D.

<u>Plane moulds</u>

mounted on cardboard & attached to face plate on H. Circular distributions can be measured as ideal rings by continuous rotation indicated thus C.
Rect. & ellipt. moulds only rotated to bring a different axis horizontal.

<u>Cylindrical shells</u>

mounted inside or outside convenient cardboard shells. Axis 11' length of measuring bench.

INTRODUCTION

Early on, Parker recognized the inadequacies of the roentgen as a practical unit for dosage and in particular for radiations other than x and gamma rays. In this regard, he was heavily influenced by one of his radiological physics colleagues Gioaccino Failla, who had questioned the value of the roentgen as early as 1937. But it remained for Parker to develop unambiguous units suitable for quantifying the dose from any ionizing radiation, from both a physical and biological standpoint. The basic concepts underlying the modern units for dose and dose equivalent were thus put forth by Parker, who in 1943 at Oak Ridge introduced the rep and the rem, the direct lineal ancestors of the modern units, gray and sievert. The first open literature publication was in 1950 in *Radiology* and was based on a paper presented before the Radiological Society of North America in late 1948.

The report on dose units to the American College of Radiology—the so-called CRUSP Report—is a lengthy tome and is not represented here because of limitations of space. The document is a veritable classic, tracing the historical development of radiation units and providing an excellent glossary of radiation quantities and units. Its clarity of language is unparalleled, particularly when considered in the context of the time. Although completed in 1955, the report was not published by the College, but has subsequently been published in its entirety in National Bureau of Standards Special Publication 625 in 1981, a publication which, ironically, Parker reviewed for the journal *Health Physics*.

His grasp of the fundamental concepts of radiological units and measurements and his ability to translate complex physical concepts into words understood by medical practitioners, biologists, and others with realtively limited backgrounds in physics, is clearly illustrated by two articles reproduced here: the 1954 article in *Radiology* "Radiation Units and Radiation Instruments" and the 1956 item in the *Proceedings of the Second National Cancer Institute Conference*, "Principles Underlying the Measurement of Radiation," both of which were originally presented in oral form.

Notes on Radiation Dosimetry

Nucleonics Dept., General Electric Company
Richland, Washington

October 4, 1949

NOTES ON RADIATION DOSIMETRY 10/4/49

By: H.M. Parker
Nucleonics Dept.,
General Electric Co.
Richland, Washington

MEANING OF DOSE

No physical definition has been used to mean

(1) energy* emitted by a radiating source : "dose emitted"

(2) energy* delivered to a biological object - "dose delivered"

(3) energy* absorbed by a biological object - "dose absorbed"

* (or physical quantities believed related to energy)

We shall say that <u>dose</u> is energy absorbed per unit ~~volume (or per unit mass)~~ mass of an irradiated biological target.

<u>INTENSITY</u> is a physically defined entity = energy flux.

e.g. parallel beam n photons, each of energy E (MEV) crossing unit cross-section (1 cm^2) in one second.

Intensity = n E MEV per cm^2 sec

or n photons per cm^2 sec

<u>EXPOSURE RATE</u> is a function also of photon energy E

$$\text{Exposure-rate} = \frac{nE(\tau + \sigma a)}{6.64 \times 10^4} \text{ r/sec}$$

$$= \frac{n \times 1 \times (0 + 3.6 \times 10^{-5})}{6.64 \times 10^4} \text{ r/sec if } E = 1.$$

$$= n \times 5.4 \times 10^{-10} \text{ r/sec}$$

<u>DOSE RATE</u> is a function of E and also of kind of tissue exposed.

For E = 1 and soft tissue (assumed water equivalent)

Dose-rate = ∼ n × 5.4 × 10^{-10} × 93 erg/gm-sec.

= ∼ n × 5 × 10^{-8} erg/gm-sec.

1. ROENTGEN-RAYS AND GAMMA-RAYS

(a) <u>Methods of measuring exposure</u>

1. <u>Photographic</u> - wide range, cheap, unreliable for low energy photons because of photoelectric effect in Ag and Br.
2. <u>Fluorescence</u> - narrow range, fatigue, dependence on energy
3. <u>Chemical effects</u> - poor for low exposure, non-linear
4. <u>Action on colloids</u> - as #3
5. <u>Color changes</u> - insensitive
6. <u>Energy absorption as heat</u> - theoretically sound - practically difficult.
7. <u>Resistivity</u> - limited range, fatigue, depends on energy
8. <u>Biological Methods</u> - limited range, statistical variations
9. <u>Ionization in gases</u> preferred physical method.
   Advantages are:
   (I) reduced to measurement of current or potential, precision over wide range (e.g. 10^{-17} amp up to microamps)
   (II) convenient range from natural-radiation (~ 0.01 mr/hr) to X-ray emission of say 1600 r/min. = working range of 10^{10}
   (III) if air is gas used - universally available
   (IV) special gases - approx. closely alleged circumstances of ionization in tissue.
10. <u>Ionization in liquids</u> - theoretical advantages offset by experimental difficulties (polarization, questionable saturation)
11. <u>Ionization in solids</u> - high efficiency conversion of energy - method being increasingly developed for biological research.

(b) <u>Original definition of the roentgen (1928)</u>

"The Roentgen shall be that quantity of X-radiation, which, when the secondary electrons are fully utilized, and the wall effect avoided, produces in one cubic centimeter of atmospheric air, at $0°C$ and 76 cm mercury pressure, such a degree of conductivity that 1 esu of charge is measured under saturation conditions". defines measurement in a particular apparatus- free-air chamber. applies only to narrow beam from a point source. All other measurements have to be made with a secondary standard - air-wall thimble chamber, (e.g. Victoreen condenser r-meter).
progressive difficulty in meeting definition conditions as energy of radiation increases.

(c) <u>Energy Absorption</u>

Absorbed photonic energy
↓
Kinetic energy of secondary electrons
↙   ↘
Ionization   Atomic or molecular excitation
↓   ↓
Increased intrinsic energy   Heat
(e.g. permanent chemical change)

(c) <u>Energy Absorption</u> - cont'd

In gases, partition between ionization and excitation is known (approximately 50:50), and we <u>assume</u> it will be the same in tissue. If the partition pattern is always the same, a measure of any one component will be a valid basis for dose, whether the significant biological factor is the increased intrinsic energy (most likely), ionization per se, excitation, or any combination of these.

The absorbed photonic energy is the easily measurable component as follows:

<u>Bragg-Gray Principle</u>

Consider a small irregular air cavity A in a medium, and a

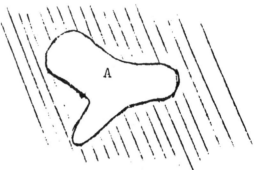

geometrically similar (smaller) volume element S of the original medium, located at the same place and with linear dimensions in the ratio ρ :1

ρ is a proportionality factor such that a secondary electron traversing the solid medium loses the same amount of energy in a short distance Δx as it would do in traversing distance $\rho \times \Delta x$ of air.

It can be shown that if the medium is irradiated,

$$\frac{\text{Energy absorption per unit volume of } A}{\text{Energy absorption per unit volume of } S} = \frac{1}{\rho}$$

Assumes (1) cavity A so small that there is negligible direct ionization by <u>photons</u> in A.
(2) cavity is deep enough in medium so that there is equilibrium between incoming photons and the secondary electrons.

Energy absorption per unit volume of A = $J_v \times W$

where J_v = number of ion pairs per cc of air

W = energy absorbed per ion pair = $\sim$ 32.5 ev

$$\therefore J_v \times W = \frac{E_v}{\rho}$$

where E_v = energy absorbed (in ev) per cc of medium

ρ = ratio of stopping powers of medium and air = known physical constant.

E_v, (required dose) is obtained from experimental ionization J_v.

Forms basis for small thimble chamber dosimetry when (1) and (2) are satisfied.

Practical limits when A is converted to a small ion chamber are:

(1) small enough air gap = few mm for Ra gamma rays.

= $\begin{cases} 0.1 \text{ mm for 200 KV X-rays.} \\ \text{or 1 mm for air at 1/10 atmos. press. etc.} \end{cases}$

(2) thick enough wall = about 6 mm graphite (or equivalent) for gamma rays.

= about 0.4 mm bakelite for 200 KV X-rays.

If wall and gas are identical in composition (except density) restriction on small size of cavity is removed.

Use this in two ways:

(1) tissue - equivalent gas in tissue wall for true 'biological' dosimetry.

(II) air in air-equivalent wall.

Choice (II) is basis of revised definition of the roentgen.

(d) <u>The Roentgen (1937)</u>

The roentgen shall be that quantity of X - or gamma-ray radiation such that the associated corpuscular emission per 0.001293 gm of air produces, in air, ions carrying 1 esu of quantity of electricity of either sign.

Note: gamma radiation now included.

<u>air</u> comes in twice - once for the air wall (the medium is air).
- once for the air-filled cavity in which the ionization is measured.

Numerically gives same value as original roentgen, but is no longer restricted to the narrow beam, point source case.

<u>Practical applications</u>

1. Measurement in air - air equivalent wall mass thickness $\underline{t}$, and appropriate mass absorption cooefft. σa.

$$\text{Exposure} = J_v (1 + \sigma_a t) \text{ roentgens}$$

where J_v is measured in esu per cc.

2. Measurements in other media (e.g. phantom)

$$\text{Exposure} = J_v (1 + \sigma_a t - \sigma_a' t') \text{ roentgens}$$

where t' = mass thickness of external radius of chamber

σ_a' = mass absorption coefft of medium.

L.H. Gray has carefully measured the ionization in a graphite chamber produced by 1 mg (Ra + decay products) at 1 cm. and filtered by 0.5 mm Pt.

$$J_o = 8.3 \pm 0.1 \text{ esu/cc/hr.}$$

Correcting to an air wall, $J^* = \underline{8.4 (\pm 0.1)}$ r/hr

This is basis for calculation of radium exposures in roentgens. Prior to revision of roentgen definition, such readings were quoted as <u>gamma-roentgens</u>. Term now obsolete as gamma ray exposures can be written in 'honest' roentgens. Note that this does not claim equivalence in effect for equal roentgen exposures of soft X-rays and hard gamma rays. One major reason is - true does (energy absorption) does not parallel the exposure.

Examples: hard gamma rays 1 roentgen $\longrightarrow$ ≈ 93 ergs/gm in soft tissue or skin.

soft X-rays 1 roentgen $\longrightarrow$ as low as 40 ergs/gm in fat, and perhaps as high as 130 ergs/gm in skin.

(e) <u>Multimillion volt Roentgen Rays</u>

1937 definition of roentgen applicable to roentgen rays or gamma rays up to a few million volts energy. Above this, nuclear properties of matter become important (e.g. pair production), and the concept of "associated corpuscular radiation" becomes more vague, because of large travel of secondary electrons in tissue.

Example: at 100 MEV at least 10 cm tissue wall needed to satisfy Bragg-Gray rules. Means one could only measure exposure at midpoint of a patient.

Avoid by going directly to an energy absorption unit.

2. OTHER RADIATIONS

(a) **Neutrons**

Fast neutrons produce effect in tissue primarily by the ionization of recoil protons - can be handled on Bragg-Gray Principle. Accurate dosimetry difficult - very small cavities needed.

Gray's energy unit for neutron radiation - "The unit dose is that quantity of neutron radiation which communicates to unit volume of water the same energy that is communicated by one roentgen of gamma radiation (i.e. about 94 ergs).

    Slow neutrons - pertinent effects   (I) capture gamma rays in H.
                                                      (II) capture gamma rays in N.
                                                      (III) reaction N^{14} (n,p) C^{14}.

Capture gammas measured conventionally. Proton recoils in a very small cavity. Practical Bragg-Gray dosimetry very dubious. Intermediate energy neutrons - dosimetry lacking.

(b) **Beta Rays**

Secondaries are electrons and indistinguishable from primary beta rays (electrons). Bragg-Gray Principle unsuitable. As gamma ray effect is due to secondary electrons, assume that beta rays are equally effective in terms of ionization by electrons. Leads to concept of "equivalent-roentgen" of which there are many.

    e.g. One roentgen gives energy absorption of 83.8 ergs/gm of air, so one equivalent is that dose of any ionizing radiation that communicates 83.8 ergs/gm to air. Then, as air and tissue are approximately alike (electronically) for hard roentgen-rays, we say crudely that 1 roentgen gives the same energy absorption per gram in tissue and in air. So, an equivalent-roentgen could be that quantity of any radiation that communicates 83.8 ergs/gm to tissue. However, for hard roentgen-rays (or gamma rays (as from Ra), the energy absorption is about 93 ergs/gm of tissue. So, another equivalent-roentgen could be that quantity of any radiation that communicates 93 ergs/gm to tissue.

Two groups of equivalent-roentgens have grown up around these numbers 83.8 and 93, and each number has an experimental range of its own - say 83 to 84 for the first, and 93 to 95 for the second.

    [See Robley D. Evans "Radioactivity Units and Standards" Nucleonics October, 1947, and in Lawrence and Hamilton's "Advances in Medical Physics", Academic Press, for an excellent discussion of this.]

The author has used "Rep" (abbreviation of roentgen equivalent physical, but pronounced as a word "rep"), and this has become popular in the AEC installations and some outside.

He has clouded the issue by starting it on an 83 erg/gm tissue basis, changing to 83 ergs/cc tissue (because logically a volume basis is somewhat preferable), and again changing it to 93 ergs/gm (because this has the widest international attraction).

Today, the rep is an energy absorption dose unit of any ionizing radiation, of 93 ergs/gm tissue. Evidently, the rep should not be the final name for an internationally accepted energy dose unit, because of those changes.

It is a useful interim form if each writer specifies 1 rep = 93 ergs/gm or whatever the basis chosen.

All physical measurements of dose can be reduced to rep when the appropriate physical constants are known.

Examples:
(I) fast neutrons - measure ionization in gases containing carbon, hydrogen and oxygen - compute the ionization in an effective tissue gas. Or try to make a tissue-gas directly.

(II) beta-rays - measure ionization in a thin-walled ion chamber - nature of gas relatively inconsequential.

Frequently, as in radioisotope therapy, dose can be calculated more accurately than it can be measured:- if isotope is uniformly distributed in an organ substantially greater than the range of beta rays in it:-

Energy absorbed/gm $\doteq$ Energy emitted/gm and this is known from radioactive constants.

(III) alpha rays — as for beta rays.

(IV) roentgen rays - normally we say 1 roentgen = 1 rep, but this is an approximation. In precise work, we could exhibit Spiers' and Lea's data as follows in terms of rep per roentgen.

	825 KeV	124 KeV	12 KeV
Muscle	1.02	1.01	0.94
Fat	0.96	0.90	0.45
Bone	1.69	2.63	9.50
Wet tissue	0.99	0.98	0.93

Obviously, the rep is primarily useful to carry over dose experience in roentgens to that for particulate radiations. The physicist would prefer a direct statement in ergs/gm, (or ergs/cc). If the radiologist converted his roentgens to energy absorption dose at the rate of 100 ergs/gm tissue, he could convert his experience with less error than arises in normal dosimetry.

3. FACTORS OF BIOLOGICAL EFFECTIVENESS

(a) Relative biological effectiveness RBE

RBE of any radiation compared to gamma rays has been measured for many biological reactions, (acute mean lethal dose, shortening of life, erythema, blood counts, etc.).

Suppose that radiation A is 5 times as effective as the standard radiation X, all other things being equal in the experiment, and both doses measured in rep.   RBE = 5.

For radiation A we shall write 1 rep = 5 rem.

(b) The Rem.

The rem [abbreviation roentgen-equivalent man (or mammal)] is that dose of any ionizing radiation which is biologically equivalent in man to 1 roentgen of hard roentgen rays, all other conditions being equal. By the last clause we mean the exposure to be given over the same part of the body, protraction and fractionation to be the same, etc.

The rem is not a formal unit, but rather a convenient shorthand notation for summing the potential damage contributions of mixed radiations.

We use a scale of relation as follows:

Roentgen rays, gamma rays 1 r	= 1 rep (approx)	= 1 rem
Beta rays	1 rep	= 1 rem
Fast neutrons	1 rep	= 10 rem
Slow neutrons	1 rep	= 5 rem
Protons	1 rep	= 10 rem
Alpha particles	1 rep	= 20 rem

Scale is especially useful in radiation protection problems. Thus, if 300 mr is the permissible weekly exposure of the whole body to roentgen rays, we may perhaps assume that the whole blood-forming system is the sensitive target, and that it can receive 300 mr per week (actual value depends on absorption and scattering of incoming radiation, and the energy dose also is a function of close proximity to bone).

Then by definition 300 mrem is the permissible weekly exposure to mixed radiations. The contributions of external roentgen and gamma rays, external neutron radiation, internal beta and alpha emitters, is measured in rep (or measured indirectly and converted to rep). Then each component is converted to rem and compared with the permissible limit 300 mrem.

One must watch out that the sensitive target is not switched. Suppose 1 r per week is acceptable on skin only. Then 1 rem is the permissible weekly skin exposure. If the actual exposures are heavily weighted with external beta-rays, the permissible skin limit can be passed long before the "whole body" limit is reached.

(c) **Limitations of the rem**

Has found most value in protection where adequate safety factors can be included to cover variations.

RBE is not a constant for a given radiation - depends on:

       (1) biological action considered.
       (2) fractionation and protraction.
       (3) dose-rate as such.
       (4) geometrical factors of the irradiation.

The above scale is meant to apply only to long-continued exposure to small daily quotas of radiation.

The values are uncertain and have been changed twice since 1943.

Certainly a different scale is needed for acute exposures (for example: atomic bombing). In routine therapy one could develop a suitable scale for perhaps two radiations used under fairly standard conditions - conceivably useful in combinations of roentgen therapy and isotope therapy.

Have already pointed out need for assumption of the relevant significant exposure.

Finally, it is conceivable that the rem to rep ratio could be infinite, e.g. if neutrons produced eye-cataracts and roentgen rays did not, for this biological effect there is no finite scale of relation.

Tentative Dose Units for Mixed Radiations

Radiology, Volume 54

February 1950

Reprinted with permission of the Radiological Society of North America.

[Reprinted from RADIOLOGY, Vol. 54, No. 2, Pages 257-261, February, 1950.]
Copyrighted 1950 by the Radiological Society of North America, Incorporated

Tentative Dose Units for Mixed Radiations[1]

H. M. PARKER, M.Sc.[2]

Richland, Wash.

THE ORIGINAL definition of the roentgen (1928) applied to x-radiation only. It was a definition of a specific measuring system, not of a fundamental unit. The development of logical systems of gamma-ray dosimetry and of the use of higher voltage x-radiation in the early 1930's made the definition inadequate, although it was obvious that the same concept, interpreted as a tissue dose, could be applied directly to radiation treatments. There was a period of confusion in which some authors confidently expressed gamma-ray dosage in roentgens, while others denied such application. This persisted until 1937, when the modified definition of the roentgen, based on the Bragg-Gray cavity principle of dosimetry, was adopted. By that time it was realized that a statement of tissue dose based on a measurement of ionization arising in a suitable tissue-wall chamber was desirable. The practical difficulties of operating such a chamber were well known and accounted for compromise suggestions such as that of L. H. Gray, in which a chamber wall of a pure, universally available material such as graphite was proposed. A further compromise to consider air as the wall material around an air cavity was incorporated in the revised definition. This had the advantage of leaving the open air chamber, at least within a restricted range of radiation energy, as one suitable realization of a measuring scheme under the new system. With this device, no numerical changes were needed in the existing applications of dose in roentgens.

The extension of radiation choice beyond that common in 1937 has again made the roentgen, as defined, inadequate for required purposes. At the present time, dosimetry in roentgens either has to be supplanted, or supplemented, by the definition of other special units useful in radiology, radiological physics, radiobiology, and allied arts, but not required in general physics. There appears at present to be a general desire to return to a fundamental unit in the science of radiation dosimetry, which shall be independent of a specific measuring system. Disregarded is the fact that it may be impossible to measure such doses in an academically absolute manner. This is a less serious objection, however, than is sometimes realized, because all existing dosimetry in roentgens is founded on the concept of instrumentation which approximates closely enough to ideal limiting definitions, and has no absolute status in science. Four feasible units have been suggested, and used by various authors:

(1) Energy absorption per unit mass (ergs per gram).
(2) Energy absorption per unit volume (ergs per cubic centimeter).
(3) Ionization per unit mass (ion pairs per gram).
(4) Ionization per unit volume (ion pairs per cubic centimeter).

PHYSICAL DOSE UNIT

The choice between the energy absorption system and the ionization system depends on whether the effect of ionizing radiation on biological materials stems entirely from the ionization produced in the material or whether it is significantly affected by that energy absorption which is dissipated in other forms such as molecular dissociation, kinetic energy of non-ionizing recoils, etc. The partition of the total

---

[1] Presented at the Thirty-fourth Annual Meeting of the Radiological Society of North America, San Francisco, Calif., Dec. 5–10, 1948.
[2] Nucleonics Department, General Electric Co., Hanford Works, Richland, Wash.

energy absorption between ionization and other modes is not constant in all cases, but is believed to be approximately so in most applications of practical interest. Although the ionization system may prove eventually to be closer to the fundamental entity, there is a wider acceptance of an energy absorption basis. While there is doubt about the role played by energy absorption, other than the ionization quota, it is safer to accept the total energy absorption basis. In laboratory practice, many of the statements of either one will depend on a physical determination of the other.

The distinction between a unit-volume basis and a unit-mass basis is less important. The former would follow Gray's *principle of equivalence* more directly, and may be more attractive for statements of dose in bone, for example. However, the unit-mass basis has lately been more generally favored. As an acceptable compromise, it is proposed to use as the basic tissue-dose unit the energy absorption stated in ergs per gram. Laboratory experiments involving radiation dose would be universally reported in these terms, which would be understood, without ambiguity, by scientists in all fields. This "cold" unit would not be well received by practising radiologists or radiobiologists, and it would not readily integrate with their years of experience in the expression of dose in roentgens. It would be certainly unnecessary, and probably unwise, to attempt to eliminate the roentgen from usage in practical radiology.

Despite the increasing application of forms of radiation which cannot be measured in terms of roentgens, it will remain true that the majority of radiologists will confine their interests to problems in which dose can be properly stated in terms of the roentgen as now defined. For this reason, it is deemed advisable to agree at an early date on a *practical* energy absorption dose unit which would correlate with prior experience in dose measurement expressed in roentgens. Many such units have appeared in the literature, as, for example, tissue roentgen, nominal roentgen, equivalent roentgen (e.r.), rhegma, roentgen equivalent, roentgen equivalent physical (rep), gram roentgen, Gray's energy unit, and so on. These hinge on one or other of two known energy absorption relationships:

(1) One roentgen of x-radiation or gamma radiation corresponds to the absorption of about 83.8 ergs per gram of air. To a relatively poor approximation, this will correspond with the absorption of 83.8 ergs per gram of wet tissue, although the actual range is known to extend from 40 to 100 ergs per gram for x-radiation in the range of 12 to 800 kv. The *gram roentgen* and the original *rep* were based on this factor.

(2) One roentgen of hard gamma radiation corresponds to the absorption of about 93 ergs per gram in pure water. This is the basis of Gray's *energy unit*.

The second figure appears to be more generally acceptable as a transitional step for translating past experience with x-ray and gamma-ray dosimetry into the more generalized dosimetry with any ionizing radiation. For practical use, an easily written and pronounceable name, preferably one beginning with "r" (as in roentgen), and forming simple, multiple, and submultiple units is most desirable.

The writer recommends, as a compromise, the adoption of a practical energy absorption dose unit called the *rep*, and defined so that "one *rep* represents an energy absorption dose in irradiated tissue of (exactly) 93 ergs per gram." The particular term *rep* has been widely used within the Manhattan Project, since 1943, and it has spread from Atomic Energy Commission installations to many of the principal centers of radiobiological research. The basis has variously been 83 ergs per gram, 83 ergs per cubic centimeter, and 93 ergs per gram, and this can cause confusion in future interpretation of data. For this reason, it is strongly recommended that authors electing to use the *rep* should include a footnote stating "1 rep = 93 ergs/gm.," or whatever form is used. The arguments for persisting with *rep* are: (1) Very little radiobiological work to date is accurate enough to make results sensitive to the difference between 83 and 93. (2) If there is any advantage in a unit beginning with the letter "*r*", selection of a new name now limits the final rational choice. (3) Such final rational choice should come by international agreement, with a more precise definition than that given above.

MANAGEMENT OF MIXED RADIATION EXPOSURES

One factor of the effective radiation dose of each component of a mixed exposure (*e.g.*, fast neutrons, slow neutrons, and gamma rays from a cyclotron or nuclear reactor) is defined by the statement of ergs/gm., or the practical derivative *rep*, for each radiation type separately. The combined effect is a function of other variables, some of which are (1) specific ionization of each radiation, (2) dose-rate of each radiation, (3) protraction and fractionation of each type, and (4) clinical circumstances such as condition of adjacent tissue. Under certain simplifying conditions, these can be reduced to a single factor, the familiar relative biological effectiveness (RBE) for each radiation type.

Two such limiting cases exist: (*a*) single acute exposure to mixed radiations; (*b*) chronic exposure to very small doses, daily or perhaps weekly, of mixed radiations. These are the only cases in which one can currently develop a plausible permissible exposure. Obviously, the first is of interest in atomic bomb attack, and the second in the everyday management of exposures in atomic energy installations.

To manipulate dose problems in the latter case especially, it was necessary to formulate an additional "unit," better described as a shorthand system than a formal unit at this time. Such a unit is the *rem*, where "one *rem* is that dose of any ionizing radiation which produces a *relevant* biological effect equal to that produced by one roentgen of high-voltage x-radiation, other exposure conditions being equal."

It is convenient to write down a *scale of relation*, which for chronic exposure is:

X-rays, gamma rays	1 r ≃ 1 *rep* =	1 *rem*
Beta rays	1 *rep* =	1 *rem*
Protons	1 *rep* =	10 *rem*
Fast neutrons	1 *rep* =	10 *rem*
Slow neutrons	1 *rep* =	5 *rem*
Alpha rays	1 *rep* =	20 *rem*

The approximate equivalence of the roentgen and *rep* has been discussed already. In the transition to *rem*, the writer has previously specified high-voltage radiation as "about 400 kv." This removes the base from the region in which the photoelectric effect gives significant contribution. The selection of gamma rays from radium in equilibrium with its products would be equally suitable. In effect, the relative biological effectiveness of "ordinary" x-rays to radium gamma rays is taken as unity, instead of the conventional 1.4. This is an example of the inaccuracies of the present system which can be controlled in radiation protection work by safety margins, but would be intolerable in therapy. In the scale of relation, fast neutrons are taken as those which produce their biological effect by generation of knock-on protons in tissue. Slow neutrons are those which produce their effect by nuclear reactions, such as capture gamma rays from hydrogen and neutron-proton reaction on nitrogen. Omitted is a range of intermediate-energy neutrons, for which no quantitative dosimetry currently exists.

Application of the scale of relation is as follows:

1. Decide on the *relevant* biological effect or the significant organ of exposure. For penetrating external radiation, this will normally be the damage to the blood-forming organs. For predominantly soft radiation, it may be damage to the skin. Internal emitters may focus the effect on a specific organ, such as the thyroid gland for radioiodine or astatine.

2. Evaluate the separate doses to the significant organ in *rep*.

3. Convert each dose to *rem* by the scale of relation.

4. Add the total dose in *rem*, and compare with the accepted standard. Thus, if the permissible whole body exposure to hard x-rays is 0.3 roentgen per week, this means, approximately, that the permissible exposure to the blood-forming organs (the assumed significant organ) is also about 0.3 *rep* per week. The related permissible exposure is 0.3 *rem* per week, and to this figure the above-determined personnel exposures are compared.

For single acute exposures, the accident cases, an entirely different scale of relation and a different permissible exposure are needed. Also, in either case, the significant organ may change depending on the predominant components of the mixed radiation. The skin is a logical significant organ if soft external radiation predominates, and the lung irradiation controls if the principal exposure comes from the inhalation of insoluble radioactive materials. For this reason, conservative radiation protection requires the evaluation of the combined dose in *rem* for each of the feasible significant organs. In many practical cases, most of the "significant organs" can be eliminated by inspection of the nature of the exposure. The major weakness of mixed exposure management has been the failure to add the various contributions in some notation equivalent to the *rem*; the permissible exposures to various irradiating agents are published separately, and there is a tendency in less conservative groups to accept close to the limit of each type, with a resulting combined exposure that may be intolerable.

Therapeutic applications of mixed radiations can in principle be controlled by a statement of dose in *rem*, where the exposure conditions are reasonably repetitive. At the present time, no profitable application is known to the writer. Much greater precision in physical dosimetry in *rep* (or directly in ergs/gm.), rather than in roentgens, is needed before the further extension to the biological equivalence unit should be attempted.

NOTE ON NOMENCLATURE

The *rep* is an abbreviation of roentgen equivalent physical. The *rem* is an abbreviation of roentgen equivalent man or mammal. The more obvious choice of *reb* (roentgen equivalent biological) is avoided because of the confusion in speech between *rep* and *reb*. Both *rep* and *rem* should be used as words, not as three spoken initials. The convenient submultiples are written as *mrep* and *mrem*, and pronounced "millirep" or "millirem." Multiples such as *kilorep* and *megarep* are permissible, but the abbreviated forms should be avoided.

SUMMARY

Energy absorption dose in irradiated tissue should be stated in ergs per gram in scientific reports. In therapeutic and other practical applications, a practical unit, the *rep*, an energy absorption dose of 93 ergs per gram, allows a reasonable transition from previous experience in dose statements in roentgens. Under simplifying conditions, the combined effect of mixed radiations can be evaluated in a common unit, the *rem*. This method incorporates a factor for relative biological effectiveness in the physical statement of dose in *rep* of each radiation type. Such a device is necessary in the management of exposures of personnel to mixed radiations. It may be useful ultimately in mixed radiation therapy, when the physical dose is more precisely stated and knowledge of biological effectiveness factors is improved.

Nucleonics Dept.
General Electric Co., Hanford Works
Richland, Wash.

SUMARIO

Unidades para Dosis Tentativas en las Radiaciones Mixtas

En las memorias científicas, la dosis de absorción de energía en el tejido irradiado debe expresarse en ergs por gramo. En las aplicaciones terapéuticas y otras de orden práctico, una unidad práctica, el *rep*, dosis de absorción de energía de 93 ergs por gramo, permite una transición bastante lógica de la costumbre anterior de expresar la dosis en roentgens. Simplificando aun más, el efecto combinado de las radiaciones mixtas, puede valuarse en una unidad común a todas, el *rem*. Este método agrega un factor referente a la efectividad biológica relativa, a la expresión física de la dosis en *reps* de cada forma de irradiación. Una fórmula de ese género resulta necesaria en la consideración de las exposiciones del personal a radiaciones mezcladas. También puede con el tiempo resultar útil en la radioterapia mixta, al expresar con mayor precisión la dosis física y acrecentarse nuestros conocimientos de la efectividad biológica.

Principles Underlying the Measurement of Radiation

From Proceedings of
The Second National Cancer Conference

December 1954

Reprinted from PROCEEDINGS OF THE SECOND NATIONAL CANCER CONFERENCE
American Cancer Society, Inc. • National Cancer Institute of the
U. S. Public Health Service • American Association for Cancer Research

PRINCIPLES UNDERLYING THE MEASUREMENT OF RADIATION

H. M. PARKER

General Electric Co.

Richland, Washington

When a piece of living tissue is exposed to ionizing radiation, the basic physical phenomena of the immediate interaction can be described. The eventual biological effects under a given set of conditions can also be described. Between these two limiting events, the description is less certain. The physical chemist carries the initiating picture to the stage of demonstrating the production of and behavior of free radicals within the irradiated piece. The biochemist discusses the reactions produced in the chemical substances of the tissue by such free radicals. If these intermediate steps were fully understood, it is to be expected that the appropriate measurement of ionizing radiation would involve a system that defined the number of ionizing events in the irradiated tissue, together with their distribution in space relative to sensitive targets and in time. In the present state of development such an approach serves only to provide an interesting and plausible account of the relative effects of different kinds of radiation delivered at different exposure rates.

The physicist properly falls back upon a reference of the initiating effect to one of the basic quantities of his professional field, namely, the energy involved in the interaction. The basic quantity to be determined becomes the *energy absorption per unit mass of tissue,* which will conventionally be measured in *ergs per gram.*

In the most familiar case, when the ionizing radiation is photonic—either

X-rays or gamma rays—the total energy available for biological action is partitioned as follows:

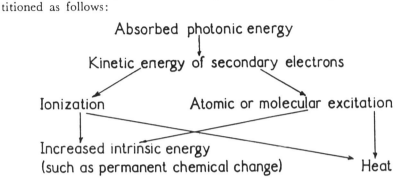

In gases, the partition between ionization and excitation is known; for such gases as nitrogen it is approximately equally divided, as is roughly shown by the average expenditure of 32 electron volts of energy to produce one ion pair compared with the 15 or 16 electron volts physically required to remove an electron from a nitrogen molecule. In the liquid state, and more specifically in tissue, the actual partition is less certain. The readily measurable components are the beginning and end points of the table of energy utilization. In principle, the measurement of the degradation of the energy to *heat* is attractive. This resolves itself into some form of calorimetry. This approach received early attention in the field, was revived some twenty years ago in the work of Stahel and others, and has received lively attention in the past few years. As a practical method it seems to be doomed to failure, because the quantities of heat involved are small. A dose of one roentgen of gamma radiation will raise the temperature of water by 2×10^{-7} °C, which is hopelessly beyond the measurement range.

The practical measurement system is therefore related to absorbed photonic energy by an argument known as the Bragg-Gray Principle of Equivalence.

Consider a small irregular air cavity A in a medium, and a geometrically similar (smaller) volume element S of the original medium, located at the same place and with linear dimensions in the ratio $\rho:1$.

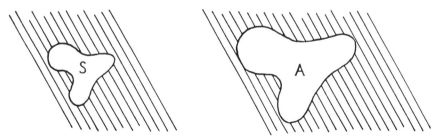

ρ is a proportionality factor such that a secondary electron traversing the solid medium loses the same amount of energy in a short distance $\Delta \chi$ as it would do in traversing distance $\rho \times \Delta\chi$ of air. It can be shown that if the medium is irradiated,

$$\frac{\text{Energy absorption per unit volume of A}}{\text{Energy absorption per unit volume of S}} = \frac{1}{\rho}$$

assumes (1) cavity A so small that there is negligible direct ionization by *photons* in A. (2) cavity is deep enough in medium so that there is equilibrium between incoming photons and the secondary electrons.

Energy absorption per unit volume of A $= J_v \times W$
where $J_v =$ number of ion pairs per cc of air
$W =$ energy absorbed per ion pair = approx. 32.5 ev

$$\therefore J_v \times W = \frac{E_v}{\rho}$$

where $E_v =$ energy absorbed (in ev) per cc of medium
$\rho =$ ratio of stopping powers of medium and air = known physical constant.
E_v, (required dose) is obtained from experimental ionization J_v.
The transition to energy absorption per unit mass rather than unit volume is elementary.

This forms the basis for measurement by converting the hypothetical cavity into a small ionization chamber when certain conditions about the chamber size and wall thickness are fulfilled.

These are:

(1) small enough air gap = few mm for Ra gamma rays.
$= \begin{cases} 0.1 \text{ mm for 200 KV X-rays.} \\ \text{or 1 mm for air at 1/10 atmos. press., etc.} \end{cases}$
(2) thick enough wall = about 6 mm. graphite (or equivalent) for gamma rays.
= about 0.4 mm bakelite for 200 KV X-rays.

If wall and gas are identical in composition (except density), the restriction on the small size of cavity is removed.

The special condition is of value in two ways:

(1) in basic "biological" dosimetry where a tissue equivalent gas in a tissue-equivalent chamber wall can be used;

(2) in practical dosimetry where air, the most convenient gas, can be used in an air-equivalent wall. This essentially is the basis of familiar dosimetry in roentgens.

The fundamental advantage is the wide range of reliability of ionization measurements in a gas. A dose rate of one roentgen per hour yields in an ion chamber of one cm volume (air at STP) a current of about 10^{-13} amp. Currents are readily measurable down to 10^{-14} amp, and under laboratory conditions to 10^{-16} or 10^{-17} amp. Equally convenient is the measurement of integrated dose in which case the change of potential of the insulated collectrode of the ion chamber is observed by an electrometer.

In visualizing the radiation mechanism and its registration by the Bragg-Gray Principle, it is convenient to describe the incident radiation as a carrier mechanism which releases the working agent, the secondary corpuscular particles, throughout the tissue. The energy absorption is sampled by examining these secondary corpuscular particles in a small cavity. That the same basic principles will apply to neutron irradiation will now be clear. The fast neutrons are an alternative carrier mechanism, which release secondary corpuscular particles in tissue. In this case, the particles are protons. The same measurement criteria will apply, although the limiting dimensions of the appropriate ion chamber will be different and the choice of "tissue-equivalent" wall in general will be more critical.

When incident slow neutrons are the carrier mechanism, the secondary corpuscular particles consist of protons generated by the N^{14} (n, p) C^{14} nuclear reaction, and electrons, themselves secondary to gamma rays generated by other nuclear reactions. Although one may technically adhere to the Bragg-Gray Principle, there are grave experimental difficulties in this case. Additionally, careful interpretation of energy absorption is needed. In the previous cases, one is led to think of the energy as being provided by the incident carrier radiation. With slow neutrons, the increment of energy from external sources may be quite negligible. That energy which appears as kinetic energy of protons and electrons is released from the tissue itself by the trigger action of nuclear reactions. It is the energy thus made available that leads to biological effect, disregarding, in our present state of knowledge, the possible additional effect of the direct disruption or excitation of molecules in the initiating action.

When the incident radiation consists of ionizing particles such as electrons (including beta rays from radioisotopes), alpha particles, or protons, the convenient picture of the carrier mechanism producing secondary corpuscular particles breaks down. In effect, the same measurement principles apply, and the same physical factors must be known to convert the ionization observation to an energy absorption. These factors, excluding the obvious geometrical and density components are:

(1) W, the energy absorbed per ion pair produced, which is different for the different particles, and which may not be entirely independent of the energy of the particles.

(2) Stopping power of tissue for the particles concerned.

The scheme of dosimetry outlined has led to extensive studies of these factors which should lead to refinement of results.

This is not the place to discuss the intriguing problems that the experimental biophysicist has in attempting to translate the basic principles into practical measurements. Neither is it possible to show how and why the practical measurement of radiation for therapy and other applications came to be developed in air-wall chambers, with dose stipulated in roentgens, nor why it is expedient to relate the two fields of experience through an arbitrary unit, the *rep*. Enough has been suggested to indicate that the most ardent supporter of the Bragg-Gray Principle will readily seek other solutions to dosimetry. In the field of radioisotope application, such escapes are frequently possible, and there is one limiting case in which the solution can be determined by a fundamental principle—one of the laws of thermodynamics:

Consider the case of a radioactive material, which emits only ionizing particles of uniform range R, and which is dispersed uniformly throughout a mass of homogeneous tissue. Any point in that tissue which is surrounded on all sides by the radioactive tissue to a distance of R or greater derives its whole ionization effect from a sphere of radius R and is essentially in an "infinite mass" of the medium. Under these conditions, the energy absorbed per unit mass is necessarily equal to the energy released per unit mass. The energy release is completely defined by the activity density (microcuries per gram) and the known energy release per disintegration. The dosimetry problem becomes a straightforward radiochemical one. This approach is frequently applicable because the distance R is on the order of 40 microns for alpha-emitters, and not more than a few millimeters for pure beta-emitters. Moreover, close approximations to the energy absorption can be written down for points not wholly surrounded by the radioactive material, such as at the surface of an organ that has concentrated the radioactive material. By further extension, the case involving an emitter of both beta and gamma rays can be reasonably well calculated from the energetics of the nuclear disintegrations.

These fine exercises of the laws of thermodynamics are inhibited only by the insistence of tissue in behaving as an inhomogeneous medium, when considered in adequate detail. Measurement is pushed to more and more detailed study of distribution of the active material by autoradiography, and refinement of the dose calculation. The ultimate solution by this or new experimental techniques is just one of the challenging problems remaining to the biophysicist.

Radiological Units

Indian Journal of Radiology

1956

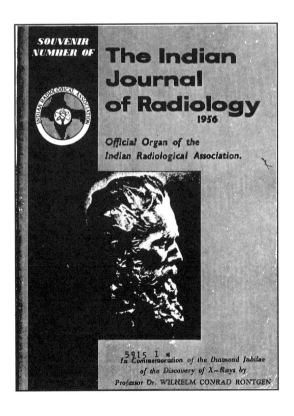

RE-PRINTED FROM INDIAN JOURNAL OF RADIOLOGY, SOUVENIR NUMBER, 1956

*RADIOLOGICAL UNITS

BY

H. M. PARKER

Radiological Sciences Department, General Electric Company, Richland, Washington

INTRODUCTION

It is fitting that the Diamond Jubilee of the discovery of X-Rays should find Prof. Roentgen honored by the almost universal use of the unit to which his name was applied.

It is not less appropriate that the ever-expanding use of radiation both in radiotherapy and in other fields of activity has led to a situation in which the *roentgen*, as a unit of dose, is rapidly losing importance; in fact, it would be more advantageous to consider it as a unit of exposure, reserving the term *dose* for what has been defined as *absorbed dose*.

THE ROENTGEN

Historically, the use of a special unit in radiology arose from the need for some physical parameter to define the radiation output of X-Ray equipment in a reproducible way, and hence to relate the observed biological effects to this common parameter.

Several factors led to the establishment of the roentgen in terms of the ionization in a defined volume of air (or more precisely, in a defined mass of air). Some of these factors were:

(1) The difficulty of measuring conventional physical parameters for radiation (such as energy flux density), when dealing with penetrating X-radiation.

(2) The existence of a saturation potential in an ion chamber, which provides a stable measuring condition for ionization.

(3) The precision of ion current measurements over a wide range, and the linear relationship between ion current and rate of production of ionization.

(4) The approximate "parallelism" between ionization in air and ionization in soft tissue for the kinds of radiation in use when the roentgen was developed.

This last factor contributed to the interpretation of roentgen as a measure of ionization produced in tissue, or of energy imparted to tissue, which is an ambiguous distortion of its definition.

---

* Work done under Contract No. W31-109-Eng. 52 to the Atomic Energy Commission.

SCOPE OF RADIATION DOSIMETRY

In the last 10 to 15 years there has been a notable extension in the range of available sources of radiation for radiotherapy. Even more striking is the catalogue of radiations to which research and operating personnel may be exposed in atomic programs, and in other branches of nuclear science. These radiations include:

PHOTONIC OR QUANTUM RADIATION

    Gamma radiation from radioisotopes.

    X-radiation at energies up to the million and thousand million volt range.

PARTICULATE RADIATION

    Electron beams up to the same high range.

    Beta radiations of radioisotopes.

    Alpha radiations of radioisotopes.

    Thermal neutrons.

    Fast neutrons, these classes of neutrons being differentiated because their interaction with tissue is fundamentally different.

    Protons.

    All the other particles of modern nuclear physics, such as positrons, mesons, and anti-protons.

Of this comprehensive catalogue, only the photonic radiations of quantum energy up to about 3 MeV can be measured in terms of roentgens. It is a striking commentary on the influence of the roentgen to observe that almost universal attempts are made to relate all other ionizing radiations to the same system through some form of *equivalent roentgen*. This is a case of the tail wagging the dog, and there is considerable evidence that the junction between tail and dog is being stressed beyond its limits of strain. If a fracture should occur at this point, practical radiotherapy could be left with the tail, with radiobiology and atomic energy applications getting the dog. Such a division would not be in the best interests of the science of radiology. Maximum integration between radiotherapy, radiobiology, and all other applications of radiation to living matter is essential to the optimum development of each of these fields. It can be achieved by a minor operation on the roentgen, the tail of the dog.

BASIC PHYSICAL DOSIMETRY

Basic knowledge of the interaction of radiation with living matter is too limited to justify dogmatic statements as to the best physical basis of dosimetry. It is agreed, however, that the most promising approach is to measure *absorbed dose* in the sense used in the I.C.R.U. Recommendations (Copenhagen 1953 — See Glossary).

The minor operation needed is now to agree that *absorbed dose shall be the only form of physical dose used in radiology*. The roentgen would no longer be a unit of dose, but could be described conveniently as a unit of exposure. The roentgen would continue to be used as a measure of output of X-Ray equipment, or of gamma-ray emitting radioisotopes, in the energy range up to 3 MeV. This would return it unequivocally to its original function, and would eliminate the innumerable ambiguities that have arisen in its broader applications.

The practical transition from roentgens to rads would be relatively simple. For conventional X-Rays incident on soft tissue, one roentgen represents an energy absorption of about 97.7 ergs per gram, which is sufficiently close to one rad to serve all clinical purposes.* Uncertainties will exist in dose at interfaces, such as between air and skin, and between tissue and bone. It is precisely in these zones that the roentgen, used as a unit of dose, is most misleading. The transition to rads will tend to promote the solution of these complex dosimetry problems.

BIOLOGICAL DOSE

When more than one radiation type is used, it is well known that biological effect per rad is a function of the type, when all other factors such as fractionation, protraction and geometric pattern of radiation are maintained constant. One parameter in this function is the specific ionization or linear energy transfer. The concept of biological dose to include such relative biological effectiveness factors is attractive. It is proposed to call biological dose, *bio-dose* when it is measured in *rems* as defined by

bio-dose in rems = $\sum$ {(absorbed dose in rads) x RBE}

where the summation is made over all radiation types with appropriate RBE values for each.

Ultimately, one may develop a more generalized biological dose which takes account of time factors and other parameters in addition to an RBE for radiation type. This would be known as Bio-dose and the term *reb* can be reserved for the unit of such a dose.

SUMMARY AND CONCLUSIONS

A consistent approach to radiological units applicable to all investigations of the effects of ionizing radiation on living matter is obtained by:

1. Differentiating between exposure, physical dose, and biological dose.

2. Exposure is a measure of the output of radiating sources; it is measured in roentgens for photonic radiation up to 3 Mev; for

---

*This conversion was formerly given as 93 ergs per gram. The change represents acceptance of W (average energy per ion pair) = 34 electron volts instead of the earlier value, 32 or 32.5 electron volts.

other radiations it is measured in the conventional physical terms, e.g. flux density, most appropriate to the radiation type.

3. Physical dose is invariably *absorbed dose* measured in rads.

4. Biological dose is arbitrarily described as bio-dose measured in *rems*. As knowledge advances, a more basic Bio-dose in *rebs* may be developed.

5. The minor distress in radiotherapy, during the transition from roentgens to rads as the practical dose unit, will be more than compensated by the long-range integration with related fields, such as radiobiology; this will arise from the more rapid assimilation of radiobiological advances through a common nomenclature.

6. The rem and reb are obvious "roentgen equivalents". The *rad* is also a "roentgen equivalent," since otherwise the fundamental unit, erg per gram, would be used for absorbed dose. Thus the roentgen is perpetuated through its vestigial remnants in all radiological dose units.

GLOSSARY OF TERMS IN RADIATION DOSIMETRY

This glossary includes a number of terms that have not been discussed in this condensation; a more comprehensive review of terminology is to be published elsewhere.

Absorbed dose:	Amount of energy imparted to matter by ionizing particles per unit mass of irradiated material at the place of interest. Expressed in ergs per gram or in rads.
Bio-dose:	Term used for doses converted to biological equivalence with respect to radiation type by the operation bio-dose in rems = Σ [(absorbed dose in rads) x RBE] summed over all the types of radiation involved.
Bio-dose:	Term reserved for the future to express "true biological dose" by converting physical dose by mathematical operators including factors for RBE, protraction and fractionation etc. It is to be expressed in rebs. The concept of Bio-dose in rebs may not become practical; if it does, it would replace bio-dose in rem.
Biological dose:	A concept of dose in which equal doses would produce equal biological effect.

4

bio-dose (rem) is a partial biological dose which takes account of spatial arrangement of ionizing events along single tracks in tissue.

Bio-dose (reb) is a true (but currently hypothetical) biological dose that takes account of spatial and temporal distribution of ionizing events and any other contributing parameter.

Bone exposure: Exposure in roentgens at a point in bone.

Bone dose: Dose in rads at a point in bone.

Dosage: That part of treatment conditions directly related to determination of dose.

Dosage-meter: Preferably not used; alternatively, any radiation measuring instrument contributing directly (absorbed dose or dose-rate) or indirectly (exposure or exposure-rate) to dose determination.

Dosage-rate: Preferably not used.

Dose: Specifically means absorbed dose; it is expressed in ergs per gram or in rads.

Generally *dose* is also used in various ways as a parameter for measuring radiation or certain of its manifestations.

Dose-meter: A device for measuring dose in rads, or dose-rate in rads per unit time.

Dose-rate: Time-rate of absorbed dose.

Dosimetry: Science of measuring dose, specifically absorbed dose or bio-dose.

Emission: Process of release of energy from a radiation source, or specifically, the magnitude of the energy release.

Emission-rate: Time-rate of emission of energy.

Exposure: Time-integrated flux density as measured in roentgens.

Exposure is a measure of a property of quantum radiation at a particular place.

Exposure-rate: Flux density as measured in roentgens per unit time. Used to measure output of sources.

Flux: Rate of flow of energy, particles, etc. can be specified as energy flux, particle flux, quantum flux, etc.

Flux density: Flux normal to unit surface.

Free air exposure:	Exposure in a region in air where the exposure of scattered radiation from objects external to the source is negligible.
Gram-rad:	Unit of integral absorbed dose equal to 100 ergs.
Integral absorbed dose:	Integration of the energy absorbed throughout a given region of interest. The unit is the gram-rad.
Integral dose:	Same as integral absorbed dose.
Intensity:	Intensity of radiation is the energy flowing through unit area perpendicular to the beam per unit time.
Irradiation:	In the general sense, the voluntary or involuntary process or act of irradiating, i.e. causing ionizing radiation to encounter a material of interest. (Specific) Irradiation could be used in lieu of *exposure* if differentiation between *exposure* and *dose* is judged to have been compromised by past practice.
Irradiation-rate:	Used instead of *exposure-rate* if irradiation is substituted for *exposure*.
K:	A numerical coefficient defining the gamma-ray output of a radioisotope. Gamma-ray exposure rate = K roentgens per hour at 1 cm from a 1 mc point source
Output:	General term for the concept of the amount of radiation emitted by a source; not a scientific term, and is variously applied to emission-rate, exposure rate at an arbitrary distance, and so on.
Physical dose:	The general concept of dose measured in physical terms as by energy absorption; specifically physical dose is expressed as absorbed dose.
Quantity of radiation:	Time-integral of intensity, or time-integrated energy flux density, or energy density.
Rad:	Unit of absorbed dose, 100 ergs per gram. The rad is an arbitrary "roentgen equivalent" replacing such terms as *rep*.
Reb:	A unit of "true biological dose"; the term is reserved in the event that such a concept should become useful.
Reference radiation:	An arbitrarily chosen radiation type used as the standard of comparison in RBE measurements: usu-

6

	ally X-Rays yielding about 100 ion pairs per micron of water.
Rem:	"Unit" of bio-dose defined by the relation bio-dose in rems = (absorbed dose in rads) x RBE
Rhm:	Roentgens per hour at 1 meter, an arbitrary compound unit useful in some applications of radio-isotopes.
Roentgen:	Considered as a unit of exposure. Its magnitude is as defined in I.C.R.U. Recommendations.
Skin exposure:	Exposure at skin surface.
Skin dose:	Dose in rads at skin surface.
Tissue exposure:	Total exposure at a place in tissue.
Tissue dose:	Dose in rads at a place in tissue.
Tumor exposure and tumor dose:	Used similarly to tissue exposure and tissue dose, generally expressing an average value throughout the tumor, unless otherwise specified.

Radiation Units

General Electric Hanford
Atomic Products Operation

March 1962

RADIATION UNITS

H. M. Parker and W. C. Roesch

Hanford Laboratories
General Electric Company
Richland, Washington

March, 1962

The purpose of this paper is to give a quick historical review of the development of radiation units together with a more detailed review of the recent refinements and developments.

A difficulty that beset radiation dosimetry through most of its history is that its basic terminology was not established until quite late. The term "dosimetry" is readily understandable as the measurement of dose or the science of dose measurement. It was the simple term "dose" that lacked appropriate definition. The relevant dictionary meaning of "dose" is "a measured quantity of a medicine to be taken at one time or in a given period." The application of the expression "dose" in this sense to an x-ray treatment must have come quite naturally to the physicians who became the first radiologists. Also, the early x-ray tubes were notoriously fickle; interest was centered on the radiation that came from the machine and was "given" to the patient -- dose is derived from the Greek word dosis, meaning "a giving". There was not a complete preoccupation with the radiation itself, however. As early as 1913, Christen pointed out that what radiologists should be concerned about is the energy absorbed from the x-ray beam by the body at a point in the body. Because of the low absorption of the x-rays, this is only a small part of the energy passing through that point. In later years, the technical problems were overcome and radiobiologists, biophysicists, and the like turned their attention to the patient and "dose" took on the meaning of "absorbed from" or "received by". It is not surprising that conflicting interpretations should have developed in a subject that translated "a giving" into "a receiving".

Another term that gave difficulty was "quantity". For the early radiologists "quantity" of radiation probably had the familiar meaning of "amount". For physicists, however, "quantity of radiation" was a special technical term; namely, the energy of the radiation per unit area. The confusion was made worse by the frequent need for the word in still a third sense, quantity in the sense of a measurable property.

Units and instrumentation are inextricably connected. Whenever a new physical agent is discovered, a measuring system is often established that uses some conveniently observable property of the agent, and from this develops a unit. For different applications, the property chosen may differ. Thus in the familiar case of electric current, units have been established based on electrostatic, magnetic, and electrochemical effects.

In the early years of radiology, measuring methods were based on many effects of the newly discovered radiation. The calorimetric method dates back to 1897. Various chemical, fluorescence, or scintillation methods were used in the first years of this century; one of the favorite systems depended on the coloration of barium platinocyanide pastilles. The photographic method, based on one of the first observed effects, was made reasonably quantitative as early as 1905. Changes in electrical resistivity and a variety of biological effects were also studied for dosimetry. The measurement of ionization in gases gained and maintained an early popularity. Each of these effects was made the basis for an appropriate unit.

The first formal statement of a unit based on ionization was due to Villard in 1908. Unit quantity of x-radiation was defined as that which by ionization liberates one electrostatic unit of electricity per cubic centimeter of air under normal conditions of temperature and pressure. This is the embryo of the roentgen unit that was not born until twenty years later.

There were several reasons for the popularity of measurements based on the ionization of gases. The existence of a saturation potential in a gas-filled chamber provided a means of making reproducible measurements by means of accumulation of electric charge or by current measurement. Either current or change of potential can be measured with accuracy and precision over a

- 4 -

wide range. The universal availability of dry air led to its adoption as a standard filling for ion chambers. Ordinarily the material of which a chamber is made has a pronounced effect on the ionization that is produced in the gas in it. Through the work of Friedrich and others, the free air chamber was developed. This is a chamber in which the ionization is independent of the material of construction of the walls. Also, through the work of Fricke, Glasser, and others, air-equivalent chambers were developed that gave the same ionization per unit volume as the free air chamber. The measurement of air ionization leads to a result that was, in effect, a sort of compromise between the two "dose" concepts. To a fair approximation, for the x-ray energies normally used in therapy in the period considered (the 1920's), air ionization is proportional to the quantity (physics) of radiation that has passed through the chamber. This relates it (approximately) to "dose" in the sense of "a giving". At the same time, for those interested in effects in tissue, the atomic composition of tissue and air are such that the ionizing effects should be roughly parallel in the two for those same energies. This relates it (approximately) to "dose" in the sense of "a receiving".

The Roentgen

The development of instrumentation that gave reproducible results and the formulation of a unit were closely related. Several proposals for ionization units were made. Finally, international agreement

on a unit was secured. At the First International Congress of
Radiology (1925) a commission, the International Commission
on Radiological Units (ICRU), was formed to deal with the matter
of units. At the second meeting of the congress in 1928, the
ICRU made the specific recommendations:

1. That an international unit of x-radiation
   be adopted.

2. That this international unit be the quantity
   of x-radiation which, when the secondary
   electrons are fully utilized and the wall
   effect of the chamber is avoided, produces
   in one cubic centimeter of atmospheric air at
   0° C and 76 cm mercury pressure, such a degree
   of conductivity that one electrostatic unit
   of charge is measured at saturation current.

3. That the international unit of x-radiation
   be called the "Roentgen" and that it be
   designated by the letter small "r".

The roentgen unit, as we shall see, has survived to the present time. The use of a common unit by scientists in all countries has contributed immeasurably to the progress of radiology. Nevertheless, the definition has contained the seeds of misunderstanding from the very start. Apparently, the intention of the ICRU was that "quantity" in their second recommendation was to mean quantity (physics) of radiation. Then it was reasoned that quantity (physics) was not measurable in general, because it involves total absorption of the energy passing through a surface. The roentgen was thus a substitute unit of quantity (physics). Critically examined, the procedure is inadmissible,

- 6 -

because two situations characterized by the same quantity (amount) in roentgens do not represent the same quantity (physics) of radiation, if the relevant absorption coefficients differ. For this reason and because the "quantity" meant is not specified, several authors have asserted that the "quantity" in the recommendation ought to be treated as quantity (amount). In that event, the quantity (measurable property) of the radiation of which the roentgen is the unit is nowhere defined.

The original definition made no reference to "dose", but almost immediately the roentgen was described as the unit of x-ray dose. If the definition refers to quantity (physics), then the dose meant must be that of "a giving".

Early Radium Dosimetry

Whereas x-ray dosimetry was largely conditioned by the need to establish a reproducible method of measuring the "output" of a variable source, radium dosimetry began with sources of fixed emission rate. It became common practice to express dose as the product of milligrams times hours of treatment. Obviously such factors as the distance and disposition of the sources had to be stipulated to complete the therapy prescription. When radon was used instead of radium element, an average strength had to be used or a more elaborate calculation of millicurie-hours effected.

The French school, under Prof. Regaud, estimated the degree of radiation as "l'émanation détruite", or millicuries destroyed.

Since a 1 mc radon source left in situ permanently contributes 133.3 mc-hours (using the old value of decay rate), the practice developed of packaging radium element in amounts of 6.66 or 13.33 mg to make the arithmetic simple.

That dose was expressed in milligram-hours or some similar unit for so long was obviously not due to lack of comprehension of the problem, but rather to the then existing technical difficulties of going at it in other ways. As early as 1919, Regaud and Ferroux pointed out the differences between "dose emitted" and "dose received", but continued to use the former. Three dose expressions were commonly recognized in this period; namely, "dose emitted" -- the energy released by the radiation source, or some function reasonably related to it; "dose delivered" -- the energy arrriving at the locus of biological interest, or some reasonably related function; and "dose absorbed" -- the energy imparted to the locus of interest, or some reasonably related function.

Innumerable biological effects were considered as bases for the dose absorbed, including that of Russ (1918), the amount of radiation necessary to kill mouse cancer, and called the "rad"; a title that has now been revived for absorbed dose measurements. Perhaps the most widely used biological frame of reference was the erythema dose, especially in the threshold erythema dose of the Memorial Hospital school. The threshold erythema dose was

- 8 -

said to be received from that amount of radiation that would produce reddening of the skin of the inner forearm in a certain percentage of patients within a certain time.

By taking advantage of the first order approximation that absorption and scattering in light elements for radium gamma-rays can be neglected, mathematical systems of evaluating "dose for external or intracavitary applicators developed. These required only a knowledge of how to integrate radiation functions, based on the inverse square law, together with a reference dose-rate at 1 cm from a unit point source of radium. Sievert, Mayneord, Failla, and Quimby were among the early contributors to this technology. One such system is that of Sievert, based on the Intensity-millicurie (Imc), a unit "intensity" defined as the intensity of gamma radiation at a distance of 1 cm for a 1 mg point source of radium filtered by 0.5 mm Pt. From the intensity unit is defined a unit of quantity, Imc-hour, usually referred to as the Sievert dose, and also frequently referred to as 1 cm.mg-el.-hr. The Sievert dose has a well-known relationship to Eve's number*, from which a translation to "roentgens" can be made. Paterson and Parker based their system of dosage on a useful set of approximations to the radiation function solutions coupled with an Eve's

---

* Eve's number is the number of ion pairs formed per second in air at standard temperature and pressure by the electrons produced in air, per cm^3 of air, by the gamma rays from a point source of 1 gm of Ra in equilibrium with its daughters and placed 1 cm away.

- 9 -

number conversion from L. H. Gray's ionization data; this system and the value of 8.4 r have been widely used since 1932. All such derivations, prior to 1937, were considered improper by some purists because the definition of the roentgen (1928) limited its field to x-rays. Actually x-rays differ from gamma rays only in the manner of their origin so the use of the roentgen for gamma radiation was quite natural.

Redefinition of the Roentgen

The definition of the roentgen was modified at the fifth International Congress of Radiology in 1937. Several changes and events of importance to dosimetry took place in the period 1928 - 1937. In this period, the knowledge of the properties and effects of radiation and of basic physics in general grew enormously. The technology for the production of radiation improved. The high vacuum x-ray tubes introduced earlier made possible much more reproducible x-ray machine outputs. The x-ray excitation voltages available increased steadily. The availability and use of radium increased.

The technology for the measurement of radiation improved. Largely through the efforts of Behnken and Taylor, the various national standards laboratories reached essential agreement on the form of free air chambers and the techniques necessary for reliable roentgen measurements in the then available energy range. Before the establishment of the roentgen, considerable work had been done on the development of thimble chambers for x-ray measurements, and commercial dosimeters based on this were available as early as 1927. The definition of the roentgen placed all these devices in the category of empirically founded secondary instruments, requiring calibration against the free-air chamber. After 1928, the extensive work on thimble chambers continued in respect to making an air-wall, choosing a satisfactory wall thickness, correcting for the absorption in that wall, and extending the energy range to include radium gamma radiation. The empiricism in thimble chamber construction offers no obstacle to their successful use, provided the use is confined to the demonstrated range of calibration against primary standards*.

The enunciation of Gray's Principle of Equivalence in 1929 profoundly influenced the course of dosimetry. The principle was established by him in connection with a study of cosmic radiation, with simplifying assumptions that might not have appeared attractive if initiated for the then normal range of radiation therapy sources. The study was basically concerned with absorption coefficients and the validity of the Klein-Nishina formula. Gray indicated that measurements of the emission of Ra (B + C) and particularly of ThC", with its strong 2.6 Mev

---

* It is a dangerous, yet common error, to consider thimble chambers as Bragg-Gray cavities. This is not so. They are chambers that have been empirically adjusted to give a response proportional to that of a free air chamber over a specified energy range.

emission, would be of particular interest. It was through this channel that his work came into the orbit of radiation therapy. The principle was presented again with supporting experimental evidence in 1936, and the results of the radium measurements appeared in 1937.

The principle expresses the energy absorption per unit volume of irradiated material in terms of the measured ionization per unit volume in a suitably specified small cavity within that material as:

$$E_V = \rho W J_V$$

where ρ is the ratio of electron stopping power in the material and the cavity gas respectively, and W is the average energy required to produce one ion pair in the cavity gas. Since it had been pointed out that W. H. Bragg had arrived at a similar proposition in 1921, the term Bragg-Gray Principle developed. Gray additionally pointed out that Fricke and Glasser independently arrived at the same principle in 1925, and it is essentially contained in a doctorate thesis by M. Bruzau. There is a limit to the size of the cavity if the Bragg-Gray principle is to apply. Gray explored this matter in one of his later papers and showed that the principle is a convenient one in the range of radium gamma-rays, but extremely taxing for x-rays generated at 200 KV or less. This accounts for many of the observed discrepancies in thimble chamber measurements, which led several authors in this period to the conclusion that there was no valid principle of small chamber dosimetry.

- 12 -

Although Gray's main interest was in the field of radium gamma ray measurement, he made an attempt to integrate this with the current x-ray dosimetry. To this end, he proposed that the roentgen be redefined in such a way as to permit its measurement with air-filled Bragg-Gray cavities surrounded by air (in practice, air-equivalent) walls. This was intended to permit measurement with the standard free air chamber in the appropriate energy range and with suitable thimble chambers at higher energies. Some such change was felt to be needed if x-ray and radium dosimetry were to be placed on a common basis. Numerous experimenters had worked with larger and larger free air chambers to evaluate the Sievert dose but found the technology not yet adequate for the purpose. Following Gray, several measurements were made of the Sievert dose with Bragg-Gray chambers (assuming a redefinition of the roentgen such as proposed by Gray). After an initial period of erratic results, good agreement was achieved between different measurements and led to the tacit agreement among most qualified experts that Bragg-Gray measurements were quite satisfactory for dosimetry.

The growing popularity of thimble chambers led to another change in dosimetry practice. The practice envisioned when the roentgen was originally defined was that a chamber would be placed in an unhindered x-ray beam and a measurement in roentgens made. Then the patient would be placed in the beam and exposed for a period determined by a correlation of previous experience with measurements made in this way. This had been and remained the common American

- 13 -

practice. Here the quantity measured in roentgens was considered a dose in the sense of "a giving". On the other hand in Britain, in particular, radiologists began to place small thimble chambers at the surface of, or inside, dummy bodies (called phantoms). In the phantom the chamber is exposed to scattered radiation that it would not receive in the unhindered beam. Phantoms were used in the belief that what was measured in this way was something that was closer to the quantity of radiological interest than that given by the original practice. It was felt to be more nearly the dose in the sense of "a receiving". The results of the measurements in phantoms were stated in roentgens. This led to discrepancies in the experience of American and British radiologists.

These changes in the technology and the practice of radiology led to pressure to revise the definition of the roentgen. At the fifth International Congress of Radiology (1937), it was redefined as follows:

> The roentgen shall be the quantity of x- or gamma-radiation such that the associated corpuscular emission per 0.001293 gram of air produces, in air, ions carrying 1 e.s.u. of quantity of electricity of either sign.

An accompanying statement was:

> The international unit of quantity or dose of x-rays or gamma-rays shall be called the "roentgen" and shall be designated by the symbol "r".

- 14 -

Gamma rays were included in the definition as well as x-rays. It was made clear that quantity meant quantity (physics). The accompanying statement made use of the word "dose" with the roentgen official. The statement seems to make "quantity" and "dose" synonomous. Omission of such words as "fully utilized", "wall effect", "conductivity", and "saturation current" removed the necessity for using a free air chamber for the primary measurement in roentgens. This was made definite by later statements in the same report that the primary standard instruments for measurement in roentgens were the free air chamber or the air-wall chamber (for harder radiations). "One cubic centimeter of atmospheric air at 0° C and 76 cm mercury pressure" was replaced by the simpler equivalent term "0.001293 gram of air".

It will pay to examine carefully the meaning of the words, given in different order than in the definition, "ions produced, in air, by the associated corpuscular emission". For one thing, particles other than electrons were included; e.g., positrons that would result from pair production at high energies or any other charged particles that might be produced (in nuclear events, for example). Figure 1.a illustrates what ions are meant in the definition of the roentgen. In a certain volume of air, secondary electrons are produced. The ionization they produce in air, regardless of whether that ionization is produced in that same volume or not, is that referred to. Figure 1.b illustrates something more akin to what is actually measured. If one collects the ionization produced in a certain volume, he

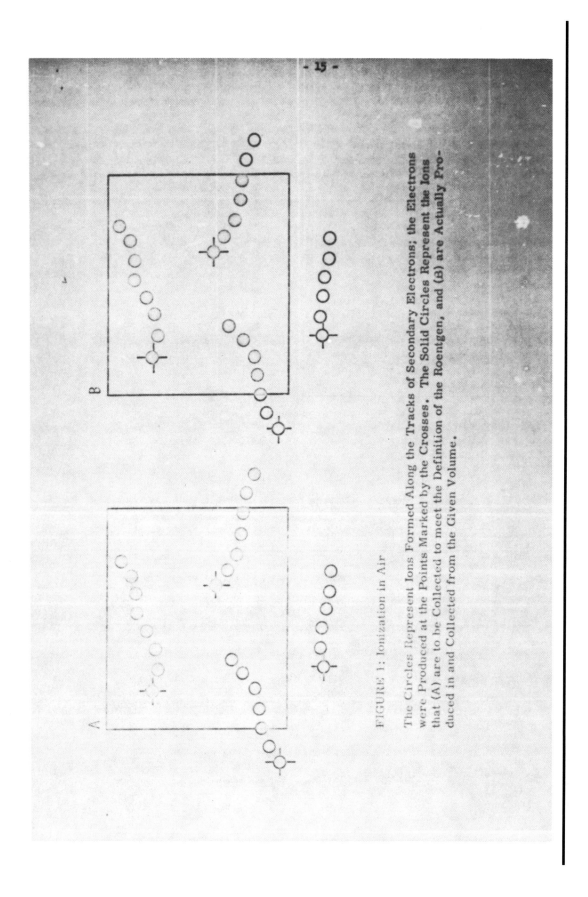

FIGURE 1: Ionization in Air

The Circles Represent Ions Formed Along the Tracks of Secondary Electrons; the Electrons were Produced at the Points Marked by the Crosses. The Solid Circles Represent the Ions that (A) are to be Collected to meet the Definition of the Roentgen, and (B) are Actually Produced in and Collected from the Given Volume.

will collect ions that were produced by secondary electrons that were produced both inside and outside that same volume. Inside a free air chamber and, effectively, inside an air-wall Bragg-Gray chamber the two types of ionization are equal because of electronic equilibrium. The words "fully utilized" in the first definition implied the necessity of establishing electronic equilibrium. This left it uncertain, however, which concept of Figure 1 was meant. In divorcing the new definition from explicit dependence on the free air chamber, it was necessary to make a choice between the concepts. The choice was made in agreement with the then commonly accepted interpretation.

At the sixth International Congress of Radiology the ICRU appended this note to the definition of the roentgen: "The unit may . . . be used for most practical purposes for quantum energies up to 3 Mev." The reason given was: "It becomes increasingly difficult to measure the dose in roentgens as the quantum energy of the x-radiation approaches very high values." It becomes difficult because the range of the secondary electrons from high energy photons is so great that it becomes impossible to achieve satisfactory electronic equilibrium. At a given point in air the ionization will be partly due to electrons that were produced in regions where the photon intensity was significantly different from that at the given point. The increased rate of energy loss by radiation by high energy secondary electrons also makes for difficulty in measurement, because the radiation so lost must not be permitted to produce ionization that is measured.

Absorbed Dose and Exposure Dose

After the redefinition of the roentgen in 1937, the continuing growth of science and technology resulted finally in another change in the basic units for radiation measurement. The developments in nuclear physics, particularly of high energy accelerators and of nuclear fission, had a profound effect by expanding the area of concern to dosimetry to include neutrons and other new radiations, high energy heavy charged particles, and artificially produced radioisotopes. These developments were accelerated by the beginning of the atomic energy program during World War II and the radiation protection and radiobiological research problems associated with it.

These new developments led rapidly to the feeling that the best way to express the dose was to give the amount of energy imparted to the matter per unit mass at the point of interest. Actually this was one of the oldest of the ideas inherent in the word "dose", but now people were forced to use it to the exclusion of other dose concepts. The case of slow neutrons can be used as one of the most illustrative (but not most important) examples of why this is so. The idea of "dose emitted" from a source is not applicable to slow neutrons because there are no sources of slow neutrons; they are produced by moderation of fast neutrons in a suitable slowing down material. The idea of "dose delivered" is just about meaningless for slow neutrons because it is so small; in the center of powerful nuclear reactors giving the highest slow neutron fluxes we know of, the energy of the slow

- 18 -

neutrons -- the dose delivered -- is only about the same as that of the photons in a beam giving 1 r/hr. The idea of "dose absorbed" makes sense; for equal energies absorbed, slow neutrons give biological effects that are comparable to those for other radiations.

The first steps toward an absorbed dose concept were the use of different "equivalent roentgens". The use of these equivalent roentgens was based on the following reasoning: The roentgen was an established, familiar unit. The unit of absorbed dose should be closely related to the roentgen. Photons that gave one roentgen would produce an energy absorption of 83 to 85 erg/gm of air (depending on the average energy loss by secondary electrons required to form an ion pair in air (W); the equivalent of the roentgen is now taken to be 86.9 erg/gm of air, corresponding to $W = 33.7$ electron volts per ion pair). They would produce 93-95 erg/gm in water or soft tissue. Therefore, an equivalent roentgen was an absorbed dose of X erg/gm. Different authors used values of X from 83 to 95.

In the atomic energy program beginning in 1942, one was faced with the practical problem of adding the doses received by a large group of workers from photons, alpha, beta, and neutron radiation. Parker was led to develop a system that was readily communicable. He used the "rep" as an equivalent roentgen, based originally on $X = 83$, but changed to $X = 93$, and "rem" as a biological "unit" obtained by multiplying the components of the dose in rep by appropriate RBE multipliers. Rep was an acronym for roentgen equivalent physical and rem for roentgen equivalent man (or mammal). The principal merit of this variation was in the communication advantage of pronounceable terms, such as rep and rem, and the simple arithmetical scale of relation.

At the first International Congress of Radiology following the war (1950), the ICRU recommended that: "For the correlation of the dose of any ionizing radiation with its biological or related effects . . . the dose be expressed in terms of the quantity of energy absorbed per unit mass . . ." In a note they explained that "energy absorbed" meant the energy imparted to the material by ionizing particles at the place of interest. They recommended that the unit erg/gm be used for this dose.

These recommendations were well received in general, but were criticized for not supplying a name for the concept considered and for not supplying a unit similar in size to the roentgen. At the 1953 Congress, the ICRU remedied these difficulties by defining the "absorbed dose" as the amount of energy imparted to matter by ionizing particles per unit mass of irradiated material at the place of interest. They also defined a unit, the "rad" to be the unit of absorbed dose. One rad is equal to 100 erg/gm. This concept and its unit received rapid and universal acceptance.

The definition of absorbed dose was bound to cause confusion in speaking about the quantity measured in roentgens. This, of course, was only part of the general confusion that had always existed about the term "dose". Now, however, a quantity had been defined that used the term. In a sense, this withdrew the sanction for the use of the term with other concepts that had been associated with it. The ICRU recognized this problem and considered the use of the term "exposure" for the quantity measured in roentgens. This was judged unsatisfactory, principally because of the widespread use of "dose" and "roentgens" in existing laws and regulations that would be very difficult to change. For this reason, the term "exposure dose" was chosen. The opening words of the definition of the roentgen (1937 version) were changed to read: "One roentgen is an exposure dose of x- or gamma radiation such that . . ." Apparently in the belief that what was measured in roentgens was well enough understood, the meaning of exposure dose was only indicated, rather than defined, as "a measure of the radiation that is based on its ability to produce ionization".

Recent Developments

To understand one of the modern trends in radiation dosimetry, it is necessary to look again at the changing practices of the radiological physicists. We have already noted that when air equivalent thimble chambers became available, the practice grew of placing these in phantoms in order to measure something more closely related to what produced the biological effect in the patient. When the Bragg-Gray principle was advanced and accepted, variations of this practice were introduced. For example, air-equivalent chambers or Bragg-Gray air-wall chambers were replaced by chambers with walls made of or simulating tissue -- sometimes soft tissue, sometimes such special tissues as fat or bone. Application of the Bragg-Gray principle to such measurements permitted the determination of the energy absorbed in the tissue; i.e., the absorbed dose. These tissue-equivalent systems assume considerable importance in the dosimetry of neutrons. Another practice that developed was the attempt to eliminate the walls of the chamber. This usually cannot be done completely because electrodes are necessary to collect the ionization current; however, some tissue-equivalent plastics that are also conducting have permitted construction of tissue-equivalent chambers in which the walls are the tissue-equivalent material itself. In general, though, the effort is to make the walls thin enough that they do not materially affect the flux of ionizing particles reaching the gas of the chamber. These thin-wall techniques are especially necessary in the dosimetry of alpha and beta rays. It is not always realized that the Bragg-Gray principle applies as well to these thin-wall chambers as to ones with thick walls. The older discussions of the Bragg-Gray principle were frequently devoted to its application to measurement of what we now call exposure dose. In this application, thick walls were required to establish the electronic equilibrium needed for the measurement of the exposure dose, but the application of the principle itself is independent of the thickness of the walls. Thin-wall chambers also came into use in dosimetry of x-ray and electron beams from high energy accelerators. The use of thin-wall chambers is essentially forced upon us in this case. The secondary electrons from high energy photons have such long ranges (of the order of

one centimeter in tissue) and the photon beam widths are usually
so narrow that it is impossible to attain electronic equilibrium.
To even come close to a relative equilibrium would require chambers
that would usually be too large relative to the system being
irradiated to give results that were of any interest.

This intense attention to the practical art of making gas ionization
measurements led to various proposals to make some ionization
quantity and the corresponding unit the basic ones for dosimetry.
As early as 1937, Failla recommended that ionization be measured
in a thin wall, vanishingly small cavity (extrapolation chamber)
placed in the irradiated material at the point of interest, and
that it be expressed in units of the same size as the roentgen
and called "tissue roentgens". In 1938, Holthusen recommended
the same quantity and called it the "ion dose". Similar recommendations were made to the ICRU by the British in 1950 and by the
Germans in 1956. The latter were based on recommendations by
Franz and Hubner who not only advocated the term "ion dose", but
suggested that the roentgen unit be generalized so that it could
legitimately be used for any ion dose. Their proposal led to
much discussion in Germany. Finally, it was adopted as the
German national standard. Schinz and Wideroe (1958) felt that
this generalization of the roentgen was not apt to obtain international approval and that change of a traditional unit was
inadvisable anyway. They suggested that a new unit, the same
as Failla's "tissue roentgen" be defined and given the name "rho".

The authors of this article feel that the attempts to define a formal ionization quantity and unit are ill advised. Adequate criticism of the attempts requires a general statement of the reasons for desiring such a quantity. We will draw on the work of Berger (1959) to frame what seems to us to be the best reason that has been advanced: Absorbed dose is concerned with an energy that exists in a biological system. There is no way of measuring it in the biological system itself. A substitute material must be placed either in the biological system (with consequent displacement of some of that system), or in a phantom and the energy imparted, or some related quantity, measured in the substituted material. The energy in the biological system and the energy in the substitute material are conceptually different things and should be treated as separate quantities.

It seems to us that this argument draws a useful distinction between what one would like to know and what one must measure to approximate to it. However, it seems to us that this distinction exists in all measurements, not just those in radiation dosimetry. Every measurement act consists of the application of an <u>instrument</u> by an <u>observer</u> to a <u>quantity</u> to be measured. The response of the instrument is related to, but is not identical with, the quantity to be measured. For example, if we want a person's temperature, we may place a mercury thermometer in the patient's mouth. We observe the height of the mercury in the thermometer. What this really gives us is an indication of the temperature of the mercury. We know that under the proper conditions this may be taken as an accurate indication of the person's temperature. We do not express the result as "mercury-degrees" or as "tissue-degrees". There is no more need to draw continual attention to the fact that measurement is a process of approximation to the desired quantity in radiation dosimetry than in any other field of science and engineering.

Some arguments for an ionization quantity say essentially that it is needed because the measurement in the substitute material may not be a good approximation to the absorbed dose. This represents a very real experimental problem. An experimenter may be able to measure the ionization in a particular situation to 1% accuracy, but be unable to translate this into absorbed dose to better than 10% accuracy. We should recognize, however, that the usefulness of his result is in general determined by the 10% uncertainty, not the 1%. An exception to this is when another experimenter would like to reproduce the original results. For this particular case, the interest is not on the dose but on the experimental techniques. It is quite adequate for this purpose to express the ionization measured in some conventional unit such as coulombs per kilogram. If desired, the old, but still legitimate unit, esu/cm^3 of air at 0° C and 76 cm mercury, could be used, the resulting numbers would be the same as those that would result for the proposed ionization units.

Adoption of any unit based on practical measurement techniques will result in problems similar to those encountered by the roentgen. Growing technology raises problems and possibilities

that were not thought of when the unit was defined. Right now
the improvements in other methods of dosimetry, scintillation,
solid state, chemical, or calorimetric devices, etc., may some
day result in some other measurement technique being more popular
than ionization methods. If this were to occur, and it might,
than an ionization unit would be very unpopular.

The proposals of ionization quantities and units provoked much
new discussion of the whole subject of the quantities and units
used in dosimetry. It became apparent that there was still
dissatisfaction with the formulation of the quantities already
in use. It was also felt that not enough attention had been
given to making sure that all the quantities and units were
related in a firm, satisfactory manner or that all the logically
necessary quantities had been defined. Still another opinion
was that the form and content of the definitions of the quantities
and units in radiation dosimetry had not kept up with developments
in other fields of science and engineering, and therefore
needed revision. These factors led the ICRU in 1958 to set
up an ad hoc committee on Concepts, Quantities, and Units in
the Field of Radiation Dosimetry. The task of this committee
was to review the fundamental concepts, quantities, and units
which are required in radiation dosimetry and to recommend a
system of concepts and a set of definitions which would be, as
far as possible, internally consistent and of sufficient generality
to cover present requirements and such future requirements as
can be foreseen. This committee met annually from 1959 to 1962

- 26 -

and finally presented its recommendations to the ICRU in 1962. The recommendations were accepted and will appear in the next report of the ICRU.

The ad hoc committee's recommendations are too lengthy to give completely here. A summary of the more important recommendations related to topics treated in this article follows:

(a) The term "fluence" was substituted for what had been called "quantity of radiation". The fluence is the quotient of the sum of the energies, exclusive of rest energies, of all the radiations that enter a small sphere, by the cross-sectional area of the sphere. The term "fluence" was essentially invented for the quantity since it appears to have no modern meaning. Its adoption should avoid many of the difficulties that have always been associated with "quantity".

(b) The name and definition of absorbed dose were left unchanged. However, in the definition of absorbed dose the term "energy imparted" appears; the committee gave this term a precise definition that substantially strengthens the definition of absorbed dose. The energy imparted to the matter in a volume is defined as the difference between the sum of the energies of all the directly and indirectly ionizing particles which have entered the volume and the sum of the energies of all those which have left it, minus the energy equivalent of any increase in rest mass that took place in nuclear or elementary particle reactions within the volume. This definition is not intended to change the meaning of absorbed dose in any way from what it had been. It was the result of a search for some way of defining very precisely what was meant by absorbed dose without resorting to a lengthy, and possibly incomplete, catalog of the different types of energy transfer to matter.

(c) The name and unit of exposure dose were left unchanged. The wording, but not the meaning, of the definitions was changed. This was done in answer to the criticism that there was no adequate definition of the quantity for which the roentgen was the unit. The exposure dose was defined as the quotient of the sum of the electrical charges on all the ions of one sign produced in air when all the electrons, liberated by photons in a volume element of air, are completely stopped in air, by the mass of the air in the volume element; the ionization arising from the absorption of bremsstrahlung emitted by the secondary electrons is not to be included. The roentgen was defined as the unit of exposure dose and as equal to 2.58×10^{-4} coulombs per kilogram of air. This was done to fit the roentgen into the framework of the MKSA systems of units; a footnote pointed out that this was equivalent to the old relation of one e.s.u. of charge per 0.001293 gram of air.

(d) A new quantity called "kerma" was defined. The kerma is the quotient of the sum of the initial kinetic energies of all the charged particles liberated by indirectly ionizing particles (photons or neutrons) in a volume element of the specified material, by the mass of the matter in the volume element. "Kerma" is an acronym for <u>k</u>inetic <u>e</u>nergy <u>r</u>eleased per unit <u>m</u>ass, with an extra "a" added for euphony. The kerma would be the energy-quantity equivalent to the exposure dose except for three things: (1) the kerma is not restricted to apply to air; (2) it is not restricted to photons; and (3) since the kerma is made up of the initial kinetic energies of the charged particles, it includes the energy that they may later radiate in bremsstrahlung. Probably one reason for introducing this quantity was to provide an energy-quantity that was similar to the exposure dose, in the spirit of the discussion of the preference for fundamental energy quantities over quantities that involved some method of measurement. Quantities identical to or similar to the kerma have, however, been used for a long time in dosimetry. For example, one finds frequently in the literature calculations of the dose produced by a given fluence; actually these are usually calculations of the kerma produced.

(e) The committee considered the possibility of widening the area of application of such units as the rad and the roentgen. They had been introduced as special units for absorbed dose and exposure dose, respectively. Special units for special quantities are not unknown in other fields, but the general practice is to permit the use of a particular unit for all quantities having the same dimensions. It was felt, however, that if the use of the rad and roentgen were permitted for other quantities, there was danger of too much confusion arising, because the names of these special units have by now become so closely associated with the corresponding quantities. For this reason, it was recommended that the present limitations on their use be retained. It was recommended that those who wish to express quantities such as absorbed dose and kerma in the same units should use units of an internationally agreed coherent system and not the special units for this purpose.

Bibliography

Report of the International Commission on Radiological Units and Measurements (ICRU), 1959, National Bureau of Standards Handbook 78, U. S. Government Printing Office, Washington, D. C.

Hine, G. J. and Brownell, G. L., Radiation Dosimetry, Academic Press, Inc., New York, 1956; especially Chapter 1, "Radiation Units and Theory of Ionization Dosimetry" by F. W. Spiers.

Jaeger, R. G., Dosimetrie und Strahlenschutz, Georg Thieme Verlag, Stuttgart, 1959.

Johns, H. E., The Physics of Radiation Therapy, Charles C. Thomas, Springfield, Illinois, 1953.

Rajewsky, B., Strahlendosis und Strahlenwirkung, Georg Thieme Verlag, Stuttgart, 1956.

Whyte, G. N., Principles of Radiation Dosimetry, John Wiley and Sons, New York, 1959.

Recent References

Berger, H., "Die praktische Anwendung der neuen Dosisbegriffe nach DIN 6809", Krebsbehandlung, Strahlenbehandlung und Strahlenforschung, special edition of Strahlentherapie 43, 1959.

Franz, H. and Hübner, W., "Zur Frage des Dosisbegriffs und der Dosiseinheiten", Strahlentherapie 102:590-595 (1957).

Schinz, H. R. and Wideröe, R. W., "Observations on the Proposed Change in the Definition of the Roentgen Unit", Fortschr. Röntgenstr. 89:486 (1958).

Units, Radiation: Historical Development

Encyclopedia of X-Rays and Gamma Rays

1963

Taken from *Encyclopedia of X-Rays and Gamma Rays*, Edited by George L. Clark, copyright 1963 by Reinhold Publishing Corporation, all rights reserved.

UNITS, RADIATION: HISTORICAL

UNITS, RADIATION: HISTORICAL DEVELOPMENT

The purpose of this article is to give a quick historical review of the development of radiation units together with a more detailed review of the recent refinements and developments.

A difficulty that beset radiation dosimetry through most of its history is that its basic terminology was not established until quite late. The term "dosimetry" is readily understandable as the measurement of dose or the science of dose measurement. It was the simple term "dose" that lacked appropriate definition. The relevant dictionary meaning of "dose" is "a measured quantity of a medicine to be taken at one time or in a given period." The application of "dose" in this sense to an X-ray treatment must have come quite naturally to the physicians who became the first radiologists. Also, the early X-ray tubes were notoriously fickle; interest was centered on the radiation that came from the machine and was "given" to the patient—*dose* is derived from the Greek word *dosis*, meaning "a giving." However, there was not a complete preoccupation with the radiation itself. As early as 1913, Christen pointed out that what radiologists should be concerned about is the energy absorbed from the X-ray beam by the body. Because of the low absorption of the X-rays, this is only a small part of the energy passing through a region of the body. In later years, the technical problems were overcome and radiobiologists, biophysicists, and the like turned their attention to the patient and "dose" took on the meaning of "absorbed from" or "received by." It is not surprising that conflicting interpretations should have developed in a subject that translated "a giving" into "a receiving."

Another term that gave difficulty was "quantity." For the early radiologists "quantity" of radiation probably had the familar meaning of "amount." For physicists, however, "quantity of radiation" was a special technical term; namely, the energy of the radiation per unit area. The confusion was made worse by the frequent need for the word in still a third sense, quantity in the sense of a measurable property.

In the early years of radiology, measuring methods were based on many effects of the newly discovered radiation. The calorimetric method dates back to 1897. Various chemical, fluorescence, or scintillation methods were used in the first years of this century; one of the favorite systems depended on the coloration of barium platinocyanide pastilles. The photographic method, based on one of the first observed effects, was made reasonably quantitative as early as 1905. Changes in electrical resistivity and a variety of biological effects were also studied for dosimetry. The measurement of ionization in gases gained and maintained an early popularity. Each of these effects was made the basis for an appropriate unit.

The first formal statement of a unit based on ionization was due to Villard in 1908. Unit quantity of X-radiation was defined as that which by ionization liberates one electrostatic unit of electricity per cubic centimeter of air under normal conditions of temperature and pressure. This is the embryo of the roentgen unit that was not born until twenty years later.

There were several reasons for the popularity of measurements based on the ionization of gases. The existence of a saturation potential in a gas-filled chamber provided a means of making reproducible measurements by means of accumulation of electric charge or by current measurement. Either current or change of potential can be measured with accuracy and precision over a wide range. The universal availability of dry air led to its adoption as a standard filling for ion chambers. Ordinarily the material of which a chamber is made has a pronounced effect on the ionization that is produced in the gas in it. Through the work of Friedrich and others, the free air chamber was developed. This is a chamber in which the ionization is independent of the material of construction of the walls. Also, through the work of Fricke, Glasser, and others, air-equivalent chambers were developed that gave the same ionization per unit volume as the free air chamber. The measurement of air ionization leads to a result that is, in effect, a sort of compromise between the two "dose" concepts. To a fair approximation, for the X-ray energies normally used in therapy in the period considered (the 1920's), air ionization is proportional to the quantity (physics) of radiation that has passed through the chamber. This relates it (approximately) to "dose" in the sense of "a giving." At the same time, for those interested in effects in tissue, the atomic composition of tissue and air are such that the ionizing effects should be roughly parallel in the two for those same energies. This relates it (approximately) to "dose" in the sense of "a receiving."

The Roentgen. The development of instrumentation that gave reproducible results and the formulation of a unit were closely related. Several proposals for ionization units were made. Finally, international agreement on a unit was secured. At the First International Congress of Radiology (1925) a commission, the International Commission on Radiological Units (ICRU), was formed to deal with the matter of units. At the second meeting of the congress in 1928, the ICRU made the specific recommendations:

1. That an international unit of X-radiation be adopted.
2. That this international unit be the quantity of X-radiation which, when the secondary electrons are fully utilized and the wall effect of the chamber is avoided, produces in one cubic centimeter of atmospheric air at 0° C and 76 cm mercury pressure, such a degree of conductivity that one electrostatic unit of charge is measured at saturation current.
3. That the international unit of X-radiation be called the "Roentgen" and that it be designated by the letter small "r."

The roentgen unit, as we shall see, has survived to the present time. The use of a common unit by scientists in all countries has contributed immeasurably to the progress of radiology. Nevertheless, the definition has contained the seeds of misunderstanding from the very start. Apparently the intention of the ICRU was that "quantity" in its second recommendation was to mean quantity (physics) of radiation. Then it was reasoned that quantity (physics) was not measurable in general, because it involves total absorption of the energy passing through a surface. The roentgen was thus a substitute unit of quantity (physics). Critically examined, the procedure is inadmissible, because two situations characterized by the same quantity (amount) in roentgens do not represent the same quantity (physics) of radiation, if the relevant absorption coefficients differ. For this reason and because the "quantity" meant is not specified, several authors have asserted that the "quantity" in the recommendation ought to be treated as quantity (amount). In that event, the quantity (measurable property) of the radiation of which the roentgen is the unit is nowhere defined.

The original definition made no reference to "dose," but almost immediately the roentgen was described as the unit of X-ray dose.

Early Radium Dosimetry. Whereas X-ray dosimetry was largely conditioned by the need to establish a reproducible method of measuring the "output" of a variable source, radium dosimetry began with sources of fixed emission rate. It became common practice to express dose as the product of milligrams times hours of treatment. Obviously such factors as the distance and disposition of the sources had to be stipulated to complete the prescription. When radon was used instead of radium element, an average strength had to be used or a more elaborate calculation of millicurie-hours effected.

That dose was expressed in milligram-hours or some similar unit for so long was obviously not due to lack of comprehension of the problem, but rather to the then existing technical difficulties of going at it in other ways. Three dose expressions were commonly recognized in this period; namely, "dose emitted"—the energy released by the radiation source, or some function reasonably related to it; "dose delivered"—the energy arriving at the locus of biological interest, or some reasonably related function; and "dose absorbed"—the energy imparted to the locus of interest, or some reasonably related function.

Many biological effects were considered as bases for the dose absorbed, including that of Russ (1918), the amount of radiation necessary to kill mouse cancer, and called the "rad"; a title that has now been revived for absorbed dose measurements. Perhaps the most widely used biological frame of reference was the erythema dose, especially in the threshold erythema dose of the Memorial Hospital school. The threshold erythema dose was said to be received from that amount of radiation that would produce reddening of the skin of the inner forearm in a certain percentage of patients within a certain time.

By taking advantage of the first-order approximation that absorption and scattering in light elements for radium gamma-rays can be neglected, mathematical systems of evaluating "dose" for external or intracavitary applicators developed. These required only a knowledge of how to integrate radiation functions, based on the inverse square law, together with a reference dose-rate at 1 cm from a unit point source of radium. Sievert, Mayneord, Failla and Quimby were among the early contributors to this technology. One such system is that of Sievert, based on the Intensity-millicurie (Imc), a unit "intensity" defined as the intensity of gamma radiation at a distance of 1 cm for a 1 mg point source of radium filtered by 0.5 mm Pt. From the intensity unit is defined a unit of quantity, Imc-hour, usually referred to as the Sievert dose, and also frequently referred to as 1 cm mg-el-hr. The Sievert dose has a well-known relationship to Eve's number,* from which a translation to "roentgens" can be made. Paterson and Parker based their system of dosage on a useful set of approximations to the radiation function solutions coupled with an Eve's number conversion from L. H. Gray's ionization data; this system and the value of 8.4 r have been widely used since 1932. All such derivations, prior to 1937, were considered improper by some purists because the definition of the roentgen (1928) limited its field to X-rays. Actually X-rays differ from gamma rays only in the manner of their origin so the use of the roentgen for gamma radiation was quite natural.

Redefinition of the Roentgen. The definition of the roentgen was modified at the Fifth International Congress of Radiology in 1937. Several changes and events of importance to dosimetry took place in the period 1928–1937. In this period, the knowledge of the properties and effects of radiation and of basic physics in general grew enormously. The technology for the production of radiation improved. The high vacuum X-ray tubes introduced earlier made possible much more reproducible X-ray machine outputs. The X-ray excitation voltages available increased steadily. The availability and use of radium increased.

The technology for the measurement of radiation improved. Largely through the efforts of Behnken and of Taylor, the various national standards laboratories reached essential agreement on the form of free air chambers and the techniques necessary for reliable roentgen measurements in the then available energy range. Before the establishment of the roentgen, considerable work had been done on the development of thimble chambers for X-ray measurements, and commercial dosimeters based on this were available as early as 1927. The definition of the roentgen placed all these devices in the category of empirically founded secondary instruments, requiring calibration against the free-air chamber. After 1928, the extensive work on thimble chambers continued in respect to making an air wall, choosing a satisfactory wall thickness, correcting for the absorption in that wall, and extending the energy range to include radium gamma radiation. The empiricism in thimble chamber construction offers no obstacle to their successful use, provided the use

---

* Eve's number is the number of ion pairs formed per second in air at standard temperature and pressure by the electrons produced in air, per cm^3 of air, by the gamma rays from a point source of 1 gm of Ra in equilibrium with its daughters and placed 1 cm away.

is confined to the demonstrated range of calibration against primary standards.*

The enunciation of Gray's Principle of Equivalence in 1929 profoundly influenced the course of dosimetry. The principle expresses the energy absorption per unit volume of irradiated material in terms of the measured ionization per unit volume in a suitably specified small cavity within that material as:

$$E_v = \rho W J_v$$

where ρ is the ratio of electron stopping power in the material and the cavity gas respectively, and W is the average energy required to produce one ion pair in the cavity gas. Since it was pointed out that W. H. Bragg had arrived at a similar proposition in 1921, the term Bragg-Gray Principle developed. There is a limit to the size of the cavity if the Bragg-Gray principle is to apply. Gray explored this matter in one of his later papers and showed that the principle is a convenient one in the range of radium gamma rays, but extremely taxing for X-rays generated at 200 kv or less. This accounts for many of the observed discrepancies in thimble chamber measurements, which led several authors in this period to the conclusion that there was no valid principle of small chamber dosimetry.

Although Gray's main interest was in the field of radium gamma-ray measurement, he made an attempt to integrate this with the then current X-ray dosimetry. To this end, he proposed that the roentgen be redefined in such a way as to permit its measurement with air-filled Bragg-Gray cavities surrounded by air (in practice, air-equivalent) walls. This was intended to permit measurement with the standard free air chamber in the appropriate energy range and with suitable thimble chambers at higher energies. Some such change was felt to be needed if X-ray and radium dosimetry were to be placed on a common basis. Numerous experimenters had worked with larger and larger free air chambers to evaluate the Sievert dose but found the technology not yet adequate for the purpose. Following Gray, several measurements were made of the Sievert dose with Bragg-Gray chambers (assuming a redefinition of the roentgen such as proposed by Gray). After an initial period of erratic results, good agreement was achieved between different measurements and led to the tacit agreement among most qualified experts that Bragg-Gray measurements were quite satisfactory for dosimetry.

The growing popularity of thimble chambers led to another change in dosimetry practice. The practice envisioned when the roentgen was originally defined was that a chamber would be placed in an unhindered X-ray beam and a measurement in roentgens made. Then the patient would be placed in the beam and exposed for a period determined by a correlation of previous experience with measurements made in this way. This had been and remained the common American practice. Here the quantity measured in roentgens was considered a dose in the sense of "a giving." On the other hand in Britain, in particular, radiologists began to place small thimble chambers at the surface of, or inside, dummy bodies (called "phantoms"). In the phantom the chamber is exposed to scattered radiation that it would not receive in the unhindered beam. Phantoms were used in the belief that what was measured in this way was something that was closer to the quantity of radiological interest than that given by the original practice. It was felt to be more nearly the dose in the sense of "a receiving." The results of the measurements in phantoms were stated in roentgens. This led to discrepancies in the experience of American and British radiologists.

These changes in the technology and the practice of radiology led to pressure to revise the definition of the roentgen. At the Fifth International Congress of Radiology (1937), it was redefined as follows:

The roentgen shall be the quantity of X- or gamma radiation such that the associated corpuscular emission per 0.001293 gram of air produces, in air, ions carrying 1 esu of quantity of electricity of either sign.

An accompanying statement was:
The international unit of quantity or dose of X-rays or gamma rays shall be called the "roentgen" and shall be designated by the symbol "r."

Gamma rays were included in the definition as well as X-rays. It was made clear that quantity meant quantity (physics). The accompanying statement made use of the word "dose" with the roentgen official. The statement seems to make "quantity" and "dose" synonymous. Omission of such words as "fully utilized," "wall effect," "conductivity," and "saturation current" removed the *necessity* for using a free air chamber for the primary measurement in roentgens. This was made definite by later statements in the same recommendations that the primary standard instruments for measurements in roentgens were the free air chamber or the air-wall chamber (for harder radiations). "One cubic centimeter of atmospheric air at 0°C and 76 cm mercury pressure" was replaced by the simpler equivalent term "0.001293 gram of air."

It will pay to examine carefully the meaning of the words, given in different order than in the definition, "ions produced, in air, by the associated corpuscular emission." Figure 1a illustrates what ions are meant in the definition of the roentgen. In a certain volume of air, secondary electrons are produced. The ionization they produce in air, regardless of whether that ionization is produced in that same volume or not, is that referred to. Figure 1b illustrates something more akin to what is actually measured. If one collects the ionization produced in a certain volume, he will collect ions

---

* It is a dangerous, yet common error, to consider thimble chambers as Bragg-Gray cavities. This is not so. They are chambers that have been empirically adjusted to give a response proportional to that of a free air chamber over a specified energy range.

that were produced by secondary electrons that were produced both inside and outside that same volume. Inside a free air chamber and, effectively, inside an air-wall Bragg-Gray chamber, the two types of ionization are equal because of electronic equilibrium. The words "fully utilized" in the first definition implied the necessity of establishing electronic equilibrium. This left it uncertain, however, which concept of Fig. 1 was meant. In divorcing the new definition from explicit dependence on the free air chamber, it was necessary to make a choice between the concepts. The choice was made in agreement with the then commonly accepted interpretation.

At the Sixth International Congress of Radiology the ICRU appended this note to the definition of the roentgen: "The unit may...be used for most practical purposes for quantum energies up to 3 mev." The reason given was: "It becomes increasingly difficult to measure the dose in roentgens as the quantum energy of the X-radiation approaches very high values." It becomes difficult because the range of the secondary electrons from high energy photons is so great that it becomes impossible to achieve satisfactory electronic equilibrium. At a given point in air the ionization will be partly due to electrons that were produced in regions where the photon intensity was significantly different from that at the given point. The increased rate of energy loss by radiation by high energy secondary electrons also makes for difficulty in measurement, because the radiation so lost must not be permitted to produce ionization that is measured.

Absorbed Dose and Exposure Dose. After the redefinition of the roentgen in 1937, the continuing growth of science and technology resulted finally in another change in the basic units for radiation measurement. The developments in nuclear physics, particularly of high-energy accelerators and of nuclear fission, had a profound effect by expanding the area of concern to dosimetry to include neutrons and other new radiations, high-energy, heavy charged particles, and artificially produced radioisotopes. These developments were accelerated by the beginning of the atomic energy program during World War II and the radiation protection and radiobiological research problems associated with it.

These new developments led rapidly to the feeling that the best way to express the dose was to give the amount of energy imparted to the matter per unit mass at the point of interest. Actually this was one of the oldest of the ideas inherent in the word "dose," but now people were forced to use it to the exclusion of other dose concepts. The case of slow neutrons can be used as one of the most illustrative (but not most important) examples of why this is so. The idea of "dose emitted" from a source is not applicable to slow neutrons because there are no sources of slow neutrons; they are produced by moderation of fast neutrons in a suitable slowing down material. The idea of "dose delivered" is just about meaningless for slow neutrons because it is so small; in the center of powerful nuclear reactors giving

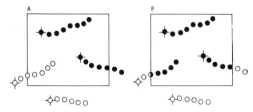

Fig. 1: Ionization in air. The circles represent ions formed along the tracks of secondary electrons; the electrons were produced at the points marked by the crosses. The solid circles represent the ions that (A) are to be collected to meet the definition of the roentgen and (B) are actually produced in and collected from the given volume.

the highest slow neutron fluxes we know of, the energy of the slow neutrons—the dose delivered—is only about the same as that of the photons in a beam giving 1 r/hour. The idea of "dose absorbed" makes sense; for equal energies absorbed, slow neutrons give biological effects that are comparable to those for other radiations.

The first steps toward an absorbed dose concept were the use of different "equivalent roentgens." The use of these equivalent roentgens was based on the following reasoning: The roentgen was an established, familiar unit. The unit of absorbed dose should be closely related to the roentgen. Photons that gave one roentgen would produce an energy absorption of 83 to 85 erg/gm of air (depending on the average energy loss by secondary electrons required to form an ion pair in air (W); the equivalent of the roentgen is now taken to be 86.9 erg/gm of air, corresponding to $W = 33.7$ electron volts per ion pair). They would produce 93-95 erg/gm in water or soft tissue. Therefore, an equivalent roentgen was an absorbed dose of X erg/gm. Different authors used values of X from 83 to 95.

In the atomic energy program beginning in 1942, one was faced with the practical problem of adding the doses received by a large group of workers from photons, alpha, beta and neutron radiation. Parker was led to develop a system that was readily communicable. He used the "rep" as an equivalent roentgen, based originally on $X = 83$, but changed to $X = 93$, and "rem" as a biological "unit" obtained by multiplying the components of the dose in rep by appropriate RBE multipliers. Rep was an acronym for *r*oentgen *e*quivalent *p*hysical and rem for *r*oentgen *e*quivalent *m*an (or mammal). The principal merit of this variation was in the communication advantage of pronounceable terms, such as rep and rem, and the simple arithmetical scale of relation.

At the First International Congress of Radiology following the war (1950), the ICRU recommended that: "For the correlation of the dose of any ionizing radiation with its biological or related effects...the dose be expressed in terms of the quantity of energy absorbed per unit mass..." In a note they explained that "energy absorbed" meant the energy imparted to the material by ionizing particles at the place of interest. They recommended that the unit erg/gm be used for this dose.

These recommendations were well received in general, but were criticized for not supplying a name for the concept considered and for not supplying a unit similar in size to the roentgen. At the 1953 Congress, the ICRU remedied these difficulties by defining the "absorbed dose" as the amount of energy imparted to matter by ionizing particles per unit mass of irradiated material at the place of interest. They also defined a unit, the "rad" to be the unit of absorbed dose. One rad is equal to 100 erg/g. This concept and its unit received rapid and universal acceptance.

The definition of absorbed dose was bound to cause confusion in speaking about the quantity measured in roentgens. This, of course, was only part of the general confusion that had always existed about the term "dose." Now, however, a quantity had been defined that used the term. In a sense, this withdrew the sanction for the use of the term with other concepts that had been associated with it. The ICRU recognized this problem and considered the use of the term "exposure" for the quantity measured in roentgens. This was judged unsatisfactory, principally because of the widespread use of "dose" and "roentgens" in existing laws and regulations. For this reason, the term "exposure dose" was chosen. The opening words of the definition of the roentgen (1937 version) were changed to read: "One roentgen is an exposure dose of X- or gamma radiation such that...." Apparently in the belief that what was measured in roentgens was well enough understood, the meaning of exposure dose was only indicated, rather than defined, as "a measure of the radiation that is based on its ability to produce ionization."

Recent Developments. We have already noted that when air equivalent thimble chambers became available, the practice grew of placing these in phantoms in order to measure something more closely related to what produced the biological effect in the patient. When the Bragg-Gray principle was advanced and accepted, variations of this practice were introduced. For example, air-equivalent chambers or Bragg-Gray air-wall chambers were replaced by chambers with walls made of or simulating tissue—sometimes soft tissue, sometimes such special tissues as fat or bone. Application of the Bragg-Gray principle to such measurements permitted the determination of the energy absorbed in the tissue; i.e., the absorbed dose. These tissue-equivalent systems assume considerable importance in the dosimetry of neutrons. Another practice that developed was the attempt to eliminate the walls of the chamber. This usually cannot be done completely because electrodes are necessary to collect the ionization current; however, some tissue-equivalent plastics that are also conducting have permitted construction of tissue-equivalent chambers in which the walls are the tissue-equivalent material itself. In general, though, the effort is to make the walls thin enough that they do not materially affect the flux of ionizing particles reaching the gas of the chamber. These thin-wall techniques are especially necessary in the dosimetry of alpha and beta rays. It is not always realized that the Bragg-Gray principle applies as well to these thin-wall chambers as to ones with thick walls. The older discussions of the Bragg-Gray principle were frequently devoted to its application to measurement of what we now call exposure dose. In this application, thick walls were required to establish the electronic equilibrium needed for the measurement of the exposure dose, but the application of the principle itself is independent of the thickness of the walls. Thin-wall chambers also came into use in dosimetry of X-ray and electron beams from high-energy accelerators. The use of thin-wall chambers is essentially forced upon us in this case. The secondary electrons from high-energy photons have such long ranges (of the order of one centimeter in tissue) and the photon beam widths are usually so narrow that it is impossible to attain electronic equilibrium. To even come close to a relative equilibrium would require chambers that would usually be too large relative to the system being irradiated to give results that were of any interest.

This intense attention to the practical art of making gas ionization measurements led to various proposals to make some ionization quantity and the corresponding unit the basic ones for dosimetry. As early as 1937, Failla recommended that ionization be measured in a thin-wall, vanishingly small cavity (extrapolation chamber) placed in the irradiated material at the point of interest, and that it be expressed in units of the same size as the roentgen and called "tissue roentgens." In 1938, Holthusen recommended the same quantity and called it the "ion dose." Similar recommendations were made to the ICRU by the British in 1950 and by the Germans in 1956. The latter were based on recommendations by Fränz and Hubner who not only advocated the term "ion dose," but suggested that the roentgen unit be generalized so that it could legitimately be used for any ion dose. Their proposal led to much discussion in Germany. Finally, it was adopted as a preliminary German national standard. Schinz and Wideröe (1958) felt that this generalization of the roentgen was not apt to obtain international approval and that change of a traditional unit was inadvisable anyway. They suggested that a new unit, the same as Failla's "tissue roentgen," be defined and given the name "rho."

Although these later proposals were well received in continental Europe, they were received less enthusiastically elsewhere. Many felt that the energy concept employed in the definition of absorbed dose ought to be used wherever possible in dosimetry because of the fundamental position it holds in the sciences. Ionization does not enjoy such a position. Also, there is a reluctance to adopt units based on one special method of measurement particularly in a period when so many other methods of measurement are being perfected.

The proposals of ionization quantities and units provoked much new discussion of the whole sub-

1107

ject of the quantities and units used in dosimetry. It became apparent that there was still dissatisfaction with the formulation of the quantities already in use. It was also felt that not enough attention had been given to making sure that all the quantities and units were related in a firm, satisfactory manner or that all the logically necessary quantities had been defined. Still another opinion was that the form and content of the definitions of the quantities and units in radiation dosimetry had not kept up with developments in other fields of science and engineering, and, therefore, needed revision. These factors led the ICRU in 1958 to set up an *ad hoc* Committee on Concepts, Quantities, and Units in the field of Radiation Dosimetry to recommend whatever changes were needed to deal with these objections. The report of the committee was presented to the ICRU in the Spring of 1962. The ICRU is expected to issue some revised definitions based on the recommendations in this report.

Note Added in Proof: The ICRU issued a new set of recommendations late in 1962 ("Radiation Quantities and Units, ICRU Report 10a, 1962," National Bureau of Standards Handbook 84, U. S. Government Printing Office, Washington D. C.) that included the work of its *ad hoc* committee. All of the definitions of radiological quantities were changed to a common pattern employing basic physical quantities in precise mathematical relations. The meanings of the definitions were not changed. The logical consistency of the definitions was assured by basing them on a mathematical analysis of the problems of dosimetry (Rossi, H. H. and W. C. Roesch, "Field Equations in Dosimetry," Radiation Research *16* 783 (1962)). No quantities other than exposure dose that were based on ionization were defined; it was recommended that the term "exposure" be used rather than "exposure dose". The name "energy fluence" was given to what we have called quantity (physics) of radiation. A new quantity, the "kerma", was introduced; it is a generalization of "exposure" and is defined by energy rather than ionization quantities.

References

1. "Report of the International Commission on Radiological Units and Measurements (I.C.R.U.), 1959," National Bureau of Standards Handbook 78, U. S. Government Printing Office, Washington, D. C.
2. Hine, G. J., and Brownell, G. L., "Radiation Dosimetry," New York, Academic Press, Inc., 1956; especially Chap. 1, "Radiation Units and Theory of Ionization Dosimetry," by F. W. Spiers.
3. Jaeger, R. G., "Dosimetrie und Strahlenschutz," Stuttgart, Germany, Georg Thieme Verlag, 1959.
4. Johns, H. E., "The Physics of Radiation Therapy," Charles C Thomas, Springfield, Illinois, 1953.
5. Rajewsky, B., "Strahlendosis und Strahlenwirkung," Stuttgart, Georg Thieme Verlag, 1956.
6. Whyte, G. N., "Principles of Radiation Dosimetry," New York, John Wiley & Sons, Inc., 1959.

H. M. PARKER
W. C. ROESCH

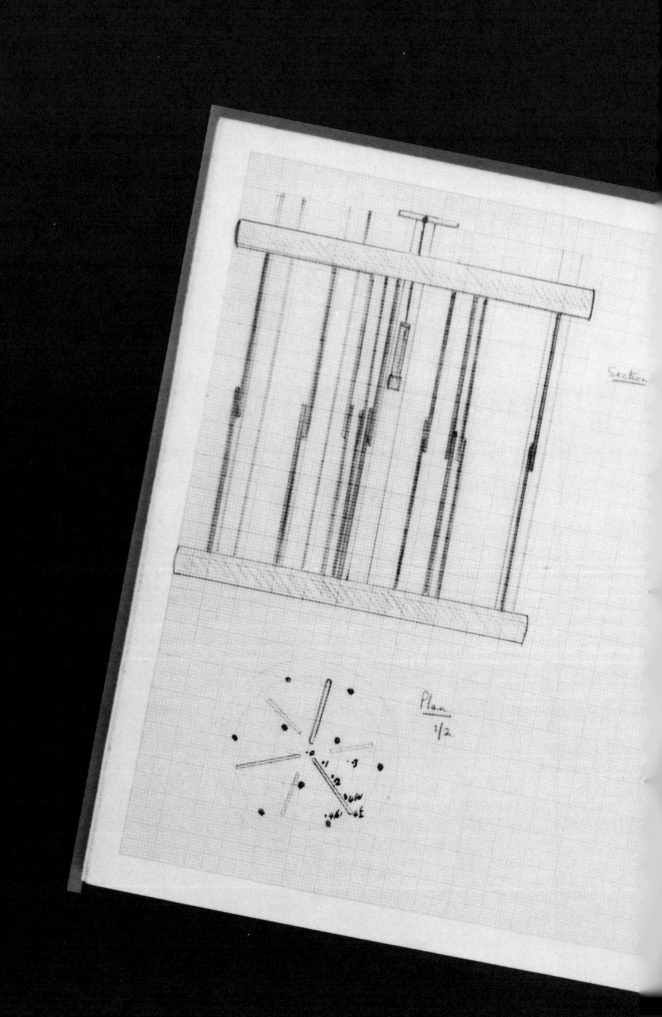

PART IV

RADIATION PROTECTION:
The Manhattan District and Beyond

INTRODUCTION

Parker's interest in radiation protection preceded his entry into the Manhattan District as evidenced by the first paper in this section, which discussed the hazards of repeated fluoroscopic x-ray exposures to infants. The principles elucidated in this article are of considerably broader application and, along with his early papers on the tolerance dose, are illustrative of his early thinking on radiation protection.

It was at Chicago, Oak Ridge, and finally and most significantly at Hanford that his radiation protection philosophy matured and was put into practice. Perhaps his greatest legacy was the sound development and implementation of strong radiation protection programs within the Manhattan District, establishing principles and practices that shaped the course of modern health physics. Much of his early work was classified, and several of his papers, now declassified, are published here for the first time. The reprint of chapter 9 from the book *Industrial Medicine on the Plutonium Project*, edited by radiologist Robert S. Stone, is of considerable historical interest and represents a great contribution to preventative medicine. Indeed, this work is a testimonial to Parker, to whom is due a great deal of the credit for the enviable radiation safety record within the Manhattan District. This chapter was coauthored with his long-time physician associate and friend Simeon Cantril and was written the day after the bombing of Hiroshima. Security requirements held up publication for several years.

This section includes numerous papers and writings on radiation protection standards, an area to which he had early devoted his attention in the Manhattan District and which he worked hard to promote. Ultimately he became concerned about the proliferation of radiation protection standards, observing the exponential growth of such standards in his keynote address "Festina Lente" at the 1971 Health Physics Society Topical Symposium "Radiation Protection Standards—Quo Vadis." With tongue in cheek in this address, he invited health physicists to ". . . determine the year at which the sole occupation of the health physicist becomes the reading of protection handbooks," projecting nearly 10,000 standards by the year 2001. His dry British humor and classical education are clearly illustrated in this work.

Several of Parker's contributions in the radiation protection area are noteworthy for their prescience, particularly those dealing with effects of reactor accidents and environmental effects. His paper on environmental effects, presented at the First International Conference on the Peaceful Uses of Atomic Energy is the first comprehensive view that distilled what might be expected in the environment. His paper on reactor hazards, coauthored with his long-time confrere, Jack Healy, is similarly a landmark work. And, although in large measure he was concerned with what might be termed radiation protection mangement or administration, he never strayed very far from science, as evidenced by his extraordinarily well written review of the radiological hazards of plutonium, in which a seemingly incredible amount of information is packed into a relatively small number of lucidly written pages.

Although this is by far the lengthiest of the five parts devoted to Parker's papers, it is also the least complete, for Herb wrote much more and made many more oral presentations than can be reproduced here. A few of those, although not reproduced here, bear mention:
- "Absorption of Plutonium Fed Chronically to Rats," coauthored with J. Katz and H. A. Kornberg, which appeared in *American Journal of Roentgenology, Radium Therapy and Nuclear Medicine,* LXXIII: 303–308 (1955), an interesting piece of biological research important to radiation protection which also appeared as an internal report in somewhat different form;
- "Tolerable Concentration of Radioiodine on Edible Plants," published January 1946 as U.S. Atomic Energy Commission Report TID-AECD-2907, a prescient work that underscored Parker's early concern for environmental effects of radiological releases;
- "General Survey of Instrumentation for Health Physics," coauthored with C. C. Gamertsfelder and P. L. Eisenacher in *Electrical Engineering* for October 1950, in which Parker bowed to his colleagues and used the term "health physics";
- pioneering consideration of the insurability of atomic energy workers, presented to the 57th Meeting of the Life Insurance Medical Directors in October 1948 and published in the *Transactions.*

Possible Hazards of Repeated Fluoroscopies in Infants

The Journal of Pediatrics

October 1942

Reprinted with permission of the C.V. Mosby Company, St. Louis, Missouri.

POSSIBLE HAZARDS OF REPEATED FLUOROSCOPIES IN INFANTS

Franz Buschke, M.D., and H. M. Parker
Seattle, Wash.

RECENTLY we became aware of the fact that apparently a number of pediatricians include a fluoroscopy in the monthly routine examinations of the infants in their care during the first and second years of life. Since we feel that such a procedure is charged with certain potential hazards, we welcome the opportunity of discussing this problem at your initiative in this group.

It is too often not realized that x-rays, in addition to being a useful diagnostic tool, accidentally and unfortunately represent at the same time a powerful drug. As Dr. Case once pointed out, x-ray diagnostic procedures are still considered by many as a kind of "glorified photography." Yet every exposure to x-rays actually delivers radiating energy to the body. The output of different diagnostic machines varies within very wide limits, depending upon the milliamperes and kilovolts used, the filtration, the distance between tube and body surface, the time of exposure, and the size of the exposed field. But even many specialized radiologists do not know the actual output per minute of their diagnostic machine.

Cowie and Scheele[1] have reported the result of a survey of radiation protection in forty-five hospitals located in twenty-four states scattered throughout the country, undertaken at the initiative of the National Cancer Institute. These hospitals were considered the type in which one might expect to find typical, if not slightly above average, equipment and practices of protection against high energy radiation. Although this survey was undertaken from the point of view of radiation protection for the personnel rather than for the patient, the inadequacy of protection during diagnostic procedures in a large number, even of these special radiologic units, shows the lack of appreciation of the amount of irradiation inadvertently delivered. It only goes to show that we can hardly expect a greater appreciation of these potential dangers where equipment used in office practice by physicians not specializing in radiology is concerned.

It is true that most of the machines today deliver a quality and quantity of radiation which, within reasonable limits of diagnostic exposure, will not cause any demonstrable gross damage. But what we have in mind are not so much the gross accidents but those potential dangers which are much more difficult to demonstrate. For the majority of diagnostic procedures used in general practice, the lack of

---

From the Tumor Institute of the Swedish Hospital.
Read before the North Pacific Pediatric Society, Seattle, Jan. 31, 1942.

appreciation of these potential hazards is regrettable but probably not a cause of major concern because these procedures in adults and older children are usually limited to fairly small portions of the body. If the small body of an infant is exposed to the same kind of radiation, a comparatively much larger volume of body is irradiated. This risk is naturally increased if such fluoroscopies are done repeatedly in infants at fairly short intervals. For this reason it seems necessary to give even more attention to details of the fluoroscopic procedure with infants in order to eliminate or minimize the possible detrimental effects.

We will meet, and as a matter of fact have repeatedly met, the objection that never has any damage been observed by those who have used this procedure throughout many years. But as we have already pointed out, the possible effects of doses under consideration here are much more difficult to demonstrate and, it is true, by their very nature can be anticipated only by implications from our general knowledge of radiation biology and from comparison of the doses delivered with those used for other therapeutic purposes. If we wait until damage is proved beyond the doubt of the unbeliever, irreparable harm may have been done.

Before entering into a discussion of these possible effects, we must determine the actual amount of radiation delivered during such fluoroscopies in order to substantiate our opinions as to their hazards. We have therefore investigated seven machines used in offices of reputable pediatricians selected at random. Not all of them do routine or even regular repeated fluoroscopies. Since, however, none of them knew the output of their machine, it is merely accidental whether those with higher outputs are actually used for repeated exposures. Table I represents a survey of this analysis, and though it is only a small and arbitrary selection, we believe that it permits us to draw some definite conclusions.

One of the machines, E, delivered at the supposed operation of 5 Ma., 35 r. per minute. The physician in charge, however, reported that the amperage is inconstant and that the amperes would change up to 40 Ma. without change of the controls. In such a case an amount of about 200 r. might be delivered to the infant's body in one minute. Wintz[2] considers 170 r. applied to the adult's ovary as sufficient to cause a temporary amenorrhea, and the dose necessary for a permanent sterilization of the adult's ovary varies around 300 r. (Martius[3]). Schinz and Slotopolski[4] caused a complete depopulation of a rabbit's testicle with only 60 r. The dose necessary for a temporary sterilization in man is considered around 200 to 300 r. (Schinz, Wintz), and we have the histologic proof of a complete atrophy without signs of restitution in a human testicle following the exposure to 600 r. (Schinz and Slotopolski). From an interpretation of available data it seems that the retardation of bone growths can be expected with doses of about 150 r. in infants and about 300 r. in small children (Ellinger[5]). In animals a marked retardation of growth was observed following doses below the epilation dose

TABLE I

X-RAY OUTPUT OF SEVEN MACHINES USED FOR FLUOROSCOPIES IN INFANTS

EQUIPMENT	NOMINAL CURRENT (MA.)	AVERAGE DOSE RATE ON SKIN (R. PER MIN.)	AVERAGE DOSE RATE 5.5 CM. DEEP (R. PER MIN.)	NOTES
A	5	6.0	1.2	
B	No milliameter	6.3	1.9	Output variable
C	5	16.0	Not measured	Output so variable that screen brilliancy changed
D	3	28.0	4.0	
E	5	35.0	Not measured	Tube current varied freely from zero to 40 Ma.; output dangerously erratic
F	5	37.5	3.8	
G Operator 1	3	23.0	4.0	Two technicians used different technique on same machine
Operator 2	5	38.0	7.0	

The dosage measurements were made in a "presswood phantom" approximately the size and shape of an infant. By this method, the dosage figures obtained are the same as if an infant were actually used for the measurements. The body of the presswood infant was 11 cm. thick (anteroposteriorly) so that the dose 5.5 cm. below the skin represents the dose in the central regions of the infant body.

In all cases the dose rate was measured with a small field of irradiation (100 cm.^{2}) and again with a large field (350 cm.^{2}). The simplified table gives the average of these for each machine. On the skin, the large-field dose is approximately 20 per cent greater than the small-field dose. Within the infant body the difference is even greater.

The percentage depth dose at 5.5 cm. ranges from 10 per cent for F to 30 per cent for B. The main factors in this variation are the distance of the x-ray tube below the fluoroscopy table, the effective filtration of the beam, and the voltage applied to the tube. The exit dose was found to vary from 0.3 to 1.6 r. per minute, while the percentage exit dose covered the wide range of from 1 to 7 per cent. From the standpoint of minimum radiation damage, it is evident that optimum conditions require: (1) large distance between tube and table; (2) screen as close to body as possible; (3) voltage and filtration chosen so that incident radiation is as penetrating as possible, consistent with good fluoroscopic differentiation; and (4) tube current the minimum for good screen visibility.

If we may generalize from the small number of machines studied, it would appear that equipment especially designed for pediatric offices ignores the important factor of tube-table distance, using twelve or thirteen inches instead of eighteen to twenty-two inches of larger machines, while ordinary radiographic equipment, used permanently or temporarily for pediatric fluoroscopy, is likely to be operated at tube current well in excess of the necessary value. We made no investigation of the voltage and filtration, although this might well be discussed in view of the incidental observation that machines A, C, and D were virtually identical except for age, whereas the outputs at the same tube current were 6, 16, and 43 r. per minute, respectively. At the voltages involved in fluoroscopy the thickness and composition of the tabletop and the choice of a mattress are prime factors in regulating the radiation output.

(Försterling quoted by Flaskamp[6]) which in human infants would mean below 300 to 400 r. Naturally these data cannot be applied to our problem without reservation, since details of technique, quality of radiation, etc., have to be taken into consideration. But they bring the dose which might possibly be applied during fluoroscopy with a machine such as the one under consideration so close to the danger zone as to justify serious concern. Although the use of such a machine today might be compared with surgery without asepsis, probably quite a number of such machines is still in use and their potential danger should be emphasized.

In another place under the direction of one of the best radiologists we found that the output differed with the operator (machine G). While the physician assumed that he was always fluoroscoping infants with 3 Ma., we found accidentally that one of the technicians when he happened to be at the control would give him 5 Ma., increasing unnecessarily the output from 23 to 38 r. This example again emphasizes that even in the best places there is not sufficient appreciation of the potential hazards, and there is a lack of attention to details with the purpose of minimizing them.

Surveying the outputs of other machines as shown in Table I we find a variation between 6 and 38 r. delivered to the posterior surface of the infant's body. For the purpose of comparison we may mention that the doses given therapeutically for inflammatory conditions, such as a mastoiditis or a furuncle, or for a thymic enlargement are around 50 to 75 to 100 r. Although there is a considerable decrease in dose while the radiation passes the infant's body (due to the soft character of this type of radiation), it seems fair to base our discussion on the lower surface dose since the most endangered organs, ovaries, testicles, spine, and extremities in small infants are quite close to the tabletop. It seems fair for this discussion to conclude from our table an average dose of 25 r. per minute. If the average rapid fluoroscopy by an experienced and well-adapted examiner takes twenty seconds, about 8.3 r. will be delivered at this rate or 100 r. during the first year of life. Actually we know from experience that some of the fluoroscopies last considerably longer.

What then are the hazards we might anticipate for doses within this range?

Four main types of radiation damage must be considered: those to the skin, the hemopoietic system, the epiphyses of growing bone, and the gonads.

The doses here under consideration are too small to anticipate any damage to the skin and to the epiphyses, particularly since they are not additive over a longer period. There is a marked recovery from each dose, if the total dose is distributed over any length of time. Consequently there is a considerable difference as to whether the dose of 100 r. is applied at once or in twelve single doses with monthly intervals. The same is probably true for the effect on the hemopoietic system. In this regard, however, one should consider the fact that the type of fluoroscopic examination under consideration here is actually comparable to a total body irradiation. If this procedure is applied in adults for therapeutic purposes (as it occasionally is in certain stages of leucemias, Hodgkin's or other generalized diseases) the dose delivered to the entire or almost the entire body varies around 5 to 15 to 30 r. With such doses in adults a very marked reduction of the white and occasionally the red count might be observed, and the procedure therefore has to be applied with the utmost caution. We have seen one case, a leucemic patient, in whom a total body irradiation of 10 r. was applied three times in ten days. This treatment was followed by an uncontrollable and fatal drop in the white count. Granted that probably the normal hemopoietic system of the infant will be more resistant than the diseased one, still the dose delivered to the entire body or large portions of it during such a fluoroscopy, as you can see from Table I, may come close to a potentially harmful dose even if it cannot be demonstrated by clinical methods at our disposal today. We also know that x-rays have an influence on the cell respiration and on the electrolytic metabolism of the cell, and thus they probably have other pharmacologic effects which cannot yet be demonstrated. Although there exists a vast amount of experimental work with regard to the physical, chemical, and physiologic-chemical changes produced by x-rays, no uniform conclusions can be drawn. But even if these effects are not yet uniformly interpreted, they should prompt us at least to consider the possibility of some metabolic disturbances if radiation is poured throughout the body.

The effect of radiation on the gonads is of two types. If a sufficient amount of irradiation is delivered, this will cause a complete and either temporary or permanent sterilization. The doses here under consideration (with the exception of those due to definite technical errors) are below this level.

Doses below this sterilizing level, however, are known which do not cause a destruction of the germ cells but do cause changes in the germ plasm; this means changes in the chromosomes and genes which are responsible for the transmission of inheritable characteristics throughout the generations to come. The doses in question here are well within the range in which such damage might be expected. Doses causing an increase in the normal mutation rate are completely additive in effects; this means the effect with regard to the number of induced mutations is the same whether a certain dose is applied at once or in intervals of days, hours, months, or years. It has been shown that the doses are even additive in succeeding generations. Duration and frequency of exposures do not influence the reaction. The result of more recent exposures simply add to the previous exposures. The number of mutations induced is in direct proportion to the total dose delivered to the gonads and is independent of the quality of radiation. The importance of these facts in their bearing on the problem under consideration is quite evident. Muller[7] has computed that if irradiation causes an increase in the mutation frequency of man similar to that in Drosophila, then it would take an average treatment of only about 30 or 40 r. per individual to make the rate at which hereditary weaknesses and lethals are produced double the natural rate. Since this effect is actually accumulated over an indefinitely long period of time, it is well possible that if germ cells received only a little secondary radiation but repeatedly at widely separated intervals, even in successive generations, the final effect would be a production of mutations as great as though the whole treatment had been given at once (Muller[7]).

The clinical manifestations of such effects which we can anticipate are of three types.

First, a general decrease of fertility in the generations to come. This is due to two different kinds of damage to the germ plasm: to chromosomal changes and to lethal gene mutations. The chromosomal changes consist of pathologic disturbances of the normal chromosome divisions. From the point of view here under discussion these changes may appear more important than the gene mutations because the chromosomal changes may cause a marked infertility even in the first generation offspring of the irradiated individual and therefore may appear somewhat closer to the general way of thought of the clinician. Lethal gene mutations consist of sudden changes of the single unit of heredity (the gene) in such a way that this particular change is incompatible with the life of its carrier. The lethal gene kills the spermovium, or the embryo, in early stages. It is now assumed that quite a number of sterile couples carry such lethal genes. Since most mutations are recessive (i.e., able to produce their characteristic effect only when an individual has received the same kind of mutated gene from both his parents) they will be lethal only when both parents are carriers. This would explain that in some of these couples either partner may be fertile if mated to another partner who is not a carrier of this gene. This type of damage will clinically manifest itself only in the third and following generations of the irradiated individual. From the point of view of quantity, this type of damage is probably the most important one since at least 80 per cent of the x-ray-induced mutations (as the spontaneous mutations) are completely lethal in their effect.

Second, an increase in the number of those persons in the general population who are afflicted with inheritable disease. This effect is due to the increased production of so-called sublethal genes and of genes which cause marked morphologic or physiologic disturbances not severe enough, however, to cause the death of the individual in the embryologic stage. They may either cause death during early life, such as from ichthyosis congenita, amaurotic idiocy, xeroderma pigmentosum with the development of skin cancers, or they may cause definite disease entities which are still compatible with life, such as diabetes, susceptibility to tuberculosis, cleft palate, certain diseases of the central nervous system, etc. The borderline between the sublethal genes and those causing marked minus variations in a viable organism is not sharp and depends upon many external circumstances. We know for instance a race of dogs with a marked inheritable tendency for the development of cleft palate causing early deaths of the puppies because of the impossibility of nursing whereas the same defect in human beings today is compatible with life following surgical intervention.

Third, only slight variations in physiologic or morphologic behavior. This group probably represents the most common but the least easily recognizable group. There are probably many physiologic changes that remain undetected because they have no outer morphologic expression at all. After all, it is partly a matter of one's philosophy of life whether he considers certain deteriorations already as pathologic. This thought was best expressed by Colwell and Ross[8] in their discussion of these problems:

"We believe that 'intelligence tests' on certain rodents have been carried out, but the normal intelligence of rabbits and guinea pigs is not usually regarded as high. And in considering the matter now under discussion, mental changes are of at least as much importance as anatomical variations. Equally, as we have indicated before, it does not help the case very much to take a collection of birth statistics and report that when the mothers had been subjected to (preconception) ovarian irradiation 'the pregnancies and the offspring were quite normal.' What may be quite 'normal' mentality and capacity in one social grade may be a marked degeneration in another. Apart from gross mental deficiencies such as idiocy or semi-idiocy, simple 'weakening' of intellect or of business capacity in the children will be much more obvious when the parents belong to the professional classes or even to the numerous class of skilled work people than when they follow some unskilled occupation. In cases like these relative values are of the highest importance.

"Parents in general wish to do the best for their children. Is subjecting the mother's ovary to radiation doing the best? The question of causing gross anatomical lesions does not enter into the discussion. If it did, there would be no discussion at all."

The fact that these slight mutations show only very indefinite manifestations does not make them less objectionable but more so since they would be harder to recognize and deal with and will lead to a general deterioration of the race. Since many of these properties probably are those of somewhat reduced efficiency of the central nervous system, it is partly a matter of how much importance one attributes to the efficiency of the brain function in general and whether he considers these variations very important or not.

How real are these dangers to the germ plasm?

To most clinicians all arguments based on genetic considerations appear somewhat as fairy tales. This is due to the fact that definite conclusions in the field of genetics can be reached only by statistical evaluations of a vast number of observations throughout many generations. Obviously therefore conclusions in regard to human pathology can be reached only in analogy to those obtained by animal and plant experiments. And the clinician has a certain instinctive hesitation to accept conclusions reached in this way on the fruit fly as applicable to the human race. The uniformity, however, of the Mendelian laws throughout the animal and plant kingdoms must be considered today as proved beyond any possible doubt. While details naturally differ, the fundamental conclusions reached on the basis of an overwhelming experimental evidence in animals and plants can safely be applied to the human race. No field of radiation biology rests on so safe an experimental basis as radiation genetics. While it is of course impossible to give even a superficial review of the experimental evidence on which our conclusions are based, we refer those who wish to convince themselves of the soundness of these inescapable conclusions to some of the elaborate reviews by competent radiation geneticists such as Muller,[7] Timoféeff-Ressovsky,[9] Hanson,[10] Oliver.[11]

On the basis of the experimental evidence which has shown that irradiation of the gonads causes a marked increase in the normal mutation rate in proportion to the dose, geneticists have warned again and again and for many years against the danger of indiscriminate exposure of the gonads to x-radiation. In a discussion of this problem in 1938, which was mainly concerned with the question of such danger from the point of view of radiation therapy, we considered this danger as exaggerated by the geneticists. We felt that, since most of these mutations are recessive in character, the occasional irradiation of the gonads would not increase the number of such recessive genes in the population at large sufficiently to increase appreciably the clinical manifestation of these genes which occurs only if both parents of an individual are carriers of these recessive characters. *This danger, however, naturally increases considerably if the gonads in a large proportion of the population are routinely exposed to irradiation.* In this way the number of mutations in the general population would considerably increase, and consequently the chances of the meeting of two recessive characters may become great enough to have practical consequences. If this routine is followed we might embark upon a gigantic experiment. The blueprint of the results which we might anticipate are in our possession in the form of the large scale experiments on Drosophila and other organisms. We might then face the paradoxical situation that while modern medicine tries by different means to eliminate defect variations, the same medicine itself produces these mutations for the generations to come on a scale much larger than nature ever attempted to do. Whether in view of the present world conditions such a change of the human race or its decreased fertility is objectionable is of course a matter of one's personal philosophy of life. The chances that the race becomes deteriorated are certainly greater than those that it will improve, because most of those mutations which are not lethal are to some extent detrimental to life, although occasional improvements were found experimentally. When changes are induced in a random fashion in a highly organized system, it should then be expected that only a very few of these random changes would happen not to be harmful (Muller).

From time to time we find in the radiologic literature reports on the offspring of mothers whose ovaries have been subjected to irradiation, and pictures are shown of apparently healthy babies born after such procedures. Since, as emphasized, most of these mutations are recessive, we will not expect any damages before the third or following generations. The decrease in fertility cannot be denied by pointing at and photographing those babies who are born. The appearance of a baby does not say anything about the further development of this child. Since we do not expect mainly gross variations but more a general but immeasurable deterioration in those who will be born, the normal appearance of a baby and even of an adult is meaningless.

CONCLUSIONS

The purpose of this discussion is not to discourage your fluoroscopies when you consider them as *clinically indicated*. We only mean to make you appreciate that fluoroscopy actually delivers radiation energy in doses which may cause certain detrimental effects and to use all possible precautions in order to minimize the risks. These safeguards we may summarize as follows:

1. The output in roentgens per minute during fluoroscopy should be known to the examiner and should be kept in mind. This means that diagnostic machines should be calibrated and should be checked at regular intervals, at least every six months.

2. The actual output should be kept as low as is compatible with the efficient diagnostic procedure. We have convinced ourselves that the screen picture with those machines in this series which gave the lowest output (6 r. per minute) are sufficient for most diagnostic purposes in fluoroscoping the small body of an infant. The milliamperes and kilovolts then should be adapted to what is really necessary. In order to keep the output as low as possible, a good adaptation of the examining physician and a good screen are prerequisites.

3. The exposure time should be kept as short as possible. No discussion of the fluoroscopic findings is permitted as long as the shutter is open. This naturally means that an explanation of the screen picture to the baby's mother during fluoroscopy is not permitted.

4. The shutter of the fluoroscope should be open only as far as is absolutely necessary for the diagnostic purpose. If ever possible, any exposure of the abdomen and of the gonads should be avoided altogether.

5. In cases in which it is doubtful as to whether the fluoroscopic examination will lead to a conclusion, it may be better to have a film exposure done with an efficient machine. This is true, for instance, for the location of foreign bodies. A rough investigation has shown us that an average film exposure of this type delivers only about $\frac{1}{10}$ r. This picture can be studied quietly. If then fluoroscopic examination becomes necessary for further localization, the field can be limited and the time

of exposure can be shorter because one approaches the problem with a more definite question in mind. We do not believe that fluoroscopic examinations give any adequate information as to the condition of the epiphyseal lines. Minor disturbances will escape fluoroscopy and need first-class film exposures. Once they are intense enough to be visible on the screen they will not need "routine" examinations.

6. Machines which cannot be calibrated adequately and are unreliable in their output, such as the one described by us as E, should be completely discarded as unsuitable.

REFERENCES

1. Cowie, D. B., and Scheele, L. A.: J. Nat. Cancer Inst. **1**: 767, 1941.
2. Wintz, H.: Strahlentherapie **37**: 407, 1930.
3. Martius, H.: Strahlentherapie **42**: 160, 1931.
4. Schinz, H. R., and Slotopolski, B.: Ergebn. d. med. Strahlenforsch. **1**: 443, 1925.
5. Ellinger, F.: Die biologischen Grundlagen der Strahlenbehandlung, Berlin, 1935, Urban & Schwarzenberg.
6. Fösterling: Quoted by Flaskamp, W.: Roentgenschaeden, Berlin, 1930, Urban & Schwarzenberg.
7. Muller, H. J.: The Science of Radiology, Springfield, 1933, p. 305.
8. Colwell, H. A., and Russ, S.: X-ray and Radium Injuries, Prevention and Treatment, London, 1934, Oxford University Press.
9. Timoféeff-Ressovsky, N. W.: Ergebn. d. med. Strahlenforsch. **5**: 129, 1931.
10. Hanson, F. B.: Physiol. Rev. **13**: 466, 1933.
11. Oliver, C. P.: Quart. Rev. Biol. **9**: 381, 1934.

The Tolerance Dose

United States Atomic Energy Commission

January 5, 1945

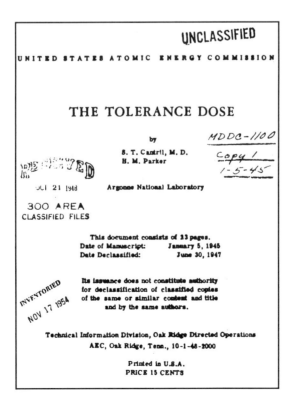

THE TOLERANCE DOSE

By S. T. Cantril, M. D. and H. M. Parker

DEFINITION OF TERMS

In the development of the science of radiotherapy, a special nomenclature has grown up, which, for the most part, is clear and unambiguous to the doctors and physicists engaged in the field. For special reasons, some of the quantities involved were defined in a different manner from that in which analogous quantities in pure physics would have been handled. The present project makes it a matter of general interest to correlate these two aspects. Some account of the terms used will therefore be given before proceeding to the main discussion.

Dose

Webster defines dose as: (a) The measured quantity of a medicine to be taken at one time or in a given period of time. (b) A definite quantity of anything regarded as having a beneficial influence. (c) Anything nauseous that one is obliged to take.

The radiotherapist presumably accepts definition (b) in considering the radiation effect on his patients, and definition (c) in considering the effect upon himself.

Webster's definitions are based on the simple picture of the swallowing or injection of a measured quantity of material. With rare exceptions, such a quantity is retained by the body for a period long in comparison with the giving of the dose. The radiation case is quite different. When the body is subjected to X radiation or gamma radiation, some part of the incident energy is converted into kinetic energy of secondary electrons. In general, the major part of the incident energy is transmitted without interaction. It is assumed that the tissue is affected only by the energy absorbed. Dose, in the sense used in radiotherapy, refers then to the energy absorbed in the tissue. In principle, dose could be measured directly in terms of energy absorption per unit volume, but the practical difficulties are great, and it is better to determine dose indirectly by ionization measurements under certain prescribed conditions.

The Roentgen

The principles of dosimetry indicate that the ionization per unit volume arising in a sufficiently small cavity in an absorbing medium subjected to X or gamma irradiation is approximately proportional to the energy absorbed per unit volume in the medium at the same point. Although this relation would in itself provide a feasible method of dosimetry, it was avoided in the setting up of international standards because it requires the use of an ionization chamber with "tissue-walls." No general agreement about such a tissue wall could be reached. It was, therefore, decided to use the ionization inside an air-wall or air-equivalent wall cavity as the standard of reference. In this system the unit dose, called the roentgen was the quantity of X or gamma radiation that liberated 1 esu of charge per unit volume of standard air in such a hypothetical air-wall ionization chamber.

The merit of the system was that over the range of wavelength used in radiotherapy, the energy absorbed per gram of soft tissue was sufficiently accurately the same as that absorbed per gram of air.

The correlation broke down when the biological material concerned was either skin, which contains enough sulphur to give an important photoelectric contribution, or bone, in which the air-walled cavity gives an entirely erroneous picture of the energy absorption. With these exceptions, it became common practice to state that the roentgen corresponds to an energy absorption of 83 ergs per gram of tissue or to the production of 1.61×10^{12} ion pairs per gram of tissue.

When neutron therapy began to be used, the air-wall chamber method of measurement broke down. It was relatively easy to make an air-equivalent material with respect to X or gamma rays since this depended only on the electron density and the artificial matching of the photoelectric effect over a wide enough range. For fast neutron irradiation, the hydrogen content of the chamber wall became the prime factor. So far the tentative methods of neutron dosimetry have used the following devices:

a) A pure carbon wall, chosen because of its reproducibility (L. H. Gray).

b) The Bakelite wall, coated with Aquadag, of the customary 100 r Victoreen condenser r-meter (the n unit of P. Aebersold).

c) Pure water (presumably as ice?). Another generally available substance, but this time with approximately the correct amount of hydrogen (L. H. Gray).

d) Basic dosimetry with different walls and chamber gases to deduce the tissue-equivalent dose (L. H. Gray).

e) Paraffin wall with ethylene gas. (E. O. Wollan).

f) The balanced double ionization chambers filled respectively with argon and a gas rich in hydrogen (Wollan, Gamertsfelder, Parker).

In the meantime, many writers have extended the roentgen as an abbreviation for 83 ergs per gram of tissue of 1.61×10^{12} ion pairs per gram to include neutron irradiation, beta radiation, and any radiation that produces ionization in tissue. This unit has been variously specified as the "tissue-roentgen," which restricts to tissue a device usefully applied to water, polystyrene, etc., as the "roentgen-equivalent" or "equivalent-roentgen," which is misleading because the biological effect is certainly not equivalent and as the "e" was used in Europe for a unit similar to, but not identical with the "r." The present writers prefer to use the expression "roentgen equivalent physical" (rep) for the time being. If such a unit ultimately proved to have any advantage over the direct statement of energy absorption in ergs per gram, it might be a more suitable vehicle for the honoring of the name Bragg than of Roentgen, since it depends on the Bragg principle of dosimetry. Frequently there is no confusion in writing "roentgen" instead of "rep" when the latter is clearly implied.

"Rem"

Another concept enters into the discussion when tolerance dose is considered in relation to varying radiations. This is the variation in effect upon similar tissues produced by an equivalent energy absorption of different types of incident radiation, i.e., alpha, beta, gamma, fast neutrons, etc. Thus, there has been formed the concept of a "biologically equivalent roentgen," namely a quantity of energy absorbed which produces an equivalent effect regardless of the character of the incident radiation. The logical term to express this concept is "roentgen equivalent biological," abbreviated "reb." This, in speech, can be so easily confused with "rep," that we have in use adopted the abbreviation "rem" which can be understood to imply "roentgen equivalent man, mouse, or mammal," depending upon the biological effect under discussion.

Dosage Rate

By convention, dose per unit time is called dosage rate. Some writers prefer dose rate. There is an alternative expression using exposure instead of dose, and exposure rate instead of dosage rate. The essential point is to distinguish between dosage rate and intensity. The former is essentially a measure of energy absorption per unit time, the latter is the energy flux. The relation between dosage rate and intensity is therefore a function of the absorbability of the radiation in question. The more recent radiological literature has carefully observed this distinction, but the older papers frequently expressed intensity in terms of r per unit time.

Tolerance Dose

Webster has "tolerance" invariably a noun, the possible relevant meanings being (a) the act of tolerating, quality of being tolerant, or (b) constitutional or acquired capacity to endure a shock or poison, etc.

Some objectors to the double noun use "toleration dose," an equally formidable double noun. Presumably, "tolerable dose" was originally intended. "Permissible exposure" would be a suitable name in the exposure nomenclature. For the present purposes, tolerance dose will be assumed to be that dose to which the body can be subjected without the production of harmful effects. It is not self-evident whether dose as used here refers to a total dose or to the elements of dose in a given period of time. This will be discussed later. The present writers will take the latter view and further specify that the given period of time shall be one day.

Tolerance Dosage Rate

Tolerance dosage rate has to be interpreted as the dosage rate that is continuously tolerable. The present writers will say that the daily tolerance dose is 0.1 r (in general). If one writes "the tolerance dose is 0.1 r per day," it is argued that this expression is dimensionally a dosage rate, and that one should write "the tolerance dosage rate is 0.1 r per day." There is here a difference in attitude which can be resolved only by a resumé of the manner in which tolerance dose has come into the literature. (Editor's note: Since this was written, the term "maximum allowable exposure" has been temporarily agreed to as the most acceptable term for "tolerance dose.")

Tolerance Dose Versus Tolerance Dosage Rate

The question of interpretation between dose and dosage rate ultimately leads to far-reaching differences of opinion concerning the permissible exposure of the body. The origin of the difficulty is closely related to the development of radiotherapy since 1928. The current International Recommendations for X-Ray and Radium Protection read as follows:

"The evidence at present available appears to suggest that under satisfactory working conditions, a person in normal health can tolerate exposure to X rays or radium gamma rays to an extent of about 0.2 international roentgen per day or 1 r per week. On the basis of continuous irradiation during a working day of seven hours, this figure corresponds to a tolerance dosage rate of 10^{-5} r per second."

It is clear that the persons responsible for these recommendations had in mind that it was immaterial whether the exposure was taken in equal daily amounts or whether it was averaged over a week. The earlier writers frequently quoted the permissible exposure per month. In all this, there was no restriction on the dosage rate other than the implication that the exposures received were such as would normally arise in X-ray or radium work. The crux of the problem is, then, the normal mode of receiving unwanted radiation. This occurred in four principal ways:

1) Fluoroscopists — The fluoroscopist was exposed to short bursts of quite intense radiation (of the order of 10^{-1} r/sec on the hands and 10^{-2} r/sec on the body).

2) X-Ray Therapy Technicians — In this case the technician was exposed to radiation for periods of about 5 to 30 minutes with intervals of the same order between treatments. Before the advent of self-protected tubes and special shielding, a typical dosage rate for these exposures would be 10^{-4} r/sec. At the present time, the exposure of X-ray technicians is so little that, in general, it adds nothing to our knowledge of tolerance.

3) **X-Ray Therapy Patients** — It is debatable whether patients should be included because their exposure is received over a period of weeks rather than years. However, in therapy, a large portion of the body receives a dose of the order of 1 per cent of that delivered to the treated part. Since it would not be uncommon to deliver 4000 r to each of two or three fields, the body can receive 80 to 120 r in a few weeks, at a dosage rate of about 10^{-2} r/sec. Such an irradiation is not without demonstrable effects, but it is believed to cause no permanent damage. A dose of this magnitude is a three-year quota of daily tolerance doses. A study of patients subjected to repeated courses of X-ray treatment would be instructive except that most of the patients so treated would die too quickly of other causes.

4) **Radium Therapists and Technicians** — These men were exposed for periods up to one hour at irregular intervals, with a background of perhaps 2×10^{-6} r/sec through the working day. The highest dosage rate normally encountered would be about 10^{-3} r/sec.

In all cases it appears that the exposures from which the present knowledge of tolerance was derived were given in relatively short bursts at dosage rates mainly in the range of 10^{-4} to 10^{-3} r/sec. This should be sufficient evidence that no special significance should be attached to a tolerance dosage rate of 10^{-5} r/sec. The inclusion of this figure in the International Recommendations was, we believe, nothing more than a recognition of its convenience as a guide when protection measurements are made with a survey meter, calibrated in r/sec. All points at which the dosage rate is permanently less than 10^{-5} r/sec can be considered safe.

On the whole, the exposure of personnel regularly employed in a radiation-hazardous occupation will be more or less evenly distributed except for week-ends.* In addition, a large body of information on repeated daily exposures of patients has been built up. For these reasons, the authors propose to restrict the meaning of tolerance dose to daily tolerance dose. This procedure is somewhat arbitrary. There is nothing magic about a period of one day, and we would manifestly be in an absurd position to claim that 0.1 r can be delivered daily with any time distribution in the day, and that 0.2 r cannot be given every alternate day or 0.5 r or 0.6 r every working week. Nevertheless, there has to be some limit to the dose-time relation, and the daily limit is convenient. No restriction need be placed on the dosage rate at which the daily general body radiation is received. Conditions should not exist in which the body can receive the daily quota of 0.1 r in less than 10 seconds, and this automatically limits the maximum dosage rate to about 10^{-2} r/sec which differs by only a factor of 10 from the dosage rates giving rise to our general knowledge of tolerance.

Statement of Tolerance Dose Levels

The tolerance dose levels which we have accepted as a working basis and which will be discussed in more detail are:

X and gamma radiation	0.1 r per day
Beta radiation (external)	0.1 rep per day
Fast neutron radiation	0.025 rep per day
Radon concentration in working rooms	1×10^{-14} curie/cc
Radium deposited in the body	0.1 microgram

---

*However, radium technicians in the larger institutions are frequently "on radium" and "off radium" in alternate months. It is fairly well established that freedom from radiation for four weeks gives the blood a chance to recover from potential damage.

MDDC - 1100

THE HISTORY OF THE TOLERANCE DOSE

A severe case of X-ray dermatitis was described in July 1896, only a few months after the discovery of X rays. It was not until 1902 that Rollins[1] attempted to formulate some idea of a tolerance dose. He suggested that "if a photographic plate is not fogged in seven minutes, the radiation is not of harmful intensity." In present day terminology this would amount to perhaps 10 to 20 r per day delivered by soft X rays. The early injuries of radiation were largely those of the skin, but the demonstration of the marked radiosensitivity of the blood forming organs (1904–1905) and of the reproductive organs of animals (1903–1904) carried some warning that more serious damage than dermatitis could be anticipated. The first organized step to insure protection from X rays was made in 1915 by Russ,[2] who read a paper on protective devices before the British Roentgen Society.

"Because of the war activity which existed then, this plan failed to bring forth important advancement. As a result of war demands, caution gave way to action and protective measures were again forgotten. Taking increased risks at this time probably was a factor which contributed to an unfortunate development in 1919 and 1921, both in this country and in Europe, when a number of prominent radiation workers died of apparent radiation injuries, particularly aplastic anemia. Unfavorable publicity developed, and definite action resulted."

The American Roentgen Ray Society formed a committee in 1920 to recommend protection measures, which were formulated and published in September 1922. The British X-Ray and Radium Protection Committee presented its first recommendation in July of 1921. The two sets of recommendations were quite similar and dealt largely with protective materials recommended for use in building X-ray and radium laboratories and apparatus.

At the first International Congress of Radiology held in London in 1925, the question of X-ray and radium protection was considered but no definite action was taken. At the second International Congress held in Stockholm in 1928, definite proposals were adopted and subsequently an International Committee on X-Ray and Radium Protection was formed. The recommendations adopted by the International X-Ray and Radium Protection Committee contained no reference, however, to a tolerance dose, merely stating that the known effects to be guarded against were: injuries to the superficial tissues, derangement of the internal organs, and changes in the blood. The report of this committee in 1931 likewise contained no statement of a tolerance dose, but in two subsequent reports (1934 and 1937) the tolerance dose is stated as 0.2 r per day.[4]

It is of interest to search for the basis on which this tolerance dose figure was established. From 1925 to 1932, various individuals published their own opinion on the tolerance dose. A somewhat detailed appraisal of the basis of these opinions is warranted here.

In 1925, Mutscheller[5] published a tolerance dose figure of .01 of an erythema dose* per month, and to quote his publication:

"Several typical good installations and fair averages were taken as a basis for calculating the dose to which the operators are now exposed during the time of one month. Thus it seems that under present conditions and standards accepted at present, it is entirely safe if an operator does not receive every 30 days a dose exceeding .01 of an erythema dose, and from the present status of our knowledge this seems to be the tolerance dose for all conditions of operating roentgen ray tubes for roentgenography, roentgenoscopy, and therapy. This dose, however, is derived from the average of a limited number only of typical examples, and is perhaps not yet sufficiently checked biologically and so it may happen that in the future this dose will have to be changed either to a larger or a smaller practical tolerance dose."

In 1928, Mutscheller[6] again published the same tolerance dose figure, and in 1934 the same figure of .01 of an erythema dose was published for "rays of higher penetration" used for therapeutic

---

*An erythema dose is one which produces a perceptible reddening of the skin.

application.[7] The erythema dose for this quality of radiation was given as 340 roentgens. Hence his tolerance dose was then 3.4 r per month 0.1 r per day.

Glocker and Kaupp[8] in 1925, and acting for the German Committee on X-Ray and Radium Protection, published the same figure as .01 of an erythema dose which they took directly from Mutscheller.

Sievert[9] published in 1925 one-tenth of an erythema dose per year as a safe dose, based again upon laboratory and hospital measurements.

Barclay and Cox[10] in 1928 published a figure of .00028 of an erythema dose as a daily exposure which could be tolerated without effect. They arrived at this figure in the following manner. Two people who had been chronically exposed without known damage were taken as a basis. One was an X-ray technician who had worked for six years and it was judged that the daily exposure which she received was .007 of an erythema dose. The second was a radiologist whose daily exposure was appraised at .0023 of an erythema dose. Barclay and Cox then arbitrarily took 1/25th of the daily exposure to which the X-ray technician was judged to have been exposed over the six-year period, namely .007 of an erythema dose, and arrived at their tolerance figure of .00028 of an erythema dose per day. It should be noted that in both examples exposure was to soft X rays and that the safety factor of 25 was purely arbitrary.

Failla[11] in 1932 published a report on the "tolerance intensity" to which his technicians had been subjected in operating a 4-gram radium installation. This was of the order of .001 of a threshold erythema dose per month (threshold erythema dose taken as 600 roentgens). The measurement of this intensity was done with photographic film by Dr. Edith Quimby. The longest term of employment had been four and a half years. No ill effects from this level of radiation were noted in the technicians. He accordingly adopted .001 of an erythema dose per month (.6 r per month or .02 r per day) as the "tolerance intensity" for gamma rays. In referring to the previously published tolerance dose levels, Failla points out that they were based upon soft X-ray radiation, having a lesser penetration than λ rays and hence less potentiality for damage to internal organs. Since protection from soft X rays is readily obtained, Failla suggests that the figure of .001 of an erythema dose per month be accepted as tolerance for both X rays and λ rays.

Kaye[12] brought together in 1928 the combined opinions on the tolerance dose and converted the figures to .001 of an erythema dose in five days as an average value.

Quality	To produce erythema
Grenz rays	100 r
100 kv	350 r
200 kv	600 r
1000 kv	1000 r
λ rays (radium)	1500 r

The dependence of erythema on quality of radiation is illustrated by the approximate values tabulated. Thus it may be seen that for soft scattered X rays (which were those considered by Mutscheller), the erythema dose may be only 1/5 of that for λ rays of radium which formed the basis of Failla's figure.

The International X-Ray and Radium Commission, acting for the International Congress of Radiology, set the tolerance dose level at 0.2 roentgen per day in both 1934 and 1937 recommendations.[4] As an outgrowth of this International Commission there was formed in the United States an Advisory Committee on X-Ray and Radium Protection which published its first proposals[13] in 1931 in the Bureau of Standards Handbook No. 15. Here the tolerance dose was set at 0.2 roentgen per day. In a later report of this American Advisory Committee (Bureau of Standards Handbook No. 20), the tolerance dose is stated as 0.1 roentgen per day, no explanation being given for this reduction in tolerance dose. In a subsequent publication in 1941 on the subject of Radiation Protection, Taylor,[14] who is Chairman of the American Committee, referred to the safety value of 0.02 roentgen per day. This latter figure, however, is not the combined opinion of the American Advisory Committee but was published independently by Taylor.

Wintz and Rump[15] in a League of Nations Publication of 1931 reviewed the various statements of tolerance dose and came to their own conclusions that the admissible dose is 10^{-5} r/sec assuming an eight-hour working day and 300 working days per year. This amounts to $\sim$.25 r per day. They qualify this dosage rate for persons remaining in proximity to sources of radiation giving off rays without intermission (radioactive preparations) by reducing it by a factor of 3 (i.e., 0.1 r per day). Both Failla and Wintz and Rump have thus specifically referred to λ rays in defining their tolerance dose. The American recommendations[18] of 0.1 r per day refer to both X and gamma rays.

The difficulties of establishing a tolerance exposure level can be ascertained from the history of its development. A clear-cut experiment on a large scale is virtually impossible to conduct because there would always arise various degrees of abnormality (in blood levels) which would have no relation to radiation exposure. In this respect Taylor concludes:

"Obviously, the determination of this tolerance dose is difficult and at best uncertain. The biologic factor differs too greatly among individuals to permit the use of a sharply defined tolerance. To be well beyond the danger limit, one must apply a generous safety factor to the result of any physical measurements."

THE BIOLOGICAL ASPECTS OF THE TOLERANCE DOSE

Experience thus far has taught that certain fundamental biologic trends will influence the allowable exposure to radiation. This information has been gained through both therapeutic and experimental work with radiation. The problem of tolerance dose is largely concerned with the radiosensitivity of tissues, and for that reason a discussion of radiosensitivity is briefly included here.

Radiosensitivity of Tissues

Radiosensitivity — By radiosensitivity is meant the relative vulnerability to radiation of a tissue living in its normal physiological environment. Although we tend to think of each tissue as having its own inherent radiosensitivity, advances in the application of radiation to medical uses have come about largely by learning to adapt the techniques of exposure to take advantage of the varying sensitivities of different tissues.

Not only may we think of different tissues as having differing radiosensitivities, but different organisms react differently to the same ionizing dose of radiation. There are also variations within strains of the same species. This is one of the obstacles in carrying over to man conclusions based on the biological effects of radiation found in lower animals.

The problem is further complicated by various biological events which can alter the radiosensitivity of a given tissue. A few examples will serve to bring out certain of these factors which are known.

Differentiation — Contrary to the usual principles of pharmacological action, the tissues which are less specialized in function tend to be the more vulnerable to radiation. The degree of specialization of a tissue is referred to as its differentiation. The highly complex cells of the nervous system are apparently less affected by ionizing rays. At the other extreme, the primitive cells of the reproductive or lymphatic system are extremely vulnerable.

Rate of Growth — In general, the more rapidly growing and active cells tend to be the more radiosensitive ones in a given tissue.

Cellular Environment — The composition of the medium or the environment of the cells comprising a tissue strongly affects the radiosensitivity. This is closely associated with the complex physiochemical alterations which must ensue within the cell when it is subjected to unnatural ionization. Whether the effect of the ionization is a direct one, taking place within the cell, or an indirect one resulting from alterations in the environment, is still largely a matter for conjecture, although evidence is accumulating to show that both mechanisms may be active. As an example of the effect of environment upon radiosensitivity, one may cite the diminished effect of radiation upon otherwise extremely radiosensitive tissues when they are subjected to a reduced oxygen supply during the time of exposure. Likewise, there is experimental evidence to show that a change in the acid-base relationship, affecting the permeability of cell membranes, can, for certain tissues, increase their radiosensitivity. Physical factors such as heat, cold, or previous radiation may alter either the growth rate or environment of the cells, and thus produce a change in their vulnerability to ionization produced in them or in the medium in which they live.

Threshold and Nonthreshold Effects — If one plots a dose-effect graph for various tissues subjected to radiation, there are, in general, two forms which the graph may take:

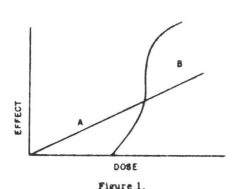

Figure 1.

Curve A illustrates the nonthreshold case, where, as the dose is increased there is a linear increase in the effect. There is no initial threshold of dose which must be exceeded before an effect is obtained. To recognize a nonthreshold effect, it must be readily observable or measurable after exposure to minimal amounts of radiation. An example is the influence of radiation upon the germ plasm of lower organisms.

Curve B illustrates a threshold effect. Here the effect is not measurable by present methods until a certain threshold of dose is exceeded. Threshold effects are not linear with dose but assume some form of S curve. The effects of radiation upon the skin and the blood-forming organs are examples. Until the dose reaches or surpasses the threshold, the first signs of skin effect (erythema) or of effect upon the blood-forming organs (as reflected in the circulating blood) are not seen.

The majority of radiation effects are thought to be of the threshold type. It may be that as more delicate indicators are found to measure effects, more of them will be seen to be of the nonthreshold type.

Reversibility of Effects — The reversibility of radiation effects is important, particularly in occupational exposure. By reversibility is meant the return of a tissue to its previously normal state after exposure is discontinued. The reversibility of any specific effect is dependent upon the reparative or regenerative properties of the tissue. Some tissues, such as skin, the blood-forming elements, membranous linings of the body cavities or glands, and peripheral nerves, are endowed with a special type of repair mechanism. Other tissues, such as brain, kidney, and lens, have no repair mechanism. In them, repair is by the formation of a scar, which does not take over the function of the original tissue which it replaces. The effects in such cases are said to be irreversible.

In order for an effect to be reversible, it must not produce injury beyond the limits of the normal capacity for regeneration. Otherwise the effect is permanent and may lead to complete destruction or exhaustion of the tissue.

Both the total dose and the total time over which it is given may affect the ability of the regenerative processes to function. If the total dose is excessive, irrespective of the time over which it is administered, regeneration and repair may be impossible. On the other hand, a total dose which will produce reversible effects if given at a rate slow enough to permit regeneration may instead result in irreversible damage to the tissues if given over a shorter period.

A tissue which has returned to apparent normal function following radiation damage may not, however, sustain repeated damage and may be unable to regenerate completely. Repeated radiation effect, initially followed by repair, will eventually exhaust the reserve for regeneration and end in death of the tissue. Hence, previously sustained radiation injury (for example to skin) which has apparently been followed by regeneration and return to normal function must be carefully observed and a repetition of of the injury avoided. Even bone marrow, in spite of its remarkable powers of recovery from radiation injury, will eventually exhaust its recuperative reserve under too frequent or too heavy doses.

General Body Effects

For purposes of simplification, the general body effects will be considered from the standpoint of (1) external radiation and (2) internal radiation. Although the effects are similar in many respects, the source and route of administration of radiation has some bearing on the tolerance dose.

The early toxic signs in man resulting from external radiation are those of (a) general lassitude and fatigue and (b) early demonstrable effects upon the leukocytes of the blood. Radium workers who are subjected to continued overexposure develop a lassitude which is out of all proportion to the physical requirements of their work. This is an effect which is real and should be cause for investigation of possible exposure. From the standpoint of protection, it is fortunate that the leukocytes of the blood furnish so readily available an index of overexposure. An extensive report on the effects of radiation on the blood and blood-forming organs has been given elsewhere (CH^{-410}) so that it will suffice here to briefly describe these early effects.

There is either a diminution in the total number of white blood cells (leukopenia) or the total number of white blood cells may remain within normal limits but there is an altered ratio in the proportion of neutrophils to lymphocytes. This latter alteration in ratio producing a lymphocytosis is more likely to appear with lesser degrees of overexposure. It is not an uncommon finding among radium workers to see an initial instability in the total white blood count for a period of months which later stabilizes on a lower level than was present before exposure began. The red blood cell elements do not participate in the early effects of continued overexposure, but a diminution in the number of red blood cells (anemia) does appear as a late effect of continued overexposure. Fatal anemias which do not respond to any form of treatment have appeared in a considerable number of radiation workers after long continued overexposure. This is a manifestation of bone marrow exhaustion in which repair has not been able to keep up with the continued insult produced by overexposure.

A regular and carefully done blood count on personnel exposed to radiation is thus indicated and serves as an early index of possible overexposure.

By internal radiation is meant radiation received through the ingestion or inhalation of radioactive substances. The earliest experience with internal radiation was obtained through studies of inhaled radon both in animals and in man. The early effects of continued inhalation of radon in excess of the tolerance limits are those upon the blood. Studies of the blood of workers concerned with the separation of radium show an eventual effect upon the blood-producing organs. The magnitude of the exposure can be correlated with the level of radioactivity of the expired air.[17] Similar studies have been made upon the workers concerned with the preparation of mesothorium.[18] The late effects of continued inhalation of excessive amounts of radon are unfortunately known for man. The high incidence of lung cancer in the Schneeberg, Sacony, and Jachymov mines of Bohemia has been directly attributed to breathing air containing radon, which gives an alpha bombardment to the lung tissues. The incidence is high. In one report,[19] out of a total of 89 deaths occurring over a period of 10 years among 400 miners, 60 autopsies were obtained and 42 of the deaths were shown to be due to primary lung cancer. This was an incidence of 9.7 ± 1.5 per year per 1000, or about 30 times the normal expectancy. Of 48 mice kept for a year in the Schneeberg mine, 28 died, 7 having developed tumors of the respiratory organs.[20] The average radon content of the air in these mines is of the order of 3×10^{-13} curie of radon per cc. On the basis of these figures, it is of interest to calculate the internal daily dose to the lungs which resulted in these findings.

We calculate the lung dosage by a combination of the methods of R. D. Evans[21] and Failla.[22] Evans has given a good picture of the lung as a series of tubes of specified length and diameter. From the total absorption of the alpha radiation from radon and its products in the first 35 μ of the tube linings, Evans has assessed the average ionization per cc throughout this lining tissue. Failla in a discussion of beta-ray effects in the lung points out that "lung" cancer in man is essentially carcinoma of the bronchus. It should therefore be correct to calculate the dose in the bronchus only. This dose is approximately 0.3 rep per day, approximately 20 times that derived as an average throughout the lung. The validity of this method might be checked by the autopsies on the Schneeberg miners, but the site of origin of the carcinoma is not given in the published reports. On the basis of ionization, and allowing for the heavy particle nature of this radiation, 0.01 rep per day would be used as the tolerance dose. This would correspond to a tolerance concentration of 10^{-13} curie per cc. This, in fact, is the concentration widely accepted in Europe and used in several states. The permissible tolerance concentration of radon recommended by the American Bureau of Standards is 10^{-14} curie per cc of air. In view of the uncertainties of the lung calculations, this additional safety factor seems to be well chosen.

The radioactive gases which in the past have been an occupational hazard were the emanations from radium and mesothorium. Our present activities bring us in contact with other new radioactive gases (xenon, argon, krypton) and a radioactive vapor (iodine). In the case of xenon, it is necessary to consider the hazard both from the standpoint of internal radiation (inhalation) and external whole body radiation from the beta and gamma rays of xenon in the atmosphere. Calculations indicate that the external radiation from radioactive xenon determines the permissible concentration in the atmosphere, which has, for the purpose of plant design, been set at 2×10^{-14} curie/cc.

The tolerance value for radioactive iodine in the atmosphere is based upon the selective absorption and deposition of iodine in the thyroid gland. Calculations on the permissible atmospheric concentration of radioactive iodine, for exposure over 24 hours a day, has been tentatively placed at 1.0×10^{-13} curie per cc.

Internal radiation from ingestion and deposit of radioactive material in the body was first encountered by Martland[23] among the radium dial painters. The history of this occupational disease is well known. About 98 to 99 per cent of the radium which was ingested through the habit of pointing the brushes in the mouth was excreted, but the remainder was deposited in the body, largely in the bones. As little as 1 to 2 micrograms has proven to be a lethal dose. Here again the continued bombardment of bone marrow results either in a fatal anemia or in the production of malignant bone tumors. The late signs of damage, anemia, or malignant bone tumors, may be evident some years after exposure has resulted in the deposit of radium. Martland does not record the early signs of radium deposit, as he had no opportunity to do so. It is possible, however, to detect radioactivity in the expired air as an early check on overexposure, and this is done routinely by the more cautious plants producing luminous dials. The alteration in blood count as an early sign of overexposure to radon was noted before. The same finding could in all probability be detected as an early sign in radium poisoning if it were looked for. There is apparently a compensatory activity of the bone marrow which attempts to keep pace with the tissue destruction produced by the continued alpha bombardment but eventually leads to bone marrow exhaustion and fatal anemia. Experience with radium poisoning in the luminous dial industry has led the Bureau of Standards to establish 0.1 microgram of radium as the limit which can be deposited in the body without resulting in later damage.[24]

The hazards of internal radiation from ingestion or inhalation and eventual deposit of radioactive materials in the body cannot be overlooked in the present undertaking. The fission products are the source of this radioactive material. Studies on the absorption, deposition, and excretion of the various fission products are being made and reports have appeared by Dr. Hamilton and his associates. Tolerance values for the limits of the fission products which could be safely taken into the body cannot be set until a complete study has been made of all the fission products and their effects are known. With the exception of the gaseous fission products (xenon and iodine), protective measures must proceed on the basis of completely eliminating this hazard.

Skin Effects

The effects of radiation on the human skin gave the first indication of any biologic effect of X rays and gamma rays. Becquerel, who carried a tube of radium in his vest pocket for demonstration purposes, developed a reaction of the underlying skin. X-ray dermatitis was in evidence within a few months after the discovery of X rays.

The greatest number of radiation injuries have been those to the skin both in X-ray and radium workers. Following the early wave of damage to the skin which came in the first 15 years of X-ray and radium experience, there were more precautions taken to prevent skin damage. With special attention given to local protection, the number of injuries to the blood-forming organs increased due to lack of complete protection. An emphasis was then placed on whole body protection; the skin injuries again assumed first place and at the present time they are still appearing in unnecessary numbers. The majority of these are physicians or dentists; a fewer number appear in radium workers.

The characteristic effect of large doses of X ray and radium upon the skin is the production of a skin erythema. In this respect, they are comparable to ultraviolet radiation except that the latent period between exposure and erythema is delayed up to ~4 weeks with single X ray and radium exposure depending on the dose. As was previously noted, there is a variation on the dose-effect ratio to produce erythema with rays of varying quality, in the direction of a larger dose for shorter wavelength. As the length of time over which the radiation is administered is increased, the dose required to produce erythema becomes larger.

The erythema is the result of a dilation of the fine capillaries, venules, and arterioles supplying the skin. The mechanism is thought to be identical with the erythema produced by ultraviolet irradiation[25] having a wavelength of ~300 $\mu\mu$, which is the effective component of sunlight. The stimulus which produces the blood vessel dilation is thought to be due to a release of a "histamine-like" substance from the living superficial layers of the skin.

The skin is composed of essentially two layers of tissue—the epidermis which consists of the epithelial cells forming the protective covering of the body and the dermis lying beneath the epidermis, which consists of the supporting connective tissue for the epidermis, and carries the nutrient vessels, accessory organs, and nerve supply. The thickness of the epidermis, or outer cellular layer, varies over the body but, in general, ranges from 0.07 to 0.12 mm. On the palmar surface of hands and fingers, it averages ~0.8 mm, on the soles of feet ~1.4 mm. In men accustomed to heavy labor which produces a compensatory thickening of the skin of the hands, the epidermis may be considerably thicker. The outermost layer of the epidermis (stratum corneum) consists of dead, hornified, flattened cells. The thickness of this outer horny layer varies over the body, being thickest on the soles (~.8 mm) and palms (~.44 mm). On the backs of the hands and over the remainder of the body it is considerably thinner, but even a thickness of 0.1 mm would give protection against natural alpha and very low energy beta radiation. As cells are lost from this outer layer by normal wear and tear, the underlying growing layer of the epidermis continues to furnish a new supply. There are no blood vessels within the epidermis— the fine capillaries nearest the surface lying directly beneath the epidermis. Thus, erythema is produced below the epidermis. Damage to the skin (whether thermal, irradiation, or chemical) is produced in the living and growing layers of the epidermis (which are shed or desquamated following the injury), or if more severe, damage may also appear in the deeper tissues resulting in necrosis, ulceration, and thrombosis (closure of the vessels).

The thickness of the outer horny layer of the skin is of importance in relation to the penetrability of alpha and low energy beta radiation. Since the alpha particles of uranium have a range in tissue of ~.05 mm, one would not expect skin injury from them. When the energy of alpha particles is greater (as those from a cyclotron), penetration in tissue becomes sufficiently great to produce skin injury.[26]

The beta-ray penetration of tissue is discussed at some length in a report by one of us (CH-930). It is sufficient here to emphasize that beta particles of the average energy associated with long-lived fission products will penetrate well below the skin, and hence there is real potentiality for injury if due caution is not exercised to avoid overexposure.

The previous discussion of erythema applies to radiation injury of a relatively acute type. The dose required to produce erythema is relatively high. As the time over which the dose is administered is lengthened to one or more years (as in occupational exposure), a considerably higher total exposure can be tolerated. Erythema may be an early sign when the dose has been acute, but it may not appear at all when the exposure is spread over a period of years. Hence, one cannot rely upon the appearance of an erythema to judge whether or not overexposure is producing damage when the exposure is prolonged over years. Other signs which should prove a warning are a loss of the normal skin ridges of the finger tips, loss of hair on the back of fingers and hand, cracking, brittleness, or ridging of the fingernails, a loss of the normal sensitivity, and an abnormal dryness of the skin.

The late evidence of injury appears as telangiectasis (a permanent capillary dilation), pigmentation, atrophy of the skin and its appendages (hair follicles, sweat, and subaceous glands), and skin thickening with the appearance of wart-like growths. The most serious late sign of damage is ulceration, which usually results from minor abrasions which fail to heal. The ulceration in many cases eventually progresses to cancer of the skin which in some 25 per cent[27] spreads beyond the hands and proves lethal. The time between onset of exposure to the recognition of skin cancer in one series was an average of nine years. The age of these victims ranged from 28 to 60 years, whereas cancer of the skin appearing without relation to radiation is an uncommon disease in early and middle life.

Stone and Larkin[28] have found that 110 n will produce a threshold pigmentation of the skin from the 4th to the 19th day, which is an effect produced by 650 to 700 r of 200 kv X rays. Thus, for fast neutron exposure there is a ratio of effect varying by a factor of 6 for the skin, which is accounted for in the tolerance dose level for fast neutron exposure of 0.01 n (0.025 rep) per day.

Unfortunately, the workers who have been damaged have no record of the exposures received over a period of years. There is no general agreement on skin tolerance among radiologists or physicists who have medical experience. On inquiry from two professors of radiology of wide experience, the answers varied from 0.5 to 5.0 r per day of X ray to the hands. In contrast to our statement under General Body Effects, many radiologists and physicists would be willing to accept 1 r per day to the hands occasionally if special work make it necessary. Past experience with fluoroscopy and radium handling leads us to assume, however, that for long-continued exposure over a period of years, a limit of 0.1 r per day exposure of the skin to radiation which will effectively penetrate below the outer horny layer (0.1 mm) is a safe limit.

Effects on Reproductive Organs

The reproductive organs may sustain damage either to the germ plasm or to the cells which carry the germ plasm (ova or sperm). The most sensitive elements in the reproductive organs are the parent cells which eventually give rise to the mature ova or sperm. Other cellular elements in the reproductive organs which are concerned with internal secretion and control the desire for and ability to consumate the sexual act are relatively radioresistant. To obtain a permanent sterilization of the female ovary requires some 400 to 600 r delivered within the ovary. Sterilization in man is produced by 800 to 1000 r in the testes. There is a threshold of dose which must be exceeded before any effect upon fertility becomes manifest.

Experiments in progress by Heston and Lorenz[29] have already shown that continued exposure of female mice to 1 r per 8 hours daily produces ovarian follicular atrophy after approximately 540 r. Experiments by Russ and Scott[30] show that continued exposure to gamma radiation 20 times in excess of tolerance leads to a reduction in fertility in mice.

The possibility of damage to the reproductive organs is the most discussed and feared hazard in the minds of most nonmedical personnel who work with radiation. The incidence of reduced fertility following occupational exposure is not accurately known, but it is not great in comparison with damage to skin. The reasons for the keen interest in the subject can best be learned from Freud. But one cannot dismiss these fears lightly as animal experimentation indicates that at least in animals, the safety factor in 0.1 r per day may not be as great as 10. Whether one can transpose these results to the human is not known — are assuming here that they are applicable.

MDDC - 1100 [13

Radiogenetic Effects

In 1927 Muller[31] demonstrated that the mutation* rate of the fruit fly could be accelerated by exposure to X rays. Bagg and Little[32] and Snell[33] have produced radiation-induced mutations in mice. Radiation increases the rate of appearance of common mutations which occur spontaneously; it produces the uncommon ones only rarely. Single exposures of 30 to 40 r will double the mutation rate in the fruit fly; doses of 500 r are required to produce mutations in mice, and these appear in far lower incidence than in the fruit fly.

Radiation-induced mutations have been found to have characteristics which bear on the practical consideration of radiogenetic changes possibly associated with occupational exposure. The most important of these is that there is a linear relationship between dose and increase in mutation rate. There is no threshold effect—the cumulation of exposure is thus additive. Furthermore, the magnitude of the effect is independent of the wavelength and dosage rate of exposure.[34] These genetic studies (largely in the fruit fly) have lead to considerable discussion concerning their applicability to man in relation to tolerance dose. There are workers in the field who advocate a further reduction in the tolerance dose for X and gamma rays because of the nonthreshold effect of X rays upon the germ plasm of the fruit fly. Further concern is added to these fears by the possibility of even greater radiogenetic damage from heavy particle radiation (neutrons, protons, etc.) than from the more familiar X and gamma rays.

Mutations, whether spontaneous or produced by radiation, are about 90 per cent lethal or sublethal. This means that the offspring does not survive the gestation or hatching period, or dies shortly thereafter. The lethal mutations are either dominant or recessive. By dominant is meant that, for the exposed parent organism, the lethal effect appears in some of its direct offspring. By recessive is meant that the effect might appear only in some succeeding generation of the radiated subject. In man it would appear in the near descendants only should cousins or near relatives intermarry. It has been calculated from the laws of genetics that some 5000 years would be required for a mutated gene to meet another mutated gene descended from the original mutation. This would indicate that we need not be too concerned about the recessive deleterious effects of mutations.

The viable mutations (about 5 per cent of all mutations, whether spontaneous or produced by radiation) are about 95 per cent deleterious ones. Of these, the majority (about 96 per cent) pertain to other than the sex chromosomes. The remaining are sex-linked mutations,† appearing in the sons of the daughters of the sperm carrying the mutated gene. The spontaneous rate of appearance of these dominant deleterious yet viable mutations (which constitute about 4 per cent of all mutations) is about 1 : 2700. It has been calculated that an accumulated dose of 300 r in the female will raise this probability to 1 : 230.[34]

Our present knowledge of radiogenetics has been gained entirely from experimental studies in the fruit fly or other lower organisms, and to a lesser extent in mice. Actually, very little is known about genetics in man, and still less about the effects which might be produced by exposure to radiation. As yet there is no convincing evidence to indicate that the present generations of radiation workers have produced offsprings which differ from those of the general population. It can be argued, however, that succeeding generations or intermarrying may bring abnormalities to light. If experimental finding in the lower organisms are accepted as valid for man, then one can only escape some degree of radiogenetic effect by avoidance of all radiation exposure, including the natural radiation (cosmic, etc.).

This concept has lead Henshaw[3] to suggest that perhaps for the nonthreshold reactions one should define rather a "tolerance injury" than a "tolerance dose."

---

*A mutation is an hereditarily transmissible abrupt alteration in germ plasm.

†Based on the number of chromosomes in the female (48) and male (47) the chance is further reduced by a factor of 1 : 24 in the female and 1 : 47 in the male.

For those readers who wish to pursue the subject further, we would refer you to Henshaw's paper.[3]

THE DEPENDENCE OF TOLERANCE DOSE ON THE NATURE OF THE RADIATION

Specific Ionization

Experimental as well as clinical studies have shown that the biologic effect produced by the absorption of a given quantity of energy is dependent in part upon the nature of the incident radiation. The effects are produced in some manner by the ionization resulting from the impinging radiation, but the distribution of ionization is not the same for all qualities of radiation. A greater ion density along the path of an ionizing particle is associated with a more pronounced cellular or biological effect. The density of ionization per unit length of path is referred to as the specific ionization of a particular quality or type of radiation. One can then compare biological effects produced by equivalent energy absorption from X rays, gamma rays, neutrons, etc. A detailed and critical analysis of this phenomenon is reviewed by Zirkle.[35] We refer to it here because it is convenient to consider tolerance dose from the point of view of the type of radiation involved rather than in terms of the specific effect on the various parts of the body as in the foregoing section. The evidence comes from the same sources and some repetition is inevitable.

X-Rays

As our knowledge of permissible exposure has been derived almost exclusively from experience in hospitals, it is best founded for X radiation. In particular, the original choice of tolerance dose was founded on two cases, both involving the exposure of operators to scattered radiation only. The wavelength of this radiation can be taken as 0.3A. Hence its penetration in tissue is about 96 per cent at 1 cm deep and 25 per cent at 10 cm deep. (These figures are derived from a consideration of the scattering contribution in a body subjected to wide beam irradiation. The corresponding figures from the absorption coefficient alone would be 80 per cent and 8 per cent, respectively.)

The tolerance dose supposedly has a safety factor of the order of 20 to 40 for radiation of this type. Modern practice involves shielding the operator by lead screens, but it is permitted that directly transmitted radiation falls on the operator, up to the limiting tolerance amount. Under these conditions, the net effect on the deeper seated organs may be different. This is especially the case with the modern high voltage machines. Four hundred kv radiation is as common now as 150 kv was when tolerance was first established. Penetration of this radiation is about 100 per cent at 1 cm and 50 per cent at 10 cm (ninety per cent and 30 per cent, respectively, from the absorption coefficient) and we might consider its effect on the hemopoietic system as approximately twice as damaging.

Little has been written about the damaging effects of X radiation generated in the range between 400 kv and 1200 kv. Such hospital installations have, in general, been well protected and have not been operated many hours per day except in recent years. In this range of voltage there is little increase in the "percentage depth dose" for depths up to about 10 cm, for wide beam irradiation. The increased penetration of the primary beam is offset by the reduced contribution from scattering. It is also known that the effect of these radiations in therapeutic dosage is quite closely the same in deep tissue as that of the well studied 200 kv radiations. As far as skin effects are concerned, there are two familiar effects both of which indicate that for equal dose as usually measured, the damage by high voltage radiation should be less severe. It is reasonable to suppose that a tolerance dose established for 200 kv radiation with a margin of safety of 10 or more will still be safe for X radiation up to 1200 kv. The widespread industrial use of 1000 kv radiation in the present war would give a fruitful source of study in a few years if the exposures were adequately recorded. Unfortunately, although it is known that many operators are being exposed to radiation in excess of 0.1 r per day, good records are the exception.

MDDC - 1100 [15

Gamma Rays

Adequate experience with gamma radiation has been restricted to that from radium and its products, and again it is derived from handling in hospitals. It has long been accepted that the danger of exposure to gamma radiation exceeds that from X radiation for two reasons: (1) X rays can be "turned off" when not required, (2) the gamma-ray penetration is higher. Wintz and Rump,[18] who gave the first consideration to gamma-ray exposure, select a dosage rate of $1/3 \times 10^{-6}$ r/sec as against 10^{-5} r/sec for X radiation. The latter figure is based on an eight hour day and the former apparently implies possible exposure to gamma radiation for 24 hours per day. In the alternative expression of tolerance dose as a daily amount, Wintz and Rump would make no distinction. On the basis of penetration (94 per cent at 1 cm, 58 per cent at 10 cm without backscatter), one might have a factor of 2 or 3 to represent the additional total body ionization.

It is commonly supposed that the effect of gamma radiation on the body is greater than that for X rays for equal surface dose solely on account of the greater penetration. (It is assumed that the action of the radiation is entirely due to the secondary electrons liberated and that change of specific ionization over this range is not important). Hence the method of expressing tolerance dose in terms of the total ionization has recently gained favor, especially abroad. The unit employed for this purpose is the "gramme-roentgen." The body exposure on this basis is simply the integral $\int D(s) \, dV$, where $D(s)$ is the dose in roentgens as a function of position s in the body, and we take the density of tissue as 1. Now $D(s) = Kn(s)$ where $n(s)$ is the number of ion pairs per cc of tissue. The integral is kN where N is the total number of ion pairs produced. One gramme-roentgen corresponds to 1.6×10^{13} ion pairs.*

Simple computations will show that there is more change in total ionization as a result of geometry than results from the change from soft X rays to gamma rays. For equal surface dose we have

	Relative total-body ionization
Soft X rays from large distance	~1
γ rays from large distance	2.5
γ rays from point source at 100 cm	1.9
γ rays from point source at 10 cm	0.6

The gramme-roentgen point of view may have more significance under conditions in which only a part of the body is exposed to radiation. At the present time there is insufficient experience to apply the method in general. It is tacitly involved in the standard protective devices for assembling radium sources wherein the hands, forearms, and head are irradiated rather liberally. Emphasis on the gramme-roentgen implies that the principal cause of damage is the general body effect. Among the present group of radium workers, the principal observed effect is skin damage, particularly to the hands. This is believed due to greater exposure of the hands, rather than to greater sensitivity, but it does indicate that the permissible exposure is not safe by much more than one order of magnitude and hence that little is to be gained by elaborating the gramme-roentgen aspect.

Exposure to gamma radiation from internal sources can be treated in terms of the ionization produced in tissue. In general, the effect would be exceeded by that of the accompanying beta radiation.

Beta Rays

There is no sound evidence on the permissible exposure of the body to beta radiation. The handling of strong sources was not common until the development of artificial radioactivity, especially by the

---

*Dose in roentgens is frequently thought of as "dose per cc," which could lead to "total dose" in the above integration. This fundamental error should be eradicated. Dose in roentgens is a measure of the ionization per cc, and is independent of the size of field or volume of tissue irradiated (increased scattering, etc. is included in the measurement of 'r').

cyclotron. (This is not strictly correct inasmuch as cathode ray tubes, e.g. Lenard tubes, have been widely used, and these are potent enough sources. With such an installation it is relatively easy to maintain adequate shielding at all times. Consequently there seems to be no record of damage by prolonged exposure to these radiations. The numerous cases of exposure, accidental or otherwise, to stray beams for short intervals is not discussed here.)

It is generally conceded that the effect of X rays and gamma rays on tissue arises from the electrons generated in tissue by these radiations. Hence, for equal ionization, beta radiation should produce the same effect,* apart from a possible correction due to specific ionization differences. This has been sufficiently well confirmed by experiments at therapeutic levels. One can therefore confidently state that an exposure to beta radiation of 0.1 rep/day will be safe. Since the external beta-ray effect is confined to the skin or to tissue within perhaps 1 cm of the skin, it is evident that beta radiation will add little or no contribution to the general body effect. On this basis, many radiotherapists would permit exposures up to 1 rep day. In view of the high percentage of superficial damage among the present workers,† one doubts whether this is entirely justified. Nevertheless, if the full spectrum of beta radiation is allowed to contribute to the ionization, the limit 0.1 rep/day imposes rather tight restrictions on the handling of beta-active materials. It can be supposed that beta rays that fail to penetrate the hornified layer will be clinically insignificant. The thickness of this layer will be taken as 0.1 mm. Thus the provisional tolerance dose for external beta radiation will be 0.1 rep/day measured in a chamber of wall thickness 10 mg/cm^2. Damage arising from the ingestion or inhalation of active material has be assessed in terms of the ionization with zero absorber.

Neutrons

The effects of prolonged exposure to low intensity beams of neutrons is unknown. Many effects of higher intensity neutron irradiation of biological materials have been compared with X-ray or gamma-ray irradiation. In this work, the neutron dose has frequently been quoted in "n" units as the reading of a particular Victoreen condenser chamber. The ionizing effect of fast neutrons in the body can be considered equivalent to 2.5 r/"n." The biological effectiveness of neutrons relative to X rays varies widely for different materials or for different conditions in the same material. According to Zirkle,' the effects on mammalian tissues show effectiveness ratios ranging from 5 to 9 (in terms of r to "n"), although there is a distinct possibility that injurious effects on mammals exist with effectiveness ratios as high as 25.

From these figures, it has become common practice to consider 0.01 n (0.025 rep) as a tolerance dose of comparable safety to 0.1 r for X rays. The most conservative would quote 0.004 n, corresponding to 0.01 rep. The figure of 0.01 n (0.025 rep) per day appears to be adequately safe, if the supposed safety factor for the X-ray dose is as high as 20.

Whereas the possible effects of prolonged fast neutron exposures have to be deduced, with some show of logic, from the known therapeutic ratios, the effects of exposure to slow neutrons is at present entirely unknown. Presumably three factors should be taken into account:

1) The production of gamma rays in the body.

2) The production of protons by the neutron reaction on some of the constituent atoms.

3) The production of new atomic nuclei.

Of these three, the first can be taken care of by ionization measurements. Whether (2) or (3) will have biological significance at the exposure levels limited by (1) remains to be investigated. In general, the body would not be subjected to slow neutron irradiation without the admixture of gamma radiation or fast neutrons. One might anticipate that the permissible exposure would be limited by the total gamma radiation or the fast neutron effect, before (2) and (3) became important.

---

* For a more detailed consideration on some of the physical aspects of the effects of beta radiation on tissue, the reader is referred to CH-930, H. M. Parker.

† Workers in industry using X ray, radium, etc.

MDDC - 1100 [17

Alpha Rays

For external radiation, the penetration of natural alpha particles is well below the thickness of the "absorbing layer." The damaging effect is assumed zero unless the intensity is such as to physically burn the part. There is now some clinical evidence to substantiate this point as uranium sheet has been worn in contact with the skin for several months. The observed damage was nil for this exposure corresponding to $\sim$ 250 rep/day, although possible late damage has to be considered.

Internally, the effects are computed in terms of the ionization produced. Since the increased biological effectiveness of neutron-produced ionization is believed due mainly to the high specific ionization along the proton tracks, it is to be expected that alpha particles would be even more effective. A tolerance dose of 0.01 rep/day would be reasonable.

Since the effects are limited to the ingestion or inhalation of alpha-ray emitters, they have been adequately treated as the special problems of bone and lung damage.

Protons and Other Particles

Accelerated particles can produce damage to the skin and superficial structures. Such exposures occur as occasional accidents. It is not appropriate to discuss these hazards in terms of tolerance dose. If it became necessary to establish a tolerance dose for protons, the figure of .025 rep/day would be proposed by analogy with the fast neutron exposure, which is essentially a proton action.

Combined Radiations

The body will, in general, be exposed to more than one type of radiation either simultaneously or at various times during a career associated with radiation work. In the absence of contrary evidence, it will be supposed that the summated daily tissue ionization, properly weighed for the specific ionization factors that have led to different tolerance doses for the heavy particle cases, should not exceed the equivalent of 0.1 rem/day. In general, one would have

$$\left[I_x + I_\gamma + I_\beta + 10 I_n + 10 I_\alpha \; 4 I_p \right] \leq 0.1 \text{ rem/day} \tag{1}$$

where the terms represent the ionization contributions of the various radiations as measured, say, in a Bakelite chamber. In a "tissue" chamber, I_n would have a weight factor of only 4, and all the terms would have the significance of "energy absorption" measurements. Slow neutron effects would contribute to I_γ and I_p in this case.

As a working policy, this formula is of the little use at the present time because the composition of the mixed beam will be unknown. If n, γ, or p terms occur, the summated ionization will have to be kept low on account of the large weighing factors. For example, with I_γ and I_n terms alone let I_γ = .09 r and I_n = .001 n. Then $I_\gamma + I_n$ = .09 + 10 x .001 = 0.1 rem. But if the beam composition is unknown, one must ascribe all the radiation to the more dangerous component. Thus,

$$I_r + I_n = 0 + 10 \times .091 = .91 \text{ rem}$$

Or the protective measures are made too stringent by a factor of over 9. Even if one could assume that the observed ionization was divided equally between the gamma and n effects one has

$$I_\gamma'' + I_n'' = .0455 + 10 \times .0455 = \sim .5 \text{ rem}$$

So that the damaging effect is still exaggerated by a factor of 3. The value of analyzing the radiation into its components is apparent, especially the determination of the upper limit of n, α, and p terms.

Some Geometrical Considerations

The estimation of the dose received by the body is normally founded on the reading at one region, e.g., the chest. Conditions can arise under which such a reading may be unreliable. The principal causes are:

1) directional radiation

2) limited or subdivided beams

Under (1), it is clear that anterior and posterior chambers on the body could give readings differing by a factor of 10. If, for example, the beam always entered the back and was always read by an anterior chamber, a serious error would be made. Such conditions can arise where an operator maintains a fixed position at a control desk and is inadvertently irradiated from an unfavorable direction.

Under (2) one has such well-known effects as the overexposure of hands and head while assembling radium source behind a lead screen. The recording chamber would, in general, be shielded by a factor of, say, 10, for equal distance. Then, since the hands might easily be much closer to the source, the net effect could be $\sim$ 150 times greater than that recorded. If the source were mixed beta and gamma, an additional factor between 10 and 50 would come in. Another aspect of limited beams arises in the shielding of equipment which requires controls to be brought to the outside, and in the use of boxes and vats with removable lids or plugs. Good design of such boxes calls for the elimination of strong narrow beams. Great attention has been paid to the blocking of fine beams in X-ray installations. This is relatively easy where the protective thicknesses involved are of the order of a few millimeters of lead. The same precautions are necessary in a large-scale project. A large shield perforated by fine holes in a regular pattern is a case in point. One is compelled to limit the permissible exposure to that corresponding to the high dosage rate in each subdivided beam, for two reasons:

a) Although it is known that a considerably higher dose can be given to a small field than to a large one, the difference is not great enough to justify a change in tolerance dose. Moreover, the clinical picture is complicated if there is a regular pattern of small fields.

b) If the exposure is measured on the person, there is the risk that the measuring device will be continuously carried in an unexposed region. For example, a man could walk by a long shield with holes on an 8-inch square lattice without exposing a film or chamber. A trivial solution of this case is the provision of a suitably sloping floor, but, in general, the possibility of patterned exposure must be excluded. These problems are concerned as much with protection as with tolerance dose and cannot be elaborated here. It is sufficient to indicate that the use of any piece of equipment capable of giving narrow emergent beams invalidates the record of pocket meters in the vicinity.

Measurement of Dose

The limitation of permissible exposure to .1 rem equivalent per day in the sense of equation 1 is futile unless steps be taken to insure that the personnel receives no more than this amount routinely. This can be done in two ways:

1) All fixed radiating sources must be shielded so that the radiation escaping is well below the tolerance level.

2) The exposure of a person in a potentially dangerous area must be recorded by a suitable device on the person.

THE LEGAL STATUS OF THE TOLERANCE DOSE

The legal status of radiation safety recommendations is discussed by Lauriston Taylor.[34] He states that the "legal status of roentgen ray safety recommendations was brought up at the outset and it is important to note that in no country do such recommendations have strictly legal recognition." The complications of this were early recognized and the British Committee, for example, felt that public opinion could be just as effective as word of law. Moreover, by not having a legal status, the recommendations could remain flexible and be readily changed to suit changing conditions. Laws, once formed, change slowly, and in the matter of roentgen ray protection, such inflexibility may lead to complications.

"The method found by the British Committee to be most effective in bringing about acceptance of the recommendation, was to put the power of inspection and approval in the hands of the National Physical Laboratory*....."

"In this country, the same conditions* did not exist, and, while recommendations have been accepted,† there is neither legal status nor the authorization for a central laboratory to put weight behind an enforcement. The charter of the National Bureau of Standards does not foster any general outside inspection of activities, nor have funds ever been authorized for the purpose. This bureau does, however, test and certify protective materials and concurs in the recommendations of the Safety Committee."

The National Bureau of Standards Handbook No. 20 on X-ray Protection[16] explicitly states in the first paragraph. "Throughout these recommendations the word 'shall' is used to indicate necessary requirements, while the word 'should' indicates advisory requirements to be applied when possible." This is included under "General Recommendations" and this is in line with Taylor's statement that the Bureau of Standards does concur in the recommendations of the Safety Committee.

Again in 1932, in discussing the work of the American Advisory Committee on X-ray and Radium Protection, Taylor[17] states that "the question of the legal status of these recommendations has been frequently raised. They have none. The Committee feels that none is needed; that legislative enactment tends to stunt development and prevent health changes. We are free to admit that our present proposals may require changes in the future as they are developed. We wish nothing to interfere with the freedom for modification. It should be pointed out, however, that lack of legal standing will probably not in any way detract from their legal value. They are a recognized set of recommendations, drawn up by qualified representatives of the art and freely distributed to those interested. A court decision involving X-ray protection would, in all probability, for lack of another source, be guided by these recommendations, and persons ignoring them may be held liable for negligence."

Certain states have set up, through their Departments of Labor and Industry, rules to cover the safety of workers engaged in luminous dial painting. New York has compromised on an allowable concentration of 10^{-10} curie of radon per liter in the working air. Massachusetts has prescribed that the whole body exposure to gamma rays shall be maintained at less than 0.1 r per eight-hour day. But insofar as we can determine, in no state are these departmental rulings or recommendations actually law on the statute books. It would seem that the time is appropriate for some government agency to establish and enforce adequate radiation protection in industry. This might well be under the direction of the United States Public Health Service, or through this agency collaborating with State Departments of Public Health. The enormous increase in the use of industrial radiography and luminous dial painting brought on by the war will continue in postwar years on a reduced but still larger scale than preceding the war.

Whether similar agencies can or should undertake to regulate radiation exposure relative to medical uses is a more complicated decision and undertaking. The continuance of damage to medical X-ray and radium workers is evidence, however, that either the medical and dental profession must put its own house in order, or accept direction in this respect from a public health agency.

In considering the radiation damage which continues, both in industrial and medical usage of radiation, we are not entirely in agreement with Lauriston Taylor that "legislative enactment tends to stunt development and prevent healthy changes." The same argument has not prevented legislation protecting the public from overexposure to lead, benzol, carbon monoxide, and a host of other toxins.

---

* A national laboratory with powers of inspection and approval.
† The recommendations of the American Advisory Committee have been accepted and published by the Bureau of Standards.

20]   MDDC - 1100

THE APPLICABILITY OF ANIMAL EXPERIMENTATION TO TOLERANCE DOSE IN MAN

In reviewing the subject of tolerance dose, it is most striking that animal experimental evidence has played practically no part in arriving at present day levels. In summary, there are only three tolerance levels which have been established and accepted as a working basis for occupational exposure:

0.1 r per day for external X and gamma radiation

1×10^{-14} curie/cc for radon in the air of working rooms

0.1 microgram of radium as the maximum allowable amount deposited in the body of a radium dial painter

Each of these levels has been established by adding a safety factor to the amount which has been known to produce lasting injury to persons so exposed. It is of interest also to note that in each case the safety factor does not exceed 10, and is more likely considerably less than 10. Human misfortune rather than animal experimentation was the basis for these levels.

The past literature is surprisingly lacking in animal exposure carried on with radiation at tolerance or near tolerance levels. This can be explained by several factors:

1) Radiologists have not given much attention to the subject as a whole. One cannot then expect to find considerable experimental work in a field which was relegated to a few committees of the various societies.

2) Experiments within tolerance or near tolerance range require long periods of time and large numbers of animals to complete. They are therefore expensive and time consuming.

3) There may have been the feeling by those engaged in radiobiology that other fields offered more fruitful paths since "tolerance" for mice, rabbits, guinea pigs, or fruit flies still was not "tolerance" for man and could never be proven to be so.

With the awakened interest in tolerance dose, one may well consider how one can apply the knowledge gained from animal experimentation to either a revision or confirmation of existing levels.

In judging that a certain level of radiation cannot be tolerated, one must look for the earliest signs of an injury which is lasting and damaging to the economy of the organism. This implies that we can establish normal standards within fairly well-defined limits and can also control other toxic agents which might either directly or indirectly produce effects confused with those resulting from radiation. These restrictions also interfere with observations on large groups of men under observation for overexposure to radiation, and at the same time are not so easily controlled. Within limits, then, we might conclude that animal experimentation could approach the problem from the standpoint of radiation effect with more chance of controlling the extraneous factors which produce alterations in leukocytic levels, weight, fertility, longevity, or whatever effects we may wish to follow.

Animal research also has the advantage of being able to sample the body elements and functions more completely by selective autopsy. It can also follow the animal to his death, either natural or radiation-induced. It is rare in clinical research to be able to observe any large group over a period of years.

The most important advantage of animal research, however, lies in the possibility of exposing the animals to known quantities and qualities of radiation and at exposure rates which are known.

What then can we expect by way of clarification of human tolerance levels from these advantages offered by experimental radiobiology?

We have already been given some insight into the variation in expected biological effect when the organism is submitted to various qualities of radiation. Some observations had previously been made from radiotherapy. More precise studies such as those now in progress by Zirkle will be of value in formulating tolerance levels for neutrons.

Animal experimentation could also be expected to give information on the relative importance of exposure rate within tolerance or near tolerance limits.

Breeding experiments with animals living in tolerance atmospheres and carried on for generations with sufficient controls would be of value in evaluating the perplexing problem of radiogenetics.

Histopathologic and metabolic studies can indicate important damage unsuspected from gross or clinical examination and perhaps eventually lead to the development of more sensitive indices than are now available to judge overexposure to radiation.

These in the main are the contributions to the subject of human tolerance dose which we might hope to gain from animal research.

The important question arises, however, as to how far we can trust animal experiments when the information must be carried over to man and his protection. Several instances in which this transposition has broken down have already come to light. Evans has found that rats will tolerate quantities of deposited radium which would be lethal to a man. Studies on the tolerance of skin in animals cannot be applicable to man because of the very great difference between the skin of man and the various laboratory animals. Nor is it expected that a group of men could submit to a daily exposure of 8 r per day of gamma radiation up to a total dose of 1350 r without serious impairment of hemopoietic function as evidence in the circulating blood, as has been done with a group of inbred mice.[29]

These observations, though at variance with animal research, do not invalidate all results obtained in animals. They indicate rather that we must look to animal research for as precise information as it can give on qualitative and quantitive effects _in the animals_. From there on it is a matter for experience and judgment to use the information in evaluating human reactions. The more precise the animal research is made, the more valuable will the information be to those interested in radiation protection.

REFERENCES

1. Rollins, W., Vacuum tube burns, Boston Med. and Surg. 146:39 (1902).

2. Russ, S., Hard and soft X rays, Arch. Roentgen Ray, London, 19:323–325 (1914–1915).

3. Henshaw, Paul S, Biologic significance of the tolerance dose in X-ray, and Radium Protection, J. Nat. Cancer Institute 1:789–805 (1941).

4. International recommendations for X-ray and radium protection, Radiology 23:682–685 (1934); Radiology 30:511–515 (1937).

5. Mutscheller, A., Physical standards of protection against roentgen ray dangers, Am. J. Roentgenol. 13:65 (1925).

6. Mutscheller, A., Safety standards of protection against X-ray dangers, Radiology 10:468 (1928).

7. Mutscheller, A., More on X-ray protection standards, Radiology 22:739 (1934).

8. Glocker, R. and E. Kaupp, Uber den Strahlenschutz und die Toleranzdosis, Strahlentherapie 20:144–152 (1935).

9. Sievert, F. M., Einige Untersuchungen uber Vorrichtungen zum Schutz gegen Rontgenstrahlen, Acta Radiol, 4:61–75 (1925).

10. Barclay, A. S. and S. Cox, The radiation risks of the roentgenologist, Am. J. Roentgenol. 19:551 (1928).

11. Failla, G., Radium protection, Radiology 19:12–21 (1932).

12. Kaye, G. W. C., Roentgenology. Its Early History. Some Basic Principles and Protection Measures, Paul B. Hoeber Inc., N. Y.

13. Advisory Committee on X-Ray and Radium Protection, X-ray Protection, Bureau Standards Handbook 15 1931.

14. Taylor, L. S., X-ray protection, J. Am. Med. Soc. 116:136–140 (1941).

15. Wintz and Rump, League of Nations Pub. CH-1054 (1931).

16. X-Ray Protection, Nat. Bureau Standards Handbook HB-20, 1936; Radium Protection, Nat. Bureau Standards Handbook H-23, 1938.

17. McClelland, W. R., Health Hazards in the Production and Handling of Radium, Dept. of Mines, Canada, Ore Dressing and Metallurgical Investigation No. 550.

18. Schlundt, H., W. McGarock, and M. Brown, Dangers in refining radioactive substances, J. Ind. Hyg. 13:117–133 (1931).

19. Peller, S., Lung cancer among mine workers in Joachimstahl, Human Biology 11:130–143 (1939).

20. Dohnert, H. R., Experimentelle Untersuchunger zur Frage des Schneeberger Lungenkresbses, Zeits, F. Krebsforschung, 47:209–239 (1938).

21. Evans, R. D., The Toxicity of Small Doses of Neutrons and Other Radiations, Unpublished.

22. Failla, G., Unpublished data.

23. Martland, H. S., Occurrence of malignancy in radioactive persons, Am. J. Cancer 15:2435–2516 (1931).

24. National Bureau of Standards Handbook H-27, 1941.

25. Lewis, Thomas, The Blood Vessels of the Human Skin and their Responses, Shaw and Sons, Ltd., London, 1926.

26. Larkin, John C., An erythematous skin reaction produced by an alpha-particle beam., Am. J. Roentgenol. 45:109 (1941).

27. Colwell, H. A. and S. Russ, X-Ray and Radium Injuries, Oxford Med. Pub., 1934.

28. Aebersold, P. C. and J. H. Lawrence, The physiological effects of neutron rays, Ann. Review of Physiology 4:35–48 (1942).

29. CH-2311.

30. Russ, S. and G. M. Scott, The biological effects of continuous gamma irradiation, with a note on protection, Brit. J. Radio. 10:619 (1937); ibid. 12:440 (1939).

31. Muller, H. J., Artificial transmutation of the gene, Science 66:84–87 (1927).

32. Little, C. E. and J. J. Bagg, The occurrence of two heritable types of abnormality among descendents of X-rayed mice, Am. J. Roentgenol. and Radium therapy 10:795 (1933).

33. Snell, G. D. and F. D. Ames, Hereditary changes in the descendents of female mice exposed to roentgen rays, Am. J. Roentgenol and Radium Therapy 41:248–255 (1939); The induction by irradiation with neutrons of hereditary changes in mice, Proc. Nat. Acad. Science, Washington 25:11–16 (1939); The production of translocations and mutations in mice by means of X rays, Am. Naturalist 68:178 (1934).

34. Mottram, J. C., Proc. Royal Soc. Med. 35:171–722 (1942).

35. CH-946.

36. Taylor, L. S., Roentgen Ray Protection, Science of Radiology, Chapter XIXX, pp 332–343, 1933.

37. Taylor, L. S., The work of the national and international committees on X-ray and Radium Protection Radiology 19: 1–4 (1932).

38. McWhirter, R., Radiosensitivity in relations to the time intensity factor, Radiology 9:287–299 (1936).

Tolerance Dose

H. M. Parker's Early Teaching Notes

1945

HMP

TOLERANCE DOSE

ALL

HMP's early teaching notes (1945?) for Hanford HI people.

Dose Units

The "Roentgen" was defined in the previous lecture. This unit of dose corresponds to the absorption of 83 ergs per gram of air. This has frequently been considered equivalent to the absorption of 83 ergs per gram or per cc of tissue. The corresponding <u>true absorption</u> varies between 42 and 89 ergs/cc in fat, and between 87 and 95 ergs/cc in muscle over the energy range 12 KEV to 830 KEV. For all except precision biological dosimetry, we ignore this difference and define a unit 'rep' (Roentgen Equivalent Physical) as follows:

Rep

That dose of any ionizing radiation which produces energy absorption of ~~83 ergs per cc~~ of tissue is defined as one rep.
93 ergs per gm

The student should note:

(1) Many writers use "per gram" instead of "per cc". Unimportant difference in soft tissue (s.g. 0.97 – 1.07). Volume method gives a truer picture of what takes place in <u>bone</u>.

(2) Some authors sponsor 92 ergs instead of 83, based on muscle. This has some justification, but we prefer to stay with 83, until there is an international unit on the lines of the rep.

Rem

One rem (Roentgen Equivalent-Man or Mammal) is that dose of any radiation which produces a biological effect equal to that produced by one roentgen of high voltage X-radiation. Note that this also is not an absolute definition. 400 KV or better is considered high voltage.

Use of rem permits additivity of composite radiation effects, according to the following scale:

X-rays and gamma rays	1 r = 1 rep =	1 rem
Beta rays	1 rep =	1 rem
Protons and fast neutrons	1 rep =	~~4 rem~~ * 10 rem
Slow neutrons	1 rep =	~~4 rem~~ 5 rem
Alpha particles	1 rep =	~~10 rem~~ 20 rem

---

* Some take 1 rep = 5 rem to preserve a 1, 5, 10 simple scale. Ratios are still uncertain. For some effects there is no <u>true</u> <u>equivalence</u>.

HMP

-2-

Tolerance Dose (or Permissible Exposure)

that dose to which the body can be subjected without the production of <u>harmful</u> effects.

Clear meaning for single shot exposure.

For repetitive small exposures, we shall call the <u>permissible daily quota</u> the Tolerance Dose.

Tolerance Dosage-Rate

that dosage-rate which is continuously tolerable. If we say that the tolerance dose is 100 mr/day, this is dimensionally a dosage-rate. Tolerance dosage-rate of 100 mr/day, equivalent to 3×10^{-6} r/sec implies the existence of a maximum permissible rate. e.g. 10^{-5} r/sec would be outside limits, or a man could not receive 36 mr (10 mr/hr) in one hour. This was never intended. There may be a true tolerance dosage-rate, but in normal exposure practice we can safely say that the <u>rate</u> is unimportant and only the total daily dose is important. This rate idea was inadvertently introduced by Wintz & Rump in a 1931 League of Nations Report. Students should read this as the distinction is important.

Exposure Rates in Past Experience

	APPROXIMATE DOSAGE-RATE
Fluoroscopists' hands	$10^{-2} - 10^{-1}$ r/sec
Fluoroscopists' body	$10^{-4} - 10^{-3}$ "
X-ray therapy - technicians	10^{-4} "
X-ray patients - general scatter	10^{-2} "
Radium therapists and technicians	$10^{-4} - 10^{-3}$ "

Main exposure experience is in the range $10^{-4} - 10^{-3}$ r/sec. Note that our restriction to 100 mr/day is equally arbitrary. We cannot define the integer N such that 100 N mr on one day, followed by (N-1) radiation-free <u>work</u> days, is hazardous. For control convenience, we take $\underline{\underline{N = 1}}$.

Brief History of Tolerance Experience

Read CH-2812 for details

Experience almost entirely with X and gamma rays.

1896   first radiation injury
1902   Rollins - first quoted limit - photographic film - perhaps 10 r/day.
Early injuries acute or sub-acute damage to skin.
1903-4 damage to reproductive organs (animal)
1904-5 damage to blood-forming organs noted
1915   Russ - first organized protective steps

HMP

-3-

1920	American Roentgen Ray Society formed committee to study protective measures
1922	This committee reported
1921	British X-ray and Radium Protection Committee report.
1925	Mutscheller: tolerance dose 0.01 erythema per month.
1925	Sievert confirmed Mutscheller's figure
1928 & 1934	Mutscheller extended work to higher voltage X-rays. Tolerance dose approximately $3\frac{1}{2}$ r per month.
1928	International Committee on X-ray and Radium Protection formed. Reported tolerance dose 200 mr per day in 1934 and again in 1937.
1928	Barclay and Cox, on a basis of two cases quoted 0.00028 of an erythema dose as a safe daily quota. Equivalent to approximately 170 mr per day. Supposed to have a safety factor of 25 fold. Typical of limited evidence on which tolerance data was founded.
1928	Kaye reviewed literature and proposed 0.001 erythema dose in a 5-day week. Equivalent to 100 mr per day.
1931	Advisory Committee on X-ray and Radium Protection published proposals in Bureau of Standards Handbook H15. Recommended 200 mr per day limit.
1931	Wintz & Rump - limit equivalent to 250 mr/day for X-rays. One third of this - or 100 mr in round figures for radium workers.
1932	Failla, for gamma rays quoted 0.001 threshold erythema dose (radium) per month. About 60 mr per work day.
1935	German Committee on X-ray and Radiation Protection accepted Mutscheller figure.
1936	Handbook HB20 set 100 mr per day. This is the recognized U.S. standard.
1941	L.S. Taylor proposed 20 mr per day.
1949-50	300 mr per week Sweden 100 mr per wk.

Translation of old values in erythema dose to roentgens is sometimes ambiguous. We have used:

a.bb. 1957   5(N-18)

Quality	To produce erythema
Grenz rays	100 r
100 KV	350 r
200 KV	600 r
1000 KV	1000 r
Gamma rays (Ra)	1500 r (<u>threshold erythema</u> = ~ 1000 r)

Current Views

Wartime animal research has fortunately caused a much more healthy respect for the 100 mr limit. Certainly a safety factor of 10 is not present. There may be a factor of say 3 in most cases. However, there is a general tendency to want to come down below 100 mr. Canada has chosen 50 mr. E. Lorenz's work on the greater susceptibility of the female genital organs (mice only) has led to some claims for a reduction to 20 mr for women workers. As a Hanford plant policy, we take 100 mr as the official limit, regulate the work so that no-one is deliberately exposed to more than 50 mr and end up with an average exposure of about 5 mr. Very few Hanford Engineer Works workers <u>habitually</u> receive more than <u>30 mr per week</u>. Safe exposure practice

HMP   -4-

in general calls for the use of a safety factor. If this exceeds 10, one can be accused of undue conservatism. The Hanford level has been held low because of the conviction of the Medical Department and H.I. heads that the current 100 mr had little safety margin. If the official level is reduced to 20 mr, we would recommend practically no change in plant policy as this new limit would be deemed to include an adequate safety margin. *

The student should read the original literature extensively.
A complete enough bibliography is given in MDDC 783.

Biologic Aspects

discussed in Lecture X, this series. See also CH-2812 and MDDC 783.

Radiation Types

Specific Ionization

Biological effect depends on the specific ionization (number of ion pairs per unit length) in the irradiated material. Although the reason is not fully understood, observations are compatible with a Hit Theory. Postulate that it takes more than one ionizing event in a particular sensitive cell region to produce significant damage. Then sparse ion tracks have a low probability of scoring multiple hits, whereas dense tracks (e.g. alpha particles) have a good chance.

Problem

Calculate the dosage-rate of 200 KV X-rays that makes the average spacing between tracks comparable with the ion spacing along a single track. At this and higher rates there is a clear physical explanation for dependence of effect upon dosage-rate. The student will find that this rate is very high. For alpha particles, it is impossibly high.

Read: R.E.Zirkle's report "Radiobiological Importance of Specific Ionization"
  also: L.H.Gray & J.Read, Nature 152, p 53, 1943

X-rays

Original tolerance data based on x-rays of effective wavelength about 0.3 Å. (Quantum energy 40 KV - tube voltage 80 KV-100 KV). For such radiation, if the dose in air is 100 mr, the dose at 1 cm below skin is about 150 mr. and the depth dose at 10 cm is about 50 mr. For 400 KV machines, corresponding figures are 120 mr and 50 mr. Systemic effect -total body- surprisingly constant over this range. No detailed information in supervoltage range - should be comparable with teleradium experience.

---

* We anticipate no multimillion volt accelerators here.

HMP   -5-

Multimillion volt X-rays show two special features:

(1) Depth of secondary electron equilibrium in tissue is several centimeters. This gives true depth doses greater than 100%. However, if we measure dose in a Bragg-Gray cavity, we choose a wall thickness that gives this electron equilibrium.

(2) At <u>very high voltages</u> there is significant pair production in bone (i.e.; in Ca primarily), and eventually in soft tissue. Fig. 1 shows the body integral dose per unit incident dose (Bragg-Gray method) for different energies.

Preliminary biological experience at 20 MEV is not greatly different from that at 200 KV.

Gamma Rays

Increased penetration compared with usual X-rays. e.g. 98 mr at 1 cm and 67 mr at 10 cm. The <u>integral dose</u> (approximately the total body ionization) is more dependent on <u>distance</u> of the source than on its radiation energy.

	Relative Integral Dose
Soft X-rays from large distance	1
Gamma rays from large distance	2.5
Gamma rays point source at 100 cm	1.9
Gamma rays point source at 10 cm	0.6

Gamma rays from internal emitters usually negligible compared with beta ray dose.

Beta Rays

Assume that beta rays produce same damage as average X-rays for equal ionization. Tolerance dose = 100 mrep/day. Damage from external sources confined to skin or tissue immediately below. Outer horny layer of skin is an inert filter - 7 mg/cm^2 thick over most of body - thicker on palms and soles. Calculation of dose from beta emitters in a specific organ is one of the main problems in radio-isotope work. Normally assume all beta energy absorbed in the relevant organ (e.g. I^{131} in thyroid) or structure (Sr^{89} in skeleton) and calculate dose in rep.

Fast Neutrons

Damage produced by proton tracks in tissue.
Tolerance taken as 0.01 n. 1 n = 2.5 rep. therefore
1 rep = 4 rem. Refined study would show dependence on energy of neutrons, e.g.

60" cyclotron	1 rep	=	3.4 rem
37" cyclotron	1 rep	=	5.5 rem
d-d neutrons	1 rep	=	8.7 rem

These values taken from L. H. Gray

HMF

-6-

Slow Neutrons

Effects due to

(1) gamma rays by interaction of neutrons on hydrogen nuclei
(2) fast protons from nuclear reactions especially $N^{14}(n,p)C^{14}$
(3) production of new atomic nuclei.

Project first took 1 rep = 1 rem, based on (1) Later work in mice showed (2) important and gave 1 rep = 4 rem. We take this in man, although effect of (2) is relatively less in a large animal.

Tolerable flux said to be 1500 n/cm^2 sec. But other calculations show up to 4000. Value uncertain.

Epi-thermal neutrons -- without producing ionization -- may produce molecular dissociation - hazard uncertain.

Alpha Rays

Dense ionizing tracks probably inconsequential on skin unless material is absorbed through surface layers. Cyclotron accelerated alpha particles do penetrate horny layer. Very hazardous internally, e.g. Ra in bone. Assume 1 rep = 10 rem.

Protons

Specific ionization between that of electrons and alpha particles. Because fast neutron damage is essentially a proton effect with neutrons used only to get the protons to the right place, we assume 1 rep = 4 rem.

Combined Radiations

All radiations added in terms of rem, because we beg the question by our definition of rem. There may not be true additivity in all cases.

Geometrical Considerations

1. Surface Dose

The European school of therapy always quotes a surface dose to include backscatter from the patient. The American school frequently uses the dose measured in air. If we take the same stand for tolerance dose, this largely reconciles the European standard of 200 mr and the American standard 100 mr for the softer X-rays on which most of the data depends. Note that monitor instruments will then give the values we look for. Pencils and badges will read high by the amount of backscatter involved. We rarely need to consider this in practice because our tolerance knowledge is insufficiently refined.

HMP

-7-

2. Directional or Narrow Beam Exposures

Pencils on the chest will give a high reading on the above basis if X-rays are incident normally on the chest. Incidence normally on the back will give a low reading (by as much as a factor of 10). We assume that radiation in the field is distributed enough to make such errors unimportant. Similarly, there is a risk of exposing a body part in a narrow beam not recorded by pencils. One guards against this by elimination of hot beams. The employee can hardly wear a battery of pencils or other meters except for special tests. In such a beam, we still take 100 mr as tolerance, although the general body effects are clearly lower in a small irradiated field. Limited beams also occur in laboratory work, especially on the hands and head in hoods. Exposures up to 1 r per day have been condoned in the past, but we try to limit this also to 100 mr. Probably 200 mr is perfectly safe on the hands each day.

Legal Status

Limits set by the National Advisory Committee on X-ray and Radium Protection are recommendations only. Apparently, no State has definite laws on tolerance limits. Some have good regulation for specific industries, e.g. Radium dial painting. We have a similar advisory rather than enforcement capacity at Hanford Engineer Works.

Policy for Specific Cases

Take handling of uranium rods as typical. We have seen that 100 mr measured in air is about 150 mr in skin for the familiar 200 KVp radiation. Skin must be able to tolerate 150 mrep beta radiation on this basis. On the palms, there is a thick horny layer which transmits about 73% of the beta radiation. For 150 mrep at the sensitive layer, we have $\frac{150}{0.73} = \sim 200$ mrep

on the surface. We permit this for holding rods in the palm, and calculate the permitted working time accordingly. e.g., bare metal and no gloves (280 mrep/hr) - time 40 minutes. Canned metal and gloves (33.5 mrep/hr) - time 6 hours.

Note that we should always be willing to adjust times to meet regular cases of such nature. We will never compromise radiation safety to oblige a group with longer time limits. Also, it is mandatory that all such special limits be documented and approved.

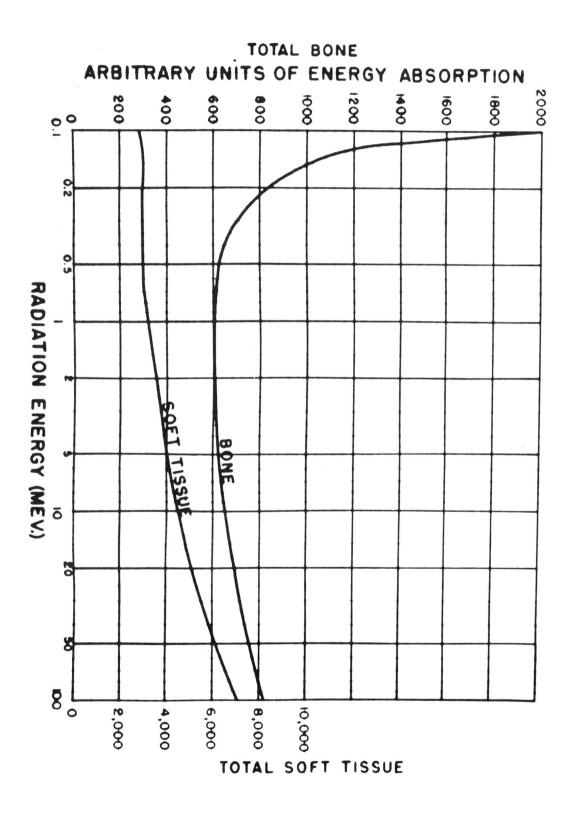

FIGURE 1

Tolerance Dose Memo Dated February 22, 1946

U.S. Atomic Energy Commission
Declassified Files

February 22, 1946

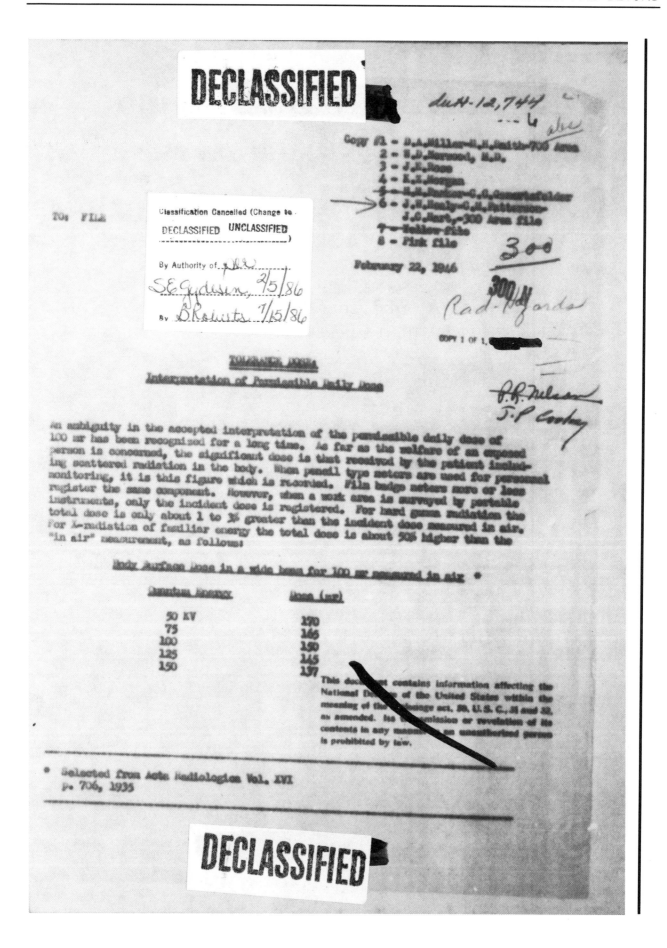

Parker-File -2-
2/22/46

The basic evidence on radiation hazard has been obtained by survey measurements, rather than by personnel monitoring. Most protection groups have carefully avoided an outright statement on this point, principally so that survey readings and pencil readings could be used interchangeably without troublesome scattering corrections. The American Standards Association "Safety Code for the Industrial Use of X-Rays" has recently made a definite commitment as follows:

"2.2.3 Dose. "Dose" is the total quantity of radiation in roentgens at a given point, measured in air, * integrated over the period of exposure.

*The expression "measured in air" has a definite meaning in radiology; namely, that the measurement is made at a given point in the radiation field without the presence of the human body."

Future practice will probably follow this policy.

The practical influence on the H. I. Section program and reports is negligible except in three specific instances:

1. **Known exposure to soft gamma-radiation or X-radiation only.**

    Detailed study of possible exposures occurring in the following cases may result in considering an observed pencil reading of over 100 mr as less than a permissible daily dose.
    (a) Fluoroscope operations - proposed pencil or badge limit 170 mr
    (b) X-Ray calibration   - limit 150 mr
    (c) X-Ray diagnosis (Kadlec Hospital) - 170 mr
    (d) Intermediate energy calibrations - to be measured

2. **Known exposure to beta radiation from uranium on the hands only.**

    It has been shown that the permissible daily dose of 100 mr really permits the exposure of the skin to 150 mr measured in the tissue for familiar X-radiation emitted at about 200 kVp (100 kV quantum energy). It will be universally agreed that the relevant sensitive portions of the skin of the hands can tolerate an equal tissue-dose arising from beta irradiation. On the palm of the hands, there is a natural absorber, the layer of passive absorption, at least 40 mg/cm^2 thick. This transmits 73% of the beta radiation from an extended source in contact with the hand. For 150 mrep at the base of the layer of passive absorption, the surface dose would be $\frac{150}{0.73}$ = 200 mrep.

    This value, which now includes backscatter and is not "measured in air" will henceforth be considered the permissible daily dose for this particular case. In the same terms, the dosage-rate of uranium metal in contact with the hand is 280 mrep/hr.

HMParker-File   2/22/46
                                 -3-

3. **Known exposure to beta radiation from other sources and on other parts of the body.**

With the exception of the palms of the hands and the soles of the feet, other skin surfaces have a layer of passive absorption of about 10 mg/cm². The absorption in this layer is a function of the beta-ray energy concerned, the dimensions of the source and the source distance. In the average case, the transmission of the layer will be about 90%, and the permissible daily dose $\frac{150}{0.90}$ = 160 mrep. Since the selection of 150 mrep (for 100 KV) is itself arbitrary, the general beta case should be handled with a limit of 150 mrep surface dose, including scatter.

Summary

For convenience in reporting, the difference between measurements "in air" and measurements "with backscatter" will normally be ignored. Special exceptions for X-ray exposures and for beta radiations, especially from uranium metal handling, will be provided. These provisions should be invoked only when the work load routinely approaches the otherwise tolerable limit. The preferred practice of restricting all planned exposure to about one-half of the permissible daily dose should be continued.

                                         H M Parker
                                         H. M. Parker
                                         Asst. Supt.
                                         H. I. Section

HMP:EWC

Status of Health and Protection at the Hanford Engineer Works

Industrial Medicine on the Plutonium Project

1951

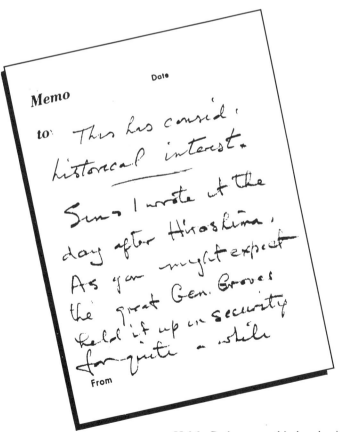

H.M. Parker gave this handwritten note to Ron Kathren about 1978 following a discussion of how this article came to be written.

Paper 9

STATUS OF HEALTH AND PROTECTION AT THE HANFORD ENGINEER WORKS*

By S. T. Cantril and H. M. Parker

[Editor's Note: The following document was written for the employees of the Hanford Engineer Works shortly after the dropping of the atomic bombs on Japan. Many of the workers at the plant had been unaware of the nature of the materials with which they were working; this document was prepared to reassure them concerning their safety on the job and to emphasize the fact that great care had been taken to safeguard their welfare. The paper is reproduced here because it is an example of a type of educational document and because it contains information, not given elsewhere in this volume, concerning the protective measures taken at the Hanford Engineer Works.]

Recent press and radio releases may cast some doubt in the minds of a few Hanford Engineer Works personnel concerning the personal safety of themselves and families. Through the magnificent cooperative effort of all the workers in the plant we have made this a safe job, and we do not want any misinformation or loose statements by the misinformed to mar either our record or the morale we have built up through confidence in the safety of our operations.

For the above reasons, it is felt that a statement is timely which will summarize our safety record and also emphasize the need for continued vigilance to maintain it.

Never before had so many people been engaged in an occupation wherein the hazard was one of radiation and radioactive substances on so large a scale. Previous experience had been limited to hospitals or scientific laboratories where x-rays, radium, and other radio-

---

*Based on Metallurgical Laboratory Report CH-3570.

active substances or cyclotrons generating neutrons were used for medical or investigative purposes. The radium industry itself, with the allied branches, was the closest counterpart to radiation as an industrial hazard. But here we were faced with a job which would require our dealing with the several more dangerous types of radiations and many heretofore unknown radioactive substances in quantities many thousand times more abundant than had been handled in the past. How is it that it has been done safely in so brief a period? This we propose to examine.

In the first place it should be understood that everyone of us has been subjected to the influence of radiation and radioactive substances from the day we were born. The earth, its soil, rocks, water, and air contain small quantities of radioactive elements —uranium, radium and its products, and others. It has been known very exactly for many years that each of us stores in certain organs of the body a fairly constant amount of these elements because we eat food or drink water which contains them. The amount of radioactive gas in the normal air has been measured the world over. The air we have always breathed contains these radioactive gases in minute quantity, by release from the natural processes of radioactive decay which are going on in nature. Furthermore, every living thing is continuously subjected to bombardment from radiations coming from the remote limits of the universe. These are the cosmic rays which penetrate all living matter —which can be measured and studied as though man himself had generated them on his own planet. So there is nothing new in man having to face radiation or radioactive substances. It was only in the quantities of them and in the existence of certain heretofore unknown radioactive substances that problems in protection had to be met.

The backlog of knowledge concerning natural radioactive substances and an experience extending over some 45 years with the effects of x-rays, radium, and other radioactive substances on living things gave us initial confidence that the job could be done safely, but would require a large expenditure of effort, money, and care to accomplish this end. Plans for the protection of health were begun three years ago, and these have grown from a modest beginning to an interlocking group of physicians, biologists, physicists, chemists, and technicians working in various centers throughout the country on the many phases of radiation protection and health which the entire project has posed. The number of these workers extends into the hundreds. The cost has reached into several million dollars. The results can be measured in the safety which we enjoy in our work.

The backlog of biological and medical knowledge on the effects of radiation also gave us quite exact limits to which radiation would have to be confined in order to be safe. Previous experience with x-rays and radium had resulted some eleven years ago in the establishing of a maximum tolerance amount of these radiations to which workers could be exposed without resulting in any permanent injury which would in any way interfere with meeting the normal strains of living. This amount of radiation was called the "tolerance dose." It was first established by international agreement of certain scientific bodies of the various nations confronted with the problem, and, with the exception of the United States, this so-called "international tolerance dose" still holds in other countries in defining the limits of radiation protection. Some nine years ago, the United States body entrusted with the definition of this standard chose to reduce it by one-half. This then is the standard of radiation protection for the majority of our radiation problems which has been adopted by this Project—one which is doubly stringent over that which has been safely used in other countries.

Previous experience in the radium industry gave us precise knowledge on the limits of protection which would have to be met in dealing with many of the radioactive solids, liquids, and gases developed in our process. Not only have these limits been defined for radium by the National Bureau of Standards, but even the maximum allowable amount of radium which a worker could take into his body without causing injury was known, defined, and could be measured in the body. These limits, which had been adopted long before our work began, have proved invaluable to us as a guide for our problems at the Hanford Engineer Works.

To be sure, we knew we would have to deal with certain radiations or radioactive substances for which no previous experience had necessitated defining a tolerance value. To this end a vast amount of biological research has been devoted by the Project, and answers have been found to these new problems, which, when adopted under industrial conditions, have proved safe. We will give some examples of these further on.

With this background, let us examine the means by which we continue to operate safely.

First of all, the design of the plant incorporated every precautionary device or material which at the time was thought necessary to do the job safely. The design phase of the work drew on the experience of those who were specialists in radiation protection. It benefited by the experience of similar but smaller plants elsewhere on the Project which were already operating and had been able to test the protective features to be used on a larger scale at Hanford Engineer Works. The more spectacular features of design included the massive concrete or steel walls surrounding operating units, the huge storage basins for radioactive materials under many feet of water, the delicate operation of 75-ton cranes by television and prisms from behind concrete walls, the high stacks used for carrying the by-product gases to great heights, and special railroad cars which carry heavy lead casks in water tanks over the desert between operating areas. These are but a few of the gigantic features of design which naturally are most impressive to an industrially minded nation and to the press and radio. There are other less visible protective features of design which are just as important, though less glamorous. To name only a few — the intricate air-conditioning systems in operating buildings and laboratories to supply air free from contamination, the hundreds of instruments stationed throughout the plant which record the efficiency of radiation protection, the special gadgets on the chemists' benches to permit them to work safely with radioactive substances, automatic locks and releases on doors to operating areas which can only be released when conditions are known to be safe and the worker can enter, coolant liquids and ventilation over the lathes of uranium-metal operators. These and hundreds of inconspicuous features of design have helped to make the job a safe one.

Having designed a plant, now can one be certain that the protection features work as designed? Fortunately radiations and radioactive substances are something which can be detected and measured with special instruments. We have therefore never needed to be in doubt as to whether radiation exists in any particular location, or whether the amount is more than can be safely tolerated. To this end a specially trained corps of physicists, engineers, technicians, and helpers has been assigned the sole task of measuring the quantities of radiation in all forms to which any worker might be exposed. Actually one in every twenty workers in the operating areas of the plant is a member of this group — known as Health Instruments. It is to this group that great credit is due for the alertness in detecting possible hazards and in measuring and recording a great mass of information which assures us that operations are proceeding safely.

All of us who enter the gates of the plant know something of their work. We all carry special badges with film in them which will detect the quantities of radiations to which we may be generally exposed. In entering all areas wherein necessary, each of us carries two "pencils" — which are delicate instruments to measure radiation. The magnitude of this program alone can only be realized when it is known that so far in 1945, over one million pencils have been measured and prepared in readiness for the next shift; more than 170,000 film badges have been processed since January 1945. To carry on the

480   INDUSTRIAL MEDICINE

mechanics of this work alone and to record all results for every worker individually has been a remarkable achievement in the monitoring of individual radiation exposure. Besides this one function of the group, it has surveyed conditions in all locations on the plant, as well as others removed from the plant site. In the separations plants alone over 31,000 surveys have been made to check on operating conditions. These are in addition to the many instruments which measure and record conditions over a 24-hr period, day in and day out. All the instruments used have to be continually checked and standardized, some 52,000 such instrument checks having been made to date. Many of us know that we are required to check the cleanliness of our hands during and before leaving work to ensure freedom from radioactive contaminants. This program alone has, by refined instruments and cooperation on the part of all of us, run into the astounding figure of some 157,000 such hand checks in a six-month period. Special clothing—coveralls, gloves, rubbers, etc.—is laundered in a special laundry and each item is checked for radioactive contamination. This amounts to some 40,000 items per month. These again are some of the more obvious and mass-production features of the protection group. Not so obvious but just as important are the many surveys and analyses made of air in the buildings, outside, and in the surrounding countryside and of all water which leaves the plant or comes from wells and rivers in the surrounding countryside, as well as the time, thought, and work which have been put into the development of instruments and analyses, both here and at other locations.

The protection program does not function because of design, instruments, and a special group of safety engineers. The understanding and cooperation of Plant Supervisors and all plant personnel has been necessary to this end. In the more technical fields, chemists, engineers, and technicians in many fields had to learn to work with new techniques and precautions for the good of themselves and others. Other workers who did not need to understand fully the complexities of the hazards were told enough about them, however, to permit them to work safely. Additional precautions were taken by studying each job before it was done, where the possibility of radiation hazard might exist. Danger zones were set up and posted as such. Before entry into these was permitted for any reason, a Special Work Permit was issued after the job had been reviewed and necessary precautions taken. Some 2,500 Special Work Permits per month have been issued, each requiring the cooperative effort of Health Instruments, Supervisory, and other groups concerned in the particular job to be done. This cooperation by so large a group of technical, skilled, and unskilled workers in an enterprise fraught with potentially great and to most of them unfamiliar hazards, has been the most gratifying phase of the work to those of us who have been closest to the protection problem. At no time have we seen signs of group fear or alarm in contemplating or doing an assigned job. To be sure, there have been a few individuals, as there are in any plant or army, who cannot assimilate the true picture of hazard conditions and who have needed reassurance and further guidance in understanding the conditions as they are and not as they have read elsewhere or imagined them to be.

What are the conditions at Hanford Engineer Works in regard to health and protection? What are the results of all the design, group, and plant-wide effort placed on personal and public safety? These questions could be answered at great length, but a briefer statement gives the full story:

It can be stated without reservation that to date no employee working in the Hanford Engineer Works has received an amount of radiation exposure which would be injurious. We have of necessity been conservative in defining our acceptable limits of exposure because of the great number of people involved and the magnitude of the operations. Not only have we adopted the U. S. tolerance dose, which is itself conservative, but we have recorded separately any exposure received by each individual and made it a point to investigate working conditions whenever there was a repetition of events which lead to an exposure of only one-half that amount which in itself is safe. In many million man-hours of work this has been necessary less than half a dozen times. In instituting safe conditions in the handling and machining of many tons of uranium metal we have adhered to a standard of dust in the air which is over three times more stringent than the one which a large amount of clinical investigation and research brought the U. S. Engineers Corps to define as a safe one. We have been able to do this without any great amount of additional expenditure. A large volume of biological research at other sites has been done to learn the safe limits in working with plutonium. In the earlier days we adopted standards of protection which were safe by a factor of ten over that which research has subsequently shown to be safe. We continue to adhere to a standard which is doubly stringent over the official limit. We find that operations can be done under those limits without impeding the work any more than if conditions were operated to permit the workers to be exposed to the maximum allowable concentrations of plutonium contamination. We do not live in a "City of Pluto," as certain elements of the press describe our village. Pluto is safely confined behind walls or barriers in the plant. What little of him as does escape is not going to relegate anyone to purgatory.

What of conditions outside the plant which cause concern to a few members of the population? It was stated before that the air outside

482 INDUSTRIAL MEDICINE

the plant was analyzed for radioactivity coming from the high stacks carrying radioactive by-products from the process. The by-products that are thus liberated are radioactive iodine and xenon, mixed with nitrous fumes and water vapor. Extensive engineering work and a study of meteorological conditions in this region preceded the erection of these stacks and their specified height. Biological work on tolerance limits of the more important radioactive gases and vapors was begun three years ago. Radioactive iodine is not unfamiliar to the field of radiobiology and medicine. Considerable research had already been done before this Project began, using radioactive iodine made by the cyclotrons. Experiments with animals and the application of these to disease in man gave us knowledge on the quantity of it which was entirely safe. We continually measure and record the quantities of these gases which do not disperse into the upper atmosphere, there to be diluted in the ocean of air. That small fraction of them which does come to ground levels is measured over the desert and surrounding countryside. The quantity in Richland air, for example, some 25 miles away from the stacks, is so small that it cannot be measured in the air itself. It is only by collecting it on delicate filters over several days and then analyzing the filters that we can find traces of them in the air of the village. The amounts are entirely innocuous and approach the levels of natural radioactivity found in the atmosphere at any location in the country. And then to check our results doubly, selected groups working in these areas closest to the stacks are individually tested to determine whether they have breathed in and deposited the radioactive iodine in the thyroid gland, where it would concentrate if it were inhaled. Many hundreds of these checks to date have not revealed a significant result. There have been times when these gases have come to the ground owing to certain weather conditions. When they do, the nitrous fume component in the mixture is the more dangerous one, and this is a hazard similar to one encountered in many more familiar industrial plants the country over. Thus far we have had no instance of injury from exposure to nitrous fumes.

There may be some concern that radioactive water released from the plant to the Columbia River will so contaminate the river that its water will be dangerous to man, fish, or fowl. It is true that the water which passes through the units to act as a coolant does become radioactive. But it is also true that the greatest part of this radioactivity is lost before the waste water is ever put back into the river. A continuous record is maintained of all water which flows from the plants to the river, and at no time has the combined waste water been in excess of that which would cause an overtolerance radiation exposure to any living thing immersed in it. The voluminous further dilution that occurs in the river brings the quantity down to one which cannot be measured below the plants. But in spite of this, the river water is periodically analyzed, with a negative result. As for the addition of heat to the river by the dumping of large quantities of coolant waters into it—there is heat added, but the dilution factor of the river itself is so vast that it requires the most delicate thermometers to register this small increase in temperature. This too is dissipated before the river flows through the limits of the reservation.

With all the precautions and protection heretofore described, the plant worker may well ask why he is submitted to periodic physical examinations, blood counts, and the many other procedures which make him feel like a guinea pig. The answer is that these are again a precautionary measure to assure him and his Supervision that the work is being done safely, and that no injury is sustained. These procedures in themselves have been no small job. In the first six months of this year over 10,000 examinations of area workers alone have been done. This has meant doing some 48,000 medical laboratory examinations in this six-month period alone. And what of the findings of all these examinations? Not one abnormal finding has been uncovered which could in any way be attributed to the hazards of radiation or to chemical toxicity of any of the materials used at the plant. We have found evidence of certain diseases common to workers or people anywhere, which have no relation to working conditions and are of the variety seen by any large hospital or clinic staff. When these are found the worker is so advised, and he is directed to his own physician for the care of them. As a result of preemployment examinations which screen out those who have conditions which make them unacceptable for work and of periodic examinations done thereafter, it can be said that the health of workers on the Hanford Engineer Works is better than that of the average community with a population of the same size as ours.

Let us look at the safety record of the entire Hanford Engineer Works. This means the major injuries due to all causes, from falling bricks to automobile accidents on the job. Thus far in 1945, the record is one of 0.89 per one million man-hours of work. How does this compare with other jobs? Here are some national figures for various industries in 1943, published by the National Safety Council:

Steel industry	7.41
Cement industry	7.79
Aircraft industry	9.91
Chemical industry	10.07
Construction industry	23.75

484 INDUSTRIAL MEDICINE

Our figure of 0.89 is low, but it is emphasized that of this none is due to injury from radiation or radioactive substances. That safety pays is an old Du Pont axiom. As Du Pont workers, the Hanford Engineer Works people can take just pride in the record to date. Only by rigidly adhering to our standards of safety and protection can we hope to maintain our present record. It is at this stage that we cannot permit any laxity in protection which might result from a familiarity with the job.

Action Taken on Spot Contamination

U.S. Atomic Energy Commission
Declassified Files

October 30, 1947

-1-

DECLASSIFIED

RECORD COPY IM-7920

Copy 1 - D.H.Lauder - G.C.Lail
2 - W.D.Norwood - P.A.Fuqua
3 - H.M.Parker - C.C.Gamertsfelder
4 - C.N.Gross - W.K.MacCready
5 - A.B.Greninger - J.B.Work
6 - R.C.Hageman
7 - J.W.Healy - R.C.Thorburn - W.Singlevich
8 - C.M.Patterson - M.L.Mickelson
9 - 300 file
10 - 700 file
11 - Pink file
→ 12 - Yellow file

FILE

October 30, 1947

This document consists of 6 pages #12 of 12 copies, Series A

Copy 1

RECORD CENTER FILE
RECEIVED
MAY 21 1956
300 AREA
CLASSIFIED FILES

ACTION TAKEN ON HOT SPOT CONTAMINATION IN THE SEPARATIONS PLANT AREAS

A. PRESENT STATUS

General

H. I. surveys of ground contamination in the T and B Plant Areas revealed the presence of many small radioactive spots on the ground. This followed an apparent increase in the general shoe contamination in the area. Similar spots of low activity would not have been distinguished from the general ground contamination by I^{131} and minor long-lived isotopes. A review of past data suggests a history of 6 months for the phenomenon, with a recent intensification. The contamination, as found, is associated with discrete particles on the ground. It is not specifically known that such particles fall to the ground. The effect may be due to droplets sufficiently small to become attached to a single particle. It is commonly supposed that the particles or droplets come from the process stacks. A less likely origin is the capricious blowing of particles from known contamination near the stacks.

Physical Form of Particles

Segregated particles have an average mass of 1 mg. The smaller ones go down to 0.1 mg., but the collection method does not include possible very small particles. The particles are normally brown in color in comparison with surrounding sand or soil. Under microscopic examination, there is no distinguishing feature indicative of the origin.

Distribution

By counting particles appearing on cleared ground in the T Plant, the rate of

DECLASSIFIED

EMP-file   DECLASSIFIED -2-   10/30/47

deposition is shown to be one per day per square foot in the prevailing wind direction from the stack. The rate is one per day per 4 square feet in the reverse direction. Particles have been found in all locations tested up to the boundaries of the 200-W and 200-E Areas. The general area distribution has not been well-patterned, and the rate of deposition may be obscured by random blowing of older particles.

Activity

The activity range is approximately 0.1 μc to 1 μc. In deposition tests, the new particles on the wind side were approximately two times as active as those found upwind.

Radiochemical Analysis

Typical analyses of T Plant particles indicate 60 - 90% Ce, and up to 15% Y. B Plant samples showed 30 - 55% Ce, 7 - 20% Sr, and 30 - 45% Y. *

Chemical Analysis

There has been popular opinion that the particles may originate from contaminated flakes of the black paint on the stacks. This is an asphalt-resin paint that should be rich in organic materials. Micro-combustion of one particle showed:

Carbon	1.5% by weight
Hydrogen as H_2O	5.3%
Silicon as SiO_2	27.9%
Iron as Fe	49.9%

In this test, bismuth would have carried with the iron. This one particle was certainly not the paint.

Spectrographic Analysis (by the Technical Department)

The first particle investigated showed high content of Pb, P, and Si. Four later samples showed no Pb, and no phosphorous above the rather high limit of sensitivity. Elements present (relative amounts) were:

	Active Particles	Control Particles
Bi	<50, <50, 100, 500	all < 50
Fe	100, 100, 100, 200	all < 10
Si	100, 100, 100, 200	100, 100, 100, 100

Ca, Mg, Na, Ba, Al - no significant difference between samples and controls.

---

* See HW-7865 dated 10/22/47

HMP-file 10/30/47 HW-7920

From the chemical and spectrographic analyses, the theory of droplet attachment to sand grains receives some support.

Age of the Particles

The age of the particles should give an indication of whether the active material comes directly from the Separations process, or whether it comes out of the stack after hold-up in ducts or the stack itself. Essential absence of short-lived isotopes (except daughters) seems to exclude direct process emission. Such emission is also improbable in view of the apparent recent development of the problem.

If process sprays contained a representative sample of the fission product mixture, the mixture found in the particles should define the age. It is conceded, however, that relative volatility and other factors alter the fission product spectrum in spray. It has been pointed out * that the low amounts of zirconium and columbium, in comparison with the quantity of strontium and yttrium which have comparable half-lives and fission yields, proves that chemical processing has altered the fission product deposition. This would not be valid if the strontium and yttrium are, respectively, Sr^{90} and the daughter Y^{90}. Isotopes of half-life in the range 50 - 70 days, such as Zr-Cb, Sr^{89}, Y^{91}, could be supposed to have been reduced to negligible activity. Proof of this isotopic assignment would give an age of about one year.

Exact age can be deduced from two isotopic ratios that would be independent of the chemical processing history. These are:

(1) Ce^{141} and Ce^{144}

If the cooking time and fission yields are exactly known, the ratio Ce^{141}/Ce^{144} precisely defines age. The radiochemical discrimination is complicated by the growth of the daughter Pr^{144}, but this can be turned to advantage. The growth can be followed through filters that exclude Ce^{141} and Ce^{144} radiations with rather small loss of the 3 MEV Pr emissions. The equilibrium emission of Ce^{144} is thus obtainable, from which the Ce^{141} radiation is obtained by difference at zero time. In practice, J.W. Healy, who proposed this method, finds that the rapid growth of Pr ($T_{1/2}$ = 17 min) necessitates difference-analysis along the growth curve. Uncertainty in counter efficiency for the different energies limits the present accuracy. Separated Ce^{141} and Ce^{144}-Pr^{144} are needed for comparison. The one particle tested to date had a minimum age of 150 days.

(2) Sr^{89} and Sr^{90}

This is a more favorable case, due to wide energy difference between the two beta-ray spectra, and to good half-life ratio. The growth of

---

* See HW-7845 dated 10/20/47

HMP-file   10/30/47

daughter Y^{90} is not troublesome, because the half-life, 62 hours, gives negligible growth during the early counts. Another separation, after one month, during which the Sr^{90}/Sr^{89} will change by a factor 1.46, should give a very accurate age, up to about 400 days. Again, test samples of Sr^{89} and Sr^{90} are required.

B. HAZARDS INVOLVED

External Radiation

There is no critical hazard from particles on the ground or on shoes. If the particles lodge on skin either by descent or by transfer from the ground, visible radiation injury will <u>certainly</u> be produced with particles of 1 μc activity, and <u>may</u> be produced at the 0.1 μc level. Reddened areas about 5 mm in diameter are to be expected, and ulceration may follow. However, it is known from "gold seed" radium therapy, that tissue is relatively resistant to radiation from highly localized ionizing sources. Such reddened areas would completely heal, and ulcerated areas would be repaired with permanent localized skin damage. If the particles lodged on clothing, visible injury might be avoided.

It seems rational to suppose that particles of mass 0.1 mg or more would be removed by normal bathing. Adherence to the skin for several hours should not produce detectable injury. Particles caught in the hair could remain for considerable periods. The normal practice of wearing of hats around a chemical plant should reduce this hazard. Contamination in droplet form would presumably be more hazardous than a shower of particles of equivalent activity. The high content of active rare earths suggests difficulty in casual removal by washing.

Internal Radiation

Inhalation of the active particles or droplets appears to be the greatest potential hazard for two reasons:

(1) The deposited material may remain in situ for a long time. The average half-life for elimination from the lungs of small particles is about 2 months. Larger particles would probably be rapidly eliminated by ciliary action and then ingested. It is clearly necessary to know whether activity is a function of size.

Again, droplets might be more hazardous, if the material is not absorbed through the lining of the lung. Also, in the lungs, there is no protective coating similar to the horny layer of skin.

(2) There is less chance of detecting inhaled particles, and, in any case, no corrective step equivalent to the washing of external contamination.

Ingested material would have a good chance of being carried through the alimentary tract with a sufficient coating of inert material to prevent serious damage.

HMP-file   10/30/47   HW-7920

C. **TESTS IN PROGRESS**

1. The current environmental pattern of particles is being established by surveys. - H.I. Survey assignment.

2. Particles are being separated and subjected to:

   (a) mass determination by microbalance
   (b) activity measurement as a function of mass
   (c) radiochemical analysis as an aid to determination of origin and age
   (d) isotopic ratio of Ce^{141}/Ce^{144} and Sr^{89}/Sr^{90} determination to fix the age
   (e) chemical analysis to establish nature of the vehicle carrying the contaminant, and hence to suggest the origin
   (f) spectrographic analysis to support (e)
   (g) microscopic examination to support (e)

   Control samples will be similarly analyzed as required.

   H.I. Methods group assignments, in collaboration with the Technical Department.

3. The rate of deposition is measured on cleared ground areas, as a function of position with respect to the stacks.

4. The rate of distribution is measured on tacky surfaces protected from as much sand blowing as possible. This supports #3, and should differentiate between falling particles and droplets. - H.I. and Site Survey assignment.

5. Collection of activity on various filtration systems, with analysis of particle size or droplet activity. The probability of inhalation can be computed from this. - H.I. Survey and Site Survey assignment.

6. Tests for Ce^{144} in the urine are being made. - H.I. Bio-Assay.

D. **TESTS PROPOSED**

1. Biological monitoring by rabbits and sheep. Sacrifice of the animals for detection of active particles in the lung. - H.I. Biology assignment

2. Collection of rain samples - differentiation between particles and droplets. - H.I. Site Survey assignment

3. Calculation of deposition pattern of standard size particles or droplets from the stack. This will check the patterns found locally and evaluate the hazard at remote points. - H.I. Industrial Hygiene assignment.

4. A literature study of exposures to point sources, both external and internal. - Major J. Brennan

5. Biological experiments on point source damage, if literature is inadequate. - H.I. Biology

FMP-file  10/30/47 HW-7920

6. Electron microscope study of very small particles, if any. - Research Laboratory (?)

7. Tests for Ce^{144} in feces - H.I. Bio-Assay.

8. Repetition of relevant tests <u>inside</u> process buildings. - H.I. Survey

9. Detailed clothing and body surface examination of H.I. employees and selected other employees (chosen to avoid general alarm). - H.I. Survey

E. FUTURE POLICY

If a finite risk of <u>damaging inhalation</u> is indicated by the tests, the wearing of respirators inside the T and B Plant exclusion areas, or a wider area if necessary, will be recommended. Alternative protection such as the use of canvas-roofed walkways may be proposed. Recommended policy will be to offer complete protection to all employees, regardless of the "scare" implications of such procedures.

If a real risk of <u>external contamination</u> is demonstrated, special procedures for hazard monitoring of all personnel in the affected area will be proposed, with removal of active spots under Operations and H.I. supervision. If this contamination is widespread, general advice on self-administered tests and decontamination steps may be offered.

In the very remote contingency that the hazard increases beyond the feasible control by these methods, cessation of the process, pending correction at the source, will be proposed.

This discussion does not include the essential programs initiated by the Production Department and the Technical Department to detect the source of the trouble, and to remedy it, except insofar as some of the H.I. tests will assist this.

Comments on and criticisms of the actions, current and proposed, will be appreciated.

H. M. Parker

HMP:swc

Radiation Exposure Data

U.S. Atomic Energy Commission
Declassified Files

November 8, 1950

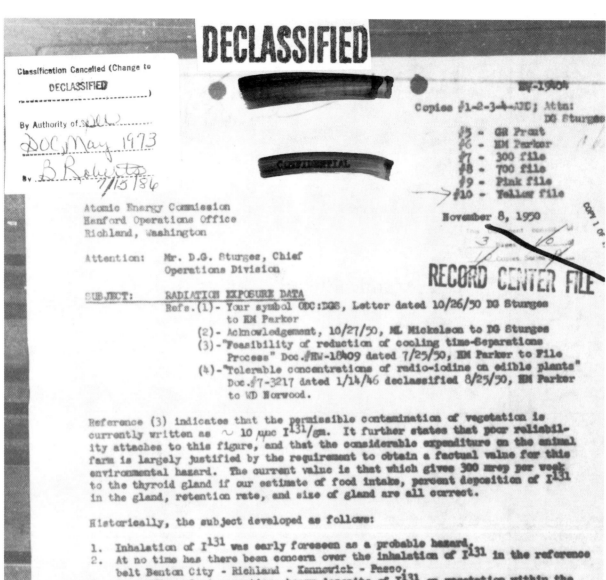

HW-19404

Copies #1-2-3-4-AEC; Attn:
D G Sturges
#5 - GR Prout
#6 - HM Parker
#7 - 300 file
#8 - 700 file
#9 - Pink file
#10 - Yellow file

November 8, 1950

Atomic Energy Commission
Hanford Operations Office
Richland, Washington

Attention: Mr. D.G. Sturges, Chief
Operations Division

SUBJECT: RADIATION EXPOSURE DATA

Refs. (1) - Your symbol ODC:DGS, Letter dated 10/26/50 DG Sturges to HM Parker
(2) - Acknowledgement, 10/27/50, ML Mickelson to DG Sturges
(3) - "Feasibility of reduction of cooling time-Separations Process" Doc.#HW-18409 dated 7/25/50, HM Parker to File
(4) - "Tolerable concentrations of radio-iodine on edible plants" Doc.#7-3217 dated 1/14/46 declassified 8/25/50, HM Parker to WD Norwood.

Reference (3) indicates that the permissible contamination of vegetation is currently written as $\sim 10\ \mu\mu c\ I^{131}/gm$. It further states that poor reliability attaches to this figure, and that the considerable expenditure on the animal farm is largely justified by the requirement to obtain a factual value for this environmental hazard. The current value is that which gives 300 mrep per week to the thyroid gland if our estimate of food intake, percent deposition of I^{131} in the gland, retention rate, and size of gland are all correct.

Historically, the subject developed as follows:

1. Inhalation of I^{131} was early foreseen as a probable hazard.
2. At no time has there been concern over the inhalation of I^{131} in the reference belt Benton City - Richland - Kennewick - Pasco.
3. Soon after plant operation, heavy deposits of I^{131} on vegetation within the reservation were noted, and protective measures were applied.
4. In the winter of 1945-46, heavy deposits developed on vegetation in the reference belt.
5. It occurred to the Health Instrument Divisions that this condition was a potential hazard for food sources. Calculations were made which represented $200\ \mu\mu c/gm$ as the permissible maintained concentration that would give 1 rep per day to the thyroid gland of farm animals. This was the then permissible amount for human exposure.

-1-

HMP-DGS -2- 11/8/50 HW-19404

6. Actual depositions at this time were as follows:

Table 1
I^{131} on Vegetation
Activity density in μμc/gm

Month	Richland	Kennewick	Pasco	Benton City	5 miles E of Stacks
Oct.1945	700	600	1200	800	7600
Nov.	600	700	1200	700	6900
Dec.	5800	12,200	7000	4100	28,300
Jan.1946	3100	3,000	1700	3200	17,800
Feb.	300	500	300	900	3,600
Mar.	300	300	200	200	1,200
April	110	240	150	70	1,100
July	80	40	60	60	1,700

7. Corrective measures were applied such that by Spring 1946, the situation was under control by the existing standards.

8. Beginning in February 1946, some furtive measurements on range animals (sheep and cattle) were made. The program was badly hampered by security requirements, despite the intent of those concerned to cooperate fully. In retrospect, these early measurements were dubious. By and large, they confirmed the calculated uptake of I^{131} within reason. For example, thyroid activity of 0.1 to 1.9 μc I^{131} was observed in animals apparently range fed on vegetation of activity density 100 to 300 μμc/gm. The expected deposition was on the order of 0.3 to 1.0 μc.

9. It is reasonable to assume that in December 1945 and January 1946 the deposition was on the order of 10 to 30 μc in the gland. Practical experience has since persuaded us that the typical animal thyroid gland is not more than half the size of that given us by the army veterinarian consulted. The estimated exposure of animals was therefore 30-100 rep per day.

These considerations led to opinion 4-B -- that range animals were heavily overexposed in terms of the revised permissible exposure. Our best guess is that the overexposure was by a factor of about 1000 in midwinter 1945-46.

10. Similarly, the estimated overexposure was by a factor ~100 in midwinter 1946-47.

11. In midwinter 1948-49, vegetation contamination was held to less than 3 μμc/gm. The same held true in 1949-50, except for temporary planned increases after the special "green run".

-2-

HMP-DOE   -3-   11/8/50   HW-19404

12. It is important to define clearly our bases for "overexposure". We are writing 10 μc/gm as the permissible limit. Correcting for reduced thyroid size, <u>if all other factors were correctly evaluated, the limit would be ~ 5 μc/gm.</u> This value we have used as the basis for the above factors. We do not believe that animals will be injured at this level or at 10 μc/gm or even at 100 μc/gm. This is speculative, or intuitive. We propose to develop the facts for the Commission, by the animal farm experiments.

If the above review fails to cover all the points raised by the comments in Document HW-18409, I would appreciate your developing the topic in the form of more specific questions.

                                            Very truly yours,

                                            ORIGINAL SIGNED
                                            BY H. M. PARKER

                                            Manager
                                            HEALTH INSTRUMENT DIVISIONS

HM Parker:swc

Health Physics, Instrumentation, and Radiation Protection

Advances In Biological and Medical Physics, Volume I

1948

Reprinted from *Health Physics* Vol. 38, pp. 957–996 by permission of the Health Physics Society.

HEALTH-PHYSICS, INSTRUMENTATION, AND RADIATION PROTECTION

By H. M. PARKER

Reprinted from
ADVANCES IN BIOLOGICAL AND MEDICAL PHYSICS
Volume I, 1948
Copyright by
ACADEMIC PRESS, INC., NEW YORK, N. Y.
All Rights Reserved

Volume 38, Number 6   June 1980

25th ANNIVERSARY ISSUE

HEALTH PHYSICS

OFFICIAL JOURNAL OF THE HEALTH PHYSICS SOCIETY

H. WADE PATTERSON
Editor-in-Chief

HARRIET J. KROOPNICK
Editorial Associate

J. RAY FIELDING	A. ALAN MOGHISSI
KENNETH R. HEID	DAVID S. MYERS
HARRY ING	GENEVIEVE S. ROESSLER
LAMAR J. JOHNSON	DAVID H. SLINEY
KENNETH R. KASE	J. NEWELL STANNARD
RONALD L. KATHREN	WILLIAM P. SWANSON
LAWRENCE H. LANZL	PAUL L. ZIEMER

Editors

J. R. A. LAKEY
News Editor

PERGAMON PRESS
NEW YORK / TORONTO / OXFORD / PARIS / FRANKFURT / SYDNEY

Health-Physics, Instrumentation, and Radiation Protection

By H. M. PARKER

CONTENTS

	Page
I. Scope of Health-Physics	226
1. Introduction	226
2. Nomenclature	226
3. Nature of Advances	227
II. Past Experience in Protection	227
1. Luminous Paint Industry	227
A. Radium Poisoning	228
B. Radon Exposure	229
C. Gamma Radiation	229
D. General Policy	230
2. X-Radiation in Hospitals	230
A. Fluoroscopy	231
B. Photofluorography	231
C. Radiography	232
D. Therapy	232
3. Radium in Hospitals	233
A. Hand Exposure of Technicians	233
B. General Exposure of Therapists	234
4. Industrial Preparation of Radium Sources	235
5. Industrial Radiography	235
6. Transportation of Radioactive Sources	235
7. Universities and Similar Institutions	236
A. Radiochemical Manipulation Hazards	236
B. General Radiation Injuries	237
8. Manhattan Project Experience	238
A. Development of Health Division	238
B. Typical Exposure Problems	239
a. Shielding	239
b. Beta Radiation	239
c. Metabolism	240
d. Waste Disposal of Fluids	240
e. Waste Disposal of Solids	241
f. Protective Clothing and Decontamination	241
g. Safety Performance	242
III. Exposure Standards	242
1. Dose Units	242
2. Tolerance Dose	244
3. Tolerance Dosage-Rate	244
4. Tolerance Dose Versus Tolerance Dosage-Rate	244
5. Foundations of Tolerance Experience	245
6. Recent Considerations	247

	Page
7. Biological Aspects of Tolerance Dose.	247
A. General Body Effects.	248
B. Internal Emitters.	248
C. Skin Effects	249
a. Erythema and the Layer of Passive Absorption.	249
b. Late Skin Damage.	250
D. Effects on Reproductive Organs.	250
E. Radiogenetic Effects	251
8. Dependence of Tolerance Dose on Nature of the Radiation.	251
A. Specific Ionization	251
B. X-Rays.	252
C. Gamma-Rays	252
D. Beta-Rays	254
E. Neutrons.	254
a. Fast Neutrons	254
b. Slow Neutrons.	255
F. Alpha-Rays.	256
G. Protons and Other Particles.	256
H. Combined Radiations.	256
9. Some Geometrical Considerations.	257
A. Conditions of Measurement of Tolerance Dose	257
B. Directional and Narrow Beam Irradiation.	257
IV. Organization and Functions of a Typical Health-Physics Group.	258
1. General.	258
2. Operational Division	258
A. Personnel Monitoring.	258
B. Area Monitoring.	259
3. Control and Development Division.	260
4. Factors Contributing to a Successful Health-Physics Organization	260
A. The Objective.	260
B. Attitude of Laboratory and Plant Personnel	261
C. Aptitude of Health Division Leadership.	261
D. Aptitude of the Health-Physics Force	261
E. Training Programs	262
F. Liaison with Operating or Technical Groups	262
G. Instrumentation.	262
H. Tolerance Limits	263
I. New Protection Services	263
J. Reports and Records.	263
K. Detailed Control	264
L. Triple Safeguard Philosophy.	264
M. Consistency of Rules and Their Enforcement.	264
N. Design.	264
V. Instruments	264
1. Status.	264
2. Calibration and Maintenance	265
3. Basic Health Instrumentation.	265
4. Fixed Instruments.	266
A. Area Monitors.	266

HEALTH-PHYSICS AND RADIATION PROTECTION 225

Page

  B. Personnel Monitors. 266
    a. Alpha Hand Counters. 266
    b. Beta-Gamma Hand Counters 267
    c. Foot Counters 267
    d. Thyroid Counters 267
 5. Equipment, Atmospheric and Miscellaneous Monitoring. 267
  A. Standard Alpha Counter. 267
  B. Simpson Proportional Counter 267
  C. Low Background Counter 268
  D. Beta-Gamma Counters 268
    a. General 268
    b. Gas Counters. 268
    c. Water Counters 268
    d. Dust Monitors. 268
  E. Special Instruments. 268
 6. Portable Survey Instruments 269
  A. Area Monitors—Alpha Types 269
    a. Poppy. 269
    b. Zeuto 269
  B. Area Monitors—Beta-Gamma Types 269
    a. Zeus. 269
    b. Lauritsen Electroscope 270
    c. Landsverek-Wollan Electrometer 270
    d. Victoreen Radiation Meter. 270
    e. Betty Snoop 270
    f. Condenser Chambers 271
    g. Portable G.M. Tube Sets. 271
    h. Sigmion. 271
    i. Victoreen Proteximeter. 271
  C. Neutron Meters 272
    a. Differential Chambers. 272
    b. BF_3 Counters. 272
 7. Portable Dust Monitors. 272
 8. Personnel Meters, Beta-Gamma 272
  A. Film Badges 272
  B. Pocket Meters. 272
  C. Neutron Meters 274
VI. Some Elementary Formulae and Calculation Methods. 274
 1. General 274
 2. Shielding 274
 3. Emission from Complex Sources. 274
 4. Emission from Sheets or Blocks 275
 5. Emission Inside Large Active Masses. 275
 6. Tolerance for Internal Emitters. 279
  A. Simple Case.. 279
  B. Other Variables 279
  C. Standard Man 282
References 282

I. Scope of Health-Physics

1. Introduction

Shortly after the discovery of X-rays and radium, the damaging properties of penetrating radiation were observed. The earliest injuries affected the surface of the body, notably the fingers and hands, and steps were taken to adjust the exposure of the individual in such a manner that these injuries would not arise in future cases. It was some time before more insidious forms of damage, frequently not accompanied by easily recognized early signs of injury, were observed. This period was one in which many people exposed to radiation were subsequently afflicted with anemia. Throughout the history of the use of penetrating radiation, steps have been taken to organize the work program in such a manner that observed injuries would not be reproduced in the future. In practically all cases, protection has been a *corrective procedure* following a series of misadventures, and in some cases there was a situation approximating a public scandal before suitable remedial measures were established. Only within the last few years has there been attention to prophylactic action against new hazards before injury experience was developed. In a general sense, Health-Physics has existed for 50 years, of which the last 25 have seen steady improvement in the techniques involved. The particular title, Health-Physics, was a product of the war-time organization within the Plutonium Project. There has been no absolute definition of the scope of the proposed new subject of Health-Physics, but in general it is concerned with the physics and biophysics involved in the interaction of radiation with the human body, with special emphasis on the protection of radiation workers against the potential hazards of their occupation. Health-Physics is, thus, a restricted aspect of Medical Physics in which the basic intention can be seen from a study of proposed alternative titles.

2. Nomenclature

While Health-Physics was used as the descriptive title at the Metallurgical Laboratory in Chicago and at the Clinton Laboratories in Oak Ridge, the equivalent organization at Hanford Engineer Works was called the "Health Instrument Section," (commonly abbreviated H. I.) for security reasons. The title "Radiation Hazard Control" has also been proposed, and this is probably the closest short description of the scope of the program. A less ambitious title, "Hazards Evaluation" has recently been introduced at the Argonne National Laboratory. The intention here is similar except that a measurement or evaluation of the hazards without any promise of effective control is implied

The pre-war title "Radiation Protection," which is understood to mean the protection of workers against radiation, is certainly not out of place, and would be more directly in line with previous experience outside the Manhattan District Project. Perhaps the only justification for regarding Health-Physics as a new subject is that the recent large-scale expansion in the use of radiation and radioactive materials has apparently established the need for a group of men engaged full-time in the considerations of radiation protection. In previous experience, medical physicists were required to give only part-time attention to the protection problems involved in their occupation. Whether Health-Physics will become a permanent branch of Medical Physics of sufficient magnitude to justify separate definition is a matter for debate; it is entirely conceivable that the subject approximates its maximum status at the present time. Despite the fact that utilization of nuclear machines and other uses involving radiation and radioactive materials will undoubtedly expand during the next few decades, it is to be expected that the methods of hazard control will become sufficiently stabilized that specialists may no longer be required to regulate safety aspects, with the exception of the consultant services of a small group of qualified experts. This would restore the relative importance of the subject to its prewar status.

3. Nature of Advances

Regardless of nomenclature, *fundamental* advances in the protection aspects of Medical Physics have been few or nonexistent. This by no means belittles the advances that have been made. It does imply that they are not of fundamental research character, but fall more into the development field or are practical problems of group organization and protection policy. This chapter will describe these phases, and will at the same time specify indirectly what is meant to be included in the expected proper field of Health-Physics.

Health-Physics is a borderline subject surrounded by Industrial Medicine, Radiobiology, Industrial Safety, Public Health, Physics, Engineering and Chemistry. The area between these boundaries is ill-defined. Whether any area at all is to remain will be largely determined by the experience of laboratories and industrial organizations with the present Health-Physics groups. The immediate prospect appears to be that such groups have thoroughly demonstrated their value and adequacy in the case of industry, whereas laboratory groups may cover the same territory by suitable extensions of the various surrounding subjects.

II. Past Experience in Protection

1. Luminous Paint Industry

Of the various modes in which radioactive materials have been used in the past, that concerned with the handling of radioactive luminous compounds has received the most careful consideration. Acceptable standards for the handling of such materials have been published in the National Bureau of Standards Handbook H-27 (1941). The known hazards, in the order of their importance, are, (1) ingestion or inhalation of solid radioactive luminous compounds; (2) inhalation of radon liberated from compounds into the air; and (3) exposure of the whole body to γ radiation from compounds.

A. RADIUM POISONING

Accumulation of small quantities of radium in the body may result in eventual damage to the blood-forming organs, or to the bone-forming cells, with malignant change as the end-result. Such injuries have been known to result fatally when the fixed radium content of the body was approximately 1 μg (Evans, 1943). The accepted limit for radium deposition in the body is 0.1 μg. The inhalation of radium dust is more hazardous than its ingestion. Approximately 100 μg administered orally, or 20 μg intravenously or by inhalation results in the deposition of 1 μg of 'fixed' radium (Rajewsky, 1939). In the past there has been inadequate attention to the possible concentration of radium dust in the air, whereas ingestion, which now must come largely from transfer from the fingers, is rather easily and routinely controlled. Emphasis on the ingestion hazard is possibly a legacy from the early catastrophe in the widely-known New Jersey poisoning case in which the material was primarily ingested by the pointing of brushes at the lips. In current plant practice, the greatest body radium content frequently is found not in dial painters, but in inspectors and assembly workers who remove small chips of the dried compounds in the course of their work. The protection of radium-dial painters against the accumulation of radium in the body depends largely on good housekeeping, which includes the provision and maintenance of a clean work location, and of suitable protective clothing. Power ventilation is used to remove radium dust as well as escaping radon. Routine breath radon analyses are used as the most sensitive index of radium accumulation. A radon concentration of 1 $\mu\mu$c/l. of air is said to correspond to a body content of about 0.1 μg Ra. However, there is no general agreement that the approximately 50% escape of radon represented by this figure always occurs. For radium of long fixation, emission as low as 7% has been quoted (Jones and Day, 1945). Despite the precautions taken, substantially all dial painters, inspectors and assemblers ingest or inhale a sizeable fraction of the tolerance amount of radium, and about 15% exceed the tolerable amount (Evans, 1943). In another report, 30% of workers of less than one year's standing and 60% of all others showed breath samples above tolerance (Morse and Kronenberg, 1943). The worker removed from radium exposure usually excretes enough radium to restore the

HEALTH-PHYSICS AND RADIATION PROTECTION

content to less than the tolerance amount. Experience of overexposure is reported from locations in which considerable care is taken to offer proper working conditions to employees. It represents the results of a philosophy in which temporary excursions above the tolerable amounts are considered proper. The efficacy of this approach depends on the frequency of the breath tests, the numerical accuracy of the results, and the extent to which temporary overexposures are harmless. In one published series, a maximum of 850% of the tolerance amount was observed. With breath analyses, every 4-6 months, a substantial portion of the radium in the body at the time of test may be permanently fixed, and accumulations of this order of magnitude would not readily be excreted down to the tolerance level (Hoecker, 1944). Conservative practice should lead to a work system in which the deposited amount *at no time* exceeds the stated tolerance value. Routine measurements of active dust concentration in the air seem to be uncommon, despite the ease with which they could be made, and the stimulus they might give to the maintenance of improved conditions.

B. RADON EXPOSURE

Evidence on the exposure of humans to radon in the atmosphere is less definite than that arising from deposits of radium in the body, but it is generally accepted that a tolerance value of 10 $\mu\mu c$ Rn/l. is safe. In European practice, 100 $\mu\mu c$ Rn/l. has been frequently used as an acceptable standard (Russ, 1943; Jones and Day, 1945). Maintenance of an acceptable air concentration in factories has been achieved by suitable design of a ventilating system. An intake-rate of about 50 cubic feet per minute per hood is the customary standard. The accepted standards of ventilation appear to insure that the room air content does not exceed 30% of the tolerable concentration. However, in a survey of 10 plants, Evans showed about 15% of the samples above tolerance. These values occurred mainly in storage or packing rooms and in offices not provided with suitable ventilation. Morse and Kronenberg reported excessive concentrations in 60% of the tests. Wartime experience in England has shown normal concentrations in luminizing rooms and radium laboratories, in the range of 20-100 $\mu\mu c$/l., with some values as high as 8000 $\mu\mu c$/l. There has been no apparent ill effect up to the present time, but the experience should ultimately be of interest in determining whether the American standards are unnecessarily stringent.

C. GAMMA RADIATION

The third hazard, namely, that arising from the external irradiation by the γ-rays from radium and its decay products, is no longer a serious consequence in any location in which a legitimate attempt is made to meet the recommended standards of protection. Control is achieved by limitation of the working amount of compound, and by housekeeping to reduce general irradiation in a room. The normal exposure appears to be about 20 mr/day.

D. GENERAL POLICY

The provision of the necessary protective mechanisms in this industry is expensive, and its continued proper operation requires careful examination. In modern plants it appears that the proper safeguards lead to conditions close to the supposed tolerance standards, an acceptable condition if those standards have been written with a reasonable margin of safety. However, there is not equal diligence in all parts of the country in maintaining such standards. Relocation of luminous paint industry in regions which have not previously had occasion to regulate such operations may lead to a relaxation of standards. Education of radium workers in the hazards of their occupation is one of the best mechanisms for promoting the preservation of good conditions. The trained worker can assist the conscientious employer in the preservation of safety and can suitably guide himself in other cases.

2. X-Radiation in Hospitals

X-ray protection has received detailed consideration, especially by the Advisory Committee on X-Ray and Radium Protection, which has published and revised at suitable intervals a unified set of safety recommendations (National Bureau of Standards Handbooks HB-15 and HB-20). Valuable additional details have been published by many authors (Braestrup, 1938, 1942; Taylor, 1944; Glasser et al., 1944). For protection considerations, the field of X-rays is subdivided into five classifications: (1) diagnostic, (2) superficial therapy, (3) deep therapy, (4) super-voltage therapy, and (5) multimillion-volt therapy. In each case, there are two primary types of possible protection failure; the one being accidental exposure to a single damaging dose or excessive dosage in a short series, and the other the eventual damage arising from prolonged exposure to low intensity radiation. Accidental exposures to physicians and technicians arise primarily from carelessness in entering the radiation beam, while accidents to patients come from omission of the proper filter, treatment at the wrong distance, treatment for an excessive period, or by an outright electrical defect in the equipment. Few, if any, of these occurrences fall outside the category of negligence on the part of physicians or technicians; their occurrence in modern usage is rare, and many are beyond the scope of Health-Physics. The hazard of principal concern is that arising from the small exposures which occur daily, and in this field there is a rather sharp differentiation between the diagnostic and therapy cases.

HEALTH-PHYSICS AND RADIATION PROTECTION 231

A. FLUOROSCOPY

In fluoroscopy, the physician must necessarily observe the fluorescent screen. The screen itself has to be backed by glass of sufficient thickness to reduce the transmission of radiaton to a safe level, and be surrounded by wings of protective material of adequate thickness. Protective aprons and gloves may be required by the physician or technician where the equivalent protection cannot be built into the equipment itself. The requirements are well known, but not always well respected. Possibly the greatest hazard in fluoroscopy arises in those cases where the physician is required to introduce his hands into the direct beam for palpation. This can be done safely when lead rubber gloves are worn. In the practice of some institutions, however, these examinations are made without protective gloves, the exposure received being held to a minimum by selection of the smallest possible irradiated field, and by fast operation. There is no unanimity of opinion on the permissible daily dose to the fingers under such conditions. The Mayo Clinic group has made careful attempts to measure the exposure which in their experience had proved to be safe (Cilley, Kirklin, and Leddy, 1934, 1935a, b). This was about 250 mr per day and as much as 700 mr per day to the finger tips. On the other hand, it has been suggested that radium exposures as low as 4 r per week have been known to produce injury (Parker, 1943). It seems evident that the restriction to 100 mr per day as required for whole body radiation is unnecessarily severe for this particular case, but whether 200, 500 or 1000 mr per day should be substituted has not been established. The continued widespread occurrence of finger damage among fluoroscopists should be sufficient to encourage the maintenance of a conservative standard. The risk of injury in fluoroscopy is greatest among those whose utilization of the equipment is sporadic. Orthopedic surgeons and veterinarians are particularly vulnerable in this sense, and reminders of the hazard (Harding, 1945) are much in order.

B. PHOTOFLUOROGRAPHY

The recent extension of photofluorography introduces additional hazards for both patient and operator in comparison with the previous routine chest film technique. The average entrance dose in photofluorography is about 1 r (Birnkrant and Henshaw, 1945), and as high as 2.5 r in some cases (Gamertsfelder, 1943). This exposure is approximately twenty times that required by direct radiography. It is preferable that radiation workers be given the advantage of the low exposure method. Repetition of a photofluorographic exposure, which appears to be required with significant frequency in some clinics, is especially out of place for a radiation worker. The

232   H. M. PARKER

increased exposure results in a corresponding 20-fold increase in the room scatter and may also lead to overexposure of the operator, if suitable restrictions on permitted approach to the patient are not maintained.

C. RADIOGRAPHY

Properly conducted radiography has none of the hazards peculiar to fluoroscopy. It can be conducted like superficial therapy. That the fingers of dentists and their assistants continue to show damage as a result of holding films in position is directly ascribable to inadequate training. The customary practice of having the patient hold his own films is presumably sound, but in one community of radiation workers a simple wooden device has been used to avoid this unnecessary exposure, and it is surprising that the same method is not universally employed.[1]

D. THERAPY

In the various branches of therapy, it is the exception to find injuries to physicians or technicians working under modern conditions of shielding with up-to-date equipment. The protection of the room in which the tube is housed is usually such that the normal work conditions give negligible exposure to persons outside the treatment room. The increasing use of electrical interlocks and similar safety devices helps to prevent such errors as operation of the tube with the lead protective doors open. Training of the operator to keep all except the patient outside the therapy room during treatment also seems to be successful. In one series of film tests (Clark and Jones, 1943), 84% of the films from a hospital read less than 100 mr per day, and 56% from a hospital where some diagnostic work was done. Two other locations, doing mainly diagnostic work showed 99% less than 100 mr per day. It should be noted that 200 mr was the accepted daily dose, and 93% of all cases came within this limit.

Protection of super-voltage therapy equipment has been for the most part at an acceptable standard, although greater possibility has existed in these cases for casual overexposure. This has sometimes arisen from the combination of the therapy tube with facilities for physical measurements, and is therefore not properly chargeable to X-ray therapy.

Multimillion-Volt Therapy, which is intended to cover the range from approximately 10 million volts up to perhaps 300 million volts, has not yet been extensively developed, but in this case an attempt is being made to establish proper safety standards before the equipment is used (Failla, 1945).

---

[1] In the same location, the "avoidance of unnecessary exposure" principle has led to a critical attitude toward the use of radiation in shoe stores, *etc*.

3. Radium in Hospitals

The protection of workers against exposure to radium, particularly in the applications to hospital practices, has been formulated by the Advisory Committee on Radium and Radium Protection, and published as National Bureau of Standards Handbook H-23 (which superseded the earlier H-18). Protection is required against, (1) local overexposure to radiation, especially upon the hands, and (2) overexposure of the entire body. Frequently overlooked in the manipulation of radium sources is the possibility of leakage which may permit the emission of radon and its accumulation in dangerous quantities in confined spaces such as radium safes (Read and Mottram, 1939). In the past, there has been a record of injury produced by rather short-time exposure to γ radiation from radium, particularly at the time when the so-called radium bombs were first developed. In present practice, the hazards are essentially confined to those which arise from long-continued exposure at low radiation intensities. The planned application of radium to patients for therapeutic reasons can easily result in damage to the patient; this is the physician's responsibility. However, it is possible for the patient to be unnecessarily injured by the introduction of leaking radium needles, or by the superficial application of radium sources with improper filtration. These in part would be considered Health-Physics problems, and methods of avoiding such injuries should logically be included in the general rules for radium protection. More generally, the injuries from hospital handling of radium occur to (a) technicians assigned to the preparation of treating units, (b) therapists and nursing staff in the operating theatre, and (c) nurses when patients are hospitalized in wards.

A. HAND EXPOSURE OF TECHNICIANS

It is especially prevalent to find that technicians tolerate a fairly close approach of the hands to radium during manipulations. In the past, there has been no specific attempt to maintain the hand exposure at the limiting value for general exposure. In fact, as recently as 1934, 5 r per day was described as the permissible exposure for the fingers (Handbook H-18). At the present time, by far the greatest amount of injury to radium technicians occurs on the hands, and there are certainly cases in which hand exposure of closer to 5 r per week than 5 r per day have resulted in skin changes which may later assume a dangerous character. The conception that a certain degree of skin injury can be tolerated and that the technician may terminate his handling of radium sources at such a stage without the expectation of subsequent deterioration in his condition has been widely sponsored (Failla, 1932; Nuttall, 1943). Although true as applied to the primary injury, it appears that radiation effects skin changes in such a way that subsequent trauma, inconsequential in a normal hand, may lead to breakdown in the irradiated skin up to several years after the termination of significant exposure. Even the professional worker finds it difficult to avoid the minor trauma that is sufficient to initiate degeneration. Precise measurements of the actual exposure of fingers of radium technicians have been infrequent. The contribution due to γ radiation is fairly well known, both by measurement with thimble chambers and by calculation from time and distance of exposure, but the additions due to β radiation are quite variable. Where sources with a minimum filtration of 0.5 mm. Pt are employed, it is rare for the secondary β-ray dose to exceed 25% of the γ-ray dose. With other sources, especially glass capillary tubing containing radon, the β-ray contribution may be extremely high.

B. GENERAL EXPOSURE OF THERAPISTS

Whereas the risk of overexposure to the radium technician may be greatest with respect to the fingers, it appears that radium therapists should consider the general body radiation as the determining factor. Such men will usually receive more radiation to the hands than to the rest of the body. It is, therefore, tacitly assumed that the limiting tolerance is higher for the hands. It has been by no means uncommon in the past decade for therapists to retire from active radium work as a result of a progressive change in the blood count, used as an index of the welfare of the individual. It has been shown (Paterson, 1943) that the average worker in the Holt Radium Institute, Manchester, shows a permanent reduction in white-cell count in comparison with other persons with similar general surroundings without radium exposure. The stage at which white-cell changes can be considered harmful is not yet known, and modern tendency has been to regulate the exposure in such a way that the average white-cell count more closely approximates the normal value. It is of interest that the best evidence for a sustained and possibly non-injurious fall and reduction in the white-cell count comes from this Institute, where considerable attention has been given to proper protection (Nuttall, 1939). This Institute has operated with 200 mr per day as the tolerable exposure, and has in fact condoned exposures closely approximating 100 mr per day. Common practice in radium institutions has permitted daily general body exposure of about 70% of the tolerance value adopted. Experience in a typical teleradium installation has shown average daily exposures between 130 and 250 mr (Wilson, 1940). This experience is extremely useful in its applications here by providing a body of evidence of exposure at the supposed present-day tolerance limit. It is not known to what extent additional exposure was tolerated in England during the war period when staff reduction was necessary.

4. *Industrial Preparation of Radium Sources*

Preparation of radium sources may be considered to begin with the mining and refining of uranium ore. The health conditions under which this has been done have not been clearly defined except in the experience of the Canadian Great Bear Lakes deposits (McClelland, 1933; Leitch, 1935). The methods of protection in the purification stages are similar to the requirements in the luminous compound industry, with the exception that as the matter is further purified the hazard is increasingly that of general body exposure to γ radiation. The further stage of subdivision of the radium into capsules for hospital use has not been governed by general regulations of the type applied to the luminous paint problem. Such work is performed by relatively few companies, and the standards of protection accepted by them appears to be variable, and in some cases less desirable than the standards proposed by the Advisory Committee on X-Ray and Radium Protection. The average radon concentration in the general laboratory of one of the most reputable companies has been quoted as 2200 $\mu\mu c/l$. (Read and Mottram, 1939). The operation of radon plants, either commercially or in hospitals, can be one of the most hazardous modes of handling active materials since it may lead to radon contamination of the air as well as hand exposure to β radiation and general body exposure. Such installations as that designed and built by J. E. Rose in the U. S. Marine Hospital, Baltimore, Md., however, demonstrate many of the best protection practices. In general, a radon plant should be located in a separate building to localize contamination.

5. *Industrial Radiography*

Radiography has had to be guided by the recommendations of the Bureau of Standards that were specifically made for hospital application. The wartime publication of rules by the American Standards Association (1945) has been a valuable advance. More could be achieved by future closer liaison between all organizations using radiation methods. Industry has developed considerable experience with supervoltage X-radiation, and the use of betatrons at 20 MV and even up to 100 MV may be common practice within a year or two. Radium experience, especially in naval ordnance, has also been extensive. The wholesale production of fission products has opened up the portentous possibility that sources of 100 or 1000 curies may be used in the future for radiography where sources of small dimensions are required.

6. *Transportation of Radioactive Sources*

The shipment of radium, radon or similar radioactive substances through the mails has been prohibited by postal regulations in the United States.

Shipment of radium up to 100 mg. in a single shipment is permitted by Railway Express under a well-established code of protection (Curtiss, 1941). These regulations were designed to prevent damage to photographic films, an immediate effect subject to damage claims. Possible damage to individuals by the transportation of radioactive sources may not become evident for a considerable length of time, and the possibility of claim and subsequent correction of malpractice becomes poor. In this instance, the public has been protected against exposure hazards by the greater sensitivity of film than of the individual. Shipment of larger quantities of radium has been performed either by special license or in some instances by courier, in which case persons in the vicinity of the source have not always been satisfactorily protected. The existing regulations have been rendered inadequate by the recent increase in shipment of active materials and by unexpected changes in its character. The establishment of revised regulations is one of the most important functions of any organization concerned with radiation protection. These will have to include special consideration of neutron sources, active materials with parent-daughter relationships, where the hazard increases during shipment, liquid samples and α-emitters where the hazard is not detectable by such simple means as Geiger counters.*

7. Universities and Similar Institutions

With the development of atomic and nuclear physics, particle accelerating equipment—such as high voltage X-ray machines, positive ion tubes, cyclotrons and linear accelerators—have become familiar features of the average University laboratory.

Pioneer work has been done with this equipment, and it is unreasonable to expect the highest standards of protection at all times. The radiation physicist was required to protect himself against neutron irradiation before the neutron beam could be used to irradiate patients or even animals. The radiochemist prepared new radioisotopes before the biochemist could determine their metabolism. That very few men have been permanently disabled by radiations from these machines and their by-products is much to the credit of the scientists responsible for these installations. Nevertheless, when the pioneer stage has been passed, there has been some reluctance to install safeguards comparable with industrial safety practice.

A. RADIOCHEMICAL MANIPULATION HAZARDS

At the low level of activation produced by exposure to most nuclear machines, no particular damage has in general occurred as a result of sam-

* *Note added in proof:* At the request of the Interstate Commerce Commission, the Bureau of Explosives, Association of American Railroads, assisted by the Subcommittee on Shipment of Radioactive Substances of the Committee on Nuclear Sciences of the National Research Council, formulated such improved regulations, effective January, 1948.

ple manipulation. This has been helped too by the fact that much of the handling has been done by graduate students not assigned to such work for prolonged periods. This has fostered the belief on the part of many of the leading radiochemists in the country that the hazards from hand exposure in the range of 200 mr per day to 2 r per day have been greatly exaggerated by the health physicists. At the present time, this is one of the major potential weaknesses in radiation protection in the country. Even when an attempt is made to maintain safe levels, the measurement of contact exposure with an instrument suitable only for remote measurements frequently gives results low by one or two orders of magnitude.

Future experience will be with sources which produce activated materials of greatly higher intensity. A change in work habit as a result of increase in source strength is one of the more difficult developments for the scientists to make, since the chemical and other associated properties remain unchanged, while only the relatively intangible radiation levels go up. The advantages of proper shielding and remote handling in future radiochemical operations cannot be too widely publicized (Levy, 1946).

B. GENERAL RADIATION INJURIES

In addition to the hazards of radiochemical manipulation of laboratory materials, the generating sources themselves are a potent source of injury, and it is recognized that they have, in many cases, been operated under conditions which depart from the ideal. Again, the number of eventual injuries which may arise from long-continued exposure at low intensity is not known, and is probably held to a minimum by the frequent turnover of personnel under University laboratory conditions. Attempts have been made to specify proper protection against new types of radiation (Aebersold, 1941; Warren, 1941). In the more immediate field of visible radiation injury occurring either in a single accident or as a fairly rapid result of overexposure, more can be stated about the customary standards achieved. There appears to be no comprehensive survey of the frequency rate of such palpable injuries in radiation laboratories, but a private survey led the writer to propose the figure of one palpable injury per 30-man-years of active employment in radiation work. Subsequent discussion has indicated that many health physicists would accept a higher rate such as one injury per 20-man-years as a representative value. Whether the rate is 1:20 or 1:50 years, it compares very unfavorably with a record such as that at the Hanford Engineer Works where over 3,000-man-years of active exposure have been compiled without a single palpable radiation injury. Original research work carried on in laboratories is of such a character that safety in handling cannot be expected to equal that in a well-regulated, established industry. However, the differences are sometimes over-emphasized, and, in the particular example cited, the first operation of Power Piles and the wholesale production of plutonium can hardly be described as well-established industries. The postwar interest in all the major radiation laboratories in maintaining radiation safety comparable with Project standards may be a deciding factor in the permanent establishment of Health-Physics.

8. Manhattan Project Experience

A. DEVELOPMENT OF THE HEALTH DIVISION

The anticipated scale of the radiation quantities involved in those parts of the Manhattan Project which were to be concerned with the development of chain-reacting units led to a fairly early recognition of the need for the establishment of an extensive Medicophysical organization for the protection of workers against deleterious effects of radiation. The Health Division, under the direction of Dr. R. S. Stone, worked along three major lines:

"(1) Adoption of pre-employment physical examinations and frequent re-examinations, particularly of those exposed to radiation.

"(2) Setting of tolerance standards for radiation doses and development of instruments measuring exposure of personnel; giving advice on shielding, *etc.*; continually measuring radiation intensities at various locations in the plants; measuring contamination of clothes, laboratory desks, waste water, the atmosphere, *etc.*

"(3) Carrying out research on the effects of direct exposure of persons and animals to various types of radiation, and on the effects of ingestion and inhalation of the various radioactive or toxic materials such as fission products, plutonium and uranium. (From Smyth, H. D., 1945)."

Paragraph (2) comes close to a definition of the Health-Physics activities on the Project. The purely medical activity of Paragraph (1) developed more and more along normal industrial lines, as the incidence of significant radiological findings proved to be essentially zero. Radiobiological research as in Paragraph (3) provided invaluable data for the progressive improvement of tolerance dose data, and in some cases there was a satisfactory integration between Radiobiology and Health-Physics.

The early work in the Metallurgical Laboratory differed neither in principle nor in magnitude from that previously practiced in several University laboratories. Since it developed from the putting together of the work of several such groups, the initial attention to radiation protection followed the standards of the parent institutions. In this period, also, the available staff for improvement in the hazard control and the necessary instrumentation for this purpose were both in short supply. However, after the suc-

HEALTH-PHYSICS AND RADIATION PROTECTION 239

cessful operation of the first chain-reacting unit in December, 1942, there was, throughout the Project, a steadily increasing interest in radiation hazards. The introduction of industrial organizations of the regular kind helped to stimulate the extension of protection rules and their more rigid observation. Those branches of the Project under such industrial control rather naturally became leaders in the establishment of functioning Health-Physics service organizations. The parent organization in the Metallurgical Laboratory continued to make valuable contributions, especially in the design and fabrication of new instruments.

B. TYPICAL EXPOSURE PROBLEMS

(a) *Shielding*. In principle, the Project program posed no new problems with regard to the irradiation of personnel by penetrating radiation. The required shielding of Pile units and of the chemical equipment involved in the Separations Process was deduced by a logical extension of existing modes of calculations of the absorption of radiation in such materials as concrete, steel and lead. Uncertainties in the calculation of the transmission through unusually large thicknesses[2] of these materials did not exceed a factor of perhaps 3, and if an entire structure was designed for an intended safety factor of 10, no great harm would have been done by finding a residual safety factor of 3 or 30. For Pile units themselves, the calculations of neutron absorption in the proposed shields represented a greater extrapolation from previous information. The early work of the theoretical and experimental physicists in this field was substantiated when the Pile structures were actually in operation. One interesting factor of the protective measures required in the new designs was the necessity for protection of workers from "sky-shine," *i.e.*, radiation scattered by free air from a primary beam which itself would not irradiate the observer. This had not been of practical consequence in earlier experience unless one considers a limited case such as the exposure of an observer standing just outside a primary X-ray beam. No new phenomenon was involved in the sky-shine calculations, which were essentially an exercise in the application of the Klein-Nishina formula. As a corrollary to shielding, the development of remote-control handling, both for laboratory tools and on the grand scale, became a fine art. Although steered by health physics, credit for this phase belonged to design engineers.

(b) *β Radiation*. The effects of β-ray exposure had to be closely considered, although, in fact, this experience was no different from that intrinsically available in the earlier radium knowledge, when the use of β-ray sources was a common practice. Reliable dosage calculations of these cases

---

[2] Greater than 3 feet of concrete or equivalent. The maximum thickness in standard tables was 10 cm. Pb, or 54 cm. concrete (Taylor, 1940).

proved to be so infrequent that the subject was essentially a new one. These β-ray exposures were expected to arise primarily on the hand, forearm and face; and typical sources of the activity were the natural uranium itself, materials activated in Piles, and the fission products. It was late in the Project history before biological data on controlled β exposures was available (Life, 1947).

(c) *Metabolism*. The metabolism of the anticipated fission products, and of uranium and plutonium, became of prime importance in the consideration of how much of each particular element could be allowed to come in contact with the individual by such methods as skin absorption, ingestion, inhalation, and introduction through open wounds. With the exception of uranium, this metabolism had to be evaluated by animal experience with meager amounts of the active materials, prior to the time that the operating Piles made them available in quantity. It was obviously desirable to have at least provisional answers on the toxicity of all these materials before this stage of plentiful supply was reached. The work of the biological organizations at the Metallurgical Laboratory, and later at Clinton Laboratories (together with the invaluable contributions of many associated laboratories, notably the Radiation Laboratory, University of California, at Berkeley, and the University of Rochester, Rochester, New York), cannot be evaluated too highly in the final assessment of the merit of the health protection of the Plutonium Project. Despite these efforts, the development of precise information on toxicity of radioactive materials at low levels over long periods of time could not be expedited by accelerated experiments, nor could the findings in animals be translated into the human case in terms of numerical tolerable dose. Fortunately, the relative deposition of active material within the animal was a good index of the corresponding deposition in the human case, and, where the toxicity was a function of the radiation effect, the health physicist could calculate the limiting intake of the noxious material to produce a specified dose in the organ most affected. This was the mode of approach to the establishment of tolerable dosage for the various fission products before any extensive body of exposure information had been accumulated.

(d) *Waste Disposal of Fluids*. As indicated in the Smyth Report, the operation of the facilities at Clinton Laboratories involved the emission of active gases, particularly radioxenon and radioiodine, into the atmosphere. It was the responsibility of the Clinton Laboratories to determine conditions of operation under which such emission would cause no hazard, not only to the operating personnel in the laboratories but to the large public area over which these gases might have been disseminated (Parker, 1944). This was successfully accomplished, and the method of approach established a satisfactory standard for the consideration of allied problems at

HEALTH-PHYSICS AND RADIATION PROTECTION

future locations faced with the problem of large-scale disposal of active gas. The second problem was the related one of disposal of active liquid wastes. This was solved by the retention in underground storage of all very active wastes and the dilution of other materials to such an extent that no possible interference with the safety of public water supplies was effected. The required standards for waste disposal of this type constitute one of the most pressing problems for future Health-Physics considerations. This is particularly true in connection with the operation of radiochemical laboratories in urban areas, where the disposal of certain types of waste to the sewer is complicated by the precipitation or absorption of the active material in the sewer pipe or its soil contents. Under unfavorable conditions this could produce a hazard at a much later date to unsuspecting persons repairing or otherwise handling such pipes. In large bodies of water, the concentration of activity in algae or in colloidal materials with its possible utilization by fish, later used for food, presents a chain of events of great consequence to the public health. Up to the present time, these problems have been bypassed by ultraconservative policy in waste disposal, but future pressure for economical disposal facilities greatly point the need for extensive research on these problems.

(e) *Waste Disposal of Solids.* All future radiation organizations will be faced with the disposal of solid contaminated items which may range from collections of waste paper used to clean contaminated equipment to large pieces of equipment which are so contaminated that their restoration to service would be more expensive than their replacement. As regards its basic nature, this problem is no new one, since it is well-known that the photographic industry has had to institute routine tests for activity of the paper used to package photographic film, as a result of the uncontrolled return for reprocessing of paper contaminated in luminous compound operations. If this source is enough to contaminate the paper stock of the country, it is easy to visualize the condition set up by the indiscriminate use of similar pieces of paper from operations which dwarf the activity of the entire radium industry. Obvious means of preventing the return of such material to the public domain are storage in special vaults, burial in underground pits, or dumping in the sea, which are currently employed. The implications of contamination of the stocks of pure metal, for example, which might occur by careless release of contaminated materials or scientific equipment require careful consideration by some Advisory Board of high caliber.

(f) *Protective Clothing and Decontamination.* Although special clothing is provided in such industries as luminous compound operations, the requirements of the Manhattan Project for deconatmination of clothing, equipment and personnel, constituted almost a new subject. There appears to be no absolute guide to the preferred method of decontamination, but information of general interest is that fission products can be most suitably removed from clothing by laundering in dilute acetic acid, and that, whenever the condition of the equipment warrants its use, nitric acid is a preferred cleanser for practically any radioactive material. For the removal of radioactive contamination from the hands or other parts of the body, a titanium dioxide paste has been found effective against fission products, while an application of potassium permanganate solution, followed by sodium bisulfite, is effective against Plutonium contamination. The use of nitric acid and hydrochloric acid on the hands has been condoned under carefully regulated conditions.

(g) *Safety Performance.* The overall safety record of the Project enterprise has been remarkable. At the Hanford Engineer Works, the largest organization concerned with virtually all the forms of activity of practical interest, the exposure record for 1946 shows, for example, that no single individual received as much as 200 mr in any one day; that only 4 individuals received more than 100 mr in any one day; and that the average daily exposure of all radiation workers was less than 5 mr per day. There has been no known instance of the ingestion, inhalation, or other mode of introduction of a damaging amount of radioactive material into any individual. This standard was achieved with a working force of less than 3% of the total payroll. The financial considerations involved have, therefore, not been out of proportion, although it has been suggested (Bale, 1946) that industrial economies might lead to a reduction in the present coverage. It is felt, rather, that in the future expansion of radiation work, the laboratories and similar institutions might be the ones which would apply somewhat less than the present rate of 3% to the protection program. It is just these institutions which most need to maintain a technical force engaged in hazard control, because of the variable nature of the basic hazards encountered in a research program. In the industrial field, it is evident that stabilization of a given operation, additions to the protective equipment, and the provision of improved instrumentation, especially in the field of automatic recording, will lead to a logical and safe reduction in the full-time protection force. However, the foreseeable change is not such as to reduce the percentage representation appreciably below 2%.

III. Exposure Standards

1. Dose Units

Discussion of tolerance dose has been obscured in some cases by ambiguity or misinterpretation of the meaning of "dose" as applied to radiation exposure. "Dose," as intended in radiobiology, is a measure of the energy absorption of radiation per unit volume or mass of tissue. In practical usage,

HEALTH-PHYSICS AND RADIATION PROTECTION 243

dose is measured in terms of the energy absorption in air which closely parallels the absorption in tissue for the particular range of X-radiation and γ radiation for which the principles of dosimetry were established (Gray, 1937a,b). The practical unit of dose, the 'roentgen,' corresponds to the absorption of 83 ergs/g. of air. This is usually considered equivalent to the absorption of 83 ergs/g. or /cc. of tissue.[3] It is convenient to use subsidiary dose units based on the energy absorption/cc. of tissue. That "dose" of any ionizing radiation which produces energy absorption of 83 ergs/cc. of tissue is defined as 1 rep. This is an abbreviation of "Roentgen Equivalent Physical," an expression which replaces such other titles as tissue roentgen, roentgen equivalent, or equivalent roentgen, which have become familiar in the literature. This unit may now be applied to X-radiation, γ radiation, α, β, proton, or neutron irradiation. It is well known that equal energy absorption arising from exposure to these various radiations will not produce the same biological effects. When the additivity of different types of radiation is to be considered, an additional unit—the "rem"— is introduced. One rem is that dose of any radiation which produces a biological effect equal to that produced by one roentgen of high-voltage X-radiation. The title "rem" is derived by abbreviation of "Roentgen Equivalent Man" or, optionally, mammal. It was selected in preference to the more obvious "roentgen equivalent biological" to avoid the confusion between "rep" and "reb" in speech. It is obvious that biological equivalence may depend on treatment conditions, protraction, fractionation, tissue exposed, and so on. In some cases, there can be no true equivalence since one type of radiation may produce damage of a character never observed with other modes of exposure. Moreover, the mere introduction of the unit "rem" does not solve the problem of writing down the numerical value of the equivalence, which remains a complex problem in experimental biology. Despite these disadvantages, the use of the expressions "rep" and "rem" has been found to simplify discussions of exposure to radiations other than X or γ radiation, and especially to the mixed radiations that have been common in the experience of the Plutonium Project. For tolerance purposes, only the relative values of "rep" and "rem" for long-continued exposure to low intensity radiation need be considered. An accepted scale of relation is as follows:

X-rays and γ-rays	1 r = 1 rep =	1 rem
β-Rays	1 rep =	1 rem
Protons and fast neutrons	1 rep =	4 rem
Slow neutrons	1 rep =	4 rem
γ Particles	1 rep =	10 rem

[3] The actual energy absorption/cc. of tissue varies between 42 and 89 ergs/cc/r in fat, and between 87 and 95 ergs/cc/r in muscle in the energy range 12 KEV to 830 KEV (Spiers, 1946).

2. Tolerance Dose

For the present purposes, tolerance dose or permissible exposure will be assumed to be that dose to which the body can be subjected without the production of harmful effects. When the exposure considered is given on a single occasion, the tolerance dose or permissible exposure is clearly defined. Health-Physics is commonly concerned with the case in which repeated small exposures occur, and it is not entirely clear whether dose in this connection should refer to the total dose or to the element of dose in a given time. A suitable convention follows the latter view, with the given time as one day.

3. Tolerance Dosage-Rate

"Tolerance dosage-rate" has to be interpreted as that dosage-rate which is continuously tolerable. It will be maintained that the daily tolerance dose is 100 mr (in the general case). If one writes that the tolerance dose is 100 mr/day, it is argued that this is dimensionally a dosage-rate, and that one should write "tolerance dosage-rate is 100 mr/day." This statement has an entirely different meaning, and implies the existence of a maximum permissible rate of about 3×10^{-6} r/sec.

4. Tolerance Dose Versus Tolerance Dosage-Rate

The question of interpretation between "dose" and "dosage-rate" as applied to the tolerance problem is fundamental in the consideration of the permissible exposure of the body. The tolerance dosage-rate of 10^{-5} r/sec. was apparently introduced by Wintz and Rump (1931), who very clearly were equally content to quote a weekly or monthly total *dose*. The same election of convenient choice between tolerable exposure of 0.2 r/day or 1 r/week and tolerance dosage-rate of 10^{-5} r/sec. is contained in the current International Recommendations. It is entirely erroneous to suppose that these sources intended a *rate* in excess of 10^{-5} r/sec. to be considered hazardous. The utility of the rate was restricted to survey measurements around fixed installations, which would be safe if all readings fell below this limit.

The practical background on permissible exposure has, of course, come from exposures at much higher rates, *viz.*,

Occupation	Approximate Dosage-Rate of Exposure r/sec
Fluoroscopists: Hands	10^{-2}–10^{-1}
Body	10^{-4}–10^{-3}
X-ray Therapy Technicians	10^{-4}
X-ray Patients: Scattered radiation	10^{-2}
Radium therapists and technicians	10^{-4}–10^{-3}

The main body of experience has been in the range 10^{-4}–10^{-3} r/sec., and the rate of 10^{-5} r/sec. should receive no special significance. On the other hand, there is no sound information relative to high rates, except for erythema and therapeutic doses (McWhirter, 1935) where no significant rate dependence exists. In the tolerance field, daily short exposures at high rates *may* be more damaging than equal doses at normal rates. Simple control policy makes it impractical to condone receipt of a daily quota of 100 mr in less than 30 secs. This is a rate of 3×10^{-3} r/sec., only a factor of 10 from the ordinary case, and significant changes in this range are unlikely. For a single major exposure, of course, the damage produced appears to be a rapid function of the rate. Most, if not all, of the variation is due to the necessary change in total exposure time (protraction) in this single shot case.

The proposed restriction to a *daily* tolerance dose of 100 mr is arbitrary and occasionally unnecessary. One cannot prove (without a long animal experiment) that 200 mr and zero on alternate days or 300 mr followed by two clear days are unacceptable. For *control purposes only* the restriction to 100 mr per day is proper in almost all cases. Deliberate exposure of $100 N$ mr balanced by a radiation-free period of $(N - 1)$ work days is foolhardy if N exceeds 10, and unwise if N exceeds 3.

5. Foundations of Tolerance Experience

The development of any extensive experience of tolerance is restricted to exposure to X-radiation and γ radiation. The following account is included primarily to emphasize the scanty and inadequate data on which tolerance figures have had to be based even in these cases. Inadequate dose measurement and statistically insignificant groups have been common.

Although the first case of X-radiation injury was described in July, 1896, the first published tolerance dose appears to be due to Rollins in 1902. His photographic plate-fogging limit was perhaps 10 r/day. Early radiation injuries were primarily confined to the skin,—but the demonstrations of the radiosensitivity of the blood-forming organs (1904–1905) and of the reproductive organs of animals (1903–1904) carried fair warning that more dangerous damage than dermatitis could be anticipated. It is important and instructive, however, to note that the first *organized* step to insure protection from X-radiation was taken in 1915 (Russ). It has been pointed out by Henshaw (1941) that the war activity at that time resulted in delay on protective measures, and undoubtedly contributed to the large group of radiation injuries, especially aplastic anemia, manifest in 1919–1921. The American Roentgen Ray Society formed a Committee in 1920 to recommend protection measures, which were formulated and published in 1922. The British X-ray and Radium Protection Committee presented its first recommendation in 1921. The two sets of recommendations were similar and dealt largely with protective materials recommended for use in building X-ray and radium laboratories and apparatus.

At the second International Radiological Congress, 1928, protection proposals were carefully considered, and subsequently an International Committee on X-ray and Radium Protection was formed. The recommendations of this Committee contained no specific reference to a tolerance dose until the reports of 1934 and 1937 which described the tolerance dose as being 0.2 roentgen per day.

Mutscheller (1925) had published a tolerance figure of 0.01 of the erythema dose per month, based on measurements in several installations in which no apparent injury to the operators was being occasioned. This figure was later substantiated (Mutscheller, 1928) and subsequently extended (Mutscheller, 1934) for rays of higher penetration. Erythema dose for this quality was 340 r and the tolerance dose, therefore, 3.4 r per month, or about 150 mr per work day. The German Committee on X-ray and Radiation Protection (Glocker and Kaupp, 1935) accepted the same figure. Laboratory and hospital measurements by Sievert (1925) led to the same statement of safe dose. Typical of the necessity to extrapolate from insufficient data was the publication by Barclay and Cox (1928) of a safe daily exposure of 0.00028 of an erythema dose[4] on the basis of determinations on two individuals. This permissible exposure, equivalent to about 170 mr per day, was believed at that time to include a safety factor of 25-fold. A reconsideration of the early data (Kaye, 1928) led to a proposal of 0.001 of an erythema dose in 5 days, closely equivalent to 100 mr per day. The first comprehensive report on tolerance exposure to γ radiation (Failla, 1932) led to a value of 0.001 of a threshold erythema dose (radium) per month, or of the order of 60 mr per work day.

An important step in the improvement of protective practice in the United States was the formation of an Advisory Committee on X-ray and Radium Protection which published proposals in 1931 in the Bureau of Standards Handbook H-15. This and subsequent handbooks have been a model of sound approach to radiation hazard control. This first handbook recommended a tolerance dose of 200 mr per day, but the revised handbook (No. 20, 1936) quoted a value of 100 mr per day with no specific explanation

[4] The translation of an early result expressed in terms of erythema dose to the equivalent roentgen dose is by no means clear-cut, but the following table includes representative values:

Quality	To Produce Erythema
Grenz rays	100 r
100 kv	350 r
200 kv	600 r
1000 kv	1000 r
γ-rays (radium)	1500 r (threshold erythema = $\sim$1000 r)

of the reduction. In an independent publication, the chairman of the Advisory Committee referred to the safety value as 20 mr per day (Taylor, 1941). A League of Nations publication (Wintz and Rump, 1931) reviewed the various statements of tolerance dose, and concluded that the permissible exposure is 10^{-5} r per second, assuming a 7-hour working day, and 300 working days per year. This is equivalent to 250 mr/day. The exposure was qualified for persons "remaining in proximity to sources of radiation giving off rays without intermission," (*i.e.*, radioactive preparations) by reducing it by a factor of 3, or in round figures, to 100 mr per day. The limit of 100 mr per day has been widely established in American practice for both X-radiation and γ radiation although the higher value of 200 mr per day still remains in the International Recommendations.

6. Recent Considerations

Recently proposed changes in tolerable exposure for X or γ radiation are a fairly general desire on the part of health physicists to establish a lower rate for multimillion-volt radiation (Failla, 1945), as a precautionary measure until more is known about this field, and a suggestion that the exposure of women employees should be less than that of men (Lorenz, 1946), in consideration of the possible irreversible effects on the reproductive organs. In both cases, provisional values of 20 mr per day have been suggested.

The stimulus of the Manhattan Project has initiated much valuable experimental work in animals exposed over long periods of time to low intensity radiations. Not much is known of the results of these experiments beyond the fact that there is a general tendency to feel that the present limit of 100 mr per day is well chosen, although the safety factor involved is by no means as large as the earlier observers intended.

7. Biological Aspects of Tolerance Dose

For completeness in developing the current approach of Health-Physics to the tolerance problem, some considerations of the biological effects of radiation on the body is required; this is not the place for an extensive description or discussion of these effects. Radiosensitivity of tissues and its dependence on differentiation, rate of growth, and cellular environment are of basic importance in the study of the effects of radiation on the tissues. Also paramount is the distinction between threshold and non-threshold effects (Henshaw, 1944). Fortunately, the majority of radiation effects are thought to be of threshold type, which simplifies the practical problems of protection at the expense of introducing many variables into the manner in which radiation exposure can be legitimately received. The reversibility of radiation effects is of particular consequence in occupational exposure. This depends upon the reparative properties of the tissue, and where recovery is large the *total* radiation dose can be materially increased. When the potential damage in question is reversible, the protraction and fractionation of the given dose are major determinants of the end result. Repeated exposure initially followed by repair may eventually exhaust the regenerative reserve and result in permanent damage. Exhaustion of bone marrow reserve is typical. The dramatic decline in the welfare of the individual following a long period of apparent normalcy is one of the most disturbing problems for the physician and health physicist to face in the investigation of permissible exposure to any type of radiation. The damaging effects of protracted exposure are customarily divided up as follows:

A. GENERAL BODY EFFECTS

The early toxic signs in man resulting from external irradiation of essentially the whole body are (a) lassitude and fatigue (Nuttall, 1943), and (b) demonstrable effects upon the leucocytes of the blood (Goodfellow, 1935; Paterson, 1943; and many others). Either diminution in the total number of white cells or an altered ratio of neutrophils to lymphocytes are considered as possible indexes of early radiation effects, although it is well known that either may be caused by many other agents. Significant changes in the red blood cell picture are said to be late effects of continued overexposure. The coagulogram has been reported to be another critical index (Kaufmann, 1946). The consensus is that regular blood counts, carefully done, serve as an adequate early index of overexposure. Technicians can arrive independently at the same answer on a blood count within 10% (Cantril, 1943). The observation of the Manchester group (Paterson, 1943) that the normal white cell count of radiation workers is maintained at a level statistically demonstrated to be significantly lower than the general norm is of prime importance. If this can be accepted, then the white count index is certainly sufficiently sensitive, provided that the subjects of the Manchester experience are found not to suffer any eventual ill effects. Unfortunately, if shortening of life is one of the residual effects, then even the large Manchester group is too small to establish this difference, nor would the results be available soon enough to be of interest to the present generation of medical physicists.

B. INTERNAL EMITTERS

General body effects arising from internal radiation have been extensively studied in the special cases arising from the ingestion or inhalation of radium, mesothorium and their decay products. The effect on the blood-producing organs has been correlated with the level of radioactivity of the expired air (McClelland, 1933; Schlundt *et al.*, 1931).

Widely discussed, too, is the effect of radon inhalation in the lungs and its possible contribution to lung cancer in the Schneeberg miners (Peller, 1939; Dohnert, 1938). If lung cancer in man is essentially carcinoma of the bronchus, it can be shown that the expected daily dose to the bronchus for men in these mines is 300 mrep per day, approximately 20 times that derived as an average dose throughout the lung. Since the exposure is almost entirely due to α radiation, the equivalent daily dose would be 3 rem. By the same calculation, the tolerable concentration of radon in the air would be 10^{-13} curie/cc. to produce a bronchial irradiation of the permitted 100 mrem per day. It is of interest that this is the concentration considered tolerable in the British and Continental philosophy, and 10-times higher than the recommended value in the United States. Specific damage to the lung or bronchus is perhaps not properly included as a general body effect, but the inhalation of radioactive gases or dust has become a prominent hazard in recent experience, and whether the effect is limited to lung damage or becomes a general body effect is determined only by the rate of absorption of the toxic material through the lung surfaces. The general effects produced by the ingestion of radioactive material are only too well known, specifically with reference to radium (Martland, 1931). Careful studies by Evans and Rajewsky and their respective co-workers have shown that damage to the individual can occur as the result of the permanent deposition of approximately 1 μg Ra in the body, and the accepted safety limit is invariably taken as 0.1 μg Ra. This represents a daily irradiation of the average bone of about 16 mrep per day, or 160 mrem per day.[5]

C. SKIN EFFECTS

(a) *Erythema and the Layer of Passive Absorption.* The characteristic effect of large doses of X-rays and γ-rays upon the skin is the production of an erythema, the result of dilation of the fine capillaries, venules and arterioles supplying the skin. The outermost layer of the epidermis, the *stratum corneum*, consists of dead, hornified, flattened cells. The thickness of this outer layer may be 0.4 mm. on the palm, 0.8 mm. on the sole, and approximately 0.05–0.1 mm. elsewhere. This layer of passive absorption (Wilhelmy, 1936) may be considered as an *inert filter* which serves to protect the sensitive levels of the skin from injury by α-particles, low-energy β radiation, and extremely soft X-radiation. Prolonged exposure of the skin to natural α-radiation, therefore, produces no erythema and probably no

---

[5] The average daily dose of the bone when the whole body receives the tolerable X-ray dose is:

    500 mrep at 100 KEV (150–200 KV therapy)
    250 mrep at 150 KEV (200–300 KV therapy)

where a volume rather than mass basis for the statement of 'rep' is used.

significant change in the skin condition.[6] On the other hand, skin injury can occur with accelerated α-particles where the rays can penetrate the horny layer (Larkin, 1941). In the average case, β radiation is well able to penetrate the passive layer and to produce severe skin injury. In very special cases, the additional thickness of the horny layer is of value in reducing the effective β-ray exposure. This might be the case, for example, if the palm only were involved. In laboratory manipulation of β sources, the exposure of the dorsal surface of the fingers usually exceeds that of the palmar surface, and here the passive layer is relatively thin.

(b) *Late Skin Damage.* The more insidious forms of damage to the skin occur in occupational exposure when the administered dose extends over one or more years. There may be at no time an erythema, and the first injury sign may be ridge changes on the finger tips, epilation, polishing around the nail beds, fissuring or ridging of the nails, and skin dryness. Later evidence of injury includes telangiectasis, pigmentation, atrophy, and thickening with the appearance of wart-like growths. Ulceration may follow minor abrasions, which heal reluctantly. This may progress to cancer of the skin which, in some 25% of the cases (Colwell and Russ, 1934), may spread beyond the hands. The widespread occurrence of skin injury has led to a diligent search for improved methods of early recognition. Dental compound impressions of the skin ridges of the finger tips have been used in one case, and have shown promising results where the picture is not obscured by ridge flattening due to other causes. Another promising approach has been the microscopic study of the capillaries of the nail bed region (Nickson and Nickson, 1946). As indicated already, uncertainties as to the permissible skin dose exists. A limit of 200 mrep per day on the surface of the skin would be generally considered acceptable.

D. EFFECTS ON REPRODUCTIVE ORGANS

This most dreaded hazard in the lay mind is usually considered subordinate to the effects on skin or on the blood-forming organs (Nuttall, 1943). Formal attempts to estimate fertility reduction in terms of the family-size of radiation workers have been obscured by the influence of other social conditions on this factor. There is no clear-cut evidence of such a reduction, and there is, in fact, some suggestion of an anxiety-stimulated overcompensation. For accidental short-term overexposure, permanent sterilization is produced by 400–600 r to the ovary, or 800–1000 r to the testes. There is a threshold dose which must be exceeded before any effect becomes evident. Early experiments in rats (Russ and Scott, 1937, 1939) show that continued exposure to γ radiation at 2 r/day led to a reduction

---

[6] Presumably α emitters applied to skin could be absorbed and so penetrate the natural filter

in fertility. Extensive work at the National Cancer Institute stimulated by the potentialities of the Manhattan Project showed damage to the ovaries of mice at levels not far above 100 mr per day. The effect on the female mouse is significantly different from that on the male, as it proves to be an irreversible change, and therefore a function of the total dose and not the manner in which it is distributed in time (Lorenz, 1946). This has formed the basis for the recommendation of reduced radiation exposure of female workers.

E. RADIOGENETIC EFFECTS

An excellent discussion of the potentialities of the genetic effect in its relation to radiation protection has been given by Henshaw (1941). Extensive experimental work has been restricted to exposure of the fruit-fly in which it has been shown that the genetic effect has no threshold, and exposure is not only cumulative in the individual but in succeeding generations. On this basis, there would be no true tolerance dose, but rather an acceptable injury-limit. Similar considerations led to doubt concerning the safety of repeated fluoroscopic exposures of children (Buschke and Parker, 1942). Fortunately, several factors indicate that fear of the genetic effect may have been over-emphasized. In the fruit-fly, single exposure of 30–40 r doubles the low natural mutation rate, whereas doses of 500 r are required to produce recognizable mutations in the mouse. Dose-effect in the human case is entirely unknown. Speculations have been made regarding the contribution of cosmic radiation to such mutations; present information appears to exclude this possibility. It is also clear that the transmission of recessive characteristics would require the irradiation of substantial portions of the population, which is not anticipated. Of fundamental importance is a recent suggestion that the genetic effect in the fruit-fly when examined closely at low dosage is not truly of the non-threshold type.* If this were substantiated, and applied also in the human case, there might be no fear whatever of genetic injury arising from prolonged exposure at very low intensity, which is a necessary corollary of radiation work. This has far-reaching consequences in the present communities of exposed persons, where, for the first time, there is a significant percentage of intermarriage of exposed persons.

8. Dependence of Tolerance Dose on Nature of the Radiation

A. SPECIFIC IONIZATION

Although the mechanism of the interaction of radiation with living matter may not be fully understood, it is certain that one determining factor is the specific ionization of the radiation in the irradiated material (Zirkle 1940; Gray and Read, 1943). This has led to the acceptance on the Project

---

* *Note added in proof:* Later developments in the experiments that prompted this statement appear to deny such a convenient escape from the problem.

of the ratio of 1 rep = 4 rem for neutron radiation, and the arbitrary extrapolation of 1 rep = 10 rem for α-particle ionization.

B. X-RAYS

The original choice of the general tolerance dose was, of course, founded on experience with low energy radiation with an effective wavelength of the order of 0.3 A. For 100 mr measured at the body surface (without backscatter), the tissue dose can be taken as 150 mr at 1 cm. deep, and 50 mr at 10 cm. deep. 400 KV radiation at the present time is as common as was 150 KV radiation when the tolerance was first established. The comparable depth doses with this radiation are 120 mr at 1 cm. and 50 mr at 10 cm. The systemic effect on the body is surprisingly constant over this range, assuming biological equality of the radiation. This is only true when the defining dose of 100 mr is measured "in air."

Little has been written about the damaging effects of radiation generated between 400 and 2000 KV. In this range the increase in percentage depth dose is not very large. It is reasonable to suppose that the tolerance dose established for 150 KV radiation with a margin of safety of 5 or better would still be entirely safe for X-radiation up to 2000 KV. Experience with teleradium sources should also be applicable in this case.

Multimillion volt X-rays show no striking increase in tissue ionization, because the increased penetration is offset by the fact that an increasing portion of the body falls in the transitional layer where equilibrium with the secondary radiation has not been established (Koch et al., 1943). The relative integral dose at different radiation energies is plotted in Fig. 1, where the incident dose is supposed to be measured in a Bragg-Gray air walled cavity. Concern has been expressed over the excessive ionization in bone at high voltage, due to the pair-production effect in calcium and phosphorus. The figure shows that up to 100 MV at least, the energy absorption in bone, is comparable with that at 150 KV. At lower energies (e.g. normal 200 KV therapy), the bone dose caused by photoelectric absorption exceeds any corresponding increase due to pair-production at high energies. The calculations are made with allowance for absorption in tissue and with a simplified picture of the body distribution of bone. On this basis, it appears that a reduction of tolerance dose with deference to bone ionization is unnecessary. Preliminary biological experience at 20 MV shows no striking difference from 200 KV experience (Quastler and Clark, 1945).

C. GAMMA-RAYS

Adequate experience with γ radiation prior to the Project was largely confined to the use of radium and its products. It has been accepted that the danger of γ radiation in practice exceeds that of "traditional" X-radiation for two reasons:

(1) X-ray equipment can be de-energized when not in use,
(2) γ-ray penetration is higher.

On the basis of 98 mr at 1 cm. and 67 mr at 10 cm. one might have a factor of 2- or 3-fold to represent the additional total body ionization, compared with the X-ray case. The expression of tolerance dose in terms of total body ionization rather than in terms of the incident dose has recently gained favor (Clarkson, 1945). The unit employed for this purpose is the gramme-roentgen. It is worthy of note that the total body ionization changes much more rapidly as a result of the position of the exposed person relative to the

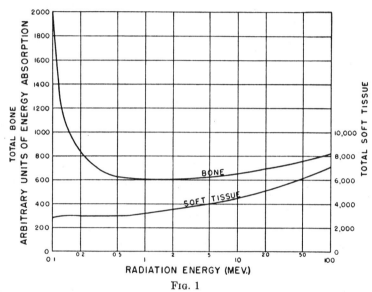

Fig. 1

Relative energy absorption in the whole skeleton and the whole soft tissue as a function of radiation energy, when the body is exposed to a remote source of radiation.

source than as a result of quality change from soft X-radiation to radium γ-rays.

	Relative Total Body Ionization[7]
Soft X-rays from large distance	1
Gamma-rays " " "	2.5
" " " point source at 100 cm.	1.9
" " " " " at 10 cm.	0.6

Although the integral dose point of view may have significance in measuring the relative potential injury for different qualities of radiation and

[7] Clarkson quotes a ratio of 5 in integral dose for exposure to stray radiation from deep therapy and diagnostic irradiations.

different geometrical depositions, it can hardly supplant the existing method of limiting the incident daily dose. This would be especially true if the damaging dose were not significantly different for blood changes, skin damage, or effects on the reproductive organs. If the existing tolerance dose does not have a large safety margin for the second and third of these effects little is gained by an elaboration of the gramme-roentgen aspect of exposure.

Recent experience in the administration of radioisotopes for therapeutic purposes has introduced the problems of permissible exposure from this source. When the distribution of the material is known, the effective dose can be readily calculated; in general, it is overshadowed by the contribution of the accompanying β radiation.

D. BETA-RAYS

There is no extensive body of evidence on the permissible exposure of the body to β radiation despite the extensive use of potent β-ray sources in the past, and related sources such as cathode ray tubes and Lenard tubes. With such installations, it is easy to maintain adequate shielding at all times, and the damage from their use has been essentially restricted to short exposure to the direct beams, with insufficient knowledge of the dose received.[8] It is assumed for the present that, for equal ionization in tissue, β radiation will produce the same effect as X-radiation of average energy. The tolerable exposure to β radiation has, therefore, been taken as 100 mrep per day. This should be conservative since the damaging effects of external β radiation must be confined to the skin or to tissue within a few mm. of that surface. The protection of the horny layer is introduced to set a limit to the lower energy of the β radiation that need be included in measurements of the skin exposure. Alternatively, one can conveniently elect to use β-ray chambers with wall thickness equivalent to that of the layer of passive absorption (approximately 7 mg./cm.^{2}).

E. NEUTRONS

(a) *Fast Neutrons.* The effects of prolonged exposure of the body to low intensity beams of fast neutrons are unknown. At higher intensity, there have been many comparisons of the radiation effects on biological materials when irradiated respectively by neutrons and X or γ radiation. The neutron dose has been frequently stated in "n" units, representing the scale deflection of a Victoreen Condenser R Chamber, which is unsuited to fast neutron dosimetry. In more recent practice, attempts have been made either to measure or to calculate the ionizing effect of the radiation in a "tissue"

[8] This applied also in a recent accidental exposure to scattered β-rays.

HEALTH-PHYSICS AND RADIATION PROTECTION

chamber (Gray and Read, 1939). For cyclotron neutrons, it is considered that 1 n is equal to 2.5 rep. Biological effectiveness ratios between 5 and 9 in terms of roentgen to "n" unit are quite common, and led to the proposed tolerance value of 0.01 n per day. This is the basis of the currently accepted relation 1 rep = 4 rem for fast neutrons.

A more refined study would make allowance for biological effectiveness dependence on energy of the incident neutrons. The reliable evidence of neutron exposure in humans rests on the erythema observations by Stone and Larkin (1942) mainly with neutrons from the 60" cyclotron. Erythema dose was 110 n for conditions approximately comparable with a 675 r erythema dose for 200 KV X-radiation or about 850 r of 400 KV radiation, the proposed standard for rep comparisons.[9] The biological equivalence ratio is, therefore, 1 rep = 3.4 rem. Gray's specific ionization calculations would give—

$$37'' \text{ cyclotron neutrons} \quad 1 \text{ rep} = 5.5 \text{ rem}^{10}$$
$$d,d \text{ neutrons} \quad 1 \text{ rep} = 8.7 \text{ rem}$$

For deeper effects, the neutrons will be slowed down in tissue with a consequent increase in biological effectiveness. On this basis, the usual value 1 rep = 4 rem is far from conservative. Moreover, there is a possibility that the effectiveness ratio may increase sharply under the extended low intensity bombardment of interest in tolerance.*

(b) *Slow Neutrons.* The present estimate of tolerance to fast neutrons has been deduced from some experimental evidence, but at the beginning of the Manhattan Project there was no experimental foundation for a slow neutron tolerance exposure. Three factors were considered:

(1) Production of γ-rays in the body by the interaction of neutrons with hydrogen nuclei,
(2) Production of fast protons by the neutron reaction on some constituent atoms, and
(3) Production of new atomic nuclei.

Although it seemed at first that the exposures would be primarily controlled by the first effect (in which case one would have 1 rep = 1 rem), it appeared later, both by revised calculations of the nuclear reactions and by animal experimentation, that the reaction $N^{14}(np)C^{14}$ played a considerable part in the total effect. This led to a provisional effectiveness ratio of 1 rep

---

[9] Lower voltage radiation introduces a photoelectric contribution at the sensitive depth in skin. Appreciably higher voltage radiation has a surface transitional layer thicker than the layer of passive absorption.

[10] Stone and Larkin observed 90 n as the erythema dose on a short series. This makes 1 rep = 4.2 rem. Their instrumentation was admittedly somewhat open to question.

* *Note added in proof:* This may now be considered proved, and a relation 1 rep = 10 rem for pile neutrons is desirable.

= 4 rem. In numerical value, the ratio must be a function of the size of the irradiated body, because the production of fast protons is a purely local phenomenon, whereas the conversion to γ radiation will be larger in the larger animal. Further refinement is hardly necessary until there is a greater body of biological observation of the slow neutron damage. The actual slow neutron flux that will produce the daily tolerable exposure of 100 mrem is said to be 1500 $n/cm.^2$ sec. This is probably a lower limit of a value which may be as much as 3 times higher.[11] The precise value depends on measurements of body ionization arising from slow neutron incidence, which have not been satisfactorily performed. Slow neutron fluxes around nuclear machines appear to be low in comparison with other hazards, and there has been no critical urgency to perform detailed physical experiments required to specify the tolerable flux in greater detail. The data on epithermal neutrons, those in an energy range between that required to produce proton tracks and the approximately thermal energy significant for the $N^{14}(n,p)C^{14}$ reaction, is incomplete. There may be some molecular disturbance in the tissue by such neutron bombardment. For measurement purposes, the epithermal flux is indicated qualitatively by surrounding slow neutron counters with hydrogenous moderators. Large counts under these conditions are corrected by alterations on the primary shielding.

F. ALPHA-RAYS

As stated before, there is no danger from the external radiation by natural α particles under normal conditions. The tolerance to accelerated particles is based on the tissue ionization with the biological effectiveness ratio of 1 rep = 10 rem. The ratio is the best guess of α-ray exposure in small organisms (Zirkle, 1946). Inside the body, the same tolerance is allowed; the practical problems are confined to the calculation of energy absorption arising from a given amount of the α emitter in the tissue or organ of interest.

G. PROTONS AND OTHER PARTICLES

Bombardment of the body by fast neutrons results in a fast proton effect in the body. Consequently, for a primary fast proton bombardment, the same biological ratio of 1 rep to 4 rem is assumed to apply. Radiological use of fast protons is anticipated when new machines yielding 125 MEV protons or better are completed (Wilson, 1946). More detailed consideration of proton effects may then be required.

H. COMBINED RADIATIONS

In the absence of evidence to the contrary, it is assumed that small contributions of different types of radiation produce additive effects when

---

[11] J. S. Mitchell (1947) has recently published a value of 1250 $n/cm.^2$ sec.

HEALTH-PHYSICS AND RADIATION PROTECTION 257

measured in the biological equivalence units of rem.[12] There is a wide field of biological experiment in the examination of this proposition. From the physical point of view, the heavy weighting of neutron, proton, and α particle bombardment in terms of biological damage makes it important to apportion the total ionization as accurately as possible to the respective causative agents. In particular, measuring devices which record mixed radiations in terms of rem are to be avoided. The rep/rem ratios are provisional, and such devices exclude the possibility of back-correction of data.

9. Geometrical Considerations

A. CONDITIONS OF MEASUREMENT OF TOLERANCE DOSE

In the field of therapeutic irradiation, there is one school which normally quotes all surface dose in terms of the measurement in an air chamber without the presence of the human body as a scattering medium; and another school which quotes the alleged skin dose which includes the air measurement supplemented by whatever scattered radiation would arrive at the surface from the patient's body. With the exception of the recent industrial regulations (American Standards Association, 1945), there has been no clear definition of the method intended in the tolerance literature. If one supposes that the European and International systems of 200 mr per day exposure include all scattered radiation, while the American system of 100 mr per day *specifically excludes it*, there would be fairly close agreement for low energy X-radiation on which the tolerance was originally founded. At higher energy, the American system becomes relatively more conservative. On this system, the readings of suitably designed monitoring instruments in exposure-areas would apply directly in tolerance considerations. On the other hand, an ionization chamber worn on the body would give a reading greater than that involved in the tolerance calculation. This difference is rarely taken into account, but it can amount to as much as 40 or 50%.

B. DIRECTIONAL AND NARROW BEAM IRRADIATION

Other principal factors which make the estimation of the dose received by the body in terms of a pocket-meter reading unreliable are: (1) directional radiation, and (2) limited or subdivided beams. With X-radiation of normal penetration, the error can be by a factor of about 10 if the meter is worn on the chest and the irradiation incident on the back. This extreme case would be evidence of considerable stupidity on the part of those responsible for protection, but similar cases of lesser degree necessarily occur.

---

[12] Although the definition of rem does not necessarily imply such additivity, the concept would be virtually useless without it.

Under item (2), one has the familiar overexposure of hands and head while assembling radium sources behind a lead shield. Another aspect arises in the shielding of equipment which requires controls to be brought to the outside, and again in the use of shielding boxes or vessels with removable lids or plugs. Proper design of such boxes calls for the elimination of intense narrow beams. Much attention has been given to this in the protection of X-ray treatment rooms. It was an important consideration in Pile design where a regular pattern of holes through the principal shield was required. Although it is recognized that a considerably higher dose can be given safely to a small field than to a large one, the difference is hardly great enough to justify change in tolerance dose. Also, if the exposure is measured on the person rather than by monitoring equipment, the recording pocket-meter may happen to spend much of its time in a shielded zone while parts of the body are in fact being irradiated. For these reasons, it is considered good policy to restrict the permissible dosage-rate of emergent beams to a necessarily safe level.

IV. Organization and Functions of a Typical Health-Physics Group

1. General

A general account of the responsibilities and functions of a Health-Physics organization has been given by Stone (1946) and by Morgan (1946a); these references should be consulted. Activities are divided into two parts: (1) Operational or Service, and (2) Control and Development.

2. Operational Division

By convention, the Operational Division has been further subdivided into (a) Personnel Monitoring, and (b) Area Monitoring or Survey.

A. PERSONNEL MONITORING

Personnel Monitoring has included routine records of the daily γ-ray exposure of each individual as recorded by pocket ionization chambers. Early equipment for this purpose was unsatisfactory; the policy of wearing 2 meters to decrease the percentage of readings totally lost by instrument defects was adopted and maintained. Mass observation of pocket meter readings has shown that well-prepared meters do not give the same reading when uniformly exposed to 100 mr (say). The readings fall on a probability curve of standard deviation corresponding to 6 mr. The chance of reading a tolerance exposure as 80 mr or less is, therefore, about 1 in 1000. In large well-protected plants this might occur about once in four years. For weekly totals, the random daily fluctuations become inconsequential. Mass preparation of data essentially accurate to within 25 mr per week is practicable.

Pocket-meter readings have been supplemented by the observation of the blackening of special dental films arranged to record β and γ radiation separately. Fast neutron exposures were recorded by the measurement of proton recoil tracks produced in the Eastman Kodak special fine grain-particle film. Slow neutron exposures have been observed by means of boron-lined pocket meters. Exposures of special parts of the body, notably the fingers, have been made either by microionization chambers or by small film packs contained in special rings. These operations have been conducted on a scale approximately 100 times that of similar pre-Project experience, and the records probably constitute the best available collection of personnel exposure data of reasonable technical accuracy.

B. AREA MONITORING

The responsibility of area monitoring has been discharged by (a) fixed instruments with recorder charts, (b) technical graduates trained in the measurement and interpretation of radiation data, and (c) technicians trained in radiation measurement. Contrary to some reports, this area monitoring was entirely similar to prewar practice. Differences were in degree only and included greater frequency, more varied instrumentation, and improved reporting. For the latter, various room or area maps have been used for data plotting. While suitable for special locations, these lack the three-dimensional feature frequently required, and well planned tabulations have proved more adaptable. Exposure standards for all radiation types were established. Special interpretation was needed for surface contamination. For α emitters, the hazard was confined to possible transfer to the body. β-Contamination required consideration both in this manner and as a direct contact hazard. Hazard limits in either case came close to the minimum conveniently detectable contamination. With the area monitoring was combined a written control system for access to a potentially hazardous area. The cooperation between operating, maintenance, technical and health groups prior to, and throughout, any maneuver involving radiation, utilized the monitor results to best advantage.

The war-time use of technical graduates for survey, when scientific talent was at a premium, was indicative that the task was more involved than would appear at first sight. Qualifications for area monitoring are given in Paragraph IV, 4, D.

As available instruments were steadily improved and made more suitable for specific applications, and as many phases of laboratory and industrial operations became stereotyped, the amount of monitoring executed competently by specially trained technicians increased. It is anticipated this procedure will be further extended in the future, but successful elimination of the graduate surveyor is improbable.

3. Control and Development Division

In the Control and Development Division is included such control items as measurement of activity on protective clothing and radio-analysis of samples of many kinds submitted by the separate monitoring groups, or collected over a wide area to maintain surveillance of the contamination conditions of air, ground, or water, at all relevant locations. Bioassay of the amount of activity excreted or exhaled by laboratory personnel is another control function. One vital responsibility is the calibration of all instruments, which, experience has taught, requires to be done with great frequency. The development field is subdivided into groups whose function is to improve methods of solution of such exposure problems as the body content of radiotoxic materials, absorption through skin, *etc*. Although the Health-Physics departments have been provided with instruments through the very successful efforts of general instrument departments, it has been found essential that they should themselves continue research and development into specific forms of instrumentation. The special biophysical requirements of radiation measurement instruments have not been well known to those most familiar with the general instrument field, and it has required a particular combination of the skills of both groups to achieve adequate results. In some cases, required radiobiological investigations have been profitably integrated with the Health-Physics activities, and form a natural part of the responsibility for the improvement in health hazards control.

4. Factors Contributing to a Successful Health-Physics Organization

A. THE OBJECTIVE

The primary difference between radiation protection in general, and that offered where this is an established Health-Physics organization, rests in the *completeness of control*. The luminous compound industry has no unsolved protection problem, yet violations of safe practice are commonplace (U. S. Dept. of Labor, 1942). In the hospital field, one clinic alone saw 80 cases of possible radiodermatitis in physicians between 1934 and 1939 (Leddy and Rigos, 1941). Another clinic saw 70 radiation injuries in 3 years (Uhlmann, 1942); 30 injuries followed treatment, and the remainder occurred in diagnostic or technical work. In a survey of 45 leading radiation hospitals, hand exposures over 20 r per occasion were noted, and skin changes in one-quarter of the radiologists observed (Cowie and Scheele, 1941). Four physicians died in 1946, as a result of X-ray accidents or complications (Editor, *J. Am. Med. Assoc.*, 1947). The Plutonium Project locations,[13] entering new levels of potential exposure for the first time, have a

[13] Metallurgical Laboratory, Argonne National Laboratory, Clinton Laboratories, and Hanford Engineer Works.

record of *no radiation injury*. What factors have permitted this advance, and can it be perpetuated in future laboratory and industrial practice? It is essential to appreciate that the success of a protection program comes only in part from the health-physics unit. Some of the factors which influence the overall program are described below:

B. ATTITUDE OF LABORATORY OR PLANT PERSONNEL

Management support is a prerequisite for success in any safety program of which radiation safety is one component. Also, full cooperation of each and every employee in protecting himself and his colleagues is required. It is common to find a junior scientist who believes he has reasonable regard for the safety of co-workers, but is willing to take chances, or, if a senior man, "calculated risks," in his own exposure. Industrial experience has shown that this fails as a safety policy; the radiation problem is no exception. Reaction of senior staff personnel to the protection program frequently conditions the standards for a whole laboratory. Where there is passive acceptance of the program, or even covert resistance, the probability of effecting detailed safety control is low. Radiation accidents befall men who know how to conduct themselves and feel that rules are for the uninitiated.

C. APTITUDE OF THE HEALTH DIVISION LEADERSHIP

In the control of a radiation project scaled up by orders of magnitude beyond previous experience, it is generally conceded that the primary health responsibility is preferably divided between a competent radiotherapist and a medical physicist. In normal laboratory or industrial practice, the special clinical skills of the radiotherapist are not required. The protection program is then largely in the hands of medical physicists, supplemented by very necessary industrial-medical examination of the health of the personnel.

D. APTITUDE OF THE HEALTH-PHYSICS FORCE

The key members of the organization are the H.I. engineers, health-physics surveyors, or radiological monitors, on whom rests the day-to-day contact between protector and protected. Requirements for a successful engineer or surveyor are:

(a) *Personality and Diplomacy*. The engineer must be able to "sell" the employee on the advantages of close hazard control. Law enforcement or policeman tactics impede the program.

(b) *Technical Skill*. He must have a sound, not necessarily profound, knowledge of the health hazards, with a knowledge of tolerance and protective policies. He needs some electronics background to standardize instruments, and critically examine their fields of performance. He must also have a general knowledge of laboratory or industrial processes involved. Where the program includes research or development, he must be competent to appreciate the objectives. He need not, and in fact should not, be a research-type himself.

(c) *Appreciation of His Responsibilities.* The engineer's occupation has a negative character, the prevention of overexposure, and he has nothing tangible to show for his day's effort. Unless he learns to appreciate the confidence that other employees develop in his recommendations, he will fail to realize the contribution that he has made. Whenever this happens, his morale and that of his contacts will suffer.

(d) *Training Ability.* The engineer is required to advise diverse members of the force—research physicists and chemists, electricians, pipe fitters, chemical operators, *etc.*—on how to execute their work safely. Special ability to indicate the necessary maneuvers simply, under these conditions, is required.

E. TRAINING PROGRAMS

Health-Physics organizations have trained all their engineers and technicians because there was no available pool with prior experience. The large industrial units have set as their objective, the training of all other senior employees in Health-Physics. This is accomplished either by formal training of personnel already assigned to other duties, or by apprenticeship to the health-physics organization. Whenever circumstances permit, assignment to health instrumentation for a period of between 6 months and 1 year has proved profitable.

F. LIAISON WITH OPERATING OR TECHNICAL GROUPS

A system of written instructions for the execution of all hazardous jobs has been formulated, and has become an important feature of radiation hazard control. At the Hanford Engineer Works, for example, a Special Work Permit is completed by all the organizations involved in a proposed job prior to its inception. This method of control was developed not by the Health-Physics group but by a general committee representing all groups. Such agreement on method of approach ensures a high standard of cooperation in the planning of each specific, hazardous operation. Where the nature of the work is technical, and especially as it borders on original research, the operation of similar formal work permits is considered less applicable. However, this difference has been overrated, and considerable success achieved in some cases with the application of permits to technical work.

G. INSTRUMENTATION

Particular stress must be laid on the use of instruments in good condition and on their frequent calibration. The manner in which instrument readings are permanently recorded can also greatly affect the overall response. For health instrumentation, printing registers are frequently superior to rate meters, although the technical reliability may be identical.

H. TOLERANCE LIMITS

With the exception of the nationally accepted value of permissible dose of X or γ radiation, there may be legitimate debate on other exposure limits. A competent Health-Physics Section must keep in touch with the main biological developments that could lead to a better understanding of such limits. Flexibility of interpretation has to be maintained, and there should be no reluctance to change exposure standards. Conservatism is essential when dealing with incompletely evaluated hazards. If X μg is the tolerable deposition of a toxic element deduced from early animal experimentation, and it later turns out that $X/10$ μg is the limit, then the health-physicist who has a number of colleagues with a body content of this element between $X/10$ and X μg occupies an infelicitous position. Final exposure limits have generally been lower than those originally proposed, and there can be little criticism of the physicist who elects to preserve an additional safety factor (up to 10) in permissible exposures. A corollary to the proper statement of standards is the necessity to measure the exposures competently. Where this hinges on a bioassay, as it does for many problems of internal deposition, the measurement of the eliminated amounts may prove to be an extremely difficult technical procedure. It is imperative that health groups establish the highest possible technical standards in these cases.

I. NEW PROTECTION SERVICES

There is danger that the protection policy becomes stereotyped perhaps with good control of general body radiation (especially γ). Method development must be continued in order to put protection against other radiations, and especially against various forms of body exposure such as inhalation of active gases, inhalation or ingestion of α-emitting dusts, β-emitting dusts, exposures to neutrons, *etc.*, on a routine basis as soon as proper procedures can be established.

J. REPORTS AND RECORDS

Good records are required to make long-range studies of small exposures. To correct any condition which may be substandard, and for legitimate protection against fraudulent claims of overexposure, the primary record should be as complete as possible. Any instance of apparent overexposure should be thoroughly investigated to ascertain its cause and prevent its repetition. Wide publicity of incidents involving imperfect control, regardless of actual exposure, is desirable.

K. DETAILED CONTROL

In the last analysis, hazard control, especially where surface contamination is the issue, depends upon attention to minute details—invariable wearing of gloves, covering of work surfaces with paper, segregation of clean from contaminated tools, *etc.* Success in this field depends largely on the personal attitude of the individual concerned. In laboratories, in particular, it may represent the determining feature between good and fair hazard control.

L. TRIPLE SAFEGUARD PHILOSOPHY

Misadventures occur when several things go wrong in sequence. All important steps should, therefore, be protected by 3 safety devices so that all would have to fail before exposure occurred. The general policy is applicable to all phases of radiation-handling from the operation of a major Power Pile down to laboratory manipulation of active solutions. Safeguards should not be substituted for vigilance—the chemist careless while wearing gloves, and the driver reckless because he has good brakes have much in common.

M. CONSISTENCY OF RULES AND ENFORCEMENT

Protection rules must be consistent, and should observe the unities of time and place. Housekeeping rules, including contamination clean-up, respirator and glove wearing, eating and smoking are particularly vulnerable in this respect. Hazard control which allows smoking in laboratories on some days, and forbids it on others, is incompatible with the best practice. Rules should *never* be enforced by a Health-Physics organization. An advisory capacity is the objective, and the degree to which advice is accepted is a measure of the group's success.

N. DESIGN

Radiation protection is relatively simple in any building initially provided with properly planned hazard control facilities. This includes suitable shielding of fixed apparatus, portable shielding for temporary sources, test stations for hands, shoes and clothing, facilities for the provision of protective clothing, and change-house facilities where needed. When the basis of operation has been properly prepared, there is a saving of time in any subsequent operation performed. Economically, this partially offsets the initial higher cost of protection planning.

V. Instruments

1. Status

One surprising feature of Health-Physics experience since 1942 has been the realization that, prior to that time, equipment needed for radiation monitoring at the levels of interest in protection was generally unreliable, improperly calibrated, or not available at all. There has been no extensive release of information on the improvements effected during the war years. From some publications (Morgan, 1946a; Jesse, 1946a), it can be deduced that considerable effort was devoted to this subject. Also, certain specific instruments have been released for general use. Brief accounts of such are included.

2. Calibration and Maintenance

No quicker way of destroying confidence in protective schemes exists than the use of instruments which do not appear to be functioning perfectly. This applies sometimes to devices such as counting rate meters which show a proper statistical fluctuation but are not as convincing to the average observer as a scaler and register that gives a definite count in the same time. More important is the elimination of loose connections, sticking needles, poor switches, tired batteries, defective tubes and like defects that arise in sensitive electronic equipment subjected to fairly rough treatment. Preventive maintenance, which could be practiced more extensively in laboratories and institutions, is required. Calibration of health instruments should be done at least once a week. This includes a direct test of the instrument response to known amounts of radiation at different scale deflections and for all ranges. Whenever possible, calibration at more than one radiation energy is advocated. Standardization, which by custom means a spot check of response at one scale-position on each range, should be practised every time a portable instrument is to be used. The provision of safe portable test sources for high intensity taxes the physicist's ingenuity. For such instruments as pocket meters, it is generally conceded that daily standardization is unnecessary, and monthly calibration adequate.

3. Basic Health Instrumentation

The primary specialty that differentiates health instrumentation from process or control instrumentation is the behavior of the radiation receptor. This has to be made to give a response related in the simplest possible manner to the pertinent biological exposure. In the simplest case one designs an ion chamber to read in roentgens; this is done empirically over the required range of wavelength and is an exercise in the construction of an electrically conducting shell with an insulated electrode, the whole having an effective atomic number equal to that of air. For more refined measurements, tissue-wall vessels with "tissue-gas" are required. As applied to fast neutron irradiation, the close imitation of the hydrogen content of tissue is indicated; for slow neutron irradiation the precise amount of all elements present has to be regulated to govern the activation. More frequently, secondary methods have to be utilized, such as the physical measurement of slow neutron flux, and its theoretical conversion to tissue dose.

The physicist is normally interested in the number of particles or photons emitted by a source, or crossing unit surface. The health-physicist searches for the energy absorption in tissue-like materials. Typical of the consequences are the difficulties of interpretation of G.M. Counter readings, especially for γ-ray counting. Very few health-physicists will accept portable counters as more than qualitative tools to locate areas of activity or contamination. Quantitative survey results come almost exclusively from ionization readings. In control work, counters, especially for α-particles, are invaluable and have been brought to standards of refinement normally unnecessary in analytical work.

4. Fixed Instruments

A. AREA MONITORS

Beta-gamma monitoring of selected work locations is accomplished by custom-built combinations of ion chambers with commercial microammeters and recorders. Where there is no size restriction, any desired sensitivity may be reached by using large chambers filled with air at atmospheric pressure. These are normally more reliable than pressure vessels and a suitable "air-wall" more easily provided. Sensitivity is approximately 1 $\mu\mu$amp = $\frac{10}{V}$ mrep/hr, where V = chamber volume (l.). It is optional whether the circuit be made to indicate the instantaneous dosage-rate or to integrate the exposure over a work period. The ideal monitor accomplishes both. Alarm systems may be coupled in, actuated either by high rate or high integrated dose.

B. PERSONNEL MONITORS

(a) *Alpha Hand Counters.* The detection of α-particle contamination on the hands is accomplished by adaptations of the standard pulse-type counters or of proportional counters with counting surface as large as the hand. A convenient modification is the two-fold counter which has flat multiwire proportional counters arranged to make contact with both sides of the hand. The collecting electrode system consists of a series of fine tungsten wires, to which approximately 2500 volts are applied, producing proportional counter-action in the flat air-filled chamber. The electronic circuit consists of an A.C. amplifier, pulse leveler, integrating circuit and vacuum tube voltmeter. Registration in this case is by counting-rate meter with a useful range from 200 to 10,000 dis/min. None of these devices records the presence of α-particle contamination on the medial and lateral surfaces of the fingers. Where some contamination on the hands is found, sound policy calls for a detailed inspection with a small proportional counter probe. The objective in this type of counting is to determine the *total* amount of toxic material present. Its local concentration is unimportant except as an aid to localized removal treatment.

(b) *Beta-Gamma Hand Counters.* Combinations of thin-walled counter tubes are normally disposed for this purpose to cover adequately all four principal hand surfaces simultaneously. As the objective is to locate contamination and remove it immediately, there is no need for quantitative measurements. The calibration of these instruments is made on the basis that contamination is spread over 3 sq. in. of the hand, to give a surface dosage-rate of 4 mrep/hr. Where the contamination can be limited to a small spot, the local surface dose may be quite high. Only if this occurred repeatedly on the same skin area would hazard arise. Protection against this is maintained by special checks for small spots. Preferred operation of the counter tubes is through a scaler and register system, with a permanent record of the results. In some cases, spot checks can be taken more conveniently on a system using 4 counting rate meters.

(c) *Foot Counters.* Groups of counters in parallel can be arranged to check for contamination on the soles of the shoes, the early detection of which reduces the tracking of active material from one location to another. These units may be coupled to the hand counters for simultaneous registration.

(d) *Thyroid Counters.* Typical of a check depending on the metabolism of a specific active material is the use of γ counters for the measurement of radioiodine in the thyroid in laboratories or other areas where this could be present. The method has been widely used in therapy with radioiodine. For Health-Physics, a suitable calibration is obtained by putting a standardized radioiodine solution in a glass model thyroid in a neck phantom. Practical sensitivity limit is about 4% of the tolerable value.

5. *Equipment, Atmospheric and Miscellaneous Monitoring*

A. STANDARD ALPHA COUNTER

The standard parallel plate chamber is used for the precision-measurement of the α-particle emission of samples of low mass, such as come from the evaporation of water samples, scrapings from contaminated surfaces, ashed tissue, *etc.* With reasonable care, the background may be reduced to 1 dis/min., and the useful range is then from 1 dis/min. to about 40,000 dis/min. Resolution of α particles occurs in the presence of up to 10^5 β counts/min.

B. SIMPSON PROPORTIONAL COUNTER

The Simpson proportional counter (Simpson, 1946) avoids the microphonic and electric disturbances of the standard type; it permits very fast counting (up to 5×10^5 dis/min.), in the presence of strong β counts (up to

5 × 10⁹ dis/min.). It is occasionally of value in Health-Physics work, but the limitation on sample size to 2 cm. diameter, and the difficulty of decontamination, make it troublesome unless the high speed and resolution are essential.

C. LOW BACKGROUND COUNTER

Of special interest is the low background counter designed to expedite the detection of minimal amounts of α emitters in the urine, tissue, *etc*. If the daily elimination of 'fixed' radium were 0.001% of the body content, as little as 0.025 µg Ra in the skeleton could be detected with ease. The background of these counters is easily maintained at 6 counts/hr, although values as low as 0.1 count/hr have been claimed. With normally pure metals for the electrodes, the natural emission would amount to 3 counts/hr (Rajewsky, 1939). Surfaces of electrolytically pure metal could eliminate most of this. Then radon and thoron and their products would introduce 0.5 counts/hr, unless inactive gas (*e.g.*, old air) were used to sweep out the counting enclosure. Radon concentration can also be appreciably reduced in a perfectly air-conditioned room.

D. BETA-GAMMA COUNTERS

(*a*) *General.* Innumerable variations on the mounting of single or multiple G.M. tubes to facilitate the contamination test of laboratory glassware, tools and other items can be postulated. Other obvious applications are:

(*b*) *Gas Counters.* Representative gas samples are collected in prepared evacuated cans, and then introduced into a previously evacuated vessel containing a counter. In some cases, introduction into an ion chamber is preferred, as in radon breath sampling. In either case, subsequent contamination of the equipment may be troublesome.

(*c*) *Water Counters.* Counters immersed in vessels of active water may be used to measure the activity, especially if readings in different sized vessels are obtained. In all such applications, contamination of the parts, and the peculiar wavelength response of some counters makes quantitative work difficult.

(*d*) *Dust Monitors.* The collection of active dust from the air on filter paper surrounding a counter tube permits the evaluation of the air contamination in terms of the rate of increase of deposited activity. A natural limit of sensitivity is set by the simultaneous deposition of decay products of radon and thoron.

E. SPECIAL INSTRUMENTS

Extrapolation chambers (Failla, 1937) are invaluable for the determination of contact dosage-rates of active sheets. Especially useful is an inverted type to measure surface activity of a liquid. The customary equipment is not portable, and the source has to be brought to it. Where this is impracticable, subsidiary standard sheets are calibrated and taken to the field. Extrapolation chambers require the best available electrometers. Both the project-improved Lindemann electrometers and the Vibrating Reed electrometer appear to have features superior to the traditional FP54 electrometer circuit. Triple coincidence counters in the standard arrangements are a valuable asset in the health instrumentation list.

6. Portable Survey Instruments

A. AREA MONITORING. ALPHA TYPES

(a) *Poppy*. For the rapid detection of α particle contamination, a proportional counter system is preferred. The typical probe operates with about 2500 volts on a fine tungsten wire collector. Its output operates an A.C. amplifier integrating circuit and vacuum tube voltmeter indicating circuit. An audio-oscillator triggered by each pulse makes the audible "pops" responsible for the popular name "Poppy." The Poppy is sensitive to mechanical and electrical disturbances, and is affected by moisture. The "geometry" is also sensitive to the operating conditions and should be tested frequently during operation. Inasmuch as the primary function of the equipment is to locate contamination that has to be removed, these variable features cause little concern. Sensitivity down to 200 dis/min. over the probe face is the usual limit.

(b) *Zeuto*. The Zeuto (Jesse, 1946b) is a convenient unit including an ion chamber with thin screen window transparent to α particles, and a circuit employing a balanced pair of miniature tubes for amplifying the ion current with a microammeter in a bridge circuit for measurement. A third tube controls positive feed back, which reduces the time constant sufficiently to permit use of a 10^{12} ohm input resistor. The instrument is sensitive to about 200 dis/min., and high readings can be accommodated by scale changing. As applied to large flat surfaces, Zeuto is at least competitive with Poppy. The latter is effective where curved surfaces and narrow strips are concerned.

B. BETA-GAMMA TYPES

(a) *Zeus*. Many successful β-γ monitors have been constructed, but apparently few are available for description. One such is the Zeus (Jesse, 1946c) a survey meter for α, β or γ work, but rather insensitive and nonuniform for α quantification. Built-in α and β shields provide a convenient method for estimating the relative intensity of β and γ components, and indicating the presence of α contamination. Zero stability is good and the time constant short. The useful range of the Zeus is from 1 to 2500 mrep/hr in three ranges, *viz.*, 1–25, 5–100 and 100–2500. Separate calibration of each range is needed. Ion chamber current is amplified by a balanced pair of miniature triodes and measured by a microammeter in a bridge circuit. The range is selected by changing the value of the input resistor, which, at maximum sensitivity is 10^{11} ohms. The wire chamber screen and metal chamber wall on some instruments will cause energy dependence with soft γ radiation.

(*b*) *Lauritsen Electroscope.* The Lauritsen electroscope (Lauritsen and Lauritsen, 1937) modified by the provision of a thin window with screens to eliminate α particles or α and β at will is an excellent tool for the small laboratory. It has to be calibrated over a wide range, because there is a peculiar intensity response due to the inadequate collection field around the fine quartz fiber system. Limiting factors in the use of the instrument are energy dependence for γ radiation, fixed sensitivity, and the inconvenience of obtaining measurements with the aid of a microscope while timing the observation with a stopwatch. Reliable determinations of dosage-rate can be made in the range from 0.1 mr/hr to 1 r/hr, provided sufficient time is available for low intensity measurements. The instrument is unsurpassed for reproducibility.

(*c*) *Landsverck-Wollan Electrometer.* Improvements on the Lauritsen electroscope have been incorporated in the Landsverck-Wollan electrometer. The new device is superior in linearity and sensitivity. A condenser-resistor circuit is used to flash a neon lamp at the beginning and end of one of two predetermined time periods, permitting the fiber to be observed at those instants, effectively making it a dual-range instrument. The timing system eliminates the need for a stopwatch to time fiber drift, when high precision is not required. The useful ranges of the instrument are 0–200 mr/hr, and 0–2 r/hr.

(*d*) *Victoreen Radiation Meter.* The Victoreen Radiation Meter[14] was primarily designed for measurements of stray radiation from X-ray installations, but can be put to wider use in the protection field. Obvious circuit changes could make the meter suitable for higher ranges, by a combination of chamber size and resistor variation. The cautious observer would wish to mount the chamber on the end of an extension rod for high level operation. It is advantageous to prepare special scales calibrated over the whole range, and to standardize frequently.

(*e*) *Betty Snoop.* The Betty Snoop is a lightweight portable meter for measuring high intensity radiation fields. Its probe contains a small plastic ion chamber, range switch, input resistors and the electrometer tube. An extension cable permits using the probe at some distance from the meter

[14] Victoreen Instrument Co., Cleveland, Ohio.

circuit. Four sensitivities of nominal maxima, 0.2, 2, 20, and 200 rep/hr are provided. The calibration curve for the three lower ranges is nearly linear, but the high range curves with a negative slope. The circuit employs a balanced pair of triodes and a microammeter in a bridge circuit, with a third tube providing positive feedback to permit use of a 10^{12} ohm input resistor on the most sensitive range. A long time constant results on the lower ranges, and attempts to reduce it by increased feedback may cause oscillation. The chamber has an easily removable cap, which permits optional γ or β registration. As the chamber is only 3 cm. in diameter, it is well suited to measurements in narrow beams or close to small sources.

(f) *Condenser Chambers*. Detachable condenser chambers of all sizes can be used as portable monitors where only an integrated dose is required. The method is cheap, reliable and versatile. Ideal construction is still similar to the Sievert chambers (1932). A series of such chambers with different wall composition can be used to calculate energy absorption in tissue. Ranges from 0.04 mr to 1000 r are easily obtained on a versatile recorder such as a Lindemann electrometer. The familiar Victoreen Condenser-R-Meter is one suitable rugged commercial application of the principle. All such meters require calibration with the specific types of radiation to be used.

(g) *Portable G. M. Tube Sets*. Since the first portable G. M. tube circuits were made, they have been used for the detection of lost radium and for health measurements. Low voltage counters (Chalmers, 1934) were particularly suited for systems weighing less than two pounds (Pallister, 1937). Innumerable circuit modifications have been used by commercial manufacturers and by private laboratories including registration by audio-signal, neon tube flashes, or integrating circuits. For health instrumentation, utility is restricted to qualitative detection rather than measurement. Good design features are, therefore, stability, light weight, and freedom from unsuspected blocking. Modern circuits use mainly self-quenching tubes, but these exhibit a temperature coefficient (Korff *et al.*, 1942), which can lead to racing counts in cold locations and no response in hot places. In survey applications, one uses a special filling of low temperature coefficient or restricts the operation to a certain temperature range.

(h) *Sigmion*. A useful Project innovation was the Sigmion, a simple chamber and D.C. amplifier set to integrate up to 20 mr and then reset. The total exposure is tallied on a register. Successful operation depends on balancing the normal reverse leak by a subsidiary active source.

(i) *Victoreen Proteximeter*. Somewhat similar in general design and function is the Victoreen Proteximeter. Dose up to 200 mr is indicated on the instrument meter. These instruments are convenient for location close to a technician. When the instrument is found to be fully discharged, there is no reset mechanism and no record of the total dose as provided by the Sigmion.

C. NEUTRON METERS

(a) *Differential Chambers.* Quantitative measurements of fast neutron exposures are made in differential ion chambers, one of which is made sensitive to neutrons by the use of a hydrogenous gas or paraffin-lined wall, while the other provides a γ-ray balance by ionization in argon. For high sensitivity, pressure vessels are required and the final equipment is cumbersome. Where rates above 50 $n/cm.^2$ sec. are involved, truly portable systems can be used. In all cases, special precautions are needed to compensate for asymmetrical γ fields. Radium-beryllium sources are widely used for calibration.

(b) *BF_3 Counters.* Slow neutrons are counted in conventional BF_3 Counters, or in boron-lined counters or ion chambers. All such slow neutron counters can be made into convenient detectors of fast neutrons by surrounding them with a 4" thick sleeve of paraffin, to moderate incident fast neutrons. Calibration is effected by comparison with the twin chamber apparatus, and is valid only for similar neutron spectra.

7. *Portable Dust Monitors*

Sampling for radioactive dust is easily accomplished by filtration or electrostatic precipitation. Standard formulae allow for the elimination of the Ra-Th contributions after two subsequent measurements. A convenient nomogram for the elimination of thoron decay product counts when α contamination is to be measured, is shown (Fig. 2). It is customary to delay the initial count until 4 hours (or better, 6) after the sample collection to reduce the radon decay product contribution to a negligible value.

8. *Personnel Meters. Beta-gamma*

A. FILM BADGES

A permanent record of integrated weekly dose is given by film badges. A typical badge contains two pieces of dental film in a silver or cadmium holder 1 mm. thick, with a window to admit β radiation. Blackening is measured on scales separately calibrated for β and γ radiation. One piece of film is the sensitive industrial radiography type, for the normal exposure, while the other is chosen with a range up to 40 rep to cover a possible gross exposure. The metal filter approximately compensates for wavelength dependence of film above the K limit of silver. However, soft quantum radiation on the window produces intense blackening which masks β contributions.

B. POCKET METERS

Pocket ionization chambers in the general form of "pencils" have been used in many forms. Typical of the commercial species is that produced by

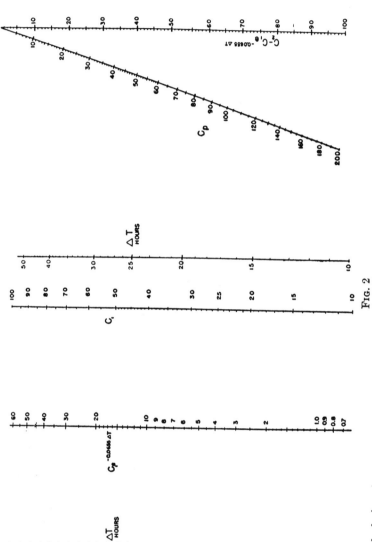

Fig. 2

Calculation of α counting rate, C_p, due to contaminant in presence of thoron decay products. C_1 = counts/min after 4 to 6 hours. C_2 = counts/min ΔT hours later. On left hand figure, straight edge through ΔT and C_1 defines $C_1 e^{-0.0655 \Delta T}$ on center scale. Subtract this value from C_2. Straight edge between $C_2 - C_1 e^{-0.0655 \Delta T}$ and ΔT on right hand figure defines C_p on sloping scale.

the Victoreen Instrument Co., Cleveland, Ohio, and used in conjunction with a simple string electrometer—the Minometer. Such pocket meters should be rugged, wavelength-independent, free from insulator leaks, and unaffected by humidity. Currently available meters fall short of these ideals. With good maintenance, reliable service can be effected, and the use of duplicate meters reduces 'lost' readings to an insignificant score. Technical improvements of the meters, especially protection against dust and moisture, and simplified manufacturing methods require study. Pocket meters with thin walls for β registration can be used, or small models can be worn on the finger.

C. NEUTRON METERS

Slow neutrons can be recorded in boron-lined chambers (Landsverck, 1947). Slow neutron doses up to 200 mrem can be read on the standard model with only 1% interference by an incident γ tolerance dose. Neutron-sensitive film can be worn for fast neutron checks. A cadmium filter over regular film is an alternate meter for slow neutrons.

VI. Some Elementary Formulae and Calculation Methods

1. General

Refined calculations of shielding, maximum permissible exposure, *etc.*, are subjects for detailed treatises on the Health-Physics specialty. Much can be done by simplified methods to indicate the order of magnitude of exposure and some typical examples are given.

2. Shielding

X-ray shielding can be calculated from standard absorption data. It is customary to quote tables for lead absorption and to refer to other materials by their "lead equivalent" (Taylor, 1941; Glasser *et al.*, 1944). The Graphic Calculator (Failla, 1945) is a convenient chart for protection calculations. Similar methods serve for the laboratory manipulation of radioactive sources. One convenient tabulation of the required shielding is shown in Table I.[15] The addition of three quantities, with due regard to sign, and multiplication by one factor permit calculation of shield thickness for all normal values of source strength, quantum energy, handling distance, daily exposure time, and customary shield material. Interpolation is unnecessary because precision in temporary shielding need not be high.

3. Emission from Complex Sources

The emission of primary γ radiation from a large radiating mass, in which the space variation of activity follows a simple power law, and which is

---

[15] Prepared by C. C. Gamertsfelder, 1943.

HEALTH-PHYSICS AND RADIATION PROTECTION

covered by a uniform filter or shield, is formally identical with the radiation problems considered by Gold (1909). The intensity can be expressed as a series of integrals of the form

$$\int_0^{\pi/2} e^{-x \sec \theta} \sin \theta \cdot \cos \theta^{(n-2)} \cdot d\theta$$

which transforms to $\int_1^\infty \dfrac{e^{-xu}}{u^n} du$

Tables of this function have been published. Approximations to the radiation from large nuclear machines, tanks of active materials, reservoirs of waste material, *etc.*, can be made on this basis. Where neutron radiation is involved, applications of diffusion theory may be needed. Many simpler cases, such as the γ radiation from thin active discs or plates are familiar in the literature (*e.g.*, Mayneord, 1932). Special cases of interest are the emission from active sheets or blocks and the emission inside extended masses of active material.

4. Emission from Sheets or Blocks

Consider a sheet emitting n particles/cc./sec., each of range R cm. in the material. The emergent integrated flux is

$$\int_0^R \frac{2\pi n r^2 \, dr}{4\pi r^2} = \frac{nR}{2} \text{ particles/cm.}^2\text{sec.}$$

If k = number of ion pairs/cm. air produced by each particle, the surface dosage-rate is: $n\,R\,k\,4.8 \times 10^{-10} \times 3600$ rep/hr. $= 8.7 \times 10^{-7}\,n\,R\,k$ rep/hr.

Radiation	k	R
α	~50,000	From tables: allow for relative stopping power
β	~75 for fast particles[a] ~100 for average particles	Range for $\dfrac{\text{maximum energy }(E)}{3}$
γ	$\dfrac{\sigma E}{W} = \dfrac{3.5 \times 10^{-5} \times E \times 10^6}{32}$ approx. $= 1.1 E$ E = energy per disinteg. in MV	Reciprocal of linear absorption coefficient in medium

[a] $k = 45$, in the physicists measure, follows the electron on its tortuous path. Our measure is a 1 cm. translation.

TABLE I

Shield Thickness for Laboratory Sources

Select column for energy required (use next higher if exact value is not given). Entry gives thickness in cm. Pb for different source strengths at 1 m. for 8 hrs./day to give 100 mr. *Add algebraically*, correction terms for other working ranges or times, and *multiply* by factor for shield material.

E.g., 500 mc of 1.8 MEV source at 50 cm. ior 4 hr./day = $(7.21 + 2.77 - 1.39) \times 1.43$ cm. Fe = 12.3 cm. Fe

Activity	0.2 Mev	0.5 Mev	0.8 Mev	1.0 Mev	1.5 Mev	2.0 Mev	2.5 Mev	3.0 Mev	4.0 Mev
10 mc	− .20	− .71	− .95	− .98	− .83	− .61	− .33	− .11	− .19
20 mc	− .14	− .36	− .27	− .11	+ .37	+ .77	+ 1.15	+ 1.40	+ 1.70
50 mc	− .07	+ .11	+ .63	+ 1.03	+ 1.95	+ 2.61	+ 3.10	+ 3.39	+ 3.69
100 mc	− .01	+ .46	+ 1.31	+ 1.90	+ 3.14	+ 3.99	+ 4.57	+ 4.90	+ 5.20
200 mc	+ .04	+ .82	+ 1.99	+ 2.77	+ 4.34	+ 5.38	+ 6.05	+ 6.40	+ 6.70
500 mc	+ .12	+ 1.28	+ 2.89	+ 3.91	+ 5.92	+ 7.21	+ 7.99	+ 8.40	+ 8.69
1 c	+ .17	+ 1.64	+ 3.57	+ 4.78	+ 7.11	+ 8.60	+ 9.47	+ 9.90	+ 10.20
2 c	+ .23	+ 1.99	+ 4.25	+ 5.64	+ 8.31	+ 9.98	+ 10.94	+ 11.41	+ 11.71
5 c	+ .30	+ 2.46	+ 5.14	+ 6.79	+ 9.89	+ 11.82	+ 12.89	+ 13.40	+ 13.70
10 c	+ .36	+ 2.81	+ 5.82	+ 7.66	+ 11.08	+ 13.20	+ 14.37	+ 14.91	+ 15.21
20 c	+ .41	+ 3.17	+ 6.50	+ 8.52	+ 12.28	+ 14.59	+ 15.84	+ 16.42	+ 16.71
50 c	+ .49	+ 3.63	+ 7.40	+ 9.67	+ 13.86	+ 16.42	+ 17.79	+ 18.41	+ 18.71
100 c	+ .54	+ 3.99	+ 8.08	+ 10.53	+ 15.05	+ 17.81	+ 19.27	+ 19.91	+ 20.21
Danger Range	plus	plus	plus	plus	plus	plus	plus	plus	plus
20 cm	+ .26	+ 1.64	+ 3.16	+ 4.02	+ 5.55	+ 6.44	+ 6.85	+ 7.00	+ 7.00
50 cm	+ .11	+ .71	+ 1.36	+ 1.73	+ 2.39	+ 2.77	+ 2.95	+ 3.01	+ 3.01
1 m	.00	.00	.00	.00	.00	.00	.00	.00	.00
2 m	− .11	− .71	− 1.36	− 1.73	− 2.39	− 2.77	− 2.95	− 3.01	− 3.01
5 m	− .26	− 1.64	− 3.16	− 4.02	− 5.55	− 6.44	− 6.85	− 7.00	− 7.00
10 m	− .37	− 2.35	− 4.52	− 5.76	− 7.94	− 9.21	− 9.80	− 10.01	− 10.01

HEALTH-PHYSICS AND RADIATION PROTECTION

Working Time	plus	plus	plus	plus	plus	plus	plus	plus	plus
1 hr. day	− .17	− 1.06	− 2.04	− 2.60	− 3.59	− 4.16	− 4.42	− 4.52	− 4.52
2	− .11	− .71	− 1.36	− 1.73	− 2.39	− 2.77	− 2.95	− 3.01	− 3.01
4	− .06	− .35	− .68	− .87	− 1.20	− 1.39	− 1.47	− 1.51	− 1.51
8	− .00	− .00	− .00	− .00	− .00	− .00	− .00	− .00	− .00
24	+ .09	+ .56	+ 1.08	+ 1.37	+ 1.89	+ 2.20	+ 2.34	+ 2.39	+ 2.39

Absorber	times	times	times	times	times	times	times	times	times
Pb	1.00	1.00	1.00	1.00	1.00	1.00	1.00	1.00	1.00
Fe	8.80	2.88	1.96	1.74	1.49	1.43	1.47	1.48	1.59
Al*	41.67	9.80	6.18	5.33	4.83	5.00	5.28	5.68	6.39
H$_2$O	106.84	21.54	13.42	11.59	10.36	11.11	11.19	12.11	12.78

* Or concrete.

Note added in proof: Source activity is quoted in millicuries or curies, where one curie is that amount of radioactive material that disintegrates at the rate of 3.4×10^{10} dis./sec. The table is computed on the further (erroneous) assumption that each disintegration yields one γ photon of the selected energy. This leads to inaccuracies whenever the disintegration scheme is complex. More accurate calculations can be made when the disintegration scheme is known. Ignored also is the increased effective transmission of shields under wide beam irradiation. The table is a useful guide for erection of *temporary* laboratory shielding.

278   H. M. PARKER

Examples

1. A uranium sheet emits 1.27×10^4 α particles/sec./g. from U^{238} or 2.38×10^5/sec./cc. There will be an equal number from the isotopic U^{234}. ∴ $n_\alpha = 4.76 \times 10^5$. Average range = 3 cm. air = 6×10^{-4} cm. U ∴ α contact dosage-rate = 12.6 rep/hr.
   Only the UX_2 β-rays will be worth consideration. $n_\beta = 2.38 \times 10^5$. Range for average energy 2.32/3 is 0.25 g. Al/cm.² = 0.34 g. U/cm.² = 0.018 cm. U
   $$\beta \text{ Dosage-rate} = 280 \text{ mrep/hr.}$$
   γ-Ray energy is 0.8 MEV, $R = 0.4$ cm. U
   $$\gamma \text{ Dosage-rate} = 7.2 \text{ mr/hr.}$$

2. A thick plastic sheet, s.g. 1.2, contains 1 mc P^{32}/cc. Range for average energy 1.7/3 is 0.14 g. Al/cm.² = 0.12 cm. in the sheet.
   Contact dosage-rate = $8.7 \times 10^{-7} \times 3.7 \times 10^7 \times 100 \times 0.12 = 385$ rep/hr.

Such calculated values can be relied upon to about ±30%. For very thin sheets, an approximate value is obtained by inserting thickness t cm. instead of R in the formula.

5. *Emission Inside Large Active Masses*

The energy absorption per cc. of the mass is evidently equal to the energy emission. When the energy absorption in the mass can be simply related to the energy absorption/cc. of air, water or tissue, the exposure in the mass can be written down immediately.

If C = concentration in μc/l. and

E = (γ energy + average β energy) in MEV per disintegration[16] the following formulae result:

$$\text{In air, dosage-rate, D.R.} = 2000 \, CE \text{ mrep/hr.}$$
$$\text{In water, dosage-rate} = 2.6 \, CE \text{ mrep/hr.}$$

or in any medium, density ρ, dosage-rate $= 2.6 \dfrac{CE}{\rho}$ mrep/hr

Persons exposed in such masses may usually be supposed to receive the β component from one hemisphere only. Also, on the ground for an active air case and at the surface of a water mass, the γ rate is usually taken as one-half the above. Tolerable concentrations can readily be established on this basis. Formally, the surface case is identical with the already considered thick sheet problem, when transitional equilibrium effects are negligible.

[16] Usual application is to β-γ emitters. Same formula applies for α emission (*e.g.*, in bone).

HEALTH-PHYSICS AND RADIATION PROTECTION 279

Examples

1. An air mass containing 1 μc Xe^{133}/l. contacts the ground.

   Average β energy = $\frac{0.33}{3}$ MEV. γ energy = 0.084 MEV [Siegel, 1946].

   A man in an airplane receives 2000 (0.055 + 0.084) = 280 mrep/hr.
   On the ground he receives 2000 (0.055 + 0.042) = ~200 mrep/hr.
   Tolerance concentration (24 hrs. daily) is: 0.021 μc/l.

2. A river contains 0.02 μcNa^{24}/l. Average β energy = $\frac{1.39}{3}$ [Siegbahn 1946], γ energy = (2.76 + 1.38). A small organism in the river receives 2.6 (0.46 + 4.14) × 0.02 = 0.24 mrep/hr.
   A man in a boat receives 1.3 × 4.14 × 0.02 = 0.11 mr/hr.
   Immersion tolerance concentration (man, 8 hrs./day) = 1.1 μc/l.

6. Tolerance for Internal Emitters

A. SIMPLE CASE

In almost all cases, the local energy absorption is governed by the particle radiation, α or β. Even a small organ is then an effective large mass and the relation D. R. = $2.6 \frac{CE}{\rho}$ mrep/hr. holds. This has been more frequently used in the form D. R. = $62 \frac{EQ}{W}$ rep/day where E = av. en. in MEV, Q = μc deposited, W = grams of tissue containing the Q μc (Cohn, 1946). Let f be the fraction of the administered dose Qa deposited in the relevant tissue. Then for a single short exposure $Q = fQa$.

For prolonged exposure $Q = f \frac{Qa}{\lambda}$ where Qa = daily dose and λ = decay constant in days^{-1}. Where the deposition is eliminated, this is assumed to be exponential with a biological decay constant λ_b. The effective decay constant is then ($\lambda + \lambda_b$). It is also necessary to distinguish between cases of prolonged exposure to a continuously maintained concentration (the main interest in tolerance dose), and to a decaying concentration, such as might follow accidental dispersal or the Bikini tests.

B. OTHER VARIABLES

Recognized formulae of radioactive transformation can be applied to the case where one active material is taken in and produces a damaging daughter. Typical of the manipulation of such formulae and their application to tolerance are the data by Morgan (1946b); some provisional tolerance values are listed in Table II. These depend on the values of f. This is usually a composite, *e.g.*, for ingestion, f = fraction absorbed from gut into blood,

x = fraction deposited from blood into tissue. For inhalation, one considers (1) lung retention from air, (2) absorption from lung to blood, and (3) de-

TABLE II

Representative Tolerance Values Calculated by Morgan (1946b)

1	2	3	4	5	6	7
Element	Grams per Curie	Assumed effective half life	Method of body intake	Organ affected and fraction of amt. in body that is in organ	Fraction taken into body that reaches organ	μc in organ to produce tolerance rate when $t = 0$
Ra^{226} pdts	1	~2 wks.	Breathing	Lungs (.5)	0.25	.013
Ra^{226} pdts	1	~10 yr.	Ingestion	Bone (.6)	0.05	.155
Pu^{239}	16	2 mo.	Breathing	Lungs (.3)	0.25	.035
Pu^{239}	16	10 yr.	Ingestion	Bone (.6)	0.0003	.42
Pu^{239}	16	10 yr.	Breathing	Bone (.6)	0.0375	.42
Natural U	1.47×10^6	2 mo.	Breathing	Lungs (.3)	0.25	.041
Enriched U	2.7×10^4	2 mo.	Breathing	Lungs (.3)	0.25	.039
U^{233}		2 mo.	Breathing	Lungs (.3)	0.25	.037
Po^{210}	2.24×10^{-4}	82 d.	Breathing	Kidneys (.05)	0.011	.010
Po^{210}	2.24×10^{-4}	82 d.	Ingestion	Kidneys (.05)	0.001	.010
Sr^{89}	3.7×10^{-5}	43 d.	Ingestion	Bone (.5)	0.075	32
$Sr^{90} \to Y^{90}$	7.74×10^{-2}	Sr—197 d. Y—2.49 d.	Ingestion	Bone (.5)	0.075	88 (20)[a]
C^{14} (graphite)	0.23	2 mo.	Breathing	Lungs (.3)	0.25	32
C^{14} (CO_2)	0.23	10 d.	Breathing	Total body (1)	0.25	2260
H^3 (water)	2.59×10^{-4}	2 d.	Breathing	Lungs (.02)	0.25	320
I^{131}	8×10^{-6}	6.3 d.	Ing. or br.	Thyroid (.2)	0.20	2.0
Na^{24}	1.13×10^{-7}	14.8 hr.	Submersion	Body		
Na^{24}	1.13×10^{-7}	14 hr.	Ing. or br.	Blood (.25)	0.25	2.2
Na^{24}	1.13×10^{-7}	11.5 hr.	Ing. or br.	Lungs (.037)	0.037	0.5
P^{32}	3.48×10^{-6}	14.3 d.	Submersion	Body		
P^{32}	3.48×10^{-6}	13 d.	Ing. or br.	Bone (.9)	0.09	39
$Ba^{140} \to La^{140}$	1.33×10^{-5}	Ba—11.75 d. La—1.51 d.	Ingestion	Bone (.6)	0.06	48
S^{35}	2.3×10^{-5}	25 d.	Ing. and br.	Skin (.2)	.05 (.1)[b]	150
Ca^{45}	6.15×10^{-5}	150 d.	Ing. or br.	Bone (.99)	.15 (.4)[b]	190

The tolerance values in columns 9, 10, 11, 12 and 13 are for continuous exposure. If the exposure is for a 40 hour week, multiply these values by 4.2.

Column 9 is the tolerance concentration rate, P, in $\mu c/sec$, to the body organ that will produce a tolerance rate of exposure after 365 days of consumption.

It should be noted that values given in column 6 depend on the chemical form and in the case of inhalation they depend upon the size particles. Until the most likely forms of these elements in a given laboratory are known, it is difficult to assign typical values of tolerance concentration in columns 10, 11, 12, and 13.

Values in column 9 can be obtained directly from equation 14 or by dividing values in column 8 by the seconds in a year.

8	9	10	11	12	13	14
μc in organ to produce an av. tolerance during year	One year tolerance concentrations rate (μc/sec)	One year tolerance concentration in air		One year tolerance concentration in water		Effective energy
		$\mu c/cc$	$\mu g/cc$	$\mu c/cc$	$\mu g/cc$	MV
.23	7.5×10^{-9}	2.0×10^{-10}	2.0×10^{-10}			$14\,\alpha\gamma$
.16	5.1×10^{-9}			4.4×10^{-6}	4.4×10^{-6}	$14\,\alpha\gamma$
.15	4.7×10^{-9}	1.3×10^{-10}	2×10^{-9}			$5.16\,\alpha$
.43	1.4×10^{-8}			2.0×10^{-3}	3.1×10^{-2}	$5.16\,\alpha$
.43	1.4×10^{-8}	2.5×10^{-9}	4.0×10^{-8}			$5.16\,\alpha$
.17	5.5×10^{-9}	1.5×10^{-10}	2.1×10^{-4}			$4.43\,\alpha$
.16	5.1×10^{-9}	1.4×10^{-10}	3.8×10^{-6}			$4.7\,\alpha$
.156	5.0×10^{-9}	1.3×10^{-10}				
.033	1.0×10^{-9}	6.4×10^{-10}	1.4×10^{-12}			$5.3\,\alpha$
.033	1.0×10^{-9}			4.5×10^{-5}	1.0×10^{-8}	$5.3\,\alpha$
190	6.0×10^{-6}			3.4×10^{-3}	1.3×10^{-7}	$0.6\,\beta$
34	1.1×10^{-6}			6.2×10^{-4}	4.8×10^{-6}	$.22, .8\,\beta$
130	4.3×10^{-6}	1.2×10^{-7}	2.6×10^{-8}			$.05\,\beta$
5.7×10^4	1.8×10^{-3}	4.8×10^{-5}	1.1×10^{-5}			$.05\,\beta$
4×10^4	1.3×10^{-3}	3.5×10^{-5}	9×10^{-9}			$.005\,\beta$
81	2.6×10^{-6}	8.5×10^{-8}	6.8×10^{-12}	5.5×10^{-4}	4.4×10^{-9}	$.2\,\beta\gamma$
		6.3×10^{-7}	7.1×10^{-14}	4.9×10^{-4}	5.5×10^{-11}	$3.3\,\beta\gamma$
960	3.0×10^{-5}	8.1×10^{-7}	9.1×10^{-14}	5.2×10^{-3}	6.0×10^{-10}	$3.3\,\beta\gamma$
258	8.2×10^{-6}	1.5×10^{-6}	1.7×10^{-13}	9.5×10^{-3}	1.1×10^{-9}	$3.3\,\beta\gamma$
		4.2×10^{-6}	1.5×10^{-11}	3.2×10^{-3}	1.1×10^{-6}	$0.5\,\beta$
750	2.4×10^{-5}	1.8×10^{-6}	6.2×10^{-12}	.011	4.0×10^{-6}	$0.5\,\beta$
170	5.3×10^{-6}			3.8×10^{-3}	5.1×10^{-8}	$.4, 2.3\,\beta\gamma$
1500	4.7×10^{-5}	3.1×10^{-6}	7.3×10^{-11}	.041	9.4×10^{-7}	$.05\,\beta$
400	1.3×10^{-5}	2.1×10^{-7}	1.3×10^{-11}	3.6×10^{-3}	2.2×10^{-7}	$0.1\,\beta$

Column 8 is the μc in the lung, bone, kidney or blood required to irradiate the organ with 3.65 roentgens of α or 36.5 roentgens of beta-gamma in a year. It is the μc in the thyroid required to irradiate it with 365 roentgens of beta-gamma in a year.

[a] The Sr-Y activity reaches a maximum after 15 days. The 88 μc is required to produce tolerance exposure rate soon after Sr reaches the bone. Only 20 μc is required to produce tolerance exposure rate on the 15th day. The 34 μc produces an average yearly tolerance dose. (See Fig. 4.)

[b] It is assumed that the fraction reaching the skin by way of the gut is 0.05 and by way of the lungs is 0.1 in the case of S^{35}. For Ca^{45} it is assumed that 0.15 reaches the bone by way of the gut and 0.4 by way of the lung.

position from blood. These factors vary for each radioactive material, its valence state, and some features of the physical and chemical form. Determination of these variables rests on prolonged radiobiological and biochemical studies, initially in animals, but in part on humans (especially for inhalation). For this reason there will continue to be discrepancies in the published tolerance values. With the termination of the war effort and the consequent restoration of channels of scientific information, the current multiplicity of values can be screened by a national committee to stabilize the best values.

C. STANDARD MAN

Computers of tolerance concentrations have used a wide range of values for body organ weights, respiratory rate, water intake, *etc*. Uncertainties in the metabolism of isotopes and ultimately the idiosyncrasies of each exposed person will always exceed the range of variation introduced in this manner. Consequently, the continued publication of tolerance values founded on irregular basic values only further confuses the issue. Agreement on uniform figures should be possible as a first step toward complete standardization. The following values adjusted* from those proposed by R. S. Stone form a logical basis for this:

Total body weight	70 kg.
Muscle	30 kg.
Skin and subcutaneous tissues	8.5 kg.
Skeleton	7 kg.
Blood	5400 g.
Heart	350 g.
Liver	1700 g.
Lungs (pair)	950 g.
Bone marrow (red)	1500 g.
Bone marrow (yellow)	1500 g.
G.I. tract (empty)	2300 g.
Kidneys (pair)	300 g.
Pancreas	65 g.
Spleen	200 g.
Testes (pair)	40 g.
Thyroid	30 g.
Respiratory rate	1 m.³/hr.
Water intake	3 l./day
Primary lung retention of small particles	25%

REFERENCES

Aebersold, P. C. 1941. *J. Applied Phys.* **12**, 345.
Am. Standards Assoc. 1945. Safety Code for the Industrial Use of X-rays. Part 1. Z 54.1.
Bale, W. F. 1946. *Occupational Med.* **2**, 1.
Barclay, A. E., and Cox, S. 1928. *Am. J. Roentgenol.* **19**, 551.
Birnkrant, M. I., and Henshaw, P. S. 1945. *Radiology* **44**, 565.
Braestrup, C. B. 1938. *Radiology* **31**, 206.
Braestrup, C. B. 1942. *Radiology* **38**, 207.
Buschke, F., and Parker, H. M. 1942. *J. Pediat.* **21**, 524.
Cantril, S. T. 1943. Private communication.

* The author is indebted to A. M. Brues and H. Lisco for this revision.

Chalmers, T. A. 1934. *Brit. J. Radiol.* **7,** 755.
Cilley, E. I. L., Kirklin, B. R., and Leddy, E. T. 1935a. *Am. J. Roentgenol. Radium Therapy* **33,** 390.
Cilley, E. I. L., Kirklin, B. R., and Leddy, E. T. 1935b. *Am. J. Roentgenol. Radium Therapy* **33,** 787.
Clark, L. H., and Jones, D. E. A. 1943. *Brit. J. Radiol.* **16,** 166.
Clarkson, J. R. 1945. *Brit. J. Radiol.* **18,** 233.
Cohn, W. E. 1946. Derivation of Equations, MDDC-95.
Colwell, H. A., and Russ, S. 1934. X-ray and Radium Injuries, Oxford Med. Publication.
Cowie, D. B., and Scheele, L. A. 1941. *J. Natl. Cancer Inst.* **1,** 767; also in *J. Am. Med. Assoc.* **117,** 588.
Curtiss, L. F. 1941. *J. Applied Phys.* **12,** 346.
Dohnert, H. R. 1938. *Z. Krebsforsch.* **47,** 209.
Editor. 1947. *J. Am. Med. Assoc.* **133,** 108.
Evans, R. D. 1943. *J. Ind. Hyg. Toxicol.* **25,** 253.
Failla, G. 1932. *Radiology* **19,** 12.
Failla, G. 1937. *Radiology* **29,** 202.
Failla, G. 1945. *Am. J. Roentgenol. Radium Therapy* **54,** 553.
Gamertsfelder, C. C. 1943. Private communication.
Glasser, O., Quimby, E. H., Taylor, L. S., and Weatherwax, J. L. 1944. Physical Foundations of Radiology, Chap. 19. Hoeber, N. Y.
Glocker, R., and Kaupp, E. 1935. *Strahlentherapie* **20,** 144.
Gold, E. 1909. *Proc. Roy. Soc. London* **82,** 62.
Goodfellow, D. R. 1935. *Brit. J. Radiol.* **8,** 669, 752.
Gray, L. H. 1937a. *Brit. J. Radiol.* **10,** 600.
Gray, L. H. 1937b. *Brit. J. Radiol.* **10,** 721.
Gray, L. H., and Read, J. 1939. *Nature* **144,** 439.
Gray, L. H., and Read, J. 1943. *Nature* **152,** 53.
Harding, D. B. 1945. *Kentucky Med. J.* **43,** 228.
Henshaw, P. S. 1941. *J. Natl. Cancer Inst.* **1,** 789.
Henshaw, P. S. 1944. In GLASSER, O., Medical Physics, p. 1356. The Year Book Publishers, Chicago.
Hoecker, F. E. 1944. *J. Ind. Hyg. Toxicol.* **26,** 289.
Jesse, W. P. 1946a. *Chem. Eng. News* **24,** 2906.
Jesse, W. P. 1946b. Survey Instrument, Mark 1, Model 10. MDDC-117.
Jesse, W. P. 1946c. Survey Instrument, Mark 1, Model 21-A. MDDC-118.
Jones, J. C., and Day, M. J. 1945. *Brit. J. Radiol.* **18,** 126.
Kaufmann, J. 1946. *Am. J. Roentgenol. Radium Therapy* **55,** 464.
Kaye, G. W. C. 1928. Roentgenology. Its Early History. Some Basic Principles and Protection Measures. Paul B. Hoeber, Inc., N. Y.
Koch, H. W., Kerst, D. W., and Morrison, P. 1943. *Radiology* **40,** 120.
Korff, S. A., Spatz, W. D. B., and Hilberry, N. 1942. *Rev. Sci. Instruments* **13,** 127.
Landsverck, O. G. 1946. New Models of Integrating γ-Ray and Slow Neutron Pocket Meters. MDDC-395.
Larkin, J. C. 1941. *Am. J. Roentgenol.* **45,** 109.
Lauritsen, C. C., and Lauritsen, T. 1937. *Rev. Sci. Instruments* **8,** 438.
Leddy, E. T., Cilley, E. I. L., and Kirklin, B. R. 1934. *Am. J. Roentgenol.* **32,** 360.
Leddy, E. T., and Rigos, F. J. 1941. *Am. J. Roentgenol.* **45,** 696.
Leitch, J. D. 1935. *Natl. Research Council of Canada, Bull.* **16.**
Levy, H. A. 1946. *Chem. Eng. News* **24,** 3168.

Life. 1947. **22,** 81.
Lorenz, E. 1946. Read at N. Am. Radiol. Soc. Meeting, Chicago, Ill. also MDDC 653, 654, 655, and 656.
Martland, H. S. 1931. *Am. J. Cancer* **15,** 2435.
Mayneord, W. V. 1932. *Brit. J. Radiol.* **5,** 677.
McClelland, W. R. 1933. Ore Dressing and Metallurgical Investigation No. 550, Dept. of Mines, Canada.
McWhirter, R. 1935. British Empire Cancer Campaign Annual Report, p. 131.
Mitchell, J. S. 1947. *Brit. J. Radiol.* **20,** 79.
Morgan, K. Z. 1946a. *Sci. Monthly* **63,** 93.
Morgan, K. Z. 1946b. Tolerance Concentration of Radioactive Substances. MDDC-240; also published 1947. *J. Phys. Colloid. Chem.* **51,** 984.
Morse, K. M., and Kronenberg, M. H. 1943. *Ind. Med.* **12,** 811.
Mutscheller, A. 1925. *Am. J. Roentgenol.* **13,** 65.
Mutscheller, A. 1928. *Radiology* **10,** 468.
Mutscheller, A. 1934. *Radiology* **22,** 739.
Natl. Bur. Standards, Handbooks HB-15, H-18, HB-20, H-23, H-27.
Nickson, M., and Nickson, J. J. 1946. N. Am. Radiol. Soc. Meeting, Chicago, Ill. Demonstration.
Nuttall, J. R. 1939. *Am. J. Roentgenol. Radium Therapy* **41,** 98.
Nuttall, J. R. 1943. *Clin. J.* **72,** 181.
Pallister, P. R. 1937. *Brit. J. Radiol.* **10,** 759.
Parker, H. M. 1943. Private communication to R. S. Stone.
Parker, H. M. 1944. Review of Air Monitoring at Clinton Laboratories. MDDC-471.
Paterson, R. 1943. *Brit. J. Radiol.* **16,** 2.
Peller, S. 1939. *Human Biol.* **11,** 130.
Quastler, H., and Clark, R. K. 1945. *Am. J. Roentgenol. Radium Therapy* **54,** 723.
Rajewsky, B. 1939. *Radiology* **32,** 57.
Read, J., and Mottram, J. C. 1939. *Brit. J. Radiol.* **12,** 54.
Rollins, W. 1902. *Boston Med. Surg. J.* **146,** 39.
Russ, S. 1914–15. *Arch. Roentgen Ray London* **19,** 323.
Russ, S. 1943. *Brit. J. Radiol.* **16,** 6.
Russ, S., and Scott, G. M. 1937. *Brit. J. Radiol.* **10,** 619.
Russ, S., and Scott, G. M. 1939. *Brit. J. Radiol.* **12,** 440.
Schlundt, H., McGaroch, W., and Brown, M. 1931. *J. Ind. Hyg.* **13,** 117.
Siegbahn, K. 1946. *Phys. Rev.* **70,** 127.
Siegel, J. M. 1946. *Rev. Modern Phys.* **18,** 513; *J. Am. Chem. Soc.* **68,** 2411.
Sievert, R. M. 1925. *Acta Radiol.* **4,** 61.
Sievert, R. M. 1932. *Acta Radiol. Suppl.* **14.**
Simpson, J. A. 1946. A Precision Alpha Proportional Counter. MDDC-80.
Smyth, H. D. 1945. Atomic Energy for Military Purposes. Princeton Univ. Press. (See Health Hazards in Index.)
Spiers, F. W. 1946. *Brit. J. Radiol.* **19,** 52.
Stone, R. S. 1946. *Proc. Am. Phil. Soc.* **90,** 11.
Stone, R. S., and Larkin, J. C. 1942. *Radiology* **39,** 608.
Taylor, L. S. 1940. *Radiology* **34,** 425.
Taylor, L. S. 1941. *J. Am. Med. Assoc.* **116,** 136.
Taylor, L. S. 1944. In GLASSER, O., Medical Physics, p. 1382. The Year Book Publishers, Chicago.
Uhlmann, E. 1942. *Radiology* **38,** 445.

U. S. Dept. Labor, Children's Bur. Publ. No. **286**. 1942. Occupational Hazards to Young Workers, No. 6. Radioactive Substances.
Warren, S. L. 1941. *J. Applied Phys.* **12**, 343.
Wilhelmy, E. 1936. *Strahlentherapie* **55**, 498.
Wilson, C. W. 1940. *Brit. J. Radiol.* **13**, 105.
Wilson, R. R. 1946. *Radiology* **47**, 487.
Wintz and Rump. 1931. League of Nations Publication.
Zirkle, R. E. 1940. *J. Cellular Comp. Physiol.* **16**, 221.
Zirkle, R. E. 1946. Radiobiological Importance of Specific Ionization. MDDC-444.

Absorption of Plutonium Fed Chronically to Rats

Medical Department (H.I. Section)
General Electric Company
Hanford Works

August 10, 1953

This paper was also published in somewhat different form in *American Journal of Roentgenology, Radium Therapy, and Nuclear Medicine*, LXXIII: 303-308 (February 1955) with Parker as senior author. Only the title page is reproduced here.

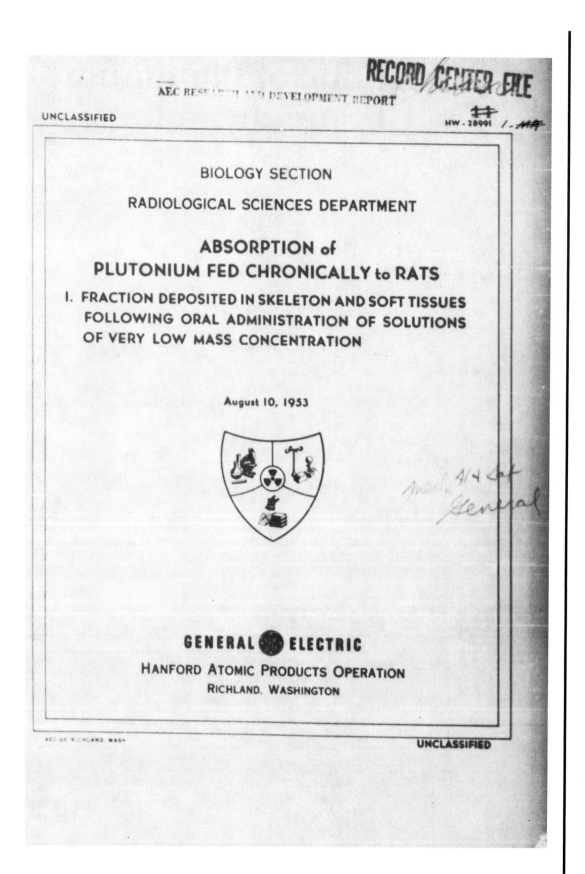

Radiation Protection in the Atomic Energy Industry

Radiology, Volume 65

December 1955

[Reprinted from RADIOLOGY, Vol. 65, No. 6, Pages 903-11, December, 1955.]
Copyrighted 1955 by the Radiological Society of North America, Incorporated

Radiation Protection in the Atomic Energy Industry

A Ten-Year Review[1]

H. M. PARKER, M.Sc., F.Inst.P.

OPERATION OF the first nuclear reactor dates back to December 1942. By 1944, production nuclear reactors for the manufacture of plutonium were in use at the Hanford Works (Richland, Wash.). The purpose of this contribution is to report on the progress made in radiation protection in such an industry in the past ten years. The scope of the problem can be sufficiently defined by recalling that the radioactivity handled is on the order of millions of times, or tens of millions, that from the available world supply of radium.

ORGANIZATION

The early radiation protection organizations were under medical leadership, a proper arrangement when the degree of direct medical participation could have been large. Experience has shown that such medical participation is minimal. Annual physical examinations and periodic blood counts have their place, but not significantly more so in the atomic energy industry than in other industries. The paramount medical contribution is the expert treatment of the occasional individual who may have been exposed to excessive radiation, externally or internally.

It was natural that leadership of the radiation protection programs should pass to a group of scientists (chiefly physicists) versed in the scientific problems involved and with enough general knowledge of biological and medical procedures to realize when and how expert assistance was needed. This led to the golden age of the health-physicist; we speculated in 1947 that the subject might have approximated its maximum status at that time. To some extent, this has been demonstrated by a tendency in industry to incorporate the radiation protection function with overall safety and health functions. In many organizations this would appear to be a wholly satisfactory development. It means, however, that the position of the industrial health physicist is limited. Somewhere there must be an atmosphere in which leading scientists in radiation protection can develop the field for future advancement. The principal difficulty involves lack of integration of the research effort with the practical field application of new methods. This difficulty arises in all branches of applied research, when the research effort is separated by organizational barriers from the intended user.

At the Hanford Works, where protection requirements apply to practically every operating task, phase of design, and construction of improved facilities, organization has developed along different lines. A few years ago, responsibility for radiation monitoring was detached from the radiation protection organization and integrated with all other responsibilities of the manufacturing organization. This change reflects the attitude that ultimately each operator must take care of his own radiation monitoring, just as he protects himself by not walking under a ladder or by not handling acid without protective equipment. The industry, however, is not yet ready for this step. The individual operator does not yet take care of his own radiation hazards; he requires a higher percentage of trained monitoring support than at any previous time.

The percentage of the radiation monitoring force in relation to the total operating force in the past ten years has been as follows:

First four years of operation............	1.5%
Three years prior to decentralization..	2.2%
Three years since decentralization.....	2.8%

---

[1] Presented as part of a Symposium on Radiation Protection at the Fortieth Annual Meeting of the Radiological Society of North America, Los Angeles, Calif., Dec. 5-10, 1954. Work done under General Electric Contract W-31-109-Eng-52 to the Atomic Energy Commission.

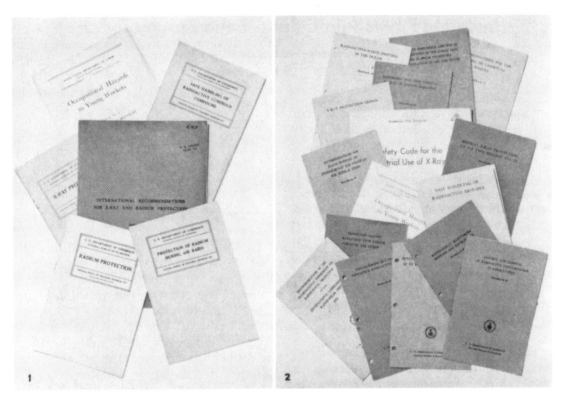

Fig. 1. Handbooks of radiation protection recommendations as of 1944.
Fig. 2. Handbooks of radiation protection recommendations as of 1954.

STANDARDS

At the beginning of the review period, the controlling standards, i.e., the recommendations of the former Advisory Committee on X-ray and Radium Protection, were few in number and generally were tailored to the experience and needs of hospital practice (Fig. 1). Today there is a library of handbooks containing the recommendations of the National Committee on Radiation Protection[2] (Fig. 2). The complexity of these standards arises largely from the many phases of hazard problems in dealing with radioisotopes; no fewer than seven handbooks (42, 48, 49, 52, 53, 56, 58) are wholly concerned with this field. Moreover, the values given in these publications do not have the comfortable assurance of ordinary engineering standards. The data of Handbook 52, on maximum permissible concentrations in air and water, are particularly vulnerable and are recognized to have at least the following defects:

1. Too few values are based on actual radiation damage experience in man.
2. Values for the bone-seeking isotopes are conventionally based on hazard ratios of the specific isotope to radium in animals, together with human data on radium. Hazard ratio changes as one proceeds from acute to chronic exposures. Specific questions are the significance of the different life span in animals and, to the severe critic, the validity of extrapolation of any biological data to man.
3. Values for other isotopes are based on simple physical concepts using uptake, deposition, and retention data from animal experimentation. Such data are suspect because of: (a) non-uniform deposition in the alleged critical organ; (b) dependence on chemical form, such as valency, pH, and presence of complexing agents; (c) dependence on physical form, such as particulate nature; (d) probable invalidity of the assumption that the same average dose is appropriate for all organs.
4. Choice of the critical organ is equivocal. Irradiation of the gastrointestinal tract has usually not

---

[2] The interest that the Atomic Energy Commission and its principal contractors have shown in the establishment of adequate standards is significantly demonstrated by the fact that almost half the membership of the subcommittees of the National Committee on Radiation Protection is derived from this source.

been considered in *Handbook 52* limits for water contamination.

For some nuclides of practical interest, the factor of disadvantage for irradiation of the bowel is as high as 1,000, and for more exotic cases as high as 40,000. Yet uncompromising switch to the intestine as the critical organ would unnecessarily hamper the great majority of practical applications of limits for water.

5. Limits currently given apply to direct application of drinking water or inhaled air to man. These limits are not always the defining ones even for such familiar isotopes as I^{131} and P^{32}. If radioiodine were released to the atmosphere so that a large area was maintained at the permissible concentration for breathing, vegetation would become so contaminated with the isotope that its consumption by range animals would be hazardous. Similarly, if P^{32} were maintained in a public water way at the formal permissible limit, its concentration in plankton and algae and its transmission to fish through food chains would be sufficient to create a hazard to a successful fisherman.

6. The whole question of limits for mixtures is in doubt. Multiple critical organs are involved, and there may be synergistic effects.

These illustrations are not criticisms of the existing standards. Rather, they indicate that the standards field is becoming increasingly complex and is not ready for formal codification. Each major application must be studied as a separate research problem.

MECHANICAL TOOLS OF THE TRADE

Among mechanical tools of the trade one can group radiation detection instruments, mechanical devices for the safe handling of radiation equipment, and protective clothing.

A full account of the advances in instrumentation would occupy more than the allotted space for this review. In brief, in ten years the field has grown from an array of laboratory-designed equipment, with a few outstanding commercial contributions, to an efficient, diversified, and highly competitive branch of the instrument industry.

For the present purposes, one can divide instrumentation into (*a*) portable dose-rate meters, (*b*) portable contamination meters, and (*c*) fixed instrumentation, including hand counters, air sampling equipment, and supporting laboratory equipment.

Under category (*a*), the 9 types of meters used in the first two years of operation at the Hanford Works became obsolete in an average period of five years. By excluding two exceptional types that are still used the obsolescence period is reduced to three years. Under category (*b*), 7 original types persisted for an average period of five years. Under category (*c*), where the more taxing requirements of battery operation, small size, low weight, and ruggedness of portable meters are not required, the average life of 10 types has been seven years.

In addition to the progress measured by the tangible yardstick of obsolescence, there has been notable advance within many of the types in the direction of improved circuit stability, reduced time constant, reduced battery weight, reduced extracameral ionization, simplified calibration, and improved decontaminability. At present, the radiation monitoring equipment field appears to be in the middle of another transition period, which will lead to the application of scintillation meters to almost all monitoring situations, with significant gains in ease of maintenance.

In the field of mechanical devices for handling radioactive materials without exposure to radiation beams, there remains no problem of interest to industry except that of cost reduction. The three questions of adequate shielding, transmission of sufficiently sensitive motion through shields, and transmission of adequate vision of an operation through shields, were all solved in the early days of the atomic energy program. For laboratory operation, there has been some improvement in the design of remote manipulators, and the use of high density lead glass has been a desirable addition to methods of seeing through shields.

Mechanical devices which also prevent the spread of radioactive contamination have been far less successful. On the laboratory scale the ingenious enclosed boxes, developed by N. B. Garden and collaborators, have frequently solved the problem. Larger scale applications have rarely, if ever, been successful for an ex-

Fig. 3. The plastic man, an impervious suit with attached exit tunnel for use in grossly contaminated areas. The suit can be left in the contaminated zone so that the offending material is not disseminated. Topologically, such a device is a cross between a barrier around contaminated equipment and a protective suit on a man.

tended period. It is of interest that, in principle, the more complex Garden boxes are remotely operated chemical plants in miniature; the sole reason the miniature plant is highly successful and the large-scale one frequently less satisfactory is that the former can be thrown away if detailed decontamination should become troublesome.

Advances in protective clothing in the ten-year period have been generally disappointing. Substantial improvements have occurred in respiratory protection, where masks were designed for specific tasks. The general question, however, of providing a comfortable garment impervious to liquid contaminants remains unsolved. This is perhaps to be expected, since the ideal solution calls for a lightweight material that will repel all liquid or moisture presented to the outside while freely passing all moisture generated on the inside.

Protective clothing development follows two divergent paths from the earlier conventional cotton coveralls, shoe-covers, etc.: in one direction the aim is a cover cheap enough to permit disposal after a single use; in the other, a material as repellent as possible to the probable contaminants and as easily cleansed as possible of such contamination as does occur.

An extreme example of personnel protection is shown by the well publicized plastic suit with access tunnel (Fig. 3). Although such suits are effective in special cases, their *general* application would be a tacit admission of the basic failure of engineering in the field. Of the choice between placing an impervious barrier around each operator and placing it around the offending equipment, the latter is obviously the ultimate objective.

In thus assuring a clean atmosphere for the operators there is nothing qualitatively different from conventional industrial hygiene control of toxic materials. Quantitatively, the problem may be greatly intensified. This is clearly seen from the representative values of maximum allowable concentrations (MAC), as follows:

Contaminant	MAC p.p.m.
Lead	~ 0.1
Typical organic solvents	~ 20
Highly toxic non-radioactive materials	$\sim 10^{-3}$
Mixed fission products	$\sim 10^{-5}$
Plutonium	$\sim 10^{-8}$
Special fission products	$\sim 10^{-11}$

The cost of reducing atmospheric contamination by ventilation rises sharply as the cleanliness requirement is increased, and reduction by more than one or two orders of magnitude becomes an absurdity. The following are representative ventilation costs, averages of a number of industrial examples:

Ventilation	Percentage of Cost* of Normal Facility
Comfort only	$\sim 10\%$
10-fold dilution	$\sim 180\%$
100-fold dilution	$\sim 1000\%$

* Includes initial cost and operating costs. Normal facility, as used here, means one in which ventila-

tion requirements are provided for human comfort only.

MENTAL TOOLS OF THE TRADE

The need for and availability of professional and semiprofessional skills represents another phase of radiation protection. Outstanding progress has been made in this field in ten years. At the beginning of the period the qualified leaders were few in number and came generally from a common background of association with radiotherapy. Today, significant contributions to hazard control are being made by scientists and engineers who have had no contact with radiology. Formal training programs have helped develop an adequate corps of men in the intermediate field of technical attainment. From this corps should come the leaders for the next decade, men who can combine an understanding and appreciation of research developments in protection with a practical interest in and application to working requirements of industries and laboratories.

At the lower echelons of control is a general movement toward development of a recognized craft of radiation monitoring, which leads to one successful method of operation (in principle perhaps not the best, but in practice probably so), provided close support of technical specialists is available. Since a high percentage of day-to-day control of hazards is founded on common sense rather than highly specialized technical experience, and since common sense is by no means a monopoly of the academically trained, it is not surprising that excellent contributions to detailed control should have come from a force of radiation monitors.

The required research and development in radiation protection have been crystallized in the ten-year period. As applied to an operation such as the Hanford Works, this includes studies in the following fields:

Field	Principle Objectives
Radiobiology	Realistic maximum permissible exposures to man, range animals, aquatic life, wildfowl, plant life, etc.
Radiation biochemistry	Uptake, deposition, and retention of radioisotopes.
Pharmacology	Toxic effects of radioisotopes; methods of treatment.
Radiological physics	Dosimetry of radiation.
Instrument development	Improved instrumentation for monitoring and radiation research.
Meteorology	Atmospheric pollution control.
Industrial hygiene	Behavior of particulate matter in the atmosphere.
Soil science	Retention of radioactive wastes by earth materials
Hydrology	Transmission of underground wastes.
Chemistry; chemical engineering	Chemical dosimetry; sampling and analytical technics.

At the Hanford Works, these activities are integrated under the arbitrary title of Radiological Sciences. To develop necessary skills, specialists in the separate fields, usually without prior radiation experience, have been introduced to the radiation protection background of the parent organization under conditions which may tempt them to be critical of the sometimes biased outlook of the earlier radiological scientists. This cross-fertilization, both here and elsewhere, has brought many novel concepts to the whole field.

OPERATING RESULTS

For administrative purposes, radiation exposure incidents can be conveniently classified as follows:

(1) *Serious Overexposure:* Cases in which the tolerance status of an individual may be affected (as used in *Handbook 59*).

(2) *Technical Overexposure:* Cases in which exposure exceeds one of the recommendations of *Handbook 59* (or other appropriate recommendations).

(3) *Potential Overexposure:* Cases in which technical overexposure could have been received under similar circumstances; these minor incidents, documented for teaching and control purposes, serve to indicate the need for revised control methods.

In all cases, it is constructive to separate the cause into (*a*) radiation beams and (*b*) radioactive contamination. Table I shows

TABLE I: AVERAGE NUMBER OF INDIVIDUALS "OVER-EXPOSED" PER YEAR

Period	"Over-Exposure" Type					
	Serious		Technical		Potential	
	Rad.	Con.	Rad.	Con.	Rad.	Con.
Early	0	0	3	1	9	24
Middle	0	0	1	1	10	21
Recent	0	0	9	8	16	32

the average annual experience, with the ten-year period divided into three equal parts.

The absence of entries in the serious over-exposure column is noteworthy. Perhaps more surprising is the failure of entries in the other columns to diminish with time. The reason is at least threefold:

(a) The total risk has increased, as the rate of production of radioactive materials has increased substantially in this period.

(b) Maintenance work, always prone to generate contamination problems, logically increases for older facilities.

(c) The frequency of incidents is controllable by procedures. Thus, improvement in control methods not shown by a decrease in the frequency of radiation incidents actually appears on the record in the form of a *decrease* in individually written radiation work permits, with a decline from an average of 50,000 permits per year in the middle period to 15,000 per year in the recent period; this represents a substantial administrative saving with no loss of control.

In conformance with national policy, employee exposures to external radiation were originally subjected to controls of a daily limit, and more recently to those of a weekly limit. Of some significance is the long-term control on the basis of an annual limit. The accepted annual exposure to gamma radiation is 15 r. The self-imposed control at this location is 3 r per year. Average annual exposure has been far below this, as shown in Table II.

Favorable scores in average annual exposure can be achieved by including enough people whose expectation of exposure is very small; to some extent, the data of Table II include such dilution. Therefore, one must present some information on the

TABLE II: AVERAGE ANNUAL EXPOSURE TO GAMMA RADIATION

Year	Annual Exposure, (r)	Percentage Exceeding 1 r	Percentage Exceeding 3 r
1944–45	0.9	0.2	0
46	0.6	0.2	0
47	0.4	0.3	0.02
48	0.3	0.2	0
49	0.2	0.08	0
50	0.1	0.06	0
51	0.1	0.4	0
52	0.2	3	0.02
53	0.2	5	0.07
54 (projected)*	~0.2	~3	~0.05

* Note added in proof: Actual figures for 1954 were 0.2, 5, and 0.2 respectively.

highest cases. This has conventionally been done at Hanford by reporting the 10 highest annual exposures to gamma radiation each year. These averages are:

1945	1.3 r
46	1.4 r
47	1.9 r
48	1.4 r
49	1.0 r
50	1.0 r
51	1.4 r
52	2.6 r
53	3.1 r
54 (projected)	4.0 r

It is evident from the data of Table II and the figures just given that control of external radiation in the atomic energy business presents no problems and can demonstrably be manipulated to levels that interpret the phrase "lowest practicable level" of *Handbook 59* to include economic incentive to avoid extremes of control.

Control of internal deposition is not so easily manipulated. Contributing reasons are: (a) difficulty of prompt signalling of an exposure; (b) greater uncertainty of deposition limits; (c) uncertainty of bio-assay interpretation as to when an assumed dep-

TABLE III: INTERNAL DEPOSITION OF RADIOISOTOPES

Period	Positive Cases per Year	Positive Cases per Year (Reduced to Consistent Detection Limit)	Cases Above Appropriate Limit
Early	0	0	0
Middle	1	1	0
Recent	30	6	2?

osition limit is being approached; (*d*) inability to remove the source in most cases.

The incidence of positive cases of deposition has risen as shown in Table III. The two entries in the last column refer to cases which in fact are probably below appropriate limits, but in which there is still some uncertainty of interpretation.[3]

Operating results with respect to release of radioactive effluents to the atmosphere or to public water supplies have been demonstrated in scientific exhibits at two previous meetings of the Radiological Society of North America. It will be sufficient to point out here that all these problems have proved to be controllable with adequate safety margins. In order to operate at the best economic level, it is necessary to maintain extensive regional monitoring programs supported by research studies.

FUTURE PROBLEMS

There are two significant administrative problems for the next decade. The first relates to *education and public relations*, and the second to *legislation and regulation*. Opinions on the released information on atomic energy hazards range all the way from the implication that the drama of the situation has been magnified by enthusiasts to the equally extreme implication that the real nature of the hazard has been deliberately de-emphasized. The public has the right to as full information from existing atomic energy programs as is consistent with prudent control of data from which the atomic potential of the nation could be inferred. The next decade will see widespread industrial application of radioisotopes and the first use of industrial atomic power. Factual information on hazards in both these fields is needed.

This contributor's point of view is suggested in the present review of ten years experience. Foreseeable hazards in external radiation can be controlled with comparative ease. Internal deposition is also controllable through the use of safety factors which may prove to be excessive when further research has crystallized appropriate limits for specific cases. The atomic energy industry started safely at the expense of considerable investment in radiation protection, and it can be safely continued.

There is an inherent fear in some quarters that it is one matter to spend government dollars on protection and quite a different matter to spend stockholders' dollars. In the present era of enlightened business management, the writer believes this fear to be ill-founded.

Regulation of radiation hazards will receive much attention in the coming period, and the manner in which it is carried out may affect the nation's rate of progress in atomic energy. Traditionally the National Committee on Radiation Protection has emphasized the advantages of control by recommendation *versus* control by statute. The recent appointment of a subcommittee to prepare model codes for state legislation presumably represents an admission that self-regulation by recommendation will not be powerful enough to assure control in a rapidly expanding field. Since the intelligent management of hazards, especially as it affects the public in terms of waste disposal, is heavily conditioned by local circumstances, it will prove extremely difficult to write generalized codes that are sufficiently restrictive of all radiation operations, without being absurdly over-restrictive of the majority.

It appears probable that unduly restrictive codes will be written. The key to progress will then hinge on the manner of interpretation and enforcement. There are good precedents in related fields, such as water pollution, for adequate and enlightened control as opposed to literal interpretation. A major disadvantage of this approach is that the large and reputable manufacturer may tend to stay within the most rigorous possible interpretation of every code clause, regardless of cost, in defense of his reputation; the marginal operator would be much less likely to

---
[3] Note added in proof: These 2 cases are now known to be below permissible limits.

accept the more expensive interpretation. To some degree, this can place the emphasis of restrictive control on the organizations most likely to maintain voluntary control at a demonstrably reasonable level for the specific case.

The fundamental problems in radiation protection remain unchanged. Typical questions and the practical corollaries are:

1. What is the basic nature of the interaction of radiation with living matter? Would knowledge of this lead to a generalized scheme of protection or to methods of moderating exposure effects?

2. What is the origin of radiation carcinogenesis? Under what conditions, if any, can a single radioactive focus in the lung, for example, induce a malignant neoplasm?

3. What is the precise significance of the genetic effect in man?

4. Can the total residual insult, say for a single radiation type, be integrated by a plausible formula to cover all cases between the extremes of a single acute exposure to a uniform chronic exposure?

5. Under what conditions are the effects of mixed radiations additive by some determinable formula? Can such a concept as the *rem* be fruitfully extended?

Applied problems of principal concern include the following:

1. How can exposures to mixed radiations be adequately integrated in the field?

2. What are the appropriate permissible limits for a wide variety of radioisotope exposures in man and other life forms?

3. How can organ exposures to low-energy emitters be determined? In particular, how can lung deposits be determined, and how can bio-assay results be reliably related to body deposition?

4. What pre- or post-exposure methods can be developed to reduce radiation injury?

The engineering problems related to protection are primarily:

1. How can radioactive materials on the industrial scale be completely contained at source?

2. How can radioactive waste be permanently contained on an economic basis?

3. While remote operation and maintenance of major equipment is feasible, initial cost is high and maintenance tends to be slow and costly. If Problem 1 cannot be solved, can effective rapid decontamination be developed to make contact maintenance attractive?

4. Again assuming that Problem 1 is not completely soluble, what standards of air cleansing, both within operating plants and for releases to the atmosphere are necessary and adequate? How can such standards be reached economically?

5. As applied to the nuclear power industry, can reactors be designed and operated with *absolute* assurance that there could be no catastrophic accident, so that they could be placed in metropolitan areas?

6. How can irradiated materials (*e.g.*, reactor fuel elements) be safely and economically transported to processing locations?

In summary, one can say that the fundamental problems of radiation protection are precisely those that have perplexed the inquiring radiologist and radiobiologist since radiation sources were first used. The new emphasis introduced by the atomic energy program is the need for solutions to these problems for mixed radiations.

In the applied problem and engineering areas, the emphasis is predominantly on the elimination of radioactive contamination and on the interpretation of contamination hazards on the body, within the body, and in many other life forms.

It seems to the writer that the next decade will see substantial reduction in the release of effluents to the atmosphere, with some progress in control of in-plant contamination. It appears unlikely that absolute success will be achieved in either field. Therefore, the chief expected burden in radiation protection will probably be in more and more refined interpretation of minimal depositions in man and other life forms. In this way, practical success in control, as opposed to unequivocal success by complete elimination of the problems at source, is the reasonable objective for the next decade.

Hanford Works
Richland, Wash.

REFERENCE

PARKER, H. M.: Health Physics, Instrumentation and Radiation Protection. In Advances in Biological and Medical Physics, Vol. I, edited by J. H. Lawrence and J. G. Hamilton. New York, Academic Press. Also issued as MDDC 783.

SUMARIO
La Protección contra la Radiación en la Industria de Energía Atómica: Estudio Decenal

Repásanse aquí los desenvolvimientos en la protección contra la radiación desde 1944, en que se emprendió la producción de reactores nucleares para la fabricación de plutonio en la Sección de Productos Atómicos de los Talleres Hanford. Obsérvase que los problemas fundamentales son precisamente los que han perturbado al radiólogo y al radiobiólogo inquisitivos desde que se usaron por primera vez focos de radiación. La nueva acentuación procede de la necesidad de encontrar soluciones a los problemas de las radiaciones mixtas.

Los riesgos predecibles de la radiación en su aplicación externa pueden dominarse con comparativa facilidad. Los depósitos internos son también dominables por medio del empleo de factores de seguridad que quizás resulten excesivos cuando nuevas investigaciones cristalicen los límites apropiados a casos específicos. En las zonas de ingeniería, se recalca predominantemente la eliminación de la contaminación radioactiva y la interpretación de los riesgos de contaminación sobre el cuerpo, dentro del cuerpo y en otras formas de vida. ¿Cómo pueden integrarse adecuadamente en el campo las exposiciones a las radiaciones mixtas? ¿Cuáles son los limites tolerables apropiados para una inmensa variedad de exposiciones a radioisótopos en el hombre y en otras formas de vida? ¿Qué métodos pre- o post-exposición pueden elaborarse que reduzcan las lesiones irradiatorias?

Opina el A. que el próximo decenio verá una reducción substancial en la liberación de efluentes a la atmósfera, con algún adelanto en el dominio de la contaminación por implantes, pero que parece improbable que se obtenga éxito absoluto en una u otra rama. Por lo tanto, la principal sobrecarga esperada en la protección contra la radiación consistirá probablemente en una interpretación cada vez más refinada de los depósitos minimos en el hombre y otras formas de vida. El éxito práctico en el dominio, en contraposición al éxito inequívoco mediante la eliminación total del problema en su foco de origen, representa el objetivo razonable para el proximo decenio.

Effects of Reactor Accidents on Plant and Community

A.M.A. Archives of Industrial Health, Volume 13

May 1956

Reprinted with permission of the Helen Dwight Reid Educational Foundation. Published by Heldref Publications, 4000 Albemarle St., N.W., Washington D.C. 20016. Copyright©1956.

Reprinted from the A. M. A. Archives of Industrial Health
May 1956, Vol. 13, pp. 441 and 442
Copyright 1956, by American Medical Association

Effects of Reactor Accidents on Plant and Community

HERBERT M. PARKER, M.Sc., F.Inst.P.
Richland, Wash.

It is inconceivable that a nuclear reactor for power purposes will ever be built so that it can be operated in such a manner as to cause a violent eruption, with consequent release to the atmosphere of considerable amounts of fission products. Although everyone agrees that reactors will not be located in communities until the risk of a major disaster is negligible, it is of interest to examine the worst possible case just to establish how certainly we must know the potential risks which have to be overcome. The degree of disaster will depend on the following factors: the power level of the reactor; the fraction of the active material that can escape in the event of disaster; the precise mechanism of escape of these offensive products; the population density around the reactor, and the meteorological conditions, particularly at the time of the incident and in the few hours following.

Calculations made on this basis are strictly "order of magnitude" and refer to the solid fuel element type of reactor where fission products will accumulate. We can perhaps say that nuclear power reactors will operate at power levels in the range of tens of thousands of kilowatts to millions of kilowatts. For the purpose of discussion, let us consider an arbitrary standard reactor, which will be 100,000 kw. of heat power and about 25,000 to 30,000 kw. of available electrical power.

The chief hazard, beyond any question, arises from the accumulation of fission products in the reactor, despite the fact that it will be loaded up with U^{235} and plutonium. It is helpful to make a comparison with the atomic bomb. A nominal bomb leaves a residue of about 10,000,000,000 curies of radioactivity one hour after its detonation. This little standard reactor will have approximately 300,000,000 curies to 400,000,000 curies at normal operation. In terms of available radioactive content this reactor contains less than the nominal atomic bomb. But the hazard from the fission products is by no means the same, as will be described later.

It is absolutely impossible for the reactor to explode in the fashion of a bomb. However, a release of fission products by a conventional type of steam explosion would be a reasonable approximation if the "impossible" happened. We can only assume for our purposes that all the fission products are released from the reactor, although it will be recalled that before reactors will be operated there will be the knowledge that no such reaction will take place and, if it does, all the products will be retained in the immediate structure.

Of importance in determining the characteristics of the contamination is the cloud height to which a release of fission products may go, and this is a function of the rate of release of energy. Mechanisms of disruption can be visualized which would cause a very abrupt release of hot gaseous fission products that would rise to something of the

---

Received for publication Oct. 6, 1955.
Director, Radiological Sciences Department, General Electric Company.
Read in the Panel Discussion on the Impact of the Atomic Energy Industry on Community Health at the 15th Annual Congress on Industrial Health.
Work done under A. E. C. Contract W31-109 Eng. 52.

order of 4000 ft. as a mushroom cloud. The condensed fission products would then move downwind and spread out by diffusion and eventually come to the ground.

In another situation, which by and large is the more probable, there would be a slower release of energy and heating up of the reactor structure leading to vaporization of the fission products. It is as though the whole unit had been set on fire, and the radioactive materials boil out and again descend downwind close to the ground, diluting themselves by turbulent diffusion. Under conditions of ultrastability of the atmosphere (and this is the most dangerous condition), the materials come out and roll along without diluting in any way by vertical motion, and thus the radioactivity at a given distance from the reactor tends to be maximized.

As the radioactive cloud goes by close to the ground, persons in the area will be subjected to external radiation from the penetrating gamma radiation and the shorter range beta rays, which just affect the skin. In addition, there will be direct beta radiation of the lung from inhaled fission products. The fission products can also be ingested and distributed to various organs in the body. As previously mentioned, the fission products from the reactor are more dangerous than from an atomic bomb, because they are longer-lived. For example, Sr^{90}, an important long-lived isotope, will accumulate to the extent of 3000 curies in about 10 days of operation of the reactor. In 500 days there would be about 150,000 curies accumulated. The longer the reactor operates, the more longer-lived and, generally speaking, the more dangerous isotopes build up. The whole composition of the radioactive isotopes in this case does not decay as rapidly as the products of the atomic bomb. In the atomic bomb the decay follows a $T^{-1.2}$ function. In the case of the fission products from a reactor the function approximates $T^{-0.3}$, where T is the time since release of the material. If these functions are carried out over a matter of days or weeks, one rapidly appreciates the relative gain in safety of the atomic bomb products. As the radioactive cloud passes by, radioactive particles will settle out, contaminating the ground, buildings, animals, and vegetation. There will be contamination exceeding 5 mc. per square foot over an area of 30 to 40 square miles. A contamination of greater than 2 mc. per square foot will cover 100 to 200 square miles, with more than 0.5 mc. per square foot for something like 300 to 500 square miles. What do these numbers mean in terms of action? They mean that, for example, 5 mc/ft.^{2} stands for "must control" for five years. At the end of this time the radiation from the ground will just equal the conventional permissible maximum as read from the National Bureau of Standards Handbook 59, "Permissible Dose from External Sources of Ionizing Radiation."

The most important defense measure, in the event of this type of disaster, is to evacuate the area in the path of the oncoming radioactive cloud. In the event of personnel contamination, the radioactive particles should be washed off the skin and clothing as soon as possible. Decontamination of buildings and equipment involves three methods: removal of radioactive material from the surface by washing; removal of the surface itself, or application of a coating to seal in the dangerous fission products. Water supplies would have to be tested for radioactivity rather carefully, but rivers would be no problem, since they would pick up one slug of activity which would be carried down and could be bypassed at any point of major use. Some basins might be troublesome where there is leaching of fission products into the water table.

Health Problems Associated with the Radiochemical Processing Industry

A.M.A. Archives of Industrial Health, Volume 13

May 1956

Reprinted with permission of the Helen Dwight Reid Educational Foundation. Published by Heldref Publications, 4000 Albemarle St., N.W., Washington D.C. 20016. Copyright©1956.

Health Problems Associated with the Radiochemical Processing Industry

HERBERT M. PARKER, M.Sc., F.Inst.P.,
Richland, Wash.

Earlier contributions at this meeting have suggested that reactors for industrial power may be of two broad types. The first, operating with solid fuel elements, would constitute a conventional power station. The second type, the homogeneous reactor, with the nuclear fuel in solution or suspension, would to some extent be a combined power station and chemical plant. We shall consider only the first case and postulate that fuel elements from a battery of power stations are returned to a central processing plant for treatment. The scope of such a plant would include separation of plutonium, recovery of U^{235} for reuse (or related products in both cases), and removal of fission-product wastes for storage or commercial application.

We are concerned with radiological health in the plant and possible effects of effluents on the environs.

Employee health protection is conveniently broken down into two categories:

1. Control of external penetrating radiation.
2. Control of radioactive contamination. Contamination is simply actual radioactive material in the wrong place. This may mean on work surfaces in the plant, in the atmosphere, in the ground, or in water supplies. Ultimately, the wrong place par excellence is deposition in the body.

Received for publication Oct. 6, 1955.

Director, Radiological Sciences Department, General Electric Company.

Work done under A. E. C. contract W-31-109-Eng.-52.

Read in the Panel Discussion on the Impact of the Atomic Energy Industry on Community Health at the 15th Annual Congress on Industrial Health.

The first category has no unsolved problems. Conventional control is by shielding, distance, or time. Of these, the first two are engineered into the plant. Some residual operations are controlled by time limits based on radiation monitoring prior to and during performance of the special work.

The second category introduces all the significant problems in the radiochemical processing industry. To date it has not been possible to design facilities that will eliminate contamination.

When contamination does occur, two classes of exposure may occur: First, the material itself may emit enough penetrating radiation to constitute a hazard from external radiation. This component can be controlled by temporary shielding, distance, or time limits, just as radiation from contained sources is controlled.

The second class of exposure arises from potential transfer to the body through inhalation, ingestion, and transmission through the skin or through wounds. Prevention of these processes is the expensive, time-consuming phase of protection.

Briefly, the defensive steps are the following:

1. Establishment and observance of contamination control areas, implemented by
    Protective clothing and equipment
    Check-out surveys of clothing and skin
    Control of removed contaminated equipment
    Decontamination of affected areas
2. Maintenance of safe breathing air, implemented by
    Ventilation zones, with air drawn from clean zones to less clean zones
    Containment of contamination-generating sources wherever possible
    Air-cleaning techniques at exit from source enclosure spaces

2

3. Personnel protection procedures, including
   Frequent monitoring of skin and clothing
   Skin decontamination
   Prompt treatment of contaminated injuries, which includes use of venous-return tourniquets, profuse water flushing, decontamination of site under medical direction, and, if necessary, small excisions around a wound
   For inhalation, nasal smears and irrigation
   Bioassay—radiochemical analysis of urine, blood, or feces to estimate possible body depositions
   External monitoring can occasionally be used to detect deposition in lung, thyroid, or other organs.

In this program there is nothing novel to conventional industrial hygiene in principle. The intensity of the effort has to be characteristically greater, because the toxic mass for radioactive contaminants may be thousands of times lower than for conventional contaminants.

There seems to be no absolute level of general contamination marking the boundary between safe and unsafe levels. The conservative practice of removing all detectable loose contamination avoids the pitfalls of having contamination gradually spread to supposedly clean areas, private or public vehicles or employee residences.

Just one incident of public contamination, in addition to being a limited health hazard, is certainly a public relations hazard to people whose morale should not be disturbed by the exaggerations that arise in such incidents.

Summing up the points on employee protection as we see them now, there are two ultimate health hazards in the business of operating radiochemical plants. First and most spectacular, there will always be, one supposes, a very few accident cases, the kind of thing Dr. Sterner mentioned so well yesterday. They are of two kinds: (1) the very severe problem of contaminated injuries and what to do where prompt medical treatment is the solution to the problem—prompt medical treatment sometimes when the radiation effects cannot be determined in full—and (2) the type of accident case which is probably inhalation of radioactive dust. Despite controls, slips can be made, and once the radioactive dust is deposited in the lungs, unless nature wills it to come out again, so far experience seems to fail to encourage success in man-induced removal.

On top of this, the long-term hazard is very likely this: that despite the best contamination control measures, there will continue to be radioactive contamination in the air of radiochemical plants, and there will be through the course of the years a very slow build-up in the bodies of virtually all the working force of some long-lived materials. This again is a problem that Dr. Sterner talked about. It is the one, I think, that has not been thoroughly analyzed, or perhaps even attempted, for parallel problems in any other industry on this scale, because one knows the final answer only some 10, 20, or 30 years from now.

Perhaps by the time the deposit becomes a hazard at a significant level—I want to make it clear we do not believe it is at or near that level at the present time—the true permissible levels for any isotope will be well known.

At the present time figures in Handbook 52 are to some extent physicists' dreams and do not yet bear the kind of factual relation to the hazard problem that we would like to have for all isotopes.

Now, as to general effects on the community, these will come in three ways: First, we are visualizing a set of power reactors, with the fuel elements shipped from the reactors to the chemical plant. In this situation shipping is an accident hazard, but normally this is a well-controlled one. Shipment of radioactive materials of any activity can be done safely, and the only problems for the community are those arising from an accident. When one ships by truck, for example, if the truck has a major crash, there is a local problem for the community; it is a problem of hazard extending some several hundred feet perhaps from the source of the "crime" in dealing with solid fuel elements.

The shipment of radioactive waste liquids might be a much more pertinent public health

3

question, since in the event of a major crash en route such material could escape and get itself into secondary complications that we can discuss under another heading in a moment. With reasonable precautions, there should be little risk in making shipments of this type from power plant to chemical plant.

So much for material coming to the plant. Now the chemical plant may send material out in two ways, in addition to planned shipments of reprocessed fuel elements. The first is by waste disposal of solid or liquid material. This disposal of solids is fairly well controlled; they are put in a box, and the box is kept in a controlled place. The volume then is the only problem. Storage of liquid material is one of the main long-range problems of the whole radiochemical processing industry, and no one knows the ideal solution. Every one dreams of a private pipeline from his radiochemical plant to some depth in the ocean from which the material could not rise because of favorable temperature gradients in the ocean, but that stage is not reached very readily. In the meantime, the requirements of waste disposal from large plants essentially determine where those large plants shall be and, I think, essentially say that they will not be in communities. It is on this one factor that probably a large share of the technology of industrial use of atomic energy may depend.

Such materials can, of course, be held in tanks for reasonable decay periods. By reasonable decay periods one is thinking in terms of something of the order of 50 years, because when one commits material to a tank one must be assured of the integrity of that tank for a given length of time; to make one that will hold one's problems for longer than that turns out to be a very expensive proposition.

Wastes of lower activity can be handled in a variety of ways with what one might call first-order safety. At least in some favored locations it has been shown by the rather close study of the actual conditions that substantial volumes of waste material can be returned into the ground, where the soil acts as a scavenger of the radioactive material and provides what amounts to a wall-less tank in the ground, ending up with a filtering-through of essentially clear water which goes to the ground water supply and ultimately gets back to the public domain. The method must be supported by a long-continued study system to make sure that there are no natural processes that can later leach out those deposits.

On a lower scale one can reasonably dispose of material to public sewers and the like, but one sets up the problem there of following any concentration of the material at points down the line from the disposal system. It is rather easy to show that many radioactive materials show an undesirable tendency to concentrate where least desired.

If one picks out specific radioelements, one can find concentration factors in nature as high as a half-million-fold, and this is a very serious determinant in estimating health hazard.

The third problem, and the second mechanism of escape from the radiochemical plant, is disposal to the atmosphere of gases or dusts. In principle this is a very easy situation to control. One knows how to remove gases from waste streams, and one knows how to remove dusts. The only difference from experience in more familiar industries is degree of cleanliness required.

Take, for example, one isotope, I^{131}, which is one that will occur in chemical plants which reprocess fuel material. Then, on a weight basis 1 part of I^{131} in 50 quadrillion parts of air constitutes the permissible limit. I noticed with interest the display down the hall of the gruesomely toxic materials of non-radioactive character that one encounters, but their MAC's are many, many zeros away from being 1 part in 50 quadrillion, and it is on this difference in degree of 1 part in 50 quadrillion *versus* something more on the order of 1 part in a million that the engineering skills, and equally so the health hazard problems, hinge.

4

Again, in release of the materials to the atmosphere the obvious hazard is not always —in fact, one might almost dare say not usually—the real hazard. The common element I^{131} is a case in point. It is not proper to use Handbook 52 figures, which are for human breathing of I^{131}, because this turns out not to be the final relevant problem. The material is released essentially as a gas or an aerosol. When traveling over the land surface, it deposits on vegetation and the like, and one must be concerned with the secondary hazards set up.

Such hazards are two in nature—the hazard to grazing animals eating vegetation contaminated with I^{131} and the additional possibilities of such occurrences being over truck-farming land where products are actually used by man or used directly as in milk from cows on contaminated pasture. One must be sure when one studies these problems to pick the critical determinant. The matter is made all the more difficult if several determinants happen to be comparable, as in this case. For I^{131} the limit established in this way is about one-thousandth part of the otherwise conventional limit.

Small fission product particles, in addition to separated materials like I^{131}, also have the habit of escaping from processing plants. For some years yet such emissions will cause concern, not because there is a known existing hazard in such things but because one cannot dismiss the problem unequivocally. More must be known about the real hazard of inhalation of very small radioactive particles. The solution will probably take at least another decade. Even so, perhaps it will come before we can make the alternative solution to have engineering that will 100% prevent escape of small radioactive particles to the atmosphere.

In the meantime, the combined problems of the escape of contaminants to the air and the disposal of liquids suggest a technology in which major radiochemical facilities are isolated from communities. With this technology there is every reason to believe that the health problems induced in the community are entirely insignificant.

Printed and Published in the United States of America

Selected Materials on Radiation Protection Criteria and Standards: Their Basis and Use

Radiation Protection Criteria and Standards

May 1960

86th Congress
2d Session
JOINT COMMITTEE PRINT

SELECTED MATERIALS ON RADIATION
PROTECTION CRITERIA AND
STANDARDS:
THEIR BASIS AND USE

JOINT COMMITTEE ON ATOMIC ENERGY
CONGRESS OF THE UNITED STATES

MAY 1960

Printed for the use of the Joint Committee on Atomic Energy

SELECTED MATERIALS ON RADIATION PROTECTION CRITERIA AND STANDARDS: THEIR BASIS AND USE

CHAPTER 1. STATEMENTS BY PRIVATE GROUPS AND INDIVIDUALS [1]

RADIATION PROTECTION CRITERIA AND STANDARDS: THEIR BASIS AND USE

(H. M. Parker, Manager, Hanford Laboratories, Hanford Atomic Products Operation, General Electric Co., Richland, Wash.)

INTRODUCTION

As a preface to a discussion of radiation protection standards, a review of the purpose and nature of standards in general seems in order to explain better some of the concepts affecting present radiation standards. Standards are developed to provide a model for judgment as to whether a given action should be taken. They are frequently the result of past experience utilized to facilitate future experience. They may be the communications of the enlightened to guide the less experienced. Finally, they are the integrating means by which there may be orderly development of a whole made up of many parts.

Of the many ways in which standards can originate, two are especially important in the radiation protection field. The first concerns moral and ethical standards which have been manifested since the recording of civilization. For example, the Hippocratic oath defines a recognized standard of conduct expected of a physician. Concern for the well-being of future generations with respect to the genetic effects of ionizing radiation leads to standards of this kind.

The second type of standards stems from the need to facilitate action to allow each individual in society to proceed in a certain direction, knowing that certain obstacles are removed. Men drive vehicles on the righthand side of the road in a given community because confusion is avoided and safety enhanced if everyone does so in that community. Similarly, radiation protection standards may be guides by which an individual may proceed with reasonable safety for himself and with the assurance of not causing unreasonable interference with other segments of his environment.

The enforcement of standards is important in considering their usefulness. Ideally, they will be accepted voluntarily through confidence in the source of the standard—the authority of knowledge—or through a clear understanding of the benefits to be derived. Thus, the voluntary restriction of freedom to drive only on the righthand side of the road is observed less because of the policeman than because this action

---

[1] See also statement of Prof. Barry Commoner attached as a supplement, p. 1239.

2   RADIATION PROTECTION CRITERIA AND STANDARDS

demonstrably affects one's own safety and ability to travel with the least confusion.

Where effects are not so obvious, where effects on the entire pattern rather than on a part are important, or where an individual or group may conclude that the burden of restriction is not equitably shared, enforcement by edict may be necessary. In the particular case of radiation protection, where the possible deleterious effects may be both latent and subtle, this type of enforcement may have an important role.

A final comment on standards in general concerns their dynamic nature. One cannot expect that standards established now will be immutable for all time hence. Just as knowledge increases, so can the quality and applicability of standards change. Unless standards are dynamic, they can become unnecessarily restrictive or completely misleading.

THE MERITS OF STANDARDS IN TECHNOLOGY-ORIENTED NATIONS

As technologies become more complex and specialized, the public must depend increasingly on the assurance that "somebody" has looked at the problems and at the hazards; and that "rules" have been established for the operation of plants and processes to provide necessary and sufficient safeguards to ensure responsiveness to the mores of the Nation. In fields involving nuclear reactions or other large-scale industrial processes, the "rules" can be the nationally recognized standards developed by agreement between major organizations, both public and private, which have both knowledge and interest in the field.

Technological standards fulfill these important purposes:
    1. They promote safety.
    2. They promote simplification by achieving agreements between suppliers and users regarding materials, equipment and procedures. This simplification should lead to broadly shared national benefits of conservation and economical performance.
    3. Well-prepared standards leave the way open for further improvements through ingenuity, new technology and materials. Standardizing bodies must be flexible enough to encourage rather than choke off initiative and technical progress.
    4. Particularly in a new field, a well-planned and well-timed system of standards can contribute greatly to the orderly development of the technology. It is easier to reach agreement *before* groups have gone too far along their separate ways.

THE ORIGIN OF STANDARDS

Formulation of standards may begin when a pressing need is felt, or preferably, when it is foreseen. Simple technological examples are agreement on a few standardized automobile wheel sizes, and the interchangeability of electrical connecting plugs. The development of such carefully planned standards in a production-minded nation is inevitable. In the relatively new field of nuclear energy utilization, many standards are today being readied for use, while the industry is still in an early stage and before practices have become deeply rooted. The combined efforts of professional technical societies, industrial associations, agencies of Federal and State Governments, labor organizations, and insuring groups are being integrated in this work.

ENFORCEMENT AND USE

The simple examples mentioned provide evident benefits to all groups. Enforcement rests comfortably on voluntary agreement. Where public safety is involved, a well-tried standard of industry may become the technical foundation for laws which require compliance and which may establish inspection requirements to aid enforcement. As an example, the boiler and pressure vessel code of the American Society of Mechanical Engineers, enforced by law in many States, has been a powerful force in avoiding boiler explosions, thus promoting public safety.

Unless a standard has the force of law—and sometimes even where it does have—its acceptance and use rest largely on its merit. The actual need for a recognized guide, the extent of experience available as background, the wisdom of the authors, the clarity of expression and the flexibility to meet changing conditions all influence the worth of such standards.

SOME SALIENT POINTS IN THE HISTORY OF RADIATION PROTECTION STANDARDS

The need for radiation protection criteria was recognized many years ago—in fact, shortly after the discovery of ionizing radiation and radioactive materials. However, many of the effects of exposure to radiation became known only as work with it progressed. During the 45 years prior to the atomic age, there were injuries and deaths resulting from the considerable amount of work with relatively high levels of radiation. These incidents showed conclusively that indiscriminate use of radiation without standards for protection cannot be tolerated. The National Committee on Radiation Protection, similarly responsible committees in other nations, and the International Commission on Radiological Protection were founded in recognition of this fact and have provided a great service through their continuing program of compiling and studying the data available, and in recommending limits for radiation.

Early experiences provide the basis upon which the knowledge of the effects of protracted radiation exposure of man is founded. The present limits for radiation protection were derived from them in a manner completely analogous to that by which many of the standards for toxic chemicals were derived. In the period up to 1940, the principal information on the effects of external radiation came from four groups of exposed individuals. (1) These were:

1. Fluoroscopists who were exposed to short bursts of quite intense radiation at rates between 4 and 400 r/hour to the body;

2. X-ray therapy technicians who were characteristically exposed to radiation for periods of 5 to 30 minutes with intervals of the same order between treatments. Before self-protected tubes and special shields were used, the exposure rate for these individuals was on the order of 400 mr/hour.

3. X-ray therapy patients in which the dose rates were on the order of 40 r/hour for a total dose to the body of 30 to 120 r from

4 RADIATION PROTECTION CRITERIA AND STANDARDS

scattered radiation and delivered over a period of weeks rather than years; and

4. Radium therapists and technicians who were exposed at rates on the order of 4r/hour for periods up to 1 hour at irregular intervals with an additional exposure from the background in the work area of perhaps 5 to 10 mr/hour through the working day.

During this period up to 1940, the limits for radiation were referred to as the "tolerance dose." This implied that there is a level of radiation exposure at which there is no lasting damage to the organism, presumably because the rate of repair is great enough and continues through the lifetime so that the damage is continually repaired.

The tolerance dose, 100 mr/day, used during the World War II period, was based on the experience with the radiation effects on people primarily occupationally exposed. Cantril and Parker (1) pointed out at this time that the basic numbers available in 1945 which were accepted as a working basis for occupational exposure were:

1. 100 mr/day for external X and gamma radiation;
2. 10^{-14} curies/cc for radon in the air of working rooms;
3. 0.1 μg of radium as the maximum allowable amount deposited in the body.

Each of these levels was established by applying a safety factor to the amount which had been observed to produce lasting injury to persons so exposed. At the time of this appraisal, the authors believed that the safety margin was considerably less than a factor of 10. It is true that these numbers were based upon limited numbers of cases so that the statistics for applying to total damage with large numbers of people were poor.

Since this time there have been some changes in the limits, primarily on the basis of further observations of early patients given radium; of epidemiological studies following the uses of radiation in diagnosis and therapy; and of extensive animal experiments. The animal experimentation has been primarily useful in providing concepts rather than in providing information directly applicable to human limits, since the quantitative response of man to radiation is different from that of most laboratory animals. Thus, the concepts of life shortening, most of our knowledge of genetic changes, values for the relative biological effectiveness of different radiations, and the knowledge of the differential effects between radium and other bone seekers is based on animal experimentation. In the final analysis, however, these concepts are applied to man and the human maximum permissible limit through the basic information which has been developed in the limited experience with the effects on man.

In the period between 1945 and 1950, the information available on the genetic effects and the increasing suspicion that not all somatic effects of radiation were of a threshold type (i.e., of a type that requires a minimum dose before manifestation) led to the change in terminology from "tolerance dose" to the present "maximum permissible dose." It was at this point that the concept of acceptable risk appeared.

I am convinced that the terminology was changed from the one form to the other specifically to underline the acceptance of a no threshold dose concept for the production of gene mutations by radiation (p. 26 et seq. of handbook 59) (2). From a language point of view, one could equally well have been changing from a "maximum permissible dose" to "tolerable dose" or "tolerance dose." The intent to change to an "acceptable risk" line of reasoning was partially obscured by an arbitrary choice of words.

The fundamental change in approach at this point has not always been sufficiently emphasized. The earlier clinical observations clearly pointed to the existence of threshold effects and perhaps conditioned observers to expect all deleterious effects to have a threshold. It was, for example, palpably impossible to produce reddening of the skin with less than some prescribed dose. If such a threshold applied to every radiation effect, an intelligent, wholly technical, search could be made to define such thresholds with precision, and persons exposed to lesser amounts would be protected in the fullest sense.

If there is no threshold dose, there is no absolutely safe dose in the same sense. The determination of permissible limits then, involves value judgments outside the areas of scientific and technical competence. It is essential that qualified observers recognize this limitation and reexamine their own role. I believe that it is equally essential that these same observers continue to play a basic role in recommending standards. This follows from the principle of better acceptance of standards having the authority of knowledge, and from one further much overlooked factor. This is that although present day standards appear to be established in detail by the considerations of physicists, biologists, and other scientists, they are basically validated by clinical experience. This experience does go beyond reproducible science and does include value judgments in accordance with the Hippocratic oath. Because some aspects of the genetic effects are beyond clinical observation, there is a needed separate voice of authority from geneticists. Then, because the decisions are necessarily based on cultural reactions to human welfare, many generations hence, the whole structure of ethical opinion must be included in some fashion.

The firm resolution of the evidence concerning the response at very low dose rate and low total dose as to whether relevant radiation effects are threshold, nonthreshold and linear, or nonthreshold and nonlinear is the most important factor, in effect, in establishing the authority of knowledge for radiation standards.

There is yet no complete definitive, scientific answer to these questions. The assumption of a nonthreshold type of response for somatic as well as genetic effects in setting limits, while plausible, is an assumption and numerical estimates of effects calculated on this basis must be treated with reservation.

It is frequently indicated that the assumption of a nonthreshold response is a "safe" assumption. This can be followed by a stipulation that permissible exposures are zero. However, there are other consequences to the Nation as a whole from eliminating radiation or radiation exposures. These consequences can be expressed in terms of a limitation in our ability to attain the many benefits which radiation can bring. In essence, then, the risk principle states that radiation exposure and the potential damage from radiation should be balanced against the benefits of radiation and the limits set at some level where the optimal benefits are attained, as compared to the losses. To establish such limits for radiation and equally carefully for any other

6 RADIATION PROTECTION CRITERIA AND STANDARDS

agent or influence on the national life, projected over an indefinite number of generations, is a herculean task.

PRESENT STANDARDS

The basis for most of the standards presently used for protection of individuals against radiation are the recommendations of the ICRP and the NCRP. As has been noted in previous hearings, these recommendations have undergone continual change both in the limits themselves, and in the basic concepts. In the early history of these organizations, radiation was utilized by relatively limited numbers of people under reasonably simple conditions. As a result, the recommendations were primarily based upon prevention of injury to occupational workers who were either adults, or at least not under 18 years of age, and who could be given special medical examinations. They were directed to individuals who had reasonable technical competence to interpret the recommendations. The present recommendations embrace the concept of dose to much larger segments of the population, including young children as well as the possible exposure of large fractions of the total population to relatively low levels of radiation.

The NCRP recommendations can be divided into two basic parts. The first comprises the recommendations for the basic exposure limits to which individuals or a population may be exposed. These include exposure limitations to the whole body or to the various organs of the body as derived from experience with radiation exposure to man and modified in terms of the concepts derived from animal experimentation. (2) Typical secondary limits associated with these basic recommendations are the calculated values of the maximum permissible body burden which will result in radiation exposures to an individual comparable with the basic exposure levels. Even further removed from the basic exposure limits are the maximum permissible concentrations in air and water which are derived from the maximum permissible body burdens with additional uncertainties added because of the necessity of incorporating the metabolic characteristics of the radionuclide into the calculation.

The second basic part of the ICRP and the NCRP recommendations consists of methods of minimizing the radiation levels. These include such recommendations as those found in the handbooks on "X-ray Protection" (3); "Safe Handling of Radioactive Isotopes" (4); "Control and Removal of Contamination in Laboratories" (5); "Recommendations for Waste Disposal of P^{32} and I^{131} for Medical Users" (6); and "Safe Design and Use of Industrial Beta-Ray Sources" (7). This type of recommendation from the NCRP is intended to provide information to the radiation user on how to be responsive to the more basic maximum permissible limits in the installation involved. These recommendations tend to become quite specific in some cases where the characteristics of facilities and their use are more or less alike, as in the installation and operation of certain types of X-ray equipment. In other cases where the variability between installations is large, no attempt is made to indicate specific criteria but rather to provide guides which may be interpreted in view of the particular conditions in the laboratory involved.

RADIATION PROTECTION CRITERIA AND STANDARDS 7

SOURCES OF EXPOSURE

Considerations involved in the derivation of maximum permissible limits depend upon the characteristics of the radiation source involved. Six different arbitrary classes of radiation exposure may be identified:

1. Background radiation

The background radiation to which man is exposed arises from the presence of cosmic radiation and radiations from naturally radioactive materials in his environment. This background includes all common types of ionizing radiations although the primary exposure to penetrating radiation is from X- and gamma radiation. The level of exposure varies from one location to another depending upon the soils, rocks, elevations, and latitude. There are areas of the world where the background radiation from external exposure alone can be close to 1 rem/year.

All foodstuffs are "contaminated" to some extent with natural radioactive materials so that there is a continual intake to the body. With modern transportation and shipment of foodstuffs from one region of the country to the other, the variations in internal exposure from natural radioactive materials may well be less than it was in former years. The recommendations for maximum permissible limits conventionally exclude background radiation. Such levels are of considerable importance, however, in providing reference points in consideration of the effects of radiation. Table I indicates some of the measured values of background radiations (8) to indicate general levels and some of the expected variations.

The conventional reason for excluding background and background variations is that nothing much can be done about this component. If, however, one is contemplating an average annual population dose of the order of 100 mrem per year, the question of whether most people live in wooden, brick, or concrete houses ceases to be wholly irrelevant.

2. Medical exposures

The gonad exposures to people resulting from medical use were estimated in 1956 (9), at 2 to 5 roentgens per 30 years with a probable dose of 3 roentgens resulting from diagnostic examinations. More recent surveys of limited, defined groups have indicated that the contribution from diagnostic radiation is on the order of 1.5 roentgens per 30 years, at least in the groups examined (10, 11). The criteria for the control of medical radiations are based primarily on the welfare of the patient where the risk versus the gain to the individual is apparent. The value judgment of the physician is the legitimate controlling factor. This is indirectly true even in mass uses of X-ray such as the chest X-ray program where many potentially serious conditions are noted in time for treatment.

8   RADIATION PROTECTION CRITERIA AND STANDARDS

TABLE I.—*Estimated background radiations*

	Soft tissue dose mrem/year	Bone marrow dose mrem/year	Lung dose mrem/year
Cosmic rays:			
Sea level	30		30
5,000 feet	50		50
External gamma:			
Out of doors	50–160	50–160	50–160
In buildings:			
Wood [1]	60–65	60–65	60–65
Brick [1]	110–125	110–125	110–125
Concrete [1]	180–230	180–230	180–230
Internal emitters:			
C-14	2	[2] 2	2
K-40	2	[2] 9	2
Ra-226		[2] 40	
Rn			[3] 70–1000
Tn			[3] 50–600

[1] Measurements in specific buildings in Sweden; includes external dose from Rn-Tn.
[2] Osteocyte dose.
[3] Variations with building types and degree of ventilation.

3. Occupational exposure

The number of persons exposed to radiation during their work is relatively limited. At present, the occupational group contributes comparatively little to the population dose although the ICRP (12) estimates that about 1.7 percent of the population could be exposed to the full occupational level before this would contribute a proportional amount to their recommended population limit of a genetic dose of 5 rems ($\sim$ 30-year dose). Thus, the criteria for occupational exposure are based primarily on the probability of somatic damage to the individual exposed. The criteria are applied by limiting the total dose which the individual can receive with subsidiary controls on the rate at which he can receive it. Even under the pessimistic assumptions which are made in setting the limits, the probability that any individual will be harmed, even late in life, seems to be remote unless he should be involved in a serious accident involving radiation or radioactive materials. The basic exposure limits of the NCRP have nothing to do with control over the occurrence of accidents. However, many of the supplemental recommendations provide practices intended to minimize the probability of such occurrences.

A possible clue to areas of importance in the control of occupational exposure is given by a survey of experience in New York State over a period of 3 years. (13) Here, a total of 19 instances of reported exposure over the standards is included. Of these, eight were determined not to involve actual exposure to people but were due to faulty technique or analysis. The remaining 11 instances involved a possible 69 individuals with both chronic and accidental acute exposure involved. A summary of these 11 instances is given in table II. Of these 11 incidents, 7 could equally well have occurred before the advent of the atomic energy program since they involved radium or radiation generating machines. One other incident involved materials which have been handled for many years in industry (uranium and thorium).

RADIATION PROTECTION CRITERIA AND STANDARDS

TABLE II.—*Summary of radiation exposure instances in New York State, 1956–58*

Type of exposure	Source	Number of people involved	Estimated dose
Chronic	Radium-radon	11	8–15 rems/13 wk.
Do	do	10	7–14 rems/13 wk.
Do	Radium	5	25–50 rems/13 wk.[1]
Do	Co^{60} and Ir^{192}	5	30–60 rems/13 wk.[1]
Acute-external	250 KVP X-ray	3	5–25 rems.
Do	50 KVP X-ray	1	300–800 rems.[2]
Do	140 KVP microwave generator	8	5–30 rems.[2]
Do	1.5 Mev particle generator	4	400–1,100 rems.
Do	Sr^{90} solution	6	0.2–4.0 rems.
Do	Tritium gas	7	2–10 rems.
Do	Uranium and thorium dust	9	0.2–5.0 rems.

[1] Primarily to hands.
[2] Presumably localized.

4. Population exposure

Exposure to the overall population results from many causes aside from medical and background radiation. Included in this category are the environmental contamination from applications of atomic energy to peaceful or military uses; the use of radioactive materials in many forms including luminous dial watches, static eliminators, thickness gages; television and other high voltage equipment, etc. In considering population exposures, it is necessary to control both the average dose to the entire population and a separate upper limit of dose of the more highly exposed individuals in the population. Control of the population exposure requires that some measurements be made of the individual components of the total dose so that proper apportionment of the maximum permissible level can be made and steps taken to eliminate the higher sources where possible. Administrative action to apportion these levels before measurements are made can be difficult since the action may become fixed and eventually work unnecessary restrictions on the overall use of radiation In proposing an average dose to a population, it is self-evident that some individuals will be exposed to a higher level than others.

The higher limits for occupational exposure are one recognition of this. Another factor is that the population dose has to provide both for probable increased susceptibility and the additional life span included when infants, presumably even from time of conception, are included. Yet a third variant is the greater range of idiosyncrasies of personal habits, bizarre food consumption patterns, and perhaps nebulous susceptibilities that are involved in a population average.

5. Special purpose exposure

Complete limitation of radiation exposure to arbitrary limits could eliminate many projects which may be worthwhile. Examples of such projects are space flight, or applications of atomic energy to aircraft propulsion and to space flight. In such cases special allowances may be required if the benefits are deemed worthy of attainment. Usually, such considerations would indicate that the benefits would be greater than those expected from normal application or that the desirable goal cannot be attained in any other manner. While basic acceptance of a risk principle makes different limits acceptable for different purposes, the determination of who has appropriate

10   RADIATION PROTECTION CRITERIA AND STANDARDS

"authority of knowledge" or other authority may be a complex question.

6. Emergency exposures

An exposure limit planned for emergency use should be differentiated from accidental exposure. An accidental exposure is one which occurs as a result of an accident and cannot be controlled although the probability of occurrence of the accident can be strongly influenced by such factors as the design and operation of the facility. A so-called planned emergency exposure can then be defined for use in the recovery from an emergency or accident. Obviously the term must be interpreted as "planned exposure limits under emergency conditions" not as "planned emergency" exposure. Emergencies can vary widely in character and extent. An extreme would be an atomic war where the criteria as to the exposure which should be taken may be based upon survival of the Nation. In this case, reasonable exposures may exceed the peacetime limits by factors of 10. In peacetime emergencies, the criteria applied in the recovery phase may depend upon the extent of the accident, the need to save lives, or the need to return vital installations to production. Here again, the criteria may be different from those used in normal applications because of the different conditions resulting from the emergency. In other words, one may wish to assume a risk which might be significant (as contrasted with the minimal risk from peacetime limits) in order to attain a rapid recovery from an emergency.

PROBLEMS IN DERIVATION AND APPLICATION

There are several basic problems involved in setting and applying adequate radiation protection limits. Perhaps chief among these is the adoption of the risk principle with incomplete knowledge of the deleterious effects of radiation in quantitative terms. This is a standard problem of calculated risk with a poor calculation. As mentioned earlier, the data are inadequate to differentiate between a threshold and a nonthreshold effect or between dose-effect linearity so that estimates of damage are invariably based on pessimistic assumptions as to the effect on the individual. Furthermore, knowledge of the benefits and the applicability of the benefits to individuals who may run the risk of being harmed are ill-defined because of the very nature of the problem. The pace of resolution of these questions by programmatic research will almost inevitably be discouragingly slow and costly, although limits of uncertainty may be reduced by well-conceived research programs. In the long run, however, the answers to the questions of risk to people will probably result from more fundamental advances in knowledge of biological systems and the effects of radiation on them. Much fundamental research in biochemistry, biophysics and biological phenomena not at all directed to radiation protection should aid considerably in shedding light on these problems.

A second potential problem with radiation is its possible widespread use and the consequent increased exposure of all segments of the population to low-levels of radiation. Technology has advanced to the point where radiation is used not only in atomic energy but in many situations encountered in normal, everyday life. Radiation is not only a useful tool to be used by more people, it is an unwanted component of other activities such as television. An indication of the present situation in terms of the average exposure to people is given in table III where some estimates of the total radiation received by the population are tabulated. The relatively small contribution of "atomic energy" as such is of significance in the present context.

A third basic problem involves the philosophy of how much enforcement is required to provide the required degree of safeguard. How rigorously should the whole field be regulated by methods representing real handicaps to the more conscientious users, in order to reduce the chance of hazard originating from some act of a less responsible individual?

This is a problem common in principle to many areas of activity. The atomic energy field perhaps may have some built-in pride in having approached safety problems with more care, foresight, and expenditure of money on safety than other fields. Yet it will not continue to be sensible to seek a degree of safety grossly out of proportion to that applied to any other activity.

TABLE III.—*30-year per capita doses in the United States from radiation sources*

Source	30-year dose	
	Gonad dose, rem	Mean marrow dose, rem
Natural radiation:		
External:		
Cosmic rays	[1] 0.8–1.4	0.8–1.4
Terrestrial radiation	[1] 1.2–3.6	1.2–3.6
Atmospheric radiation	[2] .06	.06
Internal:		
K-40	[1][2] .6	[2] .3
C-14	[1][2] .05	[2] .05
Rn-Tn	[2] .06	[2] .06
Ra		[2] .05
Subtotal	4.4	4
Man-made radiation:		
Diagnostic X-ray	[1][3] 3	[1] 1.4
X-ray therapy	.36	[4]
Internal isotopes	<.03	[4]
Luminous dials	[1][2] .03	[4]
TV	[1] <.03	[2] <.03
Atomic-energy workers	.003	[2] .003
Medical workers	.03	[4]
Subtotal	~3.5	1.4
Approximate total	~8	5–6

[1] See reference 9.
[2] See reference 8.
[3] Additional studies on limited groups have given values on the order of 1.5 rem. (See references 10 and 11.)
[4] No estimate.

A fourth basic problem concerns interpretation of certain features by the body of qualified professional opinion. Limits tend to be written for control over various time periods—any consecutive 7 days, any consecutive 13 weeks, any calendar quarter, any calendar year, and so on. There is need for a reinterpretation of the significance of these time limits.

There are limits covering specific parts of the body, the head, the hands and forearms, the feet and ankles, and so on. There are separate limits for different radiation types with some attempt to integrate these through the terminology of the "rem." Yet when this unit is

12 RADIATION PROTECTION CRITERIA AND STANDARDS

applied too broadly, it commits the user to an implicit evaluation of relative biological effectiveness, which cannot be made reliably. Then separate limits are provided for an ever-increasing battery of different radio-nuclides. Man gets so subdivided between time, space, and radio-nuclides that the basic integrating sense of standards is lost.

A fifth basic problem with radiation limits concerns their interpretation by the public. In order for radiation limits to be completely applicable, they must be accepted by the public as the most authoritative available. To accomplish this end, public awareness of the broad basic considerations involved in setting the limits is needed. If the effects of radiation against which protection is provided were dramatic and immediate, the public could visualize the need for limits and the approximate level at which those limits should be set. A popular misconception is that dire results will occur from the exceeding of a limit. The present limits are rather intended as action standards and not as effect standards. That is, standards to minimize exposure are set at such a low level that deleterious effects even late in life are expected to be minimal. The purpose of quoting absolute limits is to provide designers and operators with a firm benchmark which they can use in their practices with reasonable assurance that unacceptable risk will not result.

A contributing factor to this problem is, surprisingly enough, the ability to measure very small radiation exposures. The ability to measure the actual radiation exposure which people receive is far ahead of the ability to measure exposure for most toxic substances. Although this is an advantage in the protection of individuals from the effects of radiation, it can be a detriment if the public fears that any positive measurement of radiation will eventually result in damage.

A solution to many of these problems cannot reasonably be expected at short term. A principal contribution lies in educating the public to a better understanding of the concepts upon which radiation protection standards are based. Understanding and acceptance of the basic concepts involved without immersion in a morass of technical details, which is almost overwhelming the technically informed, is a national requirement. I believe that the present hearings will make possible a major step in that direction.

RADIATION PROTECTION CRITERIA AND STANDARDS

REFERENCES

1. Cantril, S. T. and H. M. Parker, "The Tolerance Dose," MDDC-1100, January 5, 1945.
2. "Permissible Dose from External Sources of Ionizing Radiation," National Bureau of Standards Handbook 59, September 24, 1954; Addendum—Maximum Permissible Radiation Exposures to Man. April 15, 1958.
3. "X-ray Protection," National Bureau of Standards Handbook 60, December 1, 1955.
4. "Safe Handling of Radioactive Isotopes," National Bureau of Standards Handbook 42, September 1949.
5. "Control and Removal of Radioactive Contamination in Laboratories," National Bureau of Standards Handbook 48, December 15, 1951.
6. "Recommendations for Waste Disposal of Phosphorus-32 and Iodine-131 for Medical Users," National Bureau of Standards Handbook 49, November 2, 1951.
7. "Safe Design and Use of Industrial Beta-Ray Sources," National Bureau of Standards Handbook 66, May 28, 1955.
8. "Report of the United Nations Scientific Committee on the Effects of Atomic Radiations," United Nations, New York, 1958.
9. Laughlin, J. S. and I. Pullman, "Gonadal Dose Received in the Medical Use of X-rays." Section II of "Genetic Radiation Dose Received by the Population of the United States." A report prepared for the Genetics Panel of the National Academy of Sciences Study of the Biological Effects of Atomic Radiation, November 19, 1956, Washington, D.C.
10. Norwood, W. D., J. W. Healy, E. E. Donaldson, W. C. Roesch, and C. W. Kirkland, "The Gonadal Radiation Dose Received by the People of a Small American City Due to the Diagnostic Use of Roentgen Rays." American Journal of Roent., Radium Therapy and Nuclear Med. LXXXII, 6, 1081–1097, December 1959.
11. Brown, R. F., J. Heslep, and W. Eads, "Number and Distribution of Roentgenologic Examinations for 100,000 People," Radiology, 74, 3, 353–363, March 1960.
12. "Recommendations of the International Commission on Radiological Protection," September 9, 1958, Pergamon Press 1959.
13. Kleinfeld, M., and A. P. Abrahams, "Administrative Experience With Occupational Overexposure to Radiation," Industrial Hygiene Review, New York State Department of Labor, 2, 2, 7–12, December 1959.

Statement of Herbert M. Parker, Manager, Hanford Laboratories, General Electric

Radiation Standards, Including Fallout

June 1962

RADIATION STANDARDS, INCLUDING FALLOUT

HEARINGS
BEFORE THE
SUBCOMMITTEE ON
RESEARCH, DEVELOPMENT, AND RADIATION
OF THE
JOINT COMMITTEE ON ATOMIC ENERGY
CONGRESS OF THE UNITED STATES
EIGHTY-SEVENTH CONGRESS
SECOND SESSION
ON
RADIATION STANDARDS, INCLUDING FALLOUT

JUNE 4, 5, 6, AND 7, 1962

Part 1

Printed for the use of the Joint Committee on Atomic Energy

RADIATION STANDARDS, INCLUDING FALLOUT 305

Dr. CHAMBERLAIN. Thank you, Mr. Price.
Representative PRICE. The next witness will be Dr. Herbert Parker, manager of the Hanford Laboratories.

STATEMENT OF HERBERT M. PARKER,[1] MANAGER, HANFORD LABORATORIES, GENERAL ELECTRIC CO.

Mr. PARKER. Mr. Chairman and members of the committee, the material I am reporting was assembled by a number of my associates whose help I would like to acknowledge. Perhaps my submitted paper will be acceptable for the published record and I will try to condense it here.

In the hearings on radiation protection criteria and standards conducted by the Joint Committee, Dr. Failla, whose loss its being keenly felt by all of us in the field, made some observations entitled "Giving Credit Where Credit Is Due," in which he congratulated the Joint Committee on the excellence of the hearings on radiation protection. It is a privilege to contribute to the present hearings to bring these matters up to date. Dr. Failla's testimony also emphasized the importance of the recommendations of the NCRP and the ICRP, many others at that time noted that the basis for most of the standards used for the protection of both individuals and populations at that time were recommendaions of these two bodies. The situation is essentially unchanged today. Important activities in detailed formulation of standards and regulations continue in many other bodies, AEC, FRC, some of the States, Public Health Service, American Standards Association, and others, but the bases for these standards are still predominantly the recommendations of the NCRP.

---

[1] Herbert M. Parker, manager, Hanford Laboratories, Hanford Atomic Products Operation, Richland, Wash. Place of birth: Accrington, England. Date of birth: April 13, 1910. Naturalized in the State of Washington, 1946. Marital status: Married. Education; Manchester University (England); B.S., physics, 1930; M.S., physics, 1931; fellow, Institute of Physics, 1937.
Assistant professor radiology, University of Washington, 1952 to date; honorary trustee, Northwest Scientific Association (past); technical adviser, U.S. Delegation, Peaceful Uses of Atomic Energy, Geneva, 1955; Janeway lecturer, 1955; qualified in radiological physics by the American Board of Radiology; qualified in health physics by the American Board of Health Physics.
Work history: 1932 to 1938, Holt Radium Institute, Manchester, England, physicist; 1938 to 1942, Tumor Institute, Swedish Hospital, Seattle, Wash., physicist; 1942 to 1943, Metallurgical Laboratory, University of Chicago, research associate; 1943 to 1944, Clinton Laboratories, Oak Ridge, Tenn., section head, health physics; 1944 to 1946, E. I. duPont Co., Richland, Wash., manager, health physics; 1946 to 1948, General Electric Co., Richland, Wash., assistant superintendent, medical department (in charge of radiation protection); 1948 to 1951, General Electric Co., Richland, Wash., superintendent, health instruments; 1951 to 1956, General Electric Co., Richland, Wash., director, radiological sciences; 1956 to present, General Electric Co., Richland, Wash., manager, Hanford Laboratories.
Professional societies: Fellow, American Nuclear Society; fellow, American Physical Society; associate fellow, American College of Radiology; fellow, Institute of Physics (Great Britain); fellow, AAAS; associate member, Radiological Society of North America; American Radium Society; Radiation Research Society; Atomic Industrial Forum; N.Y. Academy of Sciences; honorary member, faculty of radiologists, Great Britain; American Management Association; member, British Institute of Radiology; member, Society of Nuclear Medicine.
Committee memberships: Member, Committee on Units, Standards and Protection, American College of Radiology; International Commission on Radiological Protection, Chairman, Subcommittee on Isotopes and Waste Disposal (past); National Research Council, member, Subcommittee on Radiobiology (past); Subcommittee on Radiological Instruments (past); National Academy of Sciences, member, Committee on Waste Disposal (biological effects of radiation study) (past); chairman, Technical Advisory Panel 04 of the American Institute of Physics to the National Bureau of Standards; member, General Electric Reactor Safeguards Council (past); member, Executive Committee, NCRP; chairman, Subcommittee on Basic Radiation Protection Criteria; member, Subcommittee on Permissible Dose From External Sources (past); member, Subcommittee on Permissible Internal Dose (past); member, AEC Safety and Industrial Health Advisory Board (past); member, Committee on Radiation Protection Standards, Atomic Industrial Forum (past); member, Washington State Technical Advisory Board on Radiation Control.

As to new developments since 1960, they are possibly not too striking. In this period the philosophy and concepts and quite often the actual language of the recommendations of the NCRP and ICRP have been applied and incorporated into sections of Federal and State codes and into policies of the private industrial organizations. Noteworthy applications in this field have been made by the Federal Radiation Council and you have heard from other witnesses about the range concept which carries, as I see it, the promise of introducing much needed flexibility.

Another noteworthy development has been a statement by the NCRP on exposure limits that would apply to emergency situations. This is their Report No. 29. Although this is pitched more directly at the civil defense situation it applies quite reasonably to guide justifiable action in the event of serious emergencies of industrial origin.

Turning directly to the industrial situation, we can perhaps define three categories of industrial users of radiation. The largest single category is composed of AEC contractors who account for two-thirds of the estimated employment in the entire atomic energy field. These are involved in quite complex uses of radiation sources and need to apply their standards to a wide variety of conditions. In controlling this the AEC is assured by contract terms that the contractor proposes to maintain certain minimum radiation protection standards. These are controlled by the AEC manual chapters which are themselves based rather directly on the NCRP recommendations.

The second category of industrial users we would define as AEC licensees. This is the principal general industrial group, complying with regulations specifically formulated for the purpose and applied by the Commission.

I am told there are some 10,500 such licenses in force and compliance with standards is required through title 10 of the Code of Federal Regulations.

In this case there is a very wide spectrum of need. The standards utilized by licensees vary from minimum controls over small separate sources to situations about as complex as those characterized in the principal AEC contractors' work.

The third group would include the users of radiation sources not covered by the Atomic Energy Act. This is made up of a group of industrial firms engaged in such activities as nondestructive testing as well as those actually manufacturing radiation machines and instruments. Of course, in these kinds of work they use radiation sources such as X-ray machines, radium, and some radionuclides that escape from the AEC aegis. The principal guidance in this case is contained directly in the recommendations of the NCRP, with about half our States having State regulations which exercise some control over these situations. It is possible, of course, one should note, for one industrial firm simultaneously to fall under the jurisdiction of all three of these types of control.

In practical administrative application, almost without exception some modification or interpretation of standards and codes is necessary and must be provided before a standard becomes implemented as a working reality in the work atmosphere.

In the case of the modest user, the maximum hazard he is concerned with may be both small in itself and easily predictable. In those cases he may well elect to relate his performance directly to the language of the formal legal standards. More generally the quantity and variety of radiation sources or the very complexity of the facilities involved is such that the program has to be headed by a separate responsible radiological safety officer. Such a man will have at least some knowledge of the history and development of the basic recommendations and is in a position to guide the program by applying the underlying philosophy and intent of the standards as well as the actual terminology of the legal instruments.

The first chart (table I, p. 319) if Dr. Taylor will be good enough to show it shows some of the details which get involved in the work of the large contractors in making a tight internal system of control with the controls responsive to the legal standards but characteristically are more stringent because of the potential for significant exposure which arises in these cases and mainly because of the complexities of converting different types of exposure to a common base, which I believe is still actually the major problem for all of us.

Another important administrative aspect is the establishment in the larger industrial concerns of a work climate and an employee attitude favorable to good radiological control. We believe that written standards and procedures alone just do not give assurance that people have the understanding to enable them consistently to do the right thing.

Continuing education in the basic intent of protection standards seems to be important in this. The final success of such a radiation protection program is then, as we see it, usually more dependent on the voluntary acceptance of a way of life in dealing with radiation than upon literal conformance to some rule.

A word now about the present problem areas in the industrial application of standards. Let me say first that we see no current major problem in this area, except the one I mentioned before of the realistic adding up of all the contributions to exposure. With this limitation the possible problems are like this. First, the transfer of certain regulatory responsibilities from the AEC to the States has been going on in the period we are talking about. Despite the probable intent of all the parties to maintain reasonable uniformity within the State regulations there is opportunity for inconsistencies, gaps, and overlaps between these several codes. Some of these, although in truth they are relatively minor, have already appeared. The future or potential problem encountered by an industrial firm which may have atomic activities in several States each with different requirements and also having licensee and contractor relationships with the AEC is obvious.

Another problem is the actual format and language of the standard itself. We have our divergent viewpoints on the degree of specificity and the amount of methodology which should be contained in these standards, but broadly the industrial users join in making a very strong plea that it is important that the standards be written as performance standards and not as a specific detailing of the mechanics or interpretive methods for doing the jobs. Identifying the end point and not the method is the key issue here.

Relating to these problems of incorporating methodology into standards is the imparting of the same apparent sense of validity and weight of law in the various secondary standards that are so used as is warranted in the case of the primary or basic standards. The question of internal depositions is partly under this heading because the internal depositions may come largely from breathing and drinking. Our recommendations tend to consist of lists of permissible concentrations or radiation protection guides, in air or water. These are in themselevs somewhat secondary standards, very useful ones, as guides to prudent operation. The application of these lists in a rigorous or statutory manner instead of going back to the basic dose requirements with respect to the individual can either be very burdensome on the one hand, or actually not restrictive enough on the other hand where various biological concentrating mechanisms intervene between the initial water supplies and the consumption of food after the processing of products through a food chain.

Another significant problem not related to standards but to the way we think about them is the natural tendency among the public and perhaps to some extent even in the courts to equate the exceeding of some specific limit with injury to the recipient. Serious problems will enter into the business if radiation protection guides are erroneously used as criteria for determination of either the existence or the extent of injury. In this country radiation protection standards are not based on concepts of establishing permissible doses at levels just below the point of injury.

As I understand the present efforts of the Public Health Service, that agency is particularly cognizant of this overall problem of oversimplification of limits and tends to what I call a retrospective assessment of each case on its own merits.

As seen by industry, this approach carried to the limit won't stand up. Industry and the public which rightly attempts to judge the actions of industry must have prospective targets, not retrospective ones. Unfortunately in industry, which is technology based here, we tend to equate prospective target with a very simple go, no-go gage or the discrimination of black and white.

One almost hears a modern Decatur exclaiming, "Our numbers, may they be always in the right, but our numbers, right or wrong." The two extreme positions are not yet reconciled. If we accept the principle of acceptable risk in radiation exposure, and there is no alternative today, instead of black and white, we have only infinite gradation of gray from perhaps a black relating to significant overexposure, grading down but never reaching white. It is beyond our wits to quantify such a scale. Yet the attempt has to be made at least to define bands of gray. The three ranges as used by the Federal Radiation Council, I think, are precisely such an attempt which I have translated into fashionable color terminology with range I being Arcadian gray, range II being Achillean gray, and range III being Augean gray.

Representative Hosmer. Do you have a color chart with you?

Mr. Parker. I am not able to put precise numbers on these shades of gray but I classify Arcadian gray as pure and clean for the relevant purpose, and Augean gray containing a reference to the well-known stables of history, and the middle range, if I may clarify that, as I recall Achilles, he was pretty sound but he had a couple of weak spots one on each heel. That is the derivation of these ranges.

Public education, or in other words, doing a better job than I can do pictorially in interpreting the shades of gray we have in mind, is still vitally needed and does not come easily. We look to the Federal Radiation Council mainly to provide the Nation with authoritative judgments. Their recent Report No. 3, which we received since this was written, appears to me to be an excellent example of this as applied to the specific topic of fallout.

We see that the radiation protection standards over the past period, 20 years if you like to use that period, have served the Nation well and further aggressive research in support of establishing better principles and achieving more resourceful codification of these principles should help us to make this statement again 20 years from now.

Let us take a brief look at the numbers that we can invoke from the industrial experience. We are talking about what is looked on as an explosively expanding industry but it is neither explosive nor quite so expanding when we remember that there are said to be some 200,000 people in the present work force in atomic energy. This is small in comparison with major industries. Nevertheless, a reasonable body of experience is accumulating which points, I believe, on the whole to success in minimizing exposure through prudent design and strict enough control. We have to pick small portions of the total record to put numbers on them, and table 1 shows some of these for external exposure alone. I will not elaborate these because they are in the published record by the AEC.

Exposure records show on the whole that the vast majority of workers in the AEC complex only receive a radiation dose of less than 1 rem per year and furthermore only in a very few cases—and we count about 1 worker in 10,000—has the National Committee on Radiation Protection short-term control limit of 3 rems in 13 weeks been exceeded. This always seems to come from some kind of accident rather than from regular planning.

We attempted, since your committee announced these hearings, to make a survey to get more up-to-date information from all industry and were not able to obtain data that I would consider comprehensive. But from a fairly substantial body of representative major users covering about 30,000 people who were all actively engaged in this field, and this includes private work as well as that responsive to AEC contracts, the average annual radiation exposure for the last 2 years, that is, 1960 and 1961, seems to run at about three-tenths of a rem per person.

Thinking for a moment of the fairly standard formula for maximum accumulated dose, the one which is written as $5(N-18)$ rems with N being a number equal to the present age of the individual in years, replying to your survey and including ours, since we have recently acquired one, only two cases showed accumulated doses exceeding the formula values. If you go back to sources that we cannot always document but come from the professionals talking with each other in the field we know altogether of about 15 cases in the country in which this formula has been exceeded and we guess that if our sources were complete, this number might be doubled. So there may be about 30 situations in the whole of industry exceeding the maximum accumulated dose. I should reiterate that many of these do not represent real injury to the recipients and some, again by the numbers, will be self-correcting, since the respective values of N for these people is steadily increasing and most of them are now withdrawn from additional radiation work.

To the best of our knowledge these cases are principally due to single large accidental exposures rather than to a running steady accumulation over the years.

Coming now to the case of the experience of the small users here we could find little or no public data that would give us values needed by the committee. We examined the most recent data on licensing experience as reported by the AEC for about a 1½-year period ending last November. In that period there were some 10,000 licenses in force. There were 40 radiation incidents reported which you would classify mostly as being minor in nature. None of them, in fact, reported a very serious level of radiation exposure.

Going now to the internal deposition which is more difficult to put into numbers, the nationwide experience in this respect is just not available. We tried to get some more in our limited survey and the answers there were interesting in that they showed a nearly complete absence of significant deposition cases.

In order to have numbers that we can support better, however, let me quote the Hanford experience which was: 6,000 man-years of direct work with one of our most dangerous elements, namely, plutonium, shows us with only three employees with body burdens of plutonium approximating or exceeding the present standards. These quantities are such that none is expected to present any clinically observable symptoms and none have appeared.

At Hanford, with about 75,000 man-years of experience in working with other radionuclides, no other internal depositions have occurred except for a few minor transitory cases involving materials of short half-life.

The important aspect of the environmental radiation is what we contribute to our friends and neighbors around a plant such as we have at Hanford. Here we can conclude—again without giving wholly reliable numbers—that persons living in the vicinity of such installations receive but a small fraction of radiation from these additional sources of that acquired from natural background. In fact, the contribution there received is for the most part overshadowed by the contribution from worldwide fallout which I understand is already regarded as not being very high at this time. As to the average exposure from industrial operations which would relate to the genetically significant population around the atomic energy plants, if we tried to spread this over a few million people, we are not able to give precise numbers but we can give some evidence that it must certainly be only a small fraction of 1 millirem per person per year. You will recall the dose from natural background ordinarily falls in the range of 100 to 200 millirem. The gonad dose from fallout in our region, which is lower than that reported for the Nation at large, is about on the order of 5 millirem per year at the present time.

A slightly less favorable aspect refers to situations which may arise in the immediate vicinity of any large atomic energy plant such as ours which lead to doses several times those which currently exist from fallout. This will arise mainly with individuals with uncommon food habits or other idiosyncracies. It is very hard indeed to make reliable calculations of what these exposures may be, but using the best data available to us we have concluded for some time that in the vicinity of the Hanford project, as an example of the larger scale operations, these doses may come to about 30 percent of the appropriate limits.

Representative HOSMER. What kind of an uncommon food habit would create this situation? Like a liking for plutonium?

Mr. PARKER. No; we have not yet gotten so sold on the virtues of plutonium, although we regard it highly, as to consider it a food. We for example, Mr. Hosmer, unwittingly or unavoidably at the present time insert radioactive products into the Columbia River. This will go through various life forms including a rather noted deposition in shellfish. An uncommon food habit example might be a man who lived exclusively on shellfish rather than the normal diet.

The British situation has a community which eats a seaweed and this seaweed would have to be the one that accumulates a rather spectacular amount of radioactive debris that the British insert into the sea. This is representative——

Representative HOSMER. In other words, you cannot be a faddist in the State of Washington. That is what can be concluded.

Mr. PARKER. I think one could broaden that and say "Don't be a food faddist in any State."

Mr. RAMEY. Was there someone around Calder Hall who ate lobsters entirely as an advertisement and they had to raise their standard on his intake so they wouldn't hurt?

Mr. PARKER. I am not familiar with that specific instance.

In our case in this area where we do have uncertainty because of these individual habits things are looking up with the expanded availability of the whole body counter which is giving us a method of measuring what radioactive materials actually exist in the body. We hope, if we are asked to report to you at some subsequent time, that the data here will be very much improved.

One can get some indirect reference to the situation in industry by looking at accidents. Accidents can range all the way from minor spills of radioactive contaminants to the serious nuclear excursions, the criticality incidents up to and including loss of life. These latter are the ones that are spectacular. They are well characterized and well reported. The next chart (table II, p. 319) reveals the rate at which criticality type accidents are accruing in the United States. Within the limit of statistics of numbers like 1 and 2, one has to say that 1 and 2 are equal and the summation of this experience is that major accidents in the business is continuing at a steady rate. That situation is not conspicuously favorable nor is it conspicuously unfavorable since presumably some accidents will always occur.

I hoped to report on the feasibility and cost to industry of maintaining appropriate levels of protection, since these are important ultimately to a thriving industry. I find nothing new to report here.

I would say a good quality of protection is being achieved, though not too cheaply, and this will continue as long as the applicable base limits continue to be more or less stable. Neither do I see evidence that calls for a radical change in these limits. In some cases, in fact, as in plutonium deposition, there may even be a tendency to regard the present safety margin as more than adequate.

Finally, Mr. Chairman and gentlemen, there is a tendency to relate the careful control and work climate in this specific application of

312   RADIATION STANDARDS, INCLUDING FALLOUT

standards to safety performance as a whole. For reference, the final table (table III, p. 320) I have which I will not read gives the data for AEC contractors compared with similar experience by all industries. It is suggested that continued performance of this type should lead to a better appreciation by the general public that the Atomic Energy installations are indeed among the safest of our industrial plants.

Thank you, gentlemen.

(Mr. Parker's prepared statement follows:)

RADIATION PROTECTION STANDARDS: THE INDUSTRIAL SITUATION

(By H. M. Parker, manager, Hanford Laboratories, General Electric Co.[1])

INTRODUCTION

My name is Herbert M. Parker, and I am employed by the General Electric Co. as manager, Hanford Laboratories, Richland, Wash. The material that I am reporting was assembled by a number of my associates, including particularly A. R. Keene, L. A. Carter, J. W. Vanderbeek, and R. F. Foster, whose help I acknowledge.

I should also identify my position as chairman of the NCRP subcommittee on basic radiation protection criteria.

PRINCIPAL RADIATION PROTECTION STANDARDS APPLICABLE TO INDUSTRY

In the hearings on "Radiation Protection Criteria and Standards" conducted by the Joint Committee on Atomic Energy in 1960, one of the most knowledgeable and respected men in the the radiation protection field presented his usual thought-provoking testimony to the committee on the development and status of the bases for radiation protection standards. He also included some observations which he titled "Giving Credit Where Credit Is Due." In this last section of his testimony he offered his congratulations to the Joint Committee on the excellence of the public hearings which were conducted by the committee. He stated that these hearings have "served the purpose of clarifying the problems in the public mind and the printed reports provide an up-to-date summary of the scientific status of this field." (1)[2] It is a privilege to contribute to the hearings which the Joint Committee is conducting at this time, to bring these matters up to date.

The 1962 hearings will be missing the mature and valuable contributions of Dr. Gioacchino Failla whose well-balanced observations were an important contribution in the 1960 hearings. His loss both as a friend and as an unselfish principal contributor to the foundations of radiation protection in this country has been felt and will continue to be felt for many years by all of us.

In his testimony, Dr. Failla also gave credit to the National Committee on Radiation Protection and Measurements (NCRP) for the "introduction of many new concepts on radiation protection which are now standard practice throughout the world." (1) His testimony emphasized the importance of the recommendations of the International Commission on Radiological Protection (ICRP) in the matter of permissible limits for large populations. Many others noted that the bases for most of the standards used for the protection of individuals and populations against radiation at that time were the recommendations of the ICRP and the NCRP.

This situation is essentially unchanged today. Important activities in formulation of radiation protection standards and regulations continue in the Atomic Energy Commission, the Federal Radiation Council, some of the States, the U.S. Public Health Service, the American Standards Association, and other such agencies or bodies. The bases for such standards development and application continue to be predominantly the recommendations of the independent NCRP.

---

[1] Work done under prime contract AT(45-1)-1350 to the U.S. Atomic Energy Commission.
[2] References at end of statement.

NEW DEVELOPMENTS SINCE 1960

The philosophy, concepts, and often the actual language of the recommendations of the NCRP and ICRP have been applied, adopted, and incorporated into sections of Federal and State codes and into policies of private industrial firms. Noteworthy applications and modifications of the recommendations of the NCRP and the ICRP during the past 2-year period have been made by the Federal Radiation Council (FRC).

The FRC, since its inception has offered guidance to Federal agencies in its Staff Report No. 1 issued in May 1960,(2) and Staff Report No. 2 issued in September 1961(3). In Report No. 2 the FRC adopted a range concept in stipulating the control of radiation dose to certain critical organs of the body, which carries the promise of introducing much needed flexibility. In Report No. 2 guidance to the Federal agencies is provided in the form of allowable daily intake rates for strontium 89, strontium 90, iodine 131, and radium 226. Briefly stated, range I is the lowest of the three ranges and it spans the intake rate which is equivalent to essentially no radiation dose up to a dose equivalent to one-tenth of the so-called radiation protection guide or permissible dose. Range II extends from one-tenth of the radiation protection guide to the full radiation protection guide level; operation in this range requires a quantitative surveillance program and routine control of the releases of radionuclides to the public domain. Range III is the uppermost range and spans an order of magnitude above the radiation protection guide; operation in range III requires an evaluation program and application of additional control measures as necessary to reduce the exposure.

While the guidance offered by the FRC is for application by Federal agencies, the extension of this guidance to the industrial firm is commonplace because of the thousands of firms having a licensee or contractor relationship with the Atomic Energy Commission. A primary standard or limit for controlling radiation hazards is an expression in terms of limitation of dose to individuals or to populations at large. Federal Radiation Council Report No. 2 offers definitive guidance on a method of controlling radiation by limitation of daily rates of intakes of certain radionuclides by members of the public. This portion of Report No. 2 has the nature of a secondary standard. The incorporation of such secondary standards into a collection of Federal guides may have advantages for those engaged in activities limited to work with modest amounts of one or two radionuclides. For those activities where large quantities of radioactive materials are processed, the release, under controlled conditions, of extremly small fractions of the quantity of materials being handled requires sophisticated environmental evaluation programs. For these types of activities, rigorous application of a secondary standard may have important disadvantages. I will come back to this point later.

Since the 1960 hearings another noteworthy development has been the statement of the NCRP on exposure limits applicable to the emergency situation. These recommendations are contained in NCRP Report No. 29, "Exposure to Radiation in an Emergency" (4). They provide definitive dose and risk criteria for justifiable action in the event of serious emergencies of an individual origin as well as possible nuclear warfare. This recent guidance by the NCRP is a valuable addition to the other NCRP recommendations.

INDUSTRIAL USERS OF RADIATION PROTECTION STANDARDS

While there have been few changes in the basic radiation protection standards since the hearings in 1960, there have been many activities bearing on the generation of standards and their use and application in this formative period through which this country is now going in the area of radiation standards regulation. This is becoming particularly evident as the transfer of responsibilities for certain source, byproduct, and special nuclear materials from the Atomic Energy Commission to the States is occurring under the revision of the Atomic Energy Act.

There are perhaps three definable categories of industrial users of ionizing radiation. The largest single category is composed of AEC contractors who account for about two-thirds of the estimated employment of the entire atomic energy field (5). These contractors are frequently involved with extensive and complex uses of radiation sources and therefore often have need to apply radiation protection standards extensively to a wide variety of conditions. In its re-

314   RADIATION STANDARDS, INCLUDING FALLOUT

lationship with its contractors, the AEC is assured by contractual arrangement that the contractor proposes to maintain certain minimum radiation protection standards. The recommendations of the NCRP have been the bases for these standards as incorporated in AEC Manual chapters (6). It is my understanding that future revisions of these manual chapters will reflect more directly the specific guidance offered to Federal agencies by the Federal Radiation Council.

The second category of industrial users is composed of AEC licensees which constitute the principal industrial group complying with regulations specifically formulated and applied by the AEC. As of December 31, 1961, there were about 10,500 licenses in force (7). In issuing individual licenses the Commission requires compliance with the standards presented in title 10 of the Code of Federal Regulations (8). The standards utilized by licensees can vary from minimum controls over small individual radiation sources to extensive controls which cover as full a range of application as for the more complex atomic energy facilities.

The third group includes the users of radiation sources not covered by the Atomic Energy Act. This group of industrial firms is engaged in activities such as nondestructive testing. In the course of their work they may use radiation sources such as X-ray machines, radium, and radionuclides produced by Van de Graaff generators. The principal guidance to these firms is contained in the recommendations of the NCRP. In many States such firms may also be under regulations promulgated by State authorities such as the State department of health. It is possible, of course, for an industrial firm simultaneously to fall under the jurisdiction of all of the above types of application of standards.

ADMINISTRATIVE ASPECTS OF STANDARDS APPLICATION

The adoption of standards by regulatory agencies of the Federal or State Governments does not, in itself, insure that significant radiation exposures will not be received by workers or persons living in the vicinity of industries which handle sources of radiation. The industry or user must conduct his operations in such a manner that the standards will be easily met under normal operating circumstances with a margin to make it highly improbable that they will be exceeded under foreseeable adverse conditions. The amount of precaution which is necessary is obviously related to the size and nature of the user's business. At one extreme is the technologist who uses minute quantities of a particular radionuclide for tracer-type work and whose total supply of radioactive material is so small that it constitutes an insignificant radiation hazard, no matter how casually he may handle the material. Compliance with standards is, in this case, assured at the time the radioactive material is dispensed to the technologist.

At the opposite extreme is the large atomic energy facility which handles many tons of irradiated nuclear fuel and which must install elaborate safeguards to assure that equipment or human failure does not result in serious overexposures to perhaps hundreds of people.

Whatever the nature of the operation, the governing stature, or the contractual obligation may be, it is the common situation that there are important administrative aspects in implementing the applicable standards. Few users of ionizing radiation will find that the applicable regulatory or guiding instrument will be applied directly in his individual case. Almost without exception some modification, interpretation, selection, or emphasis will be necessary and must be provided before the standard can be implemented effectively and intelligently.

In the case of the modest user the maximum hazard may be both small and easily predictable. In such cases the user may elect to relate his performance directly to the language of the legal standards without establishing additional working limits of his own.

In the more typical situation the quantity and variety of radiation sources or the complexity of the facility is such that the radiation protection program is usually headed by a responsible radiological safety officer. This officer is responsible for assuring that the operating protection policies and practices of the installation are sound, and, with proper implementation assure that radiation exposures will remain within statutory and contractual requirements. The fully qualified radiological safety officer has expert knowledge of the history and development of the basic recommendations issued by such bodies as the NCRP and, therefore, is in a position to guide the radiological program by application of underlying philosophy and intent as well as by the terminology of the legal instruments.

The complex modes of exposure encountered in such installations often preclude a simple direct comparison to basic standards on a day-to-day basis at the operating level. It is often necessary, therefore, to set up in-plant standards and operational controls. Such local voluntary controls are responsive to the legal standards but characteristically are more stringent because of the potential for significant exposure and because of the complexities of converting different types of exposure to a common base. A few examples of the kinds of in-plant standards are—

(a) Limitations on the radiation exposure which may be received by a worker in any one administratively convenient or necessary unit of time shorter than the formal or codified time base. For example, to assure limitation of radiation dose to individuals to say 3 rems in 13 weeks, it is usually necessary to establish additional internal controls which limit dose to some fraction of 3 rems per week or per month.

(b) Requirements for the wearing of dosimeters, protective clothing, respiratory equipment, etc.

(c) Guides for the controlled release of radioactive effluents.

(d) Calibration requirements of radiation measuring instruments.

(e) Formal procedures for action in case of emergencies.

Only through the use of such in-plant administrative standards is it practical to implement the generally accepted philosophy of minimizing exposure to radiation wherever possible. Control of this type would be difficult, if not impossible, to achieve under a direct and rigid application of many basic or codified standards.

In the situation where there are no statutory or contractual requirements and the user of ionizing radiation is being guided principally by the recommendations of the NCRP, he will usually have an internal set of standards which may deviate in part but require compliance with the general intent of the recommendations of the NCRP. In such cases the principal motivating force is in the quality and the value of the guidance which is offered by the standards. The high degree of voluntary acceptance of the recommendations of the NCRP and the ICRP over the last 30 years is an outstanding example of what can be achieved by the user having high confidence in standards which are offered for voluntary acceptance and application.

Another important administrative element is the establishment of a work climate and employee attitudes favorable to good radiological control. Written standards and procedures alone just do not give assurance that people will have the understanding to enable them consistently to do the right thing. Continuing education in the basic intent of protection standards is important.

The success of a radiation protection program is, therefore, usually dependent as much or more on the voluntary acceptance of a way of life as upon literal conformance to a rule.

PROBLEM AREAS IN INDUSTRIAL APPLICATION OF RADIATION PROTECTION STANDARDS

While the development and application of radiation protection standards for control of industrial exposure are not without problems, there are perhaps no current major problems in this area. Within this framework, however, I would like to mention several areas which contain the seed of future problems.

The transfer of certain regulatory responsibilities from the Atomic Energy Commission to the States of our Nation has only recently begun. In the States where this transfer has been effected or is close at hand, State regulations generally seek to assure the level of control provided in title 10 of the Code of Federal Regulations for application to licensees. In spite of the intent of all parties to maintain reasonable uniformity within the State regulations, there is considerable opportunity for inconsistencies, gaps, and overlaps between States and between State codes and Federal codes. Some of these, although relatively minor, have appeared. The potential problems which could be encountered by an industrial firm having atomic energy activities in several States, each with differing requirements, and also having licensee relationships and contractor relationships with the Atomic Energy Commission, are obvious. Some problems of reciprocity and jurisdiction have yet to be worked out to minimize the administrative problems of industrial firms.

Another problem in the development of radiation protection standards is the format and language of the standard itself. There are divergent viewpoints on the degree of specificity and methodology which should be contained in radiation protection standards. It is important that standards be written as performance standards, or functional specifications, not as a specific detailing of mechanistic or interpretive methods to be used. The radiation conditions encountered by various users in industry differ so markedly that a standard which emphasizes method must necessarily fit poorly at one extreme or the other. Standards which define basic criteria, while permitting needed latitude in the methods employed, apply equally well to all users. Identifying end point, not method, is the key issue. The present efforts of the NCRP are concentrating on this point.

Related to the problems resulting from incorporating methodology into standards is the imparting of the same validity and weight of law into such secondary standards as is warranted in the case of primary or basic standards. As an example, in controlling the internal deposition of radioactive materials in humans the principal standard is the limitation of the amount of a radioactive material in an organ of interest. Since two principal modes of entry into the body are breathing and drinking, recommendations of the NCRP include a listing of permissible concentrations of radionuclides in air and water (9). Inclusion of such secondary standards as a guide to prudent operation is often helpful. Application of such secondary standards in a rigorous or statutory manner in lieu of assessment against primary dose standards can be unduly burdensome or expensive on the one hand, or not restrictive enough on the other hand, where biological concentrating mechanisms in the human food chain can intervene.

There is a natural tendency among the public and perhaps even in the courts to equate an exceeding of a specific permissible limit with injury to the recipient. Serious problems will result if radiation protection guides are erroneously used as criteria for determination of existence or extent of injury. Radiation protection standards in this country are not based on concepts of establishing permissible doses at levels just below the point of injury. Knowledgeable medical interpretations and decisions in the courts should provide adequate resolution of this potential problem.

As I understand the present efforts of the USPHS that agency is particularly cognizant of the problem of oversanctification of numerical limits and tends toward retrospective assessment of each case on its own merits. As seen by industry this approach, carried to the limit, would be untenable. Industry, and the public which attempts to judge its actions informally and fairly must have prospective targets. Unfortunately, a technology-based industry tends to equate prospective target with a go, no-go gage or the discrimination of black from white. One almost hears the modern Decatur exclaiming, "Our numbers, may they be always in the right, but our numbers, right or wrong." The two extreme positions are not yet reconciled.

With the principle of acceptable risk in radiation exposure, instead of black and white there is a definable black for significant overexposure, and below that, infinite gradations of gray down to but never quite reaching white. It is beyond the wit of man to quantify such a scale—there is, for example, no gray that is 10 times lighter or darker or grayer than another gray. Yet the attempt has to be made at least to define bands of gray. The three ranges as used by the Federal Radiation Council are precisely such an attempt which I would translate into color terminology as—

Range I: Arcadian gray.
Range II: Achillean gray.
Range III: Augean gray.

Public education on the acceptability of a given radiation risk or, pictorially, the interpretation of a particular gray is vitally needed. It cannot be achieved simply. Neither is it helped when prominent scientists, erroneously accepted by the public as expert in this particular field, express palpably different views on the prudence or radiation safety of actions or plans in the atomic energy field. Such differences come from socioeconomic rather than scientific interpretations. Here the Federal Radiation Council, more readily than the NCRP, could provide the Nation with authoritative or at least broadly considered value judgments.

In spite of these problems, on balance, radiation protection standards over the past 20 years do seem to have served this country well. Aggressive research in support of refined establishment of basic protection principles and standards and resourceful codification of these principles should permit this statement to be repeated 20 years hence. A very brief look at the industrial exposure experience is convincing that an effective set of controls has been in force throughout the rapid expansion of the atomic energy business in this country in the last 20 years.

RADIATION STANDARDS, INCLUDING FALLOUT 317

INDUSTRIAL EXPOSURE EXPERIENCE

Although the employment of workers by industry engaged in the handling of large sources of radiation has expanded rapidly during the past 20 years, the present work force of some 200,000 (7) is still very small in comparison with that of the major industries. Nevertheless, a substantial experience record has been accumulated which attests to the exemplary success of the operators in minimizing radiation exposure through prudent engineering design and strict control. Quantitative assessment of the performance of the larger atomic energy sites is practical in four broad areas, viz:

(1) The magnitude of the exposure from external sources received by employees during the normal course of their work assignments.

(2) The small burden of radioactive materials which may be deposited in the bodies of the workers.

(3) The magnitude of the exposure received by persons who live in the vicinity of the plant.

(4) The frequency and severity of accidents which result from loss of control.

The first three classes of exposure can be compared with pertinent limits, and the accident experience can be followed as a trend.

Compilations of exposure records for recent years (table I) indicate that the vast majority of workers employed by AEC contractors receive a radiation dose of less than 1 rem per year. Only in very few cases (about 1 worker in 10,000) has the NCRP short-term control limit of 3 rems in 13 weeks been exceeded and invariably this has resulted from some sort of an accident rather than imprudent work assignments.

We were unable to make a comprehensive survey of the radiation experience of the whole of industry. While our information from industrial users is incomplete, in replies from representative major users covering about 30,000 people, the average annual radiation exposure in 1960 and 1961 was about 0.3 rem per person.

There is a widely used formula for the control of accumulated dose for an individual over the years, which is written (10)

$$MPD = 5 \ (N-18) \ \text{rems}$$

where MPD = maximum permissible accumulated dose, and

N is a number equal to the present age of the individual in years; (the formula begins to apply after age 18, the employment of minors below this age being avoided).

Among the group replying to our survey, including ourselves, only two cases showing accumulated doses exceeding the formula values appeared. Including all the radiation accident cases known to us, the total is less than 15; if our resources had been complete, the total would probably remain below 30. It is important to reiterate that many of these do not represent real injury to the recipients and some will be self-correcting as the respective values of N increase. To the best of our knowledge these cases are principally due to single large accidental exposures. We were not able to uncover any specific case in which an employee was in excess of this limit due to radiation received chronically in the course of his work.

There is very little public data available on the radiation control experience of the small users. Examination of the most recent data on licensee experience reported by the Atomic Energy Commission for about a 1½-year period ending November 30, 1961, gives some indication of the degree of control experienced by the small user. In this 17-month period during which about 10,000 licenses were in force, the great majority of the 40 radiation incidents reported were minor in nature. None of the incidents reported resulted in a serious level of radiation exposure.

Nationwide experience on internal deposition of radionuclides in industrial workers is not readily available in published reports but in our limited survey, there was a nearly complete absence of significant deposition cases. In our own experience at Hanford in 6,000 man-years of direct work with one of the most dangerous elements, plutonium, only three employees have acquired body burdens of plutonium which approximate or exceed the present applicable standards. The quantities involved are not expected to produce clinically observable symptoms and none have appeared.[a] Additionally, in about 75,000 man-years

---

[a] There is growing evidence that the present standard for plutonium, which is based on analogy with radium depositions, may have a very considerable safety margin.

318   RADIATION STANDARDS, INCLUDING FALLOUT

of experience in working with other radionuclides at Hanford no internal depositions of radionuclides in excess of applicable permissible limits have occurred except for a few minor transitory cases involving materials with a short biological half-life.

The contribution made by industry to the radiation exposure received by the average persons living in the vicinity of the installations continues to be but a small fraction of that received from natural background and, for the most part, is overshadowed by the contribution from worldwide fallout. Although the average exposure from industry to the reproductive organs of a genetically significant population consisting of a few million people cannot be stated precisely, it must certainly be only some small fraction of 1 millirem per person per year. The dose from natural background ordinarily falls in the range of 100 to 200 millirems per year, and the gonad dose from fallout is probably on the order of 5 millirems per year at this time.

Situations may arise in the immediate vicinity of large atomic energy plants which lead to doses several times those which currently exist from fallout. Persons with uncommon food habits or other idiosyncrasies fall into this classification. Because of the relatively few individuals involved, somatic rather than genetic considerations of the significance of the exposures are appropriate. Estimates for such persons living in the vicinity of the Hanford project suggest that their doses may approximate 30 percent of the applicable limits.

The greatly expanding availability of whole body counters in the last few years has provided the technical means for measuring the radioactive body burdens in many cases with comparative ease. The present areas of uncertainty can be substantially reduced in the next few years.

Accidents associated with radiation sources range in severity from the minor spills of radioactive contaminants to serious nuclear excursions (criticality incidents) sometimes involving the loss of life. Because they are readily characterized and extensively reported, the frequency of occurrence of criticality type accidents can be watched as an indication of performance. Table II shows the experience in this country to date. Desirably there should be no accidents of this type. Realistically, some must be anticipated in an expanding technology which is heavily dependent upon new research and development. Within this framework the trend to date should not be viewed as unfavorable.

The feasibility and cost of maintaining appropriate levels of radiation protection are important factors in the ultimate development of a thriving atomic energy industry. In these phases we find nothing new to report. A good quality of protection is achievable, although not too cheaply, as long as the applicable basic limits continue to be more or less stable. There is no evidence which seems to call for a major change. Stiffening of some limits could cause considerable difficulty to the industry. It is more likely that some specific limits, for example, for plutonium deposition, may be demonstrated to have much more than ample safety margins. In any case, the organizations which provided data for our survey pointed to the need for sound basic standards unencumbered as far as possible by detailed administrative and procedural regulation. This is the avenue deemed most likely to provide the stimulus for innovation and improvement in radiation protection.

The careful control and work climate in atomic energy plants which is responsible for good radiological performance is also reflected in outstanding safety performance in other areas as well. This is evident from table III which compares the number of lost-time injuries from all types of accidents experienced by AEC contractors with similar experience by all industries. Continued performance of this type should lead to a better appreciation by the general public that atomic energy installations are among the safest of all industrial plants.

SUMMARY AND CONCLUSIONS

In brief summary, there has been no outstanding development or basic change with respect to radiation protection since 1960. The key questions identified in 1960 remain the key questions in 1962. (11) It is natural, then, that there has been no outstanding development or major change in the industrial situation with respect to radiation protection in this interval. Such minor changes as are reported are generally in the favorable direction. Exceptions are a growing concern over possible conflicts of interpretation where more than one group has real or implied authority to set standards, and most importantly an almost universally adverse reaction to such code and regulation as pinpoints specific methods and administrative procedures.

RADIATION STANDARDS, INCLUDING FALLOUT 319

The recommendations of the National Committee on Radiation Protection and Measurements and the International Commission on Radiological Protection continue to be the bases of radiation protection standards and regulations in this country. Radiation protection standards have been, and appear to be, keeping pace with the growing needs of the atomic energy industry.

Expansion of NCRP considerations to cover emergency situations on the one hand, and more amplification of the broad aspects of group and population exposures on the other are favorable trends. The introduction of the range concept by the Federal Radiation Council is regarded by many as a start in getting away from the rigidity of specific control numbers.

The industrial exposure situation continues to be characterized by good compliance with current radiation protection standards for long-term radiation control. The serious accident experience does not show any unfavorable trends, and while there is room for improvement, the accident experience of the atomic energy industry compares very favorably with other elements of industrial safety.

Some minor difficulties principally associated with implementation of standards or codification of the basic principles of good practice into regulations have appeared and very likely will continue to appear. Some problems of jurisdiction and reciprocity will be difficult to avoid in the course of the transfer of regulatory responsibilities from the Atomic Energy Commission to individual States.

Not peculiar to industry's role, but nevertheless having substantial effect on the industrial climate, is an apparent lack of public understanding in depth of nuclear energy and its associated hazards. It appears that considerable effort will have to be expended before the potential hazards associated with sources of ionizing radiation can be viewed in perspective by the layman. Comprehensive hearings such as these and others conducted in the past by the Joint Committee are major factors in increasing public understanding of this complex subject of radiation protection in the atomic energy field.

APPENDIX

SOME FACETS OF INDUSTRIAL EXPOSURES EXPERIENCE

TABLE I.—*Exposures of contractor personnel to penetrating radiation, summarized for 1959 and 1960*

Range of annual total exposure in rems	1959 (13)		1960 (7)	
	Number of workers	Percent of total number of workers	Number of workers	Percent of total number of workers
0 to 1	71,630	94.73	77,522	94.31
1 to 5	3,912	5.17	4,629	5.63
5 to 10	66	.09	41	.05
10 to 15	2	<.01	2	<.01
Above 15	1	<.01	3	<.01
Total	75,611		82,197	

TABLE II.—*U.S. criticality accident experience (12)*

Year	Number of criticality accidents	Number of fatalities	Year	Number of criticality accidents	Number of fatalities
1945	3	1	1955	1	
1946	1	1	1956	1	
1947			1957	1	
1948			1958	2	1
1949	1		1959	1	
1950			1960	1	
1951	2		1961	2	3
1952	2		1962 through April	1	
1953			Total	22	6
1954	3				

TABLE III.—*Industrial injury frequency rates in number of lost-time injuries per million man-hours* (7), *1960*

Group:
- All industries -- 6.04
- All AEC contractors -- 1.71
- AEC operating contractors (excludes construction) -------------- 1.18

REFERENCES

1. Failla, G., Statement on Radiation Protection Standards. *Selected Materials on Radiation Protection Criteria and Standards: Their Basis and Use.* Joint Committee on Atomic Energy, Congress of the United States, May 1960. U.S. Govt. Printing Office, Washington, D.C.
2. Radiation Protection Guidance for Federal Agencies; Memorandum for the President, A. S. Flemming, Federal Radiation Council, May 11, 1960. Approved by the President and Printed in the *Federal Register*, May 18, 1960.
3. Radiation Protection Guidance for Federal Agencies; Memorandum for the President, A. Ribicoff, Federal Radiation Council, September 20, 1961. Approved by the President and Printed in the *Federal Register*, September 26, 1961.
4. *Exposure to Radiation in an Emergency.* National Committee on Radiation Protection and Measurements Report No. 29, January 1962. University of Chicago.
5. *Employment in the Atomic Energy Field.* A 1960 Occupational Survey, U.S. Department of Labor Bulletin No. 1297, April 1961. U.S. Govt. Printing Office, Washington, D.C.
6. U.S. Atomic Energy Commission. AEC Manual Chapter 0524, *Permissible Levels of Radiation Exposure.* February 1, 1958. AEC Manual Chapter 0550, *Codes and Standards for Health, Safety and Fire Protection.* August 29, 1957.
7. *Annual Report to Congress for 1961.* Atomic Energy Commission, January 30, 1962. U.S. Govt. Printing Office, Washington, D.C.
8. *Code of Federal Regulations.* Title 10, Atomic Energy. Chapter 1, United States Atomic Energy Commission Regulations.
9. *Maximum Permissible Body Burdens and Maximum Permissible Concentrations of Radionuclides in Air and Water for Occupational Exposure.* National Bureau of Standards Handbook 69, June 5, 1959.
10. *Permissible Dose from External Sources of Ionizing Radiation.* National Bureau of Standards Handbook 59, September 24, 1954. Addendum, *Maximum Permissible Radiation Exposures to Man.* April 15, 1958.
11. Parker, H. M., Radiation Protection Standards: Theory and Application, *Atomic Energy Law Journal*, Vol. 2, No. 4, 334–370. Fall 1960.
12. *A Summary of Industrial Accidents in USAEC Facilities.* Division of Operational Safety, TID–5360 Suppl. 3 Revised. December 1961.
13. *Annual Report to Congress for 1960.* Atomic Energy Commission, January 20, 1961. U.S. Govt. Printing Office, Washington, D.C.

Representative PRICE. Thank you, Dr. Parker.

Dr. Parker, on page 2 you state that the basis for most of the standards used today are the recommendations of the ICRP and the NCRP and that this situation is essentially unchanged today. What then is the role of such an agency as the FRC?

Mr. PARKER. This, sir, in no way downgrades the role of that very important body. I am separating rather, and perhaps even artificially, basic standards, the broad outlook on what has to be done in the technical area. You will recall previous discussions in which it is agreed that one never completely separates the technical area from other areas of judgment.

In the technical area the standards are still essentially those put together by these two bodies. We conceive a Federal Radiation Council role as continuing to apply value judgments which are indeed no less important than the basic technical judgments and making these available to the Nation.

As I mentioned in the script, sir, I received a very recent Report No. 3, and it is a very valuable contribution in this area in making value judgments relative to fallout.

Representative PRICE. You state on page 3 that the FRC Report No. 2 carries the promise of introducing much needed flexibility. In what way does it offer flexibility?

Mr. PARKER. I think it is conceived generally in the field that through this concept of the three ranges, one gets away from some earlier objections to what was wrongly interpreted but was interpreted as a rigorous edge-of-night limit; namely, the permissible limits of the NCRP for a specific situation, such and such a number would be a limit and beyond that is bad and below it is good, has been to some extent in the past a misinterpretation of the intent. By having these three ranges, there is introduced the thought that it is perfectly reasonable to go along with a situation in which the possible exposures being received may be creeping up, provided that the looking at it, namely, the measuring devices and controls which are stipulated along with the requirements for these range applications, are in proportion to the degree of exposure that may be being received.

Representative PRICE. You further state that the FRC Report No. 2 may have advantages for those working with modest amounts of one or two radionuclides. What are some of these advantages?

Mr. PARKER. This gets back to a point that I hoped to make clear in the script, sir. My report as a whole I would like to characterize as perhaps not fairly representing the problems and situations of the very small users, since we ourselves come in contact with this rather superficially. His situation is such that he cannot have his own specialist who can study and offer professional local judgments on interpretation of cases, or say an interpretation of what the three ranges of the FRC would mean. He has to have a textbook which gives him a number. I intended this reference in that sense.

Representative PRICE. On the bottom of page 4, when you are talking about NCRP Report No. 29 and the reference to emergency exposure, was this an outgrowth of the Windscale incident?

Mr. PARKER. No, sir. Dr. Taylor is in the room and is perhaps better qualified than I to particularize this. But I believe this work was primarily started in the interest of civil defense preparation in this Nation and was finally put together in the present form. I am stipulating here that it turns out, allowing a little leeway for interpretation, to be very useful in the industrial situation, and we have already had occasion ourselves so to use it. It was not specifically prepared for that case, as I see it.

Mr. RAMEY. Did you use that in the case of one of the emergencies at Hanford?

Mr. PARKER. In our recent critical incident it was extremely valuable, sir.

Representative PRICE. You refer to transfer of responsibility from the AEC to the State for certain radioactive materials. Do you foresee about 50 different State regulations in connection with this?

Mr. PARKER. I foresee more than one. Whether it will ever get to 50—I suppose one could make 50 different variations if you tried hard. I would see perhaps a dozen variations.

Representative PRICE. What is your own feeling on the matter? What do you think should be done in this area?

Mr. PARKER. I think one has to go in this direction and do what we can to get a common understanding later. This is partly dependent on the degree to which specific items are included in the final code or the extent to which you are willing to go back to basic principles. The basic principles are certainly intended to be the same in all cases, as I see these points.

The minor items to which I refer are indeed minor. One State, Kansas, I believe, has a regulation that characterizes dose in one calendar quarter, and another State would have a limitation on any consecutive 13 weeks. This is just an internal administrative nuisance rather than any reevaluation of the hazard to man.

Representative PRICE. Do you know when the AEC plans to bring their manuals up to date with respect to plans for revisions which must be made to reflect FRC guides for Federal agencies?

Mr. PARKER. I have no date on that, sir. I just mention in the script, as you say, that we understand from the Commission that it will more directly reflect in the next revisions, and perhaps Commission representatives could be more responsive to the timing.

Representative PRICE. On page 6 you mention a third group of radiation source users. What seems to be their safety record compared to the others you indicate?

Mr. PARKER. This is the group who were not covered by the Atomic Energy Act, sir. It is very difficult again. We have no comprehensive data which allows one to testify, and one goes by impression and conversation in meetings with this group. It is characteristic that those of us with the larger enterprises who have full-time staffs in this work tend to think that our controls are more successful than others. It would be rather peculiar if that were not the case. I do believe it is in that direction. The extent to which the situation could be considered bad in this third group, I do not know, and know of no real evidence which points to a poor situation.

Representative PRICE. In the charts that you displayed and particularly in table No. 2, you listed the U.S. criticality accident experience since 1945 up until April of 1962. You give the number of criticality accidents as 22 and the number of fatalities as 6. Is this a complete and accurate picture of the accident history of the AEC?

Mr. PARKER. To the best of my knowledge, sir, this table is intended to contain the total experience on criticality incidents. There can be other accidents.

Representative PRICE. Does this include the SL-1 accident?

Mr. PARKER. Yes; that would be included.

Representative PRICE. Including that, there is a total of only six fatilities in all the years of operation of the AEC.

Mr. PARKER. I looked at a chart this morning which would add one to this. I would like, perhaps, the liberty of submitting a second look at this particular number later, so that I do not actually misquote it. It is conceivable that it may be wrong by one.

Representative PRICE. If it is, will you correct this table?

Mr. PARKER. I will do that, sir.

Representative PRICE. The figure 22, is that absolutely actual for criticality accidents in the Atomic Energy program?

Mr. PARKER. These are supposedly my associate's counting of the published number of criticality accidents.

Representative PRICE. What is the nature of these accidents? The 22?

Mr. PARKER. They were wide and varied. The early ones occurring were manipulation of weapons parts. Others came from accidents with plutonium- or uranium-bearing solutions in various vessels spread around many of the principal sites of the Commission. The so-called Y-12 incident is in here. Many of the incidents at Los Alamos are included in here.

Representative PRICE. Will you be absolutely certain before we complete the record to have the accurate figures in here?

Mr. PARKER. I will see that the figures are reviewed and accurate figures given.

(The information requested follows:)

GENERAL ELECTRIC CO.,
ATOMIC PRODUCTS DIVISION,
Washington, D.C., June 7, 1962.

Hon. MELVIN PRICE,
Chairman of the Subcommittee on Research, Development and Radiation of the Joint Committee on Atomic Energy, U.S. House of Representatives, Washington, D.C.

DEAR MR. PRICE: During the hearings on radiation protection before your committee on Wednesday, June 6, I was asked to give assurance of the completeness of a table II—U.S. Criticality Accident Experience, which appears in the appendix to my submitted material entitled "Radiation Protection Standards: The Industrial Situation."

The subject data were taken principally from reference 12 of my report which is "A Summary of Industrial Accidents in U.S. AEC Facilities," Division of Operational Safety, TID-5360 supplement 3, revised December 1961.

On page 7 of that document appears a listing of 21 criticality accidents divided into 4 categories, viz:

Metal systems in air	5
Solution systems	8
Inhomogeneous water moderated systems	5
Miscellaneous systems	3
Total	21

In transcribing these into the chronological table used in table II, an unaccountable error was indeed made. Instead of one incident each in the years of 1956 and 1960, the record should show two in 1956 and none in 1960.

I appreciate the opportunity to correct the record.

The total of 22 incidents given is believed to be correct. It adds the incident occurring at Hanford in 1962 to the previous list. This belongs in the "solutions systems" category.

As indicated in the hearing, the three fatalities in 1961 do refer to the SL-1 incident. Reference 12 does not in itself identify the number of fatalities. A separate check indicates that the total of six given in table II is correct to the best of our knowledge. I would appreciate being informed of any data which contraindicate this.

Very truly yours,

H. M. PARKER,
Manager, Hanford Laboratories.

Representative PRICE. What does this cover? It does not cover the whole area of the atomic energy program?

Mr. PARKER. This only covers a situation in which an accident occurred because a critical mass was brought together inadvertently.

Representative PRICE. It would not cover normal industrial accidents in atomic energy?

Mr. PARKER. No, sir. Only those in which a critical mass is accidentally brought together.

324 RADIATION STANDARDS, INCLUDING FALLOUT

Mr. RAMEY. You would have other accidents that involved radiation spills, radiation burns, all these other things?

Mr. PARKER. They are accidents. The only reason for presenting the criticality experience was that this is the case which is supposed to be conclusively reported so that one could use it to measure trend. I intended to use this only to show the trend of accident experience. AEC reports include these days considerable reference to so-called radiation incidents which define and elaborate a broader class. We found no way of picking these up from all sources. They come only from the licensee sources.

Representative PRICE. On page 12 you refer to the retrospective method of operation by the Public Health Service and on the other hand the prospective target needs of industry. Will you elaborate on your statement concerning needs for reconciling these differences?

Mr. PARKER. Yes, sir. Please recall that this is a personal interpretation or conceivably a misinterpretation of what I think the Public Health Service is trying to achieve in this field. I think their posture is that they would like to have things go along and then step in from time to time and say, "We examine this case now and our analysis is thus and so." This may be good or it may be that you should not have gone this far. It is to that possibility that industry is properly very sensitive. Let us assume that industry is trying to make a proper showing in radiation control, then you have to do this at the beginning of any time period and not leave oneself subject to being told after the event that this was not very wise, that you should have done it some other way.

It is this telling us in advance what we should be shooting for that I am defining as the prospective target which we need and which the public needs in order to examine our performance against these targets.

Representative PRICE. You also touched on certain inconsistencies, although relatively minor, I think you said, that have appeared in the transfer from the AEC to the States of certain regulatory responsibility. What are some of these?

Mr. PARKER. I mentioned one already. If I may refer to notes, I would have a few more. This difference on the time base between Kansas, Illinois, and New York. These States have three different time bases for measuring their external exposure. The permissible concentrations of materials put into unrestricted waters differs in minor detail between the States of New York, New Jersey, and California. The amount of material exempt from registration differs over quite a remarkable range between the States of Kansas, Minnesota, and New Jersey.

The definition of radiation area is different in the New York Code from the recommendations of title 10 Code of Federal Regulations, section 20. Surprisingly, the alleged definition of the roentgen has three different appearances as between the codes of Florida, Illinois, and Kansas.

That covers the present differences, and perhaps with the exception of this rather wide range in exemption from registration, these are administrative nuisances at the present time.

Representative PRICE. The committee staff intends to make a study of the problems involved in terminology. We have been interested in this area for some time.

You mentioned the problem of reciprocity and jurisdiction that have yet to be worked out. How would this situation affect a company such as General Electric, or others, with divisions in many States?

Mr. PARKER. I think this affects part of such companies with which I have the least acquaintance. For example, one is manufacturing radiation emitting devices, and these are to be used against different codes. One has a problem of some magnitude conceivably with the writing of certain codes which could be insuperable. This is what one has in mind in part in this industrial problem. I don't profess to be directly concerned or acquainted with this aspect.

Representative PRICE. Are there any further questions?

If not, thank you very much. You made a fine statement and contributed a great deal to the committee's knowledge on the subject.

There is a quorum call in the House. I think we will take a recess for about 10 minutes.

(The subcommittee took a short recess.)

Representative PRICE. The committee will be in order.

The next witness will be Dr. Robert Hasterlik, of the Argonne Cancer Hospital, University of Chicago.

STATEMENT OF ROBERT J. HASTERLIK, M.D.,[1] PROFESSOR OF MEDICINE, UNIVERSITY OF CHICAGO, AND ASSOCIATE DIRECTOR, ARGONNE CANCER RESEARCH HOSPITAL, CHICAGO, ILL.

Dr. HASTERLIK. Mr. Chairman, members of the subcommittee, it is a pleasure and a privilege to appear again before this subcommittee. I have been asked to discuss with you today the somatic effects of radiation and to draw attention to developments in the field that have taken place since Dr. Austin Brues appeared before you in a similar capacity 3 years ago.

I shall attempt to limit my discussion to pertinent data derived from studies of man. Over the past many years you have, I am certain, become aware of the difficulties of applying data to man derived from studies done on the small experimental animal.

---

[1] Date and place of birth: Mar. 17, 1915, Chicago, Ill.
Education: 1931–34, College of the University of Chicago, S.B., 1934; 1934–38, Rush Medical College, University of Chicago, M.D., 1938; 1938–39, fellow in pathology, Cook County Hospital, Chicago, Ill.; 1939–40, intern, Evanston Hospital, Evanston, Ill.; 1940–41, fellow in gastroenterology, Indianapolis City Hospital; and 1941–42, resident in medicine, Evanston Hospital.
Honorary societies: Phi Beta Kappa, 1934; Alpha Omega Alpha, 1938; and Scientific Research Society of America (RESA), 1950.
Military service: 1942–46, lieutenant (junior grade) to lieutenant commander, Marine Corps, U.S. Naval Reserve; active duty with the U.S. Navy; served at sea with the amphibious forces in the Pacific, and chief of medical services at the U.S. Marine Barracks, Klamath Falls, Oreg.
Certification: American Board of Internal Medicine, 1947.
Appointments: Associated with the University of Chicago since 1948; 1948–53, director, Health Division, Argonne National Laboratory; 1950–53, senior scientist, Division of Biological and Medical Research, Argonne National Laboratory; presently professor of medicine, University of Chicago, and associate director of the Argonne Cancer Research Hospital.
Other facts: Member, National Committee on Radiation Protection Subcommittee on Exposure to Radiation in an Emergency, World Health Organization Expert Advisory Panel on Radiation, U.S. delegation to the 1st International Congress on the Peaceful Uses of Atomic Energy, Illinois Legislative Commission in Atomic Energy, Illinois Radiation Protection Advisory Council; consultant, NAS–NRC Subcommittee on Toxicity of Internal Emitters of the Committee on the Pathological Effects of Atomic Radiation, U.S. Atomic Energy Commission-Department of State official scientific mission to South America, United Kingdom Medical Research Council radium toxicity program.
Home address: 5801 South Dorchester Ave., Chicago, Ill.
Office address: University of Chicago, 950 East 59th St., Chicago, Ill.

Radiation Protection Standards: Theory and Application

Atomic Energy Law Journal, Volume 2

Fall 1960

A similarly titled article with basically the same content was presented before the Euratom International Symposium on Legal and Administrative Problems of Protection in the Peaceful Uses of Atomic Energy in September 1960 and was published in the Proceedings of that meeting the following year.

RADIATION PROTECTION STANDARDS: THEORY AND APPLICATION

By Herbert M. Parker[*]

INTRODUCTION

Man, throughout history has been surrounded by hazards which subject him to varying degrees of risks. Examples of natural hazards abound. They can present themselves dramatically as in flood or storm, or subtly as in slow climatic cycles. In addition, man-made hazards develop with increasing complexity as civilization progresses. Obvious examples include atmospheric pollution, food additives, traffic, synthetic drugs, and industrial use of machinery and chemicals. Man adapts himself to his surroundings as he recognizes these hazards, and by modifying his behaviour patterns attempts to satisfy both his individual and his social concern. His attempts to do this may involve elimination or control of the source of hazard, or avoidance of exposure. Sometimes he seems to tolerate hazards with little effort at control, especially when his evaluation of personal gain or convenience ranks high compared with his estimate of the risk of personal injury.

One phenomenon, arising both from natural and man-made sources that man has found it necessary to recognize as a possible source of danger is ionizing radiation. While radiation does create problems the benefits from using ionizing radiation are conspicuous. Con-

---

[*] Manager, Hanford Laboratories — General Electric Company

This paper is similar to that presented at the International Symposium on Legal and Administrative Problems of Protection in the Peaceful Uses of Atomic Energy at Brussels, Belgium, on September 5 to 8, 1960.

Work done under prime contract AT(45-1)-1350 to the United States Atomic Energy Commission.

RADIATION PROTECTION STANDARD

sider the advances in medicine alone that can be attributed to the use of X-rays. The life span in the United States of America at the turn of the century was about 40 years; it is now about 60 years. That some part of this advance is due to the medical diagnostic and therapeutic power of X-rays is clear. A far greater benefit, often overlooked, is the contribution of X-ray crystallography to an orderly understanding of nature, making possible among other advances the rational synthesis of wonder drugs. Similarly striking benefits are apparent in industrial applications. These benefits should increase as we proceed toward the rational use of atomic power.

To gain these benefits there has been the need to control the potential hazards arising from ionizing radiation. This need continues and increasingly becomes a matter for social group action. The identification of the appropriate social groups and their optimum steps in legislative and administrative areas toward necessary and sufficient control is of major interest to this conference.

This paper traces briefly the historical background of radiation protection standards, identifies some of the technical and technical-administrative problems involved, and reviews the current status of some standards. In May and June of this year, the Joint Committee on Atomic Energy of the U.S. Congress held extensive hearings on this same topic. The record of these hearings when published will be an invaluable compilation of representative opinion.[1]

---

[1] *Selected Materials on Radiation Protection Criteria and Standards: Their Basis and Use.* Joint Committee on Atomic Energy, Congress of the United States, May 1960. U.S. Govt. Printing Office, Washington, D. C., 1960.

EARLY HISTORY OF RADIATION PROTECTION STANDARDS

When X-rays were discovered by Roentgen and radioactive substances by Becquerel, both agents were promptly applied to beneficial purposes. Almost at once, both agents produced demonstrable injury to skin. By 1905, the marked radiosensitivity of the blood-forming organs was known, and effects on the reproductive organs of animals demonstrated.

By contrast, the first record of a radiation protection "standard" was a crude one ascribed to Rollins in 1902[2] in the USA. The first organized step to ensure protection was taken by Russ[3] in 1915 in England.

By 1920 many of the early radiologists and their aides had developed cancer of the skin, particularly of the hands or face. Others had succumbed to blood diseases. In the 1920's and early 1930's there was considerable activity at the national and international level to improve and systematize protection. In the same period, investigators such as Mutscheller, Barclay and Cox, and later Failla, published tolerance-doses based on estimates of dose in hospitals in which operators continued in good health. Such tolerance-doses were conventionally quoted in fractions of an erythema dose per month or per day, or even per year at one extreme and per second at the other.

Both dose and time components of these early standards had an extraordinarily profound effect on the subject. Erythema-dose was used in the absence of a good

---
[2] Rollins, W., Vacuum Tube Burns, *New England J. Med.*, 146:39, 1902.
[3] Russ, S., Hard and Soft X-rays, *Arch. Roentgen Ray Soc. London*, 19:323 (1914-1915).

RADIATION PROTECTION STANDARD

physical unit (and long after that was corrected!). Based on clinical observations of erythema, it naturally predisposed the observer to expect other deleterious effects to follow the same pattern.

That pattern is

(a) there is obviously a threshold dose,
(b) protraction and fractionation of the dose are important,°
(c) actual dose-*rate* may be important,
(d) there is a tissue recovery effect.

It is now difficult to trace in the literature the stiffening of the time period, originally quite casually and practically chosen as one month into a rigorous one-day limitation. The observations given in the footnote undoubtedly contributed.

A quarter century later, in the face of exposure limits now set on a theory that specifically denies the existence of a time factor, this early arbitrary choice still has an influence on administrative codes.

Two events of this early period are of major significance. There is time only to refer to them. In 1927, we had the discovery of the mutagenic action of radiation by H. J. Muller. At the end of the period, Martland reported the basic work on radium deposition in the body.

Brief review of more modern times, shows that the

° In radiation therapy, protraction and fractionation refer to the spreading of a therapeutic dose over some fairly long time period (e.g. one week or one month) and to the dividing of that dose into daily or other quotas. The clinical facts on skin reaction are
(a) given equal daily doses, the reaction is almost independent of the length of time of each treatment, i.e. dose-rate is not very important.
(b) when the fractionation is every second or third day, the effect changes,
(c) for a single large dose, the reaction is less severe at low dose-rate. However, this is because the *protraction* was greater.

337

ATOMIC ENERGY LAW JOURNAL

International Commission on Radiological Protection, the National Committee on Radiation Protection (U.S.)[*] and like bodies in other nations in the period 1934 to 1940 were able to quote tolerance-doses of either 0.2 r per day or 0.1 r per day for external X-radiation or gamma radiation.

At this itme, two other numerical standards were in general use, viz:

10^{-14} curie/cc for radon in the air of working rooms.

0.1 microgram of radium as the maximum allowable amount deposited in the body.

In the United States there was an explosive growth of the scope of application resulting from the development of nuclear reactors. In a few years, radiation protection procedures for cases involving alpha, beta, neutron, and more exotic radiations were added. Means were developed to specify allowable concentrations of a host of new radionuclides. This was done on a generally satisfactory basis within the Plutonium Project of the Manhattan Project, later a part of the Atomic Energy Commission's activities.[4] In 1946, the NCRP was reorganized and began the task of improving or validating this violently enlarged scope.

RADIATION PROTECTION THEORY

The principal technical and technical-administrative factors involved in radiation protection standards can be

---

[4] Stone, R. S., et al., Industrial Medicine on the Plutonium Project, *National Nuclear Energy Series, Division IV — Vol. 20* McGraw Hill Book Co., Inc. 1951.

[*] I have not paid attention to changes in title of committees. The U.S. group has been variously the Advisory Committee on X-Ray and Radium Protection, the National Committee on Radiation Protection, and the National Committee on Radiation Protection and Measurements.

RADIATION PROTECTION STANDARD

expressed in many ways. The following classification and brief notes represents one approach:

1. *Units and Measurements*

Protection standards rely on valid measurement units for radiation dose and other quantities. Those in general use today are shown in Table I.

These units have a long history of ambiguity and conceptual change and refinement by qualified professionals. Awareness of these problems is *essential* if units and measuring methods are to be incorporated in code and regulation in a meaningful way. The 'rem', in particular, is frequently misused. Without a stipulation of the arbitrary values of relative biological effectiveness (RBE) to be used in protection, it has no meaning. Having less than the status of a unit, it is an accepted convenience symbol, albeit a most necessary and desirable one.

2. *Identification of all Relevant Biological Effects*

Complete protection presupposes a recognition of all the deleterious effects that might occur. The best that can be achieved is to complete a prudent search for all foreseeable factors. There is no reason to believe that any significant factor has been overlooked. Better understanding of some nebulous areas, particularly the aging process and low dose effect on the nervous system, is being sought.

3. *Dose-Rate*

At the experimental level, there are some biological effects responsive to dose-rate per se and others not. There is no comprehensive way to incorporate this factor

---

[5] Parker, H. M., Tentative Dose Units for Mixed Radiations, *Radiology* 54, 257, 1950.

ATOMIC ENERGY LAW JOURNAL

in a general theory, and it is probably not needed save for some uncertainty in handling very intense pulsed radiation.

4. *Protraction and Fractionation*

These factors, which are so obviously controlling ones in clinical irradiations, have varying importance down to zero for different deleterious effects. As indicated above, clinical experience with skin reactions is probably given too much weight in radiation protection.

5. *Shape of Dose-Response Curves*

It is conventional to express biological effect as a function of the dose producing it in graphical form as a dose-response curve. In Figure I, curve A shows a typical threshold response curve. B shows a response directly proportional to the applied dose. C and D are two variants of non-linear, non-threshold response. Figure 2 shows some more complex possibilities. E shows a "negative deleterious effect" or beneficial effect at low dose. This would be an absurdity for a single pure effect (such as induction of leukemia) but is at least plausible for a composite effect (such as life-shortening*). Of the nine curves in Figures 1 and 2, three (A, E and F) have an effective threshold dose. Three (F, G and H) show a deleterious effect in the absence of dose. At this point we must be careful to specify whether our dose includes natural background or not.

The precise shape of the response curve is of funda-

---

[6] Bustad, L. K., Physiological Responses of Mice to Low Levels of Irradiation, *Ph.D. Thesis, U. of Washington*, Seattle 1960.

* Recently reviewed in a doctoral thesis by L. K. Bustad of this laboratory.[6]

RADIATION PROTECTION STANDARD

mental importance to protection, especially at low dose levels, where it is most difficult to obtain experimental data. Until radiobiological theory advances, one cannot deduce this structure with confidence from measurements at higher dose.

6. *Permissible Dose*

If there is a threshold dose below which no deleterious effect of radiation occurs, there is a logical technical solution to radiation protection problems by identifying these limits and controlling exposure below them. Careful study since 1945 has led to general acceptance that some effects, notably the induction of gene mutations, occur in direct linear proportion to the radiation dose at low dose levels.

Current protection theory prudently accepts the absence of a threshold dose, although the point has not been unequivocally demonstrated. Due to some partial equating of the old term "tolerance-dose" with the threshold dose concept that term was replaced by the concept of permissible dose and its upper limit-maximum permissible dose.

There is said to be considerable misunderstanding of these terms and a desire to substitute new ones. The meanings and definitions assigned to them by NCRP Subcommittee No. 1 appear to be clear and unequivocal.[7]

7. *Value Judgments and Acceptable Risk*

Without a threshold dose there is no level of radiation without the potential of doing some harm. One is thus forced into a concept of acceptable risk. The NCRP definition of acceptable risk relates to limitation of risk to levels customarily and voluntarily accepted in ordinary

341

ATOMIC ENERGY LAW JOURNAL

occupations.⁷ Some have informally measured risk against what injury is customarily accepted in other fields. Others have pressed for a direct balancing of gains from radiation against these risks and attempts, usually unsuccessful have been made to rationalize the costs of radiation protection in terms of balancing detrimental effects against benefits.⁸

The key point is that current theory compels some type of risk-decision, expressed or implied, and necessitates value-judgments. The extent to which these judgments are to be made by scientists knowledgeable in radiation effects, or by others more broadly based in social and economic values is the central point.

8. *Differentiation Between Genetic and Somatic Effects*

Radiation-induced gene mutation in any one individual will generally be quite insignificant to him. The potential deleterious effect depends on the total load of defective genes in a population group. Within limits only the average dose is important.*

By contrast, the manifestation of somatic effects is of concern to each individual. In a large population, there is need for an *individual* limit, with somatic effects in mind, and an *average* limit, which will be the lower of two values deemed separately suitable with respect to genetic or somatic effects.

---

⁷ *NBS Handbook* 59, Permissible Dose from External Sources of Ionizing Radiation, 1954.

⁸ Report of the United Nations Scientific Committee on the Effects of Atomic Radiations, United Nations, New York 1958.

* I say "within limits" because the popular concept of gene mutation tends to bring to mind the birth of demonstrably deformed children. Every parent has the right to feel that he has not consciously and significantly increased the chance of contributing to an unfortunate birth. Thus a level of exposure that measurably increased the incidence of congenital malformation would be rejected.

RADIATION PROTECTION STANDARD

9. *Critical Organ*

The critical organ is an organ or tissue most affected by ionizing radiations from the deposition of a specified internal emitter or from external sources. The essentialness of the organ and its sensitivity to ionizing radiations are considered in determining the critical organ.

In complex radiation exposure situations, identification of the appropriate critical organ is not straightforward. This occurs if the potential effects on more than one organ system are about equal.

10. *Extent of Irradiated Tissue*

There is no valid theory for the estimation of biological effect when only a small part of a critical organ is irradiated. In practice the effect is often assumed to be equal to that of like irradiation of the whole organ unless the volume irradiated is very small. In that case arbitrary averaging over a defined "significant volume" is practised.

11. *Relative Biological Effectiveness*

Relative biological effectiveness (RBE) is a term used to compare the effectiveness of absorbed doses of radiation delivered in different ways. The concept of RBE has a limited usefulness because the biological effectiveness of any radiation depends on many factors. Thus, the RBE of two radiations cannot, in general, be expressed by a single factor but varies with many subsidiary factors such as the type and degree of biological

---

[9] Laughlin, J. S. and I. Pullman, Gonadal Dose Received in the Medical Use of X-rays. Section II of "Genetic Radiation Dose Received by the Population of the United States". *Report prepared for the Genetics Panel of the National Academy of Sciences Study of the Biological Effects of Atomic Radiation,* November 19, 1956, Washington, D. C.

damage (and hence with the absorbed dose), the absorbed dose rate, the fractionation, the oxygen tension, the pH, and the temperature.

To add up the relevant RBE dose for protection purposes, the *rem* unit or pseudo-unit is used by weighting absorbed dose in rads by RBE factors.

RBE values used in protection are arbitrary and in general today probably too high in many cases. Chosen values must be stipulated whenever the *rem* is to be used is regulation. In particular more consideration should be given to separate values for genetic and somatic effects. It is technically feasible for a given radiation type to have significantly different RBE factors for these two cases.

12. *Identity of Radiation Disease with Other Disease*

In the absence of applied dose, man develops certain diseases identical with those known to be producible by radiation. Leukemia is a familar example. The causative agent may be the natural background in part and other agents in part. Determination of the quantitative effects of low doses in epidemiological studies is then most difficult.

13. *Age Factor*

The young animal is normally more susceptible to a given radiation dose. Moreover, the uptake of internal emitters is often greater in the young. Thus there is a dual enhancement of the potential radiation damage. This is one reason why occupational limits (which are limited to selected adults) may be considerably higher

---

[10] Norwood, W. D., J. W. Healy, E. E. Donaldson, W. C. Roesch, and C. W. Kirklin, The Gonadel Radiation Dose Received by the People of a Small American City Due to the Diagnostic Use of Roentgen Rays, *Amer. J. Roetgenology.*, Radium Therapy and Nuclear Med. LXXXII, 6, 1081-1097, December 1959.

RADIATION PROTECTION STANDARD

than non-occupational ones. Some forms of radiation stress are also believed to be enhanced in the elderly or infirm.

14. *Potential Injury to Other Life Forms*

In protection concerned with radioactive waste disposal, both direct effect on other life forms and the transmission of radioactive materials through food chains must be considered.

15. *Environmental Factors*

Also for waste disposal, information on meteorology, hydrology, food consumption and other behavior patterns of an assumed standard man is needed.

16. *Accident and Emergency Procedures*

I shall omit review of these phases except to mention that irradiation reactions for acute single doses differ from those for chronic irradiation to which protection should limit us. Also, compensation features depend on knowledge of recovery and repair both from single exposures and from excessive depositions of radioactive materials.

17. *Sense of Balance*

A classification of the type just given tends to emphasize the deficiencies in the structure of knowledge. It is necessary to retain a sense of balance by recognizing that radiation has been constructively used for 60 years and that the demonstrated incidence of injury has been very low in comparison with that tolerated in many areas of private or industrial activity. As Dr. Failla recently re-

---

[11] Brown, R. F., J. Heslep, and W. Eads, Number and Distribution of Roentgenologic Examinations for 100,000 People, *Radiology*, 74, 3, 353-363, March 1960.

minded us, our information on the biological effects of radiation is by no means nil. The fact that there is no elegant single theory of the case does not prevent a sound administrative defensive program from being created.

CURRENT ATTITUDES

Essentially all responsible groups subscribe to the same qualitative articles of good practices given in Table II.

PRESENT APPLICATION

These articles have been amplified into applicable standards by a number of agencies and organizations functioning from the authority of technical knowledge or by governmental organizations functioning also by the authority of common consent.

The outstanding example of a group utilizing the authority of knowledge is the International Commission on Radiological Protection which also has the longest tenure, extending from 1928. Its members are national representatives selected by and from national professional societies. The group is purely advisory and only issues formal recommendations, not regulations.[12][13]

Individual countries holding memberships in the ICRP as well as a number of other countries then also have national authoritative and independent bodies of comparable stature. The most extensive of these is doubtless the National Committee on Radiation Protection and Measurements in the United States. A battery of about

---
[12] *Recommendations of the International Commission on Radiological Protection*, (Adopted September 9, 1958) Pergamon Press, London.
[13] Addendum to ICRP Publication No. 1 (1958 Recommendations), *Health Physics*, Vol. 2, 317-320. Pergamon Press 1960.

RADIATION PROTECTION STANDARD

fifteen handbooks is needed to encompass the recommendations of that body.

Within the United States there are also national governmental organizations which make pronouncements pertinent to radiation protection, partially from acquired knowledge and partly in fulfillment of legislative responsibilities. The Atomic Energy Commission, the Interstate Commerce Commission and the newly formed Federal Radiation Council are conspicuous examples within this grouping.

In the international sphere, examples of organizations with a concern for radiation protection are the International Atomic Energy Agency, the International Labor Organization, World Health Organization, and International Standards Organization.

Additional to national and international groups having an interest and influence in the preparation and utilization of radiation protection standards, there are at least in the United States several voluntary groups that offer professional guidance to state, city, and other political subdivisions charged with maintenance of public health and safety. These professional groups also work with industrially oriented organizations such as the American Standards Association in the preparation and voluntary acceptance of specific aspects of radiation protection.

The following categories are frequently presented in published documents from these various sources:

1. *Attitude or Philosophy*

The articles of good practice are either expressed or referred to in another source.

347

ATOMIC ENERGY LAW JOURNAL

2. *Authority and Responsibility*

This segment deals with the extent to which standards are to be applied whether application is mandatory or voluntary, what exemptions apply and the actions to be taken during routine or abnormal conditions.

3. *External Exposure Limits*

Here are often listed separately the maximum permissible whole body and partial body occupational exposure limits and the time base over which these are to be controlled.

4. *Internal Exposure Limit*

These are generally in three groupings: (a) maximum permissible body burden of a specific radionuclide, (b) a maximum permissible concentration in drinking water, and (c) a maximum permissible concentration for radionuclides in air at the work site.

5. *Exemptions*

Some groups provide for exemptions to control and regulation of very small quantities of radionuclides and for low order external exposure.

6. *Administrative Controls*

This portion of radiation protection standards deals with the identification of radiation sources, keeping of radiation exposure records, and the means and occasions requiring exposure measurement. In some cases these are very detailed and extensive.

7. *Waste Disposal*

These are normally limitations as to how, where, when, and how often radioactive material shall be released or discharged into the environment.

RADIATION PROTECTION STANDARD

8. *Non-occupational Controls*

The most common means utilized to standardize controls of radiation exposure for persons not under occupational exposure control surveillance is to arbitrarily restrict permissible exposure to a submultiple of that deemed permissible for occupational workers.

Considerations of average population dose tend to start from the other end of the scale and recommend permissible multiples of an assumed constant natural background radiation level.

Appendix A carries an abbreviated comparison of the published recommendations or regulations of the NCRP, ICRP, FRC, AEC, NY State, and British Regulations. Table III is an even more condensed form for projection here. Although a first impression is one of many differences in detail between the several proposals, the key point is that areas of agreement are far more significant. If one concentrates on different entries in any row of this tabulation, it would be difficult to find a competent panel that could *specifically* prove that one entry had greater technical validity than another. Can one not conclude that we are ready to accept a single common code responsive to the basic articles of good practice? Can one not also conclude that local situations will always be best handled by flexible interpretation of or modification of the main theme, and that codes can be written to permit if not actually promote that kind of flexibility?

SCOPE OF STANDARDS APPLICATION

Twenty years ago, radiation protection needs were fairly well limited to the medical worker, the scientific investigator, and a narrow range of industrial workers.

Today, the group occupationally concerned is still small, but markedly larger and more diversified. Table IV illustrates the industrial use in the State of New York as an example. In addition, the existence of fallout, the expanding needs of waste disposal, and the growth of radiation-producing products such as television have directed attention to population exposure problems. The broadened numerical scope has been paralleled by a broadened complexity of types and sources of radiation.

Table V lists the principal contributions to population dose in the United States of America.

The expansion of the field, especially in the non-occupational aspects, brings to an end a period in which voluntary control by industrial and other radiation-using groups was sufficient. These groups now generally look forward to the setting of sensible regulations to optimize the exploitation of atomic energy.

A striking feature of the last table is the high proportion of dose due to medical exposure. This area is most likely to be left to voluntary control, to preserve the necessary professional freedom. Yet if total population dose is at a level such that allocation of dose among various activities is to be practised, the medical quota is the most obvious target. A voluntary reduction by 10 per cent, which must be easily achievable, would compensate for all other forms of man-made radiation.

Such allocations are in the broad area of social and economic value judgments, and beyond the scope of radiation protectionists. In the United States, creation of the Federal Radiation Council has provided a new dimension that can ideally consider such allocations when

RADIATION PROTECTION STANDARD

needed. In addition, it can set reasonable graded levels of permissible dose for specialized activities of national significance. Such broad policy-setting is expected to develop into the Council's principal function. To date, the Federal Radiation Council has not been in existence long enough for the nation to appraise its full value. Its membership consists of the heads of federal government departments and agencies whose activities substantially involve radiation problems. Its chairman is Secretary Flemming, Secretary of Health, Education, and Welfare, and its statutory function is to be advisory to the President.

It is expected that over-all technical recommendations will continue to originate principally in the independent NCRP, and be accepted by the Council.

We have already pointed out that NCRP recommendations are principally technical with some infusion of value-judgments. Within the Council one would expect a much broader application of value judgments in the social, economic, and political areas from the collective experience of these distinguished department heads and their aides.

SPECIFIC APPLICATION IN THE UNITED STATES

The description of radiation protection in the United States may serve to exemplify the application of radiation protection standards. Here, all of the formal recommendations of the National Committee on Radiation Protection and Measurements are published and available at nominal cost. The publications are generally in the form of handbooks dealing with specific and practical radiation protection controls. Table VI lists the current ones by title.

ATOMIC ENERGY LAW JOURNAL

A quite comprehensive coverage of the facets of occupational exposure is provided. Except for long standing guidance on permissible concentrations in air and water, the coverage of non-occupational exposure is less complete and is currently under detailed scrutiny in committee. An independently supported study by a National Academy of Sciences committee has accelerated the assignment of limiting numbers to population gonadal dose, and has brought fresh points of view to other aspects.[14][15]

The Atomic Energy Commission, through its statutory powers has contributed materially to control by issuance of limits, licensing policies, and perhaps most of all by vigorous promotion of research into improved protection. The Federal Radiation Council recently published guidance for Federal agency use.[16][17]

Industry has used the NCRP recommendations to establish working limits which recognize the exposure conditions peculiar to a particular working situation. Operational controls are generally set with values below the maximum recommended limits. The extent to which radiation protection has been successful can be gauged by Table VII which lists some aspects of ex-

---

[14] The Biological Effects of Atomic Radiation, *Summary Reports and Report to the Public*, National Academy of Sciences — National Research Council 1956.

[15] The Biological Effects of Atomic Radiation, *Summary Reports and Report to the Public*, National Academy of Sciences — National Research Council 1960.

[16] Background Material for the Development of Radiation Protection Standards, *Staff Report No. 1 of the Federal Radiation Council*, May 13, 1960.

[17] Radiation Protection Guidance for Federal Agencies; Memorandum for the President, A. S. Flemming, Federal Radiation Council, May 11, 1960, Approved by the President and Printed in the *Federal Register*, May 18, 1960.

352

RADIATION PROTECTION STANDARD

perience of AEC contractors in the United States for the year 1958. At the Hanford Atomic Products Operation, only one-sixth of the people employed routinely received more than 10 per cent of the maximum permissible exposure. Not one person has actually accumulated one-third of his age-prorated maximum accumulated dose for external radiation. This and other AEC sites have a sound record of responsible control of radioactive effluent release to the public domain.

Radiology, which promoted practically all the early work in protection may have fallen behind to some degree. However, substantial educational efforts are now bearing fruit, and should lead to improved control of irradiations without interfering with any needed medical procedure.

Public interest in the hazards arising from fall-out has had two beneficial effects. The first is that the public has presumably been exposed already to the most extreme possible statements amplifying and deploring the hazard on the one hand and denying its existence on the other. It is probably now ready to listen to informed balanced opinion. The second is that the NAS-NRC BEAR committee, previously mentioned, has provided that balance. The national attitude to permissible gonadal dose is becoming more sophisticated through this channel.

There has been a degree of interest in federal or other central means of assessing gonadal dose for every member of the population, and possibly in establishing central control. Practical problems in making reliable measurements of gonadal dose are very complex indeed, when the effects of all sources, external and internal are considered. One has to be content to calculate most of the components of dose. Table V shows that if gonadal dose in medical procedures only were measured, practically all the man-made increment in the nation would be recorded.

Such a step would serve to draw constant attention to opportunities to reduce that dose.

All aspects of radiation that influence non-occupational groups are clearly in the domain of public health. The U.S. Public Health Service and various state services are currently equipping themselves with experience, personnel, and facilities to handle these new problems.

Not the least powerful of stimuli toward better radiation protection in this country is the critical voice of private group opinion. As an example, highly successful analyses of the complex Strontium-90 hazard situation have appeared from such sources.[18][19]

The United States is now ready to turn over much of the radiation control to legislation and action at the state level. Provided the basic uniformity of the articles of good practice is supplemented by local flexibility to meet diverse situations this would seem to be feasible and desirable.

Meanwhile, industry through its normal mechanisms of meeting its obligations is preparing comprehensive standards to assure self-control compatible with the standards of American industry generally. Activities of the Atomic Industrial Forum and the American Standards Association are examples of this. Fruition of these efforts will help to reveal radiation hazard as just one of

---

[18] The Milk We Drink, *Consumer Reports* 24, 102, March 1959.
[19] Fallout in our Food, *Consumer Reports* 25, 289, June 1960.

RADIATION PROTECTION STANDARD

innumerable controllable consequences of industrial development.

In reaching for this stage there is no better expression of the means and objective than that of Professor Wolman of Johns Hopkins University:

"The day of hand-book rule for measuring the hazard of radiation is a long way off. In the meantime, one acts upon limited knowledge. In such action, the guiding principle must be the maximum protection of people, not because of sentiment, but because society demands it."

TABLE I
RADIATION UNITS

Category	Unit	Defined*	Modified
Activity	curie	Radiology Congress 1910	IRSC - 1930 ICRU - 1950
Exposure Dose	roentgen	ICRU 1928	ICRU - 1937 ICRU - 1956
Absorbed Dose	rad	ICRU 1953	—#
"Biol" Dose (RBE Dose)	rem	ICRP 1953	ICRU - 1956**

* At the international level.
\# The *rad* is essentially a modification of the *rep*, which it superseded.
** The *rem* has been given a number of interpretations since its formulation in 1944.[5]

TABLE II
ARTICLES OF GOOD PRACTICE

1. Maintain radiation exposure at the lowest practicable level.
2. Prevent appreciable bodily injury to any individual.
3. Control maximum accumulated dose, and control the short-term increment (e.g. by quarter years).
4. Maintain significantly lower *average* limits for population groups.
5. Adequately integrate limits and controls for all different exposure circumstances.

TABLE III
CURRENT EXPOSURE LIMITS

Rem (RBE dose) Units

Application	Time Interval	NCRP	ICRP	FRC
Occupational				
Whole Body	N-Year	5(N-18)	5(N-18)	5(N-18)
	13 Weeks	3	3	3
Extremity	Year	75	75	75
	13 Weeks	25	20	25
Skin	Year	30	30	30
	13 Weeks	6	8	6

RADIATION PROTECTION STANDARD

Application	Time Interval	NCRP	ICRP	FRC
Non-Occupational				
Vicinity	Year	0.5	1.5 (adults)	0.5 (individual)
			0.5 (children)	0.17 (group average)
Population	30 Years (gonad dose)	Presently concur with ICRP.	5	5

TABLE IV

NEW YORK STATE
INDUSTRIAL RADIATION EXPERIENCE
1959

Number of Registered Radiation Producing Equipment Pieces	6,352
Number of Nuclear Reactors	8
Estimated Activity of Radioactive Materials	46,500,000 curies
Number of Industrial Installations with Radiation Sources	425
Estimated Number of Workers	5,000
Total New York State Employment	5,900,000

357

ATOMIC ENERGY LAW JOURNAL

TABLE V

THIRTY-YEAR PER CAPITA DOSES IN THE UNITED STATES FROM RADIATION SOURCES

SOURCE	30-Year Dose Gonad dose rem	Bone dose[**] rem
NATURAL RADIATION[8 9]		
EXTERNAL		
Cosmic rays	0.9	0.9
Terrestrial radiation	2.1	2.1
Atmospheric radiation	0.06	0.06
INTERNAL		
Potassium-40	0.6	0.3
Carbon-14	0.06	0.06
Radon-Thoron	0.06	0.06
Radium	-	1.2
Sub-Total	~3.8	~4.7
MAN-MADE RADIATION		
Diagnostic X-ray	[*]3[9]	1.4
X-ray therapy	0.36	no estimate
Internal isotopes	<0.03	no estimate
Luminous dials	0.03[8 9]	no estimate
Television	<0.03[8]	<0.03[8]
Atomic Energy Workers	0.003[8]	0.003[8]
Medical Workers	0.03[9]	no estimate
Sub-total	~3.5	~1.4
Approximate Total	7-8	5-7

[*] Additional studies on limited groups have given values on the order of 1.5 rem. (See references 10 and 11.)

[**] Taken here as the osteocyte dose. This principally affects the radium entry which would be 0.05 rem for mean marrow dose.

RADIATION PROTECTION STANDARD

TABLE VI

CURRENT NCRP REPORTS

NCRP Report No.	NBS Handbook No.	Date	Title
7	42	1949	Safe Handling of Radioactive Isotopes
8	48	1951	Control and Removal of Radioactive Contamination in Laboratories
9	49	1951	Recommendations for Waste Disposal of Phosphorus-32 and Iodine-131 for Medical Users
10	51	1952	Radiological Monitoring Methods and Instruments
12	53	1953	Recommendations for the Disposal of Carbon-14 Wastes
14	55	1954	Protection Against Betatron-Synchrotron Radiations up to 100 Million Electron Volts
16	58	1954	Radioactive Waste Disposal in the Ocean
17	59	1954	Permissible Dose from External Sources of Ionizing Radiation
18	60	1955	X-ray Protection
19	61	1955	Regulation of Radiation Exposure by Legislative Means

ATOMIC ENERGY LAW JOURNAL

20	63	1957	Protection Against Neutron Radiation up to 30 Million Electron Volts
21	65	1958	Safe Handling of Bodies Containing Radioactive Isotopes
22	69	1959	Maximum Permissible Body Burdens and Maximum Permissible Concentrations of Radionuclides in Air and in Water for Occupational Exposure
23	72	1960	Measurements of Neutron Flux and Spectra for Physical and Biological Application
24	73	1960	Protection Against Radiations from Sealed Gamma Sources

TABLE VII

U.S. AEC CONTRACTOR EXPERIENCE
1958

Number of Workers	124,500
Number of Recorded Exposures	66,000

Dose-Penetrating Radiation	*Number of Workers*
0-1 rem	59,400
1-5 rem	6,400
5-15 rem	181
> 15 rem	12*

* Includes one fatality.

RADIATION PROTECTION STANDARD

APPENDIX A - RADIATION PROTECTION STANDARDS: THEORY AND APPLICATION - SUMMARY TABLE

	NCRP	10 CFR 20	ICRP	NY STATE #38	BRITISH REGULATIONS	FRC(8)
PHILOSOPHY	Radiation exposures from whatever sources should be as low as practical. (1)	Agree substantially with NCRP. (2)	All doses be kept as low as practicable, and all unnecessary exposure be avoided. (3)	"reasonable and adequate protection" (4)	Do all reasonably practicable to restrict employee exposure. (5) "doses - well below those - appreciable bodily injury" (6) ICRP-1955 - "dosages as low as reasonably practicable" (7)	No man-made radiation without expectation of benefit from exposure.

OCCUPATION EXTERNAL	(9) (10)
	Whole Body
	5(N-18) rem/yr accum.
	3 rem/13 wk max.
	
	Skin
	30 rem/yr
	6 rem/13 wk (46)
	
	Extremity
	75 rem/yr
	25 rem/13 wk
		Present(11) Proposed(12)
		Whole Body Whole Body
		0.3 rem/wk 5(N-18) rem/yr
		              3 rem/qtr max.
		              1.25 rem/qtr*
		
		Skin Skin
		0.6 rem/wk 10(N-18) rem/yr
		              6 rem/qtr
		              2.5 rem/qtr*
		
		Extremity Extremity
		1.5 rem/wk 18.75 rem/qtr*
		
		*Required normally.
		
		NOTE: May exceed 7 day value to 3X for body-skin extremity if <.1X 7 day value in 13 week
			(3) (13)
			Whole Body
			5(N-18) rem/yr accum.
			3 rem/qtr max.
			
			Skin (44)
			30 rem/yr
			8 rem/qtr
			
			Extremity
			75 rem/yr
			20 rem/qtr
				(14)
				Whole Body
				300 mrem/wk
				
				Skin
				600 mrem/wk
				
				Extremity
				1500 mrem/wk
					(6)
					Whole Body
					0.3 rem/wk - advised
					5(N-18) red/yr (eyes) proposed; 3 rad/qtr to X- or gamma proposed; (100 mrem/wk, 7.5 mrad/hr - gamma, 37.5 mrad/hr - beta advised) (15)
					
					Skin
					0.6 rem/wk advised
					8 rad/qtr proposed (7) (15)
					
					Extremity
					1500 mrem/wk advised. (6)
					20 rad/qtr proposed. (16)
						Whole Body
						5(N-16) rem/yr accum.
						3 rem/13 wk max.
						
						Skin & Thyroid
						30 rem/yr
						10 rem/13 wk
						
						Extremity
						75 rem/yr
						25 rem/13 wk

361

ATOMIC ENERGY LAW JOURNAL

	NCRP	10 CFR 20 Present	10 CFR 20 Proposed	ICRP	NY STATE #38	BRITISH REGULATIONS	FRC(8)
OCCUPATIONAL INTERNAL	HB 69 for internal deposition and MPC's in air and water. (17)	Mostly values as revised from HB 52 for MPC's in air and water. (18)	Mostly values from HB 69 for deposition and MPC's in air and water. (18)	3 rem/13 wk whole body, eye, gonad; 8 rem/13 wk skin, thyroid, and 4 rem/13 wk other organs. Also 12 rem/yr whole body; 30 rem/yr skin, etc., and 15 rem/yr other, yr. - 15 rem/yr. MPC as in fn. (36) (19) (20)	300 mrem/wk in any part of the body. (14)	300 mrem/wk; or Ra - 0.1 μc; other nuclides - 0.3 rad/wk in critical organ except bone which is at 1/5 of 0.3 rad/wk - advised. (21)	0.1 μgm of Ra226 or its biological equivalent. 15 rem/yr 5 rem/13 wk
OCCUPATIONAL MIXED	Total dose within basic rules of age proration. (17)	None stated.	Not to be added, make special case. (22)	Internal values govern. Sum of external and internal. (19)	Aggregate weekly dose not exceed any appropriate weekly dose. (14)	Governed by maximum stated limit. Advise summation. (21)	None stated.
OCCUPATIONAL EXTERNAL	0.5 rem/yr (23) 125 mrem/qtr neutrons. (24)	1/10 of occupational limits with dose rates < 2 mrem/hr and 100 mrem/7 days.	125 mrem/qtr with dose rates < 2 mrem/hr and 100 mrem/7 days (26)	0.5 rem/yr - children 1.5 rem/yr - adults (19) Skin - 3 rem/yr adults (19)	1/10 occupational (27)	1/10 of occupational - advised. (21) 3 rads including 1.5 rads/yr (eyes or whole body gamma) proposed. (15)	None stated.
OCCUPATIONAL INTERNAL (Persons in vicinity)	1/10 HB 69 body burden 1/10 HB 69 MPC for 168 hr week. (23),(28)	Body burden not stated. 1/10 HB 52 MPC for 168 hr week mostly used. (25)	Body burden not stated. 1/10 HB 69 MPC for 168 hr week mostly used. (26)	Body burden not stated. 1/10 ICRP MPC for 168 hr week. Thyroid 3 rem/yr for adults.(19) (29),(36)	1/10 occupational. (27)	1/10 occupational - advised. (21)	None stated.

RADIATION PROTECTION STANDARD

	NCRP	10 CFR 20 Present	10 CFR 20 Proposed	ICRP	NY STATE #38	BRITISH REGULATIONS	FRC(8)
OCCUPATION-NON-MIXED (Persons in vicinity)	None stated. Infer summation.	None stated.	None stated.	0.5 rem/yr children. 1.5 rem/yr adults. (19)	1/10 occupational. (27)	1/10 occupational. Summation advised. (21)	None stated.
GENERAL POPULATION	Most exposed population not double background temporarily accept ICRP. (30)	None stated.	None stated.	5 rem genetic additional to background and medical over 30 year interval. (45)	None stated.	None stated.	0.5 rem/yr individually measured. 0.17 rem/yr group average calculated. 5 rem/30 yr gonad.
MEDICAL EXPOSURE	Need not be included in the exposure status of person. (17)	No limit.	Medical exposure excluded. (31)	No limit for medical diagnosis or medical therapy. (32)	Medical exposure excluded. (33)	Occupational medical exposures excluded - proposed. (34)	Not apply to purposeful exposure from the healing arts.
EMERGENCY	25 rem/once in a lifetime whole body. (11)	None stated but subsequent exposure reduction required at 1/10 rate. (15)	None stated.	25 rem/once in a lifetime - accident. 12 rem/emergency with subsequent reduction to meet age proration formula within 5 years. (13) (20)	None stated.	Occasionally up to 10X weekly limit; skin - 100 rem/decade; blood-forming - 50 rem/decade; 50 rad/age 30 to gonad and 200 rad/lifetime advised. (21)	None stated.

ATOMIC ENERGY LAW JOURNAL

	NCRP	10 CFR 20 Present	10 CFR 20 Proposed	ICRP	NY STATE #38	BRITISH REGULATIONS	FRC(8)
EXEMPTIONS	New standards not retroactive. (37)	Sources not licensed by AEC. (11)	Unchanged from present.	None stated.	Luminous dials. Equivalent natural K^{40} < 16 Kev electron TV. During transport under ICC. (38)	Sealed source < 10 mrad/hr. (38) < 5 Kev electron (39) .002 μc/gram. (38)	Natural background radiation.
CONTROLLED AREA	Defined area under access, occupancy, and working condition control. (40)	Dose rate > 5 mrem/hr and 150 mrem/7 days. (41)	Dose rate > 5 mrem/hr and 100 mrem/7 days. (42)	Where occupational exposure could exceed 1.5 rem/yr.	5 mrem/hr; 150 mrem/7 day. (16)	30 mrad/40 hr week (39)	None stated.

RADIATION PROTECTION STANDARD

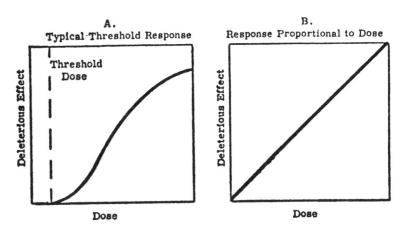

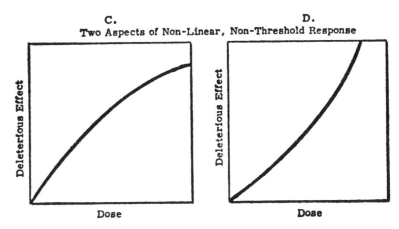

Figure 1 - Illustrative Types of Response Curves

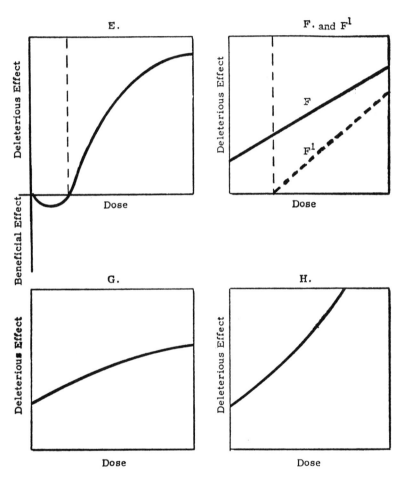

Figure 2 - Other Plausible Types of Response Curves

ATOMIC ENERGY LAW JOURNAL

REFERENCES FOR SUMMARY TABLE OF RADIATION PROTECTION STANDARDS

Although this condensed tabulation is helpful in comparing radiation protection standards from different sources, subtle shades of meaning have necessarily been omitted. The reader should refer to the defining documents listed below.

A. National Committee on Radiation Protection and Measurements (NCRP) (Published by U. S. Government Printing Office, Washington, D. C.)

— Addendum to National Bureau of Standards Handbook 59 (Permissible Dose from External Sources of Ionizing Radiation)—Maximum Permissible Radiation Exposures to Man (April 15, 1958)

— National Bureau of Standards Handbook 63 — Protection Against Neutron Radiation up to 30 Million Electron Volts (November 1957)

— National Bureau of Standards Handbook 69 — Maximum Permissible Body Burdens and Maximum Permissible Concentrations of Radionuclides in Air and in Water for Occupational Exposure (June 1959)

— Report of the Ad Hoc Committee — Somatic Radiation Dose for the General Population — appearing in the February 19, 1960 issue of SCIENCE, Volume 131 (May 6, 1959)

— Executive Committee of NCRP Report — Appearing in the July 1960 issue of Radiology, Volume 75, page 122.

B. Code of Federal Regulations — Title 10 — Atomic Energy, Part 20 — 10 CFR 20 (Published by U. S. Government Printing Office, Washington, D. C.)

— Standards for Protection Against Radiation (Original dated January 1957 — latest proposal dated May 1959)

C. International Commission on Radiological Protection (ICRP) (Published by Pergamon Press, London, England)

— Recommendations of the International Commission on Radiological Protection, Committee I — Permissible Dose for External Radiation, September 9, 1958.

— Addendum to 1958 Recommendations of the International Commission on Radiological Protection — Report on Decisions at the 1959 Meeting of the International Commission on Radiological Protection (ICRP)

— Recommendations of the International Commission on Radiological Protection, Committee II — Permissible Dose for Internal Radiation (1959)

D. New York State Department of Labor, Board of Standards and Appeals (Published by NY State — 80 Centre St., New York 13, New York)

— The Industrial Code Rule No. 38 — Radiation Protection (December 1955)

E. British Regulations (Published by Her Majesty's Stationery Office, London, England)

— Ministry of Labour Factories Acts, 1937 and 1948 — Factories (Ionising Radiations) Special Regulations (1960)

— Ministry of Labour and National Service — Precautions in the Use of Ionising Radiations in Industry (1957)

F. Federal Radiation Council (Published by the U. S. Government Printing Office, Washington, D. C.)

RADIATION PROTECTION STANDARD

— United States Federal Register, F.R. Document 60-4539 — Radiation Protection Guidance for Federal Agencies, Memorandum for the President (May 13, 1960)
— Staff Report #1, Federal Radiation Council (May 13, 1960)
(Appendix A was prepared in its entirety by J. W. Vanderbeek of this laboratory)

FOOTNOTES FOR
SUMMARY TABLE OF RADIATION PROTECTION STANDARDS

Number	Source
1	HB 59 Addendum
2	10 CFR 20
3	ICRP Recommendations, Sept. 1958
4	New York Code #38
5	British Factories Regulations (1960)
6	British Precautions (1957)
7	British Precautions (1957)
8	FRC Memo
9	HB 69
10	HB 59 Addendum
11	10 CFR 20
12	10 CFR 20
13	ICRP Recommendations, Sept. 1958
14	New York Code #38
15	British Factories Regulations (1960)
16	British Factories Regulations (1960)
17	HB 59 Addendum
18	10 CFR 20
19	ICRP Recommendations, Sept. 1958
20	ICRP Addendum, Sept. 1958
21	British Precautions (1957)
22	10 CFR 20
23	HB 59 Addendum
24	HB 63
25	10 CFR 20
26	10 CFR 20
27	New York Code #38
28	HB 69
29	ICRP Recommendations, Sept. 1958
30	Ad Hoc Committee Report (1959)
31	10 CFR 20
32	ICRP Addendum, Sept. 1958
33	New York Code #38

ATOMIC ENERGY LAW JOURNAL

34	British Factories Regulations (1960)
35	10 CFR 20
36	ICRP Recommendations — Committee II
37	HB 59 Addendum
38	British Factories Regulations (1960)
39	British Factories Regulations (1960)
40	HB 59 Addendum
41	10 CFR 20
42	10 CFR 20
43	ICRP Recommendations, Sept. 1958
44	ICRP Addendum to Sept. 1958
45	ICRP Recommendations, Sept. 1958
46	NCRP Executive Committee

Radiation Exposure Experience in a Major Atomic Energy Facility

United Nations International Conference on the Peaceful Uses of Atomic Energy

Presented August 1955
Published 1956

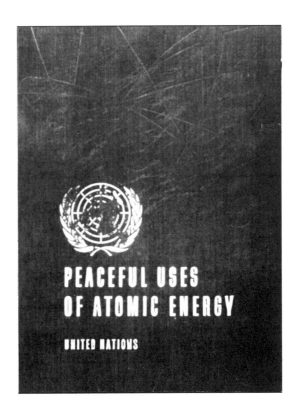

Radiation Exposure Experience in a Major Atomic Energy Facility

By H. M. Parker,* USA

The Hanford Works is one of the major facilities of the US Atomic Energy Commission. It includes a number of high-power nuclear reactors, chemical separation plants, plutonium purification facilities, associated control laboratories and major research and development laboratories. Collectively, the facilities present a good cross section of the radiation hazards that may be expected in peaceful applications of atomic energy. The magnitude of the radiation hazard can be described as the equivalent of handling some millions or tens of millions times the world's available supply of radium.

These facilities have been operated by major industrial organizations under contract to the AEC; specifically, the operator was the DuPont Company from 1944 to 1946, and the General Electric Company from 1946 to date. The radiation hazards have been managed to similar high standards as are reflected in the conventional industrial safety records of such organizations. The exposure experience should thus be broadly similar to that which may be anticipated under private industrial management.

This paper summarizes the experience in exposure of personnel in in-plant operations.

EXTERNAL EXPOSURE

Exposure to external radiation is measured by methods that are now quite conventional. The X-ray and gamma-ray components are primarily determined by pocket ionization chambers, which are worn in pairs to reduce the incidence of spurious readings due to chamber leakage. It can be shown that the lower of two readings is a valid index, unless an outright recording error is made; high accuracy is confirmed by introducing many chambers of known exposure into the system. The daily exposures so obtained are supplemented by film-badge readings integrated over two-week intervals.

The same film badges record beta-ray exposures through an unshielded aperture over the film. The interpretation of beta-ray data depends on the geometry and extent of the sources. The calibration system is adjusted to give readings equal to or higher than the actual exposure. The separation of beta and gamma exposures in the presence of soft gamma radiation is difficult. For selected operations, such as those involving routine exposure to the 17 kev X-radiation of plutonium, special badges are used.

Some operations, such as laboratory manipulations, involve significantly higher exposure risk to the hands than to the rest of the body. Measurements in these cases depend on film rings, which are essentially miniature film badges separating beta and gamma exposures. In the local experience, data of this type are influenced by such defects as pressure sensitivity of film, chemical effects, and light leaks. The recorded results undoubtedly contain many spurious high readings.

Neutron exposures are measured by means of nuclear track emulsions for fast neutrons, and by boron-lined chambers for slow neutrons. Neither system has the reliability of the gamma-ray measurements. The potential hazard for neutron exposure is low, and the incidence of significant readings has been negligible.

The average annual exposure to gamma radiation for all persons entering radiation areas is given in Table I. It is probable that the entry for 1944–45 is too high, due to initial difficulties with personnel meters and a tendency to record high if in doubt. Also, the average exposure can be held low by a conservative policy of requiring the wearing of meters in areas of low-exposure potential hazard. The data reported assume that 20% of the work in which it is forced to wear, personnel meters, has no real risk of exposure. Despite these defects, the table does show

Table I. Average Annual Exposure—Gamma Radiation

Year	Annual exposure roentgens	Percentage exceeding 1 r	Percentage exceeding 3 r
1944–45	0.9	0.2	0
1946	0.6	0.2	0
1947	0.4	0.3	0.02
1948	0.3	0.2	0
1949	0.2	0.08	0
1950	0.1	0.06	0
1951	0.1	0.4	0
1952	0.2	3	0.02
1953	0.2	5	0.07
1954	0.2	5	0.2

---

* Hanford Atomic Products Operation, General Electric Company, Richland, Washington. Prepared by H. M. Parker; including work by D. P. Ebright, W. A. McAdams, H. A. Meloeny, and other members of the radiological records and standards section, General Electric Company.

the influence of planned control of exposure. This has consisted of an attempt in the early years to reduce exposure as far as possible, and since 1950 to operate to a limit of 3 r per year. The sharp increase in the percentage of exposures above 1 r is a good index of planned relaxation. That the average annual exposure has not risen much, means that the exposure risk of many personnel groups is low.

Table II shows in more detail the number of cases of doses in excess of one roentgen. This illustrates more clearly the effects of the planned relaxation to a 3 r annual target.

Another useful index is the annual record of the ten highest individual annual exposures given in Table III for beta radiation and gamma radiation separately. The combined highest total does not necessarily represent the same cases as given separately.

Table III refers to so-called whole-body exposure. It is known that exposures of hand and forearm may considerably exceed these levels, but the data are not available in form for tabulation.

Table IV summarizes the experience in selected years with neutron exposures.

Table II. Annual Whole-Body Gamma Exposures above 1 r

Year	Number of cases*				
	>1 r	>2 r	>3 r	>4 r	>5 r
1944	0	0	0	0	0
1945	8	1	0	0	0
1946	8	2	0	0	0
1947	13	2	1	1	1 (6.1 r)
1948	10	2	0	0	0
1949	4	0	0	0	0
1950	3	0	0	0	0
1951	23	0	0	0	0
1952	179	22	1	0	0
1953	323	42	4	0	0
1954	372	68	16	3	1 (14.4 r)

* Number of cases in any column for a given year includes all the cases in the columns to the right for that year. For example, the 42 cases >2 r in 1953 include the 4 cases which were >3 r. Figures in parenthesis give the highest individual total in the years in which this exceeded 5 r.

Table III. Yearly Averages of Ten Highest Accumulated Exposures

Year	Beta radiation rads	Gamma radiation r	Beta plus gamma rads
1944*	0.9	0.06	0.9
1945	4.2	1.4	5.1
1946	7.5	1.4	7.9
1947	8.2	1.9	8.7
1948	9.2	1.4	9.7
1949	7.5	1.0	8.1
1950	5.5	1.0	5.8
1951	6.5	1.4	7.2
1952	6.9	2.6	7.5
1953	4.3	3.1	5.5
1954	4.7	4.8	6.6

* 1944 was an incomplete year.

Table IV. Neutron Exposures

	1952	1953	1954
Fast neutrons:			
Number of readings	5034	6418	6770
Number above 50 mrem	0	0	8
Slow neutrons:			
Number of readings	5971	7101	7763
Number above 50 mrem	0	0	3

Table V. Relative Hazard of Typical Operations

Type of work	Per cent above 1 r	Per cent above 3 r
Engineering, research	1.7	0.2
Reactor operations	38.5	0.8
Reactor monitoring	36.9	11.9
Reactor maintenance	10.9	0
Separations operations	4.8	0
Separations monitoring	8.7	0
Separations maintenance	2.0	0
Radiation instrument calibration	58.2	0

By comparison, approximately 3-million gamma-exposure readings are made per year. The highest recorded neutron exposures in one week have been 220 mrem for fast neutrons, and 176 mrem for slow neutrons.

In reviewing the general picture of external exposure, it is useful to know the relative contributions of different radiation types, and the relative exposure potential hazard of different tasks. For the former, beta and gamma radiations predominate. Near an operating reactor, the transmission through the shield gives approximately equal hazard for gamma radiation and neutron radiation. Both these exposures should be low. The actual exposures come in operation when the reactor is shut down. Exceptions are experimental work on reactor test holes and the operation of particle accelerators. In separations work there is a theoretical neutron hazard from (γ, n) reactions, but these are essentially eliminated by shielding for gamma radiation.

Table V gives some indication of the relative hazard for different operations, with gamma radiation as the index. (Per cent annual exposure above 1 or 3 r).

RADIOACTIVE CONTAMINATION

Experience has shown that the control of contamination is emphatically more troublesome than the control of external radiation. The performance is much less amenable to systematic presentation; only a general discussion is given here. It is compulsory to record the contamination of hands and feet on leaving a radiation area. The frequency of such contamination is shown in Table VI.

In this period, the annual total of cases requiring skin decontamination has ranged between 900 and 1700. The striking reduction in incidence of high counts in Table VI means then that the more recent desirable practice of making frequent contamination

Table VI. Hand and Shoe Contamination in Typical Years

Year	Alpha radiation		Beta-gamma radiation	
	Number of readings	Per cent of high readings	Number of readings	Per cent of high readings
1947	332,303	0.16	410,529	0.24
1948	429,163	0.29	562,166	0.23
1950	487,897	0.08	507,813	0.09
1954	563,772	0.006	720,358	0.006

checks at the work site has detected most cases before they come to the recording contamination counters. A quick pass over the body with a suitable meter additionally locates small contamination spots other than on hands or feet.

Of major importance to the future atomic energy program is the accumulation of radioelements of long half-life in the body. Such deposits must be detected by bio-assay procedures and there is considerable uncertainty in interpretation especially when the intake may have been progressive over several years. Table VII shows the experience with plutonium.

Significant features of the table are:

1. The high percentage of cases from inhalation, and the insignificance of ingestion or injection (puncture wounds).

2. Forty-five per cent of the cases not associated with known incidents, which may represent slow accumulation under apparently normal working conditions. That the number of such cases has increased sharply in the last two years supports this viewpoint, and emphasizes the need for the most sensitive bioassay procedures.

Similar data for fission products appear in Table VIII.

In both Table VII and VIII, MPL means the accepted maximum permissible limit (NBS Handbook 52) and "Partial interpretation" means that the urinary elimination has not yet been followed long enough to characterize the body content. In Table VIII, F.P. stands for a general fission product mixture.

In addition, there is a significant incidence of uranium intake in operations involving uranium dust or fume, despite generally high standards in industrial hygiene. This element is rapidly eliminated. Some specialized work with radioisotopes other than fission products has also led to demonstrable depositions. Collectively, these have led to about 3 technical overexposures per year (i.e., deposition *temporarily* exceeded the permanently permissible body burden). It is noteworthy that for no radioelement of long persistence in the body has the deposit exceeded the permissible limit.

RADIATION INCIDENTS

Untoward incidents with radiation or contamination are recorded in 4 degrees of severity.

1. Serious overexposures—those in which the tolerance status of an individual is affected (as used in NBS Handbook 59). None has occurred; one hand exposure closely approached the limit of 125 r.

2. Technical overexposures—those in which exposures exceeding any recommendation of Handbook 59 has occurred. These have averaged about 4 per year, with an additional 3 per year if the same concept is applied to contamination and internal deposition.

3. Potential overexposures—cases in which technical overexposure could have occurred under similar circumstances. These, averaging 11 or 12 per year for radiation and about 24 per year for contamination, are used for radiation safety training.

Table VIII. Confirmed Fission Product Deposition Cases in 10 Years

Per cent of MPL	Detected by routine samples	Detected after known incident	Radio-element	Route of intake
<0.5	1	40	24 Ru	Inhalation
			3 Cs	Inhalation
			14 F.P.	13 Inhalation
				1 Absorption
0.5 to 1.0	0	3	3 Ru	Inhalation
1 to 10	0	1	Sr	Inhalation
10 to 20	0	0		
20 to 50	0	1	Sr	Inhalation
50 to 100	0	0		
>100	0	0		
Partial interpretation	8	3	F.P.	Uncertain

Table VII. Confirmed Plutonium Deposition Cases in 10 Years

Per cent of MPL	First detection by routine samples	First detection after known incident	Route of intake on known incident cases				
			Inhalation	Ingestion	Absorption	Injection	Unknown
<0.5	4	29	25	1	3		
0.5 to 1.0	7	4	2		2		
1.0 to 10	13	13	8		4	1	
10 to 20	4	6	5			1	
20 to 50	0	1	1				
50 to 100	2	0					
>100	0	0					
Partial interpretation	15	2					
Total	45	55	41	1	9	2	2

EXPOSURE EXPERIENCE IN MAJOR FACILITY

Table IX. Radiation Incidents

Year	Radiation	Contamination	Miscellaneous	Total
Potential Overexposures				
1944	0	0	0	0
1945	3	32	0	35
1946	15	24	0	39
1947	10	17	0	27
1948	16	22	0	38
1949	9	27	0	36
1950	5	15	0	20
1951	8	17	0	25
1952	18	46	7	71
1953	23	38	8	69
1954	30	40	6	76
Total	137	278	21	436
Technical Overexposures				
1944	3	0	0	3
1945	5	1	0	6
1946	1	3	0	4
1947	2	0	0	2
1948	2	0	0	2
1949	0	0	0	0
1950	2	3	0	5
1951	9	4	0	13
1952	7	4	1	12
1953	12	13	1	26
1954	9	11	0	20
Total	52	39	2	93

Table X. Distribution of Radiation Incidents

Type of work	Per cent incidence
Fuel element preparation	2
Engineering, research	20
Separations operations	45
Reactor operations	20
All other	13
Reactor operations detail:	
Functioning reactor	4
Reactor shutdown	11
Auxiliary tasks	5

4. Informal incidents — minor deviations from sound control, averaging 140 cases per year, used for training and system improvement.

The incident experience is detailed in Table IX.

A significant feature of the tabulation is the preponderance of contamination incidents in the potential exposure class. If the data are translated to exposed individuals instead of incident numbers, this is even more striking. Each radiation case averages two people and each contamination case averages three.

In the technical overexposure class, radiation cases outweigh contamination cases, and this holds too for numbers of individuals; both types average two individuals per incident.

In the informal incident class, minor contamination cases predominate, with the balance made up of minor infractions of control procedures.

The relative frequency of radiation incidents in different classes of work is shown in Table X.

Of special interest to peaceful applications is the low incidence of exposure with an operating reactor. This suggests that power-producing reactors should be relatively free from radiation hazard. By contrast, separations plants will probably continue to be plagued by contamination hazards that can be held in bounds only by continuous and detailed attention to decontamination, ventilation and other controls.

In summary, the reported experience establishes that radiation hazards can be adequately controlled by methods that are economically reasonable. Evaluation of the total cost of radiation protection is a controversial field, because in addition to the obvious direct costs, one may wish to include factors for slower performance with special handling tools, time spent in planning radiation tasks, etc. Local studies on costs strongly indicate that the additional cost of operating to high standards is a relatively small increment on the necessary cost for marginal protection. This fortunately removes some of the risk of jeopardizing the future of the atomic energy industry by over-zealous competitive reduction of hazard controls. A sensible approach to the future is to accept radiation as a necessary evil of the business, and to seek engineering advances that will achieve the present or improved standards of exposure at lower cost.

Radiation Exposure from Environmental Hazards

United Nations International Conference on the Peaceful Uses of Atomic Energy

Presented August 1955
Published 1956

Radiation Exposure from Environmental Hazards

By H. M. Parker,* USA

Both nuclear reactors and radiochemical processing plants are capable of generating significant radiation hazards beyond the confines of a plant of normal size.

The expected hazards can be classified as follows:

1. HETEROGENEOUS REACTORS—AIR COOLED

The governing hazards arise from dispersion of the reactor coolant gas into the atmosphere. The predominant contaminant is A^{41}. Due to the short half-life (1.82 hours) and chemically inert behaviour of the gas, there are no appreciable complications in evaluating the hazard in terms of direct radiation from the effluent plume. Other contaminants, which may introduce more insidious hazards, are radioactive particles generated in one of three ways:

1. Direct activation of dust particles drawn through the reactor.
2. Blowing out of reactor structural materials or corrosion debris; because of the long residence of such debris in the reactor, the content of long-lived activation products may be high.
3. Escape of material from ruptured fuel elements; this introduces particles of mixed fission products and plutonium.

2. HETEROGENEOUS REACTORS—SINGLE PASS WATER COOLED

The limiting hazard develops from the release of coolant water to a river or lake. The principal contaminants are activation products, which may occur from direct throughput, or be augmented in the more dangerous long-lived components by temporary hold-up on corrosion films in the unit.

Fission products may also arise from four sources:
1. Irradiation of natural uranium in the water.
2. Uranium dust impregnated in fuel element surfaces.
3. Transmission of fission products through a fuel element skin.
4. Rupture of fuel elements.

In typical practice, some 80 radionuclides can be identified in reactor effluent, of which about 16 have separate significance in hazard calculation.

The limiting hazards for the Hanford reactors are P^{32} concentrated through biological chains, and gut-irradiation, and bone deposition from drinking water.

In all cases, an extensive radio-ecological program is necessary to validate effluent release practices.

In addition, such reactors may cause atmospheric pollution by leakage of whatever atmosphere is maintained in the reactor, and by irradiation of gas in test hole facilities. Although the escaping gases may be found to contain some unexpected components, this hazard is subordinate.

3. HETEROGENEOUS REACTORS—LIQUID RECIRCULATING SYSTEMS

In this case, the environmental hazard is normally low, being restricted to low volumes bled off for freshening the coolant, or arising from an occasional system leak.

4. HOMOGENEOUS REACTORS

From the environmental hazard viewpoint, such reactors can be treated as radiochemical processing plants.

Gross atmospheric contamination may conceivably arise from a catastrophic incident in any reactor; this phase is not considered here.

5. CHEMICAL PROCESSING PLANTS

The operation may be schematically broken into four steps, not all of which will be used in a specific plant.

Step 1. Dissolution of Reactor Fuel Elements

This step leads to periodic release of the rare gas components of the fission mixture, with Xe^{133} as the governing member, and of some volatile fission products, predominantly the radioiodines.

The controlling hazard is a function of the fuel element cooling time. Under normal operating conditions, I^{131} is the significant contaminant.

Step 2. Removal of a Specific Isotope

Either for their commercial value or because they create an in-process nuisance, one or more specific isotopes may be removed separately. Particularly if such isotopes have volatile compound forms it may be difficult to retain all the material in the system, and the release of highly active spray, evaporating to particles, is probable.

Step 3. General Chemical Separation

Regardless of the details of the chemical separation processes utilized, it has so far proved impossible to maintain all the process materials in the system.

---

* Hanford Atomic Products Operation, General Electric Company, Richland, Washington. Including work by the staff of the Biophysics Section, Radiological Sciences Department, General Electric Company.

Characteristically, fine mists or sprays of mixed fission products are released. Upon evaporation of the liquid content they form small radioactive particles of high intrinsic activity. Some of these escape through filters to the ventilation stack.

Step 4. Purification Stages

After removal of the bulk of the fission products, the product material, for example, plutonium, has to be further purified. The vented air will contain particles rich in this product material.

In Steps 2, 3 and 4 there are potential environmental hazards from the disposal of liquid radioactive wastes. This phase will not be discussed here.

It may be noted that, schematically, Step 2 of this general system is similar to the operations in an isotope factory designed for commercial recovery of a specific isotope.

PROBABLE TRENDS IN PEACEFUL APPLICATIONS

It is plausible to assume that there will be a substantial field for heterogeneous power reactors together with large central processing plants, receiving fuel elements from many sources.

Such systems place the burden of environmental hazards on the processing plants.

Past experience in this field is therefore of major interest in peaceful applications.

ATMOSPHERIC POLLUTION AROUND SEPARATIONS PLANTS

Of the available battery of contaminants, those of major concern have proved to be I^{131} in Step 1 and particles emitted in Steps 2 and 3.

The iodine problem can be wholly solved by increased cooling time. This is economically unattractive. In practice, I^{131} is removed by absorption processes. Residual hazard arises from the small fraction normally escaping, and occasionally from malfunctioning of the absorption equipment.

The primary particles escaping from the process have mass median diameter on the order of 0.3 to 0.5 micron. The maximum size is about 3 microns. The emission of 10^8 to 10^9 particles per day with activity on the order of 10^{-3} μc per particle is to be expected.

Secondary particles develop in venting systems beyond the filters by attachment to inert substances, which later flake off and escape. Such particles have diameters up to several hundred microns (or conceivably up to several centimeters) with activities up to hundreds of microcuries. These present distinctive hazards.

Potential Exposure Mechanisms

The principal exposure forms are:

1. Direct Irradiation of Persons or Animals from Highly Radioactive Ground Sources

This situation may develop from depositions of I^{131}, primary, or secondary particles. It is associated with isolated single releases of unusual amount. The intensity is greatest for emission of secondary particles, where the phenomenon is necessarily restricted to within a few kilometers of the facility.

2. Adherence of Isolated Particles to the Skin or Clothing

The significant hazard comes from the secondary particles, which in the intermediate sizes may travel up to about 15 km. Such particles on the ground can become airborne again, as in dust storms, so that isolated particles theoretically capable of producing a skin reaction may be found as far as 30 to 50 km from the source. The probability of undisturbed skin contact for the required time (of the order of days) is so vanishingly small, that this is not a pertinent hazard in practice.

Within the 15 km radius, control by radiation monitoring is adequate; there has been no experience which shows injury from contact during a normal work day.

3. Ingestion of Large Isolated Particles

Close to the process stacks, ingestion of a large particle and its retention in a convolution of the gastro-intestinal tract for days is feasible. However, normal personal hygiene makes this hazard improbable; it has not occurred in the local experience.

4. Inhalation of Small Particles

The permissible inhalation of small particles of soluble materials can be deduced from the limits given in such references as NBS Handbook 52. In these terms, the practical inhalation hazard is negligible. With the quoted emission, the average concentration at 15 km would be about 10^{-12} μc/cm^3.

For insoluble particles, there is the residual doubt as to whether a single focus, localized in the lung, may ultimately lead to malignancy. The balance of the evidence makes it improbable that this hazard has significance, at the feasible activity levels.

5. Consumption of Contaminated Vegetation

Although this is a hazard potentially originating from either I^{131} or particles, it is, in fact, limited to I^{131}. In ten years of operation the build-up of small particles on vegetation has not produced levels of significance. Areas in which large particles may fall must necessarily be withdrawn from public or grazing use for other reasons.

The hazard from deposited I^{131}, on the other hand, has required intensive control and research study. From such studies on sheep, it is determined that the permissible vegetation contamination is about 10^{-5} μc I^{131} per gram. Under equilibrium conditions, this would represent an atmospheric contamination of about 3×10^{-13} μc/cm^3. This is to be compared with the human inhalation limit of 3×10^{-10} μc/cm^3 appropriate for large populations. These values clearly define I^{131} as a ground contaminant rather than a direct air pollution hazard.

As applied to man, the secondary intake is a function of his eating pattern. Two significant routes of

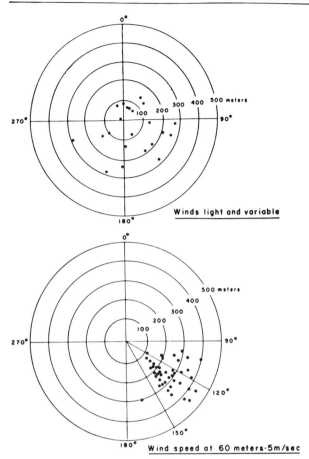

Figure 1. Distribution of puffs hitting the ground during unstable conditions

entry are consumption of fresh garden produce, and drinking of milk from cows on contaminated pasture. The limit of 10^{-5} μc per gram can be broadly applied to all cases.

6. Other Mechanisms

Once a radioactive contaminant has been released to the atmosphere it is necessary to trace its course through a wide variety of natural processes. Typical ones are direct fall into or leaching into a public water supply, uptake by soil and later by plants.

With the exception of the I^{131} case in plants, none of these processes has been found to be critical.

Sample Distribution Patterns

The actual distribution pattern of pollutants around a processing plant depends on the time-distribution of effluents, the elevation of the release point, meteorological conditions, and topography of the environs. Although these will require individual study at each site, considerable guidance can be obtained from the local patterns. The Hanford site is a good basic model because the terrain is relatively flat and barren. In site selection, if study does not assure suitability, this can be tested by controlled emissions of smoke, or better, of fluorescent particles from a high tower. An adequate picture cannot be obtained in less than one year, because there are marked seasonal variations.

Of the controlling factors listed, the stack eleva-tion is significant only close to the plant where the area will in any case be controlled. Locally, the maximum ground concentration varies approximately as the inverse square of the stack height. For conventional stacks of 60 to 100 meters, the effect of height is inconsequential beyond 5 km.

Time distribution of effluents is significant chiefly in a discontinuous process that can be interrupted at times of unfavorable atmospheric dilution. The meteorological situations can be crudely segregated into three kinds, which are functions of the vertical lapse-rate and of the wind velocity. These are:

1. *Looping*, characterized by marked vertical instability, and usually fluctuating wind direction. The stack emission loops to ground at random points as shown in Fig. 1.† Instantaneous concentration is high, but the integrated exposure comparable with that under other conditions. As a guide, it appears that a steady emission of 1 curie per day from a 60 m stack is unlikely to give (a) short-period concentration above 10^{-8} μc/cm³, (b) short-period dose above 10^{-7} μc-sec/cm³, (c) hourly average concentration above 10^{-10} μc/cm³. The maximum dose may occur in a zone 2 to 5 stack heights from the source.

2. *Coning*, the elementary picture in which the effluent forms a cone of semi-angle 5° to 7°. The point of maximum ground concentration is 8 to 12 stack heights down-wind. It implies nearly adiabatic lapse rate and strong steady wind. More realistically, the ground concentration can be estimated from the well known Sutton equations.[2]

3. *Fanning*, which occurs during temperature inversion, and usually with low wind speed. This process leads to negligible hazard close to the stack, with maximum ground concentration as much as 40 km away. It is this process that leads to generalized contamination under Hanford conditions. In local experience the standard diffusion equations, when extended to distances in excess of 20 km, yield concentrations that are not reliable to better than a factor of 5. In general, the equations underestimate the concentrations under strong inversion conditions.

Experience over ten years at the Hanford Works suggests that the environmental hazards setting appropriate protective radii are the deposition of I^{131} on vegetation, and the deposition of secondary particles on the ground. In submitting actual patterns, an existing picture will represent a partial history of past emissions. For I^{131} ($t_{\frac{1}{2}} = 8$ days), the historical period is of the order of weeks; for secondary particles, with varied half-lives from months to years, the period is long but indefinite. Existing particles may be washed into the ground, buried in dust storms,

† Figure 1 shows, in the upper figure, the number of ground contacts within 300 meters in 45 minutes in a light variable wind (∼2.2 m/s). Contacts occurred in all four quadrants. The lower figure gives the contacts within 500 meters in 51 minutes with a wind speed of 5 m/s. All the puffs fell within one quadrant; 95% of them were within a 40 degree sector, 70% within a 20 degree sector. These data, and those from which Figure 6 was developed appear in Ref. 1.

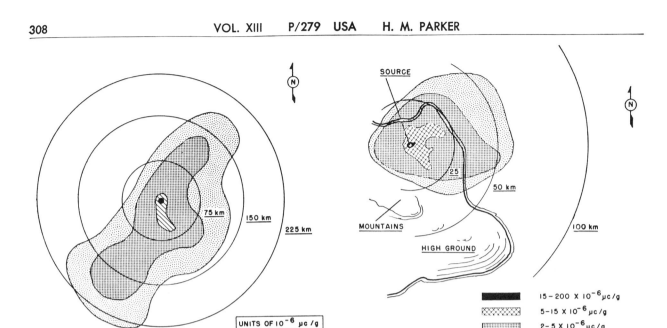

Figure 2. I^{131} deposition of I^{131} from single emission of 100 curies

Figure 3. I^{131} deposition—normal spring pattern

or transported by wind. This prevents subtraction of successive pictures to obtain short-term increment.

In the following diagrams, data for single major releases were adjusted to a 100 curie emission. Routine emission of I^{131} was adjusted to 1 curie per day. For other particle maps, the status given is approximately the worst found in 10 years.

Figure 2 relates to a single emission of 100 curies of I^{131} in a few hours, during which time inversion conditions existed near the ground with good dilution above. Effluent traveled north-east in the first half, and south-west in the second half. Wind speed was 2–3 meters per sec with low values near the time of reversal.

With a deeper inversion layer, ground concentrations 15 km and more from the stack could easily have reached 10 times the quoted values.

Figure 3 shows a typical spring condition of vegetation contamination for daily emissions of 1 curie I^{131}.

Figure 4 gives the equivalent data for the same season four years later. Figures 3 and 4 serve to measure the variability under broadly similar meteorological conditions.

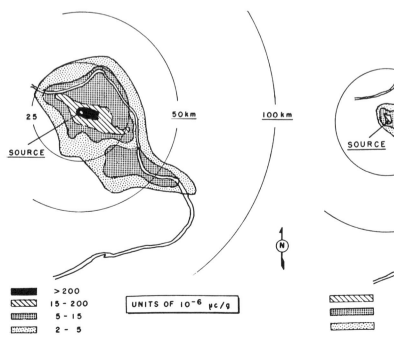

Figure 4. I^{131} deposition—spring pattern another year

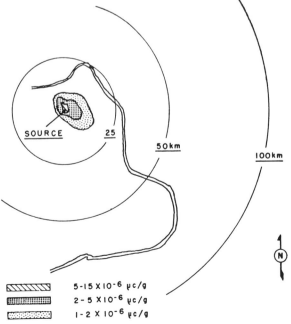

Figure 5. I^{131} deposition—normal summer pattern

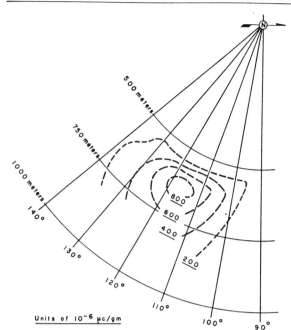

Figure 6. Local isopleths I^{131} deposition in summer

Figure 5 is a typical summer condition. High local concentration due to looping does not show on this scale. Otherwise, the remote environmental hazard is much reduced.

For comparison, isopleths of the local summer condition appear in Figure 6. These data are computed from smoke tests on a 56 m stack, assuming wind speed of 10 m/sec, constant direction, constant lapse rate of $-0.01°C/m$, and I^{131} deposition constant 2.8 cm/sec.‡

Figure 7 gives a condition resulting from the persistent inversions of the winter season. The large area of significant deposition southeast of the stack and 80 km away is typical of quiet fluid motion down a river valley and contact with vegetation on higher ground in and around a river gorge. With I^{131}, temporary depositions above the permanent limit of 10^{-5} µc/gm are tolerable, and the location of affected spots tends to change monthly. However, at 10 times the reference emission, such areas would be large and could overlap at distances up to 100 km. Since a plant boundary of this extent would be absurd, there has to be an upper limit for I^{131} emission not greatly different from the reference value.

Figure 8 gives the pattern for a single emission of primary particles, scaled to 100 curies. The unusual narrow band is real up to some 50 km. Beyond this, the conventional width to 0.1 central concentration could not be determined in the actual case, because of lack of instrument sensitivity. The actually detected width is reported. Detailed analysis of this case would provide the best practical test of Sutton-type cloud width calculations up to 50 km travel. Figure 9 shows the worst condition in ten years

‡ A deposition constant of 2.8 cm/sec corresponds with the value 10^8 µc per meter^2/hr/µc/cm^3 in the atmosphere used at Hanford since 1948. A. C. Chamberlain and R. C. Chadwick report a deposition constant of 2.5 cm/sec.^{3}

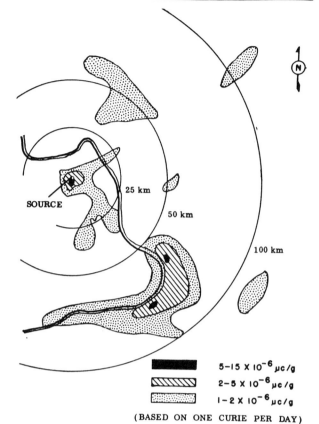

Figure 7. I^{131} deposition—remote contamination by valley drainage —winter condition

for secondary particles in the intermediate size range of about 3–100 µ.

Figure 10 is the similar condition for large secondary particles (>100 µ), which could produce damage on skin contact. Neither Fig. 9 nor 10 can be reliably scaled to a reference emission rate, because

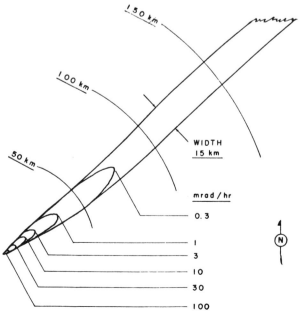

Figure 8. Ground contamination from narrow emission band of radioactive particles (scaled to 100 curies)

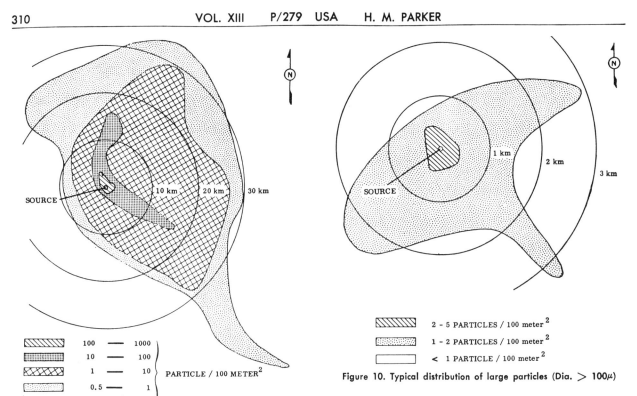

Figure 9. Typical distribution of particles in size range 3–100μ

Figure 10. Typical distribution of large particles (Dia. > 100μ)

the pictures are an integration of effects over too long a time, and suitable isokinetic sampling near the stack mouth was not always available.

In reviewing the Hanford experience, one may conclude that a comfortable boundary around a high level processing plant would have a radius on the order of 12–15 km. In areas with a distinctive prevailing wind one could make better utilization of the same area by reserving a shape modeled after the wind rose. At Hanford, for example, the required control area to the west is substantially less than it is to the east.

Reduction of the controlled area by a factor of 2 in linear dimensions would not introduce significant hazard to man; uptake by animals could be borderline and some restrictions would have to be applied from time to time.

With further reduction, the limitations on grazing animals, the potential risk of secondary particle contamination, the loss of morale in an area requiring intensive monitoring, and the potential for damage claims would make the operation unattractive.

As far as real hazard to man, with food growing and animal grazing excluded, a smaller reserved area would introduce no major hazard, provided that intensive monitoring could be practised.

In our opinion, these plants should be operated in remote areas, where land is cheap. With an ample reserved area, the public could be assured of radiation safety, and the healthy growth of peaceful applications of atomic energy promoted.

BRIEF NOTE ON LIQUID DISPOSAL HAZARDS

A comprehensive survey of this phase is reported elsewhere. Briefly, it is impracticable to create surface lakes or swamps of radioactive liquids. These sources contaminate waterfowl and are prone to create particle hazards, as the water level changes.

Highly active wastes can be retained in underground tanks for long periods; the estimated integrity of tanks is 50 years or more. The intermediate activity, large volume process wastes are most troublesome. These can be safely injected below ground at some sites. With a reservation of the size conditioned by atmospheric pollution, feasible travel time of underground sources to a public area will frequently be 50 years or more. This simplifies the environmental hazard problems. Since this disposal, where permissible, is most economical, it provides an additional incentive to reserving a substantial area around the plant.

REFERENCES

1. Barad, M. L. and Shorr, B., *Field Studies of the Diffusion of Aerosols,* Am. Ind. Hyg. Qtly. *15*: 2 (1954).
2. Sutton, O. G., *Theory of Eddy Diffusion in the Atmosphere,* Proc. Roy. Soc., *A135*: 143 (1932), and many later contributions.
3. Chamberlain, A. C. and Chadwick, R. C., A.E.R.E. HP/R 993 (1952).

Environmental Effects of a Major Reactor Disaster

United Nations International Conference on the Peaceful Uses of Atomic Energy

Presented August 1955
Published 1956

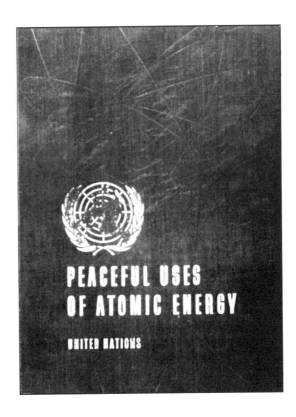

Environmental Effects of a Major Reactor Disaster

By H. M. Parker and J. W. Healy,* USA

To achieve the economic advantage of locating nuclear power reactors close to large communities, it is essential that the potential environmental radiation hazards be unequivocally eliminated.

At the present time, it seems certain that inherently safe reactors can be constructed. Even if there is still a minute possibility of serious reactor accident, release of radioactive material to the environs can be preventive by a protective envelope, of which a large steel sphere is one feasible form.

The purpose in presenting a picture of what could happen in the event that such protection were not provided is to establish reasonable criteria for the degree of safety required. It will be readily seen that high engineering standards, and the accompanying expense, are fully justified for all reactors operating at high power levels.

Since practical experience with reactor failure has been minimal, estimates must rely on theoretical considerations. The estimates are sensitive to the exact mechanism of the initiating incident, the rate of energy release, the meteorological conditions and the associated deposition rates of the active materials. The uncertainties in each phase combine to give an answer which is probably only valid as to order of magnitude. However, it has been found that the total damage, evaluated by the somewhat arbitrary method proposed later, is relatively insensitive to the detailed characteristics of the model chosen for the incident. It seems likely that total damage estimates are reliable within a factor of about three.

RELEASE OF CONTENTS

In the operation of power plants, practical reactors operating in the range of tens of thousands of kilowatts to several millions of kilowatts will be of eventual interest. In such reactors, with the resulting high specific power, the chief hazard in a major accident can be shown to arise from the fission products accumulated during the operating period. The quantities of the radioactive fission products are directly proportional to the reactor power level for a given pattern of operation, and show an increase in the longer lived components with increased time of exposure of the fuel elements. Figure 1 illustrates typical decay curves for fission product mixtures resulting from different irradiation times and shows the increase in the quantity of lingering radiation which would result from the dispersal of fission products from fuel elements which have operated for long periods of time.[1]

The curves do not include the additional fission products generated instantaneously in a runaway critical incident. This component is so low in long-lived fission products that it does not add significantly to the existing load of hazardous materials in the operating reactor.

The emphasis on the hazard of long-lived components stems both from the persistence of radiation from deposits on the ground and from the increased hazard following deposition of these isotopes in the body by inhalation or ingestion. As an example, the relative ingestion hazard for different fission product mixtures is shown in Fig. 2. The reference base is the bone-seeker hazard of the mixture characterized by one day irradiation and one hour post-incident decay.

Two basic models for release of material have been used. In one, an uncontrolled power burst is visualized to cause rapid vaporization of the fission products and to generate a high temperature radioactive

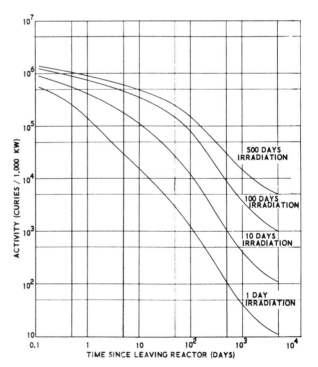

Figure 1. Fission product activity

---

* General Electric Company, Richland, Washington.

EFFECTS OF A MAJOR REACTOR DISASTER

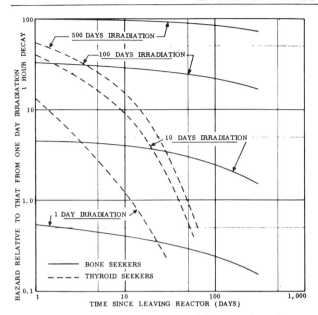

Figure 2. Relative hazard for ingestion of fission product mixtures

cloud that would rise to an elevation of several thousand feet and then travel downwind and disperse. In the other case, failure of the reactor coolant supply is visualized to lead to eventual melting of fuel elements and vaporization of fission products by the self-heating of the fission products. In this more leisurely process there is a much lower rate of heat production; a lower temperature cloud which would tend to travel close to ground level would result. An essentially similar picture would apply to a reactor consumed by fire.

In any of these models, it may be assumed that a portion of the primary escaping fission products will be retained in the reactor building structure. Only the fraction that escapes into the atmosphere generates an environmental hazard. What this release coefficient may be is best computed locally for each case. In the figures and tables:

Power level equivalent to F.P. release = (Actual Power Level) × (Release Coefficient)

In the limit, for a unit with a protective envelope of assured integrity, the release coefficient is zero, the equivalent power level is zero and environmental hazard does not occur. This is the real expected situation; however these data permit potential damage to be assessed for such pessimistic assumptions as a 1% or 10% leak from such a structure. Similarly, they may be applied to protective systems depending on large filters.

CONTAMINATION LEVELS

The concentrations of fission products downwind will also be dependent upon the meteorological conditions and the surrounding terrain. Estimates of the spread for several cases were made by the use of Sutton's theory of turbulent diffusion.[2] Basic assumptions applied to the travel of the cloud over a level plain are given in Table I.

Calculations on other conditions have shown variations in detail and pattern but little difference in over-all damage estimates. A decrease in concentration additional to that by turbulent diffusion was computed using an empirical deposition constant equivalent to a settling rate of the material in the cloud of 2.8 cm per second.

The areas involved were computed on the basis of damage levels. These limits are given in Table II.

The distance at which lethal or damaging conditions would occur is strongly dependent upon both the time and elevation of release and the meteorological conditions. Present estimates indicate lethal conditions only within the immediate vicinity of the reactor for full escape from a 100,000 kw reactor and at distances of the order of 10–50 miles for reactors in the millions of kilowatts range.

Access to land downwind will be limited for a period of time because of radiation levels from the

Table I. Assumptions Used in Calculation

	Rapid release neutral atmosphere	Slow release neutral atmosphere	Slow release inversion atmosphere
Height of release	25% - 100 meters 25% - 500 meters 25% - 1000 meters 25% - 2000 meters	70 meters	70 meters
Time of release	Instantaneous	10 hours	10 hours
Angle covered by wind during release		45 deg	30 deg
Wind speed	100 meters-5.2 m/sec 500 meters-5.7 m/sec 1000 meters-6.4 m/sec 2000 meters-7.2 m/sec	5 m/sec	5 m/sec

Table II. Damage Limits

Effect	Fallout limit	Resulting condition at boundary
Lethal		190–350 r full body* plus 800–1200 rads to lung-10 days plus 150–250 rem to bone-10 days
Significant injury to humans		60–250 r full body* plus 200–300 rads to lung-10 days plus 30–70 rem to bone-10 days
Land unusable for 5 years	5 mc/ft^3	Dose rate falls to 300 mr/week in 5 years
Land unusable for 2 years	2 mc/ft^3	Dose rate falls to 300 mr/week in 2 years
Temporary evacuation	0.5 mc/ft^3	Gamma dose of 50 r in first year (30 r in 2 months; 40 r in 6 months)
Crops confiscated	0.1 mc/ft^3	5×10^{-5} μc Sr^{90}/gram of vegetation†

* 20–30% from cloud passage; remainder from exposure to contaminated ground for 2–5 hr.

† Limit computed assuming ingestion of crop by humans for one year.

"fallout". The actual shape of these areas will again be strongly dependent upon the methods of release and dispersion with a rapid emission expected to result in a long, narrow band and a slow release resulting in a general spread over shorter distances but wider areas. Examination of typical cases and comparison with small-scale tracer tests have indicated that the areas involved are remarkably independent of the exact mechanism of release and dilution.

Average values of the contaminated areas resulting from the complete release of fission products after 100 days irradiation of the fuel are given in Fig. 3. Individual estimates for specific methods of release and rates of dilution in the atmosphere varied from these values by factors of two to three. As a rough rule of thumb, the area covered in square miles after one day decay is approximately given by the power level representative of the fraction of fission products escaping expressed in kilowatts multiplied by 2×10^{-3} and divided by the contamination level contour of interest in millicuries per square foot.

These estimates of areas are strictly orders of magnitude for average depositions with specific locations within the area possibly more or less contaminated depending upon local conditions of terrain, wind patterns and nature of the surfaces. It is to be expected that personnel contamination will also occur within these areas with immediate decontamination necessary to prevent injury from the beta radiations.

The rate of decrease of this deposited material will be primarily influenced by radioactive decay although weathering and possibly incorporation into the top soil will occur. Gamma dosage rates at several feet above the ground will decrease to 300 mr/week in about five years for the five millicurie per square foot area, in about two years for the two millicurie per square foot area, and in about one year for the 0.5 millicurie per square foot area. The specific times required will be dependent upon the weathering and upon the time of irradiation of the fuel elements.

Precipitation during the evolution of fission products will change the estimates since the rate of deposition will be increased and the areas immediately surrounding the plant will be subjected to much higher contamination levels. This, of course, increases the probability of severe damage at close approaches but tends to decrease the damage at a distance and reduces the total areas involved.

Contamination of bodies of water will occur both by direct fallout and by secondary leaching of the materials into the streams. Direct fallout could render bodies of water reasonably close to the reactor unfit for use until the material is carried away. Present indications point toward leaching into the stream slow enough to cause little trouble although monitoring downstream is advisable.

The possibility of rainout into a stream near the reactor site exists with severe contamination resulting, depending upon the rate of washout from the cloud and the stream flow characteristics. Of more concern in at least some cases is the possibility of the escape of reactor coolant containing a significant quantity of fission products. Such a mishap would cause a band of grossly contaminated water, depending again upon the flow characteristics of the stream, which could cause severe contamination of water plants or other equipment downstream and in special cases could lead to serious radiation dosages to people using the water for sanitary purposes.

DAMAGE ESTIMATES

As an illustration of the average expected cost of a major disaster, estimates were made by applying the areas of Fig. 3 to average census values[3] for several localities. The census values used are given in Table III.

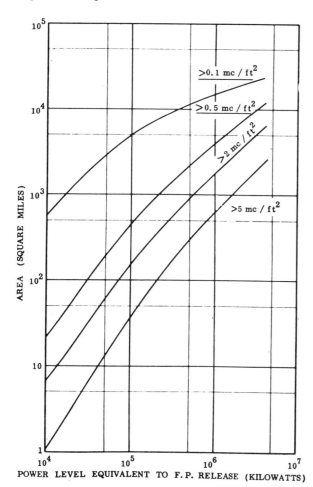

Figure 3. Average areas covered and contamination

Table III. Census Data

Area	Population density people/mile²	Property values Rural $/mile²	Property values Urban $/mile²	Crop values $/mile²
Industrial Middle Atlantic States	231	33,000	240,000	19,000
Agricultural Middle West States	154	95,000	150,000	36,000
Agricultural Western States	35	22,000	33,000	9000

It was assumed that property contaminated to greater than 2 mc/ft² would be purchased outright at market value while property contaminated to 0.5–2 mc/ft² would cost about 10% of the market value for rental or decontamination. Crops on land contaminated between 0.1 and 0.5 mc/ft² were assumed purchased for the first year with no allowance for later crops. A closer breakdown was not felt to be warranted in view of other uncertainties in the estimates. Figure 4 illustrates one series of average property damages expected for a reactor located in these three areas.

In order to illustrate the variation in contributions to the assumed damages, a breakdown by areas is given in Table IV for full release from several power reactors.

In addition it is estimated that for very large releases damage to humans could be extensive. As an example, for full release from a 1,000,000 kw reactor between 200 and 500 people could be killed in a region of population density of 200–500 people per square mile with perhaps 3000–5000 exposed to possibly damaging levels even with fairly prompt evaluation.

It may seem surprising that the estimated damage in Fig. 4 and Table IV is greater for a rich agricultural region than for an industrial area. This develops from the mode of averaging census values over a whole state. For a specific heavily industrialized zone, the property damage could be considerably higher.

To give proper perspective to these data, one must reiterate that the probability of a major reactor acci-

Table IV. Composition of Property Damage Estimates (millions of dollars)

Category	10,000 kw	100,000 kw	1,000,000 kw
Industrial area – Middle Atlantic States			
Purchase – Rural	0.2	7	50
Urban	1.2	50	380
Rent	0.4	10	70
Crops	11	90	210
Total	13	160	710
Agricultural area – Midwest States			
Purchase – Rural	0.5	20	170
Urban	0.7	30	270
Rent	0.4	10	60
Crops	20	160	450
Total	22	220	950
Agricultural area – Far West States			
Purchase – Rural	0.1	4	35
Urban	0.2	7	50
Rent	0.1	2	12
Crops	5	43	100
Total	5	56	200

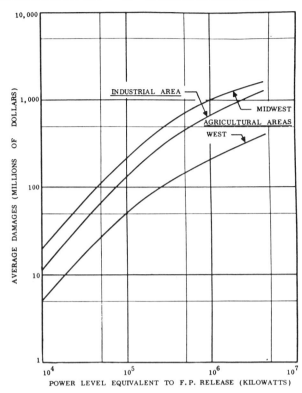

Figure 4. Estimated average damages from release of fission products

dent, although currently indeterminate, is obviously vanishingly small, and that the essential integrity of protective envelopes can be assured from conventional engineering data. The results should in no way be interpreted as a deterrent to the advancement of commercial power reactor technology. They do indicate the need for high standards of engineering and operation in all cases. More importantly, they show the high value of protective systems that are not quite perfect. Thus, for an actual power level of 100,000 kw and a protective system with a release coefficient of 0.01, the equivalent power level of 1000 kw is below the range in which it is relevant to compute estimated average environmental hazards. Over the small areas then involved, it is possible to apply decontamination methods that would be impracticable for the hypothetical large-scale disasters. The real damage in this case would therefore be even lower than is found by a plausible extrapolation of Fig. 4. This perhaps possible disaster is well within the range of disaster damage experienced from time to time in older industries.

REFERENCES

1. Healy, J. W., Pilcher, G. E. and Thompson, C. E., *Computed Fission Product Decay*, HW-33414 (1954).
2. Sutton, O. G., *Micrometeorology*, McGraw-Hill, New York (1953).
3. Hansen, N., *The World Almanac and Book of Facts for 1954*, The New York World-Telegram and The Sun, New York (1954).

Personnel Dosimetry for Radiation Accidents

"Personnel Dosimetry for Radiation Accidents"

Proceedings of a Symposium held jointly by the International Atomic Energy Agency (IAEA) and the World Health Organization in Vienna March 8-12, 1965.

Published by IAEA, Vienna, 1965.

THE HANFORD CRITICALITY ACCIDENT: DOSIMETRY TECHNIQUES, INTERPRETATIONS AND PROBLEMS

H. M. PARKER AND C. E. NEWTON, JR.
GENERAL ELECTRIC COMPANY, HANFORD ATOMIC PRODUCTS OPERATION,
RICHLAND, WASH., UNITED STATES OF AMERICA

Abstract — Résumé — Аннотация — Resumen

THE HANFORD CRITICALITY ACCIDENT: DOSIMETRY TECHNIQUES, INTERPRETATIONS AND PROBLEMS. The number and integrity of dosimetry techniques used for dose interpretations for the twenty-two personnel involved in the 1962 criticality accident occurring at the Hanford Project were unusually complete. Personnel who received excessive exposures were immediately detected and segregated by monitoring personnel using portable instrumentation for "quick-sort" procedures which rely on in vivo measurements of Na^{24} activation. The close correlation between this rapid method of dose interpretation and subsequent sophisticated laboratory procedures was noted. Primary reliance, however, was placed upon film-badge interpretations, as all persons involved were wearing film badges. An area threshold detector, located within twenty-six feet of the critical vessel, furnished data upon which the neutron spectrum and the gamma to neutron ratios were established.

Even with sophisticated and complete dosimetry techniques which were available and which are discussed, including blood and whole-body Na^{24} activity, excreta and P^{32} analyses and gold activation, many practical problems became evident. Described are methods or alternatives used to cope with or minimize actual problems including those which could have but did not arise. The "quick-sort" in vivo procedures successfully used could have become worthless if interfering external contamination were encountered. Alternatively, blood samples taken at first aid may substitute; however, in the haste of the emergency, anticoagulants may be omitted and subsequent coagulation can produce concern as to the accuracy of the dose interpretations. The film badge can include material of high neutron cross-sections introducing sufficient activity to interfere with film interpretations. The activation of the silver in the film presents a correction factor which can delay and confuse personnel who, under stress, are attempting rapid evaluations.

The support provided by fixed detectors still presents the problems that scattering, moderation and geometrical factors are not the same between exposed personnel and detector. The unidirectional nature of the heterogeneous gamma-neutron flux, although providing expected variations, presents unique complications in organ dose interpretations. Finally, the problems and delays that can be encountered in the selection and agreement of dose-related conversion factors, methods of calculation, probable applicable spectrums and other necessary supplemental data are covered.

ACCIDENT DE CRITICITÉ DE HANFORD: MÉTHODES DE DOSIMÉTRIE, ÉVALUATION ET PROBLÈMES. Toutes les méthodes de dosimétrie utiles en l'occurrence ont été employées avec une efficacité particulière pour évaluer les doses de rayonnement auxquels ont été exposés 22 membres du personnel lors de l'accident de criticité survenu à Hanford en 1962. Les personnes qui avaient reçu des doses excessives ont été immédiatement détectées et séparées par le personnel de contrôle muni d'appareils portatifs permettant de procéder à un tri rapide après mesure in vivo du radiosodium. On a constaté une bonne concordance entre le résultat de cette méthode rapide d'évaluation des doses et les méthodes plus élaborées utilisées ultérieurement en laboratoire. On a toutefois accordé une importance particulière aux indications données par les dosimètres à film, du fait que toutes les personnes irradiées en portaient. Un détecteur de zone à seuil, situé à moins de 8 m de la cuve critique, a donné des renseignements sur le spectre de neutrons et les rapports entre les rayons gamma et les neutrons ont pu être établis.

En dépit de l'utilisation des méthodes de dosimétrie complètes décrites par les auteurs, notamment la mesure de l'activité du ^{24}Na dans le sang et dans l'ensemble de l'organisme, l'analyse des excreta et du ^{32}P et l'évaluation de l'activation de l'or, de nombreux problèmes pratiques sont apparus. Les auteurs décrivent des méthodes ou des solutions de remplacement qu'ils ont employées pour résoudre ou réduire au minimum les problèmes réels ou les problèmes qui auraient pu se poser. Les méthodes de détection rapide in vivo n'auraient pas été couronnées de succès en présence d'une contamination externe. Au lieu d'appliquer ces méthodes, on pourrait faire des prélèvements de sang au moment des premiers soins, mais on risque, dans la hâte des premiers secours, d'oublier les anticoagulants; du fait de la coagulation qui en résulte, un élément d'incertitude peut intervenir quant à l'exactitude des doses déterminées. Il peut entrer dans la composition des dosimètres à film des matières ayant de fortes sections efficaces neutroniques créant une activité suffisante pour perturber l'analyse du film. L'activation de l'argent entrant dans la composition du film implique un facteur de correction qui peut retarder et compliquer la tâche du personnel qui procède, dans la précipitation, à des évaluations rapides.

En dépit de leurs avantages, les détecteurs fixes présentent encore l'inconvénient que les facteurs de diffusion, de modération et de géométrie ne sont pas les mêmes pour le personnel exposé que pour le détecteur. Si la nature unidirectionnelle du flux hétérogène de rayons gamma et de neutrons permet de prévoir les variations, elle n'en est pas moins une source de complications particulières dans l'évaluation des doses reçues par les organes. Enfin, les auteurs étudient les problèmes et les retards relatifs au choix et à la corrélation des facteurs de conversion liés aux doses, aux méthodes de calcul, aux spectres susceptibles d'être utilisés ainsi qu'à d'autres données supplémentaires nécessaires.

СЛУЧАЙ АВАРИЙНОЙ КРИТИЧНОСТИ В ХЭНФОРДЕ: МЕТОДЫ ДОЗИМЕТРИИ, ИСТОЛКОВАНИЕ ДАННЫХ, ПРОБЛЕМЫ. Необычно полно и широко применялись дозиметрические методы для интерпретации дозы, полученной персоналом в составе 22 человек во время аварийной критичности при осуществлении проекта Хэнфорд в 1962 году. Персонал, получивший избыточную дозу, был немедленно выявлен и изолирован группой радиационного контроля с помощью переносной аппаратуры с использованием метода "быстрого отбора", основанного на измерениях степени активации Na^{24} in vivo. Отмечалась значительная корреляция между этим быстродействующим методом интерпретации дозы и последующими сложными лабораторными процедурами. Основной упор, однако, делался на интерпретацию показаний пленочных дозиметров, поскольку у всех лиц были пленочные дозиметры. С помощью зонного порогового детектора, размещенного в 8 м от достигшего критичности реактора, были получены данные, на основании которых были выведены спектр нейтронов и соотношения между гамма-лучами и нейтронами.

Даже при использовании сложных и полных дозиметрических методов, которые обсуждаются, в том числе измерение активности Na^{24} в крови и во всем организме, анализ выделений и содержания P^{32} и активация золота, возникло много практических проблем. Описываются методы или альтернативные варианты, применяемые для решения или сведения к минимуму существующих проблем, в том числе таких проблем, которые могли бы возникнуть, но не возникли. Методы "быстрого отбора" in vivo, успешно примененные, могли бы оказаться бесполезными при интерферирующем внешнем загрязнении. Альтернативно могут явиться заменителем образцы крови, взятые при оказании первой помощи. Однако в условиях поспешных действий во время аварии можно опустить антикоагулянты, и последующая коагуляция может породить озабоченность в отношении точности интерпретации доз. В пленочные дозиметры может включаться материал с большим поперечным сечением нейтронов, вводящий достаточную активность для интерферирования с интерпретациями пленок. Активация серебра в пленке представляет коэффициент поправки, который может задержать и ввести в заблуждение персонал, который в напряженных условиях пытается быстро дать оценку.

Поддержка, обеспечиваемая установленными детекторами, порождает проблемы, заключающиеся в том, что коэффициенты рассеяния, замедления и геометрические факторы не одни и те же между облученным персоналом и детектором. Однонаправленная природа гетерогенного гамма-нейтронного потока, хотя и порождает предполагаемые отклонения, создает исключительные осложнения и интерпретации доз, получаемых органами. Наконец, обсуждаются проблемы и задержки, которые могут возникнуть при выборе и согласовании связанных с дозой коэффициентов преобразования, методов расчета, вероятных применяемых спектров и других необходимых дополнительных данных.

EL ACCIDENTE DE CRITICIDAD DE HANFORD: TECNICAS DOSIMETRICAS, INTERPRETACIONES Y PROBLEMAS. Para evaluar las dosis a que estuvieron expuestas las 22 personas afectadas por el accidente de criticidad que ocurrió en Hanford en 1962 se emplearon técnicas sumamente completas, tanto en lo que respecta a su número como a su eficacia. Las personas que recibieron dosis excesivas fueron inmediatamente identificadas y aisladas por personal de vigilancia radiológica provisto de instrumentos portátiles que permiten proceder a una «clasificación rápida» luego de medir in vivo la actividad del ^{24}Na. Se observó una estrecha correlación entre los resultados de este método rápido de evaluación de las dosis y los procedimientos más complejos ulteriormente aplicados en laboratorio. Sin embargo, se atribuyó una importancia primordial a las indicaciones dadas por los dosímetros de película, ya que todas las personas afectadas por el accidente llevaban

HANFORD CRITICALITY ACCIDENT

dosímetros de ese tipo. Un detector de umbral situado a menos de ; m del recipiente crítico, proporcionó datos a partir de los cuales se determinaron el espectro neutrónico y las razones rayos gamma-neutrones.

A pesar del empleo de técnicas dosimétricas refinadas y completas, entre ellas, la medición de la actividad del ^{24}Na contenido en la sangre y en todo el organismo, el análisis de los excreta de ^{32}P, y la evaluación de la actividad del oro, se plantearon múltiples problemas de orden práctico. Se describen los métodos utilizados para resolver o reducir al mínimo los problemas reales, o los que podrían haberse planteado. Los procedimientos de «clasificación rápida» in vivo podrían haber resultado inútiles en presencia de una contaminación externa. En vez de aplicar esos métodos, se podrían tomar muestras de sangre en el momento de prestar los primeros auxilios, pero se corre el riesgo de olvidar los anticoagulantes; la coagulación consiguiente puede introducir un elemento de incertidumbre en la interpretación de las dosis. El dosímetro de película puede contener materiales de elevada sección eficaz de captura neutrónica, que dan origen a actividades capaces de dificultar el examen de las películas. La activación de la plata contenida en la emulsión obliga a introducir un factor de correlación que puede retrasar los trabajos e inducir en error al personal que, bajo la tensión del momento, trata de evaluar rápidamente las dosis.

Pese a su utilidad, los detectores fijos presentan el inconveniente de que las condiciones de dispersión, moderación y de geometría no son las mismas para el personal expuesto que para el detector. La naturaleza unidireccional del flujo heterogéneo de rayos gamma y neutrones, si bien permite prever ciertas variaciones, constituye una fuente de considerables complicaciones en lo referente a la evaluación de la dosis recibida por los órganos. Por último, los autores examinan los problemas y demoras que pueden originar la selección y la correlación de los factores de conversión relativos a la dosis, los métodos de cálculo, los espectros probables y otros datos necesarios de carácter complementario.

1. INTRODUCTION

A criticality accident occurred in a chemical building where there were 22 employees on 7 April 1962 at the US Atomic Energy Commission's Hanford plant, located in the State of Washington. In contrast to several earlier criticality events at other locations in the USA where personnel dosimetry measurements were not available, especially in respect to personal film dosimetry, all of these employees were wearing film dosimeters. A formal report [1] of the dosimetry investigation, which essentially covers the period from eight hours to several weeks post incident, has been published. Where appropriate, pertinent condensations of the findings are included here. This paper describes the hitherto unreported events, uncertainties and, most important, the lessons learned from this actual test of emergency procedures. It also centres around the results of the particularly comprehensive evaluations conducted upon the three persons who received the highest doses.

2. DESCRIPTION OF THE ACCIDENT

The criticality occurred at a remotely-located plutonium waste recovery facility in a non-geometrically safe solution transfer vessel. The characteristic Cerenkov radiation was observed by the three employees in the room, one of whom was standing within three metres of the vessel. The evacuation horns sounded immediately and these three employees, together with 19 other employees from other parts of the building, evacuated the building promptly according to established procedure.

Two recording BF_3 neutron counters (one for high, one for low levels) were operating in another building nearby at the time of the accident. They indicated an initial excursion the exact magnitude of which could not be determined because the reading, which was recorded only every 30 seconds, was off-scale.

It may be expected that these readings could vary by orders of magnitude within such a time. After the initial pulse, a continuing nuclear reaction was of sufficient magnitude to keep the recorder off-scale for a period of 30 minutes. After the recorder returned to on-scale readings, the fissioning continued at a generally reduced rate until 36 hours after the incident began. The three employees in the room where the accident occurred left the room so promptly (in a matter seconds) that they were exposed only to the initial pulse.

The 22 personnel present in the building at the time of the accident were moved to a first-aid station about two kilometres away where "quick-sort" and contamination surveys were made and dosimetres were collected. None of the employees was contaminated; the "quick-sort" survey was positive for three. All involved were wearing film-badge dosimeters and, in addition, several were wearing film finger rings.

An emergency procedure, "quick-sort", had been developed [2] for use in a criticality event to detect high-level neutron exposure in the field by direct survey of the human body with a simple G-M type portable radiation survey instrument held against the stomach and abdomen area. A general relationship between body weight, count rate and dose had been developed to provide a reasonable estimate of a neutron dose received from a criticality accident. Typical sensitivity shows about 65 counts per minute for one rad to a standard man [3].

2.1. Preliminary dose estimates

On arrival at the hospital, located some 50 kilometres away, the three individuals were again carefully monitored and found to be free of contamination. The "quick-sort" procedure was repeated with essentially the same results (Table I).

All personnel assigned to the facility in which the accident occurred were thoroughly trained and, therefore, were well aware of the consequences of being exposed at short distances to a critical event. Understandably, employees A and B were in a state of anxiety; employee A having no doubts that he had received a fatal dose.

Although the "quick-sort" measurements indicated that the neutron doses were only about 32, 7 and 4 rad, radiation protection personnel had limited confidence in these preliminary estimates for three reasons:

(1) The procedure had never before been tested in practice;

(2) The estimated doses were much lower than one would expect, based on the reported distances from the source of radiation; and

(3) The magnitude of the error introduced by differences between the calibration and exposure neutron energy spectra was unknown.

In retrospect, the "quick-sort" procedure, without spectrum corrections, can be seen to have given what proved to be rapid and valid enough estimates of the neutron doses. Table II compares the results of the whole-body counter with the "quick-sort" procedure estimates.

HANFORD CRITICALITY ACCIDENT

TABLE I

"QUICK-SORT" PROCEDURE ESTIMATES

Employee	Body weight (kg)	Observed counts/min. *	"Quick-sort" dose estimates (rad)	Employees' estimated distance from source** (m)
A	70	2000	32	0.9 - 1.2
B	98	600	7	2.2 - 2.7
C	85	300	4	7.6

* Background activity negligible
** Later measurements showed that the actual distances were somewhat greater than those indicated above. These rather odd values are due to literal translation from feet to metres.

TABLE II

COMPARISON OF "QUICK-SORT" ESTIMATE AND WHOLE-BODY COUNTER EVALUATION OF NEUTRON DOSE

Employee	Whole-body counter dose estimate (rad)	Distance from critical event (m)*	"Quick-sort" dose estimate (rad)
A	23	1.8 - 2.4	32
B	9	3.0 - 3.4	7
C	3	7.9	4

* Literal translation from feet to metres

Hanford film-badge dosimeters worn by all persons present in the affected building were promptly collected and dispatched by special messenger to the film-badge processing laboratory. All dosimeters were surveyed for contamination before they were permitted in the laboratory and processing commenced within two and a half hours following the criticality exposure. This was quite rapid service considering that the film-badge processing laboratory was 40 kilometres away from the plutonium waste recovery facility and that the event occurred on a week-end when all auxiliary staffs were at a minimum. The dosimeter in use at that time consisted of a film packet, containing du Pont number 508 (sensitive) and number 1290 (insensitive) film, behind aluminium and silver absorbers and an "open window". A strip of lead tape, perforated with the payroll number, lay along one edge of the film and was normally used for film identification by local X-ray exposure coding. However, since possible excessive film darkening resulting from the exposure received during the accident might make the X-ray identification indiscernible, pinhole coding was perforated through the film packet before processing. Both the sensitive and insensitive films were developed together with a set of gamma calibration films (10, 25, 35, 50, 60, 70, 120, 150, 300, 450, 600 and 1200 r). The dose evaluations were made using the insensitive (number 1290) films, as the sensitive films turned out to be too dark to read on the densitometer.

The first estimates of gamma dose for the three involved employees were made 3.5 hours after the accident by visual comparison of their badge films with calibration films. The first estimates reported were 67, 30 and 10 r. Although these low gamma-exposure estimates tended to confirm the low neutron-dose estimates, and together indicated that no fatal or severe exposures had been received, the first "official" dose estimates were delayed until 4.5 hours after the accident, when the number 1290 film densitometer readings, rather than visual guesses, and blood sodium activation measurements were available. The estimates that were then presented to the medical personnel and to the employees involved are presented in Table III.

3. DISCUSSION AND PROBLEMS

3.1. Film badge dosimetry

Because of slow neutron activation of the aluminium and silver absorbers in the personnel dosimeter, film darkening was inconsistent with the calibration data and therefore the personnel films could not be evaluated in the usual manner. The exposure was interpreted from the densities under the unshielded portion of the dosimeter (the "open window"). This naturally introduced some uncertainties as the exposure could be as much as 15% too low because of the lack of electronic equilibrium and differing photon spectra.

The area underneath the small piece of lead tape used for identification purposes was used to compare with the area under the unshielded portion. The readings were compared with similar readings on the calibration film and they gave dose estimates that confirmed those from the "open window" readings if one assumed that there was no appreciable slow neutron activation of the lead tapes and no appreciable effect due to the degradation of the photon spectrum. Another item which became of immediate concern was the darkening of the film due to neutron activation of the materials in the emulsion. For the spectrum from this criticality event, 7% of the dose measured was attributed to this darkening. The values in rads of thermal or fast neutrons that produce the same darkening on the film as 1 r of cobalt-60 gamma rays were taken from the work of SMITH [4].

Even though it was realized that the film and ring badges of the exposed personnel could give results that would vary by factors of three or four due

HANFORD CRITICALITY ACCIDENT 573

TABLE III

DOSE ESTIMATES* 4.5 h POST INCIDENT

Employee	Gamma exposure (r) from 1290 film in Hanford film badge dosimeter	Neutron dose from blood sodium measurements (rad)
A	63 ± 10%	25 ± 10%
B	23 ± 30%	15 ± 15%
C	13 ± 50%	4 ± 30%

* Final gamma dose estimates were the same and the final neutron exposures were: employee A, 23-30 rad; employee B, 9-12 rad; employee C, approximately 3 rad.

only to differences in attenuation between what the body and badge received, it was found in this event that primary reliance for dose evaluations could still be placed on film data interpretations. Nevertheless, the need to have additional neutron spectrum measurement capability and dose range had been recognized for some time and the development work completed in early 1961. In fact, the order for replacing all the old dosimeters in use with the new Hanford personnel dosimeter had been placed in December 1961. The new dosimeters were not received from the manufacturer until July 1962; ironically, some three months subsequent to Hanford's only criticality accident. The new Hanford film badge is responsive to these needs and, although the details of the construction and the performance characteristics have been reported [3, 5], the most essential capabilities and inclusions adopted for this type of monitoring are worthy of mention here.

For instance, the inclusion of silver phosphate glass fluorod dosimeters extended gamma-dose measurement capability to 10 000 r, as compared with the 1500 r limit of the old dosimeter. The fluorods are incorporated into the film dosimeter in such a way as to be conveniently accessible and yet relatively free from contamination problems. For improved energy response, one fluorod is placed within a tantalum sleeve. A set of foils, consisting of indium, cadmium, copper and sulphur, provides a capability for measurement of neutron spectra and dose up to 2000 rad. Because of the activity induced by neutrons on the indium foil, persons grossly exposed to neutrons may be segregated by measuring the activity of the foil while it is still in the badge, using the G-M counter.

3.2. Blood

The initial blood samples from these employees were obtained at a first-aid facility near the site of the accident within 30 minutes. However, the samples were only about 2 cm³ each and the evaluators became concerned regarding the accuracy of the volume determinations, since some

TABLE IV

SODIUM-24 IN BLOOD

Employee	Sample volume (cm³)	Na²⁴ (μc/cm³ blood)	Neutron first collision dose (rad)
A	2	1.5×10^{-4}	26 ± 1.1
A	10	1.8×10^{-4}	31 ± 1.3
B	2	8.8×10^{-5}	15 ± 1.0
B	10	7.3×10^{-5}	13 ± 0.8
C	2	2.0×10^{-5}	3.4 ± 0.4
C	10	2.1×10^{-5}	3.6 ± 0.5

coagulation occurred before the measurements could take place. A second set of samples, to which heparin was added to prevent coagulation, was taken at the hospital two hours after the accident. The results [6] from both sets of samples, using the modified procedure for the second set, are shown in Table IV.

It has been recommended [7] that if the dose is believed to be more than 100 rad of neutrons, the volume of the blood sample should be about 10 cm³; however, if the dose is less than 10 rad, the volume should be 100 cm³. There is some merit in evaluating sodium activation from the blood serum only, avoiding volume errors from addition of anticoagulants.

It is expected that in cases of high exposures the subjects may be in a state of shock, requiring the administration of intravenous solutions containing sodium. If pre-empting medical considerations do not intervene, blood samples should be obtained before the administration of the salt solutions. In using blood sodium activation for a measurement of the fast neutron dose, it should be noted that mixing of blood gives an average value for exposure of the whole body and, under conditions of partial body shielding or gross asymmetry of the radiation field, results may be of limited significance.

3.3. Contamination

For personnel who received external contamination at the time of the event, the detection of the contamination itself should be used to signal involvement in the incident. The use of single-channel analyser equipment to detect promptly the sodium-24 in the body in the presence of general body or clothing contamination is technically feasible and devices of this type are reported in the literature [8]. A recently developed Hanford Laboratories' portable analyser [9] may overcome the past difficulties encountered in the use of this type of field equipment.

HANFORD CRITICALITY ACCIDENT

Generally, discriminating analysers will not be available at the right location, reliably calibrated and in operating condition. From a purely practical point of view, one will normally have to assume that a person found with contamination on his body was involved in the event and he should be segregated from the other employees. Rapid estimates of the contaminated person's dose may be made by blood sodium activation analysis. Also, in some cases it would be more helpful to try to remove the contamination from the person than to try to perform reliable dosimetry.

3.4. Whole-body counting

Personnel responsible for the operation of the whole-body counter were located promptly and the unit was in standby readiness within two hours after the accident. Counting of the involved personnel was not begun until 6.5 hours after the accident as medical procedures and observations properly took precedence. The collection of information on whole-body sodium-24 neutron dosimetry took several hours and the standard-man data were consulted for the body sodium content, the neutron dose of employee A was estimated by whole-body counting at eight hours post accident to be 23 ± 6 rad, which was in remarkable agreement with the blood sodium measurements. All 22 employees were counted and the evaluations for sodium-24 activation were negative for 19, confirming the accuracy of the "quick-sort" procedure in identifying the three exposed personnel, designated as A, B, and C.

3.5. Excreta

It was not possible to use the data from the analyses of excreta to confirm the estimations of exposure doses obtained by the other methods. The sodium-24 in urine was not found to be directly proportional to the body burden in these exposed employees [6]. Two employees, one receiving twice the dose of the other, excreted in their urine approximately the same amount of sodium-24 over a 6-d period. Additionally, it was found that the urinary rate of sodium-24 excretion was not a constant function of the body burden. The variations ranged from 4 to 11% of their body burdens during the first full day following the exposure and were certainly related, at least in part, to inability to relate the urinary excretions to the metabolic periods which they represented.

The daily faecal excretion rate of sodium-24 ranged from less than 1 to about 7% of that present in the urine. In addition to the sodium-24 detected in the faeces, trace amounts of chromium-51, zinc-65, zirconium-niobium-95 and ruthenium-103-106 were also detected. Traces of zinc-65 and chromium-51 are common in people that live near Hanford because of the presence of these nuclides in the Columbia River and it is probable that they were ingested earlier with food or drinking water [10]. The zirconium-niobium and ruthenium had probably been ingested from world-wide fall-out.

3.6. Hair

The neutron doses to localized areas and to critical organs of the employees' bodies were estimated by determining the phosphorus-32 activity

TABLE V

SUMMARY OF FIRST COLLISION NEUTRON DOSE
ESTIMATES [6]

Employee	Whole-body sodium-24	Blood sodium-24 (average)	Phosphorus-32 in hair *
A	23	28	30
B	9	14	14
C	3	3	4

* Values assume 25% of the first collision neutron dose was due to neutrons of energy greater than the S^{32} (n,p)P^{32} threshold of 2.5 MeV.

in hair sampled from selected portions of the body, together with a knowledge of the approximate neutron spectrum. Since activation in hair is a reaction at the body surface, its greatest value is in estimating personnel orientation and localized dose, especially where the radiation has not been severely degraded and where there is considerable non-uniformity of dose distribution.

In this particular incident, 1-g samples of hair from various areas of the body were used for the estimation of the dose of neutrons of energies in excess of 2.5 MeV and the agreement with other methods of estimating the neutron dose was quite good. For instance, for employee A this technique gave a first collision dose of 9.2 rad for the head, 10.5 for the chest, 6.8 for the pubic area, and 5.2 for the legs. While this method is in obvious contrast to the fast neutron average whole-body doses obtainable through blood activation analysis, for comparison purposes only, one may assign an "average" whole-body dose of about 7.5 rad to this employee due to neutrons above 2.5 MeV. Since this accounted for about 25% of the neutron dose, this method estimated the neutron dose at 30 rad, compared with about 25 rad obtained using the whole-body counting or blood sodium data (Table V).

Through the use of common low-background beta-counting techniques, as little as 0.5 dpm of phosphorus-32 can be measured. This corresponds to a first collision dose of about 0.28 rad for a 1-g hair sample.

Although it is best to use hair from uncontaminated areas of the body, some contamination of the hair from fission products does not invalidate the method. The hair is thoroughly washed before analysis and the chemical separation isolates the phosphorus-32 in a high degree of radiochemical purity.

3.7. Other materials

In order to provide support for the estimates of the applicable neutron spectrum at or in the immediate location of the exposed employee, it was found desirable to make neutron activation analyses of certain materials

HANFORD CRITICALITY ACCIDENT

TABLE VI

RADIOACTIVITY OF OTHER OBJECTS

Employee	Item	Radioisotope measured
A	Silver shield from film badge	Cu^{64}
	Ball point pen, minus tip	Cu^{64}
	Tip of ball point pen	Cu^{64}
B	Silver shield from film badge	Cu^{64}
	Nickel	Mn^{56}
	Belt buckle	Cu^{64}
	Lens of eyeglasses	Na^{24}
	Frame of eyeglasses	Au^{198}
C	Silver shield from film badge	Cu^{64}
	Pencil clip	Mn^{56}
	Button	Cu^{64}
	Watch band	Mn^{56}

such as aluminium, copper and gold. These came from the personal effects of the exposed subjects, scrap materials and tools. As for the personal items, all of the following were useful: coins, jewelry, belt buckles, metal buttons, pens and automatic pencils, wrist watches and metal from spectacle rims. Although the exact amounts of induced radionuclides in these materials could have been measured with good accuracy, such values would not have provided a measurement of the integrated neutron flux to which they were exposed. The integrated flux to which these items were exposed was determined by allowing most of the initial activity to decay and then irradiating the objects to a comparable extent in a known slow neutron flux. The details of the procedure have been reported previously [1,6]. The radioisotopes measured from various items from the involved individuals are presented in Table VI.

3.8. Dose equivalent

As noted throughout the paper, reference has been made to "first collision neutron doses". Although this is now unfashionable, these were the data available at that time. Moreover, no applicable agreed-upon substitute has yet been accepted. Prediction of the consequences of the radiation doses from the criticality accident was fraught with uncertainty. Among these uncertainties were dose rates, possible multiplicity of bursts, application of spectrum data, and a reliable index of biological damage.

TABLE VII

RECORDED EXPOSURES FOR RECORD PURPOSES

	Gamma (r)	Neutrons (rad)	Assumed conversion factors for neutrons	Total (rem) *
Employee A				
Whole body	63	23-30	2	109-123
Eyes	90	42-54	10	510-630
Employee B				
Whole body	23	9-12	2	41- 47
Eyes	28	11-14	10	138-168
Employee C				
Whole body	13	3	2	19
Eyes	13	3	10	43

* Conversion factors converting röntgens to rads and single collision doses to multiple collision doses were not used. The uncertainties associated with the geometrical factors and the extremely broad range of choices for the selection of quality factors excluded such refinements.

The heterogeneous energy spectrum of the incident radiations must have produced a very non-uniform distribution of dosage in depth, particularly where the spectrum included large, low energy components. It thus became extremely difficult to postulate the degree of probable biological insult. The exposure dose in air must have varied from 2 to 20 times greater than the midline tissue dose [11]. The use of the midline dose, or other measures such as exit dose or gram-röntgen dose, was not attempted for obtaining a definition of exposure level.

It was recognized that under the high dose and dose rate conditions, the accepted quality factor values normally used did not apply. Conversion values from rads to rems would clearly not be the same as those usually applied to repetitive small exposures commonly associated with radiation protection work. There is some evidence to support the hypothesis that the biologically effective dose may be best defined as the average dose for the bone marrow for both acute and chronic measures of injury [12]. A report [13] from dog data suggests that the RBE for mortality may not be a sensitive function of neutron energy in large species and a value of one is reasonable for bone marrow failure and a value of two for intestinal injury and its associated mortality. The composite doses were written down by converting neutron doses from rads to rems by a multiplier of 2, except for exposures of the eye lens for which a factor of 10 was applied. Table VII shows these results. They served the dual purpose of suggesting that no very striking short-term effects would be observed, and that it would be prudent to continue observation of the eye.

3.9. Dosage calculations and factors

Although the dosimetry planning for an incident of this magnitude was satisfactory, it emphasized the necessity of maintaining up-to-date agreed-upon factors and procedures, together with notes as to limits of applicability for estimating high-level radiation doses. At the referenced time, this installation had formal detailed procedures for most techniques except body sodium activation measurements using whole-body counting and phosphorus-32 in hair measurements.

Sodium-24 activity ingested from activated Hanford drinking water was routinely measured in vivo, using the Hanford whole-body counter. Therefore, no special calibrations of the whole-body counter for sodium-24 activity were needed to measure the activities involved in the incident. However, routine measurements were normally converted to a percentage of a maximum permissible body burden, and not to neutron dose. A copy of the dosimetry summary of the last criticality incident at Los Alamos was consulted and found to contain both the Oak Ridge [14] and the Los Alamos [15] dose conversion factors. These two experimentally derived factors were averaged and the resulting factor of 215 rad-kg/μc was used to estimate the neutron dose. This factor assumed that each man's sodium concentration was the same as the standard man [16] (1.5 g/kg) and that standard man was the same as the burro used in the Oak Ridge experiment.

In contrast, the detailed procedures for estimating neutron doses to the blood from fission spectrum neutrons had been established and included in formal procedures. This accounted for the rapidity with which blood analysis was accomplished. The conversion factor of 1.72×10^5 rad-cm^3/μc was the average of the values of 1.66×10^5 rad-cm^3/μc from the Oak Ridge burro experiment and 1.77×10^5 rad-cm^3/μc from measurements with sodium salt solutions exposed to the Godiva II reactor. Whole blood and blood serum sodium concentration measurements were not made; however, current emergency procedures call for such measurements to be made.

Although no formal detailed procedure was available for phosphorus-32 in hair measurements, the analytical chemistry and dosimetry evaluation techniques had been published in the literature [17] by PETERSEN et al. the preceding August. Their formula was used which gave factors of 6.44×10^6 n/cm^2 per dpm per gram sulphur and 0.0246 rad per dpm per gram sulphur. The sulphur content of the hair was found to be in reasonable agreement with their value of 47.7 mg sulphur/g of hair and was used in all calculations.

3.10. Medical treatment

Medical findings, treatment and observations have been reported elsewhere. No problems regarding such important factors as availability of physicians, hospital admittance, possible contamination or adverse reactions of personnel arose. Nevertheless, one could see how it would be possible to worsen a grave situation in the event that attending personnel are not conversant with the problem and, because of the lack of this knowledge, create unnecessary and unwarranted problems.

Although the staffs of local hospitals often may not be well versed in the handling of radiation accident cases, support can be obtained from radiologists and personnel working in hospital radioisotope laboratories. It would therefore be important for each installation where accidental radiation exposures are a possibility to have a clear-cut plan of action for the handling of accident cases worked out beforehand, particularly if there is a possibility of radioactive contamination.

4. CONCLUSION

Normal and alternate procedures should be designed around the primary purposes of emergency dosimetry: these being to:

(1) Segregate as rapidly and as knowledgeably as possible those personnel most likely to have suffered varying degrees of excessive radiation exposure.

(2) Provide physicians with timely and useful exposure data.

The first task of segregation and identification of the exposed personnel was completed rapidly and accurately. Considering the distances involved and the fact that the event occurred on a weekend when the members of the staff were not at the laboratory, it was quite an accomplishment to obtain the preliminary dose estimates within the first few hours. The one-hour delay between the preliminary dose estimates and the "official" dose estimates could have been reduced somewhat. The delay developed from uncertainties due to lack of experience in coping with such incidents and to a reluctance to believe that the doses were so low to personnel who were so close to the event. Today, as a result of the experience gained, an "official" estimate of doses can be predicted, with confidence, more quickly. The following order has been assigned to the implementation of criticality emergency procedures:

(1) "Quick-sort" sodium-24 procedures;
(2) Film badges;
(3) Blood or blood serum sodium-24 activation analysis or whole-body counting;
(4) Phosphorus-32 activation in hair.

Other procedures to enhance the accuracy of the dosimetry to help in later scientific understanding of the observations made are relegated to more subordinate positions.

Finally, even with a large and component staff available to handle the situation, back-up support in the nature of consultations with various experts was most desirable. It was especially useful to have a prepared list of readily available specialists in the area of specialized dosimetry and medical treatment. Smaller organizations would have found such preparation of paramount importance.

REFERENCES

[1] ROESCH, W. C. et al., Hlth Phys. 9 7 (1963) 757-68.
[2] WILSON, R. H., A method for immediate detection of high level neutron exposure by measurement of sodium-24 in humans, USAEC Rpt HW-73891 Rev. (July 1962).
[3] LARSON, H. W. and KEENE, A. R., "The Hanford emergency dosimetry system", these Proceedings.
[4] SMITH, R. J., Thermal and fast neutron effects on dosimeter films, Chemical Corps Nuclear Defense Laboratory, Army Chemical Center, Rpt NDL-PR-13 (1961).
[5] KOCHER, L. F., BRAMSON, P. E. and UNRUH, C. M., The new Hanford film badge dosimeter, USAEC Rpt HW-76944 (Mar. 1963).
[6] PERKINS, R. W., Personal communication, 24 Nov. 1964.
[7] HURST, G. S. and RITCHIE, R. B., Radiation accidents : dosimetric aspects of neutron and gamma ray exposures, USAEC Rpt ORNL-2748, (Nov. 1959).
[8] BALLINGER, E. R., HARRIS, P. S. et al., Nucleonics 20 10 (1962) 76-85.
[9] SHEEN, E. M., Gamma spectrometer for use in dosimetry studies, USAEC Rpt HW-84250 (Oct. 1964).
[10] PERKINS, R. W., NIELSEN, J. M., ROESCH, W. C. and McCALL, R. C., Science 132 (1960) 1895-97.
[11] GRAHN, D., Paper C-2, Book 1, Proc. Symp. Protection Against Radiation Hazards in Space, USAEC Rpt TID-7652 (1962) 275-87.
[12] GRAHN, D., SACHER, G. A. and WALTON, H. A., Rad. Res. 4 (1956) 228-42.
[13] RBE COMMITTEE, Report to the International Commissions on Radiological Protection and on Radiological Units and Measurements, Hlth Phys. 9 4 (1963) 357-84.
[14] Accidental radiation excursion at the Y-12 plant, Union Carbide Nuclear Company Y-12 Plant, Rpt Y-1234, (June 1958).
[15] HARRIS, P. S., J. occup. Med. Special Suppl. (Mar. 1961) 178-83.
[16] INTERNATIONAL COMMISSION ON RADIOLOGICAL PROTECTION, Recommendations, Brit. J. Radiol., Suppl. 6 (1954).
[17] PETERSEN, F., MITCHELL, F. E. and LANGHAM, W. H., Hlth Phys. 6 1/2 (1961) 1-5.

DISCUSSION

G. H. PALMER: The paper shows the success of your "quick sort" method, a success clearly largely influenced by the absence of contamination. At our establishment, we issue everyone on the site with an identity card embodying a small piece of indium foil. This is in addition to any dosimeter issued to people working in radioactive areas. Although our emergency plans include the "quick sort" technique described in this paper, we feel that the indium foil method is perhaps preferable because of its independence of contamination. Have you any views on the use of such foil in place of the "quick sort" arrangements used in the accident, which occured some years ago?

H. M. PARKER: I do not disagree with what you say, Dr. Palmer. In fact, in the oral presentation, I acknowledged that the quick sort method and the film badge with foils almost tie for first place. Our badges now contain indium and we would certainly expect to read these out quickly as an "indium quick sort".

There is just one factor in favour of the Na^{24} quick sort: An employee leaving hurriedly does not leave his blood behind, but he can be separated from his badge. There is an absolutely certain imprint of neutron dose in the Na^{24} method.

J. P. LOUTIT: The Na^{24} values show impressive concordance for blood and whole-body measurements. Did you use the ICRP value of 105 g for total body sodium in your calculation, or were direct estimates made?

H. M. PARKER: The concordance is due to the fact that we are simply measuring Na^{24} in both methods and this is easy to do at non-serious dose levels. We are inherently using the same body sodium value, accepting the ICRP value of 105 g for standard man. I think we also made approximate Na^{23} determinations in the blood sample later.

H. HOWELLS: In the event of a criticality accident, it is essential to make a quick sort of the personnel involved. One of the more important factors is the location of personnel relative to the site of the incident, to be found by subsequent interrogation. Could you give an indication of the effectiveness of questioning and what reliability could be placed on the information thus obtained?

H. M. PARKER: As can be seen by comparing Tables I and II, there was a change in the estimated positions of employees A and B particularly. When we had time to check the floor plans, we found that A could not have been quite as close as he thought. I might add that we did not know on the first day just what had gone critical. Employee C made a very good estimate of his distance, which he put at 7.6 metres _versus_ a later value of 7.9 metres.

The 19 others in the building were interviewed and their locations and exit routes plotted. Reference [1] contains a summary of this information.

When we later entered the facility, we found that one of the men had reported under stress that he left through a door that was found to be locked and intact. I think this confusion is to be expected of even well-trained men under the stress conditions caused by seeing a blue flash.

D. F. PETERSEN: The following is half by way of comment and half confession. The idea of distributing indium or other materials in clothing is a good one, both from the standpoint of quick sorting and for accident neutron dosimetry. However, some provision must be made for their automatic identification. In one accident in which we were involved, the brass buttons were meticulously removed from a pair of coveralls for activation analysis, but we realized afterwards that we had not even a foggy notion of their original location.

H. M. PARKER: The point is a very good one, Dr. Peterson. Mr. Larson and I have sketched out a system that might cover your point. One could make up standard buttons in various coloured plastics, with indium wire and perhaps copper and gold to give rough spectrum data, keeping the same colour for the same location on all coveralls. Perhaps four buttons back and front would do the job. The only risk remaining then would be that of confusing the whole set from one individual with another, but I believe that this would be simple to control. If mass-produced, we could all afford to use a standard system.

Critiques and Correspondence Related to Dr. Evans' "CORD" Document

Radiation Exposure of Uranium Miners

August 1967

RADIATION EXPOSURE OF URANIUM MINERS

HEARINGS
BEFORE THE
SUBCOMMITTEE ON
RESEARCH, DEVELOPMENT, AND RADIATION

JOINT COMMITTEE ON ATOMIC ENERGY
CONGRESS OF THE UNITED STATES
NINETIETH CONGRESS
FIRST SESSION
ON
RADIATION EXPOSURE OF URANIUM MINERS

PART 2
ADDITIONAL BACKUP AND REFERENCE MATERIAL TO THE
HEARINGS HELD MAY 9, 10, 23, JUNE 6, 7, 8, 9, JULY 26, 27,
AND AUGUST 8 AND 10, 1967

Printed for the use of the Joint Committee on Atomic Energy

APPENDIX 20

CRITIQUES AND CORRESPONDENCE RELATED TO DR. EVANS' "CORD" DOCUMENT (APP. 18)[1]

CRITIQUE OF URANIUM MINING HAZARDS TESTIMONY

AN INVITED FOLLOW-UP TO THE ROUND TABLE DISCUSSIONS HELD BEFORE THE SUBCOMMITTEE ON RESEARCH, DEVELOPMENT, AND RADIATION OF THE JOINT COMMITTEE ON ATOMIC ENERGY CONGRESS OF THE UNITED STATES, JULY 26 AND 27, 1967

(By Herbert M. Parker, Consultant to the Director, Pacific Northwest Laboratory Battelle Memorial Institute, Richland, Wash.)

In connection with the round table discussions on Uranium Mining Radiation Safety Standards conducted before the Subcommittee on Research and Development and Radiation of the Joint Committee on Atomic Energy, July 26 and 27, 1967, Dr. Robley D. Evans introduced a 61-page document under the acronym CORD (On the Carcinogenicity of Inhaled Radon Decay Products in Man).

Recognizing the impossibility of a prompt review by other discussants, the Chairman invited submission of critiques. This critique, under the acronym CUT, is responsive to that invitation. The CORD document contains 47 references, each of which should be considered in an authoritative review. Also, the review, if it should disagree with CORD, as it emphatically does, should be supported by an extensive literature search and bibliography. In the interval available for a timely reply, these steps had to be omitted or minimized. Details of the final posture might be modified for these reasons, and also after face-to-face discussion with Dr. Evans. However, the probability of modifying the *basic* conclusions of this review is believed to be low.

A. *Testimony of June 1967*

CORD assumes an awareness of the previous testimony by Dr. Evans, June, 1967. The relevant portions of that are briefly reviewed first. It is stated that:

"The induction of a carcinoma or sarcoma in a human by the application of ionizing radiation generally requires a continuous irradiation or a series of repeated radiation insults, distributed over a period of years or decades, and at a relatively high dose rate."

The language is even stronger in CORD:

"* * * there is no documented case of a single large radiation exposure causing a radiation-induced malignancy in man. Instead, a prolonged continuous exposure or a series of ulcer-producing single exposures distributed over a period of years seems a requirement in man."

These statements are both false. It is only necessary to cite radiation-induced leukemia in Hiroshima survivors. The medical literature is replete with reports of neoplasms following radiation therapy; these mainly involve treatments characteristically delivered over periods of weeks to a few months—not years.

Documented occurrence after a true single exposure does exist (Mitchell and Haybittle, Acta Radiol. *44*, p 345, 1955). It is relatively rare, probably because single therapeutic exposures are themselves rare, and accidental exposures severe enough to be readily carcinogenic are often lethal for other causes before a tumor could appear.

The series of reports of the United Nations Scientific Committee on the Effects of Atomic Radiation contains extensive reviews of carcinogenesis in man. The ICRP Publication #8 (The Evaluation of Risks from Radiation, 1966) contains an up-to-date review of these findings. Evidence from these sources is incontrovertible that cancer in man can be produced by single or extended doses of 100 rads or more. In the unborn child, doses as low as 5 rads have been implicated in subsequent development of malignancy.

[1] For additional comments on critiques of "CORD," see app. 27, p. 1364.

1242   RADIATION EXPOSURE OF URANIUM MINERS

Observed carcinogenic doses for lung tumors for comparison with the uranium miner experience are less easily quantified. The ICRP Publication #8 describes the bone tumor cases following ingestion of radium as unsuited to estimate the relationship between dose and cancer with any accuracy.

For the particular case of bone sarcoma, H. L. Jaffe (Tumors and Tumorous Conditions of the Bones and Joints: Lea and Febiger) accepts postradiation sarcoma as an entity induced by exposures in the range of 1500 R to 10,000 R delivered over periods from 3 months to years with tumor appearance times of 3 to 20 years, more or less.

Dr. Evans states that "the MPC for radium-226 set officially at 0.1 microcurie in 1941 on a basis of our then-studied total of about 30 cases, is well below the threshold suggested by this ever-increasing body of data."

The term "threshold suggested" is acceptable if interpreted as "tentative practical threshold suggested." It becomes progressively less appropriate if the testimony is worded to imply a "demonstrated threshold." We agree, of course, that the quoted radium limit has been one of the most effective cornerstones of radiation protection practice, and is better founded than most if not all others.

The dose-response curves as presented in Figures 4, 5, and 6 of the BJR paper referenced by Dr. Evans are graded response constructions. The response parameter (classical x-ray score) is not a definable biological response such as death or tumor occurrence. Rather, it involves an arbitrary mixture of variously weighted skeletal abnormalities which are not necessarily sequential. Doubtless it has proved of value in some aspects of the important studies of radium damage in bone. As applied by analogy to the uranium miners case, one would perhaps work out a sputum cytology score, established by some arbitrary weighting of the sputum changes so effectively studied by Dr. Saccomanno. These changes may be sequential and hence reasonably analyzed together. Dose versus graded response curves permit one to analyze a large number of cases, collectively, but their inherent limitations must be kept in mind. Thus, Figures 4. 5, and 6 deal with 270 cases, but are not relevant to establishment of a firm threshold for tumor induction.

Figures 7 and 8 deal with 360 cases in terms of age in years or years since first exposure. Inferences are drawn that life-shortening cannot be discerned at depositions below 0.5 Ci PRE. It is rather difficult to estimate life-shortening until the applicable lives have been terminated. Of the approximately 360 cases, 173 are charted with depositions above 0.01 μ Ci. Of these, 108 are presently alive and 65 are dead. Of the terminated lives, only 8 refer to depositions below 0.5 μ Ci PRE. It takes a bold observer to conclude from these insufficient data that there is a *sharp* threshold for life-shortening at 0.5 μ Ci PRE. The sample population is too small to support any conclusion other than that large radium depositions produced significant reductions in life span. This is somewhat comparable with the existing state of knowledge about radon daughter deposition in the lung.

The radium cases at issue for tumor induction total only 26 sarcomas and 9 carcinomas. These should probably be analyzed in separate dose-response curves. Thus, there would be only 26 points on the fullest observed curve. Of these cases, 10 represent ingestion of mainly ordinary radium (^{226}Ra) and 13 of mainly mesothorium (^{228}Ra). The legitimacy of grouping these together through the arbitrary device of the PRE parameter has already been indirectly questioned by Dr. Evans himself. The alpha dose patterns from the two are different in both time and geometry. The different admixture of beta and gamma radiations in the two cases may even play some role in modifying the environment of cells that are primarily assaulted by alpha radiation. Extensive studies by Dr. W. J. Bair and associates on effects of plutonium oxide particles in the lungs of dogs have been made at Hanford. Quite recently, serious consideration has been given to the possible influence of the very weak x-radiations from the plutonium. Effects from the auxiliary radiations from radium are inherently more plausible. This objection would apply also to the dosimetry models for radon daughters in the bronchi.

Dr. Evans has a consistent curve of tumor induction versus radium deposition that includes about a dozen points. It is being generous to permit the 26 sarcomas and 9 carcinomas to appear on a single dose-response curve with the lowest entity of 0.5 μ Ci PRE. If we do this, there is no guarantee that the 36th case will not show a deposit of perhaps only 0.2 μ Ci PRE. In fact, if enough data points were obtained, a whole cluster should appear at all values of PRE down to the very small normal deposit in the skeleton, due to the natural incidence of bone tumors. The dose-response curve would then be revealed as the nonthreshold curve that it truly is. To claim otherwise would be tantamount to claiming that radium-induced tumors in bone can be infallibly differentiated from natural tumors. This is not so. Peak incidence of osteogenic sarcoma in the second decade of life, and preferential occurrence in the general region of the knee perhaps justify arbitrary exclusion of some cases.

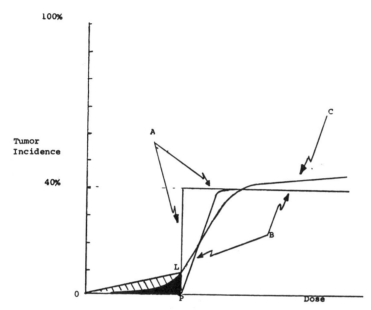

Fig. 1 Stylized dose-response curves for tumor induction.

A. An engineer's step function
B. An engineer's ramp function
C. A sigmoidal function

Dr. Evans has characterized the dose-response curve as a step function. (Curve A of Fig. 1) In this, the tumor incidence is zero at doses less than that at point P, where this corresponds to the effective dose for a long-term deposition of 0.5 μ Ci PRE. Probably, we would agree that he meant some kind of ramp function such as B. The essential difference between his position and mine rests on a single letter 'e'. For if we change 'step' to 'steep,' we would agree that a steep sigmoidal function or a S- curve such as C fits the existing positive data at doses higher than P and is consistent with the necessary inclusion of some incidence at all doses no matter how small. The question would be how to take care of tumor incidence represented in the dark shaded area OLP, where L purports to be the point on curve C for the lowest dose P for which a sufficiently reliably incidence has been determined. Proponents of the linear extrapolation concept use the straight line OL instead of the portion of curve C between O and L. The difference between the dark area OLP and the lightly shaded triangle OLP is quite small and becomes smaller with time as data points for doses between points O and P are filled in.

We have described these curves as stylized because *real* data points do not fall quite so smoothly on mathematical curves such as C. The basic issue is the same and is elaborated in the next section.

Dr. Evans goes on with the following statements:

"In the present series of hearings this Committee has been exposed primarily to the conservative, oversimplified, incorrect, linear and nonthreshold model of radiation carcinogenesis. Traditionally, this cautious model has been used by the NCRP, ICRP, AEC, IAEA, and other bodies to deduce legislated or recommended values of acceptable or permissible doses for all situations where such low values are within the domain of economic and engineering feasibility.

1244   RADIATION EXPOSURE OF URANIUM MINERS

"The time is probably overdue to use the radiobiological knowledge acquired especially in the past decade, and to admit to the radiation protection arena dose-vs.-response relationships which are nonlinear and may have a threshold.

"It is my conviction that there does exist an absolute threshold and a practical threshold for inhaled radon daughters below which these nuclides are innocuous."

It would be improper to leave this criticism of the NCRP and other activities unchallenged. Surely every professional radiation protectionist would prefer to establish a threshold limit for each application of radiation. His task would then be finished with assurance of complete protection in all occupational situations in which standards were met.

For some years I have been involved as chairman of an NCRP scientific committee engaged in reviewing such matters. The members cannot persuade themselves that a threshold model for long-term radiation damage including carcinogenesis can be justified, with the possible exception of opacity of the lens of the eye. The implication that these observers have failed to bring to bear radiobiological knowledge acquired especially in the past decade is rejected with vigor. It has to be accepted until proven otherwise that response of a given kind is a monotonically increasing function of applied dose. In the medium dose regions, the dose-response function may be experimentally measurable and it need not be linear. At very high doses it may in fact cease to be monotonic unless the terms of reference are carefully defined; obviously, very large doses may not produce a high incidence of cancer because the exposed individuals may be promptly killed by radiation of the blood-forming organs. The region of interest is that of low doses, currently below the region where definitive analysis can be achieved. It is prudent to assume that this region is governed by a linear extrapolation. This assumption is not demonstrably incorrect; it is simplified but not oversimplified; it is conservative, but not the most conservative that could be invoked. It is a responsible, prudent opinion that creates endless difficulty by transferring decisions from science to value judgment and calling for a balancing of benefit versus risk when neither can be reliably ascertained.

In 35 years of diligent study, Dr. Evans has failed to establish acceptable evidence for an absolute threshold for the tumor induction in bone by radium deposits. He should not try to persuade us by analogy that there is such a threshold for tumor induction by radon daughters in lung. He does present persuasive reasoning for some kind of practical threshold in the range of 0.1 μ Ci to 0.5 μ Ci radium.

Returning to Fig. 1, we are maintaining that the shaded area OLP is real and the triangle OLP has to substitute for it.

We agree that there is one additional factor that brings the two approaches closer together in practice. It is generally conceded that for a given type of tumor, the appearance time is prolonged as the effective dose decreases. Dr. Evans, who normally seems to eschew the use of animal data, would here have us accept a logarithmic dependence for sarcoma induction time derived from ^{90}Sr irradiations in mice. We can accept the trend but not the absolutism. Otherwise, the induction times for the present 26 sarcomas would all fall in the precise inverse order of dose.

Furthermore, the induction time of the bone sarcomas is very different from that of the carcinomas of the paranasal sinuses. The governing logarithmic formula would be different. Presumably, the relative proportions of the two will begin to change sharply with time, a feature that should ultimately be experimentally detectable.

Despite these conceptual differences, we would agree that a maximum permissible deposition of 0.1 μ Ci Ra is a very sound practical threshold point. Safety factors involved in this acceptance are (1) an obvious practical limitation on the total group of people conceivably involved with radium deposits, (2) the fact that past cases now containing 0.1 μ Ci previously had much higher deposits. Future cases will be held to a maximum deposition of 0.1 μ Ci. In full agreement with Dr. Evans (CORD: p. 22), I consider this radium limit to have far more evidence behind it than does the conventional permissible limit of 15 rems per year for most critical organs as used by the ICRP and NCRP.

B. The CORD Document

The first section of CORD is essentially a recapitulation of the analysis of lung dosimetry models as given in Preliminary Report #8, Guidance for Control of Radiation Hazards in Uranium Mining, FRC, with some interesting additions.

The section on nonlinearity in radiobiological effects and its incorporation in radiation protection philosophy differs from the usual account. The early history was heavily governed by the medical experience with x-rays and to a lesser degree with radium gamma rays. That part arising from the early luminous dial painter experience to which Dr. Evans has so brilliantly contributed advanced more or less independently of the main theme. Clinical dosimetry itself was based on the erythema dose required to produce a delayed reddening of the skin. This clearly was a threshold effect. If the dose required to produce reddening was spread out in time no erythema appeared. It was natural to assume a threshold for other delayed responses. None of these intuitive thresholds was based on definitive experience in man. Practical limits on daily, weekly, or monthly exposure were set on observations that operators at somewhat higher levels showed no apparent injury. Although radiobiological effects in various targets were known to be nonthreshold in some cases and threshold in others, it was not until the mid-1940's that serious concern developed that the relevant effects in man must be governed by a nonthreshold action. The elucidation of the genetic effect and the Boche theory of life-shortening were main contributing factors. From the 1950's on, genetic studies in animals closer to man than is the fruitfly made an acceptable risk philosophy inescapable. The plausibility that carcinogenesis is related to some form of genetic change which modifies the control mechanism in somatic cells has extended the same argument to leukemia induction and the like in man. The theory is evidently not the whole story, but it cannot be abandoned in light of present knowledge.

In the last decade, the greatest advances in radiobiology in addition to genetic studies may well have been the ability to study the survival of irradiated mammalian cells, to determine their sensitivity at different stages of the cell cycle, and to study recovery from fractionated radiations. It is important to realize that in survival curves, the effect measured is *cell death*. As early as 1944, P. S. Henshaw pointed out that radiation, in the production of biologic changes, is known to cause two effects that are biologically nearly opposite: cell death and carcinogenesis. The new biology has provided a formidable tool for understanding radiation treatment of cancer cells, but has not resolved the radiation-induction of cancer.

We agree completely with Dr. Evans' analysis of the lung cancer experience among uranium miners of the Colorado Plateau (CORD: pp. 29 and 30). The studies to date are well characterized as pathfinding. Great caution must be exercised in deducing more than that substantial exposure to radon daughter inhalation in mines of the Plateau will lead to excessive incidence of lung cancer. This reviewer places the need for a greatly intensified epidemiological program, both retrospective and prospective in aspect, second only to prompt application of best efforts to reduce the radon and radon daughter concentrations breathed by miners.

I disagree completely with the section on the effects of cigarette smoking (CORD: pp. 31, 32, 33, 24). We have both noted that great caution must be applied to interpretation of the totality of the Public Health Service data. No statistically valid conclusion whatsoever can be drawn from that very small portion of it that involves a handful of nonsmokers. I can find no evidence that either confirms or denies the existence of a synergistic relationship between cigarette smoking and exposure to radon daughters. However, the feasibility of such an effect was reiterated frequently during the hearings. It would clearly be prudent to investigate the possible effect in animals, if meaningful experiments could be devised. W. J. Bair et al. have a battery of trained smoking dogs in the Pacific Northwest Laboratory, used so far to estimate potential changes of pulmonary clearance of other radioactive particles. These pack-a-day animals can be continued indefinitely on the exposure plan which could be readily supplemented by radon daughter inhalation. Experiments along such lines should be part of the long-range attack.

The key section of CORD is that dealing with inferences from observations on human radium and mesothorium cases. (CORD; p. 36 et seq.). Six ways are claimed (CORD: p. 37) in which these observed radiation-induced malignancies are said to have a close relationship to the radiation-induced lung carcinomas seen in underground uranium miners of the Western United States.

We submit that the alleged relationships are tenuous and superficial. Collectively, they do not provide an adequate basis for relating the lung exposures in any quantitative way to the existing experience with bone tumors. A brief comment on each of the six points follows:

1. "Both are in the human species, not rodents or dogs. There is no species difference."

1246   RADIATION EXPOSURE OF URANIUM MINERS

The fact that both tissues are in the human species is irrelevant, beyond the generalization that one would always be more comfortable in using applicable human data than animal data. In the present case, the radium-induced tumors are predominantly bone sarcomas and the radon daughter induced tumors are predominantly carcinomas of the bronchial epithelium. The natural function and properties of these two tissues are so widely different and the ultimate mechanism that generates either spontaneous or radiation-induced tumors in each so little known that their comparison is futile. Many qualified observers would prefer to draw conclusions referrable to human bronchial tissue from observations of dog or other animal bronchial tissue than from human bone tissue.

2. "Both are induced by α rays, not by β, Υ, X, or neutron irradiation. There is no LET difference."

It is conceded that the presumptive primary agent in both cases is alpha radiation. There *is* a LET difference both on the average and individually along each alpha particle track. Appendix B of FRC Preliminary Report #8 shows that the linear energy transfer (LET) near the end of the range of a 6 MeV particle is about 5 times that near the beginning. Also, the rate of expenditure of energy in bone is about 50% higher than that in lung tissue. However, these LET differences are probably insignificant in terms of relative biological effectiveness; if an alpha particle intersects a relevant biological target at all, it can be assumed to deposit enough energy to initiate whatever radiation-induced changes are possible. The effect of LET differences is more likely to be reflected only in its influence on the range of the particles in tissue, and hence on the detailed geometry of the case.

The geometry is relatively uncertain in both cases. In the lung, the problems of uncertain thickness of the bronchial epithelium and uncertain thickness of the mucus blanket and the mixing of radionuclides in it have been covered in Appendix B of FRC Preliminary Report #8. In and adjacent to bone, the complications include gradual translocation of the radium, mineral exchange removal near the surface, and the desposition of new calcium layers on existing radium deposits. In addition, the diffusion of radon through the bone adds another variable, which is markedly different for deposits of ^{226}Ra and meso thorium (^{228}Ra), because of the relative half lives of the ^{222}Rn (3.825 days) from radium and the ^{220}Rn (54 sec.) from mesothorium.

As indicated in a previous section, it is by no means assured that the accompanying beta and gamma radiations do not play some role in affecting the environment of those cells that ultimately become cancerous. These contributions are different for radon daughters in the bronchial epithelium (and, in fact, in different parts of the bronchial tree) and for radium in bone and for mesothorium in bone.

3. "Both involve the irradiation of a single organ of* tissue, not a generalized whole-body irradiation."

Perhaps it is more productive to say that both involve the irradiation of a separate complex and variable tissue system, with no fruitful analogy between the two cases. If both were generalized whole body irradiations, there might well be a close relationship between them. As it is, this point argues against, not for, Dr. Evans' proposition.

4. "Both involve a nonuniform spatial distribution of absorbed alpha-ray energy within the irradiated tissue with both "hot spots" and a diffuse distribution."

The point is well taken, but again it speaks not for but against a close relationship. Intense hot spots, related to the Haversian canals, occur with radium deposition in bone. The hot spot concentration can be as high as 200 times the diffuse distribution. Permissible limit calculations by the ICRP have included a hot spot factor on the assumption that maximum damage would occur in these locations. This would be the normal expectation; the radium cases may be one outstanding exception. A review by R. E. Rowland (Studies on Bone-Seeking Radioisotopes," XIth International Congress of Radiology Progress Report, Vol. 2, p. 1530, 1965) indicates growing general acceptance that the damaging effect is more closely related to the diffuse distribution. Just what the effective dose is in the bone tumor cases is still unknown.

It is equally unknown, of course, in the lung cases. The present models show the calculated doses to be highest at regions of unusually thin bronchial epithelium in the general area of the secondary, quarternary, and segmental bronchi.

---

*The word "of" appears in the CORD document. We suspect that the word "or" was intended.

This result tends to be pleasing because the carcinomas also occur in this general area. However, no one has demonstrated that they are associated with unusually thin epithelium. Without that corollary, dose calculations do no more than establish that continued exposures (in the range of 1 WL or less, in my opinion) can provide a radiation dose capable of inducing cancer. Presumably, the actual cancers develop in a naturally susceptible tissue, adequately irradiated, and maintained in a continuing cellular environment that favors or at least condones eventual carcinogenesis. This does not have to coincide with the region of highest dose, as tacitly assumed in modeling.

Conversely, there is no a priori reason to believe that the lung tumors discriminate strongly against the hot spots, so to speak, as the bone tumors may appear to do. CORD, pp. 38-39, compares the cumulative average dose throughout the lung for a given number of WLM with the cumulative average skeletal dose for a given radium deposition. This is totally unpersuasive as a device for inferring that a lifetime exposure of, say, 500 WLM leaves a "wide margin of safety, ignorance, or relative radiosensitivity."

5. "Both involve chronic irradiation distributed over a long period of many years, not an acute exposure."

The point is correct, but hardly one establishing close relationship. Virtually all the problems relating to occupational exposure to radiation meet the same criterion.

6. "Both involve a dose rate which was initially high and which decreased with time, commonly by a factor of the order of 20 or more. For many veteran and continuing miners such a reduction in exposure is reasonably well documented, and is due to the steady improvements in mine ventilation and the consequent reduction in the WL exposure level; these mine exposures are analogous to persons with skeletal deposits of essentially pure Ra-226. For individuals who mined for a few years and then ceased underground work, the distribution in time of the dose rate may be analogous to that in persons with skeletal deposits of essentially pure MsTh, whose physical half-period of 5.7 years combined with biological excretion of the parent radionuclide results in a dose rate which decreases sharply with time after discontinuing the exposure, and may have been negligible during the final decade before the appearance of the malignancy."

The point is sound as far as the demonstrated cases are concerned, because of the general reduction of radon concentration in the mines in the last decade. The objective of both aspects is to arrive at a permissible level appropriate for continued employment. Both sets of cases present problems when viewed in this light.

Various speculations on the existing radium cases hint that the principal effective dose may be delivered in the few years following the primary ingestion. Not only is much of the radium being eliminated at that time, but the remainder may be being more or less walled off from the sensitive biological targets. As understanding of the process and its dosimetry improves, differences, if any, between pure radium cases and mesothorium cases should permit some resolution. In the radon daughter cases, there is no similar walling off. Progressive changes in ciliary action, in cell dimensions as in metaplasia, and in thickness and flow rate of the mucus layer could affect dosimetry in either the *positive* or *negative* direction.

Another significant difference is that when over-exposure is detected and the employee removed from the contaminating source, lung exposure from radon daughters will be virtually terminated within hours. Bone dose will continue indefinitely at decreasing intensity. This aspect is irrelevant to the present critique of comparisons. It is highly significant to the issue of whether a temporary WL limit that may later be lowered makes sense.

7. "Morphologically, the epithelial lining of the paranasal sinuses is the same structure of a thin ($\sim 2\,\mu$) basement membrane on which lie basal cells and their daughter ciliated columnar and goblet cells as is found in the bronchial epithelium. More concisely, the paranasal sinus tissue is respiratory tissue. The principal difference is that the columnar cells of the sinus epithelium tend to be roughly 10 μ taller (W1), i.e., 40 to 100 μ rather than 30 to 90 μ than their counterparts in the bronchial epithelium."

This attempted comparison of local tissue doses for bronchial epithelium and paranasal sinus epithelium is one of the more intriguing features of CORD. This reviewer is not well qualified to pass judgment as to whether these two tissues can be considered alike enough to permit the cross-comparison. Some indirect support for the assertion has been in a cursory literature review. J. I. Fabrikant et. al. (Brit. J. of Cancer, *18*, p. 459, 1964), in discussing tumors of the lacrimal sac describe them as closely related histologically and developmentally to

1248 RADIATION EXPOSURE OF URANIUM MINERS

similar carcinomas arising in the nasal and paranasal cavities since the lining epithelium and the respiratory epithelium develop along similar embryologic lines. This does not specifically extend the relationship to respiratory epithelium of the bronchi, but the inference is probably there.

Associates and consultants in the Pacific Northwest Laboratory tend to reject the assumption of identity or parallelism as far as tumor induction is concerned. They point to the marked differences in spontaneous incidence of carcinoma as one moves along the respiratory tract. This occurs also in the other internal regions with an epithelial lining. Spontaneous incidence is also markedly different in man and woman, frequently by a factor of 5 in various sections of the respiratory tract. The spontaneous incidence rates are generally regarded as having some sort of correlation with radiation-induced incidence rates. Certainly the immediate environment of the two tissues and the effect of possible co-factors is different. Another relevant feature is the extreme latency of radiation-induced cancers of the pharyngeal area (10 to 35 years with a mean of 25 years, according to A. W. G. Goolden, BJR, *30*, p. 626, 1957) compared with radiation-induced sarcomas of bone (3 to 22 years with a mean of 8.6 years in a series reported by A. Jones, BJR, *26*, p. 273, 1953). We do not have latency data directly applicable to the paranasal areas, except for the radium cases themselves.

In any case, to use one positive male case and negative female case as a basis for proposing a carcinogenic dose for human respiratory epithelium between 18,000 rad and 82,000 rad is not only premature, as Dr. Evans states; the assertion should not be introduced at this time lest it be inadvertently used to justify exposure of the bronchi to comparable levels. A systematic search would doubtless clearly show the induction of tumors in respiratory epithelium at substantially lower doses. Goolden refers to three sub-glottic tumors reported by Van Nieuwenhuysen 10 to 16 years after radiation therapy for benign conditions. The doses are not reliably known but they must have been of the order of a few thousand rads, certainly not 18,000 rads; these were x-irradiations, presumably some 3 to 10 times less effective per rad than alpha radiation.

The conclusion could, in fact, be refuted by the PHS data alone. I would find it impossible to believe that cumulative exposures of 1000 WLM are not responsible for a high incidence of lung cancer. This computes at about 2800 rads in unusually thin portions of the epithelium, subject to a possible margin of error by a factor of about 3 in either direction (i.e., plausible range of about 900 to 8,000 rads). I would be quite surprised if it cannot eventually be demonstrated from a combination of these cases and resonable inferences from medical exposures that doses of 2000 rads or less are clearly carcinogenic in the bronchi of the human male subject.

Despite the inadmissibility of Point #7 at this stage, Dr. Evans is to be complimented for bringing the matter up for discussion. An evaluation of all possible existing exposure data in humans that may be brought to bear on the lung inhalation problem is very much in order.

C. Summary

The main thesis of Dr. Evans' June testimony and the extensive CORD document is—

(a) to persuade that a firm threshold of 0.5 u Ci pure radium equivalent (PRE) as the long-term deposit in bone is safe against tumor induction in bone or adjacent tissue; it is also adduced that this represents a high radiation dose in tissue;

(b) to persuade that there are as many as six points of close relationship between the radium in bone cases and the radon daughter in lung cases;

(c) and hence that the permissible exposures to radon daughter inhalation in uranium mines may safely be left at a value considerably in excess of that promulgated by misguided exponents of linear extrapolation, acceptable risk and other NCRP or ICRP "mythology."

This invited critique first demonstrates that a firm threshold compatible with the limited observations made to date and with applicable radiobiological theory cannot be established.

It then considers the six points of alleged close relationship between the radium in bone cases and the radon daughter in lung cases. It discloses the generally tenuous and superficial nature of these alleged relationships.

It concludes, then, that deductions of permissible levels of radon daughter inhalation cannot properly be drawn from the radium data.

It further concludes that dosimetry for two isolated cases of paranasal sinus irradiation, while of interest because this is an effect in one form of respiratory

RADIATION EXPOSURE OF URANIUM MINERS 1249

epithelium, cannot reasonably be applied to suggest, even on a tentative basis, the existence of a high threshold dose for tumor induction in the bronchi.

It predicts that demonstrated tumor induction in bronchi at quite modest doses could be established by a thorough study of existing data.

The review supports the point of view that the dose-response curve for the radium cases is probably some approximation to a steep sigmoidal function—certainly not linear at all dose levels. By the above arguments, one must conclude that this does not lend any support to the existence of a similar steep function for the lung cancer production.

Both Dr. Evans and I agree, I believe, that the existing PHS data on the uranium mines is not complete enough to justify its matching to either a linear or sigmoidal dose-response curve. Prudent observers must currently resort to the assumption of a linear extrapolation for the low dose portion of the curve. This denies the setting of a permissible limit on wholly scientific grounds, and demands a pragmatic solution.

D. Limitation of This Critique

Before offering a brief comment on pragmatic solutions, I would like to place the foregoing critique in some kind of perspective. The critique does not profess to be a balanced analysis of the whole CORD report. I have selectively responded to those portions which in my opinion were potentially misleading in the sense of tending to persuade the Committee that there is a safe threshold level for exposure of uranium miners to radon daughter inhalation. I make the unequivocal contrary assertion that no such threshold can be established in the light of present knowledge.

I would be greatly chagrined if the tenor of this reply, due to this selectivity, led its readers to believe that I do not have the same continuing high regard for Dr. Evans' outstanding contributions to radiation protection that I have held for more than 25 years. In fact, the sense of this critique could be taken as admiration of the combination of science and intuition that led Dr. Evans to propose a radon inhalation limit of 10 p Ci per liter (more or less equivalent to 0.1 WL) in 1940, and the radium deposition limit of 0.1 u Ci in 1941. Neither of these limits seems unreasonable as a practical standard in 1967.

It is unfortunate that circumstances did not permit extensive face-to-face discussion between us. The use of such conditional terms as "if," "perhaps," and "possibly" which sometimes appear in Dr. Evans' referenced scientific papers but not in the corresponding portion of his testimony might have resolved some of the apparently sharp differences of opinion. Dr. Evans and I attempted these resolutions by telephone. Although we have differences of opinion on maximum permissible limits for radiation exposure, we rapidly and regretfully agreed that full discussion of CORD grossly exceeded our maximum permissible limit for telephone charges between Boston, Massachusetts and Richland, Washington.

I note that CORD (p. 4) quotes my statement that "far more is *written* about the subject than is *known* about the subject." Others will have to decide whether your two discussants have committed the same sin.

E. Selected Comments on Pragmatic Solutions

The concluding paragraph of CORD (p. 53) proposes that exposures of 1 to 3 WL (12 to 36 WLM per year) and a cumulative lifetime exposure of about 300 to 400 WLM should be well below the practical threshold where all workers may be exposed with only negligible risk of incurring lung cancer from radiation. This involves limitation of the working life to about 10 years at the higher levels.

My own conclusion is that there is no assured limit. Resorting to the linear extrapolation assumption, continuous exposure for a working lifetime at 1 WL would introduce a radiation risk of the same general order of magnitude as the risk (similarly *assumed* by linear extrapolation) of external irradiation at the conventionally accepted rate of 5 rems per year.

In the contractor operations for the AEC, considerable effort is properly devoted to keeping the actual exposures well below 5 rems per year whenever possible. I suggest that if history had led to the uranium mines being AEC-operated, a limit not far from 0.3 WL would have been reached. I would further suggest, by analogy with other operations, that by this time ways would have been developed to maintain such a limit without exorbitant increase in the cost of the process.

I am still hopeful that a solution of this nature will be forthcoming by a renewed imaginative attack on the removal of the contaminants.

Since the time of the round table discussion, the FRC has issued findings and recommendations on the topic under the date of Friday, July 21, 1967, approved

1250   RADIATION EXPOSURE OF URANIUM MINERS

by the President on July 27, 1967 and published in the Federal Register on August 1, 1967.

The relevant finding is: "Occupational exposure to radon daughters in underground uranium mines be controlled so that no individual miner will receive an exposure of more than six WLM in any consecutive 3-month period and no more than 12 WLM in any consecutive 12-month period. Actual exposures should be kept as far below these values as practical."

The practical reasonableness of sustained working lifetime exposure at 1 WL (i.e., 12 WLM in any consecutive 12 months) is the one feature common to the FRC recommendations, Dr. Evans' CORD document and my own review.

I fully concur with the specific FRC recommendation and with the intent to review it in one year. It is my belief that the FRC Preliminary Report #8 was a well-reasoned document contributing to the newly published guidance. Endorsement of these matters is subject to the stimulation of major efforts to—

(1) reduce radon daughter concentration in mine air, and additionally in breathed air, if necessary, by protective respirators;

(2) establish intensive retrospective studies of past and present miners, and others exposed in other radon-infested environments; and

(3) establish prospective studies of presently exposed individuals.

MASSACHUSETTS INSTITUTE OF TECHNOLOGY,
DEPARTMENT OF PHYSICS,
Cambridge, Mass., August 31, 1967.

Subject: Your "Critique of Uranium Mining Hazards Testimony (CUT)"

MR. HERBERT M. PARKER,
Pacific Northwest Laboratory,
Battelle Memorial Institute,
Richland, Wash.

DEAR HERB: Immediately after the July hearings I sent copies of my prepared supplementary JCAE testimony "On the Carcinogenicity of Inhaled Radon Decay Products" (CORD) to about 40 of our colleagues who are involved in such problems in the United States, Canada, England, France, Germany, and Austria, with a request for all the criticism they could give me on it. In the flow of responses these are some fine constructive suggestions, especially for refining the dosimetry and relative risk estimates in the sinus and bronchial epithelium,—Dame Janet Vaughan has suggested that a specific retention function for the skull rather than the usual Norris whole-skeleton retention function be used in estimating cumulative rads from terminal rad/year,—Chuck Mays has encouraged me to attempt to quantify the relative number of epithelial cells at risk in the sinuses and mastoids compared with the segmental bronchi and the distal lung,—Leo Marinelli has reminded me that Thorotrast patients who exhale about 18,000 pCi/minute of thoron for an average of 20 to 25 years have shown no increase in lung cancer in about 2,000 cases (ANL-7060, p. 42) and that measurement of the lung of one patient at ANL showed lung ThB content compatible with an average lung organ dose of 6 rads/year ($\geq$5 WL) for a typical average injection of 25 cc of Thorotrast.

I was glad to receive from John Conway an information copy of your critique because the copy which I hoped to receive from you following our telephone conversation has not shown up. It is always salutary to have a devil's advocate who will go to extremes in an all-out effort to negate a new scientific result or proposal. I, for one, am grateful to you for the large amount of time and personal effort which you spent as advocatus diaboli.

All told, it seems to me that we have a gratifyingly large area of agreement. As you said on pages 17 and 18, CUT was a selective response and not a balanced analysis. Probably most of ". . . the apparently sharp differences of opinion . . ." could have been resolved by oral discussion (or even correspondence,—hence this letter). Many of these differences involve only shades of meaning ("simplified"/"oversimplified") misinterpretations ("cautious" model), quotations out of intended context (the "arena"; "malignancy"/"carcinoma"), or peripheral topics ("no documented case"/one documented case), rather than basic matters in the main stream.

Our area of agreement on fundamental matters seems to include: (1) that the radium or mesothorium cases must be followed "until the applicable lives have been terminated" (p. 3); (2) that the dose vs. response curve for human radium cases approximates a steep sigmoidal function—certainly not linear at all dose levels (p. 17); (3) that there probably is a practical threshold for bone, sinus, or mastoid cancer, or other serious injury, by skeletally deposited radium in

RADIATION EXPOSURE OF URANIUM MINERS

humans (p. 8); (4) that the epidemiological studies among uranium miners of the Colorado Plateau have been pathfinding efforts and great caution must be exercised in drawing inferences from them (p. 9); (5) that a meticulous and intensive retrospective and prospective study of American uranium miners is of prime importance (p. 20) (I have liked the way UNSCEAR put it on page 8 of their 1964 report: "... investigations aimed at recording significant quantitative relationships between doses and observed incidence of any specific malignancy in man should be strongly encouraged and supported."; (6) that "An evaluation of all possible existing exposure data in humans that may be brought to bear on the lung inhalation problem is very much in order" (p. 16); (7) that the FRC's recommendation of 1 WL of radon daughter inhalation is comparable in safety to 5 rem/year of whole body external radiation (p. 19).

Our perimeter of partial or total disagreement may include: (1) the details of the application of human sinus carcinoma data to lung carcinoma (p. 15), (but 82,000 rads *is* a lot more comforting than if it had come out, for instance, as 82 rads in a sinus); (2) the effects of smoking cigarettes in combination with radon daughter product inhalation (p. 9), (but note on p. 33 of CORD "Within the accuracy of presently available data..."); (3) that ICRP Publication 8, submitted on 20 April 1965, is a fully up-to-date review (p. 2) of carcinogenesis in man (some things *have* happened in the last 2 years); (4) that 100 rads incontrovertibly can produce cancer in (adult?) man (p. 2); (5) that the classical x-ray score is "not a definable biological response" (p. 3) [but see Health Physics *13*, 267–278 (1967)]. I think these are mostly minor, and can be straightened out next time we get a chance to chat.

Some other items may be mere misunderstanding rather than disagreement, and several of these seem to merit further discussion. It is easy to get the impression from your selective critique of CORD and related documents that the basic point may have been missed,—that the forest was not seen because of all the trees. Because you were not present at the May and June hearings you may welcome a bit of microhistory, to fill in some gaps in the development and to set CORD in perspective. The 3 basic points are:

1. The starting point was my statement in the JCAE testimony on June 6, 1967 on page 519 of the Stenographic Transcript (hereafter, ST): "From an examination of the data available to me I find no evidence which denies that the threshold lies in the domain of 1 to 3 WL (12 to 36 WLM per year) and an integrated exposure in the domain of 300 to 400 WLM delivered at this relatively low dose rate." By "the threshold" was meant the practical threshold in which the tumor-appearance-time exceeds the residual life span, as defined on page 437 of ST. By "the data available to me" was meant the entire body of physical, medical, and epidemiological data, not only that portion of the USPHS data recorded in FRC Preliminary Report No. 8, but a much more complete and partially corrected body of detailed information on American miners, as well as published and unpublished data on the Jachymov, Schneeberg, and St. Lawrence miners which I have been collecting through correspondence and personal visits since 1938.

2. We were all aware of the many defects in this existing body of epidemiological data, and the importance of improving it by vigorous new efforts over the coming years and decades. Faced with this unavoidable delay, a question which flowed naturally from the June hearings was "Is there any other independent source of quantitative radiobiological data on humans which might be relevant to the lung cancer problem?". The bone cancer data on humans with graded chronic skeletal burdens of radium and mesothorium had been noted in June to have many parallel aspects (page 428 of ST), but the relative radiation sensitivity of bone endosteal cells and respiratory basal cells was an important missing datum in the human.

3. Early in July Shields Warren first provided the crucial news that in humans the sinus epithelium above the midturbinate level is almost identical with the pulmonary epithelium below the vocal fold, whereas the epithelium is mainly squamous between these levels. Three other pathologists soon confirmed this. We realized that our efforts over the past decade to develop physical methods for evaluating the α-ray dose to the sinus and mastoid tissues of humans might provide useful quantitative information related to the lung problem. Professor Kolenkow had already succeeded in perfecting a sound technique and had measured the α-ray spectrum from human parietal bone in several of our cases. Measurements of the radon concentration in human sinus and mastoid air spaces had also been accomplished. Measurements of the α-ray spectrum and computations of the dose rate from available human frontal sinus bones were completed during the following fortnight.

1252   RADIATION EXPOSURE OF URANIUM MINERS

I prepared CORD in order to draw together background material, to provide a perspective of the lung modelers' work, to draw attention to hot-spot doses, diffuse doses, and average organ doses in the lung and in bone, to point out the seven points of radiobiological parallelism between the sinus irradiation and the bronchial irradiation, and, in this framework, to present the experimental results available up to July 26 on the measured α-ray dosage in human sinus respiratory epithelium in one malignant and many nonmalignant cases.

You discussed in CUT the probable effectiveness of the diffuse distribution in bone. This parameter is reflected in the average organ dose in bone and in the lung. This diffuse distribution parameter was the reason for my reworking the lung models to compute the average lung organ dose, which comes to only about 0.1 rad per WLM. This α-ray dose is not above, but is well below, the average skeletal organ dose for inducing osteogenic sarcoma, or the average organ dose in the sinus epithelium, hence I said it "leaves a wide margin of safety, ignorance, or relative radiosensitivity of human organs". With respect to the hot spot doses, you and I are in good agreement that the best estimate from the lung modelers gives probable local hot sheet doses of the order of 1 rad per WLM in the bronchial epithelium, which is orders of magnitude below the measured doses associated with carcinoma in the sinus epithelium.

These experimental findings are comforting when applied to the question of whether mining exposures in the domain of 1 to 3 WL are safe—and as Laurie Taylor said in his June testimony (page 639 of ST) in the region centered around 1 WL there is no clear way of distinguishing at this time between the effect of a factor of 3 in either direction.

That microhistory got a bit detailed. In summary: (1) I said in June that nothing in the existing evidence indicated to me that 1 to 3 WL and 300 to 400 WLM was not below the practical threshold for lung cancer induction by α-irradiation, (2) we asked, can any other kind of independent radiobiological evidence be developed, (3) we found that human sinus carcinomas produced by chronic α-irradiation had many parallels, that the measured radiation insult to respiratory tissue associated with carcinoma in the sinus was orders of magnitude greater than the modelers' estimates of the radiation insult to respiratory tissue in the lung for 1 to 3 WL and about 500 WLM. I concluded that my suggestion in the June hearings, and the Federal Radiation Council's recommendation of 1 WL, would indeed involve a negligible risk. On a nonthreshold linear hypothesis, extrapolated from the origin into the domain of 10 to 100 WL and several thousand WLM, 1 WL for a lifetime mining career would be associated with a very low incidence. On a nonlinear, monotonic, steep sigmoidal response curve such as that seen so far in the radium cases, 1 to 3 WL should be "well below the practical threshold where all workers may be exposed with only negligible risk of incurring lung cancer from radiation".

I, therefore, have two reasons for agreeing with the Federal Radiation Council's selection on a risk-vs.-benefit basis of 1 WL. I believe 1 WL is safe: (1) when viewed in terms of the poor but existing Schneeberg, Jachymov, St. Lawrence, and American epidemiological evidence, and (2) when the computed average lung organ and lung hot spot doses are compared with measured α-ray doses which do and do not produce carcinoma in human sinus respiratory epithelium. You appear to have different reasons for supporting 1 WL, but I completely agree with your conclusion on page 19 of CUT that "continuous exposure for a working lifetime at 1 WL would introduce a radiation risk of the same general order of magnitude as . . . external irradiation at . . . 5 rems per year."

I hope it has been constructive to recite the theme intended from June through CORD because your own summary seems to me to be far off the mark. I totally disavow your wording of "The main thesis of Dr. Evans' . . ." given in your Summary, page 16, Section C, paragraphs (a) and (c). I hope that those who have occasion to read your CUT will realize that such expressions as "firm threshold of 0.5 μCi", "considerably in excess of", "misguided exponents of", and "ICRP 'mythodology'" are yours, not mine. I subscribe to none of them.

On page 6 of CUT you quote 3 paragraphs from my June testimony, and respond at some length to ". . . this criticism of the NCRP . . .". This section was not a criticism (I am a councillor of the NCRP and would have taken the matter up there if I felt criticism were due); it is simply a factual statement. Our policy is to be "cautious", and prudent as well, in all protection matters. But as you particularly know in such cases as iodine-131 stack effluent at Hanford, iodine-131 fallout from Soviet weapon tests, air-borne plutonium at Los Alamos, and 13-week permissible doses for test sites, economic and engineering feasibility as well as new radiobiological knowledge have to be included, and the cautious straight-line philosophy somehow got grossly violated, when risk-vs.-benefit becomes tight. The "arena" refers to the uranium mine situation and especially the JCAE hearing room, where the May 1967 hearings had involved such gross inaccuracies that the time (June 1967) was indeed "overdue" to use all the information available on human radiobiology (recall pages 489–491 of ST) in assessing the risk-vs.-benefit aspects of the mine environment. That it was indeed overdue is suggested by the fact that following these June hearings, and prior to the July round-table, at least two departments of the government altered the position taken in their May testimony, and that a unanimous agreement was then reached in the Federal Radiation Council.

Your correction (CUT p. 1 and 2) of CORD page 25, lines 5 and 6, which read "no documented case of a single large radiation exposure causing a radiation-induced malignancy in man.", is most welcome. That entire paragraph was intended to be a comment on possible species differences in the histogenesis of *carcinomas* induced by β rays, x-rays, or γ rays in rats, rabbits, and man. Do replace the word "malignancy" by "carcinoma",—to make it clear that when this phrase is taken out of context the Japanese leukemias are not under discussion. There are no known Japanese (or Marshallese) skin carcinomas attributable to radiation or fallout from nuclear devices. Now as to the "no documented case". I am delighted to be directed to the paper by Joe Mitchell and Haybittle which does seem to find the one exception. For a long time, and quite vigorously from 1947 through 1950, I sought just one such case from innumerable therapeutic radiologists in England, France, and the United States. I especially pushed Joe Mitchell on this quest, but he knew of no such case. I am a bit puckered that when Joe found the one exception I didn't get the news or a reprint of his paper. So "no documented case" becomes "one documented case", but one is enough for me, and the quest is ended. Anyway, the entire paragraph on low-LET species differences is extraneous to the evaluation of α-ray effects in man, and the simplest correction is to delete the entire paragraph after its opening sentence.

Your Fig. 1 is basically similar to my blackboard drawing (recall page 1150 of ST) showing a step, rounded off into a steep sigmoidal response, and coupled to a linear extrapolation. Our sketches seem to differ only in the slope of the linear extrapolation. Yours (line OL) goes from the origin to the *toe* of the steep sigmoidal while mine goes to the *shoulder*. Both would be in accord with conventional risk-assessment practice only if they join the sigmoid "at the considerably higher levels of exposure for which reasonably quantitative observation is possible" (ICRP Publication 8, page 1).

In your Fig. 1 the line OL, when extrapolated to the right into the high dose region, falls far below the actual incidence indicated in the high dose region. In a steep sigmoidal response curve fitted to a very large number of data points it is theoretically conceivable that the shape of the toe of the sigmoid could be determined well enough to admit use of a low level intercept extrapolation like your line OL. I know of no such case in human radiobiology, and would welcome hearing of one. For example, are there enough data on threshold erythema to define the shape of the toe of its sigmoid?

Usually the sigmoid for human responses represents a poorly defined transition region which lies between a region of negligible incidence and one of substantial incidence. If the sigmoid is steep its effective width may normally involve only a few human cases. This is why I drew the steep sigmoid which represents the practical threshold for radium cases in the bandwidth of about 1000 to 2000 cumulative rads average bone organ dose. In this bandwidth we have so far only about a dozen cases, compared with more than 80 cases above 2000 rads and several hundred below 1000 rads. Linear extrapolation from the low dose region into the high dose region for the radium cases, therefore, has to go from the origin of coordinates to the region of 1000 to 2000 rads, near the shoulder of the sigmoid, and not to its toe.

UNSCEAR, NCRP, ICRP, FRC, and many of us personally seem to use the word "threshold" to mean a number of different things, hence it tends to accumulate adjectives so we have "absolute threshold", "practical threshold", "sharp threshold," and others. In view of the necessity of following each member of any radiobiologically interesting group until its life is terminated, I wonder whether at least in human radiobiology the only threshold which is relevant and is measurable for somatic effects is the "practical threshold", that is, the looked-for effect does not appear during the lifetime of any individual in the study group. I cannot visualize how one would set up an experiment involving humans with a view to measuring an "absolute threshold", by which I suppose one would mean that the looked-for effect would not appear even if the individual lived to be 10,000 years old. Possibly "absolute threshold" concepts really are applicable only to immortal systems such as certain cell cultures or to strains of experimental animals carried through many generations of continued exposure,—such as Len Lamerton's mice. An "absolute threshold" might then be one in which cellular repair or replacement (e.g., UNSCEAR 1962, p. 64) and/or tissue and organ recovery can be demonstrated to keep pace with the rate of radiation injury. For humans, perhaps the only meaningful "threshold" is what we call the "practical threshold". Maybe we can kick this one around a bit next month when I will be visiting Richland as a member of the DBM–AEC Biomedical Program Review Committee. I hope you will be in town and that we can get together for a visit.

With best personal regards.

Cordially yours,

ROBLEY D. EVANS,
Professor of Physics.

The Dilemma of Lung Dosimetry

Health Physics Official Journal of the Health Physics Society

May 1969

Reprinted from *Health Physics* Vol. 16, pp. 553–561 by permission of the Health Physics Society.

This issue is dedicated to
DUNCAN A. HOLADAY
by

HEALTH PHYSICS

OFFICIAL JOURNAL OF THE HEALTH PHYSICS SOCIETY

Volume 16, Number 5   May 1969

KARL Z. MORGAN
Editor-in-Chief

J. A. AUXIER   C. M. PATTERSON   W. S. SNYDER
Editors

H. J. DUNSTER
News & Comments Editor

PERGAMON PRESS NEW YORK OXFORD

Health Physics Pergamon Press 1969. Vol. 16, pp. 553-561. Printed in Northern Ireland

THE DILEMMA OF LUNG DOSIMETRY*

H. M. PARKER

Battelle Memorial Institute, Pacific Northwest Laboratory, Richland, Washington 99352

(*Received* 1 *August* 1968)

Abstract—A legitimate criticism of health physics arises from the derivation of quite complex relationships for the permissible deposition of radionuclides in the body, in the absence of valid experience in man. It has to be accepted that inhalation of substantial amounts of radon and daughter products has a finite probability of causing cancer in the bronchial epithelium. Establishment of a quantitative dose–effect relationship has a formidable array of impediments. The radioactive nuclides in mine air comprise a short-lived chain existing in physical forms from free ions through small nuclei to large dust particles. Intake and primary retention of these media has to be calculated from models. Subsequent translocation up the bronchial tree is a significant factor in exposure. If the nucleus of a basal cell of bronchial epithelium is assumed to be the biological target, the penetration of alpha particles to it can be modeled in various degrees of sophistication. This leads to a numerical relationship between nature and concentration of the ambient radioactivity and the relevant tissue dose or dose-rate. The quoted number may be high or low by a factor of about three. This is essentially a permissible dose type of calculation. It is not a demonstrated dose–effect relationship. The feasibility of programs to progressively eliminate these uncertainties will be discussed in terms of refined study of mine exposees, relevance of data from radium or other human cases, and animal experimentation.

THREE-PHASE APPROACH TO DOSIMETRY

SEVERAL groups of reasons for the pursuit of excellence in lung dosimetry related to uranium mining can be advanced. The first and most compelling set of reasons is:

1. To support the assertion that substantial exposure to radon daughter inhalation leads to lung cancer (essentially begging the question by assuming that this assertion is valid).
2. To demonstrate that calculated doses for actual exposure levels in mines are in the range that might possibly lead to cancer production.
3. Hence, to derive reasonable permissible limits to concentrations of radon and daughter products in mines whose observance would virtually eliminate the occurrence of radiation-induced lung cancer in miners.

It is generally believed that this stage has been broadly reached in the recommmendations of the Federal Radiation Council (Report No. 8, Rev.—Guidance for the Control of Radiation Hazards in Uranium Mining[1]). This report establishes a proposed exposure limit of not more than 6 WLM in any consecutive 3 month period and not more than 12 WLM in any consecutive 12 month period.

The second set of reasons has to do with the refinement of information, especially with respect to low exposures continuing for a long time. A most important aspect of this would be an attempt to establish whether the lung cancer production has a threshold dose. One phase of this may be a better study of the latency effect as a function of dose. If an inverse relationship between dose and tumor appearance did, in fact, exist, a practical threshold might be identified. At the present time, information in these categories is extremely limited; it is vitally needed.

If it were easily possible to reduce the offending concentrations in mines well below the average of 1 WL, this would be the easiest solution to the uranium mining hazard problem.

---

* This paper is based on work performed under United States Atomic Energy Commission Contract AT(45-1)-1830. Presented before the Health Physics Society Symposium, Denver, Colorado, 20 June (1968).

THE DILEMMA OF LUNG DOSIMETRY

Unfortunately, efforts to get substantially below this level either by removal of radioactive material from the air or by respirators or other devices worn by the miners, appear to become potentially troublesome from competing hazards or extremely costly. If the relevant damage to man is indeed a nonthreshold effect, the ultimate balancing point will encompass issues much broader than science itself. In addition, such a solution would be a function of economics and it would not necessarily be stable with time.

The third set of reasons relates to the intellectual satisfaction of building the experience with radon inhalation into a comprehensive quantitative theory of radiation damage so that information acquired in this category could be reliably used over a broader front.

COMPARISON WITH RADIUM CASES

These sets of reasons have been separately identified to put the radon daughter lung hazard case in some perspective with the bone hazard case for radium and related products. Over a quarter century ago, the latter case was recognized, studied in the first phase and a tentative permissible exposure established. The work on the radium deposition cases, with which the name of our Symposium chairman is so closely associated, is now very deeply in the second phase. Virtually no new cases arise and the studies increasingly concentrate on very long term effects of low exposures. The attention in recent years to tumors of the paranasal sinuses is a case in point. If available cases had not been thoroughly studied this new class of tumor could well have been missed. Similarly, in the radon inhalation cases, there may be types of tumors in portions of the lung, or even elsewhere, not yet identified.

The degree of penetration of the radium cases into the third phase is somewhat argumentative. That the dose–effect relationship in man is known better for these experiences than any other is presumably self-evident. It has been used in the sense of the third set of reasons to guide permissible exposures for all other bone-seeking nuclides. Yet the interpretation of the dose effect cannot be described as unequivocal, and the rate of acquisition of further knowledge must surely be decreasing.

The relevance to the dilemma of lung dosimetry is briefly this:

1. At about the same time that Dr. Evans established a permissible deposition limit for radium, he also proposed (EVANS and GOODMAN, 1940[2]) a permissible concentration limit for radon. Had this limit (10 pCi ^{222}Rn/l.) been observed, there would probably have been no demonstrable radiation injury to uranium miners and this Symposium would not have been held.

2. With the recent activity in the field, there will probably be very few continued exposures at levels averaging more than 100 pCi/l. At this level, the generation of positive cases should virtually cease. The sophistication of the radium case analyses depends largely on examination of people whose exposure occurred in the early history of the luminous paint industry. Similar sophistication in the radon case depends on full utilization of all exposed cases to date.

3. On the whole, dose reconstruction will be decidedly more difficult for the radon products in lung case. This arises from:

(a) the nature of the decay products, which have short half lives, compared with the relative stability of radium,

(b) the variable deposition depending on distribution of three decay products over particle sizes from the molecular to the largest respirable particle,

(c) the rapid transport in the bronchial tree compared with the more or less permanent fixation of part of the radioactive material in bone; and

(d) from the difficulty of working with lung material, compared with the dimensional stability of bone.

At best, the reconstruction of the actual dose for radon daughter exposure will be difficult.

SCOPE OF THIS REPORT

It is the main purpose of this review to outline the problems of ascertaining the reliability of models relating exposure to an atmosphere containing radon and daughter products to the relevant absorbed dose in the lung. The appendix to FRC Report No. 8, Guidance for the Control of Radiation Hazards in Uranium Mining, September, 1967, contains such a review. It should suffice to refer only to some key points in that review, supplemented by the modest amount of new information that has developed subsequently.

The second main purpose is to identify the principal weaknesses of the models and their broader applications to the total problem. From this, one can define the intensity and timing of fruitful research that could be done to bring this study from Phase I through Phase II to some position within Phase III.

EXTERNAL FEATURES OF THE MODEL

The radioactive materials of interest are:

1. ^{222}Rn, the radon gas itself, entering into present day calculations only as the radionuclide cow for subsequent decay products.

2. ^{218}Po (or RaA), alpha emitter (6.00 MeV), half life 3.05 min formed in the elemental gas and thus being a free ion initially.

3. ^{214}Pb (RaB) important in this context only as the 26.8 min half life precursor of ^{214}Bi (RaC).

4. ^{214}Bi (RaC) the 19.7 min precursor of ^{214}Po (RaC'). Essentially the decay of a ^{214}Bi atom immediately signifies the release of the 7.68 MeV alpha particle of ^{214}Po.

5. ^{210}Pb, a beta emitter of 22 yr half life which is the throttle on the radon decay series. That throttle controls the release of the remaining alpha particle in the series (^{210}Po—5.3 MeV). ^{210}Pb may serve as the only useful integrator of activity retained in the system.

Materials that can be breathed in range all the way from the pure radon (which is scored as inconsequential in the model) through the free ions, aggregates of a few molecules known as nuclei, to particles of dust whose subclassification is limited primarily by the patience of the investigator or the capacity of the computer available to him. Refinements of the model (Phase II) may have to consider the fate of uranium and daughters in dust, and possible synergistic effects of ore materials such as nickel or vanadium, quartz, or other atmospheric contaminants such as diesel smoke, and tobacco smoke. The composition of the aerosol in a radon work room can be materially different from that in a mine. The detailed composition in mines can easily vary between mines and with time at any one place in a mine.

The most striking changes relate to the percentage of uncombined daughter products, which enjoy exceptional mobility and capacity for prompt deposition in the pulmonary tract. Recent measurements by CRAFT et al.[3] show uncombined activities, even in mine conditions of "moderate visible dust," ranging from zero to 73% of the total activity.

FEATURES OF THE INTAKE SYSTEM

The pulmonary deposition, retention, and translocation of these various ions, nuclei, and particles is presumably not significantly different from that for aerodynamically similar mixtures of any other composition.

The general nature of the deposition can thus be developed from the material of such monographs as that of HATCH and GROSS[4] or the report of the ICRP Task Group on Lung Dynamics.[5]

Lung cancers believed to arise from radon daughter inhalation are said to appear in the hilar region of the lung, corresponding with the smaller bronchi of the compartment (T–B) of Ref. 5. This is an area which is ciliated and covered with a moving sheet of mucus. The time relationships in the radon decay product sequence are such that small particles entering the deeper lung can be removed by the mucus escalator in time to release their most damaging radiation (the ^{214}Po disintegration) to the walls of the bronchi or bronchioles: This feature of the modeling is quite dominant compared with most applications of the generalized deposition and retention model. The thickness, rate of movement of the mucus sheath, and degree of penetration of radioactive particles within it are important factors that are not well known.

In any case, it is dangerous to assume that even the simpler quantity of primary respiratory deposition can be provided accurately by a model. In recent measurements in uranium mines, GEORGE (HASL Memo 68-11)[6] reports 37 observations of respiratory deposition in one individual ranging from 19% to 53%. In laboratory conditions the same individual scored 15–36%. While the unusually high readings possibly reflect cases in which RaA was available as free ions, separate measurements of RaC alone showed considerable variation. Measurements by the same author of deposition for different subjects, for nasal vs. oral breathing and for different tidal volumes in a standardized atmosphere show deposition rates ranging from about 18 to 51%. Thus, any respiratory model should be assumed to give "broad band" results.

THE ASSUMED BIOLOGICAL TARGET

Figure 1 has been taken from FRC Report No. 8 by permission of the FRC staff, which we gratefully acknowledge. It shows a stylized cross section of bronchial epithelium after the manner of ALTSHULER et al[7]. The lumen of the bronchus is at the top. It is lined with a viscous mucus layer of thickness a, over the continuously brushing cilia bathed in a more watery fluid of thickness b. The cilia grow from ciliated cells interspersed with goblet cells that provide the mucus secretion. All these develop from a layer of basal cells attached to the basement membrane. The integrity of the epithelial tissue depends on the continued integrity of the basal cells. Thus, the nuclei of the basal cells are taken as the relevant biological target. The central core of the dosimetry problem becomes the disposition of alpha emitting substances ^{218}Po and ^{214}Po with respect to these nuclei. Since the geometric scale is comparable with the range of the alpha particles in tissue, the real dose in the bronchial epithelium is liable to fluctuate wildly from that calculated in a stylized model.

For the segmental bronchi, Altshuler takes the elevation, d, of the nuclei above the membrane as 7 μm. The epithelium thickness $(c + d)$ can be measured in removed sections, with corrections for drying and shrinkage. For the segmental bronchi, Altshuler quotes $(c + d)$ as 29 μm in exceptionally thin locations and 56 μm as median thickness. In his model, a, b, and d are all 7 μm. Therefore, the distance between biological target and source is 36 μm minimum and 63 μm median thickness.

Figure 2 shows an actual section of bronchial epithelium (Engel[8]). If one recalls that the range of the alpha particles is 47 μm for ^{218}Po and 71 μm for ^{214}Po, it is clear that very few of the basal cells "see" any ^{218}Po at all, and that most of the cells "see" no alpha radiation whatsoever. Radiation at any tissue point also varies with time as the bronchial walls move with a sort of bellows effect. This real variability does not deter the physicist from making the most elaborate calculations of energy deposition in rigorously circular annuli.

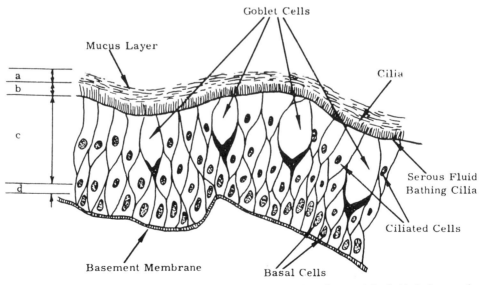

FIG. 1. Stylized cross section of bronchial epithelium after the model of Altshuler *et al.* (Reproduced from FRC Report No. 8 Rev. by permission of the staff of the Federation Radiation Council.)

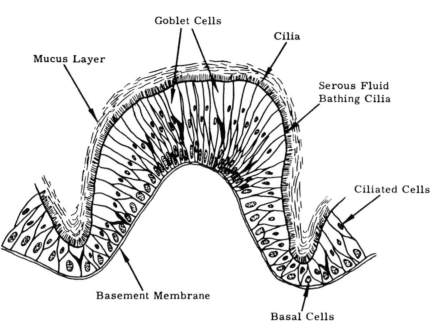

FIG. 2. Actual section of bronchial epithelium (from Engel), showing typical folding and extreme variations in thickness.
(Reproduced from FRC Report No. 8 Rev. by permission of the staff of the Federation Radiation Council.)

BRIEF COMMENTS ON SOME OF THE PUBLISHED MODELS

Early models (e.g. FAILLA, 1942[9]) considered only the radiation from radon gas itself being absorbed in the range thickness (or half the range thickness) in tissue. Since the lumen of the bronchi is less than the alpha ray range in air, maximum dose occurred in the largest tubes. This was troublesome because the then known lung tumors (from eastern European experience) did not originate there.

The modeler is inevitably pleased when he determines the maximum dose to be in the region in which tumors actually arise. Superficially, this is so in all recent models. It may be well to note that throughout the respiratory system, the environmental circumstances of target cells may vary in such a way that carcinogenesis may be quite unrelated to location of maximum absorbed dose.

BALE (1951)[10] made the first important model improvement by ascribing the principal effect to radon daughter products inhaled. CHAMBERLAIN and DYSON (1956)[11] emphasized the high mobility of free ions such as ^{218}Po as created in the radon, and the high deposition of these ions in the upper bronchial apparatus. The ICRP basis for permissible exposure to radon and daughter products is heavily weighted by a factor representing the fraction of unattached ^{218}Po atoms. The controversial matter of the free ions is frequently spoken of, and in a paper whose title includes the word "dilemma" it would perhaps be wrong to pass over the point. As suggested in Ref. (3), the percentage of uncombined activity may be higher than has been generally accepted for dusty mines. Recent models such as that of Altshuler *et al.* show less sensitivity to the free ion count than does the Chamberlain-Dyson model. Yet the difference is not such as to make the free ion count insignificant. More work needs to be done on this point. However, in our frame of reference, the issue is important as long as dosimetry is in Phase I. In Phase II, where good protection in the light of present knowledge is assumed, it is inconceivable that the free ions could not have been removed from the inspired air.

Three models have received detailed attention, that of Altshuler *et al.* and the models of

JACOBI[12] and of HAQUE and COLLINSON.[13] The Altshuler model seems preferable to that of Jacobi partly at least because each step in the modeling assumptions is clearly laid out and it is easy to repeat the calculations for one's private selection of particle distribution, lung dimension and so on. For the particular case of mouth breathing at 15 l./min in an atmosphere of the referenced composition, the absorbed dose in relevant targets of the secondary to quaternary bronchi is found to be 2.8 rads/WLM. This figure is useful for comparison with the Jacobi model. For the segmental bronchi only, where the highest figure in the Altshuler *et al.* model obtains, the value would be 3.4 rads/WLM.

The model of Haque and Collinson can possibly be preferred for two reasons. The method of determining absorbed dose from a defined geometrical distribution (Haque[14]) is unquestionably superior to the crude averaging process of the Altshuler system. As Haque shows, the differences are quite significant in the trachea and relatively unimportant in segmental bronchi. In all cases, the absorbed dose for like distributions of the alpha particle sources is lower in the Haque model. However, in total, the Haque and Collinson method gives an absorbed dose of approximately 11 rads/WLM *vs.* 3.4 for Altshuler. This markedly higher value, vis-a-vis Altshuler *et al.* appears to result mainly from a much higher component of free ions in the Haque model (fraction $f = 0.35$ for RaA *vs.* $f = 0.09$) and from the different lung model. Altshuler *et al.* used the conventional FINDEISEN–LANDAHL model[15] whereas Haque used the WEIBEL[16] model. The latter is a rigorously dichotomous model and can hardly be correct since elementary teaching shows the right lung dividing into three lobes and the left into two. This reporter cannot make a firm judgment of the relative merits of the models. He can only be surprised that there is no middle ground between the 1–2–4–8–16–32 subdivisions of Weibel and the 1–2–12–100–770 subdivisions of Findeisen–Landahl. It appears that on the whole the Weibel model should be somewhat closer to anatomical reality. The models are notably different in the size range of the segmental bronchi. Landahl calls for 100 of these tubes and a total surface area of 94 cm². Weibel necessarily has 32 tubes in this fifth generation and gets an area of 37 cm². If the models contrived to calculate equal mass depositions in these two compartments, most of the observed differences in dose would be accounted for.

In review, it would appear that the Haque and Collinson work presents sound reasons for estimating an absorbed dose higher than that due to Altshuler *et al.* In the absence of better data, one might reasonably propose the mean of the two values or about 7 rads/WLM. This would be done in the certain knowledge that the alpha particle dose to many of the basal cells must be zero. The data are not precise enough to exclude values up to (say, arbitrarily) 15 rads/WLM.

UTILIZATION OF THE DOSE RELATIONSHIP

The relationship that one WLM leads to an absorbed dose of seven rads (within a factor of 3 or so) in relevant targets in selected thin portions of the epithelium of segmental bronchi is of the nature of a calculated permissible exposure. With some arbitrarily established annual dose limit, a suitable time-integrated concentration limit for radon daughter products can be derived.

This is not the kind of dosimetry visualized in Phase II or Phase III, where a real understanding of the dose-effect relationship is contemplated. At present, it is accepted that cumulative exposures of the order of 1000 WLM lead to clearly demonstrable increases in lung cancer incidence. The model is patently compatible with this because the implied cumulative dose is nominally 7000 rads. As an alpha ray exposure, it might reasonably be assigned a Quality Factor of 3 for a dose equivalent of about 20,000 rem. Cancer incidence would be noted in many other types of lung irradiation at such a level or even at one-tenth this level. The relationship is far too tenuous to justify further application. The reasons have been catalogued before. They include:

1. Working Level (WL) as a means of expressing concentration of radon and products may not be satisfactory. It defines any combination of radon daughters in one liter of air that will result in the ultimate emission of

1.3×10^5 MeV of potential alpha energy. How much of the ambient radioactive material will be deposited in the bronchial passages is a function of the ion, nuclei and particle size distribution. Furthermore, the geometry of the case in the segmental bronchi suggests a very marginal contribution from ^{218}Po. Better definition of distribution of the several daughter products across particle size is a must. For past exposures of record, it can now be little better than a guess.

2. The quoted dose must be corrected for actual breathing pattern of each individual.

3. The lung model is seriously defective in terms of true thickness of the epithelium, thickness and rate of motion of the mucus layer, the actual radioactive dust content and its mixing in the mucus sheath, losses of calculated load due to absorption of products to the lymph nodes, effects of defects in the ciliar covering, effects of agents modifying ciliary stroking (e.g., tobacco smoke).

4. Probably the most serious single deficiency is to have developed a dosimetry that can only apply to "exceptionally thin portions of the epithelium" with no attempt to examine cases to determine the actual site of initial lesions.

CONTRIBUTORY RESEARCH

In suggesting avenues that should ultimately contribute to an adequate understanding of the dose effect relationship this review will

(a) omit those factors which primarily constitute extended epidemiological studies; the author is clearly on record as supporting this elsewhere.

(b) undoubtedly include proposed work that others may consider already resolved. This can arise because one may consider the existing data to be too imprecise, clouded by secondary variables, conceivably biased by an author's passion to achieve results supporting preconceived ideas, inadequately reported or lost in old literature (as is much of the early hospital radon experience) or escaping the reviewer's attention in recent literature.

No critical review of these points is attempted here. Rather, there are pulled together a selection of features which when added to similar lists by other reporters should form an agreed base for a long-range attack on the radon inhalation problem. Health physicists will notice what a small fraction of the proposals involves physics *per se*. This might be a clear guide for the necessity of broad interdisciplinary study.

STUDIES RESTRICTED TO EXPOSED INDIVIDUALS

This class represents the greatest urgency, once Phase I dosimetry has been achieved. The greatest human exposures are then clearly in the past. Extensive follow-up of present day miners will probably occur through the epidemiology interest. Quite irreplaceable is the opportunity to initiate lifetime follow-up of those men who previously mined for a few years on the Colorado plateau. Not only did they probably acquire relatively high annual exposures; their induction period is now reaching the interesting stage.

As stated before, present models cannot be taken too seriously until the location of the primary lesion is demonstrated to be within alpha ray range of the contaminants. The work of SACCOMANO *et al.*[17] on sputum cytology shows a progressive change in cell dimensions as they go through presumed precancerous changes. Such changes should be located in all available material. It is still conceivable that shrinking of cells in the bronchial epithelium for any cause could let alpha particles previously ineffective suddenly begin to reach target nuclei after years of exposure.

Conversely, it is commonly said that the mucus blanket tends to be unusually thick in miners. Hopefully, this should be measured as it controls feasible penetration of the radiation.

Some of the largest uranium mines in the world have been operated as open pits. It can be assumed that high concentrations of radon and daughter products occurred with relatively less dust than is common in the U.S. uranium mines. Diesel smoke and tobacco smoke were presumably absent. Hence, a valuable cross comparison with a new mix of possible synergistic agents should be possible, if cases from these areas could be studied.

Similar comparison points could come from hospital uses of radon. Also, the radium deposition cases lead to radon in the lungs essentially with the first daughter product only.

Dr. Evans has pointed out that the lining of the paranasal sinuses is a ciliated epithelium which could be judged similar to the bronchial epithelium. Thorotrast patients may also offer some experience if thoron can escape via the lungs.

While many of these factors have received sporadic attention, it is suggested here that all possible avenues be studied intensively to build up a coherent supporting background of data.

Better knowledge of actual deposition in the bronchial tree is being acquired, as in Ref. (6). Also, attempts to correlate exposure retroactively through integration by long lived products are becoming more promising as in BLACK et al.[18]. Devices to measure distribution of particle size, and distribution of radon decay products across that size spectrum in the ambient air and as actually inhaled are contributory.

From the modeler's point of view, reliable information on the thickness of the mucus sheath, its rate of motion, effect of defects on particle movement, etc. is needed. With sufficient interest generated, improved approaches to this, not necessarily related at first to uranium miners, should be forthcoming.

STUDIES IN EXPERIMENTALLY EXPOSED INDIVIDUALS

Deliberate experimentation with radon and daughter products inhalation at more than true tracer levels has little appeal. Far more encouraging is work with the ^{220}Rn (i.e. thoron) series. Atmospheres essentially stabilized with ^{212}Pb (ThB), the equivalent of RaB, are easily obtained. ^{212}Pb has the advantage of a convenient half life (10.6 hr) and a convenient gamma emission (0.239 MeV) which permits accurate scanning for location and translocation in the lung. Further down the series, the branching of the ThC' and ThC'' is favorable with ^{208}Tl giving coincidence gamma rays for precision measurements. The possibility of accurate study of such factors as particle removal via the lymph nodes is eminently superior in the thoron analog series.

ANIMAL STUDIES

A thorough discussion of suitable animal experimentation would be too time-consuming here. Two comments should suffice.

On the one side, no matter how compelling the animal results, they would be suspect for application to cancer induction in humans. This arises from the growing belief that radiation carcinogenesis requires not only some radiation dose, but some as yet undefined environmental factor. In man, such a factor could apply in the smaller bronchi but not in bronchioles, trachea, or nasal sinuses. It could apply equally to any component of the human tracheo-bronchial apparatus vis-a-vis an animal's.

Nevertheless, components of the total problem can surely be attacked with confidence through animal systems. As an example, the main thrust of the attack on synergistic interactions between radiation, diesel smoke, tobacco smoke and other irritants must be done through animals. Intercomparisons of radon daughter products with thoron daughters can be used to translate experimental thoron data in man into equivalent radon data. Most of all, because the animals can be sacrificed, there can be systematic search for progressive changes in irradiated lung tissue.

SUMMARY

Current models of the uranium miners' lung exposure problem are subject to errors involving uncertainties in:

(a) composition of the inhaled aerosol;

(b) its deposition and translocation in the tracheo-bronchial system;

(c) detailed geometry of alpha emitter location with respect to assumed biological targets. The absorbed dose to the nuclei of basal cells in the small bronchi can be written as 7 rads/WLM, good to perhaps a factor of 3.

Data that will permit development of a true dose–effect curve from human experience are scanty. Studies that might help to improve this picture are briefly considered. Improvements in mine practice should virtually eliminate further positive cases. When this is coupled with the possibility that some other environmental factor (such as synergistic combination with smoking) may be needed for tumor induction, the prospects for bringing this lung dosimetry to a stage of development comparable with that reached for bone tumors from radium are not high.

The value of direct human experience in radiation injury is judged to be sufficiently high that no effort should be spared to reach the best possible solution.

REFERENCES

1. Federal Radiation Council Report No. 8, Rev. *Guidance for the Control of Radiation Hazards in Uranium Mining*, U.S. Government Printing Office, Washington, D.C. (1967).
2. R. D. Evans and C. Goodman, *J. ind. Hyg. Toxicol.* **22**, 89 (1940).
3. B. F. Craft, J. L. Oser and F. W. Norris, *Am. ind. Hyg. Ass. J.* **27**, 154 (1966).
4. T. F. Hatch and P. Gross. *Pulmonary Deposition and Retention of Inhaled Aerosols*. Academic Press (1964).
5. ICRP Task Group on Lung Dynamics, Deposition and retention models for internal dosimetry of the human respiratory tract. *Health Phys.* **12**, 173 (1966).
6. A. C. George. HASL Technical Memorandum 68-11, May (1968).
7. B. Altshuler, N. Nelson and M. Kuschner. *Proc. Hanford Symp. Inhaled Radioactive Particles Gases*. Pergamon Press (1965); *Health Phys.* **10**, 1137 (1964).
8. S. Engel, *Lung Structure*. C. C. Thomas (1962).
9. G. Failla, CH-1347, 1942 (declassified 1948).
10. W. F. Bale, Hazards associated with radon and thoron, unpublished memo (1951).
11. A. C. Chamberlain and E. D. Dyson. AERE Report *HP/R*1737 (1955), The dose to the trachea and bronchi from the decay products of radon and thoron. *Br. J. Radiol.*, **29**, 317 (1956).
12. W. Jacobi, *Health Phys.* **10**, 1163 (1964).
13. A. K. M. M. Haque and A. J. L. Collinson, *Health Phys.* **13**, 431 (1967).
14. A. K. M. M. Haque, *Br. J. appl. Phys.* **17**, 905 (1966).
15. H. D. Landahl, *Bull. math. Biophys.*, **12**, 43 (1950)
16. E. R. Weibel, *Morphometry of the Human Lung*, Springer-Verlag, Berlin (1963).
17. G. Saccomanno, Richard P. Saunders, V. E. Archer, Oscar Auerbach, Marvin Kuschner, Patricia Beckler, *Acta cytol.* **9** (1965).
18. S. C. Black, V. E. Archer, W. C. Dixon and G. Saccomanno, *Health Phys.* **14**, 81 (1968).

Festina Lente

Proceedings Sixth Annual Health Physics
Society Topical Symposium —
Radiation Protection Standards: Quo Vadis

November 2-5, 1971

PROCEEDINGS

SIXTH ANNUAL
HEALTH PHYSICS SOCIETY
TOPICAL SYMPOSIUM

RADIATION PROTECTION STANDARDS:
QUO VADIS

COLUMBIA CHAPTER
HEALTH PHYSICS SOCIETY

Volume I

Rivershore Motor Inn — Richland, Washington
November 2-5, 1971

KEYNOTE ADDRESS

"FESTINA LENTE"*

H.M. Parker
Consultant to Battelle Northwest

The best thing to do when discussing the subject of standards is to go back and make sure that the basic definitions are in place. Rather belatedly it occurred to me long after I had accepted Ron Kathren's invitation to read the keynote address that it would not be amiss to look up the definition of "keynote." Much to my surprise, in going to Webster, which I think would be the responsible source in this case, he actually defines "keynote address" separately, calling it a noun, which I happen to think is bad grammar, which I also happen to think Webster practises fairly consistently. Here is his definition: <u>Keynote address or speech</u>, a noun he says - an address intended to present the issues of primary interest to the assembly, but often concentrated upon arousing unity and enthusiasm. Ron, I assure you, if I had read that first, I would have declined your invitation because on a topic like standards, I am sure I would find it very difficult to arouse either unity or enthusiasm. However, officially, unity and enthusiasm will be our keynote.

I would like to begin by going back to definitions and reading some earlier ones on what standards are supposed to be, their purpose and nature in general. To save time, I will simply read off these things.

---

* This material has been left in the vernacular appropriate to a keynote address and is not submitted as a formal technical contribution. - Author

Standards are: (1) Models for judgement as to whether a given action should be taken,

(2) Communications of the expert to guide those less experienced,

(3) Crystallizations of past experience utilized to facilitate future actions

(4) Means for integrating a whole made up of many parts.

Particularly in technological fields, the well-conceived standards fulfill these following purposes: They should promote safety, promote simplification, encourage further improvements, and contribute to orderly development of the technology as a whole. Of the several ways in which standards may originate and develop, two seem to be of particular relevance to the radiation protection field.

The first concerns moral and ethical standards of which the Hippocratic oath, defining the standards of conduct of a physician with respect to a patient, is a sound example. In application to radiation protection, concern for the well being of our future generations in the face of the deleterious genetic effect of ionizing radiation may lead to standards of this kind.

A standard of the second type can be called an action facilitator, a device to allow each individual in society to proceed in a certain direction, confident that certain obstacles have been removed. The rule of the road is a simple example of this. In just the same sense, there is a class of radiation protection standards designed to permit an individual to proceed with reasonable safety for himself and with the assurance of not causing unreasonable interference with the needs or work of others. These action facilitators may be either verbal rules or physical standards such as weights and measures. Both these modalities are within the purview of this Symposium.

Let me turn to the question of the application of criteria and standards and their enforcement. Ideally, a standard receives voluntary agreement and acceptance. This may be readily reached where there are evident benefits to all concerned. A simple example - the rule of the road.

We agree to be right-handed at least in one particular country. Such decisions are frequently reached by the power of a simple majority only. In other cases there is a degree of indirect enforcement. At the level of moral and ethical codes this includes the powerful forces of social ostracism. At more mundane levels it includes the powerful forces of an attack on the pocketbook, such as in the lack of insurability for non-conformity with codes such as fire or building codes. In other cases, direct enforcement is necessary through a system of laws and regulations. Unless a standard has the force of law and sometimes even where it does have, its acceptance and use rests largely on its intrinsic merit. The actual need for a recognized guide, the extent of experience available as background, the wisdom of the authors, the clarity of expression, and the flexibility to meet changing conditions all influence the worth of such standards. With this preamble on standards in general, let me proceed to radiation protection criteria or standards.

I would like to talk about the early history but there is not much time. I will just remind us that by the time that health physics was created, in other words, at the beginning of the atomic energy project, there were just three basic standards for occupational exposure in general use: the first one, the 100mR per day for external X- and gamma radiation; the second one, 10^{-14}Ci/cc for radon in the air in working rooms; and the third, the 0.1 microgram of radium as the maximum allowable amount deposited in the body. We might say for a point of historical interest that two of those three limits were mainly due to one single member of this present audience, Dr. Robley Evans of M.I.T. All these were derived from quite limited clinical observation with safety factors applied to the amount observed to have produced injury to exposed persons. The first limit, the one having to do with what we call whole body exposure, has received the most attention taken over-all. Much has been made of the point that since the time they were set in the early 1940's the limits have been progressively lowered. Obviously, the dice were loaded to make that the most likely direction from the very way they were set up so it is not so earth-shaking that they have been progressively lowered. The understanding

4

of radiation effects has indeed increased, partly through a reconsideration of those earlier human exposure cases, a field in which again Dr. Evans has been the eminent scholar for the radium cases, partly through study of additional human cases, but principally, and perhaps one could say overwhelmingly, from animal experimentation. Since these concepts have to be applied to man in quantitative terms, the validity of the translation from animal data to man becomes one of the major topics for review of the basic standards.

You will recall that Mueller unequivocally demonstrated the radiation induction of gene mutation way back in 1927. It was about the post-1945 period that enough evidence had accumulated to make it seem likely that induction of gene mutation increases in direct proportion to the radiation dose at low dose levels. It is really only in the last decade or so that the work of Russell and Russell at Oak Ridge has made it certain (I think we can use that word there) that this idea of proportionality is an oversimplification. That same period, 1945 to the mid-1950's, saw an increase in suspicion that not all somatic effects of radiation had a true threshold. If we accept these points, and many do, as we well know, there is no degree of radiation exposure low enough to be without potential change of some kind. We will generally concede that potential change will be predominately in a disadvantageous direction. That leads to the concept of acceptable risk, about which a million words have been written by this time.

In this earlier period the previous terminology of "tolerance dose" was changed to "permissible dose." The key point of that change was to underline the acceptance of a nonthreshold dose concept. The particular choice of these words to me has always seemed arbitrary. In fact, in my own slant on the language, it seems backwards. It is too bad that we did not call the limits "permissible" back in 1935 and call the same things "tolerable," but not so very good to have, in the present age.

There was a fundamental change in approach to radiation protection criteria at that general point in time. Before then, clinical observation, and we often forget that these early things were nearly always based on simple clinical observation, clinical observation had clearly pointed to the reality of threshold effects. With a threshold applied to every effect the radiation protectionist or health physicist had a clear scientific, and hence to him, wholly rational charter to set about determining all the proper thresholds for all the circumstances. And if you could do that, and at one time we felt we might, persons exposed to smaller amounts than the threshold dose could be protected in the fullest sense of that word. Without the threshold dose there is no absolutely safe dose. In the same sense again, repeating words that have been used a million times now, establishment of any quantitative standards or permissible limits involves value judgements going beyond the areas of scientific and technical competence.

It is most essential that qualified observers recognize this boundary and re-examine their own role with respect to it. The observer who unconsciously steps outside the boundary of his professional knowledge is, ipso facto, unqualified and in some cases, rather dangerously so. The nature of standards is such that value judgements will have to be made. These judgements will be more willingly accepted in proportion to their support from the authority of knowledge. This is provided by those with long work in the field, with the mentioned reservation that they must consciously accept the need to step outside scientific boundaries and be conscious of what they are doing at that time.* They should not be too surprised that in this foreign territory outside their scientific boundaries, the natives are not always friendly, nor should they be unwilling to concede that in some aspects these natives may have sounder views than their own.

In the early work, emphasis in the form of standards was on occupational exposure limits. As you well know, the AEC has often been castigated for paying inadequate attention to the environmental issues which seem to be predominant today. Actually, I think that this is a customary half truth or less. As an example - I use a local example - the extensive biological program of this site has never in twenty-five years been less than 50% involved with environmental problems. I don't think this is ex-

----
*The above background material is taken almost verbatim from the author's presentation at the J.C.A.E. Hearings on Radiation Protection Criteria and Standards, May 1960

6

ceptional for other AEC programs. It is just that the complex interrelationships in the ecological web may be difficult to codify and present to the public.

Many of us believe that the issues of acceptable risk or the attempted balancing of benefits versus risks are quite separate for the occupational case and for population exposure. That's a debatable point which I think depends on one's ethical codes to some extent, but many, I think, do feel that if a man is well-informed about the nature of his work and goes ahead with it as his career basis, it is not unreasonable that he accept some definable degree of potential deleterious effect provided a best effort is made to minimize this. And one differentiates this very sharply from the case - the population case in general - where one group seems to get most or all of the benefits while another seems to get most or all of the exposure. As we have said at previous meetings and many of you have heard me say this, that's too reminiscent to me of the old English song entitled, "She Was Poor But She Was Honest." You must know this song by now. The chorus is the relevant reference for our interest here. It reads like this: (It should be given in a rich cockney accent which I won't try to imitate) "It's the rich what gets the pleasure; it's the poor what gets the blame. It's the same the whole world over, isn't that a blooming shame." One of our distinguished British visitors should have come up and read that for me. You would have done a much better job.

But that's the way it goes. It doesn't go this way only in radiation exposure. "It's the rich what get the pleasure" generally, and we have to keep this in mind and, I think, spend more time than we have before choosing up sides with the "poor" with respect to radiation protection. One could write down a list of special considerations applying to population dose that do not apply to the occupational case. I have about seven here and in the interest of time I will leave those to the printed paper.*

----

* There were 8 points, given on pages 15 and 16 of the referenced hearings:

(a) Authority of knowledge to define limits differs from that for occupational dose;

7

(b) The national or even international character of the problem is self-evident;

(c) Means for establishing regulations and enforcement may be peculiar to this case (a 10 year prediction of the need for EPA?);

(d) For irradiation of the reproductive organs, total dose to an "effective population" is the significant parameter;

(e) It does not follow that there is no upper limit, on genetic grounds only, for exposure of an individual;

(f) For irradiation of other tissues - the somatic effects - a population average dose limit _may_ be a useful index;

(g) Nevertheless, the matter is one of acceptable risk for each individual, and this calls for specific upper limits for classes of individuals;

(h) There may plausibly be a whole series or such limits, for example, a low one for children or those with certain defined infirmities, a moderate one for definable groups of "normal" adults; the occupational limits are essentially such a case; a relatively high limit for certain tasks voluntarily assumed or otherwise in the national interest, for example, space flight; a relatively high limit for virtually all where the stress of enemy attack substantially alters the risk balancing equations.

The main thrust of what I have said so far refers to what we are now calling basic radiation protection criteria and, as you well know, my personal blinder leadeth that way. As Mr. Kathren said earlier the planning of this symposium had indeed a much broader objective in mind. Basically, I think: - how are we doing on practical day to day guidance derived from standards for operational health physics? I think that is a fair paraphrase of what the planners had in mind.

They will admit that you can't avoid the kind of philosophical "mush" that you are going to get for about half the time in the next three days but we must look also at lots of the other more detailed things, for example, instrument standards, surface contamination standards, waste disposal limits, crisis control limits, and many other phases that one could particularize. And rather curiously, some aspects are totally missing from the collection of papers. In the symposium, waste disposal for example is absent, but might become one of the prime issues that we have to deal with. Therefore, Carl Unruh, as you summarize the proceedings at the end of the sumposium, I beg you not to forget these zones of silence in your appraisal of the over-all value and significance of this conference.

The topic, as you know, went high brow and asked "Quo Vadis?" Let me answer just one aspect of that quantitatively. I've long kept, as a matter of personal interest, a record of the number of applicable protection handbooks. In 1944 and even in 1954 they were few enough that you could show legible slides of them. Since these would have been the only slides, I spared our projectionist by not having them. It is not important that you don't see the details but you can see that they made up a format in which as a slide you could actually read the title of these various things.* And in 1944 there were six of those. Really there were only five, because I threw in one on protection of radium

----

* For the present purposes printed copies of the slides were exhibited. The originals appear in "Radiation Protection in the Atomic Energy Industry," H.M. Parker. Radiology 65, 903; 1955.

during air raids which just happened to be needed at that time. In 1954 there were sixteen. In 1966, when the count read: 53, I gave up trying to make slides. You ended up taking a photograph of a bookcase by that time. In 1971, courtesy of Jack Selby, who did the counting, there are 114. If I were going to demonstrate it, I would have to do it like this by saying there are 18 of NCRP types and 15 of ICRP types and so on through this collection here which includes a rather recent addition, 19 from ANSI, which are judged by Jack to be directly related to radiation protection out of the 35 published nuclear standards.

Now, all those who work on numerical relationships spend their time extrapolating and developing accurate formulae to extend the data. I am very pleased to report that these particular data on numbers of protection handbooks versus time fit a very very simple formula. Indeed, it is strictly exponential. If you want to do the calculation yourself, the exponent is $0.148T$, where the time T is expressed in years, provided you choose the appropriate base year which is the secret of this formula. I'd have you believe that over this recorded period from 1944 to 1971 that simple formula gives the right number of protection handbooks for any year, within a few handbooks. So in view of the way we exercise this extrapolation privilege in health physics, I can extend this to provide an answer to "where are we going?" Well, here's where we are going. By the year 1981, there will be 432 protection handbooks. This doesn't count the ones that become obsolete. This is a counting of the active ones. Please remember that. In 1991, there will be 1,897 give or take a few - just a few. By the year 2001, there will be 9,662. Now as an exercise for the health physics student here I want you to take those data and determine the year at which the sole occupation of a health physicist becomes the reading of protection handbooks. What does that do to your unity and enthusiasm? "Festina lente," indeed! We had better hasten slowly in that direction. That observation is a good and sufficient reason for my title in response to "Quo Vadis?"

However, it didn't happen to be the one in mind when I was asked to give our symposium organizers a title, and they generously gave me twenty minutes to come up with a title to meet their deadline. So let me

10

attack "Quo Vadis" rather differently (the way I originally intended) by briefly citing some of the problems in radiation protection connected with standards.

The fundamental problems in radiation protection are about as follows: What is the basic nature of the interaction of radiation with living matter? Would knowledge of this lead to a generalized scheme of protection or to methods of moderating exposure effects? What is the origin of radiation carcinogenesis? Under what conditions, if any, can a single radioactive focus in the lung, for example, induce a malignant neoplasm? What is the precise significance of the genetic effect in man? Can the total residual insult, say for a single radiation type, be integrated by a plausible formula to cover all cases between the extremes of a single acute exposure all the way to a uniform chronic exposure? Under what conditions are the effects of mixed radiations additive by some determinable formula? Can such a concept as the "rem" be fruitfully extended.

Now some of the applied problems that are of principal concern and derive from these in part could be written about as follows: How can exposures to mixed radiation be adequately integrated in the field? What are the appropriate permissible limits for a wide variety of radio-isotopes exposures in man and all other life forms? How can organ exposures to short-range emitters be determined? In particular, for example, how can lung deposits be determined and how can bio-assey be reliably related to body deposition?

There are some engineering problems that go along with these. Some of these referring rather particularly to the Atomic Energy program and not the whole scope that's needed for radiation protection consideration are: - How can radioactive materials on the industrial scale be completely contained at source? How can radioactive wastes (I mentioned the deficit in waste consideration in our symposium) be permanently contained on an economic basis? What standards of air cleansing both within operating plants and for releases to the general environment are necessary and adequate? And how can such standards be reached economically? For reactors specifically - can reactors be designed and operated with absolute assurance that there could not be a catastrophic accident so that they could be placed in metropolitan areas? And lastly, how can irradiated materials - reactor fuel elements if you like, waste products too - be safely and economically transported to processing and storage locations?

Now in summary when you look at what I read as the fundamental problems of radiation protection, they are precisely the ones that exercised the first inquiring scientists who looked at the problems of the very first radiation sources that we had. Atomic Energy put some particular new emphasis on the applied problems - perhaps one of the greatest being the need for solutions for mixed radiations which I read as one of the subsets here. Now it seems to the writer that the next decade will see substantial reduction in the release of effluents to the atmosphere. There will be some progress in control of in-plant contamination. It will be unlikely that absolute success will be achieved in either of these fields. Therefore, the chief expected burden in radiation protection will probably be in more and more refined interpretation of minimal depositions in man and many other life forms. In this way practical success in control as opposed to unequivocal success by complete elimination of the problems at source is a reasonable objective for the next decade.

Now the point I want to make is not the exhortation of a literal translation of my title that we should hasten slowly but rather a stipulation that we are indeed hastening very slowly in our accrual of knowledge relevant to at least the more basic of the radiation protection standards. What I have just read to you are not today's problems; you may have sensed some minor defects in them as I read them. But this is a statement written down seventeen years ago* which also at that time said that there had been very little change in the previous decade. So that is

----

* Read at the R.S.N.A. meeting, Los Angeles, 1954, published Radiology: 65, 903, 1955.

12

to say that we have gone a quarter of a century with rather little basic accrual - some exceptions. I think the work of Russell in some aspects would certainly challenge the absoluteness of the statement, if you took it that way, but broadly speaking we are hastening slowly in this field.

Yesterday, Dr. Jack Healy handed me a copy of the old Chalk River Conference on permissible doses which is dated 1949. He had it here about a month ago for the Transuranium Conference and he said at that time that the problems that were talked about then are the problems being talked about now for permissible doses for many of the isotopes taken separately. I took time after our long board meeting yesterday to read that again last night and indeed Jack is right. All the problems are there. We are still horsing around with the same things and sometimes hope to get more elaborate formulae. But we don't have a great deal of more reliable basic data on many of these things.

Where are we going? We shouldn't just be drifting but perhaps we are drifting. A "Quo Vadis" review must call for a complete national review of the missing data-a full rational analysis of how much of this needed data can, in fact, be filled in by the proper effort, on what time scale, and at what cost. That is not a new charge, but it needs to be made again. How much of the needed data for a complete satisfactory resolution of the basic standards can, in fact, be filled in on what time scale and at what cost? And it is there, I think that we need to keynote the theme of unity and enthusiasm to have a national effort to write this down and all subscribe to the results within the normal bounds of proper differences of scientific interpretation. Let's get going on a national program dedicated to that task. As I said, the call has been made before. Meanwhile, many times the cost of making such a study (cost of making the study,not the cost of doing all the reasearch that the study would call for - no, that's much higher - but the cost of the study itself) has been frittered away and scattered with relatively futile rebuttal of attacks by about three critics of our profession.

Let's get the activity on the positive side of the ledger for a change. On some of these problems if we do, in fact, hasten slowly, irretrievable data will be lost. I will very briefly mention two. One is dear to the heart of Dr. Evans. Studies of the lungs of men who may have had uranium mining work earlier and went out of it and as they become accessible by natural or accidental death -- it's a crime that the data on the condition of the lungs of those people is not recorded. It can't be found later as some of Dr. Evan's data for radium in the skeleton can be found by exhumation. You have to have it then and there when the tissue is fresh and we sit around fiddling, letting irretrievable data slip through our fingers.

The second example I would cite is much more realistic funding - I am talking about perhaps five times as much, not doubling and not multiplying by 100, but five times as much funding for such efforts as the Mancuso Project, which we will hear about in part this afternoon. There again there are data which if not obtained at the right time, cannot be obtained later. We are still footling around in a very very feeble manner with that issue.

Going back to the point - one of Dr. Evan's favorite themes - that the only proper study of man is man, these human data are the things we must get. This is not to say we throw our biology and animal experimentation out of the window. It has a very important part but possibly subsidiary to or explicative of these other irreplaceable sources of potential information.

If I haven't overextended my welcome, I would like to close with just a few points, biased I am sure by my own views, that I consider of special interest in the program abstracts that you have before you.

Session I, I think, is an interesting if rather oddball mix on the standard-setting process. I hope that someone will relate this to the ICRP interest which is not included in the particular program and perhaps explain why some of us are tending to believe that some NCRP activities are significantly more relevant in the United States than are some ICRP activities. This is because although the science should be common to both the value judgements can be significantly different in any one

14

country should it so choose. Most value judgements are not an international issue. They are a national issue in each country - specifically a U.S. issue for most of us.

Session II concerns biological studies. It's an interesting potpourrie. It's nonsystematic. It would be impossible to be properly systematic in one afternoon anyway so I don't make this comment by way of criticism of the program. But it is grossly lacking in genetics which I think is one of the more teasing issues, because this is one that clearly affects our responsibility to generations to come rather than to anything that might happen to us if we make mistakes. Also I understand that two of the papers that were scheduled will not be read today. One of them will be read later.

Session III - special topics. I don't have much to say about that except what a welcome relief from the basic philosophy mush. We could stand something pragmatic for half a day.

Session IV. I similarly have no comment.

Session V looks like a very exciting one. I was particularly looking forward to an authoritative statement on the new interaction between the Atomic Energy Commission and the Environmental Protection Agency responsibilities. My colleagues in the operational field tell me that this is currently one of their greatest puzzles, but I have got to confess to you I have been trying to hide from them that many at the other end of the field, for example as in certain people's critiques of such documents as IOCFR 50 Appendix I, demonstrate an equal confusion. So don't despair at the operational end. You've got lots of company. Now, unfortunately, I am told that not one of the three relevant papers on that problem - AEC Contractors, AEC Licensing, and the EPA, will be presented by the original protagonist. Now I don't want to be suspicious or cynical but maybe they all chickened out and decided they didn't know the inter-relationships either. We will have to await the outcome of what will happen in the meeting.

Session VI. All shades of opinion on the benefit-risk issue from the very precise dollar per man-rem School through the Beardsley argument (which dates back to 1963) that "one life shortening is a murder" viewpoint which we shouldn't forget as a legitimate social statement. All the way through those arguments to the "you-can't-get-there-from-here" school - in other words, you can't make a balancing of benefit and risk. Only one of the seven papers - and again this is a personal view - only one of the seven papers is unequivocally correctly titled and we leave it as the second exercise to the student to determine which one.

Session VII. I have no comment.

Finally, since I was advertised as of Jack Benny vintage and, therefore, more or less an elder statesman in our particular racket of health physics, let me give a piece of advice to the relative newcomers to the field. You are all familiar with the Paolo Ucelli complex, aren't you? Well now, apparently not all of you know about the Ucelli complex. Let me refresh your memories. Ucelli, who lived 1397 to 1475, is credited with having first formulated the principles of perspective. Now you can't hope to go through a symposium on standards without having some pompous ass getting up from the audience and saying, "Let me put these things in perspective." That's the Ucelli complex, and it afflicts the older members of our profession including myself. Translated from the Italian - it means forget the distortions and exaggerations that you have heard from others and listen now to my distortions and exaggerations.

My very sincere advice to younger members of the profession is to get up and take a coffee break or something whenever the Ucelli complex rears its ugly head.

Thank you.

An Intrauterine Dosimeter

Health Physics Official Journal of the Health Physics Society

September 1972

Reprinted from *Health Physics* Vol. 23, pp. 389-390 by permission of the Health Physics Society.

Health Physics Pergamon Press 1972. Vol. 23 (Sept.), pp. 389-390. Printed in Northern Ireland

An Intrauterine Dosimeter

(*Received* 18 *October* 1971;
in revised form 14 *February* 1972)

I. Introduction

THE DEVELOPMENT of dosimetry techniques using thermoluminescent materials such as LiF, CaF and others has made practicable the measurement of low radiation dose rates sustained over long periods. With a few milligrams of these materials an accurate measurement of dose delivered over a 1-yr time span at background levels is possible. The purpose of this note is to introduce the concept of incorporation of a thermoluminescent dosimeter (TLD) into the structure of an intrauterine contraception device (IUD). Through this technique it should be possible to measure the long term integrated radiation dose delivered to a nonsurgically implanted dosimeter. The feasibility of this approach was demonstrated in a way that eliminated concern over releasing trace amounts of LiF or other thermoluminescent materials to the body. This isolation of the TLD crystals from body fluids was accomplished by first encapsulating them in thin stainless steel and then sealing the capsules within the structure of the IUD. We feel this dosimeter implantation technique should be of interest to groups studying long term radiation exposure and could be utilized through family planning clinics in many parts of the world.

II. Technical Considerations

Several types of thermouminescent materials are sensitive enough for the IUD application including all forms of LiF, CaF_2:Mn, and CaF_2:Dy. The LiF material seems most suitable for dosimetry because of its close equivalence in dose response to soft tissue. The response of LiF relative to soft tissue is constant within 10% for photons from 0.01 to 10 MeV.[1] CaF_2 has a rather large overresponse for photons with energies below about 100 keV[2] relative to ^{60}Co exposure (average photon energy 1.25 MeV) which makes the indicated exposure more difficult to relate to tissue dose.

Since the TLD is to be used to measure accumulated dose over a long period, the fading properties of LiF also make it well suited for the IUD application. CAMERON[4] lists the fading for LiF as <5% in 12 weeks. More recent studies at Battelle-Northwest have shown a fading of 6-7% per year[5] for LiF.

A second TLD which has a rather short fading time of a few hours, such as $CaSO_2$:Mn, can be used for measurements following an accidental radioactivity release or an accidental radiation exposure. In particular, $CaSO_2$:Mn is a factor of 60 or greater more sensitive to photon radiation than is LiF and exhibits a rapid fading at body temperatures. Within 24 hr of exposure greater than 50% of the induced thermoluminescence has faded. If the fading rate of $CaSO_4$:Mn at body temperature and the relative response of $CaSO_4$:Mn to LiF are carefully determined, the approximate time an accident occurred can be estimated.

Since LiF has a relatively high solubility in water (0.27 g per 100 cm^3), the dosimeters were first encapsulated in hypodermic needle tubing sealed with solid end caps by electron beam welding. The 304 stainless steel tubing wall is sufficiently thin (0.005 in.) that the effect on energy response of the system is negligible in relation to the shielding interposed by tissue when the dosimeter is in position. Because background neutron dose rates comprise such a small part (less than 1%) of the total natural radiation environment[6] either natural LiF or ^{7}LiF can be used for the IUD application. The natural LiF contains about 7% ^{6}Li isotope which has a cross section for thermal neutrons of about 1000 barn. The available ^{7}LiF contains approximately 0.01% of the ^{6}Li isotope.

Replacement of an IUD after it has been worn for a year is probably an acceptable practice. If not, since deliberate or spontaneous discontinuance of IUD use by the end of the first year averages 25%,[7] certain types of radiation studies could be accomplished by allowing unscheduled discontinuance to set the exposure period.

III. Preliminary Investigation

For this preliminary investigation, dosimeters made of natural LiF were chosen at random from a group of 200. The dimensions of the dosimeters are 1 × 1 × 6 mm. Each dosimeter weighs about 15 mg. A commercial readout instrument was used to measure the response of a number of bare dosimeters exposed to 100 mR radium gamma radiation. We considered

100 mR to be an appropriate exposure for evaluation of the system. With the reader set to reach a maximum temperature of 260°C in a 12-sec readout cycle, the net response of the dosimeters was 327 reader units or 3.27 units per mR of exposure. This net count results in a statistical uncertainty of about ±6% at the 67% confidence level for a single reading. If two dosimeters are used, of course, the statistical uncertainty improves.

Ten LiF crystals were encapsulated within thin stainless steel tubing to assure no harmful effects on the crystals themselves. Five of these encapsulated TLD's were then sealed within the stem of a double coil type IUD to demonstrate that this could be done without impairing the strength of the IUD.

IV. Conclusions

The use of LiF to measure dose at the uterine depth appears to be feasible. The preliminary results indicate a satisfactory sensitivity and accuracy for determining dose accumulated over periods as long as 1 yr. The dosimeter can be used in conjunction with a number of IUD types.

H. M. PARKER
G. W. R. ENDRES
R. G. WHEELER

Battelle
Pacific Northwest Laboratories
Battelle Boulevard
Richland, Washington 99352

References

1. J. CLUCHET and H. JOFFRE, Luminescence Dosimetry, *Proceedings of the International Conference on Luminescence Dosimetry* (Edited by F. H. ATTIX), p. 349, Stanford University (1965).
2. G. W. R. ENDRES, R. L. KATHREN and L. F. KOCHER, *Health Phys.* **18,** 665 (1970).
3. *Ibid.*
4. J. R. CAMERON, N. SUNTHARALINGAM and G. N. KENNEY, *Thermoluminescent Dosimetry*, p. 33, University of Wisconsin Press, Wisconsin (1968).
5. L. F. KOCHER, L. L. NICHOLS, D. B. SHIPLER, G. W. R. ENDRES and A. J. HAVERFIELD, Hanford Multipurpose Dosimeter, BNWL-SA-3955.
6. K. O'BRIEN and J. E. MCLAUGHLIN, Calculation of Dose and Dose Equipment Rates to Man in the Atmosphere from Galactic Cosmic Rays, HASL-228 (1970).
7. Report on Intrauterine Contraceptive Devices, Advisory Committee on Obstetrics and Gynecology, Food and Drug Administration, p. 26 (1968).

Plutonium, Industrial Hygiene, Health Physics, and Related Aspects

Uranium, Plutonium, Transplutonics
Handbook of Experimental Pharmacology
New Series

1973

Reprint from
Handbuch der experimentellen Pharmakologie
Handbook of Experimental Pharmacology
New Series

Edited by: O. Eichler, A. Farah, H. Herken, A. D. Welch

Volume XXXVI
Editors: H. C. Hodge, J. N. Stannard, J. B. Hursh

Springer-Verlag Berlin · Heidelberg · New York 1973
Printed in Germany

Plutonium, Industrial Hygiene, Health
Physics, and Related Aspects

H. M. Parker

(Not in Circulation)

Chapter 14

Plutonium. Industrial Hygiene, Health Physics and Related Aspects

H. M. PARKER

With 14 Figures

I. Introduction

Due to its unique history as a man-made element, plutonium passed very quickly from the stage of laboratory experimentation in trace amounts to manufacture as a military product in substantial quantities. If man is entitled to any satisfaction from a previous grievous experience in industrial hygiene, it would be that the tragedy of the earlier experience of the radium dial painters paved the way for the adequate protection of plutonium workers.

The broad principles of protection from the predominantly alpha-particle emitting plutonium consisted simply in isolating the material from contact with the human body by performing all operations in closed glove boxes with a secondary defense of protective clothing. From the start it was realized that the environment needed protection by the best possible filtration of exhaust air from the glove boxes.

These defenses were intended to eliminate:

1. Inhalation. Three separate hazards were recognized from the start:

Direct irradiation of the lung tissue, an effect forewarned by the experience in uranium and radium mining.

Transfer to the circulating blood and eventual residence in liver, kidney, or bone, with a quite early determination that bone deposition was one presumptive limiting factor.

Transfer from nasal and bronchial passages to the digestive tract, with secondary absorption to the bloodstream.

2. Ingestion. The prior radium experience had signalled the need for prevention of ingestion. While gross errors similar to brush-tipping would be unthinkable in this case, considerable attention was given to potential transfer to the mouth from food-handling or smoking with contaminated fingers.

3. Contamination of Intact Skin. The hazards contemplated were:

Relatively permanent deposition in a deep skin layer.

Transmission through skin to access with circulating blood.

With plutonium contamination on the *outer* surface of skin, the thickness of the horny layer is sufficient to absorb all the alpha particles. A third case, absorption into the skin, but not through it, was added later (NEWBERRY, 1964; DUNSTER, 1967).

4. Contamination of Wounds. This was an obvious concern in two aspects:

Direct access of soluble plutonium compounds to the blood stream.

Implantation of a small piece, for example, of plutonium metal with subsequent chronic leaching to the blood stream.

In retrospect, protection against these factors has been generally successful. There has been no identifiable case of plutonium transfer to the bone in sufficient quantity to demonstrably injure an individual—although this falls short of assurance that *no* late-effects case will develop. There has been a continuing history of the need to remove portions of tissue, predominantly from the hands, following wounds unresponsive to decontamination. Experience has shown that direct ingestion can be dismissed as a practical hazard (less than one case per 10 000 man-years of plutonium work). Conversely, inhalation continues as a major source of difficulty, involving one case for 40 man-years at some level of contamination, and one case per 450 man-years at 5% or more of the currently accepted occupational deposition limit in a typical major facility (HEID and JECH, 1969).

The strong attention to prevention of internal deposition by no means completes the catalog of needed defenses. Chemical manipulation of the actinide element, plutonium, frequently involves a fluoride stage. Powerful alpha-emitters react with many of the light elements to generate neutrons, and the yield from plutonium fluoride is particularly significant. The yield from plutonium oxide, which is 30 times less, must also be considered. Here, the learning experience from radium failed because concern for the alpha-neutron reaction for it was essentially confined to the deliberate mixture of radium and beryllium as a neutron source. Had the conventional compound for early radium use happened to be the fluoride, there would have been a long-unknown neutron component in its radiation effects.

Neutron generation by spontaneous fission of plutonium must also be considered. In addition, plutonium, although an alpha-emitter, does emit X-rays and gamma rays. With plutonium in industrial quantities, protection from these penetrating radiations is required. The need for these steps was correctly inferred from prior radium experience.

One unique factor with fissionable materials such as plutonium is the ability to create a critical mass, with major releases of penetrating radiation, and the creation of a host of residual radioactive nuclides. Steps to preserve subcriticality are an essential part of the health-physics protocol.

All that has been said so far inherently equates plutonium with ^{239}Pu, the particular nuclide that is predominantly produced in the initial chain reaction of natural uranium in which the ^{235}U is the fissionable component, and ^{238}U the fertile component, producing ^{239}Pu through the intermediate step of neptunium. Especially in the last decade, long term irradiation in nuclear reactors has generated substantial quantities of the higher plutonium isotopes ^{240}Pu and ^{241}Pu[1], a battery of other transuranium elements, americium, curium, californium, berkelium, and einsteinium, as well as a particularly radioactive form of plutonium, ^{238}Pu. The latter also may be generated separately for use as a heat source, now widely used in space applications, and in coming use for implanted artificial heart engines.

It is no longer possible uniquely to define the radiation hazards of plutonium. One must be cognizant of the particular complex of plutonium and other transuranium nuclides involved in any laboratory or factory process. As long as the radionuclides are successfully isolated from the body, the main changes are in recognition of the various neutron, X-ray, and gamma ray components and, of

---

[1] ^{241}Pu is primarily a beta-gamma emitter; there are some cases in which the controlling hazard of "plutonium" ceases to be due to the alpha-particle irradiation (see Appendix A).

course, in criticality control. Yet it is clear that current practices fall short of perfection of this isolation. When these complex mixtures are deposited, their proper measurement and comparison with acceptable standards, and application of corrective treatment procedures present new difficulties to radiation hygienists and physicists. Thus, in 1972, the challenges to health physicists in the plutonium field are greater than ever before. Moreover, they are not limited to the industrial hygiene aspects of inplant protection. Environmental releases, which have currently been reasonably well controlled, with the possible exception of spectacular incidents with military weapons and aborted space devices, can be expected to increase as the total usage of plutonium escalates, unless improved control of releases to the atmosphere and long term release of wastes to ground or water can be achieved. Contemplation of a projected 6 tons of plutonium per year going to waste storage by the year 2000 is a clear signal for the highest possible quality assurance of plutonium hazard control in the future.

II. Protection against Intake of Plutonium

A. Sealed Enclosures

Plutonium work, except at trivial levels, should be conducted in sealed enclosures such as glove boxes. By trivial levels, one means such things as routine bioassay samples which can be safely processed in conventional laboratory hoods. Hood operation with 1 mg Pu is undesirable or even intolerable. A switch at not more than 0.1 mg seems more appropriate because of the inherently greater risk of escape from a hood.

Conventional sealed enclosures will normally be of stainless steel with viewing windows of transparent plastic or glass. These windows may be permanently sealed or may better be gasketted to permit safer replacement. Materials passed into or out of the enclosure go through some form of airlock so that the interior space never communicates directly with the outside. In Europe, particularly, well-engineered proprietary ports are used. In the United States, it is common practice to use plastic bags with provision to seal an exiting contaminated part in an interior plastic bag. Similarly, in good practice, the manipulating gloves are attached to the box by flanges which themselves constitute a double barrier, so that a defective glove may be removed without exposure of the glove box interior to the room. The atmosphere of the box may be an inert gas, whenever the plutonium is to be handled in a pyrophoric state. Extreme precautions are needed to minimize explosions or fires in glove boxes as these are the readiest means of breaching enclosure integrity either to the operating room or to the environment. Generally, a glove box is continuously exhausted so that a breach of a glove or other part will provide an in-flow of not less than 150 linear feet of air per minute (UNRUH, 1970). Fig. 14.1 is a schematic drawing of a typical glove type hood illustrating the arrangement of machined glove ports, the box filtration system, the double airlock and its separate air control system. Fig. 14.2 shows an actual process line used for development of plutonium oxide reactor fuel. Interesting features are the absence of clutter which almost invariably appears in standard chemical hoods. This reduces the risk of casual contamination and most importantly reduces the fire risk from spontaneous combustion, a frequent cause of which is getting both oily substances and acids on the same wiping rag or paper. In this hood can be seen a narrow necked detachable metal bottle containing a carefully size-graded special sand. This provides an excellent method of quenching a small fire in a process hood. These boxes are of circa 1960 vintage, and detailed

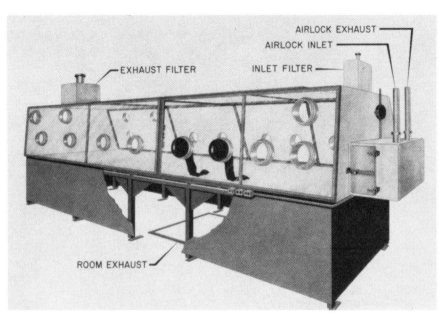

Fig. 14.1. Conventional design of process hood. Drawing of a typical glove type process hood in operations requiring protection against alpha particle contamination only. The smoothly machined circular ports permit leak-tight attachment of the long rubber gloves. The box interior will be maintained slightly below ambient air pressure by exhaust through the exhaust filter. The inlet filter prevents blowback of particulate contamination. All materials enter or leave the box through the air lock with its own air supply

improvements are being continuously made in newer installations. Some interesting examples are:

a) *Application of Human Engineering to the Spacing and Height of Glove Ports.* Fig. 14.1 and 14.2 suggest that comfortable working for long hours at these glove ports is more appropriate to the physique of a gorilla than of a man. Rather belatedly, considerable attention has been given to optimizing the placement—height, spacing, and even angling of the ports—especially where there is one designated task at an individual glove station. In addition to operator comfort, these principles contribute to safety by reducing the risk of accidents potentially leading to wounds or breakage of gloves.

b) *Replacement Absolute Filters.* The older filters were fabricated as rectangular blocks. Newer ones are cylindrical. They are gasketted into cylindrical guides so that a replacement filter ejects the old one into a plastic bag for "bagging-out" with about the same ease as replacing an injector-type razor blade.

c) *Filter Penetration Testing.* Modern boxes have installed facilities for performing the D.O.P.[2] penetration test in place. Such care eliminates gross contamination of the main exhaust ducts, which is a particular hazard if the plutonium used is in a pyrophoric form, and in all cases is a major problem if maintenance work is required on the ducts. Additionally, extra care with box filtration adds one more barrier against environmental contamination should there be trouble with the main exhaust system.

---

2 Di-octyl phthalate.

Plutonium Industrial Hygiene, Health Physics and Related Aspects

Fig. 14.2. Process line for plutonium fuels. An actual box used in the peparation of mixed uranium oxide-plutonium oxide fuel rods with Zircaloy cladding. The taller central section contains metal working equipment, such as a swage, for reducing the rods to a prescribed diameter. Note that the operator's gloves and any loose gloves balloon inward due to pressure differential. Other gloves are loosely tied off in pairs to prevent this, while some unused ports may be blanked off

Fig. 14.3 illustrates a permissible method of performing a once-only operation with a substantial quantity of plutonium, in this particular case the pressing and encapsulation of a ^{238}Pu source. A lucite glove box is prepared with complete sealing, conventional glove ports on the face, a bagging-out port on the left side, filtered exhaust on the top, and containing all the tools and equipment needed for the work. It is then placed in a standard laboratory hood (which would be a disaster area if used itself for this work) to utilize the needed services, such as argon for inert atmosphere. The plutonium material is bagged-in, the operations performed, and the finished piece bagged-out. The conventional hood services are crimped off and detached, maintaining the isolation. The whole lucite box (in the now disreputable condition shown in Fig. 14.3b) is enclosed in a steel box, welded tight, and dispatched to an approved storage area.

A well designed permanent box or process line has two fire-resistant so-called absolute filters in series. The design standard for first class filters calls for an efficiency of not less than 99.95% at rated flow for standard test conditions. These tests are made in situ by measuring the penetration of D.O.P. particles of prescribed diameter (e.g., 0.3 μm). The in situ test is essential because there is probably more risk of defect in the gasketting than in the fiter per se. The

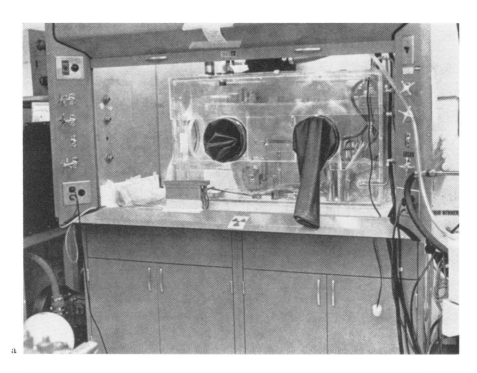

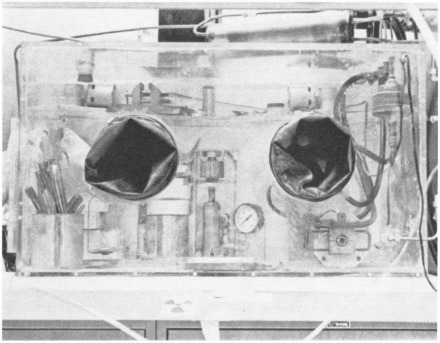

Fig. 14.3a and b. Special purpose lucite glove box. A carefully prepared sealed lucite glove box is used for special single operations not provided for in permanently installed glove type hoods. a Shows the box coupled to services in a conventional laboratory hood, and containing all the equipment for the special task. b The condition of the hood at the conclusion of the task, prior to preparation for integral removal as contaminated waste. (Courtesy: Donald W. Douglas Laboratories of the McDonnell-Douglas Corp.)

Fig. 14.4. Neutron-shielded process line. A plutonium process line with neutron shielding wall. The large viewing ports are plate glass windows with the interior space filled with water to moderate the fast neutrons from the plutonium sources. (Courtesy: Atlantic Richfield Hanford Company, Richland)

two filters in series are not necessarily claimed to achieve a compounded release as good as 0.05% of the 0.05% that penetrates the first filter. The aerosol or dust that impinges on the first filter has a spectrum of particle sizes. Those that successfully penetrate the first filter will have a new spectrum less easily removed by the second filter. Rather, the intent is to have a backup filter against the failure of the first one or its gasket mounting to the exhaust duct framework. In all major operations, an air monitor is installed after the second filter to provide a signal should both filters fail. A continuous air sample is drawn from between the two to provide warning of first filter defect. It is these practices which have led to the relatively good experience in environmental releases.

Fig. 14.4 shows a modern automated or semi-automated plutonium process line where the composition and quantity of plutonium require neutron protection for the operators. This is achieved by the use of water walls maintained in place by plate glass sheets. The walls shown are 12 inches thick. Water is effective by slowing down the fast neutrons. In some cases, soluble boron compounds are added to absorb the slow neutrons with minimal production of penetrating radiation.

It will be recognized that a 12-inch wall on a glove box would have the unwieldy effect of shortening the arms by 12 inches. In the example illustrated, the shielding wall is separated from the process line which has glove ports only for maintenance and special adjustments at times when the plutonium load is low, which, together with short exposure time in the maintenance zone, adequately limits the neutron exposure.

The complete filtration, monitoring, and air exhaust system of a plutonium plant constitutes a major integral portion of a process facility. Even in a small

Fig. 14.5. Air exhaust and filtration system of a small licensed plutonium facility. Shown during installation is a bank of eleven "absolute filters" each $2' \times 2' \times 1'$, with a total air flow of 8800 cfm. Each aperture is sealed by a plastic bag held by the gasket-lined covers with four corner screw-downs. Not installed at this stage were small bleed-off lines for continuous proportional sampling of the air contamination. (Courtesy: Donald W. Douglas Laboratories of the McDonnell-Douglas Corp.)

licensed facility the apparatus is substantial, as shown in Fig. 14.5. The dimensions can be inferred by considering that each of the eleven TV screen-shaped ports for the absolute filters measures about $30'' \times 20''$.

The almost universal practice of manipulating through rubber gloves developed in the early days when small quantities of ^{239}Pu were involved. Today, there are two sound reasons why it would be preferable to use mechanical manipulators:

1. Large masses of plutonium, especially those rich in the higher isotopes, lead to undesirable or directly damaging contact irradiation levels.

2. A substantial fraction of operator exposures occurs either through puncture wounds with the hands in the gloves or through room contamination following defects in the gloves (Fig. 14.6).

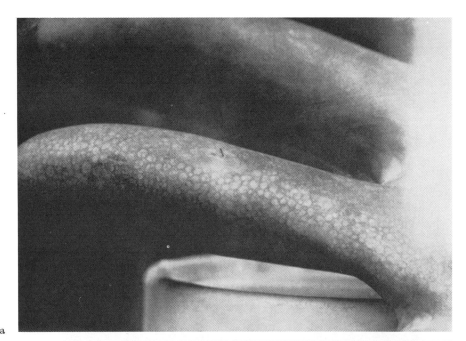

Fig. 14.6a and b. The two main causes of personnel contamination in a glove line. a A small plutonium-contaminated wire has penetrated the thick glove of the box, the thin surgical glove on the operator's hand, and the operator's finger. Essentially no contaminated air escapes, and the pin prick wound may be so small as to escape detection visually or by search for surface contamination on the finger. Yet the imbedded amount may require surgical removal. b A typical cause of room air contamination is a small glove rupture as in this glove held here for demonstration by an operator using other glove ports

The appeal of direct manual dexterity is such that a change to manipulators may be difficult to promote. Yet the experience to date, and the expected proliferation of plutonium operations related to advanced reactor fuel preparation and other tasks calls for its serious consideration.

In processing factories, design engineers prefer to develop their automatic or semi-automatic systems in long lines. Health physics pressure should be to have each work station a separate enclosure connected sequentially through air locks incorporating fire doors. While the continuous system is cheaper and simpler to operate, the penalty in the event of a fire transmitted along a line is too high a price to pay, both for internal cleanup and for potential environmental release.

B. Control in Other Working Spaces

When it is conceded that sealed enclosures are imperfect, it becomes imperative to provide secondary contamination protection throughout the operating buildings. Conventional aspects of this are:

1. Protective Clothing for the Operator

This includes rubber gloves, coveralls, work shoes or shoe covers, and ready access to full face air masks in the event of detected air contamination.

2. Continuous Air Monitoring in the Process Rooms

The customary method of measuring air concentration is to pull air preferably through a thin filter (such as a molecular pore filter) at a known rate for a known time. The alpha activity of the material on the filter face is then measured after the decay of the radon and thoron daughter products which represent a naturally occurring interference. Where a prompt warning is required there are two options:

a) Alarm response at a relatively high concentration level, sufficient to stand out above the interfering radon and thoron daughters.

b) Continuous compensation for the daughter products by reading off a beta-alpha coincidence signal from the daughters, or by a variety of other methods of compensation.

One British method uses the β-α coincidence correction, while another uses the specific signal in a CsI crystal from the 5.1 MeV alpha particle of plutonium, compensated for the degraded alpha particles above 5.1 MeV that are deduced from the total activity of alpha particles above 6 MeV originating in the radium and thorium series (IREDALE and HINDER, 1962). A French system distinguishes dusts by granulometric differences with selective filters in series (BILLARD et al., 1961).

Solid state detection is becoming increasingly effective and popular (LINDEKEN and LAKIN, 1967). A sound approach is to combine a sensitive recorder, even if its accuracy is somewhat variable, as the alarm meter, with a reliable non-alarm meter to provide a permanent record of air contamination. Electrostatic deposition of the room aerosols on aluminum plates is one method of achieving the second objective (WINKLER et al., 1969).

When contamination occurs in a process room, it is likely to be highly uneven. Opinion is sharply divided on the relative merits of having an adequate number of detection devices coupled to a central exhaust system (POMAROLA et al., 1966), where better electronics and system maintenance can be achieved, at the expense of possibly measuring at the wrong points versus the alternative of having a simple portable system on each operator with the receptor on the forehead or other point as close to the nose and mouth as possible. The data on the relatively high frequency of inhalation events would suggest that the controversy is best resolved by employing both methods. This would indeed have considerable value in contamination case follow-up if the actual intake could be determined by the monitor on the person. This is not so because of individual variations in breathing pattern. The specific respiratory cycle, the average volume intake, the nature of the inhaled aerosol or dust, and the proportion of nose and mouth breathing all affect the deposition.

A desirable feature practised in most of the principal installations is to measure the characteristics of the radioactive particles in process rooms that experience plutonium contamination. This provides an insight into the percentage of any accidental air contamination that might present a respirable hazard, by reference to standard industrial hygiene surveys of intake and retention of particles of various aerodynamic diameters. Such data are especially useful if the spectrum of sizes is shown to be fairly consistent. In the event of an accidental inhalation case, the size distribution of particles on a nose smear can perhaps be compared with the known ambient size spectrum to improve the prompt estimate of amount inhaled. The most dangerous atmosphere is one in which pure plutonium particles (often as oxide) concentrate in the size range of optimum (or more correctly "maximum" because it is the reverse of optimum for safety) respirability. Paradoxically, the safest atmosphere is an industrially dirty one in which much of the plutonium may attach to larger inert particles which do not penetrate to the lung. One is not aware of conscious attempts to promote such an atmosphere, the effort being better applied to minimizing the primary contamination.

The literature is rich in measurements of room air contamination. It is not referenced here because the recommended course for those needing it is to refer to practices in their part of the world.

3. Change Rooms

Incoming operators change into the complete protective apparel. On leaving, they discard the work clothing according to a step-off pad ritual designed to minimize accidental cross-contamination. The hands and feet are monitored for plutonium contamination; the worker may be detained for on-the-spot decontamination should the readings on these surfaces exceed prescribed levels.

4. Zoned Air Change

A process building is compartmented into zones of relative cleanliness, with reduced air pressure in the direction of higher contamination potential. The conventional plan is:

Zone 1. Offices, lunchrooms, and the "clean" side of the change rooms.

Zone 2. Corridors between change rooms and laboratory or process rooms.

A double barrier air lock is provided between Zones 1 and 2. Conceptually, the transition between Zone 1 and Zone 2 is at the point of removal of the operators' protective clothing. In practice it is at the air lock, with the clothing removal ritual used to prevent significant contamination of Zone 1. Zone 2 is normally essentially uncontaminated.

Zone 3. Laboratory and process rooms.

Zone 4. The interior of process lines and glove boxes.

In smaller facilities, there is little or no differentiation between Zone 2 and 3. The essentials of good control are to maintain the integrity of the Zone 4 to Zone 3 separation so that Zone 3 is clean except for rare and promptly recognizable accidents, and to maintain high standards in the change room procedure so that the complete cleanliness of Zone 1 is assured. A current weakness of some designs is that it is necessary to provide emergency exits from Zone 2 or even Zone 3 to the outside in the event immediate building evacuation is needed. Although these exits are normally sealed with breakable tape, it is superior practice to have each one built as a double barrier air lock.

5. Maintenance Access

Experience throughout the nuclear industry has shown that the greatest risk to operating personnel frequently arises during maintenance operations. At the same time, the risk of spreading contamination to others tends to be maximized. This is certainly true in a plutonium processing operation. Maintenance personnel should have complete protective covering, including a mask either with supplied air or with self-generating air supply. In principle, the maintenance worker is *inside* the process line or glove box; his personal protective layer should provide isolation from plutonium as nearly equivalent as possible to that of a closed glove box. In some countries, the process equipment is built in the form of a rectangle with the interior forming a sort of Zone 4A accessible only to maintenance workers clothed in the equivalent of a "frogman" suit. This type of isolation of prescribed maintenance zones seems to be a highly effective system.

6. Personnel Monitoring Stations

A major station is maintained at the change room interface. In addition, good practice calls for installed alpha-monitors at the door of each process room, used to check the hands and feet *every time* a man leaves a Zone 3. Individual monitors at the actual work station have been found useful. Periodic checks by the operator lead to prompt recognition of minor contamination usually long before it achieves a level sufficient to trigger the air monitoring devices.

Historically, the alpha sensitive surface monitors were troublesome because they required a very thin and hence fragile window over either an ion chamber or proportional counter. Development of devices such as scintillation counters has provided dependable instrumentation at reasonable cost. Here, the thin lightproof film (aluminized Mylar, for example) can be applied directly to the flat solid scintillator face. It is standard practice at each fixed station monitor to provide high- and low-activity alpha particle sources to test the proper functioning of the equipment before each test.

7. Material Transfers

All plutonium bearing material removed from process lines of glove boxes is removed through the "bagging" or equivalent devices enclosed in two separately sealed containers. The double wrapping ensures that at least the outer surface is contamination-free. Special precautions are sometimes needed to prevent damage to the package by self-heating of the sources or by radiolytic decomposition. This is especially the case if separated highly radioactive transuranium elements are involved.

Storage of removed products in special vaults is required, with provision for detection of leaks and for decontamination if necessary.

8. Waste Handling

The valuable plutonium materials for nuclear fuels, military parts, space or biomedical heat engines, or whatever, naturally get careful attention and control. It is a disconcerting fact that the small percentage that degenerates into some form of waste requires equal or even greater control with respect to contamination.

Liquid wastes are routed to prescribed storage tanks. Solid wastes such as metal chips or powder residues are collected in sealed containers and either stored or prepared for recovery. Nuisance wastes, while lower in activity, probably cause more trouble. They include large volumes of paper wipes and the like used for cleanup inside glove boxes. These are either stored in sealed containers in "permanent" repositories or further treated. Spontaneous combustion of some types of waste is a threat that must be overcome.

A principal problem is the projection of these storage demands into the future, with a proliferating nuclear industry. With material of half-life about 24 000 years for ^{239}Pu, and necessary storage time of hundreds of thousands of years, "permanence" of a surface repository creates a legitimate question in the public mind. In the United States at the present time, serious consideration is being given to storage in deep underground structures such as salt mines. European practice is in some respects ahead in this field. Compression of the solid wastes to reduce volume is a first step. A desirable first step is the separation of wastes into two categories—combustible and noncombustible. It is the latter that are then profitably compressed in a baling press. A volume reduction by a factor of 2 to 3 is conventional unless very powerful presses are used; a volume reduction by a factor of 5 would be considered good. More recently, successful incineration of the combustible plutonium wastes has been demonstrated. It is anticipated that this approach will become widespread, as it already is for conventional fission product wastes. French practice is probably the most advanced in this area. For example, at the Marcoule plant of the C.E.A. the main production incinerator fires 900 kg of waste per hour at 900° C, with an afterburner at 1 200° C. The off-gases are cleaned in three stages, and the incinerated product cooled and compacted in standard steel drums. The drum is filled up with loose gravel, sealed and taken to storage. By contrast, the incinerator for recovery of plutonium from combustible solids copes with only 1 kg per hour although it is nearly half the size of the main incinerator, reflecting the extra care needed in the plutonium case. Considerable attention has been given to the risk of fire in the loading stages, which are done in a nitrogen atmosphere. Most of the plutonium waste arrives packaged in polyvinyl chloride bags or sealed sheets for contamination control. Burning of such material releases chlorides which can adversely affect metal parts. The Marcoule installation is designed to cope with this hazard. Criticality is a serious risk in such an incinerator. It is controlled by completely emptying the system after each 300 g Pu have been treated. Accurate knowledge of the waste content of each incoming package is hard to obtain. Here it is monitored by its weak photonic radiations between 370 keV and 440 keV, with results from standard size packages good to about $\pm 30\%$. Volume reduction to 3% of the original is achieved, making chemical recovery practical. As an example of the care needed in such work, the cooling water around the final parts of the equipment is loaded with 20 g boron per liter as salts to suppress the criticality enhancement of the water blanket.

C. Control of Off-Standard Conditions

A natural order of graded response to the degree of contamination develops. These details need not be considered here. The essential feature of control is that any demonstrable level of plutonium contamination requires investigation of its origin and appropriate clean-up action.

D. Major Accidental Releases

The required action is evacuation of the operators by the best means available to them to minimize spread of contamination to other work areas or the environment. Damage control is performed by fully protected personnel (as for maintenance). The often long and laborious rehabilitation is similarly attacked.

E. Incidental Releases

Whenever plutonium contamination of the process room is detected, work is interrupted, and the affected area defined by surface monitoring equipment. For all except the most trivial cases, the operators use air masks at this stage[3]. Cleanup is begun at the periphery, with simple scrubbing with soap and water often being effective. Walls and floors of the process rooms are usually finished in special paints or other surfaces to facilitate this. Stainless steel and other parts may be treated with stronger reagents. However, the cleanup is notably different from "mopping the kitchen floor". Great care is needed to prevent further spreading of contamination. As each area is cleaned, it is covered with paper to minimize its recontamination. Also, the mopping liquid and tools must be collected and sent to plutonium waste storage.

The customary detection limit for contamination is about 500 alpha-disintegrations per minute over a 100 cm² area. Tested surfaces must be dry, for a water layer would itself be thick enough to absorb the alpha particles. Speedier identification of contamination can be done by meters sensitive to the emitted X-rays or gamma-rays, but the sensitivity is usually not competitive with the alpha-monitor. It is also clear that safe decontamination of a porous surface is virtually impossible, because the activity would be shielded from the alpha-monitor. It is a salutory reflection that a work room measuring $4\,\text{m} \times 4\,\text{m} \times 3\,\text{m}$, if uniformly contaminated, could be left with 8 μCi Pu if all surfaces were left contaminated at levels just below the detection limit. Should the material subsequently rub off and become airborne, one-five-hundredth of this amount would constitute a maximum permissible lung burden. For such reasons, it is common practice to paint the decontaminated surfaces with a sealing paint. All such surfaces subsequently require some degree of "perpetual care". When the painted surface is outside a process room (i.e., where there is normally no contamination), the sealant should be a bright yellow paint or an equivalent warning signal. British work on the physics of resuspension of contamination, and on the related hazards, has been notably thorough. Contributions of H. J. DUNSTER, I. S. JONES and S. F. POND, K. STEWART, and others, with further references to other work is conveniently grouped in a Gatlinburg symposium report (FISH, 1967).

III. Contamination of Personnel

Comprehensive reviews of the management of plutonium contaminated personnel are available from many sources.

One systematic approach to decision levels on surface contamination includes the importance of resuspension and such matters as contamination of home and

---

[3] Installations often have established limits of air contamination up to which a full face fitting mask is deemed adequate, whereas at higher levels, a mask with supplied air (piped in or chemically generated) is mandatory. The switch point is usually 10 or 20 MPC_a.

environment from residual contamination of workers (HEALY, 1971). Each proposed limit has a symbol identifying the mechanism leading to the lowest value according to the code

        A — skin absorption,
        H — inhalation from clothing or skin,
        R — resuspension,

with other codes for gamma emitters. For the alpha-emitting plutonium isotopes, the signal is H for contamination of skin or clothing, and often R for transfers to home. Interestingly, the signals for ^{241}Am and all the higher transuranium elements switch to A for skin and transfers, remaining H for worker clothing. To some extent, this is based on the known eagerness of plutonium to form complexes, compared with americium and higher transplutonium elements. HEALY (private communication) reports that the switch is perhaps an artifact due to current paucity of knowledge about the behavior of the higher elements. His decision levels are meant to be a basis for organized discussion of realistic action levels.

Here, the highlights of systems will be described. This account will relate primarily to the Hanford system for convenience, with no implication that other systems are not equally useful. In particular, the British writings on the topic are universally highly regarded. As an example, a report of specific considerations of protection against airborne particulates by HOLLIDAY et al. (1969, 1970) is a model of conciseness and clarity. Indeed, there is no major group that has engaged in plutonium handling that has not made some contribution worthy of adoption by the others, and of eventual incorporation into an international protocol. Since the best results in this phase of industrial hygiene often hinge on apparently minor details, it is recommended that newcomers to the field both review the literature in depth and personally inspect operating methods at a plutonium site.

A. Skin Contamination

The extent of the contaminated area is defined and surrounding areas protected from contamination spread. Loose material responds to removal with gauze and mild detergents. More obstinate cases are then scrubbed with soap and water, including quite abrasive soaps if necessary. Recalcitrant cases are treated chemically. One scheme uses a 10% solution of the chelating agent Na_4 EDTA (ethylene diamine tetraacetic acid). This is contraindicated if there is a break in the skin in the affected part (WILSON and SILKER, 1960). The customary last resort is the use of saturated potassium permanganate solution followed by a 4% solution of sodium bisulfite. In effect, this chemically removes the outer layer of skin. An alternative method is to apply a sticky tape or a plastic spray-on film, which upon removal brings some contamination (and possibly superficial skin debris) with it. This dry method has the merit of reducing the volume of contaminated wastes.

The simpler steps are commonly accepted as non-medical treatments. Persistent cases with abrasion of substantial skin areas are preferably regarded as medical cases (JECH et al., 1969a). National customs vary widely in this respect; health physicists who go beyond the accepted boundaries acquire dangerous liability.

B. Eye, Nose, or Mouth Contamination

Irrigation with water or a normal saline solution is the usual limit of decontamination.

C. Wounds

For this purpose, in addition to major wounds, cuts, punctures and severe abrasion are included in the category. The conventional surface counter is no longer effective for detection of the source. Instead, a wound counter consisting of a thin slice of sodium iodide crystal with suitable electronics as a scintillation counter is used (ROESCH and BAUM, 1958). The chosen slice thickness, usually 1 mm, is an optimizing compromise between absorbing the low energy L X-rays of ^{239}Pu (13.6, 17.2, and 20.2 keV) to give a strong signal, and rejecting as much background signal from external natural radiation as possible by minimizing the crystal mass. The minimum detection amount for ^{239}Pu in a 10-minute counting period is about 0.1 nCi. For more complex mixtures of plutonium it is more realistic to claim a limit of 0.1 nCi to 0.6 nCi depending on isotope mix and depth of deposition. Precise location of the imbedded contamination can be achieved (except as to depth) by using a thin metal diaphragm as a collimator; the signal from the soft X-ray occurs only when the hole is over the plutonium. Reasonably fast and accurate localization contributes materially to the physician's confidence when he is called on to make excisions close to a part that would damage finger dexterity, for example.

There are two reasons for gaining the highest possible removal from a wound. There may be a possibility of radiation damage at the site, with possible malignancy. Alternatively, residual material may be transmitted to the bone, probably the more cogent of the two reasons. There is a reluctance to leave at the wound site an appreciable fraction of a potentially transmissible maximum permissible body burden.

For related reasons, there is a continuing effort to improve the sensitivity of detection of contamination in wounds and their speedy location. In one method employed by WRIGHT LANGHAM et al., at Los Alamos, a dual scintillation counter circuit is employed. A thin sodium iodide crystal to read the L X-ray signal is backed by a thick cesium iodide crystal which accurately reads the ambient radiation background signal and subtracts the appropriate background from the thin crystal signal.

In one development, a lithium-drifted silicon detector is used to measure the depth of a contamination source (SWINTH, 1968, PALMER et al., 1968). Depth discrimination is possible because a silicon or germanium detector can resolve the three L X-ray lines at 13.6, 17.2, and 20.2 keV. The relative intensity of the lines varies with source depth. For a point source, discrimination can be achieved to within ± 0.2 mm. The technique is ideal where an identifiable plutonium piece has been imbedded. For routine use it suffers two disadvantages:

a) Discrimination is gained at the expense of sensitivity.

b) Puncture wounds typically have contamination along the whole track.

In another application at Hanford, the hand is examined by 26 thin slice NaI crystals above and below the hand. The sensitivity of this version is between 0.6 nCi and 1.2 nCi plutonium depending on isotope mix, average depth, and dispersion of the deposition. In a series of 200 routine examinations, no previously undiscovered depositions were found, whereas all known depositions, run as a blind test, were correctly identified (ANDERSEN, 1972).

D. Lung Contamination

Modern technology leads to more and more plutonium work in powder form, frequently as an insoluble oxide. This increases the inherent risk of inhalation and of the deposition of insoluble particles in the lung. The biological evidence of the long retention time of such particles, the experimental production of neoplasms of the animal lung, the notable concentration of the particles in lymph nodes, and the difficulty of removing known deposits, all lead to some concern about the reliability of present maximum permissible lung burdens. There is no health physics area in which full attention to the concept of lowest practical exposure is more relevant.

This concern is matched by continuing improvements in the early detection of airborne contamination, improvements in means of measuring lung burdens more accurately, and improvements in the very limited steps to effect removal of lung burdens.

E. Detection of Airborne Contamination

The use of continuously operating air samples has been referred to in Sec. II.B.2. Within the last decade, the sensitivity of such devices has increased by about one order of magnitude. Unless there is a breakthrough in technology, further gains will be increasingly difficult. Good recognition at the present time involves one accepted maximum permissible concentration (MPC_a) in one hour, when there are no interferences from conflicting radionuclides other than the natural radon and thoron daughters. Since the occupational limits are predicated on a 40-hour week, there is some safety margin here. Yet this is offset by the risk of appreciably higher concentration at the breathing zone. As the detection limit is driven downwards, reliable measurement is reduced because of the random nature of the particle size distribution. Circumstances are reported in which it takes 2000 m^3 of air to get a representative sample accurate to 0.1 MPC_a (SHERWOOD and STEVENS, 1965). Man himself, breathing 40 to 50 m^3 in a workweek, is not a representative sampler at that concentration level. The sample activity may be almost totally contained on one single active particle, and it may be such a particle that the operator inhales.

1. Detection of Contamination by Nasal Smears

Whenever a high air concentration is observed, or if there is any possibility of leakage from a ruptured glove box, nasal smear samples should be taken. With commonly employed materials such as plutonium oxide powder, there is believed to be a reasonable correlation between the nasal smear and the lung deposition. In 30 oxide inhalation cases, the following relationships, each good within a factor of ± 5 were found (HEID and JECH, 1969):

Burden per 1 nCi nasal smear	
Initial pulmonary	1 nCi
Pulmonary at 7 days	0.6 nCi
Initial lung	3 nCi
Ultimate systemic	0.1 nCi

With other aerosols, there is no assurance that such relationships apply. In fact, with very fine particles there seems to be a finite possibility of lung deposition without a nasal smear signal.

2. Detection of Lung Burdens by External Means

The whole body counter is a versatile tool for the detection of various radionuclide burdens in the body. It is not at its best when applied to plutonium.

The characteristic X-rays of ^{239}Pu, used as signal generator in the wound and hand counters, are ineffective here because of their poor penetration of the chest walls (WAITE et al., 1970). Reliance is placed on measuring the 60 keV radiation of ^{241}Am, present as a daughter product of ^{241}Pu. Conventional counters for this purpose employ thin NaI crystals of large cross-section (LIDÉN and McCALL, 1962). A double crystal system using a thin CsI crystal for the soft radiation and a thicker NaI in anticoincidence improves the background by 60% permitting measurement without a steel shielding room (LAURER and EISENBUD, 1969). Several other systems are in the development stage; for example, at Los Alamos. An alternative detector, the Maushart proportional counter, has enjoyed considerable popularity in European practice (KIEFER and MAUSHART, 1962). The Hanford crystal counter employs four crystals approximately 9.5 mm thick and 127 mm diameter. It provides a detection capability of 0.15 nCi to 0.6 nCi ^{241}Am, depending on chest wall thickness. In the absence of more specific data, it is provisionally assumed that the plutonium burden is 15 times this. However, the true burden is a function of both the plutonium isotopic mix and the age of the sample since separation and in-growth of the daughter ^{241}Am. In a plant or laboratory using a constant source of plutonium, it is desirable to reserve a sample for actual measurement of these relative amounts at any subsequent time. Failing that, the activity in the nose smear may be enough to permit analysis. By coupling these data with phantom measurements simulating the build of the contaminated individual, amounts comparable with the maximum permissible lung burden or better are measurable. This at least provides a reasonable signal for attempted therapy of substantial depositions. One would like to have a system that reveals chronic buildup of minor depositions at the level of, say, 5% of the MPLB. This is impossible with present sensitivities. In any case, over time, the chronic burden would probably involve an aggregate of many plutonium mixes, with various unknown ages of americium growth. In addition, there is the possibility of biological separation of americium from plutonium in the lung.

3. Detection in Feces

An inhalation event will normally be accompanied by deposition in the upper respiratory passages, part of which by ciliary action will be transferred to the digestive tract. Most of this material will appear in the feces, especially for relatively insoluble material such as the oxide. Even for soluble materials, the fecal elimination will be initially some six times higher in feces than urine. After an inhalation incident, fecal samples are usually analyzed; if only one is done, the second-day sample is expected to give the peak value (LISTER, 1966, 1968). A representative schedule for feces sampling at Hanford is as follows (HEID and JECH, 1968; HEID et al., 1968):

Schedule 1. Obtain five daily fecal samples within the first seven days post intake.

If any of the following circumstances arise:

A. Nasal smears exceed 500 d/m.
B. Nasal smears $>100 <500$ d/m and exposure duration >5 min.
C. Nasal smears $>5 <100$ d/m and exposure duration >50 min.
D. Exposure to fumes from a fire.
E. Air sample results exceed 2×10^{-10} µCi/cm^3 for an 8-hour period and exposure duration >1 hour.

Schedule 2. Obtain one fecal sample on the second day post intake.

Plutonium Industrial Hygiene, Health Physics and Related Aspects 631

If any of the following circumstances arise:
A. Nasal smears are positive but do not meet criteria in Schedule 1.
B. Any other person in the same incident meets the criteria in Schedule 1.
C. Air sample results exceed 2×10^{-11} µCi/cm³ for an 8-hour period.
D. Widespread skin contamination in a dry form or facial contamination > 1000 d/m.
E. Clothing contamination in a dry form > 5000 d/m.
F. Possible plutonium inhalation is suspected for other reasons.

In Schedule 1, a correspondence, again within a factor of ± 5, is claimed between 5 times the nasal smear activity and the integrated fecal excretion in the first five days post intake, but some non-conforming cases are known. Nevertheless, the combination of nasal smear, lung count, and fecal examination does seem to provide a reasonable basis for early identification of substantial lung burdens that would prompt removal attempts.

4. Detection in Urine [4]

Internal contamination by plutonium at the levels of interest cannot be detected by the weak photonic irradiation emitted, in most cases. Thus, the initial health physics effort when plutonium was first used concentrated on means of estimating body deposition from radiochemical analysis of available excreta. Parallel work with animals at that time was exclusively with soluble plutonium compounds, and it became clear that the activity of urine provided a reasonable basis for detection of body burdens. The program was placed on a sound quantitative basis following the work of LANGHAM (1956), who introduced known amounts of plutonium into a number of selected people with terminal illnesses, judged not likely to materially affect the normal excretion rate. LANGHAM also measured fecal elimination rate. Although the fecal signal is stronger in the first few days after intake, the urine signal is thereafter higher, and in any case tends to be more consistent. Combining this with the greater acceptability of urine sampling by both worker and analytical technicians, urine analysis became the universal method of choice. With the later growing use of plutonium in oxide form and the prevalence of intake by the inhalation route, the relative significance of fecal sampling has perhaps steadily increased. Yet urine analysis remains as the major tool for routine detection and attempted quantification of all "systemic burdens" of plutonium; in fact, insoluble plutonium deposits in lung and lymph nodes are arbitrarily excluded from the terminology of systemic burden, and are sought either by direct measurements of emitted photonic radiation or deduced from pertubations of the estimated systemic burden over time. Regularly scheduled urine analyses have successfully pointed to increments of systemic burden from transfer from both lung and contaminated wound sites (see Appendix A). These observations indirectly support the interpretability of urine samples for standard cases not obfuscated in part by such long-term accretions from specific compartments.

Routine analysis for plutonium in urine has thus become standard practice in all major plutonium operations. Several methods are available of more or less comparable effectiveness. The particular method used at Hanford is briefly as follows:

Home collection: Voidings one-half hour before retiring and one-half hour after rising on two successive days closely approximate a total 24-hour sample, and minimize inconvenience to the workers.

---
[4] The systematic ordering of this chapter would have called for notes on urine analysis after 3 or 4 of the previous sections. Instead, they have been grouped together here.

Analytical method: Chemical extraction with TTA (thenoyl trifluoracetone) followed by electrodeposition on a stainless steel plate is used. The plutonium activity is measured by counting alpha particle tracks on nuclear track emulsion (SCHWENDIMAN et al., 1951).

Sensitivity: The sample counting detection limit is 0.05 dis/min. For soluble plutonium in a single intake the sample activity can be related to the initial intake by the now classical LANGHAM equation (LANGHAM, 1956).

$$\frac{\text{Daily excretion rate}}{\text{Initial intake}} = 0.002\, t^{-0.74}$$

where t is the time in days after intake. Thus the sample sensitivity converts to a sliding scale of inferred body burden. For ^{239}Pu only, the maximum permissible body burden (MPBB) is taken as 0.04 µCi. Detection levels become:

Time post single intake	Fraction of MPBB
30 days	0.005
90 days	0.01
1 year	0.03
5 years	0.08

For known accident cases, a single voiding, say, 2 hours after the event provides an adequate preliminary sample analyzed by less efficient rush procedures to signal the need for removal therapy for gross intakes.

In the more general case of chronic minor intakes over a working career it is impossible to convert the bio-assay reading to known values of MPBB, because the excretion rate becomes the sum of a series of power functions of time, where each time is unknown. However, an upper limit can be set by assuming that each incremental intake occurred the day after the previous sample was taken. On this basis, routine schedules for frequency of bio-assay of personnel, depending on the anticipated severity of their chronic intakes, can be set. A frequency of sampling scaled to detect 1% of the MPBB is recommended (WILSON, 1967). This equates to a monthly or quarterly schedule in the regular plutonium plant staff. Some installations plan schedules responding only to 5% MPBB or more. An annual sample for occasional workers gives 3% MPBB under the upper limit assumption. Those with access to a plutonium facility but never engaged in plutonium work should prudently be sampled at 5 year intervals. The pessimistic limit for that group is 8% MPBB and the more probable limit (based on median time exposure) is 6% MPBB. Naturally, these long-interval samples are expected to show no contamination. The complete overview of bioassay results from all interval classes from one month to five years does provide one of the better indices of the effectiveness of the defenses against plutonium contamination (WILSON, 1967).

The British group has challenged the rationale of a comprehensive program of this type, "provided that the environmental control and monitoring are adequate" (BEACH et al., 1966). Depending on the precise interpretation given to the word "adequate" one can say that this begs the question; obviously a no-defect containment system would lead to zero deposits. More generally, their arguments merit consideration because the only time at which treatment has worthwhile effect is shortly after the intake. This group recommends five base-line samples before commencing routine plutonium work and a series of five samples at 6-month intervals. This indeed tends to eliminate panic reaction if a single result on a regular sampling program throws a high reading due to cross-contamination. Clearly the total sampling programs under the two regimes are about the same size. Adherents of the monthly or bimonthly mode expect that trends indicating the beginning of off-standard conditions that escape the monitoring network will be suggested earlier in their system. Demonstration of this in a given system would require statistical analysis of many sample results. The British attitude, while not explicitly based on the assumption that exposures are nowadays predominantly due to inhalation of insoluble oxides, seems to lean strongly that way. For such cases, in which the urine signal is very slowly accumulated and probably not well formulated in models, their points are very well taken. BEACH and DOLPHIN (1964) have given revised formulae for fractional excretion of soluble plutonium compounds incorporating a more rapid component for the first few days. These are:

$$\text{Urine:} \quad \frac{\text{Daily excretion}}{\text{Initial intake}} = 0.0041\, e^{-0.67\,T} + 0.0016\, T^{-0.68}$$

$$\text{Feces:} \quad \frac{\text{Daily excretion}}{\text{Initial intake}} = 0.004\, e^{-0.46\,T} + 0.004\, T^{-1}$$

where T is the elapsed time in days. These equations have been referenced by the ICRP Report of Committee IV (1968), although it is not clear whether they wholly endorse them as a substitute for the LANGHAM equations. BEACH and DOLPHIN (1964) also presented an expression for urinary excretion which includes a factor for slow mobilization of plutonium initially deposited in the lung, as follows:

$$\text{Urinary excretion rate} = Q(b-1)\, 0.0016 \int_{1}^{T} \tau^{-b} (\tau - T)^{0.68} d\tau$$

where Q (in μCi) is the amount of inhaled plutonium eventually transferred across lung epithelium. The power function τ^{-b} is perhaps an improvement on the Healy method of the next paragraph. b was arbitrarily taken as 1.177, corresponding with 50% transfer from the lung in 25 days, 75% in $3\frac{1}{2}$ years, and 84% in 50 years.

When the intake, either acute, accidental or chronic, involves relatively insoluble material, the Langham equations break down. HEALY (1957) has presented a mathematical compartment model for the case of lung deposition with consideration of the subsequent time variation of activity in urine, feces, and blood, all of which can be measured to get a more or less empirical fit to the data. The Healy model is based on an assumed constant elimination rate from the supply pool. More recently, HEID et al. (1970a and b), have applied the ICRP lung model, including some recent tentative modifications (ICRP, 1971) to incorporate multiple mobilization half-lives up to as long as 10000 days. Inferred lung burden is still probably not highly reliable, but it is the best that can be done at levels below the sensitivity of the lung counter. The present state of knowledge does not indicate a need for treatment of these cases. Rather, they comprise an information bank for eventual determination of the true deposition at autopsy, and for a concurrent search for possible malignant changes or precursor changes through the mechanism of the U.S. Transuranium Registry (see Sec. V, p. 635).

The uncertainties of measurement of lung body burden are well recognized by those active in the field. The perturbation of bioassay results when DTPA has been administered is well known. A comparison has been made between the interpretations of three major U.S. locations—Los Alamos, Oak Ridge, and Hanford—based on six well-documented Hanford cases (BRAMSON et al., 1972).

Los Alamos uses a computer code (LAWRENCE, 1962) and Oak Ridge a more complex code accounting for statistical variations in following the Langham equations (SNYDER et al., 1967); Hanford still uses manual calculation. In general, the Los Alamos and Hanford interpretations are in close agreement. The authors wisely point out that this is no proof that they are superior to the Oak Ridge results. Some of the points may not be resolved until enough cases come to autopsy through the Transuranium Registry. These intercomparisons, in any case, relate only to data manipulation. More realistic intercomparisons are being organized to test lung counter signals. These will include the circulation of loaded chest phantoms, and ultimately the circulation of contaminated individuals, if possible. With no reflection on the three laboratories cooperating above, there are locations where the reports of sensitivity and detection limits seem to be a product of local enthusiasm. Considerable effort can be wasted at other locations by prematurely installing a new system that will not match the originator's claim. Cross comparison is desirable not just for lung counters, but for all systems used in plutonium hazard evaluation, where the needed results press the limits of resolution.

IV. Treatment of Plutonium Contamination Cases

The field of health physics stops at the boundary at which information on the nature and extent of body depositions is transmitted to the physician to permit an informed basis for his actions. Current approaches will be superficially included here. A comprehensive summary of the state of knowledge to 1967, although naturally weighted heavily by U.S. contributions, can be found in the Proceedings of a Symposium on Diagnosis and Treatment of Deposited Radionuclides held at Richland, Wash., the Hanford atomic site (KORNBERG and NORWOOD, 1968)[5]. There is no agreed protocol. In fact, it is proper to leave an area for the physician's judgment and it is socially desirable in some borderline cases for the patient himself to participate in a decision which balances some present impairment and discomfort against a very remote risk of future radiation injury.

As a guide, for wound cases one finds a readiness to use DTPA locally to remove contamination. Excision is usually not employed if the residual amount is assuredly less than 1 nCi Pu. Conversely, if excision is started it may be continued until there is no measurable residue (0.1 nCi to 0.6 nCi).

In the few cases in which a substantial fraction of a permissible systemic burden of soluble plutonium is known to be in the body from any mode of entry, there is no hesitation to use DTPA to enhance removal via the urine. National practice has to be responsive to the respective drug codes. In the United States, the Federal Drug Administration regards DTPA as an experimental drug requiring special blood tests prior to administration. This poses problems because the key feature of effective chelating therapy is to begin treatment at the earliest opportunity, ideally within about 30 minutes after plutonium intake. It is this factor, above all others, that makes it desirable to have a battery of special tests for fast recognition of the general magnitude of an accidental intake. Currently, such steps as urine analysis are fast if completed in two hours, unless the signal is strong enough for detection of plutonium or americium by the photonic radiations. Similarly, for lung counting, it is essential first to be sure that there is no external contamination on the patient, which might completely vitiate the lung counter

---

[5] See also Chap. 8, Sec. III. F, and Chap. 10, Sec. IX of this volume. (Editor.)

reading, and then to move him from the accident location to a central testing laboratory. This accounts in part for the strong interest in nasal smears that can be immediately read.

For lung deposits, especially with insoluble materials, there is more controversy. The extreme range covers nebulized DTPA for any known deposit to total rejection of the value of the method (SCHULTE and WHIPPLE, 1968). Other methods discussed, but not standardized, are mild lung irritants such as 10% saline solution to stimulate coughing, which is probably more useful to give a confirmatory positive sputum sample than to remove plutonium, and the use of lavage.

Lavage is a procedure not to be undertaken lightly. The expected mortality is comparable with that in anesthesia ($\sim$1 in 4000 cases). The incubation or latent period for lung cancer may be of the order of 10 to 40 years. Lung lavage for a person over age 50 may well be a poor tradeoff of competing risks. For a person age 20 to 30 after one week for thorough investigation of the deposit, lung lavage for 50 times the MPLB would appear to be a good risk. Others might make this determination at $10 \times$ MPLB.

V. The U.S. Transuranium Registry

During the summer of 1968 the Atomic Energy Commission authorized the establishment, under the directorship of the Hanford Environmental Health Foundation, Richland, Washington, of the U.S. Transuranium Registry to serve as a national focal point for the accumulation, evaluation, interpretation and dissemination of data on the uptake, retention, distribution and biological effects of plutonium and other transuranium elements in occupationally exposed workers. The purpose is to identify hazards to the worker that may exist from internal depositions of plutonium and other transuranium elements, which exhibit long latent periods between the deposition within the body and the possible demonstration of adverse effects. The number of workers potentially exposed has been estimated at 20000 in facilities currently cooperating with the Registry. With the expected expansion of the nuclear industry, this number will continue to increase.

The tissues and organs from some recent autopsy cases of exposed plutonium workers indicate the health physics estimates of internal systemic depositions to be conservative, ranging from 1 to 13 times the depositions extrapolated from the autopsy samples when the amount in the lung is not included (NORWOOD et al., 1972). The periodic estimates of employee systemic depositions generally made by health physics personnel from urinalyses data, do not reflect the insoluble plutonium deposited in the lungs. Recent cases analyzed by the Registry show that these lung depositions can be significant and have been as much as ten times greater than the estimated systemic burden.

The distribution of plutonium found at autopsy is greatly different from the original assumptions made by the National Council on Radiation Protection and Measurements (NCRP, 1953, 1959) and the International Commission on Radiological Protection (ICRP, 1960, 1964, 1968). The autopsy data indicate the percentage of the total skeletal burden to range from 0 to 77 with a mean of 32, while the liver ranged from 0 to 80 with a mean of 20. This compares with the assumption used in determining the maximum permissible body burden that 80% or more goes to the skeleton and 7% to the liver. In studies of 12 of the 14 cases that have come to autopsy so far, the ratio of *concentration* in skeleton to that n liver ranged from 0.05 to 5.2 with a mean value of 1.0. The Registry is now concerned that its bone samples are too small (usually ∼100 g), and that extrapolation to the total skeletal burden in 10 000 g is unrealistic. This is particularly troublesome because the ratio of plutonium concentrations in various representative bones is far more variable than anticipated. For example, in five cases, the concentration ratios rib/sternum were 30, 2.6, 1.25, 0.49, and 0.19. Rib/vertebrae ratios in 3 cases were 0.74, 0.30, and 0.25.

Existing models have assumed that the contribution to total body burden from muscle, fat, skin, and other organs not included in the standard "search list" of lung, lymph nodes, skeleton, liver, spleen, and kidney is negligible. By scaling up from the lowest measured concentration in such tissues in one case, the Registry came to a value of 27%, considerably higher than either liver or lungs in the specific case. This triggered a search for deposits in muscle in two available cases, which scaled up to 2.5% and 4.0% body burden, respectively. The Registry prudently notes the anomaly of scaling from less than 50 g to the total muscle taken as 28 000 g in standard man. Equally prudently, it is preparing to undertake major sampling of muscle (as well as bone) in future. Ideally, of course, total body sampling of a substantial number of higher exposure cases is needed. This runs into two social problems. First, the families of those enlightened workers who have willed their tissues to the Registry are making an irreplaceable contribution to science. Yet the American ethic has historically been sensitive to major destruction of the willed body, and the Registry must properly respect this sensitivity. What is needed most, and unfortunately does not currently seem probable, is a major increase in the analytical sensitivity of nondestructive testing for the photonic radiations of plutonium. Second, the detractors of the nuclear energy establishment could easily imply that certain organizations do not wish to cooperate with the Registry, because it may be shown that their practices are less than perfect, or more despicably, they may be accused of wanting to avoid evidence supporting compensation. It can be stated categorically that none of the cases to date shows any evidence of causation between the plutonium burdens and death. It would be absurd, however, to expect any valid epidemiological conclusions from 14 retrospective studies. Unfortunately for those seeking the truth, the nuclear detractors cannot be said to have confined themselves to *valid* epidemiological conclusions in other contexts.

It seems abundantly clear that the complete story of plutonium in man cannot be written without the enlightened cooperation of families of plutonium workers, management of plutonium operations, and dedicated physicians and scientists. Given this support, the results to date clearly indicate that the Registry could eventually become a key factor in establishing rational protocols for the management of plutonium contaminated individuals, comparable with the understanding of radium cases developed over more than three decades by R. D. Evans (1933, 1943, 1972).

VI. Protection against Criticality

The need to protect against an accidental nuclear chain reaction—criticality—is unique to operation with the fissionable elements, including plutonium. The mystique of this form of protection has been vested in a coterie of nuclear physicists and engineers; the detailed rites are foreign to the industrial hygienist or health physicist. It is well that this should be so, because a half-informed practitioner is an invitation to disaster.

In basic principle, though, the methods of control are simple, and two in number.

Plutonium Industrial Hygiene, Health Physics and Related Aspects 637

A. Safe Geometry

Whenever plutonium operations can be conducted in specific vessels and pipes, it is possible to keep their dimensions below the size that could allow criticality including all feasible accidental alterations to the vessel's geometry. This is the preferred modus operandi whenever applicable.

B. Administrative Controls

The control principle is to permit not more than 45% of the critical mass appropriate to the particular configurations and materials used to be present at one time. Accidental double-batching then preserves a small safety margin. For example, a plutonium-water system has a minimum critical mass of about 500 g, from which one conventional batch limit of 220 g is derived.

The art of criticality protection is to derive from basic experimentation and theory the correct critical mass for all proposed plutonium isotope mixtures and chemical and physical forms to be used. Similar precautions extend to transportation and storage of plutonium-bearing materials.

An excellent summary of such work appears in Chap. 27 of the AEC Plutonium Handbook, together with 68 references (CLAYTON and REARDON, 1967). The approaches in many countries are well described, although at an earlier date, in a European Nuclear Energy Agency report of a Karlsruhe Symposium (ENEA, 1961). When the claim is made that the principal hazard from plutonium is intake in the body, and especially inhalation, the hazard from radiation generated by plutonium criticality is being overlooked. This is the one circumstance that has already led to at least one fatality (SHIPMAN et al., 1961).

VII. Protection from External Radiations

A. Nature of the Radiations

As indicated in the introduction, with the increasing interest in and use of isotopes other than ^{239}Pu, protection from plutonium radiations has progressed rapidly from simple protection against alpha radiation to a rather complex situation involving shielding against X-rays, gamma rays, and neutrons at levels otherwise potentially damaging or even lethal to the operators. As a graphic example of the change, let it be assumed that a mythical country with a dynasty of King Pluto's created a coinage from plutonium metal, equal in size to a U.S. silver dollar. A coin collector holds one commemorating Pluto the 238th in his left hand, and one commemorating Pluto the 239th in his right hand. The coins are physically indistinguishable except for the markings, ^{238}Pu and ^{239}Pu. The right hand receives 1 rem/hour from X-rays and gamma rays, with an insignificant 3×10^{-5} rem/hour from neutrons. Meanwhile the left hand receives about 330 rem/hour from X-rays and gamma rays, plus 3 rem/hour from neutrons. Once the feasible doses from plutonium surfaces are recognized, there are no particularly unique features to providing effective shielding, testing the emitted radiation with survey meters, and supplementing the protection with personnel meters. The relevant factors appear in Chap. 25 of the AEC Plutonium Handbook (LARSON, 1967). However, they have been modified by newer experiments and calculations (FAUST et al., 1971), the major change being a reduction from 960 rads/hour to 330 rads/hour for the surface dose of pure ^{238}Pu. In addition, the changes in dose with changing composition of the plutonium mix have been codified, so that designers can incorporate appropriate shielding into proposed processing facilities for handling mixed UO_2-PuO_2 nuclear fuels and the like.

The conventional starting fuel for present-day commercial nuclear reactors is uranium oxide enriched in ^{235}U. As the ^{235}U is consumed by fission, some of the ^{238}U, the main constituent of natural uranium, is converted to ^{239}Np by neutron absorption, and this decays with a half-life of 2.35 days to ^{239}Pu. This plutonium in situ continues to be bombarded with neutrons. Some of it undergoes fission but some captures another neutron to yield ^{240}Pu, which in turn is partially boosted to ^{241}Pu and ^{242}Pu. Meanwhile, through other secondary chains and radioactive decay, lesser amounts of ^{236}Pu and ^{238}Pu are created. The plutonium isotopic content of the fuel thus changes continuously with fuel exposure, and is conditioned also by the details of the fuel disposition, choice of neutron moderator (whether graphite, light water, or heavy water) and other factors. In the context of the reactor industry, fuel exposure, which is a measure of the integrated neutron bombardment per unit mass of uranium, is stated indirectly in the units of megawatt-days of power per metric ton of uranium (MWd/MT or MWd/MTU, a confusing nomenclature because the two M's have different connotations).

Table 14.1 shows representative isotopic compositions that would arise after three degrees of exposure, which have arbitrarily been taken as:

  Brief exposure   about 140 MWd/MT,
  Moderate exposure about 13 000 MWd/MT,
  Heavy exposure   about 33 000 MWd/MT.

Table 14.1. Plutonium compositions in reactor fuel

Pu isotope	Weight percent of the isotope		
	Brief reactor exposure	Moderate reactor exposure	Heavy reactor exposure
236	—	9.10^{-7}	6.10^{-6}
238	—	0.3	2.0
239	~99.44	79.0	55.5
240	~0.5	14.8	25.3
241	—	4.9	12.1
242	—	1.0	5.1
Total	~100	~100	~100

Upon reprocessing the spent fuels, the plutoniums as a family are separated from the residual uranium and the americium and higher transplutonium elements present. If the plutonium were then prepared as a metal block the surface doserate will change with time due to the ingrowth of ^{237}U and ^{241}Am, the two radioactive daughters of ^{241}Pu, in the following pattern.

These figures apply to the metallic form. The reconstitution for further fuel use is now generally done in the form of plutonium oxide, for which the doserates are approximately one-half (usually specified as 1/1.93), due to the lower density of the oxide form.

Some of the protection considerations, in brief form, are:

Most of the plutonium isotopes and the immediate daughter products are alpha-emitters. Among the beta-emitters, such as ^{241}Pu, none has a vigorous gamma emission comparable with those that provide the strong gamma radiation in the radium decay series, used in clinical radiation therapy.

Plutonium Industrial Hygiene, Health Physics and Related Aspects 639

The sources of X-rays and gamma rays to be considered are:

a) Gamma rays coincident with alpha decay: For the plutonium isotopes, these are weak, except for ^{238}Pu. They are the principal radiations of ^{241}Am and ^{237}U. It is the 60 keV emission of ^{241}Am that facilitates its recognition in the lung counter.

b) X-rays from internal conversion: These are the principal photonic radiations of ^{238}Pu, ^{239}Pu, ^{240}Pu, and ^{242}Pu. The conversion X-rays give 98% of the very high surface dose-rate of ^{238}Pu, and 92% of the modest surface dose-rate of ^{239}Pu. They have the merit of falling in the energy range of 17 keV (^{239}Pu spans 13.6 to 20.2 keV, which is practically equivalent). Thus they are easily shielded. The percent transmission is essentially zero through a typical glove box wall, and 30% through a typical viewing port. An additional lead glass sheet or even ordinary glass completes the shielding.

Similarly, their transmission to the critical organs, except possibly the testes, is insignificant. In any case, it is inexcusable not to have this radiation confined to the glove box, except at the glove ports, which are themselves shielded when in use by the hands and forearms at the expense of irradiation of those parts.

c) Gamma rays from spontaneous fission: These are of relatively low intensity in all cases.

d) Gamma rays from fission products generated by spontaneous fission:

While the energies of these radiations are high, the total yield from most conventional plutonium mixes is quite low, being largely controlled by the insignificant contribution from ^{239}Pu. The high energy implies that thick shields are required for its elimination. This would be considered only for such special operations as prolonged manipulation of a fabricated ^{238}Pu piece, for which the dose-rate at the unshielded body might be of the order of several millirads per hour.

e) Gamma rays secondary to neutrons: Conceptually, both the neutrons from spontaneous fission and those generated by the alpha-neutron reaction in light elements, as in PuF, will give rise to gamma rays as they are absorbed in intervening materials, and will also induce activation products, some of which will continue to emit gamma radiation. These are not normally classed as practical hazards. If the levels were appreciably higher, the gradually increasing induced radioactivity of all glove boxes, shipping containers, etc., would have to be taken into account.

The activity of plutonium samples containing ^{241}Pu is not constant because of the in-growth of the daughter products ^{241}Am and ^{237}U, both of which make notable contributions to the surface dose rate, when, as is the usual case, there is little ^{238}Pu present.

The ^{237}U activity rises rapidly in the first few days, and then levels off after two months to a steady plateau, being governed by the relationship $(1-e^{-0.102t})$, where t is the post-purification time in days. The growth of the ^{241}Am is virtually linear with time at least for several years. As measured by the X-ray and gamma-ray emissions, ^{241}Am overtakes the ^{237}U after 3 months (Note: the 4 months cross-over shown in many curves still repeated in the literature is based on superseded 1958 data), and rapidly becomes the leading component, e.g., 40 times the ^{237}U component after one year. Operators must be constantly aware of this changing pattern whenever radiations from these daughters are not fully shielded. Some illustrative numbers have been included above (Table 14.2).

Another variable hazard is presented by spreading a plutonium source. Since most of the photonic radiation is in the low energy range, most of it is self-shielded in a compact source. The effect at a distance then depends on the solid angle

Table 14.2. Contact photonic dose-rate (rads/hour)

Time after purification-days	Brief reactor exposure	Moderate reactor exposure	Heavy reactor exposure
0	1	2.2	10.3
100	1	2.6	11.2
1000	1	5.9	19.1

subtended by the source at a point on the body. When it is spread, the effective solid angle increases. Poor housekeeping in glove boxes can lead to a thin layer being spread over the box surfaces, materially increasing the dose to unshielded part of the body.

The sources of neutron radiation to be considered are:

a) Neutrons from Spontaneous Fission

The dose rate per unit flux density is reported as 0.115 mrem/hr. ROESCH has derived a formula for the neutron dose rate at the surface of a plutonium sphere from which the coin examples were calculated (ROESCH, 1958). ^{238}Pu, ^{240}Pu, and ^{242}Pu are the isotopes of concern, in relative contributions of 3.3:1:1.7. In practical terms of reprocessing irradiated fuel elements, the neutron dose-rate of plutonium surfaces rises from virtually zero for short-term irradiation to 700 mrem/hr after reactor exposure to 10000 MW days per ton.

b) Neutrons from α-n Reactions

With plutonium fluoride in a safe batch quantity, the neutron dose-rate close to the source is of the order of 5000 mrem/hr. More realistically at a glove box face it is about 20 mrem/hr, intolerable without shielding. Thus glove boxes for fluoride extractions and even for all PuO_2 operations need neutron shielding in the form of hydrogenous materials. Such materials as water, pressed wood, or plastics are suitable. Special formulations incorporating boron are convenient. The hydrogenous materials slow the fast neutrons (initially 1 to 10 MeV) which are then rapidly absorbed by the boron.

The practical absorption half-value thicknesses of these materials are in the range of 5 to 10 cm. Thus a useful water wall tends to be a cumbersome attachment to a glove box. Nevertheless, water may be the material of choice because of the fire propagation propensities of some of the other materials. It also serves conveniently for the viewing ports. The possibility of leakage of such systems and a resultant drastic change in criticality parameters has to be considered at all times.

B. Survey Instrumentation

For the photonic radiations conventional gamma-ray survey meters are adequate. However, if at any step the 17 keV X-radiation is not contained, special meters responsive at this low energy are needed. The neutron dose is generally obtained by moderation in specially sized hydrogenous cylinders with measurement of the moderated neutrons in a BF_3 counter or thermal neutron sensitive scintillation counter. The moderator size is empirically adjusted to cause a response that can be interpreted as a dose-equivalent rate in rems/hr. The original form of these meters utilized two coaxial cylinders of paraffin, the inner one functioning essentially as a neutron flux meter, and the combination as a dosimeter (DE PANGHER, 1959). A successful commercial instrument achieves the same result more conveniently by judicious placement of partial boron shielding with the moderator (TRACERLAB, 1968).

C. Personnel Monitoring

Records of personnel exposure throughout the nuclear industry have been customarily obtained from standard film badges worn on the outside of the protective clothing. Suitably chosen filters over the film flatten the response curve so that there is approximately equal sensitivity to gamma rays over the range 50 keV to 1.25 MeV (UNRUH et al., 1963). Routine calibration can thereafter be done with one calibration source, optionally radium, ^{137}Cs or ^{60}Co. Most film badges were not intended to measure such soft X-rays as the 17 keV internal conversion groups, but they suffice if special filter systems and calibration are made with a suitable fluorescent X-ray source (LARSON et al., 1954). Separation of the dose for the soft component needs to be continued because only about 35% of its total contribution is properly assigned as a whole body dose.

The neutron component has normally been measured by nuclear track film, a tedious procedure when relatively low doses are to be evaluated.

In recent years, measurement of both photonic and neutronic components has been improved by thermoluminescent dosimetry (TLD) (ATTIX, 1967; KORBA and HAY, 1970).

A representative receptor as used at Hanford includes five LiF chips and special filters to provide dosimetry for both photons and neutrons (KOCHER et al., 1971). Two of the five LiF chips are used for the photon dosimetry giving a measure of the penetrating component and the non-penetrating component (17 keV) or soft component. The neutron dosimetry is provided by three chips including chips of ^{6}LiF and ^{7}LiF. These systems work on albedo neutrons from the body. Their response is dependent on the neutron energy spectrum, hence it is necessary to calibrate such systems with neutron spectra closely duplicating field exposure conditions. With such calibration, this type of TLD system is probably the best available for personnel neutron dosimetry, although the current performance capabilities of these systems do not fully meet the desired levels of performance. Some groups use dose-rate monitoring instruments and time-keeping to calculate a neutron dose for entry in the personnel exposure records in lieu of using personnel dosimeters. AEC Workshop Reports review the current status of personnel neutron dosimeter recent development (UNRUH, 1969, 1971).

Plutonium fluoride neutron sources are commonly used for dosimeter calibration when a plutonium fluoride component may be encountered in the plutonium work. Plutonium beryllium sources may be used when only plutonium metal or plutonium oxides are present.

The ratio of gamma to neutron personnel dose-equivalents in plutonium facilities may vary widely depending on the plutonium material form, dispersal and local shielding. Gamma to neutron dose-equivalent ratios varying from 1:1 to 1:5 are not uncommon. Neutron dosimetry is clearly an important part of a modern radiation protection program in plutonium operations.

Measurement of the dose to hands and forearm is particularly important with close work with "hot" sources. Film in wrist badges and in small finger rings has long been used, but the finger rings particularly have a poor performance record. Substitution of TLD offers improvement (ENDRES, 1968). The TLD finger ring dosimeters are preferred over the film system because they are more resistant to environmental damage such as might arise from light leaks, heat and humidity. The lower energy dependence properties of the TLD material also offer the opportunity for better dose interpretations. Generally, though, the records of finger doses lack precision; it is impossible to have the finger rings at the maximum dose-rate position, and the neutron component is usually ignored. Presumably, some sort of neutron dose estimate could have been made with nuclear track films. The obvious application of combined ^{6}LiF and ^{7}LiF chips encounters a problem. It is effective for the whole body dose because it actually measures slow neutrons moderated in the body. There is not enough moderating mass in the hands to achieve reliable results.

Operator protection is often maintained by calculating or measuring both photonic and neutronic doses in various geometries and then establishing daily time limits for specific operations. Custom has allowed hand and finger doses to come much closer to permissible limits than is the case for whole body exposure. Special limits for hands and forearms are also more generous. Recent recommendations of the NCRP suggest some degree of interest in reducing such doses (NCRP, 1971). While there is legitimate room for a range of opinion, the intuitive feeling that if some late effects occur on the hands they will be correctable, and by no means comparable with possible whole body effects such as leukemia from external radiation or bone tumor from deposition of plutonium is not the whole story. The conservative view is that impairment of function of the hands is troublesome and cosmetically displeasing. In addition, surgery or electrocoagulation used to treat finger lesions is exceptionally painful because nature designed fingers to be sensitive. Demonstrably deleterious effects on the hands are not being found to date. The extreme escalation of surface dose in modern plutonium sources and its wide variability set the stage for future problems if handling practices are not correspondingly improved. Once the need for, and feasibility of, remote manipulation are accepted, there could be a major reduction in hand exposures to levels more responsive to the recommended concept of lowest practicable level of exposure in all operations.

D. Protection against Criticality

Meters, preferably neutron meters but sometimes gamma meters, set to give an audible alarm, provided with redundant circuitry to minimize failures, and tested on a regular schedule are mandatory. A fast pair of legs to respond to an alarm is the other necessary equipment.

Health physics personnel responding to a criticality incident should have a Geiger counter or other equally sensitive meter that can be held against the stomach of the evacuated personnel. A strong signal from the ^{24}Na activity, induced by slow neutrons moderated in the body exposed to a fast neutron flux, fortunately arises at levels well below those at which serious radiation injury will result (WILSON, 1962). Such a quick-sort is invaluable to direct medical attention to critical cases. Almost more importantly, negative results promptly relieve the anxiety of the unexposed personnel. The knowledge that criticality accidents have caused painful deaths is foremost in the mind of one who sees the "Blue-flash" of a criticality incident until he can be confidently reassured that his exposure was clearly sub-lethal.

Re-entry to an evacuated area is made under full protective clothing and air masks, and carrying high-reading dose-rate survey meters as well as those of conventional range. It is absolutely essential that neutron meters be used, because the most insidious situation is to have a criticality event that continues as a nuclear reactor without recognition.

VIII. Summary of Plutonium Internal Deposition Experience

USAEC experience from 1957–1966 reported cases in excess of 25% of the permissible body burden (Ross, 1968). There were 136 cases, rising to 203 through 1970 (personal communication). Twenty-nine of these exceeded the limit (37 through 1970). About 65% represented inhalation, and 25% wounds, the remainder being uncertain. Chelating agents were used in about 25% of the cases; wound excision covered about 18%. Ross recognizes the uncertainties in the measurements. The experience with plutonium is not conspicuously good, even measured against full utilization of accepted maximum permissible burdens. It is less satisfactory when measured against admittedly arbitrary standards of "lowest practicable level" achieved today in many other areas of radiation protection. On the other hand, as measured by the conventional standards of industrial hygiene, the performance to date has been outstanding. With the exception of criticality accidents, injury has been limited to removal of small amounts of tissue around wound sites, and the trauma of occasional lung lavage. There has been no significant deleterious effect of chelating agents, such as DTPA, used to promote removal of plutonium from the body. Above all, there seems to be no identifiable case of malignancy of bone, lung, or other critical organs and tissues. None of this is cause for complacency in view of the anticipated long latent period for tumor induction, and the lack of proof that these carcinogenic effects show a real or practical dose threshold.

Hanford experience[6] for 1946 to 1967 was as follows (JECH et al., 1969b; HEID and JECH, 1969):

Injury cases potentially contaminated	230	(310)
Cases containing measurable plutonium	136	(162)
Cases surgically decontaminated	86	(100)
Deposition above 5% MPBB	15	(16)
Deposition above 50% MPBB	5	(5)
Plutonium inhalation cases	1140	(1501)
Chelating agent used	12	(13)
Deposition above 5% MPBB	98	(101)
Deposition above 50% MPBB	12	(14)

Figures in parentheses represent a private communication updating through 1971. They demonstrate an encouraging stabilization of the cases above 5% MPBB in the face of considerably improved measurement capability since 1967.

Of the 136 cases reported by Ross, 132 could be identified as to year of origin. Sixty-three of these, or about 48%, had occurred in the last 3 years of his reporting decade. He was concerned that this might indicate an accelerating trend. Actually, 67 cases have been added in the remaining four years of record through 1970. Thus the USAEC data continue to show a fairly high incidence rate, but not an explosively expanding one at this time. An incidence rate based

---

[6] For easier comparison with the total AEC experience which is cut off at 25 percent MPBB the Hanford cases with depositions equal to or greater than this amount are—injury cases, 9 through 1967 and 10 through 1971; inhalation cases, 29 and 31 respectively.

644   H. M. Parker:

on the number of injuries per kilogram of plutonium used, or alternatively, per man-hour of plutonium operations, would inherently seem to have more interest, but it cannot be derived from published data. In any case, such an index might be expected to be a function of the form of plutonium used, with an expectation of higher incidence rate for inhalation with plutonium in oxide form as for nuclear fuels, and higher incidence rate for wounds with plutonium in metal form as for space heaters and other applications.

If one uses the percentage of cases exceeding the maximum permissible annual dose commitment as a crude index of severity, it will be seen that the severity rate has fallen from 21% in the prime reporting decade to about 12% in the added report period. Possibly the number of cases is too low and their occurrence too sporadic to justify a rigorous conclusion that the severity rate has materially fallen.

AEC Licensees'[7] experience to 1966 shows one plutonium case and one americium case above 25% MPBB, but the workload must have been relatively light in this field (Roeder, 1968).

Autopsy determinations of lung burden among Hanford plutonium workers, non-radiation workers, and residents found results compatible with air concentrations typical of plutonium facilities for the first group, and accountable from world wide fallout for the other two groups (Newton et al., 1968).

A recent updating of the above material extends the cases to 1971, with a total of 356 autopsy cases (Nelson et al., 1971). Some observations on these results are included in Appendix A, which also describes the learning experience from a number of specific plutonium contamination cases, some of which have been followed for almost 20 years. This experience is mainly in confirming or refining metabolic pathways in man. The absence of deleterious effects to date precludes analysis of dose-effect relationships in man.

IX. Environmental Protection for Plutonium[8]

As indicated in Chap. 15, environmental considerations for plutonium have not had the worldwide attention given to such nuclides as ^{90}Sr and ^{131}I. Many factors may have contributed to this—the origin of plutonium production as a highly classified operation, long continuation of certain security measures that conditioned scientists to minimize publication (whereas the contrary policy might have won public confidence), the fact that plutonium in the environment is re-concentrated to a much lesser degree than such well-studied nuclides as ^{32}P, ^{90}Sr, and ^{131}I, and the fact that committed atomic explosions have provided a long term artificial background of plutonium. Current worldwide attention to environmental problems has stimulated research plans to fill in gaps for the transuranium elements as a class. Meanwhile, competent reviews of environmental studies, even though they omit significant masses of classified data that would now be releasable, show a modest to reasonable breadth beginning with 1948 (Price, 1971, 1972). The referenced 1971 report addresses itself specifically to the transuranium elements in one section. In addition, it has the merit of providing abstracts of other papers not specifically analyzed in the text. Other

---

7 In the United States the U.S. Atomic Energy Commission controls the use of reactor-produced nuclides through a regulatory and licensing system. An "AEC licensee", as used here, is an individual or organization permitted to use specified nuclides under these controls.
8 This section takes up in some detail those aspects of environmental protection most pertinent to the nuclear energy industry. More general aspects are considered in the following Chapter (15). (Editor.)

Plutonium Industrial Hygiene, Health Physics and Related Aspects 645

useful source material includes reports of three AEC Radioecology Symposia (1st, SCHULTZ and KLEMENTS, 1963; 2nd, NELSON, EVANS et al., 1969; 3rd, Oak Ridge—in press).

The worldwide surface contamination peaks between latitudes 40° N and 50° N at about 1.5–2 nCi ^{239}Pu/m^2 (HARLEY, 1971, and earlier reports). Separation of plutonium from soil samples at these low levels is tedious and subject to variable yield. Good chemistry in one series gave a recovery from soil of 67.9% with a standard deviation of 32.2%, decidedly inferior to air filter yields (79.9% ±15.5%) (DE BORTOLI, 1967). It is now standard practice to spike soil samples with trace ^{236}Pu so that the yield can be experimentally found, making its variability less important. One concern with this technique, which applies to a similar application to urine analysis, is that there is no absolute assurance that all chemical and physical forms of plutonium will be extracted in the same ratio as the tracer ^{236}Pu. With the vigorous chemistry resorted to in some forms of soil analysis this may not be a major factor. Various U.S. methods of analysis for plutonium in soil were reviewed at a recent AEC Environmental Plutonium Symposium at Los Alamos. (Proceedings of Environmental Plutonium Symposium, 1971.) Wide variability of sample recovery by laboratories recognized to be highly skilled in such work was noted. In addition to the chemical problem, the problems of standardizing the method of taking a soil sample, and particularly the inherent randomness when the total contaminant may be present as a few specific particles were noted.

Representative air contamination in mid-latitudes, circa 1965–1966, was 0.1 pCi/1 000 m^3. Measured lung burdens were ~0.45 pCi/kg (MAGNO et al., 1967), ~0.5 pCi/kg (NEWTON et al., 1968) and ~0.15 pCi/kg (SMORODINTSEVA et al., translation 1969). These are compatible with the air concentrations, except that the Russian air concentration seemed low by an order of magnitude. They probably fairly represent lung contamination throughout the northern mid-latitudes at that period of time. (See Appendix A, Part 1.)

Environmental sampling and some degree of ecological study have been applied to the U.S. Nevada test site, and the atomic bomb incidents at Palomares and Thule. As an example, the Danish studies at Thule represented a sound ecological approach, which revealed no notable surprises, found no conditions of major concern, and correctly reserved judgment on some topics to be pursued later (AARKROG, 1971).

A major plutonium fire at Rocky Flats in 1969 led to concerned outside scientists measuring plutonium in soil samples, and questioning the adequacy of the protective and monitoring systems. There was eventual agreement on the soil sampling competency. It was determined, however, that the particular samples arose from spread from a temporarily substandard storage of contaminated oils and solvents in drums. It is no surprise to health physicists to find that the most frequent cause of environmental contamination is neither the operation of the obviously most dangerous facility, nor the controlled storage of the obviously most dangerous major wastes. Rather, as in this case, it comes from an intermediate class of wastes, temporarily held in seemingly adequate control, but with failure to recognize and correct a deteriorating storage condition in time. Examples from other sites would include such aberrations as a contaminated waste ditch, safely operated for years and then abandoned without adequate immobilization of radioactive sediments and restriction of access by wildlife. PRICE (1971) has pointed out the large number of animals which are attracted to food, shelter, and water around a pond in the advanced stages of plant community succession, with increased potential for contamination spread. Despite the popular trend to improve the "quality of life" by making industrial facilities attractive, the use of surface disposal ponds for dilute radioactive wastes is one case in which ecological common sense demands that the facility be established and maintained in a simple, early succession stage, or even sterilized.

Review of the Rocky Flats case led to publication of the following data applying to the total experience to date (HAMMOND, 1971):

Controlled releases	airborne effluents	42 mCi
	liquid effluents	90 mCi
Accidental releases	1957 fire	$\sim$0.06 Ci
	1969 fire	$\sim$0.0002 Ci
Waste-contaminated soil	transference	$\sim$2.6 Ci

The almost incredibly low releases from the Rocky Flats fires stimulated a review of all reported release incidents and development of theory to assist accident analysis (HUNT, 1971a, b). The crux of the matter is that the greatest mass of plutonium is attached to large particles that do not escape from the vicinity of the accident. Such large particles would in any case not contribute to an inhalation hazard. However, one cannot dismiss the possibility that they would subsequently disintegrate by weathering to a transportable respirable size.

Soil samples from the environs of the Hanford plant show evidence of local contamination in the vicinity of the plutonium working facilities, dubious results at two miles from the stack, which is still 10 miles within the controlled boundary, and none off-site (CORLEY et al., 1971). When the surface activity is plotted against distance from the stack, there is randomness in the data with no clear relationship between activity and distance. By contrast, had there been steady minor stack releases averaged over years of operation, the reliable meteorologcal data at this site would predict a fall of over two orders of magnitude over the range covered, perhaps somewhat modified by redistribution. Because of the random errors in sampling, probably due to real variations in point to point settling of discrete particles, it is difficult to discern that distance at which plant discharges demonstrably add to existing fallout. The question was ingeniously resolved by measuring the ratio of plutonium to ^{137}Cs in the samples. This ratio is distinctly higher for the local particles whereas the fallout particles are relatively rich in ^{137}Cs and other nuclides. The results clearly show that only the immediate inner area has a local infestation. In other words, living within two miles of this major plant throughout its operating career would not have added significantly to the body burden of plutonium (however, see Appendix A, Part 1). As a point of perspective, the highest plutonium alpha activity close to the Hanford plutonium facility was 7.5% of the natural alpha activity of the uranium and thorium series in typical soil.

Wind transference of activity was seen to be the major cause of difficulty at Rocky Flats. Hanford is by no means deficient in wind. The possibility of resuspension was foreseen over 20 years ago, and studies were made in preparation for an untoward event (HEALY and FUQUAY, 1958, 1959). The work is a fascinating exercise in atmospheric physics, but the parameters have never yielded fully to quantification. Despite the one experience at Rocky Flats, the transference is usually not severe, because, as it were, the fine radioactive particles hide behind larger inert particle neighbors on the ground surface or adhere to them. They are essentially in a quiescent boundary layer. Thus the problem is principally one of movement when a whole layer is picked up in a dust storm. There is some evidence suggestive of increased air concentrations during dust storms, but the extreme randomness of sampling for particulate matter under such conditions contraindicates a sound statistical analysis unless many samples are taken. To relate the increased air contamination to augmented inhalation risk requires knowledge of the particle size of the actual contaminated particles. Elementary theory, as given above suggests that when the surface layer is resuspended, much of the activity may be attached to larger non-respirable particles. A major restudy of resuspension problems is currently underway (HEALY, unpublished). With vegetative cover, the problem of resuspension is trivial.

The demonstrably low releases of plutonium to the atmosphere point clearly to the health physics control of the problem at its source. The solution is as follows:

Installation and maintenance of the highest grade filtration on all exhaust ducts.

Selection of fire resistant filters and frames in duplicate. A fire presents probably the greatest risk of breaching the exhaust.

Continuous sampling between the two filters so that failure or reduced efficiency of the first is promptly detected.

Alarm type sampling on the exit side of the second filter.

Air-locking of all entrances and exits including the emergency exits.

It is less easy to list the elements of a sound waste management program. The high level wastes must have first class retention with capability to protect against a tank leakage. The third line of defense results in material being deposited in the ground. Retention on sediments is remarkably good so that movement into ground water is minimal. Nevertheless, uncontained deposits can never be dismissed because of the long half-lives involved. The general renaissance of interest in man's obligation not to prejudice the environmental future has included a most vigorous consideration of the ultimate issues for transuranium wastes. Calculations of waste composition 30 million years from now must surely be man's most forward look to date (BELL and DILLON, 1971). It is of passing interest to note that reprocessing wastes (not plutonium plant wastes) have the transuranium elements, plutonium and americium, as the key offenders between ages 300 and 30000 years. Curiously, the learning experience between plutonium and radium that started plutonium work on the right foot, comes full circle. From about 50000 to 500000 years the key offender in nuclear wastes will be none other than ^{226}Ra. Even more surprisingly, the most-unwanted title thereafter passes until the 30 millionth year to none of the heavy element decay chains but to a radioiodine, ^{129}I. Meanwhile, improved engineered storage can and will be developed for a more modest time span—of the order of 100 years.

At the present time, the most cogent environmental risk in the plutonium industry (excluding nuclear weapons incidents) probably involves substandard handling and transportation of the secondary wastes, the discarded equipment, the contaminated paper, and so on. At this writing, this aspect is receiving renewed attention now that the transuranium elements, as industrial products and wastes, are here to stay in more than one sense. As an example of the care taken in disposal of solid wastes, one can cite the example of a licensed laboratory close to the Hanford Works, which is using ^{238}Pu in the development of heat sources for space applications, artificial hearts, and the like. Casual waste from the glove box operations is collected in small cans, which are separately sealed, and then sealed in larger cans. A number of these cans are then placed in a stainless steel box, which is welded hermetically tight and tested by helium leakage. This box is then transported to a major burial site within the Hanford reservation with virtually no risk of escape of contamination en route.

The questions of safe packaging, intra-plant shipment, off-site shipment of completed pieces such as PuO_2 fuels, and off-site shipment of wastes to central repositories have not been covered in this review. Considerable talent has gone into engineered safety to protect packages from damage under severe impact conditions and when subjected to severe fires. The International Atomic Energy Agency has been productive in establishing uniform regulations for safe transport of all radioactive materials. The experience to date has been good (D. E. PATTERSON and V. P. DE FATTA, 1962). Yet, with a markedly expanding frequency of shipment, the assurance of absolute perfection is absent.

All the major nations involved with plutonium wastes are actively strengthening their plans for ultimate storage. In the United States their commitment to a disused salt mine, previously considered only for disposal of the highly radioactive mixed fission product wastes, is under active discussion. In West Germany, plans are well advanced for retention of virtually all that nation's nuclear wastes in the Aase Salt Mine, near Braunschweig. Other nations less favored by underground formations that offer effective isolation from the biosphere are seeking the best approaches open to them.

Lastly, as a matter of practicality, use of nuclear materials at all will involve some residual releases to the environment, occasionally from accidents but steadily from a diverse battery of individually trivial sources. Every contaminated plutonium worker is a source of environmental pollution, and he cannot be kept in engineered storage. More pertinently, ground disposal of so-called decontaminated equipment, which is obviously still contaminated in lesser degree, will have to continue. Criteria governing this will probably hinge on intuitive arguments that increments of one or a few percent addition to worldwide plutonium fallout are inconsequential, or more generally that similar increments to the natural radioactivity (radium etc.) of soils, plants, and animals are acceptable. As in all radiation protection problems in which there may be no threshold of absolute safety, the acceptable solutions are not wholly contained in science, but include social, economic, and political vectors.

The issues of plutonium contamination of the environment are far from resolved. The current experience does not suggest substantial present risks to man or any other life form; neither is it comprehensive enough to dismiss the risks as insignificant. Importantly, there is a trend to improve present practices, and in some cases to recover and relocate some past releases. More importantly, there is a responsiveness to the expected exponential increase of plutonium availability, coupled with higher sensitivity to environmental impacts. With these signals, it is unlikely that plutonium misuse, short of nuclear war, will ever rank with man's major mistakes.

Acknowledgments. The author is greatly indebted to C. M. UNRUH and K. R. HEID, Battelle-Northwest Laboratory, for collaboration in selecting source material and for manuscript review. Former colleagues in this Laboratory, B. V. ANDERSEN, L. A. CARTER, J. P. CORLEY, J. J. FUQUAY, H. V. LARSON, C. E. NEWTON, JR., J. M. NIELSEN, and J. M. SELBY, were most helpful in providing material, as were also Dr. P. A. FUQUA and Dr. J. A. NORCROSS, Hanford Environmental Health Foundation; Mr. M. L. SMITH, Donald W. Douglas Laboratories; and Mr. W. C. SCHMIDT, Atlantic Richfield Company, all of Richland, Wash.

Work performed under AEC Contract AT-45-1-1830.

Plutonium Industrial Hygiene, Health Physics and Related Aspects 649

Appendix A

Some Learning Experiences from Autopsy Data and Selected Personnel Contamination Cases

Part 1. Autopsy Data

The data from 356 autopsy cases should provide a wealth of information on the outcome of casual incidental intake and substantial accidental intake of plutonium (NELSON et al., 1971). However, as the authors point out, there are several defects in the data which have gradually been reduced by learning experience.

Had NELSON et al., been measuring the concentration of plutonium in moon rocks, they would doubtless have had access to the most sensitive analytical methods feasible. As it was, they used good, but not outstanding methods, which steadily improved. The minimum sample activity which has a 95% probability of being distinguishable from zero was about 1 pCi prior to 1954, and about 0.02 pCi thereafter. The early data were difficult to interpret in an overall sense because so many readings were at or below the minimum detectable level. Usually, in any sampling program, positive results not far above minimum detectable level have wide error limits. These limits are not quoted in the report. Neither is there a record of analytical yield which might have been obtained at greater expense by salting the samples with trace amounts of ^{236}Pu. Differentiation of fallout plutonium from industrial plutonium could have been made by isotopic analysis. At an approximate cost of $ 400 per sample that, too, was beyond the resources of this program. A great improvement was made circa 1963 by radically increasing the size of the autopsy samples.

The cases were divided into four categories as follows:

Hanford Plutonium Workers	71
Others having had Hanford work experience	89
Local residents of more than 3 years	146
Individuals not residing locally	50

The second and third categories may be polluted by accidental inclusions of a former atomic energy worker, since the cross-checking of autopsy cases against former employment is not complete.

As a result of these problems, together with the risk of cross-contamination in the laboratory, the total mass of data is relatively uninformative, beyond the fact that the plutonium workers clearly have higher burdens on the average. Only two of the workers were known to have positive burdens from routine bioassay samples. Thus there is at least presumptive evidence of generalized deposition below conventional detection methods during employment.

With such data, a so-called inferential average concentration for each class can be obtained as follows:

If p_i is the ith person with a positive result P_i fCi/g and n_i is the ith person with no detectable burden at the detection limit N_i fCi/g, the inferential average concentration lies between:

$$\frac{\Sigma P_i}{\Sigma p_i + \Sigma n_i} \text{ f Ci/g and } \frac{\Sigma P_i + \Sigma N_i}{\Sigma p_i + \Sigma n_i} \text{ f Ci/g}.$$

For the whole data, these averages are shown in Table 14.3.

Table 14.3. So-called inferential average concentrations (f Ci/g)

Class		Organ or tissue		
		lung	liver	bone
I	Pu workers	4.4–5.4	2.3–2.9	3.5–7.7
II	Other work experience	0.4–1.5	0.5–1.3	1.7–6.9
III	Local residents	0.2–1.0	0.2–1.1	0.6–4.5
IV	Non-residents	0.1–1.0	0.2–1.3	0.2–4.2

The data on activity of tracheo-bronchial lymph nodes has been disregarded because of the small sample size. However, it is clear that some of these concentrations were the highest of any recorded. This may further complicate the lung data where node separation was not performed completely.

Such inferential averages are not informative except for the plutonium workers. As a first step in clarifying the data, the average sample weight, rounded to grams, is shown in Table 14.4, for Class I and Class III. Class II is quite similar to Class I as is Class IV to Class III.

Table 14.4. Average sample weights (g) by 5-year intervals

Interval	Lung		Liver		Bone	
	Class I	Class III	Class I	Class III	Class I	Class III
1951–1955	34 (6)	37 (6)	30 (5)	33 (6)	7 (1)	9 (2)
1956–1960	41	41	33	25	7	5
1961–1965	84	78	57	95	10	11
1966–1970	437	367	199	215	146	66

The numbers in parentheses are hypothetical weights made more comparable to subsequent data by accounting for the lower analytical sensitivity prior to 1954.

It is now clear that more interpretable data should arise by discarding the first decade. This notably increases the percentage of positive samples as in Table 14.5.

Table 14.5. Percentage of positive samples by class and decade

Class	Decade	Lung	Liver	Bone
I	1951–1960	26	48	4
	1961–1970	71	55	12
II	1951–1960	30	23	7
	1961–1970	50	47	10
III	1951–1960	17	20	2
	1961–1970	46	38	11
IV	1951–1960	20	0	7
	1961–1970	31	38	4

The recalculated inferential averages appear in Table 14.6.

Table 14.6. Inferential average concentrations (f Ci/g) 1961–1970

Class	Organ or tissue		
	Lung	Liver	Bone
I	5.8–5.9 (1.8–1.9)	2.2–2.3 (1.2–1.3)	5.4–7.3 (0.2–2.2)
II	0.4–0.5 (0.4–0.5)	0.4–0.5 (0.3–0.5)	3.6–6.7 (0.4–3.5)
III	0.3–0.4 (0.2–0.3)	0.2–0.4 (0.2–0.4)	1.0–3.6 (0.4–3.0)
IV	0.1–0.3 (0.1–0.3)	0.2–0.4 (0.1–0.4)	1–3.6 (0.4–3.7)

In some cases, the range of the inferential average is sufficiently closed up to persuade one that the average is meaningful. Unfortunately, this is not necessarily the case. Without resorting to complex statistical analysis, one can see that the average in a large enough class should not be materially affected by eliminating the extremes, i.e., the highest and lowest concentrations in the class. The numbers in parentheses in Table 14.6 are so derived. Many of the entries are so weighted by a single case that the apparent averages change significantly. The date for bone are seen to be relatively uninstructive. Even the apparently stable value for lungs of plutonium workers is grossly distorted by a single case.

The ratio of concentrations in lung, liver, and bone for a given class should be more or less constant if the plutonium is being acquired through the same mechanism. This might be expected to be inhalation, supplemented by some minor intake through cuts and abrasions for the plutonium workers. The data are not incompatible with this assumption.

Some tentative conclusions can be drawn from the lung data. There is other evidence that atmospheric contamination for local residents is insignificantly greater than that due to plutonium fallout. Thus Classes III and IV can be added to give a total of 103 cases showing an average lung concentration of (0.27 ± 0.07) fCi/g (or 0.27 pCi/kg). This is probably the best available figure for lung burdens in mid-latitude over the last decade.

Only 42% of the 103 cases showed positive results; treatment of the below detection limit data is plausible, but arbitrary at best. Therefore, it is more reasonable to quote an average concentration of (0.3 ± 0.1) fCi/g. Presumably, the lung deposition occurs as discrete particles possibly with a wide range of activity. A small sample of lung may show quite non-representative activity. The actual observations cover two decades of activity. If the non-worker lung cases are reduced to those with a sample size of about 100 g or more, 47 cases are left (34 female and 13 male), and about equally divided above and below age 55. In this limited class, the women had three times the activity of the men, and the older group twice the activity of the younger. It is difficult to assign statistical significance to these variations, but they should be watched with interest as the sub-classes expand.

If 0.3 fCi/g is a reasonable number for the unexposed population, it would suggest that Hanford front-line plutonium workers have an additional lung burden of about 1.6 fCi/g on the average, from 40 cases, and that other Hanford workers have a real but small increment of about one-half of the fallout background. The total lung burden equates to 1.6 pCi for the typical plutonium worker, orders of magnitude below that detectable by external means. Only the accumulation of massive autopsy data can prove or deny this tentative conclusion.

The biological consequences of these deposits is uncertain, but at most it must be very small relative to other available insults. In terms of dose, the apparent population lung dose equates to 0.3 mrem per year, which is trivial compared with natural variations in background radiation. Yet if the "hot-spot" theory of particulate contamination has merit, it cannot be stated categorically that no detriment will arise by combination of this insult with some other cancer-promoting condition.

Part 2. Selected Deposition Cases
General Interpretation

Hanford methods of interpretation of deposition data have been given by HEID et al. (1970a, b) and JECH et al. (1971, 1972). The prompt actions which permit rational medical treatment have been briefly described in Sec. III. In this appendix, the interest is in what can be learned about plutonium behavior by long term follow-up. Only rarely do the late observations contribute to the welfare of a particular contaminated worker (see Cases 1 and 6). Plutonium has so often (and wrongly) been described as the most toxic of the radioactive elements that it is natural that an affected worker should become concerned about his prognosis when long-continued tests are made. NORWOOD and FUQUA (1969) refer to a case in which therapy was discontinued due to the emotional make-up of an individual. The relentless search for additional scientific data, especially when therapy is not involved, is an improper assignment of priorities.

Fig. 14.7—14.11 are taken from the Hanford papers and internal protocols with the consent of the authors and of the Battelle-Northwest Laboratories.

Fig. 14.7 illustrates the fit of urinary excretion rate to the Langham equation over a 60-day period, for a known soluble deposition which can be calculated. An *essential* parameter is knowledge of the date of intake. Interpretation of a positive urinary excretion in a routine bio-assay program is inevitably uncertain when the specific cause cannot be identified. Since the excretion rate drops by a factor of 8 in the first month, there are two sound reasons for preferring a monthly sampling interval over the 6-month interval advocated by others:

a) There is a higher probability of obtaining a reliable demonstrably positive sample.

b) The ultimate estimate of intake date cannot be wrong by more than one month.

Fig. 14.8 shows the fit of excretion data to the Healy model (1957) of mobilization of plutonium from initially insoluble plutonium. For example, the derived value of Q_o could be a measure of a lung burden of plutonium oxide (minus that initially cleared by ciliary action).

Fig. 14.9 shows a 7-year follow-up of urine activity originating from a wound source of plutonium nitrate, with the wound cleaned to eliminate a continuing source. After some 200 days, the scatter of the individual low activity samples is perhaps more impressive than their fit to the curve. The visual impression can be improved by grouping the data in sets of 3 or even 5 if the interval between samples is small compared with the time post intake. Computer analysis of such data is surprisingly consistent. It can be set to give the best position of a straight

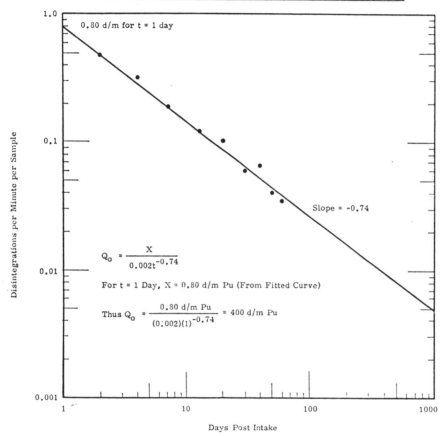

Fig. 14.7. Calculation of initial systemic burden, Q_0, from application of the Langham model to urinary excretion of initially soluble plutonium with a known intake date

line known to have the power slope $t^{-0.74}$ or to determine the best fit to any power slope, which might give values between about -0.72 and -0.78 for the power exponent. In the subject case, the best fit gave -0.76, which changes the intercept at day 1 from 120 dis/min to 130 dis/min.

Fig. 14.10 demonstrates the problems of interpretation when DTPA is used to chelate the systemic burden. The enhancement ratio of urinary excretion was 100 on treatment days. In other individuals it is variously between 30 and 100. A value of 50 is conventionally used to estimate the systemic burden during treatment. JECH et al., (1971, 1972), point out that the estimate of deposition is likely to be too high for some time after treatment. Here it was calculated as 18 nCi for the period 25 to 120 days post treatment (130 to 220 days post intake). It reverted to 10 nCi in the period 200–9120 days post intake, confirmed later by a continuing LANGHAM fit over 8 years. A cluster of points

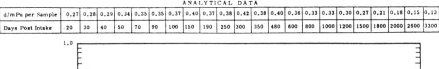

d/m Pu per Sample	0.27	0.28	0.29	0.34	0.35	0.35	0.37	0.40	0.37	0.38	0.42	0.38	0.40	0.36	0.33	0.33	0.30	0.27	0.21	0.18	0.15	0.13
Days Post Intake	20	30	40	50	70	90	100	150	190	250	300	350	480	600	800	1000	1200	1500	1800	2000	2600	3300

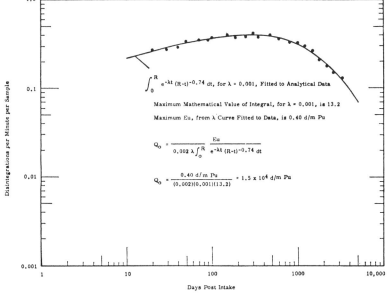

Fig. 14.8. Application of the Healy model to calculation of systemic burden slowly acquired from initial deposition of nominally insoluble plutonium in a specific compartment such as the lung

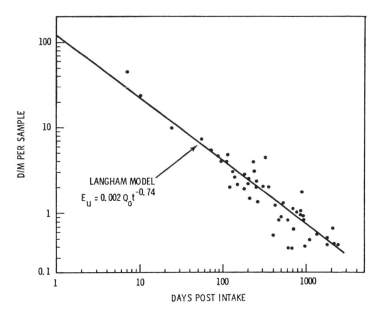

Fig. 14.9. Fit of urinary excretion data to the Langham model following plutonium nitrate intake from a wound

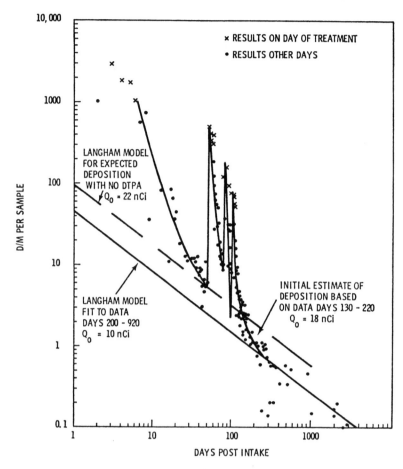

Fig. 14.10. Distortion of urinary excretion resulting from a plutonium nitrate intake, treated sporadically with DTPA over the first $3^1/_2$ months. At the end of 7 months, the estimated systemic burden is nearly twice the finally accepted value

is seen well below the fitted line at about 120 days. A review of all Hanford data shows this to be the usual occurrence (HEID, personal communication). If so, the temporary suppression merits study as a possible DTPA effect, and for practical analysis, these data points should be rejected. The LANGHAM line for 22 n Ci burden refers to the calculated deposition had DTPA not been used. Thus DTPA prevented 60% of the potential burden, and this is shown to be typical.

Fig. 14.11 shows the fit of urinary excretion data arising from a plutonium oxide inhalation case to a combined Langham model for transportable plutonium and a Healy model for the non-transportable fraction slowly mobilized from the lung. Oxide inhalation cases treated with DTPA present serious difficulties in modeling, due in part to loss of interpretability during the treatment period, and uncertainty with respect to time of return to an undisturbed excretion curve (JECH et al., 1971, 1972) (see also Case 7).

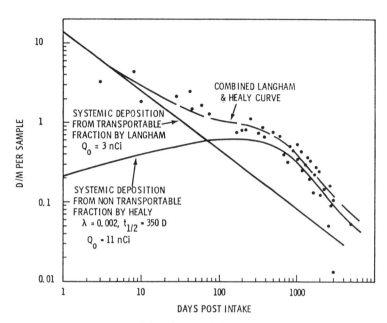

Fig. 14.11. Application of combined Langham and Healy equations to the transportable and non-transportable fractions of the ultimate systemic burden, following inhalation of plutonium oxide particles. With the low activity of the samples, the curve-fitting over the period 50—2000 days post intake is reasonably good

Refinement of Body Burden

The conventional maximum permissible body burden of plutonium (0.04 µCi) was originally based on ^{239}Pu only. With more mature plutonium, ^{240}Pu is the first to grow in. Both are customarily measured together (e.g., in urine) by their alpha-emissions, and they have the same permissible limits. With even more mature plutonium, all the isotopes from ^{238}Pu to ^{242}Pu are present. For practical purposes, the alpha count is proportional to the body burden for the 4 alpha-emitters. ^{241}Pu has to be accounted for by other means. Hanford refers to a normal plutonium, which by weight is 93.5% ^{239}Pu, 6.0% ^{240}Pu, and 0.5% ^{241}Pu. The measured alpha MPBB of this composition is 81% due to ^{239}Pu and 19% due to ^{240}Pu. In addition, the beta radiation of ^{241}Pu contributes 35% of a MPBB. It will be clear that mixtures much richer in ^{241}Pu will need correction for the unmeasured ^{241}Pu. For the heavy exposure plutonium of Table 14.1 (p. 638) at the MPBB, ^{238}Pu contributes 79% of the alpha activity, while ^{241}Pu constitutes an unmeasured contribution of 14 × MPBB. The picture is further complicated in the body by the growth of the alpha-emitting ^{241}Am. The biological effect of the beta and gamma radiations from ^{241}Pu may be sharply different from that of the alpha-emitting plutoniums.

Specific Cases of Interest

Case 1. Late Detection of Minor Wound

An operator received a puncture wound in 1952 while using a steel brush on plutonium in a glove box. Hand contamination of 0.003 µCi plutonium removed easily from the external skin surface. No bioassay sampling was requested at the time and by chance one week later the employee transferred to another assignment which did not routinely require bioassay. The first indication of intake was obtained three years later when the employee again changed work and a routine bioassay examination gave positive results.

Under questioning the employee stated that the area around the wound had itched for several months and that approximately six months after the incident he noticed two small black spots which showed up near the spot where the skin had been punctured. When these spots disappeared, the itching stopped. An examination using a plutonium wound counter in 1956 ($3^1/_2$ years after the injury was received) indicated 0.02 µCi remaining at the wound site. Excision of tissue reduced the contamination at the wound site to 0.0001 µCi.

Based on an evaluation utilizing urine data, the employee had an internal deposition estimated at 0.048 µCi of plutonium initially, and now about 0.04 µCi. Excretion data 20 years after the event still follow a Langham curve.

The incident emphasizes the need to investigate even minor puncture wounds. Had this wound deposition been detected promptly, practically all the systemic burden could have been avoided by minor surgery.

Case 2. Plutonium Metal Fragment in Wound

A technician was compressing a plutonium specimen inside a glove box when an explosion occurred. Fragments of the plutonium specimen were ejected from the jaws of the press, some of which inflicted a puncture wound in the arm of the technician. Efforts to decontaminate skin surfaces were effective except at the wound site where 0.05 µCi activity remained. A wound counter measuring the 17 keV radiation of plutonium showed initial activity of 36 µCi, with 8.5, 2.6, 2.8 and 0.1 in four stages of excision. These are uncorrected for tissue absorption. The third excision removed a plutonium metal flake (Fig. 14.12) which contained 7100 µCi (LARSON et al., 1968). Activity in excised tissue, other than the flake totaled 3.2 µCi. Faced with readings which went down and then up again, the authors concluded that the counter malfunctioned due to the extremely high activity of the metal flake.

As an exercise in health physics, it can be shown that the results could well have been real. The removed piece has a surface area of 16.8 mm²; from the known activity of 7100 µCi, a mass of 112 mg and hence a thickness of 0.3 mm can be derived. The initial depth of the flake was about 19 mm. The 17 keV radiation escapes only from a range thickness of the metal. Therefore, the appar-

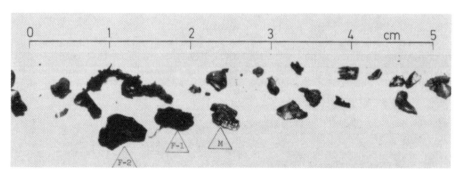

Fig. 14.12. Plutonium metal fragments, ejected in an explosion. The piece F-1 was protruding from the injured arm. F-2 is the deeply embedded flake removed by surgery

ent activity from the face would be ~ 190 µCi, and from the edge only ~ 19 µCi. The shielding factor for tissue depth is about 0.15. Thus the wound counter would see between 30 µCi and 3 µCi, depending on the orientation of the flake. That orientation could have changed with the probing at excision. The initial counts and counts after excision are compatible with this explanation. The interest in the exercise is the practical application of plutonium metal dosimetry.

The patient was given chelation therapy for 50 weeks, during which the systemic burden was estimated to be about 0.16 µCi with a DTPA enhancement ratio of 30. Experience with other cases suggests that this would overestimate the burden. The current burden, 8 years later, is estimated to be about 0.1 µCi. In addition, 0.06 µCi remained at the wound site. Measurements are continuing after 8 years.

Case 3. Plutonium Oxide Inhalation from Improperly Adjusted Mask

An employee was changing the primary filter from a glove box when a spread of plutonium occurred. The filter area had previously been completely enclosed in a plastic greenhouse or tent arrangement. The employee was exposed to the contaminated atmosphere inside the greenhouse for an estimated 15 minutes. Upon exit from the greenhouse, a personal survey indicated extensive skin and nasal contamination up to 0.02 µCi plutonium, ascribed to an ill-fitting mask. A lung examination detected 0.003 µCi of ^{241}Am on the day of the incident, 0.001 µCi on the following day, and 0.0005 µCi a week later. Sputum samples collected immediately following the incident contained 3×10^{-5} µCi Pu/ml. Fecal samples collected during the first 5 days post intake contained about 0.2 µCi plutonium. The employee was administered DTPA on eight separate occasions. From current evaluations, the employee incurred an initial deep lung burden of 0.018 µCi of plutonium (based on an initial burden of 0.03 µCi clearing to 0.018 µCi in 7 days) and a projected ultimate systemic burden of 0.016 µCi or 40% of the MPBB, with bone as reference.

This was a 1968 case reflecting, in the primary records, increasing sophistication in using auxiliary methods to refine the analysis. A nasal smear removed 20 nCi, enough to permit an accurate isotopic analysis which read:

	Atom percent	Percent of total alpha	Wt percent Pu only
^{238}Pu	0.0261	5.6	0.026
^{239}Pu	92.2	72.4	92.2
^{240}Pu	6.88	19.4	6.91
^{241}Pu	0.782	—	0.79
^{242}Pu	0.078	<0.1	~0.079
^{241}Am	0.0521	2.6	—

The initial ratio of $\frac{\text{Pu alpha}}{\text{Am alpha}} = 37.5$.

The particle size in the smear had activity mean aerodynamic diameter (AMAD) of 7.5 µm. Its solubility in normal saline was less than 0.9% in 2 hours. Smears from the mask had AMAD of 4.5 µm, the size used to fit lung model data. The true AMAD of the original aerosol could have been smaller.

Feces samples in the first five days yielded 165 n Ci. Adding the nasal smear as another mode of early clearance, conventional lung models for 4.5 μm particles gave

Initial pulmonary burden	26 nCi,
Pulmonary burden at 7 days	16 nCi,
Total systemic deposition	3.5 nCi.

Delayed feces samples, too few for good evaluation, gave a tentative initial pulmonary burden of 15–18 nCi.

Lung counting of the ^{241}Am radiation, using equations derived by NELSON (1968) from the ICRP lung model (MORROW, 1966) gave these results:

	nCi ^{241}Am	nCi Pu (alpha-emitter)
Initial pulmonary burden	0.75	30
Pulmonary burden at 7 days	0.45	18
Lung burden at 2.7 yrs	0.6	8

Fig. 14.13 shows the measured lung burden of ^{241}Am and the inferred plutonium burden. After the rapid fall in 7 days, the americium count slowly rises, due to growth of new americium offset by removal of americium and the parent ^{241}Pu. The inferred plutonium burden in the lung fell slowly. Such calculations can only be done when the original isotopic composition is known, and the continually changing Pu/Am ratio can be established. Determination of the systemic burden by urine analysis continues to show 10.2 nCi after 3 years. The model projects an ultimate burden of about 16 nCi. However, there is a continuing discrepancy between the present systemic burden deduced from the lung model (4 nCi) and the straightforward excretion data.

The case is particularly instructive in showing the general reliability of crude initial guides, at least to the extent of signalling rational treatment. The comparison as given by HEID et al. (1970a, b) is:

	Evaluation technique	Systemic deposition nCi Pu	Initial pulmonary deposition nCi Pu
Prompt	nasal smear guide	2.1	21
	initial in vivo guide		
	for assumed Puα/Amα = 15	1.3	13
	for true Puα/Amα = 40	3.3	33
	initial urine — Langham model (corrected for DTPA influence)	1.3–6	—
Interim	fecal — lung model		
	early ^{241}Am and Puα/Amα data	3.2	24
	final Pu data	3.3	25
	in vivo — lung model early data	4	33
	urine — Langham early data	2–10	—
Formal	in vivo — lung model	4	33
	urine — Langham	10	—
	urine — Healy	7.5	—

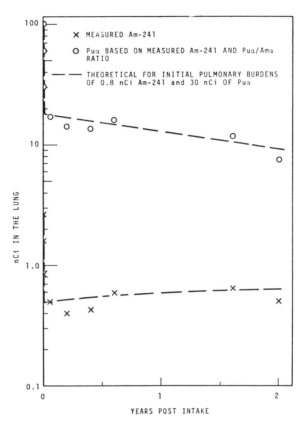

Fig. 14.13. Measured ^{241}Am lung burden and calculated Puα lung burden after plutonium oxide inhalation (Case No 3). The precipitous drop in the first seven days is virtually a vertical line on this scale. The main interest is the good fit of the derived falling plutonium burden to a theoretical model in the presence of an increasing burden of ^{241}Am

Case 4. Lung Burden Unrelated to a Specific Accident and below Detection Limit of Routine Bioassay

From retrospective study of the employee's work schedule it is believed that the inhalation occurred while he was performing work which involved welding on special plutonium fuel elements. Routine bioassay sampling did not detect any plutonium in urine samples collected during or immediately following this work. A routine lung examination some 8 years later detected 0.004 μCi of ^{241}Am and approximately 0.008 μCi of ^{239}Pu in the thorax region. The latter measurement is somewhat questionable since it is close to the minimum detectable for the equipment used. A Pu/Am ratio of 2 is in reasonable agreement with the ratio currently being observed in his fecal samples. Based on evaluation of these data, this employee incurred an initial intake to the deep lung of 0.1 μCi and received a systemic burden estimated at 0.001 μCi of plutonium or 2.5% of the MPBB, with bone as reference. The transference from lung to systemic burden is not compatible with existing models. The lung burden is close to existing detection limits. The employee was sent to two other major laboratories to check the lung burden, with these results:

	nCi ^{241}Am	nCi Pu
Hanford	4.5 ± 0.6	8 (dubious)
Lab A	5.7 ± 0.9	<4
Lab B	4.1 ± 0.5	<13

This is a fair measure of the present state-of-the art.

Case 5. Repetitive Depositions of Plutonium Compounds

The immediate incident involved an employee who was bagging material out of a glove box using a plastic bag which ruptured in the process. The employee remained in the contaminated atmosphere for an estimated 15 minutes after the rupture occurred; however, he was wearing a mask during most of this period. At the completion of work, personal surveys indicated nasal smears of 0.015 µCi plutonium and a lung examination indicated 0.004 µCi of ^{241}Am. A week later this had decreased to 0.0004 µCi. Utilizing a Pu/Am ratio of 22 and particle size of 7 microns as determined for a sample of the material, one can calculate an initial deep lung burden of 0.009 µCi of plutonium. Evaluation of the urine excreta data indicates the intake resulted in a systemic burden of 0.013 µCi plutonium or 33% of the MPBB with bone as reference. In addition, it was determined that there was a systemic burden of 0.105 µCi ^{241}Pu, which is 12% of the relevant MPBB. In total, then, the case had a 50% MPBB. Nine DTPA treatments were administered.

Two items are of interest in this case:

a) Fecal excretion provided a more definitive signal than did urinary excretion, because of high activity in the early samples.

b) The individual was involved in six separate incidents with plutonium in different solubility forms over a span of 7 years. Fortunately, all except the one reported above gave very low depositions. Otherwise, the interpretation of an additional burden would have been troublesome. Carrying this to the limit of more or less continuous chronic exposure at very low levels (as implied in Part I of Appendix A), it would be difficult to detect a systemic burden, and virtually impossible to quantify it.

The case also raises a fundamental social problem. The experience of this individual was so at variance with the norm over 50 000 man-years of plutonium work that he can properly be classed as incident-prone. Should such a man be removed from hazardous work for his own protection or the protection of fellow workers? In a large institution in which he could be found useful employment in non-radiation work, one would tend to answer in the affirmative. If displacement of an older worker with no other special training would lead to probable unemployment [9], the balancing of benefits versus risks is a difficult task. If the risk were confined to the individual, he could properly be made a party to discussing his options. Should the risks to fellow employees be judged to be significant, management would be compelled to weight that factor heavily.

Case 6. Undetected Wound in Experimental Laboratory

An employee had worked with several special enrichments of plutonium during November of 1966. No unusual conditions were reported during this work

---

9 Although one knows of no social studies bearing on the point, it appears that the incident-prone worker is not necessarily a careless irresponsible type, but is rather the willing worker always pressing to do more than his fair share. Unemployment for him is a devastating morale breaker.

nor was the employee involved in any known radiation incident. However, a result of 2×10^{-6} µCi plutonium was reported for a routine urine sample collected on January 1 of the following year. At first it was assumed that the intake had occurred through inhalation, the usual mode of intake when first detected by routine bioassay sampling. However, the urinary excretion pattern observed during the next few weeks did not follow the expected pattern. Consequently, a search was made for contaminated wounds on 2/21/67, using a plutonium wound counter. A wound was found in the employee's hand which contained 0.01 µCi of plutonium. This was reduced through surgical excision of tissue at the wound site to 0.0004 µCi. Based on isotopic analysis of the material removed from the wound, 11/9/66 was estimated as the most probable date of intake, since the employee had been handling material of this composition at that time. Sixteen DTPA treatments were administered. Evaluation of the systemic burden, utilizing urinary excretion data, was 0.02 µCi of plutonium, or 50% of the MPBB, with bone as reference. Had this employee not been working in a Critical Mass Laboratory where a complete, accurate inventory of plutonium compositions with their times of use is kept, it would have been difficult to assign a deposition date to support the evaluations. The 8.9 nCi removed from the wound was a sufficient source to measure the 4.6% abundance of ^{240}Pu. This compares with the precisely known $4.62 \pm 0.3\%$ ^{240}Pu content of the material used only for a few days.

The material was a dry nitrate so that the mobilization data are representative of transfer from a soluble deposit in fleshy tissue. As a result of this and a few other nearly-missed cases, the practice is now encouraged of supplementing bioassay samples with routine photonic radiation counts of hands and lungs. On the whole, there is always a good chance of suspecting inhalation from signals of air contamination. The small puncture wounds can clearly occur with so little surface contamination that it is not detectable by the conventional alpha-counting hand counters.

Case 7. Puncture Wound with DTPA Treatment

An employee was removing stainless steel piping from a solvent extraction hood when he received a puncture wound in his right hand. Personal surveys upon exit from the work area detected external skin contamination of 0.005 µCi of plutonium. A blood sample collected at that time indicated 0.0002 µCi plutonium/ml. An examination of the wound site using a sodium iodide plutonium wound counter detected 0.12 µCi plutonium which was reduced by surgical excision to 0.01 µCi. The wound was flushed with DTPA before closure and a total of 47 DTPA treatments were administered. Based on evaluation utilizing urine excreta data the employee received a systemic burden estimated at 0.015 µCi plutonium as alpha emitters or 40% of the MPBB, with bone as reference. In addition, a systemic burden of 1.2 µCi ^{241}Pu is derived, or 130% of the formal MPBB. This is a relatively recent case illustrating growing attention to the usually ignored ^{241}Pu burden. The excretion pattern is shown in Fig. 14.14, which has some data points averaged to reduce the clutter. As pointed out by JECH et al. (1971, 1972) the DTPA enhancement ratio was 60–80 although the apparent ratio between days of treatment and other days was only about 5–10. Thus the DTPA was decidedly more effective than would have been judged at the time. This case maintained elevated excretion rates for 80 days post-treatment, and then shows the marked suppression below the eventual Langham fit. There was a typical elimination of 60% of the otherwise expected systemic burden.

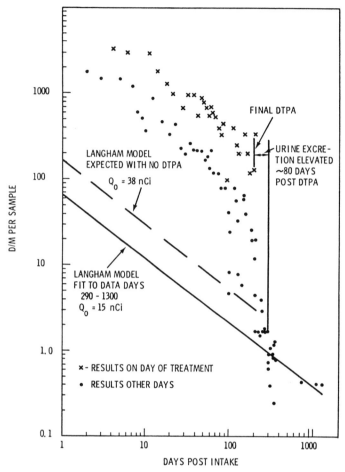

Fig. 14.14. Prolonged elevation of urinary excretion of a soluble salt (with no supplementation from the initial wound site), following intensive DTPA therapy. Some of the data points have been grouped together for clarity

Case 8. Long Term Reliability of Langham Excretion Formula

A simple case of deposition of soluble plutonium occurred in 1952. It was studied with frequent bioassay samples from January 1, 1953, to July 26, 1961. The best fit to a Langham curve of exponent -0.74 showed an estimated initial systemic deposition of 26 nCi. In early 1959, DTPA was administered over 12 consecutive days during which the enhancement ratio averaged 40. This demonstrates the effectiveness of late chelation in terms of enhancement ratio. As a practical therapy for a 7-year deposition, it is questionable because the elimination rate is still trivial. In the subject case, the enhanced elimination was 0.65 nCi of the then existing deposition of 23.5 nCi (a 10% reduction from the original burden through 6% elimination by urine and 48% by feces according to Langham equations), or less than 0.3% removal by the late chelation.

The principal interest of the case is that it has now been followed for 20 years, and the power function of $t^{-0.74}$ is still unchanged. This does not prove that the calculated deposition is correct because this involves also the numerical constant 0.002 in the Langham equation. There is a preliminary suggestion that deposits are lower than calculated but this can only be ascertained by total body analysis of well-documented cases at autopsy. It is remarkable enough that Langham was able to derive so powerful a tool from his limited observations.

In another Hanford case of a lung burden of soluble plutonium, the elimination rate was successfully reconciled to a combined Langham and Healy model over the first 1 000 days. Now after 4 500 days with the lung burden essentially removed the total systemic burden of only 26 nCi is faithfully following the $t^{-0.74}$ function.

References [10]

AARKROG, A.: Radioecological investigations of plutonium in an arctic marine environment. Hlth Phys. 20, No 1, 31–48 (1971).
ANDERSEN, B. V.: Battelle Northwest Laboratories—private communication (1972).
ATTIX, F. H. (ed.): Luminescence dosimetry. Int. Conf. on Luminescence Dosimetry, USAEC, 1-514 CONF-650637 (1967).
BEACH, S. A., DOLPHIN, G. W.: Determination of plutonium body burdens from measurements of daily urine excretion. In: Assessment of radioactivity in man, vol. II, p. 603–615. Vienna: I.A.E.A. 1964.
BEACH, S. A., DOLPHIN, G. W., DUNCAN, K. P., DUNSTER, H. J.: A basis for routine urine sampling of workers exposed to plutonium-239. Hlth Phys. 12, 1671–1682 (1966).
BELL, M. J., DILLON, R. S.: The long-term hazard of radioactive wastes produced by the enriched uranium, Pu-^{238}U, and ^{233}U-Th fuel cycles. USAEC-ORNL-TM-3548 (1971).
BILLARD, F., MIRIBEL, J., PRADEL, J.: Controle de la contamination atmospherique par "Double Mandarin." French CEA. CEA-2085 (1961).
BORTOLI, M. C. DE: Radiochemical determination of plutonium in soil and other environmental samples. Analyt. Chem. 39, 375–377 (1967).
BRAMSON, P. E., JECH, J. J., ANDERSEN, B. V.: Comparison of manual and computer evaluation of systemic plutonium burden. Hlth Phys. 22, 911–922 (1972).
CLAYTON, E. D., REARDON, W. A.: Nuclear safety and criticality of plutonium. 875–919 in AEC Plutonium Handbook. Ed. O. J. WICK. New York: Gordon & Breach Science Publ. 1967.
CORLEY, J. P., ROBERTSON, D. M., BRAUER, F. P.: Plutonium in surface soil in the Hanford Plant environs. Proceedings of Environmental Plutonium Symposium, LA-4756, 85–88 (1971).
DUNSTER, H. J.: The concept of derived working limits for surface contamination. Surface contamination (B. R. FISH, ed.), p. 139–147. Oxford: Pergamon Press 1967.
ENDRES, G. W. R.: Characteristics of teflon-LiF thermoluminescent dosimeters. BNWL-741 (1968).
ENEA: Criticality control in chemical and metallurgical plants. Karlsruhe Symposium, p. 1–622 (1961).
EVANS, R. D.: Radium poisoning: a review of present knowledge. Amer. J. publ. Hlth 23, 1017–1023 (1933).
EVANS, R. D.: Protection of radium dial workers and radiologists from injury by radium. J. industr. Hyg. 25, 253–269 (1943).
EVANS, R. D.: Radiogenic effects in man. In: Radiobiology of plutonium (STOVER, B. J., and JEE, W. S. S., eds.), p. 431–468. Salt Lake City: The J.W. Press, Univ. of Utah 1972.
FAUST, L. G., BRACKENBUSH, L. W., SMITH, R. C., NICHOLS, L. L., BRITE, D. W.: Radiation dose rates from UO_2-PuO_2 thermal reactor fuels. BNWL-SA-3661 (1971).
FISH, B. R. (ed.): Surface contamination. Proceedings of a symposium held at Gatlinburg, Tennessee. Oxford: Pergamon Press 1967.
HAMMOND, S. E.: Industrial-type operations as a source of environmental plutonium. Proceedings of Environmental Plutonium Symposium, LA-4756, 25–35 (1971).
HARLEY, J. H.: Worldwide plutonium fallout from weapons tests. Proceedings of Environmental Plutonium Symposium, LA-4756, 13–19 (1971).
HEALY, J. W.: Estimation of plutonium lung burden by urine analysis. Amer. industr. Hyg. Ass. Quart. 18, 261–266 (1957).
HEALY, J. W.: Surface contamination: decision levels. USAEC-LA-4558-MS (1971).

---

[10] See note in References for Chap. 8 regarding availability of governmental and other "in house" documents (Editor).

Plutonium Industrial Hygiene, Health Physics and Related Aspects

HEALY, J. W., FUQUAY, J. J.: Wind pickup of radioactive particles from the ground. Proceedings of the 2nd Internat. Conference on Peaceful Uses of Atomic Energy 18, Paper P/391, 291–295 (1958).
HEALY, J. W., FUQUAY, J. J.: Wind pickup of radioactive particles from the ground. Progress in nuclear energy series XII, vol. 1 — Health physics. Oxford: Pergamon Press 1959.
HEID, K. R., HENLE, R. C., SELBY, J. M.: Prompt mitigatory action after accidental exposure to radionuclides; in the monograph No 2 on nuclear medicine and biology "Diagnosis and treatment of deposited radionuclides" (eds. KORNBERG, H. A., and NORWOOD, W. D.), p. 593–599. Amsterdam: Excerpta Medica Foundation 1968.
HEID, K. R., JECH, J. J.: Assessing the probable severity of plutonium inhalation cases, p, 1–17. BNWL-SA-1595 (1968).
HEID, K. R., JECH, J. J.: Prompt handling of cases involving accidental exposure to plutonium. BNWL-SA-2062 (1969).
HEID, K. R., JECH, J. J., ANDERSEN, B. V.: Interpretation of data on internal contamination. BNWL-SA-3486 (1970a). Also in the Proceedings of the Seminar on Radiation Protection Problems Relating to Transuranium Elements, Karlsruhe (W. Germany), Sept. 21–25, 1970. EUR 4612 d-f-e (1970b).
HEID, K. R., JECH, J. J., ANDERSEN, B. V.: Interpretation of human urinary excretion of plutonium for cases treated with DTPA. BNWL-SA-4071 (1971) and in Hlth Phys. **22**, 787–792 (1972).
HOLLIDAY, B., DOLPHIN, G. W., DUNSTER, H. J.: Radiological protection of workers exposed to airborne plutonium particulate, p. 1–14, AHSB(RP)R 96 U.K.A.E. (1969). Also read as Paper No 1 of the 2nd Congr. of I.R.P.A. Brighton, England (1970).
HUNT, D. C.: Restricted release of plutonium—Part 1—Observational data. Nuclear Safety **12**, 85–89 (1971a).
HUNT, D. C.: Restricted release of plutonium—Part 2—Theory. Nuclear Safety **12**, 203–216 (1971b).
I.C.R.P. Publication 2: Report of Committee II on permissible dose for internal radiation. Oxford: Pargamon Press 1960.
I.C.R.P. Publication 6: Recommendations of the Internat. Commission on Radiological Protection (as amended 1959 and revised 1962). Oxford: Pergamon Press 1964.
I.C.R.P. Publication 10: Report of Committee IV on evaluation of radiation dose to body tissues from internal contamination due to occupational exposure. Oxford: Pergamon Press 1968.
ICRP Publication 10A: The assessment of internal contamination resulting from recurrent or prolonged uptakes. Oxford: Pergamon Press 1971.
IREDALE, P., HINDER, G.: The detection of airborne plutonium hazards. AERE-R-3783 (1962).
JECH, J. J., ANDERSEN, B. V., HEID, K. R.: Interpretation of human urinary excretion of plutonium for cases treated with DTPA. BNWL-SA-4071 (1971) and in Hlth Phys. **22**, 787–792 (1972).
JECH, J. J., HEID, K. R., CROOK, G. H.: Interaction between health physicists and industrial physicians immediately following radiation incidents. Read at Western Industrial Health Conference, San Francisco. BNWL-SA-2663 (1969a).
JECH, J. J., HEID, K. R., LARSON, H. V.: Prompt assessments and mitigatory action after accidental intake of plutonium. IAEA Symposium on Handling of Radiation Accidents, p. 77–93, IAEA (1969b).
KIEFER, H., MAUSHART, R.: Determination of Plutonium-239 body burden using gamma spectrometry with proportional counters. Proceedings of the symposium on whole-body counting, p. 289–293. Vienna: IAEA 1962.
KOCHER, L. F., ENDRES, G. W. R., NICHOLS, L. L., SHIPLER, D. B., HAVERFIELD, A. J.: The Hanford thermoluminescent multipurpose dosimeter. BNWL-SA-3955 (1971).
KORBA, A., HAY, J. E.: A thermoluminescent personnel neutron dosimeter. Hlth Phys. **18**, 581–584 (1970).
KORNBERG, H. A., NORWOOD, W. D. (eds.): Monographs on nuclear medicine and biology, No 2—Diagnosis and treatment of deposited radionuclides. Amsterdam: Excerpta Medica Foundation 1968.
LANGHAM, W. H.: Determination of internally deposited radioactive isotopes from excretion analyses. Amer. industr. Hyg. Ass. Quart. **17**, 305–318 (1956).
LARSON, H. V.: Factors in controlling personnel exposure to radiations from external sources, p. 845–857—AEC Plutonium Handbook, ed. O. J. WICK. New York: Gordon and Breach Science Publ. 1967.
LARSON, H. V., MYERS, I. T., ROESCH, W. C.: A wide beam fluorescent X-ray soucre. HW-31781 (1954) declassified (1956).

LARSON, H. V., NEWTON, C. E., JR., BAUMGARTNER, W. V., HEID, K. R., CROOK, G. H.: The management of an extensive plutonium wound and the evaluation of the residual internal deposition of plutonium. Phys. in Med. Biol. 13, 45–53 (1968).

LAURER, G. R., EISENBUD, M.: In vivo measurement of nuclides emitting soft penetrating radiations. New York University Medical Center, AD690243 (1969).

LAWRENCE, J. N. P.: Hlth Phys. 8, 61 (1962).

LIDÉN, K. V. H., McCALL, R. C.: Low-energy photon detectors for whole-body counting. Proceedings of symposium on whole-body counting, p. 145–166. Vienna: IAEA 1962.

LINDEKEN, C. L., LAKIN, R. W.: Improvements in the solidstate plutonium-alpha air monitor. UCRL-50228 (1967).

LISTER, B. A. J.: Early assessment of the seriousness of lung contamination by insoluble alpha and low energy beta emitting materials after inhalation exposure. UKAEA Report AERE-R-5292 (1966). Also in Proceedings of the First Internat. Congr. of Radiation Protection, Rome 1966, vol. 2, p. 1191–1198. Oxford: Pergamon Press 1968.

MAGNO, P. J., KAUFFMAN, P. E., SCHLEIEN, B.: Plutonium in environmental and biological media. Hlth Phys. 13, 1325–1330 (1967).

MORROW, P. E. (CHAIRMAN): ICRP Task Group on lung dynamics: Deposition and retention models for internal dosimetry of the human repiratory tract. Hlth Phys. 12, 173–207 (1966).

N.C.R.P. Report 11: Maximum permissible amounts of radioisotopes in the human body and maximum permissible concentrations in air and water (N.B.S. Handbook 52) (1953).

N.C.R.P. Report 22: Maximum permissible body burdens and maximum permissible concentrations of radionuclides in air and in water for occupational exposure (N.B.S. Handbook 69) (1959).

N.C.R.P. Report 39: Basic radiation protection criteria. National Council on Radiation Protection and Mesurements (1971).

NELSON, D. J., EVANS, F. C., (eds.): Symposium on radioecology. Proc. 2nd Nat. Symp. Ann Arbor, Michigan. CONF-670503 (1969).

NELSON, I. C.: Theoretical excretion of plutonium in urine based on the new ICRP lung model. In: Monographs on nuclear medicine and biology, No 2—Diagnosis and treatment of deposited radionuclides, p. 266–278. Amsterdam: Excerpta medica Foundation 1968.

NELSON, I. C., HEID, K. R., FUQUA, P. A., MAHONEY, T. D.: Plutonium in autopsy tissue samples. BNWL-SA-4077 (1971).

NEWBERRY, G. R.: Measurement and assessment of skin doses from skin contamination. In: Radiation and skin, AHSB (RP), R 39, 44–67 (1964).

NEWTON, C. E., JR., LARSON, H. V., HEID, K. R., NELSON, I. C., FUQUA, P. A., NORWOOD, W. D., MARKS, S., MAHONEY, T. D.: Tissue analysis for plutonium at autopsy. In: KORNBERG, H. A., W. D. NORWOOD (eds.) Monographs on nuclear biology and medicine series, No. 2 p. 460–470. Amsterdam: Excerpta Medica Foundation 1968.

NORWOOD, W. D., FUQUA, P. A.: Medical care for accidental deposition of plutonium (^{239}Pu) within the body. IAEA-SM-119/25 in Proceedings of a symposium—Handling of radiation accidents, p. 147–162. Vienna: IAEA 1969.

NORWOOD, W. D., NORCROSS, J. A., NEWTON, C. E., JR., HYLTON, D. B., LA GERQUIST, C.: Preliminary autopsy findings in U.S. Transuranium Registry Cases. Read. May, 1972 and to be published in the Proceedings of the 12th Annual Hanford Biology Symposium (Radionuclide Carcinogenesis) (1972).

PALMER, H. E., WOGMAN, N. A., COOPER, J. S.: The determination of the depth and amount of ^{238}Pu in wounds with Si(Li) detectors. Proc. of the symposium on Diagnosis and Treatment of Deposited Radionuclides, Richland, Washington, p. 164–170 (May 1967). Amsterdam: Excerpta Medica Foundation 1968.

PANGHER, J. DE: Double moderator neutron dosimeter. Nucl. Instr. and Methods 5, 61–64 (1959).

PATTERSON, D. E., DE FATTA, V. P.: A summary of incidents involving USAEC shipments of radioactive material 1957–1961. U.S.A.E.C. TID-16764 (1962) (and a series of updating supplements such as TID-16764 Suppl. 1, not all by the same two authors).

POMAROLA, J., RISSELIN, A., FELIERS, P.: Assessment of individual risk during atmospheric contamination by plutonium. CEA Fontenay-aux-Roses, NP-18254 (1966).

PRICE, K. R.: A critical review of biological accumulation, discrimination and uptake of radionuclides important to waste management practices. (1943–1971), p. 1–67. BNWL-B-148 (1971).

PRICE, K. R.: A review of transuranic elements in soils, plants, and animals. (This is the transuranium portion of the above in-house document.) To be published in Journal of Environmental Quality (1972).

Proceedings of environmental plutonium symposium, Los Alamos Scientific Laboratory. USAEC LA-4756, 19 papers, 1–119 (1971).

ROEDER, J. R.: A statistical summary of USAEC licensees' internal exposure experience. (1957–1966), p. 435–450. In: KORNBERG, H. A., and W. D. NORWOOD, Monographs on nuclear medicine and biology, No 2. Amsterdam: Excerpta Medica Foundation 1968.
ROESCH, W. C.: Surface dose from plutonium. Proceedings of the Second United Nations Conference on the Peaceful Uses of Atomic Energy, Geneva 23, 339–345 (1958). Geneva: United Nations 1958.
ROESCH, W. C., BAUM, J. W.: Detection of plutonium in wounds. In: Proceedings of the Second United Nations International Conference on the Peaceful Uses of Atomic Energy, Geneva 23, 142–143 (1958). Geneva: United Nations 1958.
Ross, D. M.: A statistical summary of U.S. AEC contractors' internal exposure experience (1957–1966), p. 427–434. In: KORNBERG, H. A., and W. D. NORWOOD (eds.), Monographs on nuclear medicine and biology, No 2. Amsterdam: Excerpta Medica Foundation 1968.
SCHULTE, H. F., WHIPPLE, H. O.: Chelating agents in plutonium deposition—a minority view, p. 587–592. In: KORNBERG, H. A., and W. D. NORWOOD, Monographs on nuclear medicine and Biology, No 2 (1968).
SCHULTZ, V., KLEMENT, A. W., JR. (eds.): Radioecology. Proc. 1st Nat. Symposium on Radioecology, Colorado State Univ. New York: Reinhold Publishing Corp., and Am. Inst. Biol. Sci. Wash. D.C. 1963.
SCHWENDIMAN, L. C., HEALY, J. W., REID, D. L.: The application of nuclear track emulsions to the analysis of urine for very low level plutonium. USAEC Report HW-22680 (1951).
SHERWOOD, R. J., STEVENS, D. C.: Some observations on the nature and particle size of airborne plutonium in the Radiochemical Laboratories, Harwell. Ann. occup. Hyg. 8, 93–108 (1965). Also as AERE-R-4672 (1963, declassified 1964).
SHIPMAN, T. L., LUSBAUGH, C. C., PETERSEN, D. F., LANGHAM, W. H., HARRIS, P. S.: Acute radiation death resulting from an accidental nuclear critical excursion. Special Suppl. J. occup. Med. 3, 147–192 (1961).
SMORODINTSEVA, G. I.: Study of uptake of airborne Pu-239 by the human organism. U.N. Scientific Committee Document A/AC.82/G/L.1301, HASL Translation (Nov. 1969).
SNYDER, W. S., FORD, M. R., WARNER, G.: Proceedings of the 13th Annual Bioassay and Analytical Chemistry Meeting (Oct. 1967).
SWINTH, K. L.: Wound counting with solid state detectors. Pacific Northwest Laboratory Annual Report for 1968 to the USAEC Division of Biology and Medicine. Part 3, Instrumentation, p. 1–6, BNWL-1051 (1968).
Tracerlab (now Trapelo West) Technical Bulletin S-6 (1968). (Ascribes design basis to ANDERSON, I. O., and J. BRAUN: A neutron REM counter AE-132 and further work by UCLRC. Evaluation of a neutron REM dosimeter. Hazards Control Quarterly Report No 18. UCLRL-12167.)
UNRUH, C. M.: AEC workshop on personnel neutron dosimetry. BNWL-1340 (1969).
UNRUH, C. M.: Transuranium nuclides—a manual of good practice. BNWL-SA-3075, 99 pps. (1970). (A revised draft in press will be available from C. M. UNRUH, Battelle Northwest Laboratories, Richland, Wash. USA.)
UNRUH, C. M.: Second AEC workshop on personnel neutron dosimetry. BNWL-1616 (1971).
UNRUH, C. M., KOCHER, L. F., BRAMSON P. E.: The new Hanford film badge dosimeter. HW-76944 (1963).
WAITE, D. A., ANDERSEN, B. V., BRAMSON, P. E.: A projection chest phantom for in vivo dosimetry. A.I.H.A. Journ. 31, 322–326 (1970).
WILSON, R. H.: A method for immediate detection of high level neutron exposure by measurement of sodium-24 in humans. HW-73891 Rev. (1962).
WILSON, R. H.: Controlling and evaluating plutonium deposition in humans, p. 831–844 AEC Plutonium Handbook, ed. O. J. WICK. New York: Gordon and Breach Science Publ. 1967.
WILSON, R. H., SILKER, W. B.: Plutonium contaminated injury case study and associated use of Na_4 EDTA as a decontaminating agent. USAEC Report HW-66309 (1960).
WINKLER, R., HOETZL, H., SANSONI, B.: Rapid identification and activity determination of long-lived alpha emitters in air. CONF-690540, 184–196 (1969).

Unwanted Radiation Exposures — The Meaning for Man

Lecture Notes

March 5, 1975

UNWANTED RADIATION EXPOSURES - The Meaning for Man

H. M. Parker   3/5/75

OUTLINE

1. Early protection considerations

   a. medically related
   b. industrial - radium and uranium

2. Status as of early 1940's

   Tolerance dose
   Threshold concepts
   Developing problem of genetic effects

3. Changes in permissible limits 1940 - 1957

   NAS - Genetics Report 1956
   Acceptable Risk and ALAP
   Environmental limits v. Occupational limits

4. Late somatic effects

   Leukemia
   Other cancers
   Atomic Bomb Casualty Commission results
   Medical irradiation results
   Non-threshold concept
   Linear concept

5. Current reports on protection "philosophy"

   UNSCEAR
   BEIR
   NCRP #39 and #43
   WASH-1400

6. Some numerical consequences of literal acceptance of the present linear hypothesis

   Natural radiation - 3500 cancer deaths/yr in U.S.
   Unwanted medical - 2500   "   "   " " "
   Fallout   - 140   "   "   " " "

Outline
H.M. Parker   -2-   3/5/75

    Occupational exposure - ∼ 30 cancer deaths/yr in U.S.
    Nuclear power (1970) - ∼ 0.1   "   "   " " "
    Nuclear power (2000) - < 35   "   "   " " "

All these values may be zero. It seems "most likely" that they will be lower than these numbers by a factor of 5(?), 10 (?), 100(?). No one knows.

For comparison, there are 365,000 cancer deaths/yr in the U.S. It is said that 75% or more of these (i.e., > 1/4 million) are related to some form of environmental contamination. (WHO Report).

Recommended Reading:

A. Essential

    Ref. 1. NAS-NRC BEIR Report. The Effects on Populations of Exposure to low levels of ionizing radiation, Nov. 1972

    Ref. 2 UNSCEAR Report. Ionizing Radiation: Levels and Effects, Vol. 1. Levels, Vol. II, Effects, United Nations 1972.

    Ref. 3 NCRP Report #39. Basic Radiation Protection Criteria, NCRP 1971.

    Ref. 4 NCRP Report #43. Review of the Current State of Radiation Protection Philosophy, NCRP 1975.

B. Additional

    Ref. 5 Permissible Dose from External Sources of Radiation. NCRP Report #17, published as NBS Handbook #59, 1954.

    Ref. 6 NAS-NRC Report on the Biological Effects of Atomic Radiations, 1956.

    Ref. 7 Various reports of the ICRP. #8 and 14 each contain material on the Evaluation of Risks from Radiation.

Principles of Standard Setting: The Radiation Experience

Third Life Sciences Symposium, Los Alamos

Presented October 1975
Published 1976 in CONF 751022

Principles of Standards Setting: The Radiation Experience

H. M. Parker

ABSTRACT. The basis and use of standards generally are repeated from material used in the Joint Committee on Atomic Energy Hearings on Radiation Protection Standards. A brief history of radiation protection standards is related to this base. The recent history emphasizes the delayed production of leukemia and other cancers as a principal risk. A linear relationship between such effects and dose is accepted as a prudent policy. What really happens at low dose and low dose rate is unknown and may remain so.

Whether linear or not, it is likely (but also unproven) that the deleterious effects are nonthreshold. Standards setting then becomes a public responsibility not soluble on purely technical grounds. It is implied that hazards of numerous environmental pollutants from alternative energy sources operate in the same way. To apply equal diligence to all pollutants as is now given to radiation would cost astronomical sums. Steps are outlined to reduce some of the excesses that have troubled the radiation case. The most useful step is probably to avoid the "Father knows best" approach that has encouraged some lack of credibility.

It was reported in the local paper this last weekend that the director of the Washington Public Power Supply System was bemoaning the need to observe 1624 nuclear application standards of the Nuclear Regulatory Commission, compared with fewer than 100 when he applied for a similar reactor license in 1970. Standards documents on radiation protection increased exponentially from 6 in 1946 to 114 in 1971, a rise that projects to an absurd 9662 in the year 2001. At the start of the nuclear energy program, we had but three basic standards: one for external radiation, one for radium deposition, and one for radon inhalation.

Obviously, we have had a proliferation of standards, yet equally obviously the above measures are not internally consistent. We are not always talking about the same kinds of standards. I have no definitive classification to offer. As often as not, we refer to a *basic* standard when it is important enough to work on oneself, whereas the *applied* or *derived* standard is snobbishly left as a simple exercise for the juniors. To the extent that these classifications have value in the present symposium discussions, I shall consider a radiation standard *basic* if it sets an acceptable exposure or exposure rate for a broad class of conditions. An *applied* or *derived* standard would have specific application to particular circumstances. It might refer to a required thickness of lead around a radiation device, to an exclusion radius for a nuclear power station, or to an air concentration above which protective devices are to be used.

At the risk of being dull and perhaps elementary, it may be useful to repeat 15-year-old words from the Joint Atomic Energy Committee Hearings on Radiation Protection Standards: their basis and use:[1]

INTRODUCTION

As a preface to a discussion of radiation protection standards, a review of the purpose and nature of standards in general seems in order to explain better some of the concepts affecting present radiation standards. Standards are developed to provide a model for judgment as to whether a given action should be taken. They are frequently the result of past experience utilized to facilitate future experience. They may be the communications of the enlightened to guide the less experienced. Finally, they are the integrating means by which there may be orderly development of a whole made up of many parts.

Of the many ways in which standards can originate, two are especially important in the radiation protection field. The first concerns moral and ethical standards which have been manifested since the recording of civilization. For example, the Hippocratic oath defines a recognized standard of conduct expected of a physician. Concern for the well-being of future generations with respect to the genetic effects of ionizing radiation leads to standards of this kind.

The second type of standards stems from the need to facilitate action to allow each individual in society to proceed in a certain direction, knowing that certain obstacles are removed. Men drive vehicles on the right-hand side of the road in a given community because confusion is avoided and safety enhanced if everyone does so in that community. Similarly, radiation protection standards may be guides by which an individual may proceed with reasonable safety for himself and with the assurance of not causing unreasonable interference with other segments of his environment.

The enforcement of standards is important in considering their usefulness. Ideally, they will be accepted voluntarily through confidence in the source of the standard—the authority of knowledge—or through a clear understanding of the benefits to be derived. Thus, the voluntary restriction of freedom to drive only on the right-hand side of the road is

---

Parker: HMP Associates, Inc., Richland, Washington.

observed less because of the policeman than because this action demonstrably affects one's own safety and ability to travel with the least confusion.

Where effects are not so obvious, where effects on the entire pattern rather than on a part are important, or where an individual or group may conclude that the burden of restriction is not equitably shared, enforcement by edict may be necessary. In the particular case of radiation protection, where the possible deleterious effects may be both latent and subtle, this type of enforcement may have an important role.

A final comment on standards in general concerns their dynamic nature. One cannot expect that standards established now will be immutable for all time hence. Just as knowledge increases, so can the quality and applicability of standards change. Unless standards are dynamic, they can become unnecessarily restrictive or completely misleading.

THE MERITS OF STANDARDS IN TECHNOLOGY-ORIENTED NATIONS

As technologies become more complex and specialized, the public must depend increasingly on the assurance that "somebody" has looked at the problems and at the hazards; and that "rules" have been established for the operation of plants and processes to provide necessary and sufficient safeguards to ensure responsiveness to the mores of the Nation. In fields involving nuclear reactions or other large-scale industrial processes, the "rules" can be the nationally recognized standards developed by agreement between major organizations, both public and private, which have both knowledge and interest in the field.

Technological standards fulfill these important purposes:

1. They promote safety.

2. They promote simplification by achieving agreements between suppliers and users regarding materials, equipment and procedures. This simplification should lead to broadly shared national benefits of conservation and economical performance.

3. Well-prepared standards leave the way open for further improvements through ingenuity, new technology and materials. Standardizing bodies must be flexible enough to encourage rather than choke off initiative and technical progress.

4. Particularly in a new field, a well-planned and well-timed system of standards can contribute greatly to the orderly development of the technology. It is easier to reach agreement *before* groups have gone too far along their separate ways.

THE ORIGIN OF STANDARDS

Formulation of standards may begin when a pressing need is felt, or, preferably, when it is foreseen. Simple technological examples are agreement on a few standardized automobile wheel sizes, and the interchangeability of electrical connecting plugs. The development of such carefully planned standards in a production-minded nation is inevitable. In the relatively new field of nuclear energy utilization, many standards are today being readied for use, while the industry is still in an early stage and before practices have become deeply rooted. The combined efforts of professional technical societies, industrial associations, agencies of Federal and State Governments, labor organizations, and insuring groups are being integrated in this work.

ENFORCEMENT AND USE

The simple examples mentioned provide evident benefits to all groups. Enforcement rests comfortably on voluntary agreement. Where public safety is involved, a well-tried standard of industry may become the technical foundation for laws which require compliance and which may establish inspection requirements to aid enforcement. As an example, the boiler and pressure vessel code of the American Society of Mechanical Engineers, enforced by law in many States, has been a powerful force in avoiding boiler explosions, thus promoting public safety.

Unless a standard has the force of law—and sometimes even where it does have—its acceptance and use rest largely on its merit. The actual need for a recognized guide, the extent of experience available as background, the wisdom of the authors, the clarity of expression and the flexibility to meet changing conditions all influence the worth of such standards.

At this stage, we should address ourselves to a history of radiation protection standards, which, in the interest of time, will have to be condensed into note form:

1. Early considerations were largely medically oriented and based on obvious threshold observations for the effects. The principle was simply the measurement of how much radiation had not seemed to injure people, and often dividing by 10 to be extra safe.

It was a well-intentioned voluntary control system, but mostly a coterie group talking to itself. It solved *all* the short-term hazards, *most* of the intermediate hazards, and *none* of the long-term hazards. There was no way of *telescoping time* to be aware of the long-term effects, much less learn how to minimize them.

2. In the early 1940s, genetic effects received intensive study under the atomic energy program. The basic scientific facts had been set forth by H. J. Muller in 1927. For the first time it seemed likely that there was a nonthreshold effect at the radiation levels of interest. With this in mind, there was a much-heralded change in terminology from *tolerance dose* to *permissible dose* to emphasize that there is no level of exposure which is absolutely safe. This was a semantic misdemeanor, as now seen in retrospect, because the sense of exposure being tolerable, although always undesirable, is the essence of the modern construction.

3. The period from the late 1940s to the late 1950s was characterized by:

    a. An important NAS Genetics Report, 1956, which underlined the nonthreshold and linear nature of the effect and offered some arbitrary judgment levels on acceptable exposure.

    A later modification that retracted the linear element in the light of positive evidence of repair of genetic damage has been relatively overlooked.

    b. The principle of acceptable risk was widely promulgated. Such a principle is needed to meet any nonthreshold argument of effect versus dose—not just the *linear* argument. This distinction is often overlooked.

    Acceptable risk can be looked on as a version of informed consent for occupational levels to exceed public levels.

    It was in this period that the need to maintain all exposures at the *lowest practicable level* was formulated. This, under the acronym ALAP or in variations like ALARA (as low as reasonably achievable), is a currently fashionable aspect of regulation.

   c. The growth of environmental limits appropriate for man and all other life forms occurred in this period. Population dose was used as a criterion of integrated injury to the gene pool.

4. In the more recent period, studies have concentrated on the late somatic effects of radiation, especially leukemia and other cancers. The evidence depends on the Atomic Bomb Casualty Commission results from Japan and various medical irradiations (treatment of spondylitis with X rays, irradiation of infant thymus glands, etc.). Today, there is heavy emphasis on the induction of late cancers as the dominant radiation risk.

5. Current popular wisdom leans toward a linear extrapolation of effect versus dose for the late somatic effects. Meanwhile, the simple theory for genetic effects from which this linear concept came has been replaced by more sophisticated arguments including both dose and dose-rate effects.

The linear extrapolation is considered prudent by most people, overly conservative by many, and not necessarily conservative enough by a few, who advance possibilities that there is supralinearity. No one knows the real answer at the low levels of dose applying in occupational exposure and even more so for environmental levels. This does not prevent a current wave of calculation of alleged deaths, or lives saved by reductions of dose, or alleged economic balances of benefits versus risks by assigning cash values to man-rem of exposure.

6. The main line of the arguments can be found in the following four references:

Ionizing Radiation: Levels and Effects, 2 Vols., A Report of the U. N. Scientific Committee on the Effects of Atomic Radiation, United Nations, New York, 1972.

The Effects on Populations of Exposure to Low Levels of Ionizing Radiation, Report of the Advisory Committee on the Biological Effects of Ionizing Radiations, National Academy of Sciences, National Research Council, Washington, D. C., 1972 (the BEIR Report).

NCRP Report No. 39, Basic Radiation Protection Criteria, National Commission on Radiation Protection, Washington, D. C., 1971.

NCRP Report No. 43, Review of the Current State of Radiation Protection Philosophy, National Commission on Radiation Protection, Washington, D. C., 1975.

7. Some numerical consequences from the BEIR Report are:

U. S. cancer deaths per year	
Natural radiation	3500
Unwanted medical radiation	2500
Fallout	140
Nuclear power (1970)	0.1
Nuclear power (projected to year 2000)	<35

(End of brief history)

Compared with many deleterious agents or competing energy sources, ionizing radiation offers a particularly difficult case, because the hazards are presented in three forms:

 1. Electromagnetic radiation—X rays and gamma rays.
 2. Ionizing particles—alpha particles, beta rays, and many others.
 3. Radioactive depositions from inhalation or ingestion or transmission through intact skin or wounds. All this more nearly parallels the issues with chemically toxic agents.

Unifying theories have been attempted, especially for dosimetry units such as the rad and rem. Some 30 or 40 years ago, biophysicists believed they could solve the basic problems (by Hit Theory, etc.) with which their biologist colleagues had struggled so ineffectively. Biology seems to have had the last laugh by refusing to cooperate with the simplifying approaches of the physicists. Nature denies a comprehensive theory of the case, and experimentation at the low exposure levels sinks into a quagmire of marginal statistics.

We have referred to the impossibility of telescoping time in human experimentation. In the United States the Atomic Energy Commission attempted to solve this by a major program of animal experimentation. Such countries as Great Britain, unable to afford this, concentrated on study of scattered and often indirect medical experience. There is a strong recent tendency (e.g., BEIR Report) to use human data, and minimize inferences from animal data, when the relevant issue is believed to be late cancer.

Nature has provided obstacles understanding of these effects by making them identical with natural disease, which complicates recognition above a statistical background, and by making cancer a complicated disease which may need two or more interacting factors to cause its expression. Any other carcinogenic process is likely to present the same problems of hazard analysis.

The cost of all the radiation hazard studies is obscured by the combination of arbitrary assignment of study costs to various alternative purposes and of administrative reticence. As a conservative estimate, direct investigations amount to at least $100 million per year, and applications at least an order of magnitude greater. The integrated expenditure must be of the order of 20 billion dollars.* What have we learned from it? Certainly much needed radiobiological knowledge, but equally certainly no resolution of the risk assessment to which all responsible observers will subscribe, nor is there an accepted mechanism to discount what seem to be irresponsible claims in some cases. An upper limit to risk is set by the health experience of those who have worked with radiation since the 1940s. To date, that experience is not clearly influenced at all by radiation effects, but one can claim that the

---

*Erratum: This number was erroneously read as 2 billion in the meeting, evoking a comment that it was not very much.

latency of cancer would still be obscuring effects. Another decade or so should clarify that issue.

Meanwhile, we can say that there are 365,000 cancer deaths per year in the United States. A *WHO* publication asserts that 75% or more of these have some environmental causation factor. That is about a quarter million deaths versus the one-tenth death (1970) assigned to radiation from nuclear power. At the assumed 10^8 per year direct research, an equal study of all other environmental pollution agents on a "lives lost" basis puts the annual cost in the 10^{14} range. Without accepting such numbers literally, it is clear that radiation protection costs are out of proportion to the balanced national need to control agents dangerous to man or his environment.

Does the radiation experience offer any guidelines for attacking the broad parallel problems from alternative energy sources? Perhaps a few, but clearly no easy answers.

First, the case could be solved and safely extrapolated to low exposures if a generalized theory of the deleterious actions of the agent could be developed. This must be attempted for any new agent. Then a broader umbrella should be developed for classes of agents, if it exists technically. Short of these unifications, the problems of each new agent might be looked at as follows:

1. Make an early attempt to determine whether the relevant risks have a threshold or not. In doing this, develop measurement systems that are unambiguous and acceptable to all. If animal data are going to be discarded because the only relevant subject is man himself, review early whether animal research is needed at all.

2. If comprehensive theory for the particular agent is compatible with a hazard threshold or even a practical threshold for low exposure (due to latency period inversely related to exposure), there is an acceptable technological standard to be found.

3. If not, the eventual standard involves value judgments. The efforts of the scientific fraternity will *never* be sufficient in themselves. The key element is to identify the plausible effects as early as possible and to promote informed public discussions as early as possible.

A clear fault in the radiation case has been paternalism. The "Father knows best" or "Big Brother will help you" attitude was partly necessary for radiation because of the legitimate national security aspects of atomic energy. It was continued far too long by the AEC and did much harm to credibility. That much of it was either accidental or well-intentioned seems probable. I have served on the NCRP, and my name is connected in part with both NCRP Report Nos. 39 and 43. I know, for sure, that No. 43 was not in any way intended as a restatement of how right one was in No. 39. Yet in rereading it, I could not blame a critic for drawing such a conclusion. Both AEC (now ERDA and NRC) and NCRP must take more lessons yet from Caesar's wife.

Now, with a proliferation of controlling federal agencies, and some signs of the Big Brothers squabbling among themselves, we play increasingly into the hands of a few dissenters, who get more public attention than all the fine efforts of the agencies. The stakes for future energy sources are too high to allow that kind of anomaly to infect the studies of the new standards problems.

REFERENCE

1. Hearings Before the Joint Committee on Atomic Energy, *Selected Materials on Radiation Protection Criteria and Standards, Their Basis and Use,* pp. 1-3, Superintendent of Documents, U. S. Government Printing Office, Washington, May 1960.

DISCUSSION BY ATTENDEES

Richmond: I hope that all the big brothers are listening. We have time for a few questions, and while we are finding the first person to volunteer, I would like to make a comment. The $2 billion doesn't really bother me because I didn't pay it, but it will yield much information on mechanisms that can be applied to areas other than radiation. Seriously, you mention 1975 dollars, and you can compare the number with something in real life right off, and that is the size of our national program to pay for pneumoconiosis which is a billion dollars a year. So 2 years' output to pay for black-lung disease equals that entire 2 × 10^9 dollar outlay which goes back over some 30 years.

Parker: Then, if people were listening to what we are saying, I think that might be all right, but I am suggesting that they are listening to people who are saying our work is useless, and that is rather disastrous.

McClellan: You noted the extent to which the experimental animal data with regard to somatic effects has essentially been ignored in favor of heavy dependence on human data. It has been suggested by some individuals that this relates in part to some inadequacies in the experimental animal data. Do you feel that there are critical experimental animal studies that have not been done, that could be performed, that would provide the necessary cross-linking between experimental animals and man, and that would lead well-intentioned scientists who were involved in making these decisions to accept the experimental animal data?

Parker: That is a very tough question, Roger, and you know my field is not animal experimentation but physics; the observation of animal experimentation is done by you and others. Specifically in answer to your question, of course, there are things that could have been done with animals. A lovely program was proposed that would cost more money than we talked about in these other outrageous sums to remove animals from any source of radiation and see what happens there, and it is perfectly possible technically to maintain animals in an atmosphere where they get a hundredth of the present natural radiation. It is not cheap to do that because you not only put them in cages that cut out the cosmic rays but you also feed them food that has been relieved of the natural radioactive elements, ^{40}K and so on. It would make a vital experiment since it would give you one point that would not be an extrapolation point but would give you a crack at getting some incidence data at a tenth of the present natural limit; experiments like that have not been done, and I am not saying they should be done for the following reasons. The difficulty of recognizing the effect at those very low levels is known to all, and I think the statisticians should have the last word on that kind of thing. Would the experiment yield meaningful data? As to the general thrust of your question, no, I think good animal data exist that, in my opinion, should have had better weight in the collective judgment reflected in the BEIR Report, but, Roger, that is a personal opinion and others may have contrary views. But it is rather surprising, if you take time to read these four documents, to note the number of very distinguished scientists of this country who seem to be writing on both sides of that question. Some of the distinguished contributors to the report seem to say that it was deficient when they write a review. It is very unusual and, I think, relevant to our studies. Is that a reasonable answer to what you were asking, Roger?

Nellor: If we accept the UN figures of a quarter of a million for nonradiation-related and environmental cancer deaths, are we going to have to exploit an annual death rate to impress people that monies should be allocated to these nonradiation areas? What do we cite as salable for greater monies? Is it a yearly figure, and is there some other justification?

Parker: I hope we have a better justification. What we are trying to do is to protect all people in the United States from any unnecessary radiation, while continuing to use radiation for beneficial purposes, and I would throw out any counting of cancer deaths because I don't think a single one of those numbers is realistic. Let me say again, that is a very personal viewpoint, but I think there is an ethical posture, and I don't sit up here as a prig as the only one who wants to be ethical. I think we all feel this way, and the issue is how in the world do you do it since, at some stage, there is a conflict where the money runs out. You can't go ahead putting more money into radiation protection, and I would start by saying we have put too much money in relative to other hazards. I am not going to play any more until you have looked at some of these other things that I think are much more disadvantageous to our population. I would answer it that way.

Schneiderman: My boss said earlier this month that approximately 20% of our Cancer Institute budget, $600 million, is going into problems of environmental cancer, so that does give you $120 million a year.

Parker: It is already higher than I said. Yes, I would agree with that. My original notes didn't say $100 million, but I am in rather a predicament; if Dr. Liverman is not willing to tell me how much goes in, I am very hard pressed to get the total information.

Schneiderman: What I meant was that more money is going into total environmental cancer research than is going into radiation protection.

Parker: Yes, but not the amount that would be the result of slavish adherence to the relative number of deaths in the various experiences. Would I recommend it? No, because I don't understand where that quarter of a million environmentally related deaths comes from. It is not my field.

Plutonium — Health Implications for Man

Official Journal of the Health Physics Society

October 1975

Reprinted from *Health Physics* Vol. 29, pp. 672–632 by permission of the Health Physics Society.

PLUTONIUM—HEALTH IMPLICATIONS FOR MAN: CLOSING REMARKS AND FINAL DISCUSSION

H. M. PARKER
H.M.P. Associates, Richland, Washington

(*Received* 10 *February* 1975)

FOR THE benefit of Gilbert and Sullivan fans, I intend to play the role of the Duke of Plaza-Toro, who led his regiment from behind, and give my concluding remarks first and then move into the discussion. In contrast to the well known TV program, *Issues and Answers*, sessions such as this at the end of symposia usually are more properly described as "*Issues and No Answers.*" There is no reason to believe that we can do better, but I do hope that we at least keep the possibility of generating some answers in mind.

For purposes of this closing session, I have broken my remarks into seven topics which I will introduce as we come to them; when I am through, we will select some of the topics for further discussion and discard others because we will not have time to cover them all.

(1) INTEGRATED NUCLEAR FACILITIES

This topic arose in Commissioner Larson's opening of the Symposium and it seems to be a proposal to avoid transport. If the proposal is made solely for the purpose of the potential diversion problem, I believe that it is not within the scope of this symposium. However, if it is dictated by the health problems of transporting plutonium it is very much within our scope. This position is augmented by Dr. Pigford's paper in which he emphasized the differences in quantities and types of plutonium which could arise from different systems. Interestingly, he left out the one system that avoids all of these, the Canadian Candu system. This should be placed into perspective with the other items in that excellent paper.

(2) PLUTONIUM—THE MOST TOXIC ELEMENT KNOWN TO MAN

My questions here are—is it?—and where is the evidence for that statement? We started, in the early days of the Plutonium project, offering occupational protection to a group of people who could not possibly take the time to learn even the meager facts available at that time. As a part of this, we drew an analogy with the preproject disastrous experience with radium. At this time we started preaching that plutonium is *one* of the most toxic elements known to man. Somewhere between then and now the word "one" has disappeared. Despite Roy Thompson's well-expressed thought about the psychological effect of this teaching, I would do the same thing under the same circumstances. But, with 30 yr experience how do we stand today? I believe that this is a question we should examine.

(3) RELATIVE WEIGHT OF HUMAN DATA VS ANIMAL DATA

Generally in this country we are now tied to the somatic aspects of radiation effects as presented in the BEIR report. (We are eliminating the genetic aspects in this particular part of the discussion.) That report omits much of the information from radiobiology and tends to distort the meager human data to fit a predetermined intention to use the linear hypothesis. As a result, we seem to be all strongly accepting this postulate. Should we not, then, be following the same approach with plutonium? Specifically, should not our conclusions for this element be almost solely based upon human data? Wouldn't this approach say, again for the somatic case, that plutonium at occupational levels, and even more so for environmental levels, deserves respect but is a relatively comfortable thing compared with many other accepted toxic agents in the scheme of things in this country?

(4) WASTE MANAGEMENT

We heard Harvey Soule discuss the elaborate studies for control of high-level wastes to provide interim retention for about a century and ultimate disposal with proper precautions adequate for something like a million-year period. However, nothing was said about the probably greater hazard of the release of copious volumes of low-level liquid wastes which seem to present the highest potential for population exposures. If we established a sweepstakes for the most hazardous element, despite the emphasis of this symposium, it is rather interesting to note that there is no way that plutonium can appear in the win, place, or show position. There are many other issues that must be considered in that sweepstakes.

I believe that management of the high-level wastes we have heard about is a case for which the lowest practicable level is more applicable than, say, the program given by Dr. Gamertsfelder. The cost values that arise from the studies are incredible. The most complex of all the systems studied by the AEC to date make a relatively trivial increase in the cost of nuclear power—an increase that I believe any reasonable person would accept. There is no question of economic balancing. Man just has to do the very best he can with these high-level wastes.

(5) INDIFFERENCE LEVEL

By indifference level I mean the throwaway level or the level at which wastes can be disposed of without concern. H. Soule has made an attempt to provide this that I think most people believe is wrong. However to my knowledge, no one has done any better so that it is not all that wrong. The problem is that we must arrive at an indifference value which will be good for the entire lifetime of the nuclear fission program which is currently estimated at several hundred years. We need a level that reasonable people would accept with that as a premise. I feel very strongly that the nuclear business needs, and deserves, a firm figure from us to avoid what would otherwise be a major retrofit problem.

(6) ISOTOPIC SEPARATION

This is, perhaps, a diversionary item. I have heard many people talk about the biological isotopic separation of ^{238}Pu and ^{239}Pu. This is a gross misuse of the language. We have heard several examples which could be misinterpreted. Pat Durbin gave us one where the retention following intramuscular injection was up to 80% for ^{238}Pu and only a few per cent for ^{239}Pu. This, in my language, is a straightforward difference in chemical and mass characteristics. The other example shows clearly in the Hanford experiments on inhalation of plutonium oxide. Here the difference is apparently due to the intense radiation from ^{238}Pu which changes the very local chemistry and grossly alters properties such as solubility. This leads to very different practical results between the ^{238}Pu and ^{239}Pu. We must express clearly to others that neither mode is properly called isotopic separation.

(7) MAXIMUM PERMISSIBLE BODY BURDEN FOR PLUTONIUM

There is one topic that I will not refer to here and that is the hot particle problem. I say that in the sense that the opposition side to this argument is not represented and we do not want to appear to be arriving at a consensus without their views represented.

I would like to refer to Roy Thompson's data on tumor incidence in various animal species at various levels of alleged rads. I use this term advisedly since this dosimetry can be very misleading. First we should convert those two selections of Roy's into some form which has a body burden scale or something close to a dose in the toxicological sense. If those redrawn data, by any chance, showed tumor formation at, or near, the presently allowed burdens, surely we should lower the limits and do it now—not put it off for years. In the early days one could call for an experiment and within a few months get an answer that pragmatists, which is a proper term for health physicists and radiation protectionists, could use to improve an existing limit or create a new one. Of course, cancer induction with the latency period is now a substantive issue in setting the limit and we can't study that in 2 months. However, can't we get together and forecast when enough evidence will have accrued in humans to confirm or deny the animal data at some agreed level of probability? Or can we look again at latency, which now appears to be mentioned only casually, as a function of dose and deduce some formulation to see if there is a level of dose below which there is what has been called a "practical threshold"—i.e. a level of dose where the expectation is that the person dies from some other cause? At least, let us develop an orderly plan to achieve what is needed, publish that plan and then go after it in the sense of a dedicated project, a sense that sometimes seems to have been lacking in the otherwise very fine endeavors of radiobiologists and many other contributors to our joint field.

That concludes my closing remarks. We have had two specific questions from members of the symposium and I believe that we should answer these before our discussion. The first, from Dr. Stannard, was: Indication of the possible role of gonads was based on Los Alamos data on man which was shown to be in error. What do the animal data show? It would not take much to be of interest because of the long time scale for effects to show.

RICHMOND, C. R.: I can address the question but there are undoubtedly people in the audience who will want to add to this. There were some data reported on human gonads as part of the Los Alamos tissue program. They were found to be in error as a result of a procedural change in using ^{242}Pu as the tracer instead of ^{236}Pu. Harry Schulte may wish to elaborate on this. Many people are now interested in obtaining the best values for the concentration of plutonium in gonadal tissue. There is a body of information on species ranging from rodents through dogs and rabbits and, I believe, pigs. A number of people have been exchanging information and we have been looking at this question at Los Alamos. It appears that the fraction of the amount injected that gets into gonadal tissue is about 10^{-4}. There is a spread in the data but it does give a general feeling that the concentration developed in soft tissue is low. The real question, of course, is what we can now infer about the effects of high LET radiation from the data from Hiroshima and Nagasaki or other sources.

SCHULTE, H. F.: The problem with the tissue analyses arose from a small amount of ^{239}Pu in the ^{242}Pu tracer. The quantity was insignificant for other tissues but at the low levels in the gonads, it did make a difference. There is still a problem here even with absolutely pure tracers because the low quantities present in the gonads results in a large uncertainty in the final results. I think that the only way that we are going to get good human data is to pool gonadal tissue from a number of individuals.

GRAHN, D.: I would like to comment that I don't think that you are going to get any information with regard to the effects of radiation on germinal tissue from the Hiroshima-Nagasaki data. The sample sizes are not adequate and, even if they were, there is not a good estimate of the RBE for the neutron component. Data I have seen on somatic genetic effects from lymphocyte cultures from persons exposed at graded doses show a very distinct difference between the Hiroshima and Nagasaki exposures for the high LET component although they are coming out somewhere in the expected range. There is a little bit of human data but it seems consistent with what we are thinking.

MAYS, C. W.: I believe that it would be well to get this gonadal question settled as well as we can as soon as possible. A number of us have data and I believe that it would be useful to pool all of these data and have some individual, such as Chet Richmond, put this together and get it into the literature. I think that a high priority should be given to assembling this material and getting the results in a form where people can look at them.

PARKER, H. M.: We have only to inquire whether Chet is willing to accept the responsibility on behalf of the profession.

RICHMOND, C. R.: I think that this needs to be done but it does require more discussion. I would not like to get involved in an operation requiring many teletypes and facsimiles. I believe that it could be done by a few people getting together for a short period of time.

PARKER, H. M.: Perhaps we can use this as an assembly of knowledgeable people with Chet appointed unofficially as an obvious leader. Let me move to the second question from BRYCE RICH: Larger particles are less likely to be retained in the lower respiratory tract and are also, from the data presented, less carcinogenic. The smaller particles are not as well retained and, because of their low activity, are also less likely to produce cancer. My question was whether that leaves us with the conclusion that the respirable, retained particles are the most likely to produce cancers?

RICHMOND, C. R.: You are correct. The larger particles have a smaller probability of being retained. I don't think, however, that we should confuse another issue as to whether a large particle is more carcinogenic than a smaller one. The question we discussed yesterday was whether the non-uniform distribution was more or less harmful than the situation where you have a high degree of uniformity. It is difficult to interpret the problem in terms of whether a little particle or a big particle is carcinogenic. The issue is whether, for a given total activity, it is sparing biologically to have it aggregated into fewer but individually larger particles. There are many parameters involved and you can induce lung cancer with particles if you have enough of them and a high enough dose.

PARKER, H. M.: Since our time is getting short and we can't discuss all seven topics which I outlined, let's arbitrarily see if anyone has a comment on how to arrive at a rational indifference level.

MAYS, C. W.: The thing that I object to is the concept of an indifference level because it is something that cannot be proven or disproven. We have all carefully avoided the real question which is the degree of risk we are willing to consider permissible. I have estimated the risk at the permissible level for radium and I come with a figure in the domain of 1%. If we are talking about adjusting the permissible level for plutonium, we must decide what we consider permissible in terms of hazard. An important comment at this conference was that it is hard to make good decisions on data that we know are in error. Harry Schulte presented results that indicate that the Langham equations overestimate the body burden. Perhaps the time has come when this equation should be revised so that the results correspond to reality. It can be argued that it is conservative to overpredict the body burden. However, if you use these results to conclude that plutonium is not very effective in causing damage to people, you are not being conservative. On another point, apparently the difference between ^{238}Pu and ^{239}Pu lies in the amount of self-irradiation of the particle and the high specific activity of ^{238}Pu causes it to solubilize more quickly. Perhaps there is simply a time factor change of, say, 300-fold and, with the ^{239}Pu that is being released into the environment, there could be the same type of change in ^{239}Pu. This could result in different dynamics in the future from the aging of the plutonium now in the biosphere.

PARKER, H. M.: Thank you, Chuck. I believe you've attempted to answer all except question number one instead of number five, which is still with us. I agree with what you say about particles but I don't believe that all of the facts are known. The position you take is similar to that of Dr. Evans from the parallel data from earlier radium and mesothorium experience. Is that fair, Bob?

EVANS, R. D.: I would want to talk with Chuck about his risk estimate at 0.1 μCi radium burden; I assume that he means that there are not enough human cases at exactly that level to disprove a 1% risk. This is an important point and not everyone would agree with your 1% estimate.

PARKER, H. M.: Thank you, Bob. In your point, Chuck, you came to what we are all struggling with—a reasonable man's interpretation of as low as practicable and the level cannot be zero. We cannot wait forever to do something about it. If we decide that the proper thing to do is to expose no one to the risk of injury from plutonium, then we have a problem. We should have been out of the business before the business started, which is a bit difficult. From now on if we really must *not* let any plutonium into the public domain, we have a very real problem. Some of us couldn't go home at night which is awkward, to say the least. I suppose that a part of the problem could be relieved by the standard Mexican prison arrangement, but I do not seriously envision the United States going to that extreme. The issue is that small amounts could get into larger volumes of water and other materials, when projected to the year 2000, with an anticipated multiplication of nuclear power by a factor of approximately 100, it becomes a horrendous problem if we go to something like one-tenth of the level proposed earlier by Harvey Soule. His level is one that is reasonably capable of measurement with present equipment. If no one else will propose a better way for determining an indifference level, it would have to be a level below which we couldn't reasonably find out whether or not you were meeting it. This is as logical as some of the ways of controlling releases. However, we do owe the business a number and rather soon, say within a few years.

Who would like to try the question of how we should weigh the balance between what we may achieve on animal data relative to plutonium and human data—Roy?

THOMPSON, R. C.: I think that I am repeating what I said before but I will continue to insist that there are no human data on plutonium effects. We have done too good a job on controlling this "most toxic element known to man" for the past 30 yr and we are not likely to see any human effects. We can play the BEIR game by applying numbers that have no experimentally demonstrated relevance to plutonium toxicity in humans by applying those numbers to humans and saying that they must bear some relationship to plutonium exposure. However, the only data we have is in animals and we should use those data for all they are worth. I think that they are worth considerably more than the indirectly derived human data.

HEALY, J. W.: I would like to comment on what Roy said about the lack of human data. I believe that he really means that no effects have been noted. To the contrary, there is a great deal of data which show no effects. Sometimes an experiment which shows zero is more valuable than one which is designed with high levels and *does* show effects. This is one of the questions that many people have on the program Chet described with the particles since very few tumors have resulted. I feel that it will be a great success if it shows that tumors are *not* induced.

PARKER, H. M.: One of the present problems is that we have reason to expect, from other data, that tumors can, indeed, be induced in the exposed human. The only thing we can do is to observe all of the people who have been exposed and make sure that none escape a first-class examination. It is unfortunate that the funding for those programs doing their best to ultimately arrive at such data, including the Transuranium Registry, is so small. Dr. Norwood, who heads this project is here and, perhaps, he would give us his views on what we might expect to learn and what we can do to get more prominence on this only way of acquiring any human data and answering that part of the problem.

NORWOOD, W. D.: I think that a great deal can be learned and that we are learning a great deal. Roy and I agree that, hopefully, we will find that we are not damaging people at the levels in those we are examining. I certainly agree that this is good information to see. However, if we have a case with several times the permissible burden and he comes down with a bone sarcoma, I believe that we would say that the probability is greater than 50% that there is a causal relationship. We are learning a great deal on many of the autopsies. In the usual autopsy we are taking about ten primary organs or samples and in most of them we are getting whole organs. In some cases, where the deposition is expected to be about one-half a body burden or greater, we try to get 20 or 30 different organ samples. One of our routine samples is the gonads and we are finding low deposition in those organs. With some of the newer techniques we will try to find out whether the plutonium is located in the connective tissue or in more important areas. Since I indicated earlier that we were anxious to get people who permit the use of a whole body so that we can find out what is in each bone, we have had a volunteer from the audience. I think if we let it be known what can be done, we can improve our programs.

PARKER, H. M.: I would like to see someone venture some positive numbers on forecasts. If we take the parallel Mancusso study, which involves the health of radiation workers, sometime in the future we will be able to look at the data and say that the incidence of cancer from a given dose of radiation cannot exceed a specific number plus or minus a fairly broad spread. This should be documented ahead of time, say by decades, so that we will know when we can arrive at a fairly firm position. What I am talking about with plutonium is the same thing, but realizing that this is probably more difficult for two reasons. The first is that there will inevitably be fewer cases than for general radiation workers since the first class also contains the second class and there have to be fewer exposed to plutonium than to all radiation. The second is that the chances are that the incidence may in the long run, be very low. At least let's document these numbers and let the world know when we think we will have them. I consider it very important in the design of experiments to define first what the feasible objective is to be, conditioned always by the actual incidence rate. If the incidence rate is zero, we will go on a long time, but this will be fine if no deaths can be ascribed to this alleged toxic agent.

BISTLINE, R. W.: We have been discussing biological effects but the only biological effect that has been pointed out is cancer. Is this the only effect that we are going to look for in human populations?

PARKER, H. M.: I suppose not. We have talked about the genetic possibility and I think that the chance of observing this in humans is essentially zero. The emphasis on cancer is the same as that developed, for example, from the BEIR report where this is essentially taken as the major probable risk. With respect to plutonium in man this is only an assumption since, to my knowledge, there have not, as yet, been any cancers in man attributed to this material. Certainly it is proper to look for other effects and a total program would, indeed, do that.

NORWOOD, W. D.: If we have a few cases of a rare disease, such as cancer of the bone, I think that we can answer that question. If the clustering effect that Voelz talked about yesterday occurs we should be able to pick that up but nobody can say when that will occur. The other thing we are looking for is whether we have more of a certain type of cancer or more life-shortening in our present sample of about 6000. We hope to get 15,000 or more eventually. With that size sample, about a 10% difference at the 90% level in the number of deaths is the best we can say.

PARKER, H. M.: We have only a few minutes left in our *Issues and No Answers* program. Does anyone wish to make a comment in that time?

BONNELL, J. A.: There has been considerable discussion in England over the past few months on this question of risk and an attempt has been made to put it into perspective by comparing risks of cancer from radiation or plutonium with risks of everyday life. These are the risks accepted by the community without question such as the risk from widespread prescription of the contraceptive pill. There is a risk assumed by the community, for example, a certain level of blood alcohol which increases the hazard in driving. One of the figures that seems to come out of the latest discussion at a meeting of the Medical Research Council was that all of these activities had a risk of less than about one in 10^5. This risk was perfectly acceptable and nobody seemed to bother about it. But when you got somewhere below that, say about one in 10^4, then there was a certain amount of public concern. I mention this for the purpose of putting a number on what might be regarded as acceptable risk.

PARKER, H. M.: That is a good start. We do have some authors in this country who have written extensively on these risks—Dr. Chauncey Starr and many others. They arrived at, more or less, the same numbers because they used the same data. There could be a course of action on the indifference level, for example, in which the hazards are made comparable with the so-called Acts of God which I believe is on the order of one in 10^6. That might be a level that most reasonable people would accept. However, we do have some people in this country who are not willing to accept any number near that level. This is indeed a problem. It is, however, a problem that we, as scientists, can never solve nor should we. It is a societal problem and all that we can do is to place our concepts of the real risk before the public, honestly and forthrightly, and hope that there is some form of public examination so that we have agreement on these risks sometime before we get to the nuclear economy of the year 2000.

I will use this as a closing comment since we have used up our time and will take this opportunity to express, on behalf of all of you, our thanks to the organizing group for a well-conceived symposium and most cordial hospitality.

Statement on Proposed Rule-Making on Environmental Radiation Standards for the Uranium Fuel Cycle

Public Hearing Testimony

March 8–9, 1976

Page 1.

Statement on Proposed Rule-Making on Environmental Radiation Standards for the Uranium Fuel Cycle

Intended for presentation at an EPA-sponsored public hearing, March 8-9, 1976

by H. M. Parker

1. Credentials

   My name is Herbert M. Parker. I have been connected with some radiation protection activities of the atomic energy program since 1942, serving the interests of five contractors variously at the University of Chicago, Clinton Laboratories, Oak Ridge, and the Hanford Operations, Richland, until my retirement in 1971. I am currently a consultant to Battelle Memorial Institute on its ERDA and other programs at Richland. I have been a member of the National Council on Radiation Protection and Measurement since 1946.

   My statement is made at the invitation of ERDA, but is wholly my own, in no way reflecting either ERDA influence or a position necessarily subscribed to by ERDA. I acknowledge; however, that I have had ERDA approval to request a limited amount of calculation checking by my former Battelle associates.

2. General Comment on the EPA Draft Environmental Statement for the Proposed Rule-Making Action, May 1975

   I was one of the originators of the principle of "lowest practicable level" of exposure, which can now be abbreviated either as ALAP or ALARA. I continue to be a strong advocate of the principle. I regard the subject DES as a reasonable first attempt to establish an ALAP approach to standards across the entire fuel cycle. Therefore, I support the broad intent of the proposed rule-making. Few, if any, of the specifics can be accepted without extensive exchange of views between informed agencies and individuals. A national effort to define

Page 2

the best collective judgment on these standards is needed.

When one considers the time consumed in crystallizing an acceptable position on ALAP for light water reactor operations alone, it seems likely that the appropriate interactions may take perhaps as much as five (5) years for equally intensive ALAP study across the entire fuel cycle. I consider it axiomatic that implementation dates for protective measures (e.g. substantial removal of a particular long-lived effluent) would only apply when the national posture has been agreed upon. The projected exponential expansion of nuclear power means that the main thrust of postulated health effects increases rapidly toward the end of the century so that a five-year delay in improving protection now is about equivalent to a four-month failure to respond correctly in 1995.

The particular version of dose commitment selected by EPA is one that heavily emphasizes the potential risks of a few long lived radionuclides. This is a positive contribution as the early warning allows time for the development and commercial proving and costing of removal methods, all of which usually takes far more time, effort and money than is foreseen by rule-making organizations.

3. Specific Comments on the Draft Environmental Statement and Supporting Documents

A balanced or comprehensive review of the material is not attempted. Instead, a selection of half a dozen points of personal interest is presented. The selection has in mind the purported saving of approximately 1000 health effects by the EPA approach; this results from superior retention of classes of controlled long-lived materials as clearly set forth in Table 10, page 82 of the DES. It leaves the tritium and ^{14}C problems for later resolution.

Page 3

3.1 **National Planning for updating best evidence on health effects of enviornmental radiation levels**

Positions concerning the prudence of linear non-threshold extrapolation of health effects to low doses and dose rates on the one hand, and the danger of gross distortions of cost-effectiveness, if the real hazards are substantially lower on the other hand, have been repeated ad nauseam. There may be some accommodation between the protagonists of these positions, but the evidence for it seems to this observer to be less visible than the fervor of each side to emphasize that it was right the first time. The issue is too important to have the final answer rest with any one agency.

It is my opinion that we need a national forum to perform the following functions:

(1) obtain a consensus on the types of evidence of health effects acceptable for standards making for the various types of potential exposure.

(2) provide orderly prediction of when human data can be expected to yield positive results that can be tested against health effects postulates.

(3) perform a formal review of these data at regular intervals (e.g. every five years) and make consensus publication of the best evidence.

As a minimum, these matters should have broad national attention, preferably supplemented by learned international opinion either as in UNSCEAR reports or through specific international symposia.

Under (1), one has in mind such issues as balance of animal data versus human data, sub-classifications of

Page 4

radiation types, low dose and low dose rate factors, agreed biological end-points, uniform use of a recognized risk calculation protocol, and so on.

Under (2), Appendix A presents an order-of-magnitude version of what could be done with health and mortality studies of populations of atomic workers to validate or modify BEIR Report type predictions. Most of the definable populations offer problems. The atomic worker class has a fairly low radiation exposure, and omits all in-utero and juvenile exposures that rank high in total health effects.

The prediction method should be helpful in a reasonable time for populations exposed mainly to measurable amounts of penetrating radiation, and using late somatic injuries as the index. For direct human genetic effects, and for internal exposure to transuranic elements where the first human lethal health effect is yet to be recognized, the study may be unproductive. Yet it should be attempted. At least, it might help to throw earlier light on claims that health effects are 10^4, 10^5 or 10^6 times as frequent as predicted by "establishment" methods.

3.2 Current Alternatives to the BEIR Report

The Reactor Safety Study (Rasmussen Report) WASH-1400, Appendix VI, October 1975, presents a study of health effects by a distinguished group comparable in stature with the BEIR committee, and in fact, including five (5) common members. For low LET radiation, this study proposes a curvilinear non-threshold dose-effect relationship, as widely found in animal studies. For health effect impact studies, this is approximated by a dose-effectiveness factor as presented in Table VI 9-7 taken from page 9-36 of WASH-1400 Appendix VI.

Page 5

TABLE VI 9-7 DOSE-EFFECTIVNESS FACTORS

Total Dose (rem)	Dose Rate (rem per day)		
	<1	1-10	>10
<10	0.2	0.2	0.2
10-25	0.2	0.4	0.4
25-300	0.2	0.4	1.0

Most of the human exposures considered in the BEIR Report occurred at high dose and high dose rate. All of the postulated environmental exposures will be at low dose and dose rate, with a dose-effectiveness factor of 0.2. Acceptance of this approach would reduce the predicted health effects (somatic component) by a factor close to 5.

Also, the subject report uses only absolute risk estimates, as recommended by UNSCEAR analyses, and extends the elevated incidence plateau for 30 years. The BEIR Report uses some form of averaging over absolute risk and relative risk estimates, and over a 30-year plateau and a lifetime plateau. The BEIR estimates are roughly twice those of the absolute risk-30 year plateau method.

Collectively, then, this responsible report, for equal doses, predicts health effects about one-tenth of those postulated in the BEIR Report for low LET radiation and somatic effects.

These observations do not say that the Rasmussen Study data are *better* than the BEIR Report data (although they do have benefit of hindsight of the BEIR Report). They do say that reasonable groups can present health effects data differing by at least an order of magnitude. This confirms my plea for a national program. To go on to cost-effectiveness

Page 6

arguments when such order of magnitude discrepancies exist is not persuasive, and is an unreasonable burden on the nuclear industry.

3.3 Retention of ^{85}Kr

NCRP Report #44 July 1975 presents projections of ^{85}Kr releases which could result in a widespread skin dose of 3 mrem/yr by the year 2000, and a whole body dose of 0.02 mrem/yr. Our attempts to reconstruct the EPA handling of these releases suggest that there is not as much as a factor of 2 difference in the dose predictions. The thrust of the NCRP study is that skin would be the critical organ with respect to skin cancer. The implication of the NCRP report is that one may wish to control the releases vis-a-vis skin effects before the year 2000, but not on a crash basis. Through the statement that the dose to gonads and whole body are orders of magnitude lower, it is implied that their health effects are insignificant.

By contrast, the EPA accepts a threshold effect for skin cancer, and computes health effects for other cancers and genetic changes.

I consider that ^{85}Kr release is an international problem. The appropriate U.S. agency or agencies should initiate moves toward international agreement now. If U.S. industry successfully removed all its ^{85}Kr with no reciprocity, the U.S. dose might fall by only some 20-25%, a cost-ineffectual step.

Conversely, if one decides not to further the international approach, it is questionable whether the world population should be included at all in the DES projections.

3.4 Radioiodines

EPA has been a leader in drawing attention to the long-term problem of ^{129}I release. My former colleague, J. K. Soldat, has recently reviewed the environmental dose relationships.[1] To the extent that the EPA model can be reconstructed, it appears that EPA dose estimates are compatible with this study within a factor of 3 or so. Soldat emphasizes the peculiar function of dose versus age--peculiar in the sense that it differs sharply and not monotonically with age from the familiar ^{131}I pattern. Refinement of the EPA data when ^{129}I becomes a major issue should include an age-distribution factor.

For ^{131}I, WASH-1400 notes that ^{131}I irradiation of the human thyroid is only about one-fiftieth as effective per rad as is X-ray irradiation. Values more in the range of one-tenth to one-twentieth seem to be observed in animals. WASH-1400 therefore applies a conservative quality factor of 0.1 for ^{131}I irradiation.

Presumably, a corresponding factor needs to be developed for ^{129}I. Should it prove to be comparable with that for ^{131}I in humans, ^{129}I would virtually be removed from consideration for the 20th century. Again, the need to achieve a national consensus on such matters is paramount.

3.5 Systematic presentation of source term, model dispersion and health effects data

At the present time, it is extremely difficult to follow the EPA calculational steps from source term through the dosimetry modeling sequences to health effects calculation. Parts of the work appeared in different documents, and

---

[1] J. K. Soldat, "Radiation Doses from Iodine-129 in the Environment," Health Physics, 30, 61-70. Jan 1976.

Page 8

there is some evidence that different modelers have used different methods. We note this not to be critical because the labor involved is well known to us. Rather as a constructive suggestion, we urge that the next round of publication should present these data systematically and uniformly so that the processes can be reviewed carefully by qualified people. Mr. J. K. Soldat prepared a few pages of notes for my use, which show in detail the kind of discrepancy that is confusing to qualified reviewers such as he is. Unfortunately, Mr. Soldat developed flu before he could polish his comments. If the material would be helpful to the EPA staff, we would be glad to provide it--not for formal publication--provided the Chairman rules that such a procedure is proper.

3.6 Unusual and Temporary Conditions

According to the DES page 68, the proposed standards are designed to govern regulation of the industry, under normal operation and provide a variance to be exercised by NRC to accommodate unusual and temporary conditions. This is a most unusual and open-ended form of standard setting, quite at variance with the approach of most bodies essaying to present basic inclusive standards. The issue leads to endless bickering as to how unusual is unusual and how temporary is temporary. Unfortunately, EPA's laudable attempts to promote higher and higher decontamination factors for effluents would progressively aggravate that situation. For example, if the DF is intended to be 1000 and the removal device malfunctions to a DF of 1 for just one day per year--a reasonable unusual and temporary condition--the annual emission would be 3.4 times the normal value. In the case of reactor safety, the public has clearly demonstrated greater concern about accidental releases than conventional ones, a concern partially answered by WASH-1400.

We would expect a satisfactory environmental statement to include an extension of WASH-1400 methodology to all phases of the nuclear fuel cycle. This amounts to a probability calculation of all accident-caused health effects over the same time frame as the normal operation health effects calculation. At some level of assumed DF's the anticipated accident-related effects will overwhelm the normal ones. Cost-effectiveness would then direct one's attention to accident reduction rather than to further decontamination of normal effluents.

The answer as to which is the controlling factor is almost obviously unusual and temporary conditions for the reactor phase, perhaps the opposite for mining and milling, and perhaps worthy of intensive study for fuel reprocessing and fuel fabrication.

Of the 1020 health effects to be saved by the EPA approach, it appears from our reconstruction that about 980 come from ^{85}Kr removal, and about 40 from ^{129}I removal. Both of these nuclides are ones for which the "real" health effects may be small, because the quoted values are the product of very low doses and very large populations, where the uncertainty is greatest. Each of them has the potential for clear lethality if, and perhaps only if, it has been removed from the effluent stream.

For ^{85}Kr, the retention in gas cylinders has an obvious risk to the operator from a cylinder rupture or leak. Long-term retention of ^{85}Kr in terminal storage along with truly long-lived fission products or actinides would appear to be one of the more foolish concepts in waste management.

Page 10

In the case of ^{129}I, highly efficient devices have long been used to remove ^{131}I from stack effluents. The ^{129}I naturally goes along with the ^{131}I. When the latter decays for several months, the residual activity allows one to regenerate the collector. The fact that this may release the ^{129}I whose continuing presence has been forgotten can lead to disastrous local exposures.

4. Minor points

A nearly endless stream of minor points and not-so-minor points of principle can be raised, as indeed could occur for any set of documentation as complex as that supporting the proposed rule-making. Two such points are briefly entered for the record.

4.1 Man-made doses versus natural radioactivity

The DES, page 22, calls for different treatment of avoidable man-made doses and doses due to natural radioactivity. A follower of ALAP would want to minimize all doses regardless of origin. The proposed separation is arbitrary and unacceptable in two ways:

(1) A man-made dose from a uranium tailings pile is neither more nor less natural than the dose from a building constructed of radioactive granite or marble. Both represent natural radioactivity in man-made configurations, and both are involuntary dose burdens on the public. If one is controlled, the other should be correspondingly controlled.

(2) If one wishes to reduce the national health effect from radiation there may be a time at which nuclear industry emissions have been suppressed to such a degree that the most cost-effective steps (excluding interference with medical usage) could be such things as:

Page 11

- prohibition of eating Brazil nuts, which are thoriated.
- control by treatment or prohibition of drinking water made radioactive by uranium or thorium daughters.
- removal of natural radon and thoron from the atmosphere.
- decontamination of areas of high natural radioactivity or restriction on living there. (This would be of special interest if the USA had areas of notable activity such as those in India or Brazil, and especially if we could have the natural nuclear reactor that developed in the present Oklo mine some 1.7 billion years ago.)
- shielding against cosmic rays at high elevations.
- limitation of high altitude flights.

Decision making for standards is probably more conditioned by whether a particular exposure is voluntary or involuntary than whether it is based on natural radioactivity.

4.2 Occupational Exposure versus Public Exposure

The DES, page 17, properly defines the EPA's authority limit not to regulate the occupational exposure of workers. A concerned public is interested in minimizing the total national avoidable exposure. A satisfactory environmental statement which is proposing removal of ^{85}Kr and ^{129}I to reduce environmental exposures must demonstrate that the net effect is not an increase due to the additional process and storage operations.

Page 12

The accident probability aspect has been referred to under topic 3.6. To this must be added an analysis of occupational exposure from the additional normal operations.

Reasonable observers will expect to see some initial difficulties in the interrelationships between agencies. If the needed cooperation is not evident after a due interval, I presume that an informed public would properly seek modified legislation.

5. Summary of Principal Points

The documentation supporting the proposed EPA rule-making is a most reasonable first step in applying ALAP principles to the entire nuclear fuel cycle. Due to the arbitrary nature of each component decision in an ALAP approach, and to their multitudinous interactions, I expect it to be several years (arbitrarily about 5) before the required national wisdom will have been brought to bear to reach a sound solution.

The thrust of the EPA proposal is mainly to eliminate about 1000 potential health effects by superior removal of such effluents as ^{85}Kr, ^{129}I, and the transuranic elements. In practice, 96% of the health effects seem to be related to ^{85}Kr. We suggest that international agreement on control of this worldwide contaminant is of paramount importance.

Major uncertainties in health effects predictions are well understood by EPA. We strongly believe that no one agency can ultimately be responsible for selecting a final health effects format from a series of plausible formats whose predictions are shown by examples in the text to differ by one or two orders of magnitude.

Page 13

The constructive feature of the statement is a <u>positive call for a national forum to achieve a consensus on use of health effects predictions</u>. A three function charge is defined for such a forum.

APPENDIX A

PREDICTIVE CAPABILITY FOR HEALTH EFFECT FREQUENCY

Some classes of health effects, not the particular selection used by EPA in the draft environmental statement, can be tested either now or at definable times in the future. Atomic workers who have been subjected to measured external exposures, since 1944 in some cases, have been studied in AEC Health and Mortality Studies. They form one of the favorable classes for analysis of excess cancer deaths. The relevant predictions appear on page 170 of the BEIR Report:

	Calculation of the excess annual number of cancer deaths for individuals exposed from 20 to 65 years of age			
	Absolute Risk Model		Relative Risk Model	
Exposure Conditions	Excess Deaths Due to:		Excess Deaths Due to:	
	Leukemia	All Other Cancer	Leukemia	All Other Cancer
U. S. Population 0.1 rem/yr	195	(a) 721 (b) 808	436	(a) 1,444 (b) 1,793
10^6 people 5 rem/yr	81	(a) 300 (b) 336	181	(a) 601 (b) 746

(a) refers to a 30-year plateau
(b) refers to a lifetime plateau

We choose two arbitrary populations of workers,
    (1) 1000 persons at 5 rem/yr, and
    (2) 10,000 persons at 1 rem/yr.

While there is no population quite so clear cut, the examples are reasonable simulations of available exposees. Dr. Ethel S. Gilbert of the Battelle Northwest staff has calculated the odds on confirming

Page 15

health effects if they occur at frequencies equal to the BEIR Report, 5 times as severely, or one-fifth as severely. We give the results for the absolute risk model, which seems to be almost universally preferred by statisticians, and for our best interpretation of the average of absolute and relative risk used in the body of the BEIR Report. For all practical purposes, the calculated excess death rate is about doubled by that method.

A necessary delay in recognition for cancers other than leukemia is the allowed 15 year latent period. Leukemia recognition alone should be better with a latent period of two years, but the advantage is almost offset by the smaller incidence. Perhaps the best study would have to follow each cancer type separately.

The present exercise is illustrative only to show that positive results are conceivable. The statistical basis is not submitted for the record because I have not asked Dr. Gilbert to polish it to publication standards. It can be made informally available to EPA staff.

Summary
Population: 1000 persons at 5 rem/year
Column I - absolute risk basis
Column II - absolute and relative risk basis

Cancer Death Frequency	Feasible Observations	
	I	II
1/5 BEIR	Nothing	Nothing
BEIR	Nothing	Some signal after 30 years
5X BEIR	Reasonable answer after 30 years. Suggestive answer after 20 years.	Probably answered in less than 20 years

Page 16

Population: 10,000 persons at 1 rem/year

Cancer Death Frequency	Feasible Observations	
	I	II
1/5 BEIR	Nothing	Nothing
BEIR	Nothing	Nothing
5X BEIR	Fair chance of answer after 30 years.	Good answer in 20 years. Some evidence sooner.

Note that the times, 20 or 30 years in the above refer to times <u>after</u> the allowed latency period (15 years). So, if such real classes existed from 1944, answers would be available in 1979 and 1989, or a decade earlier for leukemia alone.

Protection Programs of the Plutonium Project

Health Physics, Official Journal of the Health Physics Society

October 1981

Volume 41, Number 4   October 1981

HEALTH PHYSICS
OFFICIAL JOURNAL OF THE HEALTH PHYSICS SOCIETY

H. WADE PATTERSON
Editor-in-Chief

GORGIANA M. ROTTER
Editorial Associate

THOMAS R. CRITES	MARGARET A. REILLY
KENNETH R. HEID	JOHN RUNDO
HARRY ING	DAVID H. SLINEY
LAMAR J. JOHNSON	WILLIAM P. SWANSON
KENNETH R. KASE	RALPH H. THOMAS
LAWRENCE H. LANZL	DAVID A. WAITE
THOMAS L. PITCHFORD	PAUL L. ZIEMER

Editors

J. R. A. LAKEY
News Editor

PERGAMON PRESS
NEW YORK TORONTO OXFORD PARIS FRANKFURT SYDNEY

INVITED PAPER DELIVERED AT THE 25th ANNUAL MEETING OF THE HEALTH PHYSICS SOCIETY

PROTECTION PROGRAMS OF THE PLUTONIUM PROJECT

H. M. PARKER

WHEN I was a struggling student, I used to earn a few extra bucks (or more correctly, pounds, shillings and pence) in the summer by checking the scores in University Entrance Examinations. This provided the wherewithal to lure a certain young lady named Margaret to the movies a little more often, that being the only darkened location respectably available in those prehistoric college days. Well, my reflexes were unusually quick in those days (which has nothing to do with the movies) and I always completed more checks than the other 99 checkers did. This gave me time to read the examination answers of interest, and to compile a rather remarkable collection of school-boy howlers. English essay answers were especially intriguing; you would be surprised at how many schoolpersons believed that Queen Elizabeth I was "first a bastard and then a princess." My favorite was the Essay on Dogs which began:

"There are, of course, three sizes of dogs—the large, the small, and the middlesized."

If you examine technical literature, I believe that you will find that a quarter to a third contains an equally pompous and tautological assertion.

However, the point of these introductory comments is that— "There are, of course, three sizes of standards—the large, the small, and the middlesized." I mean that there are three important phases to the development of standards:

(1) The setting of an appropriate pattern for achieving some desired result.

(2) The reduction of that pattern to accepted codes and standards.

(3) The regulatory process by which standards are enforced, or indefinitely elaborated and reinforced.

I have little patience with the third category, and will not consider it further, unless you choose to ask for my solution for the United States' regulatory malaise.

The thrust of my remarks will be random anecdotal comments reflecting the health physics pattern-setting of the early years of the Plutonium Project. The conventional histories of this period, mostly written by people who were not there, are full of misconceptions, perhaps unavoidably because of poor documentation. They more or less portray the original glorious Chicago Seven riding off into the sunset, after establishing a whole new profession. I rather think that the truth is nearer to a praiseworthy but prosaic response to more complex radiation protection problems than had been handled before, with assistance from a large supporting cast, especially those well removed from the higher halls of Academia. Some key features were:

● The potential presence of dose levels some 6 orders of magnitude higher than we were accustomed to.

● A much greater mix of external radiations than was usual.

● The possibility of radioactive contamination externally and internally both in the workplace and in the environment from a wide battery of fission and activation products.

● The need to transfer protection knowledge to large groups of people of diverse educational backgrounds.

● The perceived need to document personnel exposure to provide, ultimately, a reliable dose-effect relationship.

Existing standards were few in number (three, to all intents and purposes)* and the primary effort was to extend the sense of such standards to the new situations. To aid this, Dr. S. T. Cantril and I, who had worked together for 4 yr at the Swedish Hospital just up the hill from here, reviewed all the available literature on the Tolerance Dose (Ca45). "Available" did not include obscure references from the earliest days. We simply dared not go after some sources lest it stimulate curiosity about our interest. This is just one problem of security to which I shall return later.

I find that I was recently honored by having two papers included in the Health Physics Society 25th Anniversary Issue. If I had been asked for my choice, it would have included the tolerance dose review which is a classical example of the needed interplay between medicine, biology, and the physical sciences. Because Sim(eon Cantrill) and I were both European trained our story was a bit different from Dr. Taylor's with more emphasis on European reports and especially the League of Nations 1931 Report by Wintz and Rump. Should the BEIR III Report ever escape from its committee's womb, a comparison of it with our reference should define what we have learned about radiation risks between 1942 and 1980 or 1981, or 1982, or whatever.

A subject close to my interest was the need to reduce exposures from all radiation types to a common scale of relation (Pa50). As you know, this was done in early 1944 by introducing the concept of the rep and the rem, later changed by others to the rad and rem. I loosely said that *rem* stood for Roentgen Equivalent Man, Mouse or Mammal. I have seen learned analyses of the meaning of those terms extending over several pages. Those few of you who were in the early Oak Ridge group will smile with me upon recalling that the rem was originally the *reb* (*roentgen equivalent biological*). By great good fortune, I had a horrible cold in the nose when I first explained the system to the staff. It took some time to discover that the blank stares were due to the virtually identical sound of "rep" and "reb", under those conditions. And that is the sole reason why we hastily created a *rem*.

Personnel dosimetry combined ionization chamber measurements with film badges, fortunately now superseded by TLD. Neither modality was new, but it was not the normal practice to measure doses to individuals before the project.† The Victoreen Instrument Co. had suitable chambers for X-rays, but Mr. Terry has confirmed that fewer than 20 of the Minometers were in use pre-war. Those ionization chambers were modified to suit the generally harder radiation of the project, and Victoreen supplied these cooperatively and competently for many years. They are notoriously fickle with respect to insulation leakage; if dropped, they almost inevitably read a false fullscale dose. We found that by wearing two we could get a reasonable value by accepting the lower reading, but "double drops" were always a pain in the neck. This policy required proof that the lower reading could *never* be too low (plus or minus a small error factor), an exercise that is very much more difficult than you would suppose.

Film badges formed a backup to cover that contingency. Phillips (Eindhoven) had developed a pretty good film system *circa*. 1935. I had tried it for about a year in England (1936), but abandoned it because smart-alecky interns would expose the badge to various sources, and engage in the range of other sophomoric responses with which you are familiar. Plutonium Project results were better, because of the disciplined security aspects of the work. Dr. E. O. Wollan and N. Goldstein developed a superior badge with silver shielding for energy dependence sup-

---

*Two of those (for radium and radon) were credited to our co-speaker, Dr. Robley Evans.

†Protection in my first medical physics work was watched by monthly blood counts. In fact, I am Mr. Z in D. R. Goodfellow's "Leucocytic Variations in Radium Workers." Br. J. Rad. **VIII** 752–780 (1935).

pression, and we added an "open window" for crude beta ray measurements.*

The actual working of a good system depended, in my case, first at Oak Ridge and then at Hanford on a key pragmatic individual and a battery of dedicated technicians. Who but a Jim Hart could organize a *SWAK* test to maintain quality control over insulation leakage, and *schedule* 120 *technicians to cover* 10 *badge locations* around the clock? Curiously enough, one obvious control could not be used because of security. Our badges were integrated with the security photographs, and we got a flat "no" to having fictitious badges† for blind testing.

A useful summary of the personnel dosimeter work is given by John Auxier in the Ron Kathren-Paul Ziemer compilation of radiation protection history (Ka80).

For measurements of ambient radiation and radioactive contamination, the early health physics groups at Chicago, Oak Ridge, and Hanford developed about 20 new instruments. Collectively, they covered all the relevant radiations—photons, β-particles, α-particles, neutrons and all the modalities of interaction—external, internal, concentrations in air, water, food and so on. They have been described in the Kathren—Ziemer book, in various papers in the Health Physics Anniversary Issue (He80) and elewhere. In looking up material for today, I was surprised to discover an old paper by Karl Morgan and me (Mo46) which was intended to be part of a *Manhattan Project Handbook* CL-697. Written as a secret document for quick internal use, it contains rather blunt statements about the performance of some instruments, which I am sure we would have "tactfulized" had Karl and I had access to it at declassification.

I will make only three or four comments about the instrument program.

(1) It *did* set the permanent pattern of what had to be measured as personnel doses, internal emissions, fixed area monitoring, portable surveys, environmental radiations in air, water, and the ecological chains. Thus, it was a *standard setting achievement of the large dog size*.

(2) It was a peculiarly British type enterprise in that one muddled through, with a quickly conceived model that was barely good enough for the task, while developing the next generation. Current critics should remember that we were designing and building instruments before we had the devices that would emit the various radiations. We did not even know if those devices would really work, in some cases. Few of the original instruments survive, but I would note that the CP Meter or Cutie Pie was one excellent one that stood the test of time. I note that Karl and I ascribe its development to G. O. Ballou, which now makes the third or fourth different account of its origin.‡

(3) There has been much speculation about the code involved in the "funny" names. There was no code. We gave instruments mnemonic names to facilitate the *pattern-setting* of teaching their use to countless beginners. Surely, Fish-Pole is more informative than Meter, Survey, remote extension, Mark 2 and Poppy (simply because it made a popping sound) more relevant then Meter, Surface Contamination-Alpha Particle and so on. With these clues you can easily rationalize the name Chang and Eng for the first practical fast neutron meter, which was a Siamese Twin pair of ion chambers, one containing CH_4 (CHang) and the other argon (Eng). The advanced student could now even identify the allusion in the Devizes water contamination meter.

I have mentioned some security problems. I suspect that the modern health physicist

---

*Few of you will remember that the silver shields of the Hanford badge were riveted together each week to reduce tampering. Later models had a magnetic latch for the same purpose. For these reasons and the close attention to wearing badges by the security tie-in, Hanford film badge data are probably closer to being what they purport to be than are most other systems.

†Of course, the intended badges were not fictitious, but we needed real photographs of fictitous persons!

‡Generalized moral: build a better mousetrap, and everyone will claim to have invented it.

does not appreciate the problem of getting data when one cannot dare to reveal any public interest in radiation. I will use two anecdotal examples from Hanford.

(1) When we first discovered that ^{131}I released to the atmosphere refused to behave as a gas, but preferred to "plate out" on vegetation, we started a chain that led to reduction of the permissible concentration in air by a factor of 6000. Jack Healy and I constituted the first remote environmental monitoring team. Every Saturday we traveled 100-200 miles mostly on back roads. At each unobserved spot one of us dashed out, sliced off a vegetation sample, crammed it into a plastic bag for later tests in the laboratory, and retreated rapidly to the car.

(2) I, the most dismal of bowlers (unless you include the game of cricket), played on our department's team to the great disadvantage of their scores. But at each shot I chose a different ball and fastidiously wiped the finger holes with Kleenex. At games' end I had a fine collection of wipe samples to test whether hand contamination was detectable in the public domain.

On the day after the Japanese atomic bombing, Dr. Cantril and I wrote an account of the radiation protection at the plant (Ca51) which had to be accepted on faith by the vast majority of the workers. The document had to pass through classification. In this special case, it ended up on the desk of a certain general whose contempt for scientists is (a) well-publicized, and (b) matched only by our reciprocal contempt. I do not have an official release date, but it was at least 2 yr later, too late to build the confidence that was needed.

Full and understanding communication is one of the most important components of a sound standards-setting program. I have harped on the obstructions by security, not because they were wrong, but they did *produce a mind-set of caution*. We were, after all, charged with protecting a major secret. In later years, the AEC was accused of hiding the facts of radiation from the public. Perhaps they were slow in some cases, but please reflect on some of the forced absurdities (the foliage collection, the bowling balls, etc.) and develop some charity toward those who had been conditioned toward the *clam concept* by that earlier period.

In this anecdotal way, I have pictured the project as setting the patterns for good protection, *the large dog in my analogy*. The middle-sized dog was the conversion of these patterns to approved recommendations of the NCRP and others. That began in late 1946 with the restructuring of NCRP after the wartime hiatus. I served on Scientific Committee #1 under Dr. Failla. While officially concerned only with external radiation it was effectively the "philosophy arm" or basic protection criteria arm. The important features of the Plutonium Project work were rapidly incorporated. These included:

● The additivity of potential deleterious effects through the rep and rem system, later changed by ICRP to rad and rem.

● The first statement of the desirability of holding all exposure to the lowest practicable level.

● The creation of specific limits for exposure of body parts, e.g. the hands and forearms.

● The extension of the time base of measurement from 1 day to 1 week (and later to much longer periods).* This greatly simplified personnel dosimetry.

The work of this committee was essentially completed in 1948 but was not published until 1954 (NBS54), partly due to pressure of work, but mainly, I believe, to the intolerably cumbersome consensus process. As a result, much of this essentially U.S. protection structure is commonly thought to have originated with ICRP, to whom our thinking was transferred by Dr. Failla.

Other concurrent activities of NCRP were the elegant detailing of internal emitter standards under Karl Morgan of NCRP Scientific

---

*In retrospect, my influence tended to be counterproductive. I was raised in an institution celebrated for its superior work in studying protraction and fractionation in radiation therapy. I was well aware of the severe reactions produced at high dose and high dose rate, and carried these over to protection. The statement on p. 970 of (Me80), as an example, is not one that can be supported today.

Committee 2 and a series of topical guides. My assignment to prepare a recommendation for safe handling of radioisotopes (*NBS Handbook* 42) was completed in about a year. It is always the philosophy end that gets us, not the applications, perhaps because we always have *non-philosophers discussing philosopy* as we had this morning in an earlier program.

In the period following 1954, the NCRP kept actively in touch with developments in protection. When the National Academy of Sciences published its genetic report in 1956 (NAS56) it was only a matter of months before NCRP responded by adopting my age-proration formula.

Maximum Accumulated Dose = 5 (N-18) rem where N is the age in years. It should have been 4(N-18) except that we had a misconception that a scale of 1, 5, 10 better emphasizes numerical uncertainty. However, I was pleased with the accumulated dose concept. It was some degree of penance for my previous obsession with quite short time limits. Unfortunately, I was horrified to see the formula used to justify catch-up dose if the 5 rem is not used up in the current year. How that can be reconciled with an ALAP or ALARA concept, I have always failed to see.

Later, I participated with Lauriston Taylor in preparing the NCRP Basic Radiation Protection Criteria document (NCRP71). This document was disappointing in that it came out when the total picture of late somatic effects seemed to be coming into focus. The report does actually include projected rates for leukemia and for total cancer which are quite reasonable, but it does not address them in the forthright way that some of us would have preferred.

This brings the account up to 1972 which is where Dr. Taylor asked me to quit. Let me say that I am trying to continue my interest in basic criteria by serving on Dr. Cassaret's current committee; that I have felt that UNSCEAR 1977, BEIR I and some of the ICRP papers have seemed to be more progressive than our own; but that watching the labor pains of BEIR III has made me less disappointed that we did not claim to have sound risk rates in our documents of a decade ago.

My own instinct would be to go for a courageous risk basis knowing that *evidence* as differentiated from logical deduction and induction will never accrue at occupational levels, and of course never, never at environmental levels. I would seek an improvement of the rem values.* Remember that its original definition underlines the word *relevant* in the phrase "one rem is that dose of any ionizing radiation which produces a *relevant* biological effect equal to that produced by one roentgen of high-voltage X-radiation, other exposure conditions being equal." I *submit that we have learned in the last two decades* that the *relevant* biological effects at low dose are

● late cancers
● genetic effects, perhaps especially those reflecting seriously handicapping disease and haven't adjusted our rem values to suit that.

Finally, I submit that the quality factors for these two components are *not the same* for occupational exposure and environmental exposure, and that I do not know an unequivocal way of summing them for *either* case.

---

*Multiplying the rem by 100 and calling it the Sievert is not an improvement in my ledger.

REFERENCES

Ca45 Cantril S. T. and Parker H. M., 1945, "The Tolerance Dose". Manhattan District Report MDDC-1100, Oak Ridge Associated Universities, Oak Ridge, TN.

Ca51 Cantril S. T. and Parker H. M., "Status of Health and Protection at the Hanford Engineer Works", Based on Metallurgical Laboratory Rep. CH-3570; Also appearing as Paper 9 in *Industrial Medicine on the Plutonium Project* (Edited by R. S. Stone), 1959 (New York: McGraw-Hill).

He80 *Health Physics*, 1980, Vol. 38, No. 6.

Ka80 Kathren R. L. and Ziemer P. L. (Eds), 1980, *Health Physics: A Backward Glance*, pp. 167-169 (New York: Pergamon Press).

Mo46 Morgan K. Z. and Parker H. M., 1946, "Health Physics Beta and Gamma Radiation Instruments", DUH-7875, 12, March, Declassification date illegible.

NAS56 National Academy of Science—National Research Council, 1956, "Report of the Committee on Genetic Effects of Atomic Radiation", in: *Biological Effects of Atomic Radiation, Summary Reports*, pp. 3–31 (Washington, DC: NAS–NRC).

NBS54 National Bureau of Standards, 1954, "Permissible Dose from External Sources of Ionizing Radiation", *National Bureau of Standards Handbook* 59; also known as *NCRP Report* 17.

NCRP71 National Council on Radiation Protection and Measurements, 1971, "Basic Radiation Protection Criteria. *NCRP Report* 3 (Washington, DC: NCRP).

Pa50 Parker H. M., 1950, "Tentative Dose Units for Mixed Radiations". *Radiol.* **54**, 257–261.

The Squares of the Natural Numbers in Radiation Protection

The Lauriston S. Taylor Lecture Series in Radiation Protection and Measurements

December 1, 1977

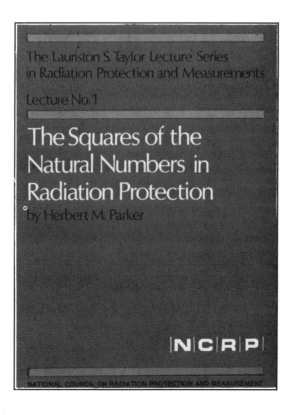

The Squares of the Natural Numbers in Radiation Protection*

by H. M. PARKER

Introduction

I am most honored by the invitation to deliver the First Annual Lauriston S. Taylor Lecture on Radiation Protection and Measurement before the NCRP. I first met Dr. Taylor in 1934 at the Zurich meeting of the International Congress of Radiology, when he was already well known, not so much for radiation protection as we now understand it, but for his contributions to x-ray dosimetry and its international standardization. One of the personal satisfactions arising from my emigration from England to the United States was the opportunity to be more closely associated with Dr. Taylor than I otherwise would have been through the triennial meeting schedule of the ICR.

As things turned out, I began a commitment to radiation protection in the U. S. atomic bomb project in 1942, while Dr. Taylor was helping to defend my homeland in scientific

* Presented March 17, 1977 in the John Wesley Powell Auditorium of the Cosmos Club, Washington, D.C.

areas not related to ionizing radiation. When the earlier activities of Dr. Taylor's Advisory Committee on X-Ray and Radium Protection were renewed after World War II and expanded in late 1946 or early 1947 to form the "old" NCRP (National *Committee* on Radiation Protection), I renewed an association that has not been interrupted since.

It was natural for the NCRP to ask for a history of radiation protection to honor Lauriston S. Taylor because his long career seems to many of us to be exclusively devoted to that one topic. Before essaying such a history, I would like to recall that during Dr. Taylor's long tenure at the National Bureau of Standards, his principal activities were in leadership of a broad range of atomic physics research with relatively little connection to radiation protection. For many years, I was a member of an Advisory Committee to the NBS to review these programs, and it gave me the opportunity to appreciate Dr. Taylor's work in these diverse fields. It also happened to cover a time in which the NBS activity in radiation protection was perceived by its Director to bear some of the characteristics of a foreign body. This was because it required exercise of judgment in biological or biomedical areas. I watched with interest the mounting of an immune rejection of this invading tissue, resulting in its expulsion. Very fortunately, this expulsion led to the formation of the "new" NCRP (National Council on Radiation Protection and Measurements) as a quasi-governmental agency, or federally chartered organization, if one prefers that terminology. I have waited with equal interest for realization by the NBS that most of its standards involving such things as human comfort in the home or factory, or noise insults at work or in the environment involve a physiological, biological, or biomedical factor not contained in the hard science familiar to the NBS. Unique dis-

14

criminatory treatment of radiation issues in this case and in environmental pollution control has been a major factor, and a deleterious one, in the convoluted history of radiation protection.

Radiation protection is an intrinsically dull subject; the better it is conducted, the duller the result. Eye-catching titles such as *Poisoned Power, Death of All Children,* or *The Careless Atom* were hitherto reserved for studies that emphasize the negative aspects of radiation usage. The introduction of the squares of the natural numbers in my title is a valiant but probably disappointing attempt to get attention. I use the squares in the same sense that a square section of wood is said to be applied to the flank of the mule to get his attention. However, let me hasten to add that I draw no comparison, expressed or implied, between the attributes of the mule and the attributes of the present distinguished audience.

9^2

This history begins with the *square of nine*, which when subtracted from 1977 leaves 1896. Within the required accuracy of radiation protection, that is the date of the discovery of x rays by Roentgen, and the discovery of radioactivity by Becquerel. The midpoint of the year 1896 was precisely the time of the first report of x-ray dermatitis. There is a record of systematic collection of information on injuries by 1898. One of the earliest references to a protection standard is that of Rollins [1] in 1902, based on the blackening of photographic film. While the early demonstrable injuries were superficial, it was as long ago as 1904 that the marked sensitivity of the blood-forming organs and the reproductive organs was recognized. The first organized step to insure protection from x rays is usually attributed to Russ

The Lecturer: Herbert M. Parker

[2] and the British Roentgen Society in 1915. A separate essay on the influence of war on radiation usage and protection could be written. World War I is generally charged with having diverted attention from improvement of protection to speed up military x-ray diagnosis and the like. Miller and Jablon [3] found no statistically significant problems among military technicians in World War II. Shipyard radiography is said to have involved undue exposure but I know of no definitive report on that. Returning to World War I, the situation is said to have contributed to many fatalities then ascribed to aplastic anemia. In the superb work of Court Brown and Doll [4, 5] on the British spondylitis patients, the authors note that diagnoses of aplastic anemia that can be checked by preserved clinical material were correct about 25% of the time, and should have been leukemia more than half of the time. Probably the even earlier World War I cases were more grossly misdiagnosed leukemia cases. The key position of leukemia in risk estimates might have come sooner had these mistakes been correctable.

In the early 1920's professional societies in both the United States and England developed recommendations for protective shielding and the like*, at least partly stimulated by the wartime results.

7^2

Subtraction of 7^2 from 1977 leaves 1928, a key year in the history of protection standards. It was then that the International Committee on X-Ray and Radium Protection (essentially the ICRP) was formed under the auspices of the Inter-

* This portion of the history is abbreviated from *The Tolerance Dose*, S. T. Cantril and H. M. Parker, Jan. 1945, declassified 1947, Doc. MDDC-1100. Another account is also given in *Radiation Protection Standards*, L. S. Taylor, CRC Press, 1971.

national Congress of Radiology. This group provided the first generally accepted guidelines on radiation protection — working hours, general recommendations (ventilation, etc.), x-ray shielding, electrical precautions, and radium handling. No mention whatever of a tolerance dose appears, which is all the more remarkable in view of the lucid proposal of limits (10^{-5} R/sec, equivalent to about 0.2R/day) in the Wintz and Rump, League of Nations Report of 1931 [6]. It was not until 1934 that such a limit was proffered by the ICRP and not until 1937 that gamma radiation was included in it.

The 7^2-derived date is even more significant for the NCRP because it was the year in which the international effort stimulated the first coordination between three U. S. radiation societies to establish an Advisory Committee on X-ray and Radium Protection, under the chairmanship of Lauriston S. Taylor, and centralized at the National Bureau of Standards. As we all know, Dr. Taylor has retained that chairmanship through the already described maturation to the NCRP. During the 7^2 years of that tenure, Dr. Taylor has organized the publication of 7^2 NCRP reports.*

As early as the 1934 Report No. 2 on radium protection [7], the following peculiar limit appears: — "The safe general radiation to the whole body is taken as 0.1R/day for X-rays, and may be used as a guide in radium protection." It was peculiar in two ways:

1. It used 0.1R/day, when the rest of the world used 0.2R/day.

* There are 50 reports of record. The most reasonable approach to this small numerical discrepancy is to maintain that for any sensible protection purpose, 50 is a sufficiently close approximation to 7^2. For the purist, one can say that there was a gap in World War II, when Dr. L. F. Curtiss acted as Chairman, but that undershoots by one, so one can select a report of one's choice (e.g. the only venture into comments on the legislative issue) [11] and exclude that one.

18

2. It half-heartedly applied it to radium gamma rays at a time when only a very few renegades, including this speaker, were willing to ascribe a meaning to "gamma roentgens" that was only formally accepted in definitions in 1937, upon acceptance of the Bragg-Gray Principle. The revised 1936 report on x-ray protection [8] used the same limit. Report No. 4 on radium protection [9] (1938) legitimately used the same limit. It became the U. S. national standard with no very good reasons for deviating from the international 0.2R/day on the one hand, or from a separate recommendation of 0.02R/day by none other than Dr. Taylor [10].

6^2

The 6^2 subtraction brings one to 1941. The country was at war, NCRP activities at a standstill, and an unprecedented scientific effort mounted to develop nuclear weapons. The so-called Plutonium Project* fortunately began with a realization of the horrendous radiation problems that might arise. The Health Division staff had two experienced radiotherapists and one experienced radiological physicist. The tone of the protection work was set by this all-too-small group while they busily trained others for the future. One of the three segments of the activity was known as health physics, and eventually a virtually new profession of full-time radiation protectionists came from it.

* The Plutonium Project was that part of the total Manhattan Project concerned with the production of plutonium in reactors, and its chemical separation as a weapons material. I have singled it out as the component expected to have the principal explosive expansion of radiation problems (roughly a million fold in radioactivity), and perhaps unconsciously because I was part of it. In so doing, I leave out Dr. Stafford Warren and Dr. Hymer Friedell, both of whom served in the military uniform of the U. S. Corps of Engineers. Dr. Friedell's extensive contributions to the NCRP are well known to this audience.

ARTICLE IV-30

Audience

In the stress of war work, few scientists were concerned with defining their specific contributions. In addition, the radiation work was necessarily classified; it never entered the peer review system. Despite some useful summaries by Dr. R. S. Stone [12] and others, a definitive history of these radiation protection practices has not yet been written. After 6^2 years, only about one-third of the central figures are alive. More unfortunately, the memories of the remainder seem, at times, to cloud the contributions of their former colleagues, and to vivify and amplify their own past efforts. I am persuaded that some of these actions were important enough to justify a national effort to reconstruct them. This is less to credit the originators than to review where their approaches had to be made on the most tenuous grounds and should not be so cast in concrete. The intertwined stories of dose-equivalents, dose-effect curves, and the choice of relevant effects is a case in point. The Plutonium Project had to have a system to add up exposures from all types of ionizing radiation; this was done early in 1944 through the scale of relation of roentgen, rep, and rem [13], which was later formalized as the roentgen, rad and rem in the unfortunate terminology of RBE-dose, and still later as dose-equivalent, with qualifying factors. An examination of the origins of the rem scales would show that they were based on limited data on relative biological effectiveness, ranging from the Stone and Larkin irradiations of patients with fast neutrons to the Mottram and Gray observations of growth rate of the bean root. No accommodation has been made over the years that the quality factors should relate to some combination of late somatic effects and genetic effects.

A case can be made that the concept of some residual injury was always contained in the defining reports, and

21

that it is implicit in the term "tolerance dose." That was certainly true in the classical Wintz and Rump report [6] which states " . . . the tolerance dose is never a harmless one . . . " However, it is at variance with the explanation in NCRP Report No. 17 for the change to "permissible dose" terminology, namely that "The concept of a tolerance dose involves the assumption that if the dose is lower than a certain value—the threshold value—no injury results" [14]. The issue is a semantic one, because in common English parlance, tolerance implies a capacity to accept a pain or detriment, whereas the medical usage specifically refers to ability to endure effects without showing unfavorable effects. Regardless of the semantics, the course of radiation protection at one stage was largely determined by the practical actions in the Plutonium Project, which were responsive to the statement: "The majority of radiation effects are thought to be of the threshold type. It may be that as more delicate indicators are found to measure effects, more of them will be seen to be of the non-threshold type" [15]. This prophetic quotation was intended to refer to somatic effects only.

Linearity was separately accepted for basic genetic effects, but no practical account was taken of it. To Henshaw [16] belongs the credit for suggesting a "tolerance injury" approach for the genetic effect in 1941. This concept can be considered as the nucleus of the lowest practicable level of exposure concept, which we would now recognize under the acronyms ALAP (as low as practicable) or ALARA (as low as reasonably achievable). The viewpoint was introduced by the NCRP about 1948, widely used in the Plutonium Project since that time, but not publicly documented by the NCRP until 1954 [14], by which time Dr. Failla had gained the acceptance of the ICRP for the same approach.

22

Cancer was well accepted as a late result of radium deposition, and was also believed to occur from exposure to radon in uranium mines. The risk that plutonium might produce bone cancer was at once recognized in the Plutonium Project. (Parenthetically, it has not yet done so in any demonstrable case in man, despite its success in winning the annual award as the most dangerous element known to man.) Yet the recognition of the probably non-threshold development of a wide variety of late cancers seems, in retrospect, to have been slow in coming. There was suggestive evidence from early radiotherapy, and considerable likelihood that leukemia was a specific end product. Jablon* has pointed out that one such report is dated as early as 1906 [17]. It was 1944 before a firm association of leukemia with exposed radiologists was made [19]. The effects of prenatal irradiation were noted in 1958 [20]. The general scope of radiogenic cancers was first measurable in the Japanese atomic bomb survivors, beginning in 1960 and with better evidence still steadily accumulating [21].

4^2

The year of record becomes *1961*. By that time, the daily limit of 0.1R had been superseded by a weekly limit of 0.3 rem, and we had reached an occupational accumulated dose formula of 5 (N–18)rem, where N is the number of years of age. This was the result of a major National Academy of Sciences study of the genetic effects in 1956, which accepted linearity [22]. The function of the formula was to

---

* In providing references to some of the early effects, I have relied heavily on the fine chapter on Radiation as an Environmental Factor by Seymour Jablon in *Persons at High Risk of Cancer* [18]. Also, in the case of obviously-difficult-to-obtain references such as Ref. [17], I have not professed to have read them.

Dr. and Mrs. Lauriston S. Taylor in the audience

tend to minimize the exposure of young people and to limit the integrated dose during the child-bearing period to a value acceptable to geneticists. The derivation came from accepting 30 years as the average age of all parents, and allowing individuals to receive not more than a total accumulated dose to the reproductive cells of 50 roentgens up to this age, and not more than 50R additional up to age 40. The 5(N–18) formulation gives 60 rem at age 30. Internal discussion with geneticists on the NCRP committees supported this approximation.*

A 1960 revision of the BEAR Report recognized that genetic effects are reduced at low dose rates [23]. This very important modification has been frequently overlooked; it inherently makes linearity for the somatic case less probable.

In this same time frame, the existence of the environment and the question of population dose came to the fore. The NAS proposed a limit for average population dose to the gonads of 10R in the first 30 years of life, excluding background, but including medical, fall-out, industrial, and all other factors. In January 1957, the NCRP, in an addendum to NCRP Report No. 17, quoted a population dose that

---

* In retrospect, it would have been better to promote the formula:
maximum accumulated dose = 4(N-18)rem

It dates from a time when I was a crusader for simplification of numerical coefficients in protection because most of our numbers imply more accurate knowledge than we have. At one time, I had a hang-up on a 1, 5, 10 scale, after I failed to get support for the far more rational 1, $10^{1/2}$, 10, $10^{3/2}$...scale, rounded off to 1, 3, 10, 30.... In 1957 when the accumulated dose formula was introduced, differences between gonadal doses and so-called whole body doses were not recorded, including the radical differences for men and women workers, in some cases. The NAS report dosimetry in roentgens only was not very refined for the times. Also overlooked was that the NAS allowance included medical exposures whereas the NCRP did not. That made 3rem difference, which we now know should have been about 1rem for 30 years.

rather curiously was withdrawn in April 1958. The first NCRP attempt toward a population dose for somatic effects was written in 1958–59, just before the views on genetic repair were accepted [24]. It accepted linearity as a postulate – not a matter of proof. Later reports have generally advanced linearity more stridently, often supporting it by an inference of cancer deriving from a gene aberration in a somatic cell, a process which can certainly be repaired in some cases.

This, too, was a period of intense national activity partly stimulated by concern for world-wide fall-out. Congressional hearings beginning in 1957, the Federal Radiation Council, 1959, and the USPHS National Advisory Committee on Radiation, 1958, were parts of this. In this selective and anecdotal history it is impossible to cover these areas. Nor is it possible to do justice to the important contributions of the ICRP.* Fortunately, Dr. Taylor's monograph, "Radiation Protection Standards," performs that function up to 1971 [25]. It seems appropriate that the First Annual Lauriston S. Taylor Lecture should be assumed to contain the whole of that source material. The vital role of the ICRU

---

*Having been given the opportunity to add a few footnotes to the lecture prepared for oral presentation, it is naturally tempting to fill in some of the gaps referred to here, but that could not be done adequately in much less than the 110 pages used by Dr. Taylor in Ref. 25. However, I will specifically reference two ICRP documents which were highly contributory to protection advancement. These were one of the first balanced essays on risk estimates [26] in 1965, and what is in part a 1969 updating of similar data [27]. The ICRP Reports appear in two series, one in brown covers which are formal pronouncements, and one in blue covers which are of the nature of ad hoc reports to a formal committee. Both the referenced documents are in the blue series. I find them more stimulating than the usually sterile 'official' documents. This may be a personal bias against the despoliation of a report by its gang-raping by multiple reviewers.

in promoting sound units and measurement principles should also be covered; Dr. Taylor was a major contributor [28]. Reluctantly, condensation also requires omission of the major work on internal emitters or other partial body irradiations. After the pioneering work of Martland and Evans led to a permissible limit for radium [29, 30], the Plutonium Project began a new dimension of accounting for hundreds of potential internal emitters, beginning with the work of Waldo Cohn, and ultimately leading to major national and international studies with which the name of K. Z. Morgan is so closely associated [31, 32].

N^2

The section heading of a generalized squared number signals a review of some phases over an extended time. What has been presented so far is a partial history of dose limits and their origin. The best limits in the world could be accompanied by deplorable protection practices. In other words, the history of radiation protection is vastly broader than the history of standards. I suspect that many of us equate the NCRP and ICRP with dose-setting standards; that is a misconception. Utilization of the first 7^2 reports of the NCRP illustrates the point. I have divided the reports into three types by function—more or less basic limits or criteria, measurement methods, and practical guides. Function is not quite so clear-cut in some cases, so my figures are an average of two interpretations. I had to form a fourth class labeled "None of the above," but neither of these factors affects the thrust of the comment. The relevant numbers were:*

* Mr. Wil Ney, who kindly provided the sales data, will note that round-off has lost 907 copies.

Report Type	Copies Sold
Basic Limits	242,000
Measurement Methods	94,000
Practical Guides	940,000
"None of the Above"	9,000
	1,285,000

Thus, basic limits are less than 20% of the interest, as reflected by document sales. If the reports are placed in rank order of sales, the ten best sellers amount to 703,000 copies. But the first four, Nos. 26, 18, 33, and 6, (349,000 copies) are all in the field of medical x-rays or gamma rays. NCRP started as a supportive instrumentality for radiology, and has substantially remained so. Conversely, the professional time spent on radiation protection is overwhelmingly on the nuclear energy side. In effect, the health physicists and others developed their own practical guidelines from the start; these guidelines were often written into NCRP reports when they were already becoming obsolescent. Available government money, favoring atomic energy, may be a principal reason for this imbalance. If so, and if healthy integration of all phases of radiation protection has been impeded, and if integration would have been helpful, there may be logical demands on government resources to adjust the balance.

In previous reviews of radiation protection performance, the components have been discussed under the topics [33, 34]:

Organization
Standards
Mechanical Tools of the Trade (i.e., instruments, etc.)
Mental Tools of the Trade (i.e., training)
Operating Results
Future Problems

There is time for no more than a comment on some of

these. Dr. Dade Moeller is currently conducting a survey among the "Elder Statesmen" of health physicists, part of which relates to major research findings since World War II. He has graciously allowed me to review his data. In what I would have grouped under Standards, the consensus of outstanding features is:

Metabolism of the radionuclides, especially the actinides-

Atomic Bomb Casualty Commission studies and application to somatic effects on large populations-

Genetic effects of ionizing radiation, especially the Russell work on mice-

Late cancer in patients-

Lung cancer in uranium miners-

As mechanical tools of the trade, one finds:

Thermoluminescent dosimetry, both for personnel dosimeters and for environmental monitoring-

Conversion of monitoring equipment to solid state circuitry-

Specific gamma-ray spectroscopy both for environmental measurements and for whole-body counting-

Incredibly, not a single respondent mentioned any single development that actually reduced exposure to any individual, whether in the occupational or environmental context. Has there really been no advancement in real radiation protection in 6^2 years? Or, do most of us really equate protection with standards and instruments? Or, is it a quirk of the survey question including the word "research" while the useful advances are reduction to good practice, or at most development, not research?

3^2

The operative year is now 1969. This date can be taken as the time of an explosive growth of the writings of the

Herbert M. Parker

nuclear critics. I omitted this phase from the presented lecture because such a topic did not seem appropriate for an occasion designed to honor the retiring president of the NCRP. Yet, a history of radiation protection without some reference to it would be incomplete. I shall be brief.

The flood of documentation included the critic's views and proponent's refutations. One of the things many found disturbing was the scant use made of the conventional outlets for scientific writings. The archival journals were essentially ignored and a bibliography of this material would have to include a curious assortment of hearing records, statements, comments, symposia transcripts, and independently published documents. Some of the writings were highly speculative in nature and this was not overlooked by those in opposition. Unfortunately, some of this material also included allegations of selective use of data and charges of subjective bias.

Good work and reasonable speculation doubtless appear on both sides of the nuclear controversies. Some better method of progressively approaching scientific truth rather than personal polarization is much needed. Absolute certainty of the effects of very low levels of radiation is denied us by the statistics of the case. Unbiased analysis of cancer induction in available exposed groups fails to show demonstrable effect, and will probably continue to do so for the rest of this century whether the true measure of risk is 10 times as severe as BEIR Report predictions, or one-tenth as severe, or nothing at all.

There is one positive result of the controversy. Much more attention has been focused on these issues since 1969, sometimes with apparent waste of time and money, and yet it has led (in my opinion) to a lower level of environmental exposure in the United States than would otherwise have applied.

$(1^2 + 2^2)*$

It is necessary to resort to the sum of squares to come precisely to the year 1972. That banner year saw the publication of an updated UNSCEAR report on levels and effects [35] and the National Academy of Sciences BEIR Report [36]. The former report gives risk estimates for late somatic effects and genetic effects, with adequate caveats about the dangers of extending the results to low dose and dose-rate. The latter gives similar estimates with advocacy of the linear hypothesis as the "only workable approach to numerical estimation of the risk in a population" (ibid. p. 89). By contrast, the NCRP Report No. 39 [37] can be accused of shirking the risk estimate issue. The more recent review of so-called philosophy, NCRP Report No. 43 [38] reads more as a justification of NCRP reports than a bold attempt to make improved risk estimates. As a major contributor to No. 39 and a minor contributor to No. 43, I know that neither of these effects was intended.

The key question is "which of these reports is most influential?" Is there any doubt that the answer is the BEIR Report?

0^2

Mathematics was fundamentally changed when the position of zero in the scale of things was recognized, interest-

---

* A member of the audience criticized my use of the sum of squares here, claiming, as is technically correct, that if I use a series which includes 1^2 I can come to any date. I apologize for this peccadillo, employed to get exactly to 1972. Alternatively, I could plead that the paragraph was headed by an honest 2^2 when I wrote it late in 1976.

ingly about as late in that history as the quantification of late cancer was in the radiation story. Here, I use it to bring the history up to the minute by literally selecting a manuscript from the mail as these notes were written. It was an advisory from the American College of Radiology concerning a move by the EPA apparently to assume jurisdiction in a medical radiation issue normally assigned to the Bureau of Radiological Health. It can be taken as symbolic of a growing overlap of regulation between those agencies, ERDA, NRC, and others. Perhaps one of my successors in this Lecture Series will choose regulation as the theme. Time has squeezed it from this review.

$(42,426)^2$

From zero, the text goes to a very large integer. The 9^2 paragraph acknowledged the discovery of x rays and radium, and the 6^2 paragraph began the nuclear age. $(42,426)^2$ which comes to 1,799,965,476, is very close to the date 1800 million B. P. assigned to the first known uranium fission chain reaction in the Oklo mine in Gabon, Africa. [39] This reaction ran intermittently for perhaps 600,000 years, and produced as much nuclear energy as one of our modern reactors would do in about 10 years. Meanwhile, the research of Richard Leakey and his group has pushed the known presence of man in Africa back some 1.8 million years. Hominid history goes back about 35 million years, and the primates span 70 million or more years. If there were some younger Oklo's and some older recognizable ancestors of man, perhaps there was an overlap. That pre-man actually witnessed his first nuclear chain reaction in Africa some $(10,000)^2$ years ago rather than at Stagg Field, Chicago, some 6^2 years ago is no more outrageous a speculation than many seen in radiation papers of the last decade.* For the purposes of this section we could equally well have picked any square up to $(67,823)^2$ which is our notation for the age of the earth. The point is that living organisms have been subjected to radiation since the beginning of life. Whether that exposure has been beneficial or deleterious is, I submit, not known to us. I shall look forward to that Lauriston S. Taylor Lecture which first determines whether we are here today despite these radiation insults, or *solely because of them.*

$(-3)^2$ OR $(-4)^2$

In conclusion (those happy words for the audience), I now reveal the true reason for the lecture title. The squares of the negative numbers give the same values as squares of the positive numbers and they can be added to 1977 just as comfortably as being subtracted. Therefore, I have created a logical framework to give the future history before it happens—and to do it in a negative frame of reference. In other words, I take advantage of the privilege of presenting protection history to use the day which is my last day of active membership in NCRP to advance two points of personal concern.

In meeting to honor Dr. Taylor and especially his contributions to both NCRP and ICRP, and in welcoming a new

---

* As a non-technical aside, we here indicated that although the lecture was dedicated to the NCRP, we reserved the rights to the movie suggested by this scenario. In a country that enjoyed EARTHQUAKE and the TOWERING INFERNO, it should certainly be a hit. By chance, this talk was made within two weeks of the extraordinary reception of ROOTS on TV. If the 200-year story of one segment of our population commands such attention, how much more should the origin of the whole family of man $(10,000)^2$ years ago grip us. And in view of the format of the lecture, was there ever a more logical title than SQUARE ROOTS?

President to carry on the tasks for NCRP, it behooves us to enhance the NCRP potential, whenever we can. My "negative" remarks are meant to bear on this positive purpose.

It is a shock to discover that if one selected from the history of radiation standards, the five most influential documents, the list might not include more than one, or at most two, from the NCRP and ICRP combined. That impact is disappointing.* At least two steps must be taken:

---

* A reaction to this statement was that if the situation is so bad, perhaps the NCRP has failed and has no real function. No such conclusion had occurred to me. In refuting it, several points have to be made.

1. In the history section with squares of positive numbers, I have given a brief anecdotal history which is biassed no more than that each of us must naturally see things through our own screen. In going to negative numbers, it should be clear that I am going to a personal viewpoint which emphasizes my biasses.

2. My N^2 section underlines the powerful force that NCRP publications have been. It also shows that basic standard setting is only 20% of the effort.

3. The 4^2 section notes, mainly by reference to Dr. Taylor's monograph [25], the great contributions of NCRP, ICRP and ICRU since their inception.

4. In the terminology of the table set out in the N^2 section, my interest has been much more with the Basic Limits than with the Practical Guides. I do indeed wish that NCRP had been responsible for more influential documents on Basic Limits.

5 On the other hand, NCRP could have performed valuable service with *never* a basic document. Much of the foundation work is expensive research beyond the NCRP scope. If we had always converted basic documents such as the NAS Genetics Report to protection practice as rapidly as we did in that case, we would indeed have a notable record.

6. I am very strongly devoted to the topic given as Step 1 below. I see it as the best way to repair the schisms of the age of controversy and the growing problems of overlapping, and potentially conflicting regulation.

7. Lastly, I have referred both here and in an earlier section to the most *influential* documents. In history, that which is most influential is not necessarily *right*. For some of the documents, we are not likely to know the answer much before the year $(45)^2$.

"In conclusion (those happy words for the audience)..."

1. NCRP and ICRP must play a more vital role in the setting of the *scientific* basis for risk estimation. There will have to be at least a national forum in which all risk data are subjected to peer scrutiny in public — not in committee rooms. There must be a national commitment for all to use the same data and update it at prescribed intervals. This is not a consensus process; it is of the nature of a science court procedure (but not a formal adversary procedure).

   This science phase is the only one actually requiring close interaction of NCRP and ICRP. Application of the agreed risk estimates is a *social* exercise, and the relevant society is essentially the U. S. population. Just how these societal judgments should best be made is yet to be determined. We must face the fact that the role of NCRP in it may be minimal, although that is not a foregone conclusion.

2. Whatever recommendations NCRP does make *must* be done far more expeditiously than has occurred in the last decade. That every member's voice should be heard equally is idealistic but palpably unattainable. If one had a council of 10 members, the possible number of interactions in dialogue form is 28,501. With a council of 65 members, it spirals to about 5×10^{30}.* While only a fraction of these will ever be exercised, the complexity of a consensus tends to be more nearly proportional to these numbers than to the number of members. Surely we can construct some mechanism to get speedy reliable conclusions from selectively chosen subgroups, and thus restore the position of timely leadership to NCRP Reports.

*Being in each case of the mathematical form $[(3^N + 1)/2] - 2^N$, where N is the number of council members.

In conclusion, I again thank the incoming President and the officers and my many friends in the NCRP for inviting me to speak to you, and I publicly wish to express the pleasure that I have derived from long professional and personal acquaintance with Lauriston S. Taylor. I wish him well for the future.

References

1. W. Rollins, "Vacuum Tube Burns," *Boston Med. Surg. J.*, 1902.
2. S. Russ, "Hard or Soft X-rays," *Brit. J. Radiol.* **11**:110, 1915.
3. R. W. Miller and S. Jablon, "A Search for Late Radiation Effects among men who served as X-ray technologists in the U. S. Army during World War II," *Radiology* **96**:269–274, 1970
4. W. M. Court Brown and R. Doll, *Leukemia and Aplastic Anemia in Patients Irradiated for Ankylosing Spondylitis* (HMSO, London, 1957).
5. W. M. Court Brown and R. Doll, "Mortality from Cancer and Other Causes after Radiotherapy for Ankylosing Spondylitis," *Brit. Med. J.* **2**:1327–1332, 1965.
6. H. Wintz and W. Rump, *Protective Measures against Dangers Resulting from the Use of Radium, Roentgen and Ultraviolet Rays*, League of Nations, C. H. 1054 (League of Nations, Geneva, 1931).
7. NCRP Report No. 2, *Radium Protection*, published as National Bureau of Standards Handbook 18 (U.S. Government Printing Office, Washington, 1934).
8. NCRP Report No. 3, *X-Ray Protection*, published as National Bureau of Standards Handbook 20 (U.S. Government Printing Office, Washington, 1936).
9. NCRP Report No. 4, *Radium Protection*, published as National Bureau of Standards Handbook 23 (U.S. Government Printing Office, Washington, 1938).
10. L. S. Taylor, "X-Ray Protection," *J. Am. Med. Soc.*, **116**:136–140, 1931.
11. NCRP Report No. 19, *Regulation of Radiation Exposure by Legislative Means*, published as National Bureau of Standards Handbook 61 (U.S. Government Printing Office, Washington, 1955).
12. R. S. Stone (Ed), *Industrial Medicine on the Plutonium Project* (McGraw-Hill, New York, 1951).
13. H. M. Parker, "Tentative Dose Units for Mixed Radiations," *Radiol.* **54**:257–261, 1950.

14. NCRP Report No. 17, *Permissible Dose from External Sources of Ionizing Radiation,* published as National Bureau of Standards Handbook 59 (U.S. Government Printing Office, Washington, 1954).
15. S. T. Cantril and H. M. Parker, *The Tolerance Dose,* Manhattan District Report MDDC-1100 (Oak Ridge Associated Universities, Oak Ridge, Tennessee, 1945).
16. P. S. Henshaw, "Biologic Significance of the Tolerance Dose in X-ray and Radium Protection," *J. Nat. Canc. Inst.* **1**:789–805, 1941.
17. K. Ziegler, *Experimentelle und Klinische Untersuchungen uber die Histogenese der Myeloischen Leukamie* (Gustave Fischer, Jena, 1906).
18. J. F. Fraumeni, Jr. (Ed.), *Persons at High Risk of Cancer* (Academic Press, New York, 1975) (The Jablon chapter "Radiation" is pp 151–165).
19. H. C. March, "Leukemia in Radiologists," *Radiol.* **43**:275–278, 1944.
20. A. Stewart, J. Webb, and D. A. Hewitt, "Survey of Childhood Malignancies," *Br. Med. J.* **1**:1495–1508, 1958.
21. S. Jablon and H. Kato, "Studies of the Mortality of A'Bomb Survivors, 5. Radiation Dose and Mortality, 1950–1970." *Radiat. Res.* **50**:649–698, 1972, and in many reports of the Atomic Bomb Casulaty Commisson, now called the Radiation Effects Research Foundation, Hiroshima and Nagasaki, Japan.
22. *The Biological Effects of Atomic Radiation: A Report to the Public* (National Academy of Sciences–National Research Council, Washington, 1956) or, in more detail, "Report of the Committee on Genetic Effects of Atomic Radiation" pages 3–31 in *The Biological Effects of Atomic Radiation: Summary Reports* (National Academy of Sciences–National Research Council, Washington, 1956).
23. "Report of the Committee on Genetic Effects" pages 3–24 in *The Biological Effects of Atomic Radiation: Summary Reports 1960* (National Academy of Sciences–National Research Council, Washington, 1960).
24. "Somatic Radiation Dose for the General Population," Report of the Ad Hoc Committee of the NCRP, *Science,* **131**:482, 1960.
25. L. S. Taylor, *Radiation Protection Standards* (CRC Press, Cleveland, Ohio, 1971).
26. *The Evaluation of Risks from Radiation*, ICRP Publication 8 (Pergamon Press, Oxford, 1965).
27. *Radiosensitivity and Spatial Distribution of Dose,* ICRP Publication 14 (Pergamon Press, Oxford, 1969).

28. L. S. Taylor, "History of the International Commission on Radiological Units and Measurements (ICRU)," *Health Physics*, **1**:306–314, 1958.
29. H. S. Martland, "Occurrence of Malignancy in Radioactive Persons," *Am. J. Cancer* **15**:2435–2516, 1931.
30. NCRP Report No. 15, *Safe Handling of Radioactive Luminous Compounds*, published as National Bureau of Standards Handbook 27 (U.S. Government Printing Office, Washington, 1941).
31. NCRP Report No. 22, *Maximum Permissible Body Burdens and Maximum Permissible Concentrations of Radionuclides in Air and Water for Occupational Exposure*, published as National Bureau of Standards Handbook 69 (U.S. Government Printing Office, Washington, 1959) and *Addendum 1 to National Bureau of Standards Handbook 69* (U.S. Government Printing Office, Washington, 1963).
32. *Report of Committee II on Permissible Dose for Internal Radiation (1959)*, ICRP Publication 2 (Pergamon Press, Oxford, 1950.)
33. H. M. Parker, "Radiation Protection in the Atomic Energy Industry," *Radiol.* **65**:903–911, 1955.
34. H. M. Parker, "Occupational Radiation Protection," presented at the *Plenary Session of the 20th Anniversary Meeting of the Health Physics Society*, Buffalo, New York, 1975.
35. *Ionizing Radiation: Levels and Effects*. A Report of the United Nations Scientific Committee on the Effects of Atomic Radiation (United Nations, New York, 1972).
36. *The Effects on Populations of Exposure to Low Levels of Ionizing Radiation*. Report of the Advisory Committee on the Biological Effects of Ionizing Radiations (National Academy of Sciences-National Research Council, Washington, 1972).
37. NCRP Report No. 39, *Basic Radiation Protection Criteria* (National Council on Radiation Protection and Measurements, Washington, 1971).
38. NCRP Report No. 43, *Review of the Current State of Radiation Protection Philosophy* (National Council on Radiation Protection and Measurements, Washington, 1975).
39. C. Fréjacques and R. Hageman, "Conclusions from Studies of the OKLO Natural Reactors" pages 678–685 in *Proceedings of the International Symposium on the Management of Wastes from the LWR Fuel Cycle*: Report No. ERDA CONF-76-0701 (Energy Research and Development Administration, Washington, 1976).

Warren K. Sinclair presenting commemorative certificate to the Lecturer

X-Ray Measurement and Protection 1913-64

Health Physics Official Journal of the Health Physics Society

June 1983

Reprinted from *Health Physics* Vol. 44, pp. 709-710 by permission of the Health Physics Society.

Volume 44, Number 6   June 1983

Beginning our second quarter-century of publication

HEALTH PHYSICS

OFFICIAL JOURNAL OF THE HEALTH PHYSICS SOCIETY

GENEVIEVE S. ROESSLER
Editor-in-Chief

BETH MAGURA
Editorial Associate

Editors

W. ROBERT CASEY	MARGARET A. REILLY
THOMAS R. CRITES	JOHN RUNDO
LEO G. FAUST	WILLIAM J. SILVER
RICHARD V. OSBORNE	DAVID H. SLINEY
DENNIS C. PARZYCK	RALPH H. THOMAS
THOMAS L. PITCHFORD	ROBERT G. THOMAS
WILLIAM S. PROPERZIO	DAVID A. WAITE

J. R. A. LAKEY	JOHN W. POSTON
News Editor	Book Editor

PERGAMON PRESS
NEW YORK TORONTO OXFORD PARIS FRANKFURT SYDNEY

BOOK REVIEWS

X-ray Measurements and Protection 1913–64, by Lauriston S. Taylor. U.S. Dept of Commerce, National Bureau of Standards Special Publication 625, December 1981.

DESPITE its broad title, this work is actually a chronicle of the National Bureau of Standards effort in this area as seen through the eyes of the author. Dr. Taylor was central to the work of the Bureau in this field, and thus cannot serve as an unbiased historical reporter, although he can and does provide valuable observations and information. Indeed, in the Preface he himself emphasizes that he is not a trained historian, a caveat supported by the style and context of the text. The format is similar to that of Dr. Taylor's 1979 tome "Organization for Radiation Protection". Both are anecdotal—almost autobiographical in part. They reproduce various relevant documents and letters. Use of unpublished and otherwise unavailable letters is a practice with some negative aspects.

The outstanding contribution of the Bureau was undoubtedly the development of the parallel plate ionization chamber for measurement of medium voltage X-rays, and its international comparison and standardization. This is covered well, but all too briefly. As the interest in higher energy X-rays or γ-rays increased, attempts were made to increase ionization chamber size or to resort to pressure chambers. In other countries, resort to Bragg–Gray cavity ionization methods occurred earlier. This part of the history is not made clear in the present text.*

Curiously, the document includes a 30-page interruption to cover a historical background of dosimetry prepared by one of use in 1955 (HMP) for the American College of Radiology. Some sense of chiding the Bureau for not going toward Bragg–Gray methods may perhaps be detected in this. If Dr. Taylor could have found four or five other such summaries, we would have had an excellent historical reference base.

One area not covered here (it was in the 1979 tome) is the real relationship of NBS to the protection handbooks issued first by NBS and later by NCRP. Most health physicists appear to believe that the early ones represented NBS work. They were the product of outside advisory committees chaired by Dr. Taylor. After World War II, during which NBS was in low gear, its revival in the dosimetry field (1946) was faced with the upstart health physics started in the Manhattan Project. A major interest there was unification of dose across all types of radiation, the rad/rem system as it later became. This depended on radiobiological data, an area foreign to NBS expertise. This led to the NCRP being established as a non-profit corporation chartered by Congress.

It is not surprising, with this background, that NBS never did address itself diligently to the advancement of health physics. The present report covers competent but pedestrian work on film badge standardization, work on radioactivity of fission products and activation products, and some contributions to radiological instrumentation.

The health physicist, and even more so, the medical physicist, will find some items of interest throughout the text. Appendix B is a useful compilation of 684 publications by the NBS staff. Above all, because this is a government publication at the bargain price of $9, it has to be a "best buy".

H. M. PARKER
R. L. KATHREN

Battelle
Pacific Northwest Labs
P. O. Box 999
Richland, WA 99352

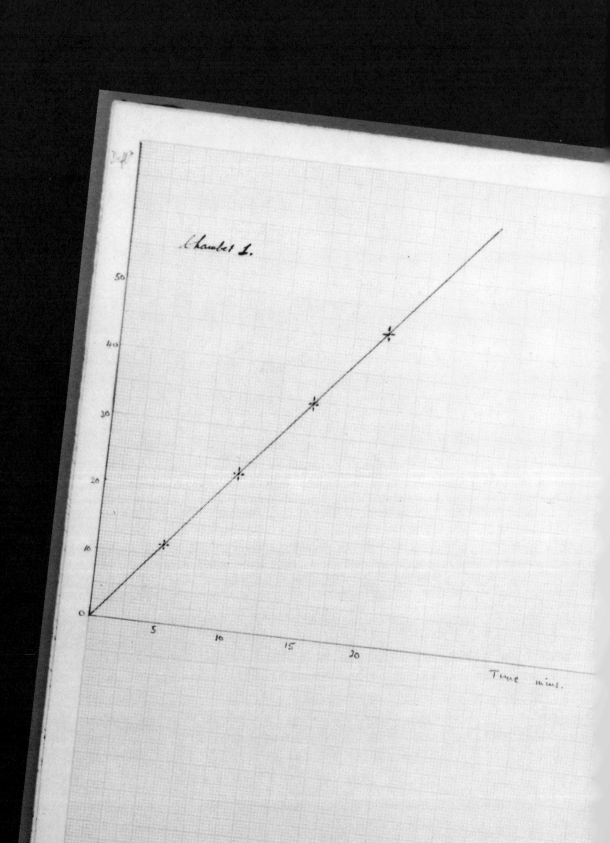

PART V

RADIOACTIVE WASTE MANAGEMENT

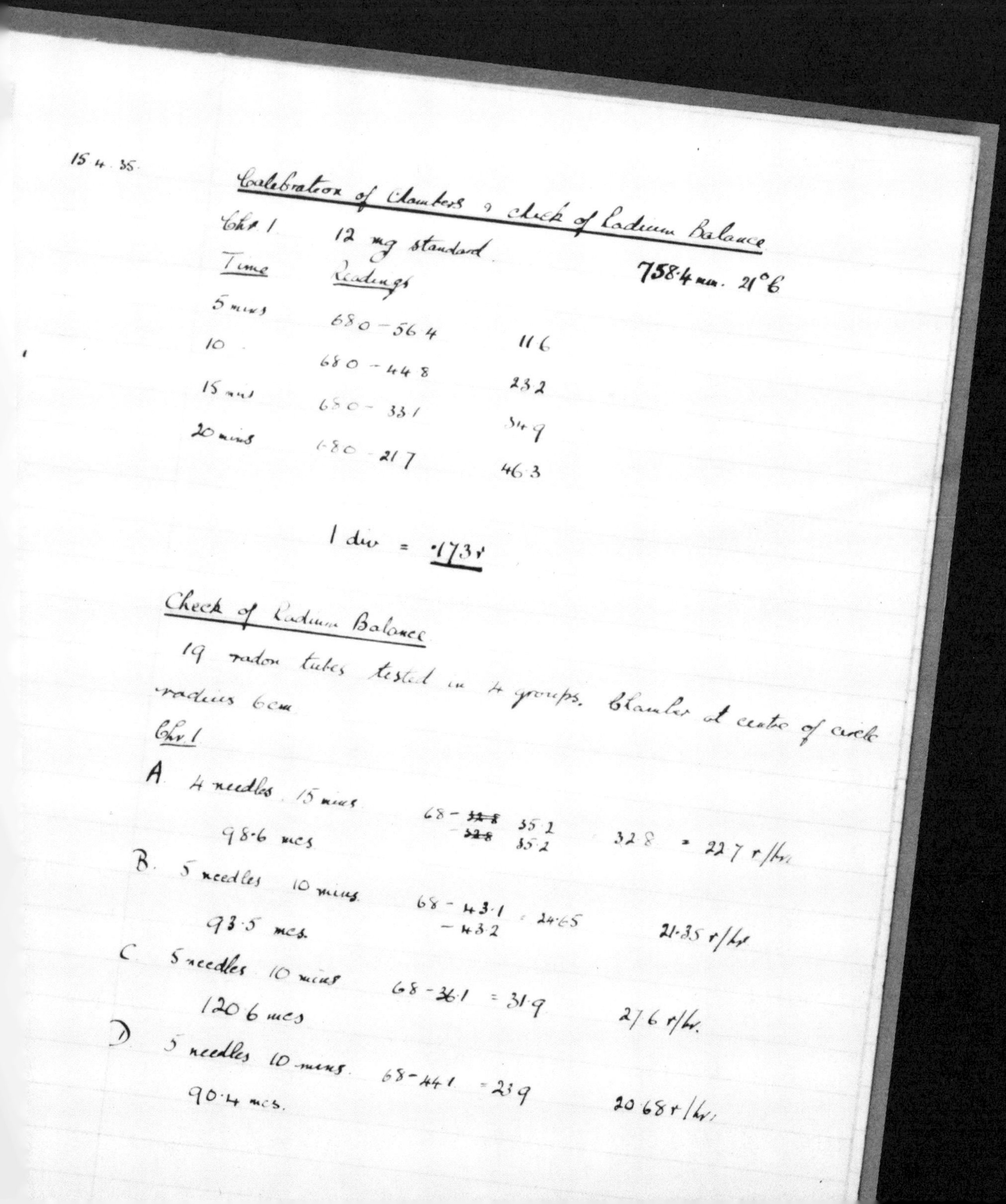

INTRODUCTION

The management of radioactive wastes was one of Parker's special interests. Early on he recognized that we as a society would have to come to grips with the radioactive waste problem, as evidenced by his paper "Speculations on Long-Range Waste Disposal Hazards." Although he published relatively few papers on radioactive wastes — only five are included in this section — he has exerted considerable influence on others who were involved on a full-time basis, as evidenced by the first two items in this Part, both hitherto unpublished documents.

Speculations on Long-Range Waste Disposal Hazards

U.S. Atomic Energy Commission Files

January 26, 1948

HW-8674
Technology-Hanford

SPECULATIONS ON LONG-RANGE WASTE DISPOSAL HAZARDS

By: H. M. Parker

Date: January 26, 1948

Medical Department (H.I. Section)

General Electric Company

Hanford Works

HW-8674
Technology-Hanford

-2-

DISTRIBUTION OF DOCUMENT HW-8674

Internal Distribution

#1 - D. H. Lauder - G. G. Lail

#2 - C. N. Gross - W. K. MacCready

#3 - A. B. Greninger - J. B. Work

#4 - H. M. Parker - C. C. Gamertsfelder

#5 - #6 - #7 - Atomic Energy Commission; Attn: R. C. Hageman

#8 - F. P. Seymour - R. E. Brown

#9 - Chicago Patent Group

#10 - 700 file

#11 - 300 file

#12 - Pink file

#13 - Yellow file

External Distribution

#14, 15, 16   - Atomic Energy Commission, Washington

#17, 18   - General Electric Company

#19   - Patent Advisor

#20, 21, 22, 23, 24   - Research Division, Oak Ridge

HMP-file  HW-8674
-3- 1/26/48

SPECULATIONS ON LONG-RANGE WASTE DISPOSAL HAZARDS

The operating policies of the Hanford Works call for the retention in buried tanks of all strongly radioactive wastes from the Separations Plants. The provisional studies during the war indicated that it was feasible to discharge less active wastes to ground to avoid absurd costs on tank storage, evaporation equipment, or equivalent. For approximately one year the H.I.Section has studied the disposition of underground wastes for the following purposes:

(1) To determine whether there is a significant risk of pollution of potable water sources, from the past practices,

(2) To devise corrective steps if such risk exists,

(3) To ascertain whether additional radioactive liquors can be similarly released, and thus to effect considerable economies.

Anticipated Disposal Situation

The salient facts from well drilling to date are:

(1) Plutonium contamination is very readily held by Hanford subsoils. The contamination has nowhere penetrated to a depth of more than 25 feet from the point of entry, although it has traveled laterally to about 200 feet.

(2) Fission product contamination is moderately well held by subsoil. It has penetrated downward as much as 100 feet in two years in one case. Lateral penetration up to nearly 300 feet has occurred.

(3) When the two species of contaminant are introduced together, they are separated out. Presence of fission products does not measurably alter the soil retention of plutonium.

(4) In the one special case of uranium waste disposal, rapid percolation to ground waters was noted.

These results confirm laboratory tests on soil columns. Other data from soil columns can be used with reasonable security, until checked in nature. Two such important findings are:

(1) Leaching of a plutonium soil column by copious volumes of water is relatively insignificant.

(2) Variation of pH of the disposed liquid has rather little effect on plutonium retention. There is more variation for fission products, which are, in any case, less tightly bound.

-3-

HMP-file  1/26/48 HW-8874

-4-

These facts are applied to make the following propositions:

(1) Disposal of approximately neutralized wastes to ground is a very local problem.

(2) Penetration of plutonium contamination will not substantially exceed 25 feet in two years.

(3) There may be lateral spread in local strata up to say 300 feet. There is a risk of surface run-off in sloping terrain.

(4) Penetration of fission products should go at a rate not substantially exceeding 100 feet in two years. This is less certain than other forecasts. The following factors were considered:

   (a) The downward progress of the contamination front should not accelerate in uniform soil.

   (b) The coarser nature of some lower strata may lead to some acceleration.

   (c) In all locations examined in 200-West Area there is a deep layer of fine sand, silt and clay, known to be a good sorber. This should offset (b) in part.

   (d) The contamination is apparently advancing in the form of a distorted spheroid, as the retention of the core reaches an equilibrium value. In this case, the front would advance at a rate which is a function of the cube root of waste volume. The progression would then go as the cube root of time.

   At present, there is only a static picture of the Hanford disposal. Approximately one year will be needed to get a fair dynamic picture. Possibly, data on other chemical waste disposal could answer the problem. It is important to appreciate that knowledge of water travel in soil does not answer this point. What the water does at Hanford is irrelevant, if the toxic matter is retained.

(5) If a spheroid of contamination is approaching the water table, the waste disposal can be terminated, and the contamination will be essentially fixed. Present rainfall, or even a ten-fold enhanced precipitation, would not be an effective leaching agent.

(6) The spheroids will include only a small fraction of the beds on the high terraces near the present Separations Plants. The whole process could be repeated several times from new local disposal points. This gives capacity for hundreds of years of plutonium disposal, and for perhaps 50 years of fission product disposal.

(7) The present use of superficial cribs is advantageous in comparison with the previously tried deep dry wells. Likewise, maintenance of position on the high terraces (about 250 feet to the water table) is preferred.

(8) The above propositions might be affected if the disposal rate at any one

-4-

HMP-file 1/26/48 HW-8674

point had to be increased to about 1000 gpm from the current 100 gpm.

(2) Disposal of large volumes of inactive water adjacent to the waste disposal points might also affect retentivity.

Dismal View of Disposal

Suppose that the relevant above anticipations prove invalid, and that, for any reason whatever, all the radioactivity currently in the ground (outside of tanks) is suddenly released and discharged into the Columbia River. Table I shows the time of river flow that would be needed to cope with this situation.

TABLE I
Release of all Cribbed Waste to the Columbia River

	Plutonium	Fission Products of half life >16 days
Total discharge to ground	1 kilogram	100 curies
Permissible concentration in drinking water	10^{-8} gm per liter	0.5 µc per liter
Minimum dilution water required	2.65×10^{10} gallons	5.3×10^{7} gallons
Dilution time in Columbia River, at minimum flow and full mixing	27 hours	4 minutes
Dilution time in Columbia River, at minimum flow and assumed 10% mixing	12 days	35 minutes

It appears that such a release is inconceivable in nature within the 12 days required for safety, under the most pessimistic assumptions.

Evidently, the Hanford Works has not set the stage for a <u>potential regional disaster</u>, by its planned disposal policies. Nor will this conclusion be materially altered by many more years of operation.

Major Disasters

Suppose that <u>all</u> the buried tanks are disrupted, and the material carried immediately

EMP-file 1/26/48 HW-8674

to the Columbia River. The hazard picture is given in Table 2.

TABLE 2

Release of Buried Tank Wastes to the Columbia River

	Plutonium	Fission Products $T_{\frac{1}{2}} > 16$ days	Uranium
Available quantities*	~ 30 kg.	5×10^7 curies	500 tons
Permissible concentration in drinking water	10^{-8} gm per liter	0.5 μc per liter	300 μg per liter**
Minimum dilution water required	8×10^{11} gallons	2.65×10^{13} gals.	4×10^{11} gals.
Dilution time in Columbia River, at minimum flow and full mixing	27 days	1230 days	19 days
Dilution time at average river flow and full mixing	12 days	384 days	6 days

In this calculation, full mixing is assumed, because a disaster of the magnitude considered could presumably be met by large-scale efforts to minimize channeling in the river. Alternatively, a channel in the right place might help to bypass the wastes beyond water intakes of local cities. In any case, rapid release of buried wastes would be intolerable, with respect to fission product pollution.

---

* Plutonium and fission products are calculated in terms of the total amount in the tanks. An unknown fraction has precipitated as insoluble sludge, which would probably not reach the river even under disaster conditions. If it did reach the river, it would not give rise to an ingestion hazard. The uranium waste has been more carefully studied (see reports by J.B.Work), and the quantity in the supernates is quoted.

** One finds no generally accepted data for this tolerance. Figured for heavy metal chemical toxicity by analogy with lead, it would be 100 μg per liter. A safe value deduced from mice experiments is 1 mg per liter. We have taken an arbitrary intermediate value.

HMP-file  1/26/48 HW-8674

-7-

Common sense indicates that the postulated catastrophe is out of the question. Already enough is known about adsorption of the radioactivity in soil to ensure that almost all the activity would be held up locally. It would take years for the pollution to reach the river by underground routes. Ten years has been quoted as the travel time of underground water, and the contamination front should move more slowly than this.

Suppose conservatively that the travel time is 5 years. Then the fission product activity is reduced by a factor of about 10.* The dilution time would be 38 days at average flow. One fails to visualize total release in less time than this. Increase beyond five years is comparatively unhelpful in producing further decay. A second five-year period would gain a factor of only 2. It may be noted incidentally that there is little interest in knowing the travel time accurately, if it exceeds a few years. Hold-up time, of course, has no effect on the plutonium and uranium hazards.

With 5 years or more for method development, it may be feasible to recover or divert a fraction of the escaped wastes, and thus alleviate the river pollution at the critical time.

There is a remote contingency that the contamination would run off to the Yakima River. In this event, one section of the river, including the dam and irrigation take-off below the Horn, would be intolerably contaminated. Below the junction of the Yakima and Columbia, the status quo would be restored.

Other Considerations

The current storage amounts have been considered above. After several years, the problem will be magnified by a factor of perhaps 10. By this time, additional knowledge of underground travel should permit a more precise appraisal of the existing hazard.

One has also not considered the accumulation of river pollution in algae or mud, with the subsequent scouring of the river beds at times of early summer high flow. Heroic, but feasible, methods of removing such contamination exist.

Pollution of the water table may be presumed to establish contamination which could pass <u>under</u> the two rivers and be used as well water at a more remote point. However, if the activity is diluted in ground water, a tolerable condition is rather quickly obtained.

Reconsider the major disaster under which all the buried tank activity gets into the water table after a 5 year travel time, and does not escape to the rivers. Take the simplified picture in which a circular disc of uniformly active water

---

* Decay computed from the curves of H. Kamack, Doc. No. 3-3332, on the simplifying assumption that retention in buried tanks has proceeded uniformly for 1000 days.

HMP-file 1/26/48 HW-8674

-8-

is produced. Suppose the water table to hold 10 quarts per c. ft.

TABLE 3

Pollution radii of water table

	Plutonium	Fission Products	Uranium
Available quantities	30 kg.	5×10^6 curies	500 tons
Minimum dilution water for tolerance	8×10^{11} gals.	2.65×10^{12} gals.	4×10^{11} gals.
Pollution radius for 4 yard pollution depth	18 miles	32 miles	12 miles
Pollution radius for 40 yard pollution depth	6 miles	10 miles	4 miles

In these speculations, one ignores the fact that if all three contaminants occur together, a further dilution (up to 3-fold) is required. In general, there will be a predominant hazard at any one point.

These pollution radii confine the serious problem to a comparatively local area. The lateral extent of contamination is troublesome if the pollution depth is low. In such cases, water withdrawn from wells can safely come from a much greater depth with an adequately cased well. Frequent and close control of water sampling is presupposed.

In a real case, one can reasonably assume 90% of the activity to be retained by adsorption in the high terraces. The resultant radii are reduced by a factor of $\sqrt{10} = 3$ approximately.

Reasonably Pessimistic Viewpoints

It may develop that the planned ground disposal system forms an equilibrium column, and therefore a transmissible channel sooner than anticipated. Then the output rate to the rivers would eventually equal the input rate. It is assumed that the hydraulic gradients are such that the time of underground travel is at least 5 years. The plutonium would then be the determining hazard. The quantity involved is currently ~ 5 gm per month. Suppose this to be increased to 50 gm per month by Plant expansion, or disposal of other wastes. Minimum permissible dilution will be in 1.4×10^9 gallons. The average monthly flow in the Columbia River is 2×10^{12} gallons. There would not be a critical condition.

HMP-file  1/26/48 HW-8874

-9-

Another rational assumption is to postulate a leak in one storage tank, say halfway down a tank in which 50% of the initial activity has settled out as sludge. This would release quantities of the order of 200 gm Pu, 3×10^5 curies fission products, and 20 tons of uranium. These can be cleared by the river in no more than 8 days at minimum flow.

Necessity to Continue or Improve Present Practice

The above reasoning purports to show that there would be no critical hazard if, for example, all second cycle wastes were returned to the river through a filter bed. If so, one must defend the intention to build additional second cycle storage tanks, and consider also why even hotter wastes cannot be economically dumped into the ground. The answer is three-fold:

(1) A great time-wise extrapolation from existing data has been made.

(2) Conservative practice would never condone the <u>planned</u> pollution of a national asset such as the Columbia River with a long-lived contaminant such as plutonium. The contractor and the Atomic Energy Commission share a major responsibility in the proper protection of the public interest in this field. Once the river becomes contaminated with alpha-emitters, the present careful watch for beginning pollution from unsuspected sources will be impossible.

(3) If the rivers or water table sources were contaminated at any level approaching the presumed "tolerable" concentrations, an extensive survey and water sampling program would be needed over a very wide area. This would be expensive enough to represent a significant fraction of the cost saved on tanks. Coupled with the greatly lowered public morale in the Columbia Basin, it would be a poor business risk.

Summary

Present disposal procedures may be continued, in the light of present knowledge, with the assurance of safety for a period of perhaps 50 years. Projection of the problem to future geological ages, as proposed by some authors, appears to be irrelevant in view of the technological progress in corrective measures that can be anticipated. Major <u>foreseeable</u> disasters would not seriously jeopardize the health of communities dependent on the river. In the worst case, radical curtailment of the use of river water may be required.

Conclusion

Currently planned geological studies are developing good data on the disposal system. While the continuation of these studies is deemed essential, there is no cause for hysteria, or for the radical expansion of the proposed program. This conclusion would require further study if:

(1) disposal points off the high terraces around the Separations Plants were required

(2) substantial changes (factor of 10) in disposal volumes were considered

HMP-f8bc   1/26/48

(3) certain major changes in chemical or physical status of wastes occurred. (For example, ultra-fine toxic metal particles may not be retained).

Some of the unresolved problems should be managed by a first-class soil chemist. The addition of such a man to the staff, or the association with a consultant, is recommended.

HMP:swc

H. M. Parker

Hazards of Waste Storage at Hanford Works

U.S. Atomic Energy Commission Files

August 2, 1949

~~UNCLASSIFIED~~

HW-14058

RECORD CENT FILE

FILES:

RECEIVED
MAY 22 1956
300 AREA
CLASSIFIED FILES

Copies #1,2,3 - Atomic Energy Comm.-
              attn: W.K. Crane
#4,5,6,7,8,9,10 - Atomic Energy Comm.-
                  Washington, D.C.
#11 - GR Prout
#12 - AB Greninger
#13 - CN Gross
#14 - HA Kornberg
#15 - CC Gamertsfelder
#16 - HM Parker
#17 - 300 files
#18 - 700 files
#19 - Pink file
#20 - Yellow file

This document consists of
56 pages # 20 of
20 copies, Series

R. H. Wilson 3746
12715 JUL 10 '62

HAZARDS OF WASTE STORAGE AT HANFORD WORKS

PART I

Date: August 2, 1948

by: H.M. Parker, Manager
    Health Instrument Divisions

Classification Cancelled (Change to
UNCLASSIFIED
By Authority of B.C. Feldman
teletype dated 1/29/74
By L. Pope 2/1/74

General Electric Co.

Hanford Works

UNCLASSIFIED

HW-14058

HAZARDS OF WASTE STORAGE AT HANFORD WORKS

PART I

INTRODUCTION

Speculations on the hazards involved in the accidental dissemination of waste materials at Hanford Works are required for all foreseeable circumstances including enemy attack or sabotage. Due to delay in obtaining data on the potential effects of bombing on waste storage tanks, the report has been held up beyond the anticipated date. It is therefore being prepared in two parts of which this first part is concerned only with potential disaster to the nuclear reactors. From the nature of the process, these are major waste storage units.

Separations process waste stored in buried tanks or otherwise will be considered later. Preliminary speculations on this topic have been given already [1], and it is not anticipated that much improvement in this phase will be effected.

In all cases, it is not expected that numerical conclusions will be valid to a factor of two, but they should be reliable within a factor of 10. This is the estimated accuracy of the excellent British reports of the Plant Location Panel, which have been extensively used in this compilation.

This report is a joint effort of the staff of the Health Instrument Divisions. In particular, F. G. Tabb is responsible for the calculations of fission yield, and Dr. H. A. Kornberg and Dr. R. F. Foster for the principal general biological and aquatic biological data, respectively.

1. Storage of Radioactive Wastes in a Hanford Nuclear Reactor

In an assumed average operation of 200 days at 275 MW, a Hanford reactor will contain

Waste $\begin{cases} 584 \text{ megacuries of fission products} \\ 220 \text{ metric tons of uranium} \end{cases}$

Products $\begin{cases} 200 \text{ megacuries of neptunium} \\ 2.7 \times 10^4 \text{ grams of plutonium} \end{cases}$

By logarithmic classes of half lives, the fission products are divided thus:—

$T_{\frac{1}{2}}$	0 - 1 day	1 - 10 days	10 - 100 days	> 100 days
Megacuries f.p.	443	70	70	1
Mass of f.ps. (gms)	46	274	1408	2442

UNCLASSIFIED

The total mass of fission products approximates 9 lbs; fission product release from an atomic bomb is said to be 2 lbs., with a presumed activity of 10,000 megacuries.

The decay of the total fission products is specified by

Time (hrs)	0	1	2	5	10	24	48	72	120	240	480	720
Megacuries	584	398	339	277	230	176	143	124	105	80	56	43

This is compatible with decay laws

$$\text{Activity} = \frac{\text{constant}}{(\text{time})^{0.2}} \text{ for the first 5 hours}$$

$$\text{Activity} = \frac{\text{constant}}{(\text{time})^{0.3}} \text{ thereafter.}$$

The decay of atomic bomb fission products follows

$$\text{Activity} = \frac{\text{constant}}{(\text{time})^{1.2}}$$

After a few hours, the release of all the activity from the nuclear reactor would exceed that from an atomic bomb at the equivalent time. The hazard is at all times greater in the present case, because of the higher relative content of long-lived fission products.

The initial activity of certain key fission products is

Isotope	Megacuries
Sr^{89}	6.85
Sr^{90}	0.09
Ru^{103}	6.9
Ru^{106}	1.49
$Te^{131} (\rightarrow I^{131})$	6.45
I^{131}	6.45
I^{133}	10.35
I^{135}	13.6
Cs^{135}	0.0001
Cs^{137}	0.0008
Ce^{141}	11.9
Ce^{144}	0.0008

2. Circumstances Producing Uncontrolled Release

These have been discussed in the A.E.C. Reactor Safeguard Committee report "Review of Certain Hanford Operations"[2]. The reactivity of the Hanford reactors is adequately controlled under all foreseeable conditions except the following:

1. **Planned shutdown without addition of poison**

    After the decay of the natural poison Xe^{135}, the unit could become reactive. It will be assumed that such an operating error is inconceivable.

2. **Loss of water in one tube** *

    If accompanied by failure of the safety devices, the metal in the affected tube would melt and be ejected with an accompanying minor steam explosion. The management of the ejected 0.45 megacuries of fission products would be a serious local problem. If the reactor were then shut down, access for addition of poison might be denied, if the final disposition of the activity prohibited the rapid dumping of shielding material upon it. An explosive run-away is not visualized in such a case.

3. **Total loss of water**

    The loss of cooling water increases the reactivity of the unit by about 850 in-hours, which can be held by the vertical safety rods (1700 in-hours) or in emergency by the third safety device (1800 in-hours). An uncontrolled condition requires the concurrent failure of both safety mechanisms. This is visualized to be possible under three circumstances:-

    (a) Earthquake
    There is no history of earthquakes in this region of sufficient intensity to disrupt the water supply to the reactors and to damage the safety mechanisms. Nevertheless, there are records both in the western slopes of the Cascade Mountains and in Montana, of earth movements not significantly below that required to produce such damage. The relatively short recorded history of the Pacific Northwest casts legitimate doubt on the permanent stability of the units against earthquake. This is the one plausible case in which _all_ the reactors could be affected simultaneously. This extreme case has not been developed in this report.

    (b) Enemy bombing
    The Division of Biology and Medicine reports information from the National Military Establishment to the effect that although the reactors are considered relatively invulnerable to bombing attack, the slab construction of the main shield leaves some possibility of disruption of the shield. Evidently an enemy attack could, by chance, demolish the vulnerable safety mechanisms, and damage the water supply. Advance warning of a few seconds only would permit safety to be assured by insertion of all rods. Subsequent bombing, even with disruption of the unit should not lead to a runaway condition.

---

* The PLP reports adequately discuss the case in which the ejected metal from one tube burns.

-5- UNCLASSIFIED   HW-14058

Bombing should hold a low priority in the foreseeable hazard pattern.

(c) <u>Sabotage</u>
Sabotage by an enemy agent or by an unbalanced person intrinsically loyal to the United States appears to be the outstanding potential hazard. With two or more persons in collusion, the risk of detection would be negligible. There is no point in elaborating here on the feasible sabotage methods.

The general picture is that of rupture of the water supply, together with lateral motion of the reactor sufficient to close the thimbles to the vertical rods. The fluid of the third safety device would fail to enter the thimbles, or the piping to the thimbles would be broken. Deliberate tampering with the safety mechanisms is an alternate to the lateral motion.

3. The Uncontrolled Nuclear Reactor

The approximately 6 cubic meters of water in the pile can be raised to the boiling point in about 7 seconds at normal operating power. The water in central tubes would boil in about 3-4 seconds. The power level of the reactor increases rapidly; the excess reactivity above critical is said to be $K_e = 0.025$ without water. On elementary reactor theory, the power increase after loss of water is

$$\frac{\text{Power after t secs}}{\text{Steady power}} = e^{20 K_e t}$$

The 250 tons of uranium in the structure is boiled by about 10^{11} cals. By integration, this will occur in 13 secs. The actual course of events is the boiling of the central uranium in a fraction of this time, with possibly the termination of the reaction. In any case, the sequence of events is so rapid that there can be no substitute control. The real rate may be affected by the unknown temperature coefficient, release of stored energy from the graphite, release of Xe^{135} as the uranium melts.

To determine the environmental hazard, we take two simplified pictures.

(a) the 2×10^{11} cal. explosion

    basis: all the uranium just boils = 1.16×10^{11} cal.
an equal amount of heat goes into the graphite, aluminum and other parts.

    or: some of the stored energy of graphite (calculated as 2.6×10^{11} cal. at the present time) is released, thus terminating the nuclear reaction at a lower integrated flux.

(b) the 2×10^{10} cal. explosion

    basis: the Reactor Safeguard Committee states that "it does not seem possible for the metal temperature much to exceed 10,000° C at the time when the course of the accident itself disrupts the chain reaction. This temperature will be reached in a power burst of about 7.5×10^7 KW seconds".

    This is equivalent to 1.8×10^{10} cal., which is insufficient to boil the uranium. Presumably the given figure refers to the power burst required to raise the central temperature of an unflattened pile to about 10,000° C. Alternatively, very different physical constants have been used for the metal. This report uses:-

        Specific heat of uranium = 0.028 cal/gm at any temp.
        Melting point = 1150° C
        Heat of fusion = 13 cal/gm
        Boiling point = 3900° C
        Heat of vaporization = 391 cal/gm

Analysis of the 2×10^{11} calorie explosion

Primary Effect

All the uranium boils and approximately 1.5×10^{11} cal. goes into the hot gas bubble. This is equivalent energetically to an explosion of 150 tons of high explosive. The nuclear reactor will be explosively shattered, although this is probably minimized by the heavy shielding. More probably the hot gases will be ejected violently from each of the reactor tubes. In either case we shall abandon the physical model and proceed to calculate the behavior of the gas cloud as if it were a regular high explosive detonation, following O.G. Sutton's method[3].

Behavior of the gas cloud

The instantaneously generated heat in the cloud is about 1.5×10^{11} cal. The initial temperature is probably $4000°$ to $8000°$ K.*

From $PV = RT$

the volume V after the initial rapid expansion to atmospheric pressure is

$$V = \frac{8.3 \times 10^7 \times 150 \times 10^6 \times (4 \text{ or } 8) \times 10^3}{25 \times 10^6} \text{ cm}^3 = 2 \text{ or } 4 \times 10^{12} \text{ cm}.$$

The initial radius = 80 to 100 meters

Note that the mean molecular weight of the explosion gases is taken as 25 for the normal products of explosion, and bears no relation to much higher value in the real model.

The velocity of ascent from the original position after expansion is given by

$$V^2 = \frac{1.5\, g\, Q}{T_a\, C_p\, \rho\, \pi^{3/2}\, c^2\, z^m}$$

where V = velocity of ascent
g = acceleration due to gravity = 9.8 meters/sec^2
T_a = temperature of surrounding air = $\sim 300°$ K
C_p = specific heat of gases produced = $\sim 1/3$
ρ = mean density of air over the range = ~ 1000 gm/meter^3
C = generalized diffusion coefficient =
         ~ 0.2 (meter)$^{1/8}$ or less.
m = index of turbulence = 1.75
Z = height in meters above original expansion location

$$\therefore V^2 = \sim \frac{2 \times 10^7}{Z^{1.75}}$$

---

* Sutton's formal calculation will always give $3000°$ C as the temperature. We take a higher value because $3000°$ C would not start a cloud of uranium gas.

Z meters	Height above ground (meters)	Vel. of ascent meter/sec
200	~ 300	90
1000	~ 1100	23
2000	~ 2100	12
4000	~ 4100	6

The average wind speed in the Hanford region is

Height (meters)	Wind speed (meters/sec)
1000	5
2000	7
3000	9
4000	11

The cloud is pictured to rise until the velocity of ascent is comparable with the horizontal wind velocity (perhaps one half of it). The subject cloud rises to 3000 to 4000 meters, say conservatively 3000 meters. The main uncertainty in the calculation is the appropriate value of C, which is a function of Z.

$$C = (0.17 - 0.042 \log_{10} Z) \, (\text{meter})^{1/8}$$

Perhaps an average value of C between Z = 0 and Z = 4000 is applicable. This would send the cloud higher, and we take the conservative low value.

Activity in the cloud

The cloud contains M = 584 megacuries of fission products. It moves horizontally with velocity 9 meters/sec. Under average conditions of turbulence, the maximum concentration at the ground is

$$\frac{0.073 M}{r_0^3} \text{ curie/meter}^3, \text{ where } r_0 \text{ meters is the distance from the cloud center to ground.}$$

Over level terrain, max. conc. f.p. = 1.6 μc/liter
similarly max. conc. Pu = 7×10^{-5} μg/liter
max. conc. N_p = 0.5 μc/liter
max. conc. U = 0.6 μg/liter

These concentrations occur at a time, t secs. after the explosion (strictly after the ascent) given by

$$t = \frac{1}{U} \left(\frac{2 r_0^2}{3 c^2} \right)^{1/m}$$

where U = horizontal wind velocity (meter/sec)

If C = 0.2 (ground value), t = 5200 sec. = ~ 1½ hrs.

If C = 0.02 (at 3000 meters), t = 73,000 sec. = ~ 20 hrs.

The probable value is about 5 hrs.

HW-14058

The corresponding distance from the reactor is ∼ 100 miles. Additionally, the activity after 5 hrs. has decayed to approximately one half the initial value.

As a man stands at a point 100 miles from the reactor, the cloud substantially contributes its activity to him in 11 minutes.* The time variation of concentration at the ground is shown in Figure 1.

If the average energy of the fission products is 1.7 MEV, the integrated dose (neglecting scatter) is:

$$\frac{310 \text{ (}\mu c/liter\text{) sec.} \times 1.7 \text{ MEV}}{2 \times 3600 \text{ sec.}} \text{ rep} = 0.075 \text{ rep}$$

The variation of activity outwards from the center of the moving cloud at this time is

Radius (meters)	0	1000	2000	3000	4000
Concentration ($\mu c/liter$)	2.9	2.5	1.3	0.6	0.2

The gamma-ray component comes from a range of ∼ 300 meters which includes zones of higher concentration than at the ground. This is a small factor, and we can write

$$\text{Dose} = 0.1 \text{ rep}$$

The width of the cloud in Sutton's arbitrary definition is terminated at points where the activity is one-tenth of the central activity. The subject cloud has a width of 7500 meters at 100 miles from the reactor. It will be noted that significant concentrations can occur at points well beyond the conventional boundaries of a radioactive cloud. Also, in a real case, the effective width at 100 miles will be greatly affected by wind shear. We shall elaborate the effect of meteorological variables for a much more dangerous cloud later.

Inhalation Figures

Assume 17 liters/min with 50% retention

Fission products - deposit 45 μc - initial dose-rate 40 mrep/day.
no significant hazard if material is absorbed from lung.

Plutonium - deposit 4×10^{-3} μg - no hazard.

Neptunium - deposit 40 μc - initial dose-rate 20 mrep/day. formation of Pu insignificant.

Uranium - deposit 40 μg - innocuous.

---

* Neglected throughout is the normally lower wind speed near the ground; exposure time could be about twice the calculated value.

Mountainous Terrain

If the ground elevation rises away from the reactor, a more significant hazard can be encountered at these intermediate distances. In Hanford terrain this is insignificant for the present high cloud. For reference, the central fission product concentration is given as a function of distance, together with the width of the cloud, from which the concentration at any distance and any elevation above the base line can be estimated.

Distance (miles)	0	10	20	40	60	80	100
Central concentration (μc/liter)	2×10^5	3000	270	45	15	7	3
First order width (meters)	180	1000	1750	3400	4800	6000	7500
Second order width	?	1400	2500	4800	6800		
Third order width		1750	3000	5900			

First order width = width to points of 1/10 central activity.
Second order width = width to points of $1/10^2$ central activity, etc.

Second and third order widths probably have poor physical significance, but their use should lead to <u>overestimates</u> of exposure.

> Example: A man on Rattlesnake Mountain is $\sim$ 1000 meters above the base-line of the reactor = 2000 meters below the drift-line of the cloud, and 14 miles away.
> The central concentration is 750 μc/liter, and the subject point is beyond the fourth order semi-width (= 1250 meters below drift-line). The concentration is below 0.075 μc/liter.

Fall-out

The fall-out of particles under gravity will be inconsequential with an initial high cloud. For example: a particle of specific gravity 5 and diameter 6 microns will fall only 70 meters in the postulated 100 miles horizontal travel

Rain-out

In the Hanford area, a rain originating above 3000 meters is so exceptional that one can disregard the risk of rain-out of activity within say 40 miles of the Plant. If rain-out did occur, it could be a potential major hazard. Sutton estimates from Leicester data that one-eight of the soot content of a cloud can be washed out by rain. We invoke two additional factors:-

(1) Typical British rain is obviously a more formidable scrubbing agent than the Hanford variety.

(2) The British figure is probably weighted by removal of the heavier particles. Most of the fission product activity will be concentrated on very small particles, for which scrubber efficiency is low.

HW-14058

We shall assume a **maximum** rain-out of 5%. Initially, a deposition of

$$\frac{0.05 \times 584 \times 10^6}{\pi (90)^2} = \sim 1000 \text{ curies/sq.meter}$$

is feasible. At 10 miles, the maximum deposition is ~ 30 curies/sq.meter; at 40 miles it is ~ 6 curies/sq.meter. Such a deposit gives a radiation dose-rate of the order of 100 r/hr. Also deposited would be 400 μg Pu/sq.meter. We conclude that a rain-out could produce intolerable conditions at sporadic locations up to 100 miles or more from the reactor.

Analysis of the 2×10^{10} calorie explosion

Primary Effect

Computed for a flattened Hanford reactor, and this time assuming all the heat goes into uranium heating, the distribution in arbitrary classes will be:-

        5 central tons boil
     55 tons at the boiling point
    40 tons at $\sim 3700°$ C
    40 tons at $\sim 2500°$ C
    55 tons just melt
    55 tons just fail to melt

Behavior of the cloud

       Heat in cloud = $\sim 3 \times 10^9$ calories
 Estimated temperature = $\sim 4000°$ C
      Expanded radius = 20 meters
     Velocity of ascent = 13 meter/sec at 200 meters
                     = 3.3 meter/sec at 1000 meters

Cloud rises to ~ 800 meters.

Activity in the cloud

The cloud contains ~ 20 megacuries of fission products, 4.5 metric tons of uranium, 6 megacuries of Np, and 800 gms of plutonium. Under average conditions of turbulence, the maximum concentration at the ground is $\sim 3 \mu c$ f.p./liter or 1.8 μc/liter, corrected for decay. The cloud moves horizontally with velocity ~ 4 meter/sec, and touches the ground at times between 40 minutes and 5 hours, depending on the assumed average value of the diffusion coefficient C. A reasonable value is C = 0.1 (meter)$^{1/8}$. The time of maximum concentration at the ground is then 1½ hours after the ascent, and the distance is 13 miles. Figure 2 shows the time variation of concentration at this point. The integrated exposure is 550 (μc/liter)sec., and the estimated dose = <u>0.25 rep</u>.

Inhalation Figures

- Fission products — deposit 80 μc — initial dose-rate 70 mrep/day
  no significant hazard anticipated
- Plutonium — deposit 3×10^{-3} μg
- Neptunium — deposit 40 μc
- Uranium — deposit 20 μg.

The ratio of these deposits differs from the previous case for two reasons:

(1) different relative concentration in the <u>central</u> reactor tubes
(2) different <u>decay time of fission products</u>.

For general interest, the structure of an instantaneously generated cloud has been calculated with the following assumptions:

Height of center line = 800 meters
Velocity = 4 meter/sec.
C_x (downwind) = C_y (crosswind) = 0.2 $(meter)^{1/8}$
C_y (vertical) = 0.12 $(meter)^{1/8}$
Total activity, Q = 1 curie
Concentration, X in μc/liter.

time, t in secs.

The curves of equal activity are, of course, true ellipses in the vertical plane and circles in the horizontal plane. (Figure 3).

Mountainous Terrain

In this case, persons on either Rattlesnake Mountain (14 miles) or on Saddle Mountain (9 miles) could intersect the axis of the cloud. A representative time is 4000 sec. Q = $20 \times 10^6 \times 0.65$ (for decay) = 13×10^6 curies,

$$X = 13 \times 10^6 \times 3.7 \times 10^{-7} \text{ μc/liter} = \underline{5 \text{ μc/liter}}.$$

The integrated exposure is $\sim$ 2000 (μc/liter) sec.
Appreciable contamination of the mountain slopes would occur. The peculiar air currents over such ridges distort the polluted air stream, and contamination could occur on the far slopes.

Rain-out

The nominal maximum deposition is $\sim$ 800 curies/sq. meter. The value falls to $\sim$ 1 curie/sq. meter over the upper Wahluke Slope (5 miles).

Release of Activity from residual molten metal

The "junior atom bomb" cloud so far pictured in the 2×10^{10} calorie explosion is the minor hazard component. Left at the site is a mass of molten uranium, assumed not to be sucked up as the small hot bubble ascends.

HW-14058

The boiling points of the fission products are:

Element	B.P. °C	Element	B.P. °C	Element	B.P. °C
As	615 (subl.)	Cb	3300	Xe	gas
Se	688	Mo	3700	Cs	670
Br	58.8	43	> 2500 ?	Ba	1140
Kr	gas	Ru	> 2700 (vol.?)	La	1800
Rb	700	Rh	> 2500	Ce	1400
Sr	1150	Sb	1380	Pr	> 1000
Y	2500	Te	1390	Nd	> 900
Zr	> 2900	I	184	61	> 1000 ?

We now calculate that 440 megacuries of fission products will evaporate from the arbitrary metal classes above. This initiates a cloud whose initial height is comparable with that of smoke from a hot fire. At distances not too close to the reactor, the height will be inconsequential. Consider an average case [4]

where $C = 0.2$, $m = 1.75$, wind velo. $U = 5$ meter/sec

$$X \text{ (}\mu\text{c/liter)} = \frac{2 \times 4.4 \times 10^{11}}{\pi \times 0.008 \times (5t)} \; 2.62 \times e^{-r^2 / 0.04 \times (5t)^{1.75}}$$

where 2 is a ground reflection factor, and other terms are obvious.

$X = 120 \;\mu\text{c/liter at 11 miles } (= 1 \text{ hr})$

$ = 80 \;\mu\text{c/liter, corrected for decay}$

Integrated dose as cloud passes = $\sim$ <u>4 rep</u>

During an inversion (Sutton's bad case) assume $C = 0.1$, $m = 1.5$, and

$U = 3$ meter/sec.

$X = \sim 3 \times 10^4 \;\mu\text{c/liter at 11 miles}$

Integrated dose = $\sim$ <u>1200 rep</u>

Whereas Sutton's equations are well known to give reliable results for the average or zero-lapse rate case, there is reason to doubt the applicability to the marked inversion or bad case, in which the constants are rather dubiously related to physical reality, and wind shear is neglected. We choose an arbitrary model as alternate.

Assume the initial cloud has the dimensions of the reactor.

Concentration = $2.5 \times 10^8 \;\mu\text{c/liter}$.

P.E. Church's[5] Hanford meteorological data shows dilution during inversion = 10^4 at 1.2 miles.

∴ Conc. = 2.5×10^4 μc/liter

Sutton's data shows dilution = 20 between 1.2 and 11 miles for the bad case.

∴ X = 1.25×10^3 μc/liter at 11 miles.
= 750 μc/liter, corrected for decay.

Integrated dose = ∼ 60 rep. (cloud size is different now)

Let us weight these two cases, and write

Integrated dose in inversion = ∼ <u>400 rep at 11 miles</u> *

This approximates the human lethal dose, and in view of the uncertainties in such calculations, it fully substantiates the conclusion of the Reactor Safeguard Committee that the whole Wahluke Slope (radius ∼10 miles around the reactors) is <u>unsuited for settlement</u>.

This is done on the basis of the external radiation alone. Presumably far worse is the inhalation hazard. The computed deposition is ∼ 100 - 200 millicuries! The corresponding initial lung dose-rate is a few thousand rep per day. The highly speculative integrated dose in one month is a few tens of thousands of rep. Such a cloud is <u>conceivably lethal</u> at distances up to 50 or 100 miles, and <u>almost certainly damaging</u> in this range.

<u>Deposition of active material from an active cloud contiguous with ground</u>

With the limited library facilities at this site, we were unable to develop theoretical data on the feasible deposition on the ground as a cloud of particulate matter sweeps over it.

We resort to three "wild" methods from the Hanford experience, together with industrial pollution data.

(1) <u>Emission of long-lived fission products</u>
For the first three years of operation at Hanford Works, the concentration of fission products (excluding Xe, Kr and I) in the stack of a Separations plant was ∼ 10^{-8} to 10^{-7} μc/liter. Feasible average concentration at a point 6 miles away cannot exceed 10^{-13} to 10^{-12} μc/liter at ground level. There is agreement between the data of Church and Sutton on this. It assumes a usually 'bad' case, and the wind in one direction continuously.

Take 10^{-13} μc/liter as the representative concentration. The vegetation contamination at this time and place was ∼ 0.01 μc/kg. (excluding I). The vegetative cover approximates 2 kg/sq.meter. The practical deposition rate D

---

* This would agree with a Sutton calculation for initial cloud height =

∼ 100 meters and accurate <u>reflection factor</u>

for cloud concentration

$$1 \ \mu c/liter = \frac{10^{13} \times 0.01 \times 2}{3 \times 365 \times 24} \ \mu c/sq.meter/hr.exposure.$$

$$\therefore D = 7 \times 10^6 \ \mu c/sq.meter/hr.$$

This agrees with the calculation at 2 miles.

(2) **Emission of I^{131}**

By similar methods, $D = \sim 10^4 \ \mu c/sq.meter/hr$ at 6 miles

and $D = \sim 10^6 \ \mu c/sq.meter/hr$ at 30 miles.

assuming that the observer who samples the vegetation effectively looks at the average deposition within the last two weeks, as a result of radioactive decay.

(3) **Emission of discrete active particles**

Here the calculation hinges on a count of the number of radioactive particles per liter of air, and the number deposited per unit area of a horizontal catching frame in the same time, both observations being by radioautography. In a typical case, about 2 miles from the stack $D = \sim 10^4$ particles/sq.meter/hr per unit concentration.

(4) **Atmospheric Pollution**

In measurements in New York City, the average dust concentration in the air was 1.7 mg/10 c.meter, and the deposition rate was $\sim$50 tons per sq.mile/month.

Hence $D = 1.2 \times 10^5 \ \mu g/sq.meter/hr$ for 1 $\mu g/liter$

The concentration 1.7 mg/10 c.meter is known to correspond with 0.3 million particles per c. feet. The range of concentration in many principal cities is 0.2 to 10 million particles per c. ft. in short time samples. The reported deposition rate (longer term samples) ranges from 25 to 190 tons/sq. mile/month, (Washington, D.C., and St. Louis, Missouri, respectively). These ranges are compatible with the value $D = \sim 10^5$.

We shall assume that $D = \sim 10^5 \ \mu c/sq.meter/hr$ for cloud concentration of 1 uc/liter. This figure may be in error by a factor as large as 100. On the whole, the agreement between the quoted results for a fine aerosol on vegetation, iodine vapor on vegetation, radioactive particles on prepared surfaces, and for normal industrial pollution are surprisingly consistent.

Typical depositions will be

(a) from the 2×10^{11} calorie explosion cloud $\sim 10^4 \ \mu c/sq.meter$ at 100 miles

(b) from the 2×10^{10} calorie explosion high cloud $\sim 10^4$ μc per sq.meter at 13 miles

(c) from the 2×10^{10} calorie explosion low cloud ~ 50 curies per sq.meter at 11 miles

Fate of the molten metal

The residual 95 tons of uranium now contains 51 megacuries of fission products. Some writers have visualized such material running down to the river, and producing gross contamination. This appears most unlikely. If it is ejected violently it will have large surface area. If it cascades from the reactor most of it will be trapped in the pits around the unit, even if these are filled with discrete pieces of the structure after the explosion. We shall consider the naive model of the molten metal cascading from one face of the reactor at an average temperature of $2800°$ K. The initial area will be a square of side 12 meters, and we assume a comparable area maintained. By Stefan's Law, the rate of radiative cooling is:

Temperature °K	2800	2500	2000	1500
dT/dt (°C per sec)	27	17	7	2

The estimated cooling time is $\sim 3 - 5$ minutes, insufficient to permit significant travel of the fluid.* The cooled metal presents a local hazard, but the contained fission products can escape only by surface oxidation and weathering. This can be controlled.

4. Accidents leading to disruption of a reactor accompanied by fire.

In any accident which leads to exposure of molten metal to the atmosphere, it is believed probable that the metal would ignite. Consider the case in which the reactor is laid open by the primary event, and the 250 tons of uranium and 2200 tons of graphite burn. The rate of combustion will be so much a function of the local conditions that the combustion time cannot be computed. If the graphite is initially dispersed into particles of dia. 1 cm, and the initial temperature is $2800°$, the combustion time is ~ 4000 secs. It is reasonable to assume that the postulated fire will burn uniformly for at least two hours. The maximum environmental hazard will then be independent of the actual time of combustion, and can be computed from Sutton's equations for diffusion of smoke from a continuous point source.

---

* Experimentally, molten metal (mercury) flows toward the river at 30 ft./min. in a layer of comparable thickness.

Environmental hazard of combustion of all reactor material

The relevant equation is (4)

$$X \, (\mu c/liter) = \frac{2 \, (\text{for reflection}) \times 1000 \, (cc/liter) \times Q \times e^{\frac{-z^2}{C_z^2 \, x^m}}}{\pi \, C_y \, C_z \, U \, x^m}$$

where X = ground concentration at distance x meters

Q = emission rate = $\sim 8 \times 10^4$ curies/sec (for 2 hours)

C_y, C_z, m, as used before

U = wind velocity = 4.5 meter/sec.

z = height of source above ground (meters)

The appropriate value of z is unknown, probably of the order of 100 meters.

If $z < 50$ meters, X will differ from the value for $z = 0$ by less than 10% for $x > 4000$ meters. As we are primarily interested in the hazard at great distances, we shall take $z = 0$.

Atmospheric Stability

	Average case	Bad case
C_y =	$0.2 \, (\text{meter})^{1/8}$	$0.15 \, (\text{meter})^{1/8}$
C_z =	$0.12 \, (\text{meter})^{1/8}$	$0.10 \, (\text{meter})^{1/8}$
m =	1.75	1.5

Choice of wind speed

Although we calculate formally for $z = 0$, we picture z as ~ 100 meters. The wind speeds at 7 foot and 200 foot above ground are tabulated below for the Hanford meteorological station.

PERCENT FREQUENCY OF WIND SPEEDS

Wind Speed Class Intervals (mph) at 7-ft. Level

	0-3	4-9	10-14	15-19	20-24	25-29	30-39	40 or >
1st Quarter	53	37	6	3	1	L	0	0
2nd Quarter	27	55	13	4	1	L	0	0
3rd Quarter	32	51	13	3	1	L	0	0
4th Quarter	55	33	8	3	1	L	0	0
Annual Mean	42	44	10	3	1	L	0.	0

Wind Speed Class Intervals (mph) at 200-ft. Level

	0-3	4-9	10-14	15-19	20-24	25-29	30-39	40 or >
1st Quarter	27	32	21	11	5	2	2	L
2nd Quarter	8	41	23	15	8	3	2	L
3rd Quarter	11	42	21	14	8	3	1	L
4th Quarter	25	36	18	10	6	4	1	L
Annual Mean	18	38	21	13	7	3	2	L

These wind speeds are averaged over periods of one hour, and apply well to the present case. A wind speed of 10 mph (= 4.5 meter/sec) is a reasonable average for the estimated height of the source. X is inversely proportional to U, so that the most conservative case (say 2 mph wind) can be computed by multiplying the given values of X by 5.

Concentration of fission products (uncorrected for decay)

Distance (miles)		3	6	12	19	30	60
Conc. μc per liter	Average case	160	48	14	7	3	0.8
	Bad case	2080	750	270	144	64	16
	Bad case $Z = 100$	1	60	80	80	40	16

First order width of cloud (meters)

Distance (miles)	3	6	12	19	30	60
Average case	1070	1970	3600	5150	8000	15,000
Bad case	280	470	800	1080	1600	2,700

Integrated external dose

Assume an average available energy of 1.7 MEV per disintegration (chosen to agree with the Reactor Safeguard Committee's power calculation of dose). The integrated dose computed for a hemisphere without scatter correction is:-

Integrated Dose in rep

Distance (miles)		3	6	12	19	30	60
No decay	Average case	135	40	12	6	2.5	0.7
	Bad, Z = 0	1780	640	230	125	55	15
	Bad, Z = 100	1	50	70	70	35	15
Decay at $U = 4.5$ m/sec	Average case	110	30	8	3.5	1.4	0.3
	Bad, Z = 0	1470	460	150	75	30	7
	Bad, Z = 100	1	35	45	40	20	7

Values for no decay are included to facilitate recalculation for wind speeds other than 10 mph. For speed x miles per hour,

$$\text{integrated dose} = \frac{10}{x} \text{ (appropriate value in table)} \times \text{(decay factor)}$$

This cloud gives lethal doses up to $\sim$ 6 miles in the bad case if there is no initial rise, and gives troublesome doses up to about 50 miles.

Inhalation figures

The retained amounts in the lung (50% retention) are:

Distance (miles)		3	6	12	19	30	60
Deposit (mc)	Average	130	36	10	4	1.7	0.4
	Bad, Z = 0	1,760	560	180	90	36	8
	Bad, Z = 100	1	45	50	50	22	8
Initial dose-rate rep/day	Average	6,700	1,850	510	210	90	20
	Bad, Z = 0	90,000	29,000	9,300	4,600	1,800	400
	Bad, Z = 100	45	2,300	2,800	2,500	1,100	400
Total dose first month (rep)	Average	52,000	14,000	4,000	1,600	700	150
	Bad, Z = 0	700,000	220,000	72,000	35,000	14,000	3,000
	Bad, Z = 100	350	18,000	21,000	20,000	9,000	3,000

The crude total dose calculation assumes a rapid elimination of a further 25% of the inhaled amount, with radioactive decay of the rest. Neglected are such factors as the failure to retain gas (e.g., Xe, Kr), and the rapid transmission of some radioisotopes through the lung wall. For the bad case, plutonium deposition in the lung is 35, 12, 7, 3, 0.7 μg at 6, 12, 19, 30 and 60 miles, respectively. This will ultimately be damaging at $\sim$ 30 miles. Neptunium deposition is approximately one third that for fission products.

Deposition on the ground*

If $D = 10^5$ μc/sq.meter/hr per unit concentration, the ground contamination (fission products) is:-

Distance (miles)		3	6	12	19	30	60
Ground Contamination curies/sq. meter	Average	30	10	7	1.5	0.5	0.2
	Bad, Z= 0	400	150	55	30	12	3
	Bad, Z= 100	0.2	12	16	16	7	3
Initial Dose-rate roentgens per hour	Average	200	70	50	10	3	1
	Bad, Z= 0?	3000	1000	400	200	80	20
	Bad, Z= 100	1.5	80	120	120	50	20

All these numbers are freely rounded off. For the dose rate from ground contamination, we have assumed 1 curie/sq.meter. $\longrightarrow$ 7 r/hr. This is greatly affected by the self-absorption of vegetation. A general conclusion is that the ground would be temporarily dangerous at distances up to $\sim$ 50 miles.

Rain-out

Estimated maximum rain-out is 170, 40, 20, 8, 3 curies/sq.meter at 6, 12, 19, 30, and 60 miles, respectively.

---

*Calculations ignore weakening of cloud by deposition.

HW-14058

Mountainous terrain

The foregoing calculations for inversion conditions are extremely sensitive to the assumed initial height of the smoke plume. The actual concentration at ground level on a true horizontal plane is conjectural. However, the initial rise cannot exceed a few hundreds of meters. The local terrain (Figure 4) is such that the plume could intersect the ground at distances $\sim$ 6 miles. The given values for the bad case with $\underline{Z} = 0$ would apply. Some dilution must occur in the primary ascent, but this will be an incalculable function of the rate of combustion and the geometry of the fire. A factor of 10 can perhaps be applied. Conditions on the Wahluke Slope can then be computed from the above data.

5. Circumstances leading to Columbia River Contamination

So far it has been considered that all the available energy will be used up as heat. Undoubtedly, some will go into physical disruption of the unit, and ejection of all sizes of particles from chunks down to submicron size.

The most naive picture shows all the reactor material spread uniformly in a circle of radius x feet about the reactor P.

Let p (feet) = perp. dist. from P to the far bank of the river

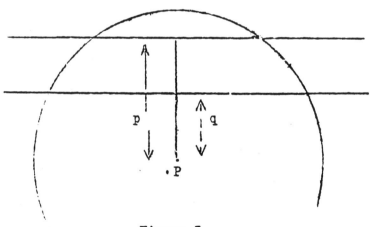

Figure 5

and q (feet) = perp. dist. from P to the near bank of the river.

The area of the circle between the parallel chords is:

$$A = p\sqrt{x^2 - p^2} + x^2 \sin^{-1} \frac{p}{x} - q\sqrt{x^2 - q^2} - x^2 \sin^{-1} \frac{q}{x}$$

The ratio of this to the whole area of the circle is $A / \pi x^2$

This has a maximum value when $\dfrac{d\left(\dfrac{A}{\pi x^2}\right)}{dx} = 0$.

i.e.: if $x = \sqrt{p^2 + q^2}$

$$\dfrac{A}{\pi x^2} (\text{max}) = \dfrac{1}{\pi}\left(\sin^{-1} p/x - \sin^{-1} q/x\right)$$

Representative values for each Hanford reactor are:

$p = 3200$ and $q = 2000$, whence

Maximum fraction falling in river = $\underline{0.145}$

We now elaborate the model to the Jericho Case, in which we picture the reactor shield walls to fall down, and the upper hemisphere of an assumed spherical reactor to be initially projected radially from the center with equal force in all directions. The radial elements are assumed to break up after the initial velocity is established. We further assume that 10% of the maximum possible energy 2×10^{11} calories goes into energy of motion. On this simple model, the maximum range is $\sim$ 1000 meters, comparable with the worst value of x in the previous case. The trajectories of all particles and the ground pattern of contamination can be computed. This pattern can be approximately fitted to a curve of the type

Contamination per unit area at radius $\underline{r}$ = $A e^{-b r^2}$

in which the constants are related by:-

Total activity in half the reactor = $\displaystyle\int_0^\infty A e^{-b^2 r^2} \times 2\pi r\, dr = \dfrac{\pi A}{b}$

The fraction of activity thrown into the river can be calculated for various "reasonable" values of $\underline{b}$. Such integrations show between 2.5% and 7.0% in the river.

We postulate that in a 2×10^{11} calorie explosion most of the fission products will ascend in the primary hot bubble. Conversely, the 2×10^{10} calorie explosion has insufficient energy to give maximum coverage of the river. * We shall assume that not more than 1% of the total activity is directly thrown into the river. This is $\sim$ 5.8 megacuries of fission products, and proportionate amounts of other activities. The active particles will fall at high

---

* These calculations are admittedly crude. A firmer opinion is being sought from competent experts in explosives.

velocity on the simple model. If highly pulverized, much could escape upwards. We shall assume uniform mixing of the full quota throughout the depth of the river in a strip approximately 4000 feet long.*

   Width of river   = 1200 feet
   Effective length   = ~ 4000 feet
   Average flow   = 120,000 c.ft/sec.
   Average velocity   = ~ 3 mph = 4.4 ft/sec.
   Volume contaminated = 8.2×10^7 c.ft. = 2.3×10^9 liters.

Concentration:-

   Fission products - ~ 1.4 mc/liter
   Plutonium   - ~ 0.1 µg/liter
   Neptunium   - ~ 0.5 mc/liter
   Uranium   - ~ 0.5 mg/liter

Alternate method of contaminating the river

This refers back to the case of burning of all the reactor material. Under inversion conditions, it is feasible for the smoke plume to drift toward the river, and then be caught in the peculiar air stream which flows down ** the Columbia River and is directed over the bend of the river by the White Bluffs on the far side for about 20 miles. Take the simple model in which the plume is centered over the river at a height of 100 meters (reactor is 25 meters above the river). Assume that the deposition in the river is the same as that calculated for the ground.

   Deposition rate = 12, 16, 16 curies/sq.meter at 6, 12, and 19 miles.

The average deposition across the river (360 meters) can be calculated from the known cross river pattern of the cloud.

   $\dfrac{\text{Average deposition}}{\text{Axial deposition}}$ = 0.7, 0.8, 0.9 at 6, 12, and 19 miles.

   Total contamination = ~ 0.8 (6x14 + 7x16) curie-miles/sq.meter x
   5.8×10^5 sq.meter/mile = ~ __90 megacuries__

---

*Most of the fission product material is expected to be oxidized. These oxides are normally insoluble, but at low concentration the solubility product is exceeded, and the material is postulated to hydrolyse.

**or __up__ the river

Colonel Holzman[6] estimates that oxides in the hot bubble of the atomic bomb cloud agglomerate to particles of diameter $0.1\,\mu$. Assume this happens in the cooler fire considered. Assume that all the particles in the present cloud are normal inactive particles uniformly coated with an active layer $0.1\,\mu$ thick.

Fission product oxides have specific gravities between 3.9 and 6.9.

Assume mean coating density = 5 gm/cm^3

mean carrier material density = 2.5 gm/cm^3

For any coated particle of final radius r microns.

$$\frac{\text{Active mass}}{\text{Total mass}} = \frac{0.6\,(r - 0.1)}{r^2 + 0.3\,r - 0.3} \quad \text{to 2nd order}$$

$$= \frac{0.6}{r + 0.3} \quad \text{to 1st order (good if } r > 1\,\mu)$$

The distribution of activity as a function of particle size is now calculated from the Alamagordo data, assuming same distribution applicable here.

Diameter μ	Relative number of Particles	Estimated Relative Mass	Active mass / Total mass	Relative Activity
0. - 3.3	98.3	98	0.75	73
3.3 - 4.7	0.94	60	0.26	15
4.7 - 6.6	0.55	110	0.18	20
6.6 - 9.4	0.10	50	0.14	7
9.4 - 13.2	0.07	74	0.11	8
13.2 - 18.7	0.04	120	0.08	9
18.7 - 26.5	0.008	83	0.05	4
Coarse grains	--	--	v.low	--

The rate of fall of particles in air (Bagnold's data corrected to dens.5)[7] is:-

Diameter μ	Velocity meter/sec.	Fall in 5 hrs. meters	Percent Cloud * content falling
9.4	2×10^{-2}	300	75
6.6	5×10^{-3}	100	50
3.3	8×10^{-4}	16	8
~1.	2×10^{-4}	4	2

---

*Calculated from vertical thickness of plume, allowing for reflection at the water surface, but not for vertical pattern of concentration. Cloud moves downriver at 4 mph.

Activity in river is:-

Class μ	Percent Total Activity	Percent Class Falling	Percent Activity in River
> 9.4	15.4	75	11.5
6.6 -9.4	5.1	50	2.5
3.3 -6.6	26	8	2.0
1 -3.3	~ 40 ?	2	2.0
		Total	18

Eighteen percent for the mean decay time is $\sim$ 50 megacuries.

In comparison of the two methods, the first should be adjusted to the lower wind speed of the second, giving $\sim$ 200 megacuries. The agreement is satisfactory. The value 50 megacuries appears more probable, as the higher value is incompatible with the observed rate of disappearance of fine smoke near the ground.

This activity descends in a peculiar pattern distorted by the motion of the river. The initial mixing in the river is probably very low. However, we shall assume 50 megacuries uniformly dispersed in $\sim$ 15 miles of river. The average concentration is 1.1 mc/liter. This is the same as that calculated for direct dumping in the river, but the potential nuisance is greater by a factor of 20 due to the longer contaminated strip.

6. Speculations on biological effect on man

Permissible exposure -- ingestion toxicity
Known to this writer is no significant improvement on the early calculations of W.Cohn[8] with respect to the real hazard of fission product ingestion. Cohn's methods are adopted, with the current more conservative radiation limits, and occasional use of newer biological constants (Morgan, Brues, etc.) where these seem well-founded.

Fission Products

(a) Lethal dose, LD 50 in 30 days
Single dose of 1000 rep delivered to the gut in 24 hours from a single ingestion is given by $\sim$ 30 mc, because the radiation at the gut wall will be approximately one-half that at the center of the filled tube.

Lethal concentration in water = $\sim$ 30 mc/liter (single drink)
= $\sim$ 10 mc/liter (contamination persists one/day)

If 30 mc fission products are ingested, the bone marrow deposition may be ~ 1.2 mc, and the initial dose-rate ~ 40 rep/day. The integrated dose is ~ 500 rep. While probably not lethal in 30 days, this is presumably damaging. There is a rather close balance between the gut radiation and the skeletal deposition as the principal damaging agent. *

(b) <u>Permissible Emergency Concentration for 1 day</u>

Gut damage - assume 100 rep (comparable with fluoroscopy) is insignificant.

Permissible emergency concentration = ~ 3 mc/liter (single drink)

= ~ 1 mc/liter (for 1 day)

Deposition in bone marrow - as the fission products age, the relative concentration of dangerous long-lived products, especially those prone to deposit in bone increases. Assume 100 rep (integrated dose) to the bone marrow is tolerable. Permissible emergency concentration (30 day material) = ~ 2 mc/liter (single drink)

= ~ 0.7 mc/liter (for 1 day)

(c) <u>Permissible emergency concentration for 1 year</u>

The values according to K.Z. Morgan[9] can be adjusted with the new radiation exposure limit to -

Permissible emergency concentration = 1 uc/liter (for 1 year)

This is higher by a factor ~ 1000 than the planned waste disposal limit of an atomic energy installation.

---

* Appropriate values of LD 50/30 are rather poorly known. In the animal, experimentation has concentrated on other modes of entry, or on the transmission to other organs of orally administered material. One value can be deduced from the ingestion of Y^{91} by rats [14]. Animals fed 1 - 6 mc lived until eventually killed by adenocarcinoma of the colon. Take the mass of gut plus contents as 5% body weight (standard 1 year rat = 280 gm.). Estimated dose = 6000 rep.

Flaskamp[15] reports that an X-ray dose of 140% SED (= ~ 850 r) to the colon does not produce irreparable damage. The writer estimates that in man, the LD 50/30 for gut irradiation is 2000 - 3000 rep.

We adhere to the 1000 rep of the tabulations, because the higher figure would probably lead to intolerable depositions in other organs, and while not a true LD 50/30 it may well be a representative lethal dose.

Plutonium

(a) <u>Lethal dose LD 50 in 30 days</u>

Gut as limiting organ receives 1000 rem (= 50 rep for alpha radiation) from ~ 5 mg Pu.

Lethal concentration = ~ 5 mg Pu/liter (single drink)

Note that concentrations of the order of 1/3 of this may deposit enough plutonium in the skeleton to be <u>ultimately</u> lethal.

(b) <u>Permissible Emergency Concentration for 1 day</u>

The skeletal deposition is of the order of 0.05% of the ingested dose, and the permissible skeletal deposition is ~ 0.1 ug Pu. *

∴ Permissible emergency concentration = ~ 200 µg/liter (single drink)

= ~ 60 µg/liter (for 1 day)

(c) <u>Permissible Emergency Concentration for 1 year</u>

Permissible emergency concentration = ~ 1.5×10^{-2} µg/liter.

Neptunium

Lethal concentration = ~ 100 mc/liter (single drink)
= ~ 35 mc/liter (one day)

Permissible Emergency concentration = ~ 10 mc/liter (single drink)
= ~ 3 mc/liter (one day)
= ~ 1 µc/liter (for one year)

Apparently the neptunium need only be considered separately in a circumstance in which primary plutonium is absorbed out, for example, by seepage through soil, and neptunium is transmitted. The neptunium may then become a carrier of plutonium to the skeleton. No case is currently visualized in which this effect would be significant.

Uranium

Lethal concentration (radiation effect) = ~ 6×10^5 mg U/liter (single drink)

but chemical toxicity probably intervenes at ~ 700 mg U/liter if the material is in soluble form.

---

* Hamilton[16] quotes oral absorption as 0.007% with 65% going to bone. i.e. Skeletal deposition = 0.0045%. Morgan writes skeletal deposition = 0.03%. We consider this critical value to be inadequately known, especially for low concentrations. A definitive experiment on this is in progress at Hanford Works. Meanwhile we use the more conservative value of 0.05%.

Uranium - continued

Permissible emergency concentration (soluble U compounds)
= ~ 5 mg U/liter   (single drink)
= 1.7 mg/liter   (one day)
= 250 µg/liter   (for one year)*

Inhalation toxicity

The inhalation toxicity depends on

(1) retention of particles - we assume arbitrarily a primary retention of 50%, with rapid removal of one-half of this which then presents an ingestion hazard. The residuum is supposed to be eliminated with a biological half-life of 2 months;
(2) transmission through lung wall - bromine, rubidium, strontium, ruthenium, iodine, cesium and barium, will be assumed rapidly soluble, and all other active materials totally insoluble;
(3) hazard of discrete particles - this will be neglected here, although it is a conceivable limiting hazard.

Fission Products

For simplicity, the composition at 2 hours after the explosion will be used throughout. Of the calculated depositions, the percent distribution is:

Lung insoluble	62%
Lung soluble	
bone-seekers	14%
iodine	12%
others	5%
Gas - escapes	7%

Lethal deposition (LD 50 in 30 days) = ~ 20 mc.
(assumes all deposit stays for one day and gives 1000 rep to the lung tissue - this is conservative).

---

* These values for chemical toxicity of uranium are taken from Rochester Reports (Neumann, et al).

Fission Products - continued

Permissible emergency deposition P.E.D. = 0.3 mc.*

Such a deposit may send 36 uc radioactive iodine to the thyroid gland, where it produces 63 rep/day initially, but a total of ~ 80 rep due to rapid decay of all except I^{131} and I^{133}.

Similarly, the 54 uc bone seekers, if 50% is retained in bone, gives ~ 0.2 rep/day, for a total of ~ 3 rep.

---

* Approximate calculation as follows:

0.3 mc resides in lung for 1 day; dose in 1 day

$$= \frac{62 \times \text{Energy per dis (MEV)} \times \text{deposited amount (mc)}}{\text{Lung mass (kg)}} = \frac{62 \times 1 \times 0.3}{1.2} = \underline{15.5} \text{ rep}$$

Thereafter, 25% of the deposit has been eliminated (ciliary action, etc.) and ~ 60% of the remainder is lung insoluble. The integrated dose in 2 months is

$$\int_1^{60} \frac{\text{initial daily dose}}{t^{0.3}} e^{-\frac{.693 t}{60}} dt = \sim 17 \times \text{initial daily dose}$$

∴ additional integrated dose $= \dfrac{17 \times 62 \times 1 \times 0.5 \times 0.6 \times 0.3}{1.2} = \underline{79} \text{ rep}$

Total dose = 96.5 rep = ~ 100 rep.

Note that this is conservative on two counts:-
(1) the revised value of 1 rep is 93 ergs/gm[10] instead of the earlier 83 ergs/gm used here.
(2) Following Cohn, the average energy per disintegration is taken as 1 MEV, unless otherwise specified. From Wigner and Way [11] [12] we have

Time After Fission Days	Average beta energy per beta disintegration MEV	Average gamma energy per beta disintegration MEV
10	0.35	0.7
100	0.3	0.45

Our use of curie has the customary questionable meaning of 3.7×10^{10} beta disintegrations per sec.

In body tissues, not all the gamma energy will be absorbed. About 30% will be absorbed in gut or lung. We are concerned with times in the general order of 10 to 200 days on the above time scale.

The average energy is therefore about (0.35 + 0.21) to (0.3 + .15) MEV = ~ $\underline{0.5 \text{ MEV}}$

Plutonium

Lethal deposition (LD 50/30) = ~ 2 mg ?

P.E.D. = ~ 5 μg

This is not much higher than the normal permissible deposition ~ 2 μg.

Neptunium

Lethal deposition (LD 50/30) = ~ 50 mc

P.E.D. = ~ 1.5 mc

Uranium

Lethal deposition (LD 50/30) = 270 gm.

P.E.D. = 650 mg.

Application of speculative exposure limits

We have drawn a series of pictures of the potential air, ground, and river contamination arising from a reactor explosion. Along with these should be considered the somewhat different pictures of the P.L.P. reports. Enough data has been included so that the reader who prefers other postulates can deduce the corresponding contamination. In applying the equally speculative exposure limits we shall choose the bad cases, but not necessarily the worst cases for elaboration. For example, if the major damage could arise from a particular explosion on a day of minimum wind velocity, with the wind in just the right direction to put the cloud over the river, with a severe rain-out at that time, and the river at minimum flow, it will be considered inconceivable that all these disadvantages should arise simultaneously. An exception will be made for cloud deposition in the river, where incomplete mixing of contaminants with the water is physically probable.

(1). **Immediate Effects of River contamination on man**
Table 6-1 exhibits the predicted contamination in drinking water from the Columbia River, subjected to representative water treatment. The following assumptions are involved:

(a) Sand filter beds remove 90% of the plutonium and no other activities.*
(b) Ground filtration (recharge systems) removes 100% Pu and 90%

---

* This is conservative because sand is said to remove fission products by the following schedule: (13)

Time of Contact	Percent Removal
1 min.	8
15 min.	20
1 hr.	37
6 hrs.	64
24 hrs.	78

HW-14058

TABLE 6-1

Comparison of Concentrations in Water Systems with damaging concentrations

Material	LD-50 1 drink	LD-50 1 day	P.E.C. 1 drink	P.E.C. 1 day	Concentration in River (Blow-Up)
FP	30 mc/L	10 mc/L	3 mc/L	1 mc/L	1.4 mc/L
Pu	5 mg/L	1.7 mg/L	0.2 mg/L	0.07 mg/L	0.1 μg/L
Np	100 mc/L	35 mc/L	10 mc/L	3 mc/L	0.5 mc/L
U	700 mg/L	230 mg/L	5 mg/L	1.7 mg/L	0.5 mg/L

Concentration in Water Distn. System

Material	Direct (no filtration)	Sand Filter Beds	Ground Filtration
FP	1.4 mc/L	1.4 mc/L	0.14 mc/L
Pu	0.1 μg/L	0.01 μg/L	0.00 μg/L
Np	0.5 mc/L	0.5 mc/L	0.005 mc/L
U	0.5 mg/L	0.5 mg/L	0.05 mg/L

For humans drinking 1 Liter, fraction they will receive of the

Material	LD-50 Direct	P.E.C. Direct	LD-50 Sand Filter	P.E.C. Sand Filter	LD-50 Ground Filter	P.E.C. Ground Filter
FP	0.047	0.47	0.047	0.47	0.0047	0.047
Pu	2×10^{-5}	5×10^{-4}	2×10^{-6}	5×10^{-5}	0	0
Np	5×10^{-3}	5×10^{-2}	5×10^{-3}	5×10^{-2}	5×10^{-5}	5×10^{-4}
U	7×10^{-4}	0.1	7×10^{-4}	0.1	7×10^{-5}	0.01

For humans drinking 3 Liter, fraction they will receive of the

Material	LD-50 Direct	P.E.C. Direct	LD-50 Sand Filter	P.E.C. Sand Filter	LD-50 Ground Filter	P.E.C. Ground Filter
FP	0.14	1.4	0.14	1.4	0.0014	0.14
Pu	6×10^{-5}	1.4×10^{-3}	6×10^{-6}	1.4×10^{-4}	0	0
Np	1.5×10^{-2}	0.15	1.5×10^{-2}	0.15	1.5×10^{-4}	1.5×10^{-3}
U	2.1×10^{-3}	0.3	2.1×10^{-3}	0.3	2.1×10^{-4}	0.03

HW-14058

fission products, neptunium and uranium.*

The feasible eventual slow leaching of activity from the filter beds is not considered here. The two types of explosion considered give initial strips of contaminated water respectively < 1 mile and 15 miles long. With an adequate warning system, currently under development by the Disaster Control Committee, municipal systems could be by-passed with a generous margin for expansion of the strips by mixing and diffusion.

The tabulation predicts that all exposures will be below the relevant LD 50/30 by a safe margin. Borderline will be the ingestion of fission products and uranium relative to the permissible emergency concentration. This assumes that a storage tank has been refilled during the time of passage of the contaminated strip. Not only is this unlikely, but also the holdup will introduce a useful decay factor for fission products ($\sim$ 3-fold in one day).

If the reactor burns up and the river contamination is acquired from the contacting cloud, the mixing in the river may be far from uniform. At low flow, when the possible dilution is least, water intakes will tend to be near the river surface, and the surface-contaminated layer could be directly drawn into the system. If the water activity is assumed 100 times greater than before, the dose is well into the lethal range. At 10 times greater, the P.E.C. is clearly exceeded, and injury would be anticipated.

Listed in Table 6-2 are communities, populations, sources of sanitary water, and the fractional LD 50's and P.E.C. they will receive if no effort is made to have their pumping stations turned off temporarily. (This table has been abstracted from as complete a listing as could be provided by State officials, of all towns between Hanford Works and the mouth of the Columbia river, using the Columbia River for sanitary water). The concluding figures are based on drinking contaminated water from the various systems for one day. They may be uniformly divided by 3 to find fractions of LD 50's and P.E.C. for people taking but one large drink (1 liter) before being warned. Longer periods are not considered since a supply showing contamination in excess of 1 μc/liter of fission product, 1.5×10^{-2} μc/liter of Pu, 1 μc/liter of neptunium, or 0.25 mg/liter of uranium, a few weeks after the emergency should be decontaminated or abandoned. Table 6-2 includes the decay factor of fission products travelling downstream. Without such a factor, the exposures in all towns except Richland, for the burn-up case (calculated as 10 times the previous concentration) would exceed the LD 50/30. In view of the many uncertainties in computation, we can only conclude that a <u>critical</u> condition <u>may</u> arise in the water supply of any downstream community using the Columbia River.

---

*This is possibly optimistic. Anions (Br, I) and Group I cations (Rb, Cs) will be poorly retained. Of these, only I^{131}, I^{133}, and Cs^{137}, are present in sufficient yield and long enough half-life to merit examination.

UNCLASSIFIED

HW-14058

TABLE 6-2

Community	Population	Source of Sanitary Water	Blow-up, 1 days drinking, receiving fraction of		Burn-up, 1 days drinking, receiving fraction of	
			LD-50	P.E.C.	LD-50	P.E.C.
Richland, Wn.	25,000	2-Y W + 1-C W	0.0002	0.002	0.002	0.02
Kennewick Irrigation Proj.	400	C	0.06	0.6	0.6	6.
Pasco, Wn.	8,000	CF	0.06	0.6	0.6	6.
McNary, Ore.	1,500	CF	0.04	0.4	0.4	4.
Arlington, Ore.	600	CF	0.02*	0.2*	0.2*	2.

Key:

YW = Yakima River water stored in earth basins to raise table for wells

C = Columbia River water, no filtration.

CF = Columbia River water, sand filtered.

* $\frac{1}{2}$ calculated value due to dilution from tributaries.

(2) **Immediate Effects of river contamination on lower forms** (17, 18, 19, 20, 21, 22)

River Population

(1) Fish
During the summer run, there are about 300,000 adult salmon and trout in the Columbia River, based on counts at Bonneville Dam and estimated fishing mortality. With the normal 20% survival of potential spawn of adult salmon, and assuming half the young fish in the Lower Columbia, there will be 85 million young migrant salmon.

Other migratory fish (smelt and shad) are in the Lower Columbia, and sports fish, (bass, whitefish and sturgeon) are found throughout the section below the Hanford Works. Coarse fish are abundant. From test samples, the relative populations can be set at

Adult salmon and trout	300,000
Young salmon and trout	85,000,000
Other fish	1,500,000,000
Total	~ 1,600,000,000

This represents 1000 fish per linear foot of river.

(1-b) Other aquatic vertebrates
There is no important form. Lamprey eels have been included as fish.

(2) Micro-organisms
Near the Hanford Works there is 5 lbs. of algae and associated microscopic forms per 100 sq. ft. of productive bottom. If such bottom averages 500 feet from either bank from Hanford Works to the mouth, there is 40,000 tons of algae in the river. Planktonic organisms are a special case, because they will float downriver with the contaminated water. The mass involved for the "blow-up" case is ~ 1 ton, and for "burn-up" is ~ 17 tons. The mass is approximately 9 parts phytoplankton to one part zooplankton.

(3) Insect larvae and associated forms
At Hanford, larvae, snails, etc., approximate 2.7 lbs. per 100 sq. ft. productive bottom. The estimated total mass is ~ 22,000 tons.

Concentration factors

(1) Fish   (21)
From data of Prosser, et al, the concentration of fission mixture (disregarding differences in age of mixture, etc.) is estimated as:

Tissue	Concentration Factors	
	Blow-up ~ $\frac{1}{2}$ hr. exposure	Burn-up ~ 10 hrs. exposure
Intestine	10	100
Bone	0.2	1.5
Viscera	1	2
Muscle	1	2

UNCLASSIFIED

HW-14058

Hypothetical values for plutonium and uranium are derived as follows:

Some will be ingested with food, and perhaps 0.1% Pu and 10% U absorbed from gut. For an average fish (200 gm), 10 grams of food is estimated to contain < 0.01 µg Pu, which deposits < 10 µµg Pu ($= \sim 1$ mµg/kg). Uranium deposit is ~ 5 ug. These amounts are small compared with those in the water. Ion exchange across gill membranes should equilibrate the specific activity in the fish with that in water (0.1 µg Pu/kg and 0.5 mg U/kg, respectively).

(2) Micro-organisms

The estimated concentration factor for fission products in bottom forms is ~ 1000 for a 10-hour exposure, and ~ 100 for one-half hour, based on the fecal activity of Prosser's goldfish, and Hanford data on concentration factors of other radio-elements. Plutonium and uranium concentration factors are arbitrarily taken as 2 in both cases.

Plankton travelling with the contamination is estimated to have a concentration factor (F.P.) of ~ 1000. Arbitrarily, the factor for plutonium or uranium is written as 10.

(3) Insect larvae and associated forms

By generous extrapolation of Hanford data, the average concentration is written as 50 for $\frac{1}{2}$ hour (blow-up) and 500 for 10 hours (burn-up).

The concentration of available contamination is exhibited in Table 6-3, in comparison with crudely estimated damaging doses. The $LD_{50/30}$ is that estimated to give 3000 rep to the gut in fish.

In the above picture, damage is confined to:

(1) Fish in Case III for fission products
(2) Larvae in Cases II and III for fission products
(3) Borderline chemical toxicity of uranium, possibly of all forms.

(3) Delayed effects on aquatic forms -- food chains

Plankton of the Columbia River consists mainly of diatoms. These and the filamentous algae of the river bottom concentrate radioisotopes from the water and subsequently provide contaminated food for higher forms. The diatoms are probably the most efficient concentrators of activity in the river, other things being equal. Here they are unequal because they float freely in zones of maximum activity, and travel with the contaminated water. Possible intake into unfiltered water systems has to be considered. An important distinction is that, if not captured enroute, they escape in about 5 days from the biological system of the river. All other forms will tend to set up a biological chain of activity transfer in the river.

Bottom forms include the algae, protozoa and insect larvae. All these participate in the food chain. For the customary Hanford

TABLE 6-3

Concentration of Radioactivity in aquatic organisms

HW-14058

Contaminant	Damaging exposures				Conc. in River	Specific Activity (rounded-off)						
	LD 50			PEC		Fish				Algae	Plankton	Larvae
	Fish	Algae	Plankton	Larvae	Fish		Viscera	Bone	Muscle	Gut			
Fission Products mc/kg	100	10,000	10,000	500	(viscera)	I 1.4	2	0.5	2	15	140	2800	70
						II 1.4	3	2	3	140	1400	2800	700
						III 14	<u>30</u>	<u>20</u>	<u>70</u>	1400	<u>14,000</u>	<u>28,000</u>	<u>7,000</u>
Plutonium mg/kg	15	1,500	1,500	75	1	I 10^{-4}	$<10^{-3}$	$<10^{-3}$	$<10^{-3}$	$<10^{-3}$	$<10^{-3}$	$<10^{-3}$	$<10^{-3}$
						II 10^{-4}	$<10^{-3}$	$<10^{-3}$	$<10^{-3}$	$<10^{-2}$	$<10^{-3}$	$<10^{-3}$	$<10^{-3}$
						III 10^{-3}	$<10^{-2}$	$<10^{-2}$	$<10^{-2}$	<0.1	$<10^{-2}$	$<10^{-2}$	$<10^{-2}$
Neptunium mc/kg	300	30,000	30,000	1500	20	I 0.5	0.5	0.1	0.5	<5	5	5	3
						II 0.5	1	<2	1	5	5	5	3
						III 5	10	10?	10	50?	50	50	25
Uranium mg/kg	100	200	200	200	1	I 0.5	Chemical toxicity basis						
						II 0.5	includes concentration factors						
						III <u>5</u>							

NOTES: All values highly speculative, except for fission products. Underlined values are those Uranium calculated as for lead poisoning. possibly above a safe limit.
Case I - Blow-up of reactor
Case II - Burn-up
Case III- Burn-up, and non-uniform mixing of deposits in river.

effluents, aquatic insect larvae, such as of caddis flies, which feed on plankton, or midge larvae that feed on algae and plankton, acquire about the same specific activity as their food. Forms such as the may fly which are partly carnivorous are less active by a factor of two. Carnivors, such as crayfish, acquire one-fifth the activity of plankton feeders. Similarly, the carnivorous game fish are less active than carp or suckers.

In application to the current contamination, it follows that coarse fish, escaping direct injury by ingestion of fission products could easily acquire a lethal dose from contaminated algae. Such forms as adult bass or trout would probably avoid this fate. Young salmon or trout, still feeding on micro-organisms would be likely candidates for extinction.

No attempt has been made to calculate the distribution pattern of activity in the water. Known from experience with reactor-cooling effluents is that mixing from a small source is incomplete 20 miles downriver, but approximately complete at 40 miles. Bottom forms probably have access to the explosion contaminants 10 to 20 miles below the reactor. Their concentrating action will unbalance the concentration. If all the algae in the river were exposed to the active water, <u>all</u> the activity could be removed:-

$$
\begin{aligned}
\text{Mass of algae} &= 40{,}000 \text{ tons} = 36 \times 10^6 \text{ kg.} \\
\text{Activity of water} &= 1.4 \text{ mc/kg} \\
\text{Concentration factor} &= \sim 1000 \\
\text{Activity of algae} &= \underline{50} \text{ megacuries}
\end{aligned}
$$

A <u>plausible</u> distribution of the initial 50 megacuries from the burn-up case is:-

Consumer	Activity curies	Location
Algae	7,500,000	10-200 miles downriver
Fish	1,000,000	throughout the river
Larvae	2,000,000	10-200 miles downriver
Plankton	45,000	carried to sea
Irrigation	60,000	distributed
Water supplies	30,000	as listed
Animals & birds	500	distributed
Balance	$\sim$ 40,000,000	carried to sea

Relatively superficial contamination appears feasible in the burn-out case. This materially alters the pattern in the river. Contamination of algae, etc. would be displaced perhaps 50 miles to 100 miles downriver. Mixing would certainly occur in the rapids at the Dalles, if not before. Nearer the Plant, fish prone to rise to the surface for insects would enter the very active water. Scum on the surface would presumably accumulate activity by adsorption. Such scum is a favorite food of carp.

(4) Effects on farm animals and wildlife of the Columbia River

Assumptions made in this case are:
(1) 1000 rep to gut (single exposure) is a conservative LD-50
(2) the contaminant, on the average, is diluted in the ratio of the daily intake of water to the normal gut volume
(3) only fission products need be considered
(4) only Case III of Table 6-3 is significant
(5) if the permissible emergency concentration is taken as one-tenth the conservative LD 50, the range between P.E.C. and LD50 will be a plausible injury-range.
(6) no corrections for decay are made.

The relevant data is assembled in Table 6-4.

The potential affected population of domestic animals is taken as 20% of that of counties adjoining the Columbia River[23]. Gut retention times are as estimated by Morrison[24]. Other animal physiology data comes from Dukes[25].

See Table 6-4

The tabulation suggests that ∼ 300,000 domestic animals, and ∼ 900,000 wild animals may receive exposures between the estimated P.E.C. and the LD50, and may be injured. For the more probable Case I or Case II, there is no injury. The general speculation is that injury by drinking will be moderate or non-existent. Reservations with respect to aquatic birds eating contaminated algae have to be made.

(5) Effects of river contamination on irrigated crops

Data on the translocation of fission products, plutonium, and uranium, in plants of various types is scarce in the literature. L. Jacobson and R. Overstreet[26][27] quote 0.1 mc/kg soil as the specific activity causing damage. Such an activity is available for Case II, and is readily exceeded in Case III. Plutonium and uranium hazards appear to be insignificant.

Fission products such as Zr-Cb and the rare earths predominantly go to the roots, while Sr has significant translocation to the leaves. Quantitative transformation of the data to the field case is difficult. The laboratory tests were made with plants grown for 24 hours in suspensions of bentonite clay in water. Initially, the activity is concentrated on the clay, and there is competition for the cations between roots and clay. Presumably then, the picture in irrigation is an initial deposition on soil up to some saturation point, with later

HW-14058

TABLE 6-4

Damage to Animals drinking contaminated river water

Animal	Population thousands	Daily Water-intake liters	Gut Retention days	Gut Volume liters	LD50 mc/L	P.E.C. mc/L	Fraction received of	
							LD50	P.E.C.
Horses	6	40	2	200	80	8	0.2	2
Cows	60	60	2	350	100	10	0.14	1.4
Swine	11	5	1	33	200	20	0.07	0.7
Sheep	60	5	2	45	140	14	0.1	1.0
Goats	0.5	5	2	45	140	14	0.1	1.0
Turkeys	5	0.3	1	0.8	90	9	0.15	1.5
Hens	160	0.2	1	0.5	80	8	0.2	2.0
Game Animals	10	5	2	40	130	13	0.1	1.0
Beavers	0.2	0.5	1	1.5	100	10	0.14	1.4
Muskrats	5	0.1	1	0.5	170	17	0.08	0.8
Otters	0.05	0.5	1	2	130	13	0.1	1.0
Other fur bearers	10	0.2	1	0.1-5	100	10	0.14	1.4
Canada Geese	5	0.8	1	1	40	4	0.35	3.5
Ducks	20	0.4	1	0.5	50	5	0.3	3.0
Other aquatic birds	50	0.03-0.3	1	0.1-0.8	50	5	0.3	3.0
Pheasants	10	0.1	1	0.4	130	13	0.1	1.0
Quail	100	0.01	1	0.05	170	17	0.08	0.8
Song birds	500	0.01	1	0.04	130	13	0.1	1.0

translocation to the plants. The laboratory results, converted to specific activity, are:-

Element	Concentration factor				
	A			B	
	Leaves	Roots		Leaves	Roots
Y	0.5	33		0.2	14
Ce	0.1	33		0.06	14
Zr - Cb	1.4	40		0.6	17
Sr	8	13		3.4	5.5

A = 0.1 mg bentonite/ml.

B = 0.6 mg/ml.

In Case III, the specific activity of Sr^{89} in water is 14 x 0.011 mc/liter = 0.15 mc/liter. In the leaves will be found ∼ 5 x 0.15 mc/kg = ∼ 0.75 mc/kg. A vegetable consumption of 0.3 kg/day gives a daily ingestion of ∼ 0.2 mc Sr^{89}/day. Over one year, the average daily intake (allowing for decay) is ∼ 10 μc. The daily intake which gives the permissible daily exposure after one year is ∼ 2 μc. In the event of disaster, the translocation into irrigated crops may require intensive study. We may consider here irrigation projects, either now in existence or contemplated for the future, which would be made unavailable for the production of crops following the accident if the irrigation water could not be turned off in time to avert contamination by radioactive materials.

The Pasco Irrigation Project covers 7,750 acres, with 5,552 acres to be irrigated in the near future.

Below the confluence of the Snake River in the Columbia River Bottom irrigation territory, approximately 1500 acres are being irrigated in small tracts, the largest of which is 700 acres on Blalock Island, located between Umatilla and Wallula. For much of other irrigation required near the Columbia River, tributaries are used which, in rare instances, may be increased supplementing with Columbia River water.

Between the Dalles and the mouth of the Columbia River, small acreages are located which are not irrigated but are subject to inundation by flood. These farms frequently have dikes which would prevent contamination in some measure and consist of plots varying from 10 to 100 acres.

In the distant future, it is contemplated that possibly 280,000 acres will be irrigated by the Columbia River located below Kennewick and through Umatilla, and will include some of the Horseheaven area. Source of the water will probably come from the McNary Dam with Hidden Valley to be irrigated first.

The crops grown between here and the Dalles are almost entirely pasture for stock with little truck farming. No permanent crops are being anticipated until the irrigation projects presently being installed are

HW-14058

complete. It may take several years before crops of this type, including orchards, are extensively planted.

In summary, a total of slightly more than 7,000 acres will shortly be irrigated by water from the Columbia River, and perhaps 290,000 acres in the distant future. Whether ground and crops would be contaminated and rendered useless for a protracted period in the event of a catastrophe is dependent on how quickly irrigation operators could be made aware of the hazard and stop drawing water for perhaps a day or two until the bulk of radioactivity has progressed downstream. This should be possible if the communication system is efficient.

Speculations on the biological hazard of the radioactive cloud

The calculations for reactor burn-up with initial height 100 meters in an atmospheric inversion will be accepted as the worst case reasonably expected. The relevant figures are summarized below:

Distance (miles)	3	6	12	19	30	60	100	200
Cloud dose, external (rep)	1	35	45	40	20	7	--	--
Lung deposit, F.P. + Np. (mc)	1	65	75	80	35	14	6	2
Lung deposit Pu (μg)	--	6	8	8	4	2	1	0.5
Ground Contam. (curie/m^2)	0.2	12	16	16	7	3	1.5	--
Ground dose-rate (r/hr)	1.5	80	120	120	50	20	10	5
Rain-out (curie/m^2)	large	170	40	20	8	3	--	--
Cloud width (meters)	280	470	800	1080	1600	2700	--	--

Uranium effects can be neglected, and the neptunium grouped with fission products. The significant initial hazard is <u>inhalation toxicity</u>.

Values above the estimated LD-50 persist to distances of the order of 50 miles. Values above the P.E.D. extend to about 200 miles. At face value, the burn-up could cause substantial injury to the population of Walla Walla, or borderline injury to a strip of the population in Spokane. Such factors as the improbability of an inversion persisting over the required length of time, the weakening by deposition enroute, and the wind shear effect, mitigate this risk.* Nevertheless potential inhalation hazards exist in a triangular area of base about 1 to 2 miles and length $\sim$ 30 to 80 miles, with apex at the reactor. The unpredictable hazardous area will be determined by the wind direction.

For the State of Washington average population, the triangular area would include only $\sim$ 30 people. For the given location, the average target is $\sim$ 500 people. Worse still is the concentration of population in cities including Yakima valley towns, Richland, Kennewick, and Pasco. The three latter are the prime targets, due to level intervening terrain and prevailing wind direction. The probability of scoring a hit on such a center is $\sim$ 1 in 10.** As many as 10,000 people could then be affected. Fortunately, just these towns are the most likely to have effective evacuation schemes operative in the 3 to 5 hours available.

---

* On the other hand, an additional factor of 2, due to low wind speed is not unreasonable.

** Weighted according to average wind rose.

UNCLASSIFIED HW-14058

The secondary hazard to man in this triangular zone comes from irradiation from contaminated ground. Accurately enough this can be taken initially as ~ 100 r/hr at distances of 6 to 30 miles from the reactor. The rounded-off integrated doses (assumes no leaching) are:

Time after deposition	1 hr	2 hr.	5 hr.	1 day	3 days	1 week	1 month
Dose-rate (r/hr)	100	80	70	40	30	20	10
Integrated dose (r)	100	200	400	1300	3000	5000	15,000

The inhabitants of this zone have been potentially killed by the inhalation hazard, unless previously evacuated. We are concerned only with subsequent re-entry for rescue, etc. Obviously, re-entry times must initially be kept to the order of one hour. Even at one month, entry for more than a few hours would be undesirable. There is a hidden safety factor of 2 because the figures refer to the axial dose-rate, which is about twice the average across the width of the cloud.

An additional hazard to man is the consumption of vegetation that has received surface contamination. In the most dangerous triangle, the specific activity on plants will be ~ 5 curies/kg, which is perhaps on the order of 50-1000 times the permissible limit for human consumption. Regardless of the atmospheric turbulence, dangerous depositions on vegetation could occur up to at least 60 miles. The hazard is not elaborated because it is avoidable by control. Local water supplies - reservoirs and ponds - will be contaminated to the extent of ~ 1-10 mc/liter. This hazard can be traced from the discussion of the river, except that it could persist longer if the supply is not dumped. Control is feasible with a good monitor service.

The hazard to domestic animals and wildlife will exceed that to man, because of the impracticability of removing any other than perhaps bloodstock. Approximately 13,000 domestic animals, and 80,000 others can be expected to perish. As many more could move into the triangle later and be killed by radiation from the affected ground or by eating contaminated vegetation.

The calculations presented are not sufficiently precise to justify ramifications of the discussion for less concentrated clouds of greater width.

Direct injury to plants is a further sequel to the ground deposition. Much of the target area is waste desert land or very low-grade grazing land. Sporadically located are valuable irrigated orchards. To the east and south are extensive dry-farmed wheat areas. It is presumed that these will be rendered useless for at least one growth season.

ARTICLE V-2   765

 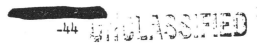 HW-14058

SUMMARY AND CONCLUSIONS

	Page Reference
This is a preliminary draft, diffuse and un-edited in order to make the speculations available immediately. Comments and criticisms directed to the principal author will be appreciated, and helpful, in preparation of the final draft.	
Part I treats the Hanford nuclear reactor as a waste storage unit, and relates the foreseen circumstances under which these wastes may be disseminated to produce environmental hazards outside the Hanford reservation. Disaster requires simultaneous loss of cooling water, and failure of safety mechanisms by earthquake, bombing or sabotage. Simple physical and meteorological pictures of two explosions of heat release 2×10^{10} calories* and 2×10^{11} calories**, respectively, are drawn. The feasible most dangerous cases are probably in this range.	2 4 6 7 ** 11 *
Considered next is the more probably case in which primary explosion is followed by burning of reactor contents. This is the <u>most hazardous predicted event</u>. The behavior of the resultant radioactive cloud under average and adverse meteorological conditions is defined for	16
(1) activity in air at ground level up to 60 miles	18
(2) width of cloud	18
(3) integrated external dose	19
(4) lung deposition of presumed aerosols	19
(5) deposition of activity on the ground	20
(6) rain-out of activity	20
(7) effects of elevation (mountainous terrain)	21

#(4) is the critical immediate hazard, whereas #(5) or #(6) leads to persistent environmental hazard.

Two models leading to gross contamination of the Columbia River are developed:

(1) Blow-up, in which the primary explosion ejects active material directly to the river. This model is highly speculative.	21
(2) Burn-up, in which the active cloud deposits activity in the river.	23
The estimated specific activity (fission products) in the river is 1-2 mc/liter in either case. The contaminated strip is ∼ 0.8 miles in (1) and ∼ 15 miles in (2).	23, 25
Speculations on biological effect in man present quite tentative values for a lethal dose (LD-50) and a permissible emergency concentration or deposition for ingestion and inhalation,* respectively, for fission products, plutonium, neptunium, and uranium. These	25 * 28

Summary and Conclusions - contd.

are adequate to define a general damage range, when the current uncertainty of the physical pictures is considered. (We anticipate criticism and wide divergence in values of LD-50 and P.E.C. or P.E.D. From submitted comments, it is hoped to present weighted values for the final draft).

Comparison of the damage range with calculated exposures shows:

	Page Reference
(1) effects of river contamination on man - borderline damage for unfiltered water systems; certain damage if active water is unfavorably channelized.	30
(2) immediate effects of river contamination on lower forms - fish damaged in the worst case by fission products. Insect larvae generally damaged by fission products. Dubious chemical toxicity in the worst case.	34
(3) delayed effects on aquatic forms - food chains with the activity initially concentrated by algae or planktonic forms enhances the damage to higher forms, for fission products only.	35
(4) effects on animals watering at the river - widespread damage in the worst case only. Probably damage to aquatic birds in an average case.	38
(5) effects on irrigated crops - primary damage seems feasible. Residual long-lived activity in food plants would have to be checked.	38

From the possible radioactive clouds:-

(1) the principal hazard is <u>inhalation</u> of fission products and possibly plutonium, rather than the external radiation from the cloud.* Dangerous doses occur up to about 60 miles from the reactor.	42
(2) the secondary hazard, from deposition of activity on the ground, denies unrestricted access over a similar range for at least days, and possibly months.	43
(3) the same secondary hazard is lethal to animals by external radiation.	

---

* This has generally been inadequately discussed in most earlier discussions. (2) (28) (29). Plutonium inhalation has been considered by Jane Hall (30), and iodine inhalation by C. C. Gamertsfelder (31).

Summary and Conclusions - contd

(4) ingestion of contaminated food aggravates the damage to animals.

(5) growing plants may be injured by the deposits. Where possible, the affected populations are speculatively estimated.   43

The Appendix includes a few notes of interest not included in the appropriate text location.

APPENDIX

1. **Note on additional activity generated in the power burst**

   From the Reactor Safeguard Committee Report.

   $$\text{F.P. activity after disruption} = \frac{0.1 \times \text{previous steady power in KW}}{(\text{time(sec) after accident})^{0.2}}$$

   $$\text{Additional activity during run-away burst} = \frac{0.02 \times (\text{integrated burst in KW sec})}{(\text{time(sec) after accident})^{1.2}}$$

   Steady power = 2.75×10^5 KW

   For the 2×10^{11} calorie explosion, integrated burst = 8.3×10^8 KW sec.

   $$\therefore \frac{\text{Additional activity}}{\text{Stored activity}} = \frac{600}{\text{time (sec)}}$$

   Our calculations refer generally to times in excess of one hour, so the additional activity is negligible in approximate calculations.

HW-14058

APPENDIX (Continued)

2. Note on activity in the graphite moderator

Hanford graphite has specific activity = 0.032 curies/kg
of which 0.02 curies/kg is the gamma-ray activity,
and 0.002 curies/kg is C^{14}.

The total activity in a unit = 6.4×10^4 curies
This is negligible in comparison with stored fission product activity.

Predominant gamma emitters are: Eu^{152}, Sm^{153}, Fe^{54}, Fe^{59}, and unidentified rare earths.

3. Note on iodine deposition on vegetation

For stable iodine, Thomas and Hill[32] report a concentration of 0.8 μg I_2 per liter in air to give a deposit of 67 mg I_2/kg in alfalfa leaves in $\sim$ 19 hours. In two hours, we can assume the concentration to have been $\sim$ 7 mg I_2/kg. The same ratio holds for radioiodine, if the early radiation does not damage the plant's transfer mechanism. Therefore, 0.8 μc I^{131} per liter for 2 hours should give $\sim$ 7 mc I^{131}/kg in leaves.

In the triangular danger zone of the "bad" case from burn-up, the predicted 80 μc F.P./liter gives by our empirical deposition rule, 16 curies/sq.meter = $\sim$ 8 curies/kg of vegetation (assumes all the activity caught in a half-grown field of alfalfa). Of the atmospheric contamination, 1.1% is due to I^{131}. By rule of thumb we predict vegetation contamination of $\sim$ 90 mc I^{131}/kg, about 12 times the Thomas-Hill value. Take the lower value and assume animals feed on such vegetation. The initial thyroid irradiation will be $\sim$ 4000 rep/day. The I^{131} component alone will be hazardous. It will require 130 days decay time to reach the contamination level considered as the permissible permanent limit at Hanford (0.01 μc I^{131}/kg). The total radioiodine deposit will be 8 times that of I^{131} alone. By crude weighting of half-lives and disintegration energy, the total deposit is worth about twice the I^{131} in hazard.

Note that the agreement between 'D' for iodine and 'D' aerosols or particles is presumably fortuitous, and arises from the concentration factors for iodine reported by Thomas and Hill. The copious Hanford data on iodine deposition on vegetation has not yet been reconciled with the Thomas-Hill data. We would anticipate higher depositions than are quoted above. By scaling up the normal Separations Plant emission of I^{131} and the observed deposition at distances up to 50 miles, Gamertsfelder [31] predicts maximum deposition on the order of 400 mc I^{131}/kg. His assumptions on the time in which observed deposits are laid down leads to values about 4 times higher than those of the writer. This would give 100 mc/kg in agreement with the 90 mc/kg predicted in this report.

There appears to be an error of 1000 between the script and tables of the Thomas-Hill report. The script value has been used above.

APPENDIX (Continued)

4. **Note on leaching of activity from the ground**

Wherever contamination is laid down by an active cloud (burn-up) or by the physical ejection of reactor material (blow-up), there exists the possibility of leaching of such activity into potable water sources. In all except the blow-up case, this effect is probably low compared with the primary contamination of water. It can be crudely estimated from the data given for ground filtration water systems. Low-level, long-continued water contamination of this type will be discussed in Part II. The general Hanford operating experience has been that whenever an area is accidentally contaminated, it is virtually impossible to remove the contamination by copious washing (exception: uranium). Conversely, if one wants the activity to remain in place, some fraction of it will be transported by one mechanism or another. This gloomy philosophy should be applied by the reader to the subject case.

5. **Note on the acute toxicity of inhaled plutonium**

For the rat, Abrams, et al[33] report

Exposure µg Pu/rat	8	24	70	210	500
Median Survival time (days)	203	154	67	24	26
Lung dose (rep)	2600	7200	15,000	29,000	72,000
Ratio Lung dose/bone dose	8	10	26	93	86

Taking the lung as 9% body weight in the rat, and 14% in man, and assuming all other factors equal, the comparable depositions in man would be 380 times greater. The "true" LD-50/30 is of the order of 80 mg, but 3 mg is a lethal dose. This should be compared with our current guess of 2 mg as the lethal deposition.

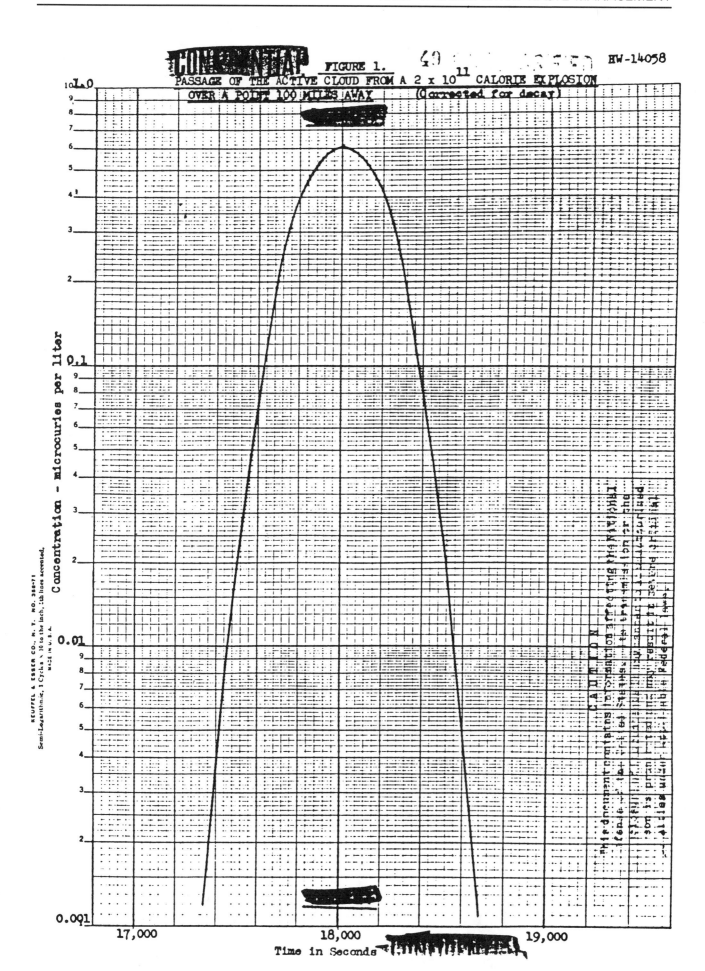

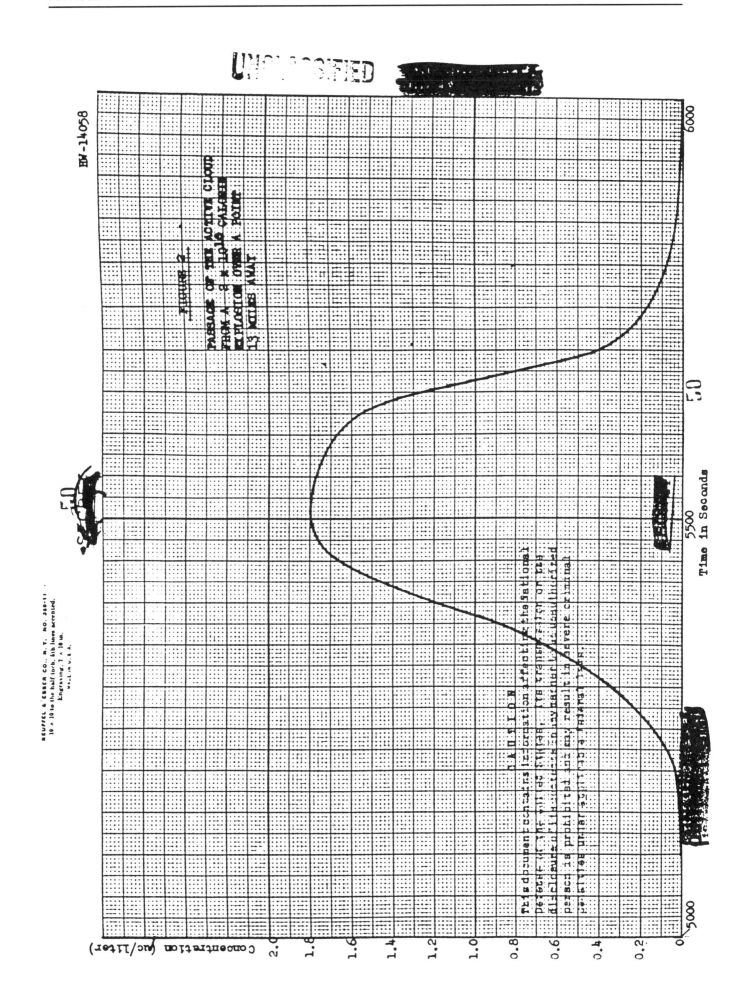

PART V: RADIOACTIVE WASTE MANAGEMENT

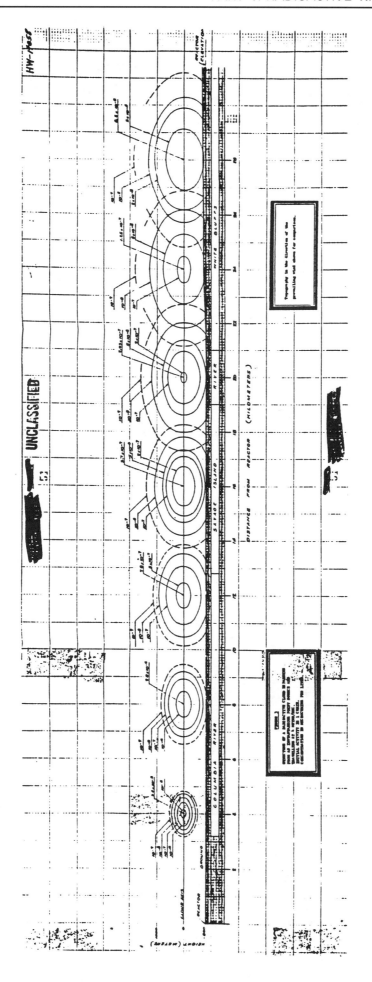

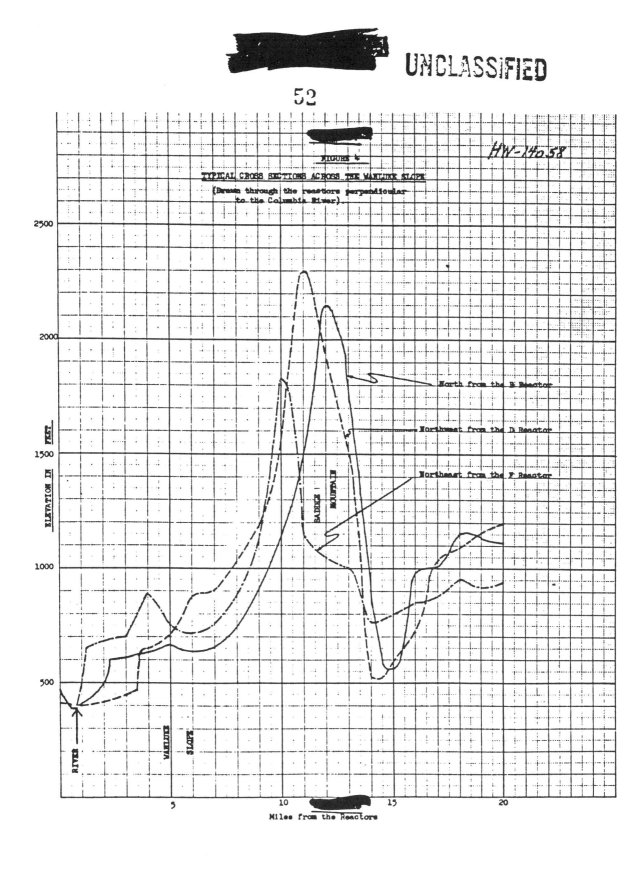

HW-14058

BIBLIOGRAPHY

1. Parker, H. M. - "Speculations on long-range waste disposal hazards"
   Doc. HW-8674, 1/26/48

2. Reactor Safeguard Committee (Teller, E., Benedict, M., Kennedy, J., Wheeler, J., Wolman, A.) - "Review of certain Hanford Operations"
   Doc. GEH-14,040, 11/4/48

3. Sutton, O. G. - "The atomic bomb trial as an experiment in convection"
   Weather (Gt. Britain) p. 105 April 1947

4. Ministry of Supply, Dept. of Atomic Energy, Plant Location Panel
   "Memorandum on the diffusion of matter from a fission plant"
   PLP-8

5. Church, P. E. - "Hanford Works Meteorology Section Report" - Jan. 1943 to July 1944

6. Holzman, Col. B. - unpublished data for Dr. J. Hirschfelders "Atomic Weapons Handbook"

7. Bagnold, R. A. - "Physics of blown sand and desert dunes"
   Morrow & Co., N. Y., 1943

8. Cohn, W. E. - "Tolerances for fission products" in the Project Handbook
   CL 697, Nov. 1944. Revised May 1945

9. Morgan, K. Z. - "Tolerance concentration of radioactive substances"
   MDDC 240, also in J. Phys. Colloid Chem. $\underline{51}$, 984, 1947

10. Parker, H. M. - "Tentative dose units for mixed radiations", read at N. A. Radiol. Soc. Meeting, San Francisco Dec. 1948 (to be published in "Radiology")

11. Wigner, E. P. and Way, K. - "Summary and correlation of data on the rate of decay of fission products", CC-3032, 7/2/45

12. Way, K. and Wigner, E. P. - "The rate of decay of fission products"
    MonP-303, 7/16/47

13. Johnson, K. D. B., and Wilkinson, J. - "The absorption of fission products from sea-water, on sea-bed and shore", AERE/C/R-294, 1948

14. Lisco, H., Finkel, M. P., and Brues, A. M. - "Carcinogenic properties of radioactive fission products and of plutonium", Radiology $\underline{49}$, 361, 1947

15. Flaskamp, W. - "Rontgenschaeden"

16. Hamilton, J. G. - "Metabolism of fission products and the heaviest elements"
    Radiology $\underline{49}$, 325, 1947

17. U.S. Commissioner of Fish - "Bonneville Dam and protection of the Columbia River Fisheries", Report - Doc. #87 to 75th Congress, 1937

18. Rich, H. W. - "The salmon runs of the Columbia River in 1938", U. S. Fish and Wildlife Service, Fishing Bull. #37, 1942

19. Biennial Report of the Fish Commission of the State of Oregon for 1949

20. Pritchart, A. L. - "Efficiency of natural propogation of Pacific Salmon", Can. Fish. Cult. Vol. 1 #2, 1947

21. Prosser, C. L., Pervinsek, W., Arnold, J., Svihla, G. and Tompkins, P. C. "Accumulation and distribution of radioactive strontium, barium-lanthanum, fission mixture and sodium in goldfish", CH 3233, 1945

22. Bonham, K., Seymour, A. H., Donaldson, L. R. and Welander, A. D. "Lethal effects of X-rays on marine microplankton organisms" Science $\underline{106}$, 2750, 1947

23. U.S. Dept. of Commerce, Bureau of Census Report Vol. 1 part 32, 1945

24. Morrison, F. B. - "Feeds and Feeding", Morrison Publ. Co., Ithaca, N.Y. 1937

25. Dukes, H. H. - "Physiology of domestic animals", Comstock Publ. Co., 6th Ed., 1947

26. Jacobson, L. and Overstreet, R. - "Progress report on plant studies: the fixation and absorption of fission products by plants", MDDC 571 1944

27. Jacobson, L. and Overstreet, R. - "The uptake by plants of plutonium and some products of nuclear fission adsorbed on soil colloids" MDDC 1006, June 1947

28. Wheeler, J. A. - "Hazard following explosion at Site W", CH 474, 1943

29. Compton, A. H. - "Hazard from radiation in event of various failures" CH 504, 1944

30. Hall, J. H. - "Hazards in event of product being dispersed", Doc. 7-635 1944

31. Gamertsfelder, C. C. - "Effects on surrounding areas caused by the operations of the Hanford Engineer Works", Doc. 7-5934, 1947

32. Thomas, M. D. and Hill, G. R., - "The absorption of iodine and hydriodic acid by alfalfa", private memorandum within AEC. Official publication unknown

33. Abrams, R., Siebert, H. G. and Forker, L. - "Acute toxicity of intubated plutonium", Doc. CH 3875, 7/19/48

Other sources not specified in the text are:

34. Sutton, O. G. - "Theory of eddy diffusion in the atmosphere", Proc. Roy Soc. A $\underline{135}$, 143, 1932

35. Sutton, O. G. - "The problem of diffusion in the lower atmosphere" Q.J.R.M.S. 73, 257, 1947

36. Sutton, O. G. - "The theoretical distribution of airborne pollution from factory chimneys", Q.J.R.M.S. 73, 426, 1947

37. Sutton, O. G. - "The diffusive properties of the lower atmosphere" MRP-59, US. AEC issue April 1949

38. Dobson, G. M. B. - "Some meteorological aspects of atmospheric pollution" Q.J.R.M.S. 74, 133, 1948

39. Marley, W. G. - "Note on possible disaster to a water cooled pile", PLP 16

40. Durst, C. S. - "Notes on the amount of atmospheric pollution which would be breathed over a long period", PLP 18

41. Sutton, O. G. - "Revised calculations on the danger from air-borne products of an atomic pile explosion", PLP 20

42. Plant Location Panel - "The concentration of pollution in mountainous country", PLP 22

43. Dallavalle, J. M. - "Micrometrics", Pitman Publ. Co., N.Y., 2nd Ed., 1948

ARTICLE V-2

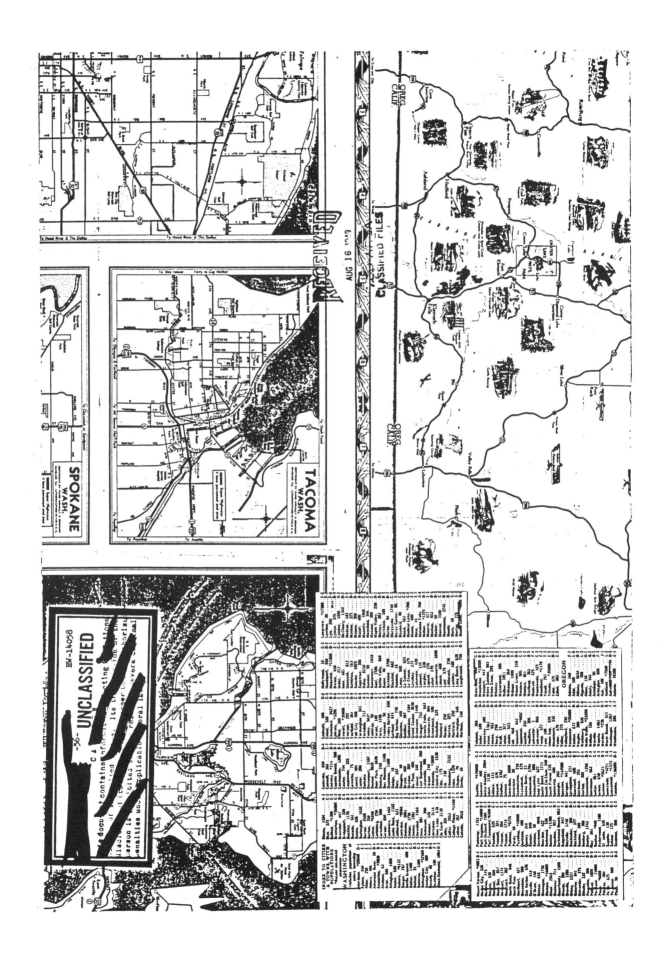

PART V: RADIOACTIVE WASTE MANAGEMENT

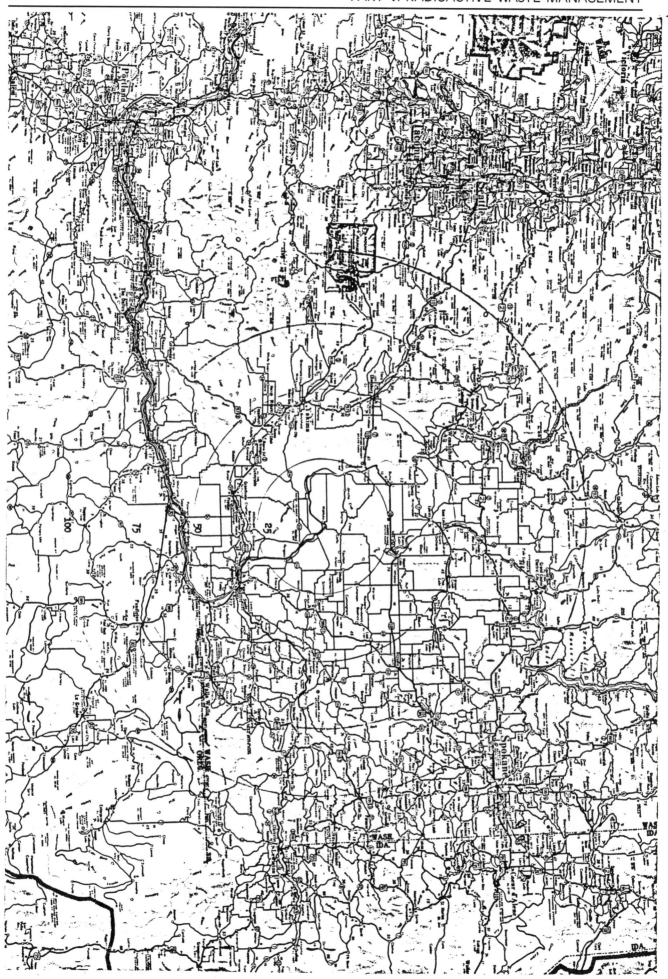

North American Experience in the Release of Low-Level Waste to Rivers and Lakes

International Conference on the Peaceful Uses of Atomic Energy

1964

P/287 United States of America

North American experience in the release of low-level waste to rivers and lakes

By H. M. Parker,* R. F. Foster,* I. L. Ophel,** F. L. Parker*** and W. C. Reinig****

The release of liquid effluents from atomic energy installations in North America has always been carried out under the philosophy that the quantities of radioactive materials present in effluents which eventually reach surface streams and lakes should be kept at a minimum. There is, however, a practical limit to the cleanliness of any effluent. Complete exclusion of all radionuclides from the waste water that leaves any plant is neither economically feasible nor would this accomplish a rational objective in view of the amounts of naturally-occurring radionuclides already present in surface waters.

All North American installations have controlled their releases of radioactive wastes so that exposure of people from this source has been considerably less than the limits recommended by the International Commission on Radiological Protection (ICRP) and other authoritative bodies. Better evaluations of this exposure have been possible during the past five years with the availability of new technology and additional knowledge of the fate of the radionuclides in the environment. The combined experience of the several sites embraces such a broad spectrum of nuclides and variety of circumstances that it appears highly unlikely that any substantially different or unique problems will be encountered in the environs of future reactors or fuel reprocessing plants of contemporary design.

THE ADDITION OF RADIONUCLIDES TO STREAMS

The experience reported in this paper is dominantly that of the large government-owned atomic energy sites represented by the authors. The facilities at these sites differ in basic design and consequently the characteristics of their liquid wastes are vastly different. Waste treatment and local topography and hydrography further modify the kinds and quantities of radionuclides that reach surface waters. The marked differences in the rates of addition of various nuclides to rivers by the major atomic energy sites

---

* Hanford Atomic Products Operation, Richland, Washington.
** Chalk River Nuclear Laboratories, Chalk River, Ontario, Canada.
*** Oak Ridge National Laboratory, Oak Ridge, Tennessee.
**** du Pont, Savannah River Plant, Aiken, South Carolina.

and by some typical nuclear power stations are shown in Table 1. Nuclides other than those shown in the table have also been detected downstream from some of the major sites, but the ones listed are of greatest interest because of abundance or biological importance. The amount of water available for dilution in the receiving rivers is also shown in Table 1.

Chalk River Nuclear Laboratories

Most of the low-level waste released to the Ottawa River from the Chalk River Nuclear Laboratories (CRNL) consists of neutron-activation products. These activation products are formed in the light-water system that cools the NRX reactor [1] and passes via delay tanks to the process sewer that discharges directly into the river. Some activation products are also present in the water of the fuel storage bays and in liquid wastes generated in other parts of the establishment that are routed to ground-disposal pits. Such wastes may eventually reach the Ottawa River after seepage through the ground and drainage through Perch Lake [2]. Tritium is produced in the heavy water that is used to moderate the NRU and NRX reactors and the portion that reaches the waste pits soon enters the ground water and moves with it toward Perch Lake. In 1963 a weak "front" of the tritium inventory in the ground water had reached Perch Lake [3].

Most of the low-level fission-product waste is sent to the waste pits where the nuclides are effectively detained by the soil. The principal source of fission products shown in Table 1 as entering the Ottawa River is the process sewer. Strontium-90 from a comparatively large disposal in a special pit in 1954 is now seeping into Perch Lake at the rate of about one curie per year. In 1963, world-wide fallout contributed much greater amounts of radionuclides to the Ottawa River than the CRNL facilities; e.g., about 0.1 curie of Sr^{90} and 100 curies of H^3 from fallout were transported by the river each day.

Hanford Plant

The Hanford facilities release larger quantities of radioactive materials to surface waters than any of the other North American sites. Nearly all of this consists of short-lived neutron-activation products

Table 1. Quantities of low-level radioactive wastes added to streams
(curies/d)

	Nuclide	Half-life	Chalk R.	Hanford	Oak Ridge	Savannah R.	Nuclear Power
Activation products	Cu^{64}	13 h	—	200[a]–1 000[b]	—[d]	—	—
	Na^{24}	15 h	(Trace) T-10	200–1 000	—	—	—
	As^{76}	26 h	T- 5	50– 300	—	—	—
	Np^{239}	2.3 d	—	200–1 000	—	2[c]	—
	P^{32}	14 d	0.01–0.1	20– 70	—	—	—
	Cr^{51}	28 d	—	600–2 000	—	3.5	T
	Fe^{59}	45 d	—	T	—	—	T
	Co^{58}	71 d	—	T	—	—	T
	S^{35}	87 d	0.001	T	—	0.4	—
	Zn^{65}	250 d	0.002	30– 100	—	0.09	—
	Co^{60}	5.3 yr	T- 0.05	1– 2	0.04 –0.2	0.01	T
	H^3	12 y	0.1–20	T	—	205	—
Fission products	I^{131}	8 d	—	1– 3	0.001–0.01	0.1	T
	Ba^{140}	13 d	—	T	—	0.1	T
	Nb^{95}	35 d	—	—	0.002–0.2	—	—
	Sr^{89}	50 d	—	T	—	0.09	—
	Zr^{95}	65 d	—	T	0.001–0.1	0.1	—
	Ce^{144}	285 d	0.005–0.015	—	0.003–0.1	0.1	—
	Ru^{106}	1 yr	0.002	—	1 –5	0.03	—
	Sr^{90}	28 yr	0.001–0.005	0.1	0.02 –0.2	0.03	—[e]
	Cs^{137}	30 yr	0.002–0.02	T	0.01 –0.2	0.3	—
	Total beta	(Exclusive of H^3)	0.05 –15	2000	1 –6	(0.5 in river)	10^{-5}–0.01
	Receiving stream		Ottawa R.	Columbia R.	Clinch R.	Savannah R.	Various
	Flow 10^{10} l d	Range	3–13	10–75	0.6–2	1.4–8	1–10
		avg.	6	27	1	2.5	
	Measurement point		Ottawa R. process sewer Perch Lake	Columbia R. (Pasco)	White Oak Creek	Storage-basin discharge	Various waste streams

[a]–[b] Where a substantial variation is reported, both the low a and high b values are listed.
[c] Where a yearly average is reported, or there is little variation, only one value is listed.
[d] (—) Indicates the nuclide is not reported. It may be present but in amounts that are trivial in relation to other nuclides encountered.
[e] Fallout contributed from 0.1 to 1 curie of Sr^{90} per day to large rivers of North America in 1963.

associated with the single-pass cooling system of the production reactors [4]. The comparatively small amounts of fission products indicated in Table 1 result from the fissioning of natural uranium present in the Columbia River water used for cooling and from rare defects in the cladding of fuel elements.

Low and intermediate level wastes from the chemical plants that process the fuel elements are discharged to the ground some 16 km from the Columbia River, and most of the fission products are held in the 60–90 m of soil above the water table [4]. The only nuclides which have been found at appreciable distances from the points of discharge are H^3* and Ru^{106}. Both of these were detected in test wells quite near the Columbia River in 1963, but neither has been detected in water samples taken from the river.

---

* H^3 is sometimes formed when U^{235} fissions [5]. When irradiated fuel elements are dissolved, a part of the H^3 is carried by the liquid-waste streams [6].

Oak Ridge National Laboratory

The radioisotope production center is the initial source of most of the radionuclides that ultimately enter the Clinch River from the Oak Ridge National Laboratory (ORNL), but various waste handling operations occur throughout the site which contribute some low-level wastes. The wastes listed for Oak Ridge in Table 1 are transported to the Clinch River by White Oak Creek which receives the effluent from the process waste water treatment plant and seepage from waste pits [7]. The major source of the Cs^{137} and Sr^{90} is the treatment plant effluent, but most of the Co^{60} and Ru^{106} comes from the waste pits. The ranges in rates of discharge shown in Table 1 are for the five-year period ending December 1963, but the minimum values are characteristic of the releases during 1963 [8].

Savannah River Plant

Nearly all the radionuclides that enter the Savannah River from this site come from basins in which the fuel elements discharged from the reactors are stored and disassembled. Activation products from the surfaces of the fuel elements and any fission products that may escape through defects in the cladding enter the basin water. Removal of the spent fuel elements also results in the transfer of tritium to the basins. This tritium is formed in the heavy water that is used to moderate and cool the production reactors. The basin water is continually replenished and the waste is discharged to small natural streams that flow 16-21 km through the reservation before they enter the Savannah River. In one case the discharge is to an artificial lake, called Par Pond. The outlet of this impoundment flows about 29 km through the reservation before entering the Savannah River.

The release rates shown in Table 1 for the Savannah River site do not include contributions from the fuel processing plants because the amounts added from this source are insignificant. All of the low-level liquid waste from the fuel processing plants is sent to seepage basins, except for one very small discharge into a surface stream. Liquids placed in these basins must seep through a minimum of 150 m of earth in one case and 700 m in another case, before entering Four Mile Creek. Tritium is the only nuclide associated with seepage basins that has been detected in Four Mile Creek [9].

Nuclear power stations

The quantities of radioactive materials discharged from nuclear power reactors of contemporary design are so small that individual nuclides cannot easily be identified in the bulk effluent leaving the plants. Gross measurements of the beta activity are ordinarily made and reported, however, and such results for three typical installations that have operated for more than one year are included in Table 1. The installations included are: (a) the Dresden Nuclear Power Station of the Commonwealth Edison Company, a boiling water reactor located near Chicago, Illinois, (b) the Yankee Station of the Yankee Atomic Electric Company, a pressurized water reactor located in the northwest corner of Massachusetts, and (c) the Indian Point Station of the Consolidated Edison Company of New York, a pressurized water, thorium converter reactor, located near New York City. The radionuclides that appear to be most common in some process liquids and may sometimes be present in the gross effluent have been indicated as T in Table 1 [10].

The low release rates characteristic of nuclear power stations (10^{-5} to 10^{-2} curies per day) are achieved because the reactor water is recirculated and passed through clean-up equipment, because of the cleanliness of the steam generated, and because the spent fuel is processed at other locations. As shown in Table 1, such releases are insignificant in comparison with those from some of the major sites. Additional perspective of the very small quantities of radionuclides released by the power stations can be gained from a comparison with natural and fallout nuclides present in the receiving streams. The Illinois River, for example, carries about 50 times more radioactive material from natural sources (K^{40}, radium + daughters, and uranium) than is added by the Dresden Station, and the quantity of Sr^{90} in the river from fallout was several thousand fold greater in 1963 than that added by the station [10].

FATE OF THE RADIONUCLIDES

Implicit in a sophisticated evaluation of radiation exposure that may accrue from the release of low-level wastes is a knowledge of the rates and mechanisms of transport of the radionuclides away from the points of release and of the places where the nuclides may accumulate. During the past five years, substantial progress has been made in tracking some of the radionuclides downstream from the major sites, in defining the rates at which certain nuclides are depleted from the water, and in measuring the accumulation in bottom sediments and biota [11, 12, 13]. These studies have been stimulated and materially aided by rapid advances in the technology of measuring individual radionuclides. Gamma-ray spectroscopy together with new analytical procedures for beta-emitters such as Sr^{90} have made it possible to analyse large numbers of samples rapidly and economically and thus to undertake programs which would have been impractical a decade ago [14, 15, 16].

All the major sites have made observations which show that significant amounts of the radionuclides which enter surface waters are retained in the aquatic system within a few kilometers of the point of entry. Most of the retention is in bottom sediments, especially if a lake or impoundment is involved. Beyond the initial area of deposition, depletion of the nuclides from the water proceeds much more slowly and may not again be significant until the physical-chemical conditions are modified, possibly in the estuary.

At the Chalk River site, Perch Lake (40 hectares) provides a unique opportunity to study the fate of some radionuclides in a fresh water community and observations were begun in 1956 [17]. About 25% of the Sr^{90} that has entered the lake has been retained within the lake ecosystem. Of this amount, 90% is present in the bottom materials, and less than 1% is held in the dense aquatic vegetation that covers almost one-third of the lake area. Data are now being accumulated on the S^{35} and Co^{60} that enter and leave Perch Lake and preliminary evaluations suggest that 50–75% of the S^{35} and 80–90% of the Co^{60} remain in the lake. Contrasting with the removal in Perch Lake, observations in the Ottawa River showed no significant depletion of Sr^{90} or Cs^{137} with distance downstream and barely detectable amounts of artificial radionuclides have been found in the bottom sediments of the Ottawa River within 15 km of CRNL [11, 18]. Most of the burden of these nuclides in the river water in 1962 and 1963 came from world-wide fallout that entered the river upstream from the plant.

In the vicinity of the Hanford production reactors, the Columbia River is swift and clear and there is little opportunity for deposition of suspended solids until the current is slowed by the backwaters of McNary Dam. Nevertheless, some studies have indicated that, besides the loss from radioactive decay, substantial depletion of several nuclides occurs in the first 50 km below the reactors and before an area of appreciable sedimentation is reached [19]. For example, about 30% of the P^{32}, 15% of the Cr^{51}, and 40% of the Zn^{65} is depleted from the river in this first 50 km reach. Below this point, the river flows more slowly and passes through a series of impoundments. While additional depletion occurs, the rate of depletion in relation to distance appears to be significantly slower. In a 350 km reach of the lower Columbia River some 10–40% of the P^{32}, about 30% of the Cr^{51}, and about 50% of the Zn^{65} was depleted [20]. The nuclides that are dominant in the sediments behind the dams are Cr^{51}, Co^{60}, and Zn^{65}.

At the Oak Ridge National Laboratory, significant deposition of Ru^{106}, Sr^{90}, Cs^{137}, Co^{60}, and other nuclides occurs in White Oak Lake (8 hectares). However, the radioactive materials carried into the Clinch River with the overflow from the lake are essentially all swept downstream. With the use of mass balance techniques, it has been possible, within 10%, to account for the quantities of radio-nuclides discharged from White Oak Dam in the water of the Clinch River passing a sampling station some 25 km downstream [21]. The physical form in which the nuclides are transported is quite variable, however. About 90% of the Cs^{137} is carried on particles whereas 80–90% of the Sr^{90}, Co^{60}, and Ru^{106} is in solution.*

Studies of the Clinch River have also included extensive sampling (coring) of the sediments in order to measure the quantities of nuclides deposited on the bottom. Although the presence of nuclides that are dominant in the river water can easily be measured in the bottom sediments, only about 3% of the total inventory of radioactive materials discharged to the river system has been retained in the stream bed. A negligible percentage is incorporated into the biomass [22].

At the Savannah River plant, a significant fraction of the radio-nuclides released from the reactor areas is removed by organic and inorganic sediments in the small streams and their contiguous swampy areas and in Par Pond (1 100 hectares) before the waste enters the Savannah River. Only 4% of the cesium and 30% of the strontium, released from the reactor areas in 1963, remained in the water of the Savannah River sampled 16 km below the plant. During 1963, the concentrations of most long-lived nuclides in the Savannah River were nearly as great upstream, because of world-wide fallout, as they were below the atomic energy plant. Tritium was the notable exception. Suspended silts and clays play an important role in the transport of the radionuclides down the turbid Savannah River. Some deposition of the sediments occurs in locations where little or no current exists, but much greater deposition takes place in the tidal intrusion zone of Savannah Harbor, some 220 km below the atomic energy plant.

The radioactive materials contained in the bottom sediments downstream from atomic energy sites would not create significant radiological problems should the deposits be moved to flood-plain land. A practical evaluation of such a situation was carried out in 1963 when sediments were dredged from the Clinch River, just below the mouth of White Oak Creek, in order to create a ship channel. Extensive radiation monitoring of this operation showed that direct radiation from the exposed piles of dredged sediments presented no immediate or long-range radiation hazard to people or animals, and that there was no need to prevent beef cattle from grazing on vegetation expected to grow on the exposed sediments.

Studies on the dispersion and transport of low-level wastes in rivers adjacent to the major sites have, in every case, involved complicated hydraulic systems. The flows of these rivers are no longer natural but are controlled by dams. Control of the flow generally results in more water being available for dilution of wastes at times of the year when runoff is ordinarily at a minimum. On the other hand, the daily release of water from dams that produce hydro-electric power may fluctuate widely. The recently constructed Melton Hill Dam on the Clinch River, about 4 km above the confluence of White Oak Creek, is an extreme example. Beginning in 1964, the flow of the Clinch River will essentially be stopped on week-ends during the summer low flow period, but may reach 4×10^7 l/min at certain hours on week days [23]. Over a year's time the quantity of water available for dilution of the nuclides will not change, but the flushing action is expected to alter the mode of transport and higher concentrations will exist near the mouth of White Oak Creek for brief periods of time.

In addition to the variations caused by uneven discharge from dams, thermal stratification in slow-moving streams and impoundments has resulted in anomalous distributions of the radionuclides during the summer. In the Ottawa River, radionuclides from CRNL are sometimes transported a short distance upstream from the outfall by cool water at the bottom of the river that flows in the opposite direction to the warmer surface current. Stratification in

---

* Nuclides that were not separated from water samples with particles larger than 0.7 μ in diameter were considered to be in solution.

an impoundment of the Clinch River, about 100 km downstream from the Oak Ridge site, causes the dilute radioactive waste being carried by the cool river water to flow underneath the warm surface layers [21].

Studies of the complicated hydrographic conditions which exist in the rivers that receive the radioactive wastes have been materially aided in recent years by the use of fluorescent dyes, such as Rhodamine B [24] which can be measured in concentrations as low as one part in 10^{11}. The dyes have distinct advantages over radioactive tracers in situations where the chronic release of low-level wastes interferes with the measurement of very low concentrations of individual radionuclides. Discrete tracer-type tests with such dyes have provided information on the time required for water masses to travel from the atomic energy facilities to various points of use downstream, on the rate of transverse mixing, and on the extent of longitudinal dispersion.

The feasibility of accounting for the fate of radionuclides from sensitive analyses of environmental samples and the preparation of material balances has now been demonstrated. A logical next step is the construction of improved mathematical models for the prediction of conditions that may develop downstream from significant sources of radioactive waste.

RADIATION EXPOSURE EVALUATION

Of greater significance than the presence of low concentrations of radionuclides in waste streams is the magnitude of the radiation exposure that people living in the vicinity of atomic energy plants may receive as a consequence of the releases. Human exposure from water-borne waste occurs only at locations where the water is used or where the nuclides are somehow incorporated into foods. A number of different pathways that conceivably could provide such exposure was suggested in 1959 [25]. During the past five years, several studies have been carried out at the major sites that have confirmed the potential significance of some of these pathways and, as predicted, the insignificance of others.

The exposure pathways for low-level radioactive wastes in surface waters that have received the most attention are: (*a*) drinking of the water; (*b*) consumption of fish caught from the rivers; (*c*) consumption of farm crops and milk derived from farms irrigated with the river water; (*d*) consumption of marine products harvested near the mouths of the rivers, and (*e*) swimming or other recreational use of the water. Pathways *a* and *b* exist downstream from all the major sites, but irrigation is practiced only below the Hanford site. Nuclides from the Hanford reactors are measurable in shellfish near the mouth of the Columbia River, but comparable situations do not exist below the other North American sites. Direct exposure from river water while swimming or boating is negligible (less than 25 mrem/yr) at all locations.

The environmental surveillance programs at all the major sites have gained in sophistication as the relative importance of the several different exposure pathways has been clarified and as new analytical techniques, especially gamma-ray spectroscopy, have become available. In the past, calculations of human exposure have been based almost entirely on estimates of the rates of intake of radionuclides derived from measured concentrations in foods and water and from broad surveys of dietary habits. Significant refinements in exposure estimates are now emerging, however, as more exact information is developed on water use and the habits of individuals. The advent of wholebody counters capable of detecting accumulated or recently ingested gamma-emitters provides a vastly superior method for determining the actual body burden of some nuclides. This technique is now being used at Hanford and Oak Ridge as an aid to evaluating exposures from environmental sources. The results have usually shown lower body burdens than predicted from dietary assumptions.

Recent estimates of the exposures received by people from low-level wastes discharged into North American rivers are summarized in Table 2, where they are expressed as percentages of appropriate limits for particular organs.* Although these estimates include the contribution from all radionuclides in the water, only those few nuclides contributing the greatest part of the dose are shown. Similarly, only the exposure pathways which might lead to the largest exposures are listed. The reference limit (to individuals or to populations, and the critical organ) was chosen to show the most restrictive case. The estimates were made on the basis of actual measurements of the amounts of the radionuclides present in the receiving water, or in fish or crops.

Downstream from CRNL, water is pumped from the Ottawa River to supply the city of Pembroke and the Sr^{90} in the water is the nuclide of greatest interest [18]. Most of the Sr^{90} present in 1963 was from fallout rather than from plant operations, however, and the calculated dose to the bone of Pembroke residents was about average for persons drinking the water of other rivers of Ontario or Quebec. Fish and clams within a few kilometers of the CRNL process sewer contain P^{32} and Zn^{65}, but the concentrations are so low and use of the fish is so limited that the potential human exposure from this source is also negligible.

In the Columbia River below the Hanford reactors, P^{32} is clearly of greatest importance because of its accumulation in local fish and, to a lesser

---
* These limits do not represent thresholds above which discernible radiation damage begins, but rather are levels only a few times greater than natural background where there is a negligible probability that an individual would be adversely affected should he receive such exposures throughout his lifetime.

Table 2. Significance of exposure from various sources

Site	Nuclides of greatest interest	Mode of exposure	Critical organ	Type of person receiving greatest exposure	Percent of limit [a]	Year	Reference limit
Chalk R.	Sr^{90}	Drinking water	Bone	Pembroke resident	< 1[b]		ICRP, population at large
	P^{32}	Fish	Bone	Fisherman	< 0.1	1963	ICRP, Group B (c)[c]
Hanford	P^{32}	Fish and irrigated crops	Bone	Fisherman, farmer	< 40		ICRP, Group B (c)
	$As^{76}+Np^{239}+Cr^{51}$	Drinking water	G.I. tract	Pasco resident	< 8	1963	ICRP, population at large
	I^{131}	Drinking water	Thyroid	Pasco child	< 6		FRC[d], exposed population
Oak Ridge	Sr^{90}	Fish	Bone	Fisherman	< 30[f]		ICRP, Group B (c)
	Sr^{90}	Drinking water	Bone	Clinch R. resident[e]	< 5	1961	ICRP, Group B (c)
	Ru^{106}	Drinking water	G.I. tract	Clinch R. resident[e]	< 5		ICRP, Group B (c)
Savannah R.	H^{3}	Drinking water	Whole body	Savannah R. resident[e]	< 4		ICRP, genetic apportionment
	I^{131}	Drinking water	Thyroid	Savannah R. resident[e]	< 0.3	1963	ICRP, population at large
	Sr^{90}	Fish + water	Bone	Fisherman	< 1[b]		ICRP, population at large
Power station	Co^{60}	Fish	G.I. tract	Fisherman	< 0.01		ICRP, Group B (c)

[a] Excludes atmospheric pathways, but includes contributions to the same organ from other man-made isotopes present in the water.
[b] Most of the Sr^{90} contributing to this exposure was from fallout and not plant operations.
[c] Recommendation adopted 9 September, 1958. Group B (c) is "members of the public living in the neighborhood of controlled areas".
[d] Federal Radiation Council (US) Recommendations (September 1961).
[e] A hypothetical person who drinks untreated river water. No such person has been found.
[f] Assumes that whole fish, including bones, is eaten. If only the flesh is eaten, the estimate is < 6 per cent.

extent, in produce from irrigated farms [20]. The individuals who eat the largest numbers of local fish may ingest P^{32} in amounts approaching 40% of the annual limit. Zinc-65 follows the same exposure pathways as P^{32} and can also be measured in marine shellfish beyond the mouth of the river. Despite the fact that Zn^{65} is present in the Columbia River in approximately the same concentration as P^{32}, its radiological significance is an order of magnitude below that of the P^{32}. None of the other nuclides present in Columbia River water are concentrated to an important extent by the fish, but Co^{58}, Co^{60}, and Cs^{137} are usually detected. Salmon and other fish that obtain most of their food from the ocean contain negligible amounts of radionuclides of Hanford origin and they contribute essentially no exposure to the people who eat them.

Residents of the cities of Kennewick, Pasco and Richland drink water pumped from the Columbia River below the Hanford reactors. For the year 1963, the dose to the G.I. tract of Pasco residents was less than 40 mrem, and was contributed mostly by short-lived As^{76} and Np^{239}. Chromium-51, the most abundant radionuclide in the water, makes only a minor contribution to dose because of its very low energy release. Iodine-131 from the reactors is also measurable in the drinking water, but the dose to the thyroid of a small child from this source was not greater than 35 mrem in 1963. This thyroid dose was less than half that estimated to have resulted from the presence of fallout I^{131} in the commercial milk supplies of this region during the previous year.

In the Clinch River, below the Oak Ridge plants, Sr^{90} is the nuclide of greatest exposure potential [8]. The most important source of Sr^{90} to people may be local fish if they are consumed in large quantities, and if the bones are eaten as well as the flesh. If the bones are not eaten, then drinking of the water becomes an equal or more important source of Sr^{90}. Ruthenium-106 is also measurable in Clinch River fish, but drinking of the river water is a more significant source of this nuclide to humans.

At the Savannah River plant, H^3 is the radionuclide that reaches the river in greatest amounts. Tritium is not concentrated in food organisms and drinking water is the greatest potential source of this nuclide to people. Potential exposures from the I^{131} and Sr^{90} carried by the Savannah River Plant are not significant but have been included in Table 2 for comparative purposes.

Since it has been recognized as a fission product, tritium has received special attention not only at Savannah River, but at the other sites as well. Because no effective way of removing it from liquid wastes has yet been developed, appreciable inventories of H^3 now exist in the ground at Hanford and at Chalk River. These inventories do not pose significant radiological problems. The applicable limits for drinking water would not be exceeded even if the entire inventories were to enter the rivers in a single day.

A speculative value for exposure from nuclear power stations has been included in Table 2 for comparison with the experience at the major sites. For this purpose, Co^{60} was chosen as the probable nuclide of greatest interest because it has been identified as one of the most common nuclides in reactor water and because it is concentrated by fish. On the basis of data from the Columbia River [20], the potential dose to people from Co^{60} in fish caught below nuclear power stations of contemporary design would be substantially less than 1 mrem/yr. A dose of this low order is completely overshadowed by the exposure received from natural background and several other sources of radiation.

CONCLUSION

The radioactive materials that have entered the rivers adjacent to the major North American atomic energy sites have received continuous attention so that their fate in the environment and the several important exposure pathways could be evaluated. A gratifying outcome of this work is the confirmation that the total dose to the public from multiple environmental sources has been well within appropriate limits at all sites. The general tenor of the environmental studies indicates sound and continuous control of the radioactive effluents from these sites.

In comparison with the quantities of radioactive materials released from the major sites, the releases from the nuclear power stations have been exceedingly small and the dose to the public has been negligible. Methods are thus available which can provide the necessary continued reliable control of liquid effluents associated with the expected growth of the nuclear power industry in North America, and such effluents need not be a deterrent to that growth.

REFERENCES

1. Merritt, W. F., and Patrick, P., Atomic Energy of Canada Limited, Document AECL-1177 (1963).
2. Ophel, I. L., and Fraser, C. D., Atomic Energy of Canada, Limited, Document CRHP-709 (1956).
3. Parsons, P. J., Atomic Energy of Canada Limited, Document AECL-1739 (1963).
4. Foster, R. F., Junkins, R. L., and Linderoth, C. E., J. Water Pollution Control Fed., *33*, 511–529 (1961).
5. Albenesius, E. L., Physics Review Letters, *3*, 274 (1959).
6. Haney, W. A., Nuclear Safety 5 (4) (1964), in press.
7. Browder, F. N., in *Industrial Radioactive Waste Disposal 1* (Hearings before Special Subcommittee on Radiation, Joint Committee on Atomic Energy, Eighty-Sixth Congress of the US), Washington, D.C., 461–514 (1959).
8. Cowser, K. E., Snyder, W. S., and Cook, M. J., in *Transport of Radionuclides in Fresh Water Systems* (Kornegay, B. H., et al., editors) Document TID-7664, Dept. of Commerce, OTS, Washington, D.C., 17–38 (1963).
9. Reichert, S. O., Presentation to Geological Soc. of Amer. (1962) (unpublished).
10. Unpublished data supplied by the Atomic Power Equipment Department, General Electric Company, Consolidated Edison Company of New York, and Yankee Atomic Electric Company.
11. Ophel, I. L., in *Disposal of Radioactive Wastes 2*, International Atomic Energy Agency, Vienna, 323–328 (1960).
12. Kornegay, B. H., Vaughan, W. A., Jamison, D. K., and Morgan, J. M., Jr. (editors), Document TID-7664, OTS, Washington, D.C., 1–406 (1963).
13. Harvey, R. S., Health Physics, *10*, 234–247 (1964).
14. Boni, A. L., Health Physics, *9*, 1035–1045 (1963).
15. Butler, F. E., Health Physics, *8*, 273–277 (1962).
16. Butler, F. E., Analyt. Chem., *33*, 409–414 (1961).
17. Ophel, I. L., in *Radioecology* (Schultz, V., and Klement, A. W., editors) Reinhold Publishing Corp., New York, 213–216 (1963).
18. Guthrie, J. E., Atomic Energy of Canada Limited, Document AECL-1765 (1963).
19. Nielsen, J. M., in *Transport of Radionuclides in Fresh Water Systems* (Kornegay, B. H., et al., editors), Document TID-7664, 91-105 (1963).
20. Wilson, R. H. (editor), Document HW-76526 Hanford Operation (1963).
21. Parker, F. L., in *Transport of Radionuclides in Fresh Water Systems* (Kornegay, B. H., et al., editors), Document TID-7664, 161–180 (1963).
22. Nelson, D. J., *ibid.*, 193–201.
23. Morton, R. J. (editor), Oak Ridge National Laboratory, Document ORNL-3409 (1963).
24. Merritt, W. F., Health Physics, *10*, 195–201 (1964).
25. Parker, H. M., in *Industrial Radioactive Waste Disposal 3* (Hearings before Special Subcommittee on Radiation, Joint Committee on Atomic Energy—Eighty-Sixth Congress of the US), Washington, D.C. 2359–2372 (1959).

ABSTRACT - RÉSUMÉ - АННОТАЦИЯ - RESUMEN

A/287 Etats-Unis d'Amérique

Expérience nord-américaine concernant la décharge de déchets de faible activité dans des rivières et des lacs

par H. M. Parker et al.

Le rejet contrôlé d'effluents contenant des quantités résiduelles de radionucléides est devenu virtuellement inévitable il y a presque vingt ans, lorsque l'homme a commencé à produire des substances radioactives en quantités appréciables. En Amérique du Nord, les principales sources ont été les grandes installations d'énergie atomique de Hanford, Oak Ridge et Savannah River aux Etats-Unis et de Chalk River au Canada. Les genres et quantités de déchets radioactifs provenant de ces installations sont toutefois très différents à cause des différences fondamentales dans les installations principales.

Les évacuations ont varié de quelques curies pour les produits de fission à plusieurs milliers de curies par jour pour des produits d'activation par les neutrons. En comparaison, les quelque douze réacteurs de puissance actuellement en fonctionnement en Amérique du Nord rejettent, au plus, quelques millicuries d'effluents liquides par jour.

L'attention portée à la surveillance de routine dans chacune des installations est fonction de l'importance relative des évacuations et de la possibilité de radioexposition d'êtres humains ou d'autres formes de vie. De plus, la nature variée des déchets radioactifs, ainsi que la grande diversité des caractéristiques physiques, chimiques et biologiques des eaux de surface au voisinage des différentes installations a permis des recherches qui apportent une connaissance étendue sur les conséquences de l'introduction de matières radioactives dans des milieux aquatiques et sur les relations écologiques existantes.

Le mémoire est consacré à la grande expérience acquise au cours des cinq dernières années sur ces questions; il met l'acent sur plusieurs facteurs d'importance primordiale dans la gestion des déchets et qui ont pris une importance plus grande. Ces facteurs sont les suivants:

a) Progrès importants dans l'évaluation de la dose d'irradiation à laquelle le public peut être exposé du fait des multiples sources environnantes;

b) Identification des voies particulières pouvant conduire à une radioexposition significative à partir d'endroits où sont rejetés des déchets liquides;

c) Evaluation des quantités de radionucléides dans l'eau, dans les sédiments et dans la flore et la faune par des méthodes de bilan matières;

d) Améliorations notables du point de vue des instruments, y compris les anthroporadiamètres;

e) Importance des variables locales, telles que la qualité de l'eau, et de la nature des sédiments sur le sort final des radionucléides.

Bien qu'une grande variété de produits de fission et d'activation par les neutrons pénètrent dans les lacs et les rivières, un petit nombre d'entre eux seulement a jusqu'ici mérité une étude détaillée à cause de leur potentiel actif pour les êtres humains qui pourraient y être exposés.

La plupart des produits de fission, y compris le strontium 90, le ruthénium 106, le cérium 144 et l'iode 131 ne sont pas transmis par des chaînes alimentaires aquatiques de telle manière que la chair de poissons présenterait un risque plus grand pour l'homme que le fait de boire l'eau dans laquelle les poissons ont vécu. Quelques produits d'activation, en particulier le phosphore 32 et le zinc 65, constituent des exceptions importantes. Le tritium pourrait se révéler être le nucléide présentant le plus de danger dans l'eau potable à proximité d'usines de traitement du combustible.

Un résultat rassurant des études concernant la radioexposition provenant de sources multiples dans l'environnement est la confirmation du fait que la dose totale à laquelle le public est exposé s'est révélée bien au-dessous des limites de sécurité dans toutes les installations. Les conclusions générales des études de l'environnement indiquent un contrôle sûr et continu des effluents en question. Les études devraient fournir une méthodologie destinée à assurer un contrôle sûr et ininterrompu des effluents liquides qui doivent résulter de l'expansion prévue de l'industrie de l'énergie d'origine nucléaire en Amérique du Nord.

A/287 США

Сброс малоактивных отходов в реки и озера Северной Америки

М. Паркер *et al.*

Около 20 лет назад, когда началось производство радиоактивных веществ в значительных количествах, возникла необходимость контролируемого сброса жидких отходов, содержащих остаточные количества радиоактивных изотопов. В Северной Америке основными источниками таких отходов являются крупные атомные центры Ханфорд, Окридж и Саванна-Ривер (США) и Чок-Ривер (Канада). Однако ввиду различия в оборудовании этих центров количества и виды радиоактивных отходов из этих центров будут отличаться.

Объем сбрасываемых продуктов колебался от нескольких кюри для продуктов деления до нескольких тысяч кюри для продуктов нейтронного облучения в день, тогда как для более 12 действующих энергетических реакторов в Северной Америке предельный объем жидких отходов составляет несколько милликюри в день.

Степень повседневного контроля для каждого центра выбиралась с учетом величины сбрасываемой активности и возможности облучения людей и других живых организмов. Кроме того, различный характер радиоактивных отходов в сочетании с широким диапазоном физических, химических и биологических свойств, присущих поверхностным водам в различных центрах, позволил провести исследования, которые дали новые данные о последствиях сброса радиоактивных материалов в водную среду и об основных экологических соотношениях.

Рассматриваются значительные достижения, достигнутые за последние пять лет в этих исследованиях, и подчеркиваются некоторые факты, имеющие важнейшее значение для контроля отходов, которые находятся в центре внимания. К этим факторам относятся следующие:

a. Существенные усовершенствования методов оценки дозы облучения населения от всех источников среды.

Radioactive Waste Management in Selected Foreign Countries

Nuclear Technology

December 1974

Reprinted with permission of the American Nuclear Society.

RADIOACTIVE WASTE MANAGEMENT IN SELECTED FOREIGN COUNTRIES

H. M. PARKER *HMP Associates, Inc., Richland, Washington 99353*

Received May 6, 1974
Accepted for Publication July 19, 1974

KEYWORDS: *radioactive waste management, international control*

A panel of the National Academy of Science Committee on Radioactive Waste Management (CRWM) has been reviewing the status of waste management in various foreign countries. The objective has been to look mainly for differences between U.S. and foreign practices to identify policies or procedures that might improve technical practices or achieve equal results at lower cost. This paper is an informal summary of the findings that are expected to be reported by the CRWM in the next few months.

The panel on foreign activities of the National Academy of Sciences Committee on Radioactive Waste Management (CRWM) is preparing a report on waste management in other countries which is expected to be published in a few months. The report will take the interesting form of personal statements by the various panel members on the specific countries, some 12 or 13 in all, with which they are familiar, together with a more formal summary report, edited by the panel and the CRWM as a whole. This report is an informal "unofficial" condensation of that summary.

The panel report will make no claim to be comprehensive, and in practice it reduces to considerations of the wastes from chemical reprocessing of nuclear fuels. Within this context, the policies and practices of the selected nations are broadly similar, and relatively well-known to each other. The International Atomic Energy Agency (IAEA) has contributed substantially to this, as described by Jacobs[1] in another paper of this special issue.

Another agency, the Nuclear Energy Agency of the Organization for Economic Cooperation and Development (OECD-NEA) has made a number of interesting contributions for Western Europe. These include planning of major symposia, often in collaboration with IAEA, and the organization of several internationally cooperative sea dumpings in the North Atlantic Ocean. A document, "Radioactive Waste Management Practices in Western Europe" OECD-NEA (1971), is of exceptional interest. The tenor of the report is that waste is currently being handled in a manner that competently protects people from adverse radiation effects, but there are topics of continuing concern. In brief, these are as follows:

1. Storage of high-level wastes in liquid form is an interim solution.

2. Long-lived alpha wastes represent a virtually permanent hazard, with inhalation the critical risk factor.

3. No man-made structure can be guaranteed to provide containment on geological time scales. Therefore, deep geological formations offer better possibilities. Some criteria are given for suitability of bedded salt deposits and salt domes.

4. The alternative for long-lived alpha wastes of disposal in packaged form on the deep ocean bed, especially where the sea bed is sinking, is not rejected.

5. The issues of long-term buildup of ^{85}Kr and of tritium in the atmosphere are analyzed.

6. Lastly, vigorous attention is directed to the problems of ultimate decommissioning of nuclear facilities in the first half of the next century.

Essentially, these topics define the future strategy for Western Europe. However, the implied policy is unofficial, as the report is a scientist's report, neither seasoned nor sterilized, as the case may be, by the bureaucratic process.

In summary of worldwide trends, there is more or less general interest in and acceptance of the following steps:

1. *For High-Level Liquid Wastes.* Reduce to a relatively nonleachable solid, such as a glass, within a few years of generation of the waste. Then either store in a retrievable form in an engineered storage facility, or commit to ultimate disposal, most likely in a geological formation. The only arguable issue is the need for retrievability in the ultimate disposal mode.

2. *For Intermediate-Level Liquid Wastes.* Convert to relatively high-level solids (although generally these can be handled more easily than the high heat producing solids of the above section). The voluminous liquid portion becomes either an innocuous liquid, or at worst a low-level liquid waste.

3. *For Low-Level Large-Volume Liquid Wastes.* Continue to release these to the environment, but work toward reducing the released activities and concentrations. How powerful this effort will be depends on national interpretations of the concept of "lowest practicable level." It is probable that the U.S. working levels will be among the lowest attained, as they already are for nuclear power reactor releases.

It is quite likely, although difficult to prove, that this category of wastes will lead to more environmental exposure than all the rest combined. One could also point out that to a first approximation, if release levels are successfully reduced by a factor of 100 while the source term increases by a factor of ~100 to the year 2000, one has effectively stood still.

4. *For Miscellaneous Solid Wastes Incidental to Processing.* Package and retain these more effectively in fewer and better defined locations. Particular attention is now being given to wastes that contain plutonium or other transuranium elements, since time reduces the activity so slowly. A key question, still unresolved, is the proper *indifference* level, i.e., the degree of contamination below which isolation is not required.

5. *For Gaseous Wastes Incidental to Processing.* Prepare to remove more effectively those contaminants that will accumulate in the biosphere, especially ^{85}Kr and tritium. (The U.S. concern for the very much longer lived ^{129}I was not noted elsewhere.) Reference to typical environmental impact statements for U.S. reactors shows removal efficiencies for noble gases of ~99.998%.

To add two more nines before the eight will truly tax technology.

In the remainder of this paper, a few of the highlights of observations in other countries will be presented. There is no significance to the *order* in which these countries are mentioned.

1. *France.* Until about 1970, French authors wrote confidently about the safety of waste management practices at all levels. Emphasis has now shifted to problems of the future, which are treated as global or international matters. This is a consequence of projection of the fission product and transuranium wastes to the year 2000, and the impracticality of amplifying the storage at Marcoule, for example, by a factor of 50 to 100.

The panel was particularly impressed by the French handling of wastes other than the main stream high-level wastes. The segregation of solid wastes and their incineration or compaction is superior to general U.S. practice. The extensive use of bitumen for encasement was noted. Particularly effective is the process for bituminization of low to medium activity liquid wastes by a combination of bitumen and emulsifier. In the first step, ~80% of the water is removed as the remaining sludge is incorporated in the bitumen at 90°C. Removal of the remainder is done by heating to 130°C. The process owes much of its success to well-engineered equipment that was first developed for the plastics industry. An IAEA Technical Report, "Bituminization of Radioactive Wastes" (1970), is an excellent reference[3] for this topic.

2. *Federal Republic of Germany.* The program here was found to benefit from the relatively late start in nuclear energy, because it avoids the risk of locking into methods that are obsolescent. The West German research at Karlsruhe and Jülich, with basic research at university institutes, is well-conceived, timely, and coordinated to minimize duplication. It is a model that deserves U.S. study.

The most interesting feature is the practical use of a salt mine for centralized permanent disposal. This is the Asse Mine in a salt anticline formation 25 km from Braunschweig. Used as a source of salt and potash since about the turn of the century, the facility has 145 underground rooms providing 3.5×10^6 m^3 of storage space. Conventional drums for low activity solid waste (no liquid wastes may enter the mine) are to all intents and purposes simply "warehoused" there. In a separate chamber, intermediate-level solid wastes are accumulated with enough salt shielding to protect the operators. The operation of chief interest is the preparation for high-level wastes because of the analogy with the U.S. interest in a salt formation repository. Experimental work on heat transfer and strain induction with simulated hot sources is being done. By 1976 or 1977, borosilicate glass blocks will be introduced from the experimental reprocessing plant, VERA, at Karlsruhe. Each block will contain 250 000 Ci so that the operation will be a full scale test of ultimate disposal. The eventual German plan does not appear to be formally documented; it is expected to involve a national reprocessing plant built on the site of a salt dome or anticline other than Asse, to eliminate transportation of the final waste products.

Other notable features of the German program are the fundamental research on irradiated glasses and the search for more durable low solubility media—so-called glass ceramics or true ceramics.

3. *The U.S.S.R.* A point of interest here is the practice of injection of liquid wastes into permeable zones at depths of 1000 to 1500 m and pressures up to 50 atm. It seems likely that in the U.S. a combination of the appeal of early solidification, some pressures toward potential retrievability, and the controversial question of earthquake induction by high pressure injection will combine to make the approach undesirable.

4. *The United Kingdom.* This country has a well-conceived and integrated waste management program that successfully maintains all aspects below the limits recommended by the International Commission on Radiological Protection (ICRP). The British response to the lowest practicable level concept is probably not to press for levels as low as those now applied to reactor releases in the U.S. In this sense, the excellently studied releases to the Irish Sea would probably not be continued in the U.S. scene.

The main line wastes have been well taken care of in high quality stainless-steel tanks, but a projection of the size of the tank farms in the year 2000, plus other factors, have encouraged a firm decision to proceed with solidification (the British FINGAL glass process is a good one) and interim storage in engineered surface facilities. It appears that these facilities, originally conceived as air cooled will definitely be water cooled. As shown in the proceedings of this symposium, the U.S. Atomic Energy Commission (USAEC) is studying both cooling modalities, with perhaps some inferential preference for a passive air-cooling design.

Two British points are of interest:

a. The British are adamant on the need for reliable retrievability under any form of ultimate disposal.

b. The author was particularly impressed by the superior interrelationship between the disciplines of chemical engineering and radiation protection. In a detailed search for the reasons, a logically better organizational structure was not evident. Rather, the success seems to be some product of the more compact geography, a national pragmatic attitude (vulgarly known as muddling through), and the availability of highly qualified personnel. It appears that considerable benefit would accrue to U.S. programs if radiation protection and environmental science skills were integrated into the top planning levels of USAEC waste management.

5. *Canada.* High-level waste storage in Canada is unique because, in the CANDU system, there is no chemical reprocessing, and the fission products are left stored in a favorable ceramic form, namely the original nuclear fuel pieces. The analogy between their storage and the U.S. retrievable surface storage will be well covered in the symposium paper by Morgan.[4]

6. *Japan.* The Japanese practices are representative of what a progressively industrialized island nation with minimal land resources must do in its nuclear energy program. Either ocean dumping or export of high-level wastes to other countries becomes important. In addition to the forthcoming panel report, an excellent report was very recently published by the OECD-NEA.[5] Japan became a full member of the NEA in 1972.

7. *Other Countries and Areas.* The remaining countries and areas considered include India, Pakistan, Italy, Belgium, the Netherlands, and possibly Scandinavia. Time does not permit their coverage, and on the whole, inferences that might suggest improvements in U.S. practices (the principal objective of the panel study) are less effective.

The intended panel report will also include comments on ocean disposal, transuranium waste problems, potential world-wide contaminants, bituminization of wastes, plastics in waste packaging, public opinion, and perceived risks.

Six recommendations to the USAEC will be derived from the study; the principal ones can be sensed from the content of this brief summary.

REFERENCES

1. DONALD G. JACOBS, "Management of Radioactive Wastes—Programs of the International Atomic Energy Agency," *Nucl. Technol.*, **24**, 300 (1974).

2. "Radioactive Waste Management Practices in Western Europe," Organization for Economic Cooperation and Development-Nuclear Energy Agency, Paris (1971).

3. "Bituminization of Radioactive Wastes," IAEA-116, International Atomic Energy Agency, Vienna, Austria (1970).

4. W.W. MORGAN, "The Management of CANDU Fuel," *Nucl. Technol.*, **24**, 409 (1974).

5. Y. NAGAI, "Radioactive Waste Management Practices in Japan," Organization for Economic Cooperation and Development-Nuclear Energy Agency, Paris (1974).

Nuclear Waste Disposal

"Science of the Future"
Encyclopedia Britannica Yearbook

1978

Reprinted by permission from the *1978 Yearbook of Science and the Future*, copyright 1977, Encyclopedia Britannica, Inc., Chicago, Illinois.

Nuclear Waste Disposal
by H. M. Parker

Unlike household garbage, radioactive wastes from atomic power plants cannot be merely discarded and forgotten. Expected increases in the number of nuclear reactors by AD 2000 demand secure methods of waste storage.

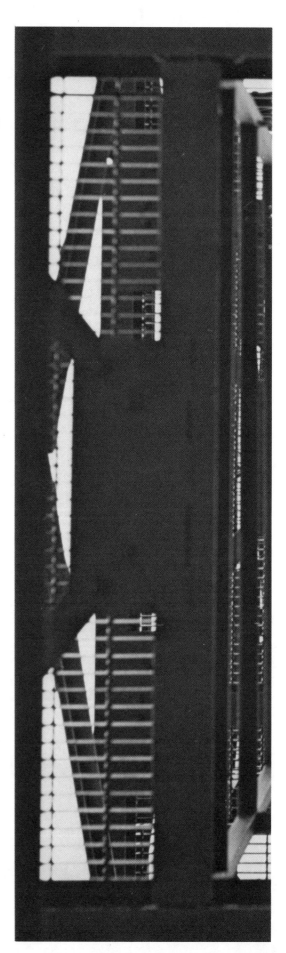

The production of electricity from coal, gas, or nuclear fission is accomplished identically after each of these sources of energy is used to generate heat. Before that stage two important differences distinguish nuclear power from its counterparts. On the favorable side, the fission of one gram of uranium-235, which is the fissionable content of 140 grams of natural uranium, provides as much energy as the burning of three tons of coal or 700 gallons of fuel oil. On the unfavorable side, the fission process generates life-threatening fission products whose radioactivity exceeds that of any sources previously known and whose toxicity may persist for thousands of years. A modern nuclear reactor contains radioactivity that is equivalent to about 2,000 tons of radium, which is several million times the amount of this element in commerce.

Fuel for nuclear power originates as uranium ore. Radioactive itself, this material contains other radioactive elements that must be removed prior to its use as fuel. In fact, all stages of the nuclear fuel cycle—from milling ore to reprocessing spent fuel—may generate some undesirable radioactive components. It is during the intense magnification of radioactivity in the reactor, however, that the major high-level radioactive wastes are produced. Until their decay, these wastes must be segregated from the biosphere to the greatest extent possible.

In weighing the hazards of radioactive waste, the main concern of society is not the possibility of massive radiation injury. That result is much less likely from accidents in waste storage than from a catastrophic accident to a nuclear reactor, a case that has been extensively analyzed. More relevant is the potential for small quantities of radioactive substances to escape into the air or water sources and eventually enter the body. There is no question that radioactivity can produce cancer and genetic mutations in man, although the degree of harm that can arise from low levels of exposure is still uncertain. Inadequate processing, storage, transport, or disposal can disperse radioactivity

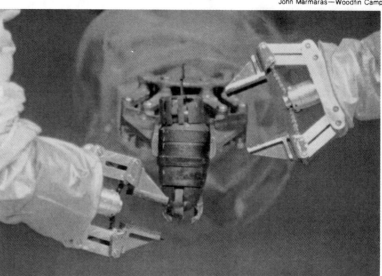

John Marmaras—Woodfin Camp

(Overleaf) Suspended 30 feet aloft, a crash-shielded model of a shipping cask for radioactive material awaits testing within a drop tower at the U.S. Oak Ridge National Laboratory. Manipulators (above) operated by remote control from behind special radiation-protective glass are often used to transport radioactive matter within the confines of a single room or chamber.

H. M. PARKER *is President, HMP Associates Inc., Richland, Washington.*

(Overleaf) Photograph courtesy, Union Carbide Corporation, Nuclear Division, Oak Ridge National Laboratory. Illustrations by Dave Beckes

into the environment. The biological consequences of such events must be predicted through a chain of prudent assumptions.

To assess future problems of commercial waste management from a nuclear economy, one must use past experience in handling radioactive wastes. For the United States that experience lies in its 30-year-old nuclear weapons program. Recently the comprehensiveness of that country's existing or proposed waste-management systems has come under increasingly severe public and technical scrutiny, with considerable justification. In 1973 reported leakages of high-level liquid wastes from short-term storage tanks of the Hanford Works, a federal plutonium production plant near Richland, Washington, did not inspire public confidence. Progressive addition of nuclides to the list of significant hazards has deflated claims of scientific omniscience; tritium was first recognized as a fission product in 1959, iodine-129 as a very long-lived biological hazard only about a decade ago, and carbon-14 as a product of impure uranium oxide fuels a few years later. In addition, several years of apparent vacillation by the U.S. government in the choice of a long-term underground repository or retrievable surface storage facility has not helped.

Nevertheless, over the past 30 years control of nuclear wastes has become more effective than control of such age-old chemical wastes as mercury and arsenic. The allocation of another 30 years to develop more secure methods, such as a well-tested geological repository, is not unreasonable. However, public concern demands that such time be used decisively in the pursuit of excellence, with no unreasonable burden left to future generations.

Composition and classification of radioactive wastes

The most significant radioactive wastes from the nuclear program arise through two separate mechanisms. The principal one is the neutron-induced fission process that occurs in uranium-235 and in a few other

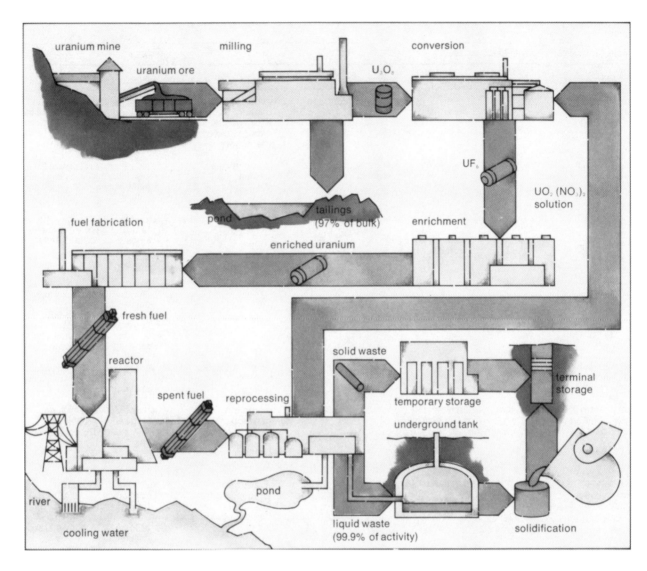

fissionable materials. In the course of liberating heat to generate power and neutrons to sustain the fission process, fissioning uranium nuclei divide into pairs of smaller nuclei called fission products. The species so created are regular elements of the periodic table but with unstable nuclear configurations. In their search for stability, they decay into other nuclear species, often several times in rapid sequence, emitting electrons, or beta (β) rays, and frequently energetic photons called gamma (γ) rays. This decay provides the intense radioactivity of the fission products and causes the chemical composition of fission-product wastes to change continually with time.

The second mechanism involved in waste production is the modification of the nuclear identity of fuel and fission products by direct neutron absorption. Characterized by a chain of successive absorptions and nuclear decays, this process is especially important in the creation of transuranic species from unspent uranium fuel. Many transuranic nuclides emit energetic helium nuclei, or alpha (α) particles, and are

All stages of the nuclear fuel cycle may generate radioactive wastes. Whereas milling of uranium ore produces 97% of waste bulk, more than 99% of the potentially hazardous activity of nuclear wastes is contained in liquid waste produced during reprocessing of spent fuel. Bulky and unwieldy, it is liquid waste that creates the majority of immediate storage problems.

Table I. Significant Radioactive Nuclides				
nuclide and half-life	principal emissions	source	manner of release	biological significance
tritium (hydrogen-3), 12.3 years	β	fission product and neutron activation of heavy water	isotopically mixed with ordinary water in liquid wastes and in vapor released to the atmosphere	incorporation in water enables it to pervade all life forms including genetic material
carbon-14, 5,730 years	β	neutron-absorption product	carried in liquid wastes	only recently recognized as a significant hazard; long half-life and high biological activity
krypton-85, 10.7 years	β and γ	noble gas fission product	into the atmosphere	worldwide distribution; hazard to skin; intake from the lungs to blood and fatty substances
strontium-90, 29 years; cesium-137, 30.1 years	both β emitters; cesium-137 also a γ emitter	fission products	carried in liquid wastes	strontium is a chemical analog of calcium and has a strong affinity for bone, decays to strongly radioactive yttrium-90; cesium, a potassium analog, is broadly distributed and retained in soft tissue*
technetium-99, 213,000 years	β	fission product	carried in liquid wastes	very long half-life; migrates easily through soil sediments
ruthenium-103, 39.6 days; ruthenium-106, 369 days	both are β emitters; ruthenium-103 also a γ emitter	fission products	escape into atmosphere from stacks of fuel-reprocessing plants; carried in liquid wastes	short half-lives preclude long-range problems; ruthenium-106 decays to a short-lived β emitter
iodine-129, 15.9 million years; iodine-131, 8 days	β and γ	fission products	escape into atmosphere from stacks of fuel-reprocessing plants	nuclides deposit on vegetation, transfer to milk through cattle; affinity for thyroid gland; long half-life of iodine-129 allows steady accumulation in the environment; iodine-131 is a key consideration in accidental releases
plutonium-239, 24,400 years; also plutonium-238, 87.8 years, and other isotopes	α	neutron-absorption product	carried in liquid wastes; when used in oxide form as a nuclear fuel, it tends to escape confinement as a fine dust	affinity for bone; dust retained in lungs and lymph nodes; demonstrated proclivity to produce cancer in animals; undoubtedly carcinogenic in man, although no case has yet been demonstrated
transuranic nuclides, including neptunium, americium, curium, and higher elements (including plutonium, described separately above); several species have very long half-lives	α, β	neutron-absorption products and decay products of higher transuranic nuclides	carried in liquid wastes	may show a radioactive toxicity similar to that of plutonium; these nuclides are members of the actinide series and are often identified by this more inclusive term, which also encompasses some relatively innocuous members

*Note: taken together, radioactive strontium and cesium nuclides generate a considerable fraction of the heat of high-level wastes during the first decades; their relatively long half-lives, high abundance, and biological compatibility materially influence waste-management policies.

regarded collectively as biologically dangerous. They also decay into other alpha emitters, thus prolonging the hazard.

Through a similar process of neutron absorption, called neutron activation, various normally stable elements present in the reactor as fuel cladding, moderators, control elements, coolants, and impurities can be rendered radioactive. For the most part, these activated materials are less significant and often not as persistent as fission products. Nitrogen in air coolant, for instance, is activated to a powerful gamma emitter, fortunately of very short life. Argon in air coolant gives argon-41, whose rate of production may set limits on safe reactor operation. Irradiation of cladding material often leads to the production of cobalt-60, with a significant five-year half-life. The hydrogen isotope deuterium (hydrogen-2), a constituent of the heavy water used as a neutron moderator, is activated to tritium (hydrogen-3), an unstable isotope with a half-life of about 12 years. Other activation products have much longer, troublesome half-lives: for example, carbon-14 (half-life, 5,730 years) and nickel-59 (half-life, 80,000 years). Important characteristics of several species of radioactive nuclides found in nuclear wastes are delineated in table I.

Although it is possible to consider the composition of radioactive waste in terms of elemental mass abundance, it is often the case that rapidly decaying nuclides will have negligible mass in proportion to their activity. In addition, mass abundance depends upon many variables, including the kind of fuel used and the length of time the fuel is allowed to fission and cool. Hence it is usually of more interest to discuss waste composition in terms of its activity. Activity is the rate of disintegration of a radioactive species; it is usually expressed in curies (Ci), with 1 Ci = 3.7×10^{10} disintegrations per second. (The becquerel, or Bq, equivalent to 1 disintegration per second, is not yet in common use.) For example, for uranium fuel allowed to fission for 33,000 megawatt-days in a pressurized water reactor and cooled 30 days, the ten most active nuclides account approximately equally for 76% of the total activity. By contrast, after ten years of cooling, the four most active nuclides are responsible for 90% of the total activity. Per metric ton of nuclear fuel, the total activity as a function of cooling time will decrease from a maximum of 138 megacuries (MCi) at the time of discharge from the reactor to 10.8 MCi after 30 days of cooling, to 2.2 MCi after 1 year, and to 0.04 MCi after 100 years.

Both mass abundance and activity, however, are only secondary indices of the hazards of radioactive wastes. One needs to estimate the radiation dose to people and other life forms for any given release. Complete specification of the hazard over the lifetime of nuclear wastes is difficult or speculative. For wastes that present themselves through water sources, one method of specification is through the use of a toxicity index, defined as $\log_{10}$ of the volume of water (in cubic meters) that is required to dilute the wastes to conventionally accepted maximum permissible concentrations. A useful simplification, it nevertheless ignores all the problems of transfers through the ecological web

Cask-encased element of spent nuclear fuel is lowered into an unloading pool. The fuel element is then removed and transferred underwater to an adjoining storage pool to await reprocessing.

and the obvious fact that widespread contamination at such concentrations would be unacceptable.

More than 99% of the total activity of nuclear waste is first encountered in liquid form, remaining as a by-product after spent nuclear fuel is chemically dissolved and all reusable fuel and fission products are reclaimed. Bulky and unwieldy, liquid creates storage problems that require immediate attention. Liquid wastes are commonly categorized by level of activity as low, medium or intermediate, or high, but agreement on the activity ranges of each has never been achieved. In the U.S., high-level wastes are practically defined as those arising from the first cycle of solvent extraction during fuel reprocessing or equivalent. Each site tends to have its own terms, which may be source-oriented or treatment-oriented, and only secondarily related to waste activity or relative hazard.

At the Hanford Works, for example, the limits arbitrarily and consistently have been: low level, below 5×10^{-5} microcuries/milliliter (μCi/ml); intermediate level, 5×10^{-5} to 10^2 μCi/ml; high level, above 10^2 μCi/ml. The International Atomic Energy Agency (IAEA), by comparison, offers six categories ranging from a throw-away level for wastes with activities below 10^{-6} μCi/ml to a level requiring long-term storage with cooling for wastes with activities above 10^4 μCi/ml. These categories are not consistently used, nor are three similarly conceived ones for gaseous wastes. However, four categories of solid wastes have formed the basis for orderly transportation regulations in many countries. As applied to storage, arbitrary terminology is customary.

The Hanford Works: a study in short-term storage

Built during World War II, the Hanford Works was the original complex of fuel fabrication plant, nuclear reactors, and chemical separations plants for the production of plutonium through the irradiation of uranium-238. It occupies 570 square miles adjacent to the Columbia River in a region of southeastern Washington that experiences an average annual rainfall of 6.5 inches. Its separations plants and waste-storage facilities were built near the center of the reservation on a plateau that offers 200–300 feet of dry sediments above the water table.

During chemical processing, noble-gas fission products are released into the air along with radioactive iodine nuclides, some ruthenium, and some entrained fission products and actinides. All except the noble gases, however, are retained with high efficiency on chemical absorbers and filters. The main liquid wastes generated during processing are stored in large underground tanks.

A total of 152 large underground storage tanks exist at Hanford for high-level radioactive wastes. The older tanks are of single-wall carbon steel and concrete. Currently, all high-heat liquid waste is stored only in newer double-wall tanks; low-heat waste and the salt cake from a waste-solidification program are stored in older single-wall tanks. The tanks are in four sizes: 55,000 gallons, 500,000 gallons, 750,000 gallons, and 1,000,000 gallons.

Lowell J. Georgia—Photo Researchers, Inc.

Houses built to gauge the effects of radiation from uranium ore tailings are studied by scientists at Colorado State University. Presently hundreds of millions of tons of tailings from uranium mills stand in outdoor mounds in the western U.S.

Since the Hanford plant was built, 20 confirmed tank leaks have occurred, all of which have been from single-wall tanks. Each of the leaks has been assigned to one of four causes: corrosion of carbon-steel tank liners, cracking or mechanical failure of the steel liner, thermal expansion due to local overheating, or buckling due to other causes. These tank leaks have varied in size from very small, *i.e.*, about 1,500 gallons, to 115,000 gallons for the notorious leak that occurred in 1973. Total volume from the 20 leaks is 464,000 gallons. Current government policy for these wastes requires their solidification to salt cake, which will continue to be held in the tanks. Although the probable release from this condition is very low, such storage will not be as secure as solidification to a high-quality glass.

As the remaining liquid waste is converted to salt cake, additional tank leaks are expected. In the future, an increasing fraction of liquid waste will be stored in double-wall tanks, where a leak in the inner tank can be detected and corrective action taken without escape of liquid from the outer tank to the ground. Detection systems currently under development have the objective of locating a leak before it exceeds 1,500 gallons. These systems plus the use of double-wall tanks should virtually guarantee that a leak as large as 115,000 gallons will not occur again under normal operations.

None of the leakage from underground tanks at Hanford has been proved a hazard to the groundwater located some 200–300 feet below the surface. Essentially all of it has remained fixed in sediments beneath or adjacent to the tanks and is expected to remain in this status, short of almost cataclysmic changes in climate or hydrology.

The effectiveness of this retention was determined incidentally from monitoring the movement of deliberately released intermediate-level wastes. These wastes contained about 40,000 Ci of cesium-137, about one-fifth to one-sixth the activity of the eventual accidental leaks. An independent review by the U.S. National Academy of Sciences (NAS) in

1966 looked unfavorably on these releases, and they have since been reduced by a factor of 1,000.

When the planned releases of large volumes of intermediate-level wastes were made, extensive programs were mounted to study the consequences. For all short-term and medium-term purposes, the results of the studies were favorable. Dry sediments below a disposal point were found to absorb liquids like a sponge and hold them in place as effectively as a tank. Such sites now hold 5.3×10^7 gallons containing 60,000 Ci of active wastes. This method of disposal, called the specific retention method, is no longer approved.

Additionally, it was found that larger volumes would essentially saturate the sediments and proceed downward to the water table. The soil columns were found to function as ion-exchange media, releasing elements already in the soil while retaining in their place most of the active materials, usually quite close to the point of entrance. Each radioactive element has a characteristic retention curve—sometimes more than one for different chemical forms. By chance, most of the more mobile forms are short-lived. Hence, it would have been possible to permit such disposal until the significant nuclides just approached the water table. At one stage in the history of these releases, a prescribed level of actual contamination of the groundwater was permitted before changing disposal points, a poor policy.

The best studied case of this type occurred between 1952 and 1956 when 40 million gallons of waste containing about 750,000 Ci of mixed fission products including strontium-90 and cesium-137 were released into underground timbered cribs. Disposal was stopped when instruments in test wells indicated strontium and cesium at accepted concentrations 18 meters above the water table. In 1966 advanced monitoring instrumentation determined that downward movement of the liquid and of its activity was greater than expected. The groundwater became locally contaminated with strontium-90, as well as with ruthenium, technetium, and tritium.

Despite the indisputable failure of waste management to avoid contamination, the episode was not as serious as it appeared. Subsequent testing established that more than 99.9% of the dangerous activity of strontium and cesium would continue to be retained within 15 meters of the cribs. Moreover, it would be an oversimplification to expect all waste constituents to migrate with equal swiftness through saturated sediments to the Columbia River. Over the past 20 years it has become increasingly clear that some constituents—e.g., tritium, ruthenium, and technetium—do move rather freely through the ground. In fact, these have been used as tracers to follow movement of the contaminated groundwater. According to present estimates, travel time for such highly mobile species could be as low as 15–30 years, and, indeed, tritium has already been detected in the banks of the river as far as 11 miles from the release point. For strontium, cesium, and other slow-moving nuclides, however, travel time is calculated to be in excess of 1,000 years. In addition, analytic and computer models of water-table

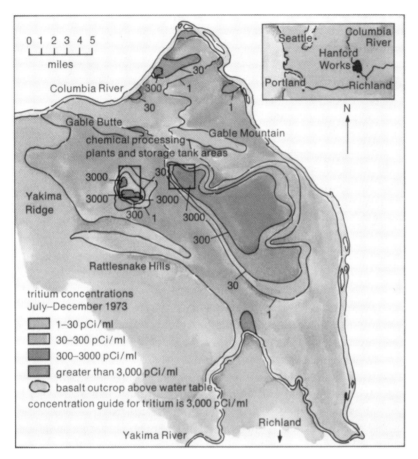

contours have indicated that by the middle of the 21st century travel time to the river can be doubled by a concerted reduction of further liquid input to the ground.

A separate case at the Hanford Works precipitated yet another round of controversy over disposal practices. The release of relatively large amounts of plutonium wastes, variously estimated at 25-70 kilograms, into a single crib led to public fears that the material could become sufficiently concentrated to undergo a spontaneous chain reaction. Although this eventuality was shown to be scientifically unfounded, the main mass of plutonium at the site was ordered removed. This situation would not arise in commercial reprocessing.

In retrospect, the leaks at the Hanford plant and the use of its unique dry sediments for waste retention probably did significant damage to its overall waste-management image. Molded by such lessons of experience, the future of waste management at Hanford has been projected to include the following guidelines: (1) High-level wastes will continue to be evaporated to salt cakes stored in the large tanks; whether they should later be converted to a less leachable solid for shipment to a federal repository or basalt-cavern burial at Hanford is undecided. (2) Intermediate-level wastes will not be released until they have been treated to reduce their activity to releasable amounts. (3) Solid wastes will be controlled with emphasis on careful packaging and

Studies over the past 20 years have made it clear that certain of the more short-lived components of liquid wastes, particularly tritium, ruthenium, and technetium, migrate rapidly through saturated sediments. Monitoring of test wells at the Hanford Works during the latter half of 1973 revealed a pattern of widespread groundwater contamination with tritium from liquid wastes that had been intentionally released to the ground. Tritium was also detected in the banks of the Columbia River as far as 11 miles from the release point.

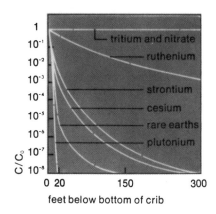

Extensive investigations at the Hanford Works established that soil below its underground liquid-waste storage cribs functions as ion-exchange media, releasing elements already in the soil while retaining most of the active wastes. Shown above are retention curves for several radioactive species monitored during their downward migration through typical Hanford sediments; C/C_0 is the ratio of nuclide concentration at a given depth to its initial concentration. Fortunately most of the activity of the more hazardous, long-lived species is effectively trapped within 15 meters of the cribs.

isolation against leaching. (4) Such controversial situations as the relatively large accumulation and storage of plutonium wastes at one location will be corrected.

The Savannah River Plant: a comparative study

The Savannah River Plant (SRP) is the second of the major plutonium production centers in the U.S. First operated in 1953, it is a complex of nuclear reactors and separation plants in a reservation occupying 300 square miles in western South Carolina.

From the viewpoint of waste management, three important characteristics differentiate the SRP from the Hanford Works. First, its later start permitted evaluation of Hanford experience. Du Pont & Co., responsible for the SRP, had the advantage of having designed, built, and operated the Hanford facilities. Second, in contrast to the graphite moderators used in the production reactors at Hanford, heavy water is employed for the purpose at the SRP. For this and other reasons, tritium is a more significant waste nuclide at the SRP than elsewhere. Third, climate and groundwater conditions are strikingly different from those prevailing at Hanford. Safety margins for retention of wastes in dry sediments do not exist at the Savannah plant.

High-level liquid wastes are retained at the SRP's reprocessing plants in large carbon-steel tanks that have progressive improvements in design over earlier Hanford models. The first design was essentially a cup-and-saucer arrangement in which any leakage from the tank could be held in the saucer. Newer versions are complete double-walled tanks with internal cooling coils and a central column that greatly stiffens the entire structure.

The reprocessing of highly radioactive wastes involves storage and settling of sludge in one tank, leaving predominantly radiocesium in the remaining liquid, or supernate; transfer to a sludge collection tank; concentration of the liquid in an evaporator; and accumulation of salt in a salt storage tank. The active materials produced in the first cycle of solvent extraction are termed high-activity wastes, to use SRP terminology. Later-cycle wastes are called low-activity wastes, but these are what might normally be called intermediate wastes. These too must be stored, evaporated, and reduced to salt cake. The primary difference between the two types is that the latter do not require continuous cooling and can be handled in simpler tanks.

Tanks at the Savannah River Plant range in volume from 750,000 gallons to 1,300,000 gallons with a total capacity of 30 million gallons, currently about two-thirds occupied. Leakage (about 1% of the amount at Hanford) has been encountered in the tanks with no serious consequences to date. The newer tanks, short of some unforeseen catastrophic mode of destruction, should have high integrity for several decades. There now seems to be a greater chance of accidental spilling during transfer or by overfilling than by tank leakage. Nevertheless, despite the reduced mobility of sludges and salt cakes compared with liquids, storage cannot be accepted as final at this location.

Table II. U.S. Commercial Wastes in AD 2000 (based on a 500-Gw installed nuclear capacity)

waste type	annual addition volume × 10³ cu m	annual addition activity MCi	accumulated to AD 2000 volume × 10³ cu m	accumulated to AD 2000 activity MCi
ore tailings (mining and milling)	19,000	0.2	260,000	3
nontransuranic wastes (miscellaneous low-level wastes: 90% from reactors, 10% from rest of cycle)	200	1	2,000	3
low-level transuranic wastes (emissions: α greater than 10^{-9} Ci/g, low β, γ)	6	5	60	40
intermediate-level transuranic wastes (emissions: α greater than 10^{-9} Ci/g, high γ)	2	1	22	5
high-level solidified waste (in glass or ceramics)	0.4	1,100	2.5	6,700
noble gases (particularly krypton-85)	in cylinders	60	—	440
iodine (iodine-129 from stacks)	0.01	0.0003	100	30
tritiated water (from coolant activation)	60	0.3	660	2
tritium (fission product)	0.1	4	1	30
carbon-14 (from nitrogen-14 in fuels)	—	0.005	—	0.05
cladding wastes	0.5	50	4	200

Current expectation is that the material will eventually be solidified and transferred to a federal repository, if the necessary technological and political decisions can be made. The Savannah River Plant staff have long considered a method of storing the wastes some 1,500 feet below the SRP site in bedrock. This concept has been independently examined by a committee of the NAS, which concluded that the project may be feasible but only after extensive exploratory work. Between 350 and 1,000 feet below the site lies the Tuscaloosa aquifer, a main water resource for the region. It seems unlikely that convincing proof of continued safety of the aquifer over geologic time can be developed.

Meanwhile, the stored wastes are causing no radiation exposure except to SRP staff involved in the processing and transfer. Unavoidable processing releases to the atmosphere and to plant streams are also causing some exposure, but comparable with local variations in natural background radiation.

Courtesy, Southern California Edison Company

Technicians work to remove spent fuel elements from reactor core at the San Onofre Nuclear Generating Station in California. About one-third of the reactor's 157 fuel assemblies are replaced during refueling every 16 months.

Terminal storage

Until recently, major projections of waste accumulation in the U.S. assumed an installed nuclear power of 1,200 Gw by AD 2000. In mid-1975 this figure was revised downward to 850 Gw, and the current best guess is taken to be 500 Gw. Precision in the power forecast is not needed, however, because the choice of reactor and fuel types materially alters waste composition. Estimated volumes of wastes are also speculative, being based upon the expectation of improving management practices. Hence, any forecast of the magnitude of the waste problem, such as depicted in table II, cannot be accepted as final.

The ore tailings shown in the table are expected to be returned in bulk to disused mines. They release small amounts of radon-222, a noble gas whose decay products include polonium-218 and polonium-214, both strong alpha emitters with a preference for deposition in the human lung. Currently more than a hundred million tons of tailings from uranium mills stand in outdoor mounds in the western U.S.

Miscellaneous low-level beta- and gamma-emitting wastes are generated throughout the nuclear fuel cycle and will probably be buried in surface pits. Tritiated water arising from reactor operations is also expected to be sent to disposal sites. All other waste types in the table must be afforded high-level storage.

Conventional high-level wastes may require separation from the biosphere for at least a million years. Retention of liquids or relatively immobile sludges for such times is out of the question. Their reduction to insoluble solids has been studied for nearly 20 years and, as was mentioned above, satisfactory methods, such as incorporation in special glass compounds, now exist. One key question is how and where to store these solids.

In the U.S., where the responsibility for long-term disposition of commercial wastes has been inherited by the Energy Research and Development Administration (ERDA) from the defunct Atomic Energy Commission (AEC), federal regulation requires solidification by the waste producer within five years and transfer to a federal repository within ten years. Yet, as of 1977, this repository does not exist, and proposals for it have not progressed beyond the review stage.

As originally conceived, the federal repository was to have been a salt mine, as recommended to the AEC by the NAS in 1957. In 1970 the AEC proposed a demonstration salt-mine facility near Lyons, Kansas. Specific review of the proposal, however, revealed several preexisting man-made flaws in the integrity of the site; among the worst was the distinct possibility of the destructive intrusion of water from mining operations in the vicinity.

The AEC then proposed a temporizing method of retrievable surface storage. In one suggested version, solidified wastes would be sealed in large steel canisters, which in turn would be placed in concrete cylinders and erected on barren land. The need for extensive surveillance and the vulnerability of the stored wastes to war, sabotage, neglect, and natural disasters are obvious and have overshadowed

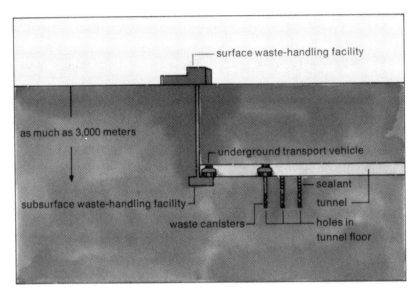
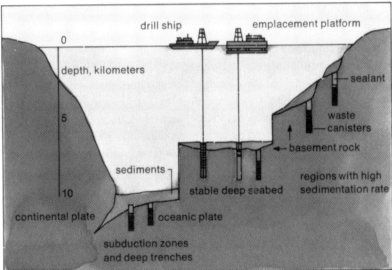
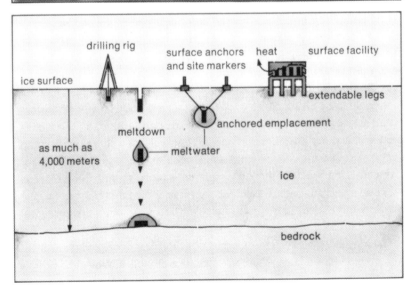

Depicted are several concepts for geological disposal of nuclear wastes (top to bottom): solid waste emplacement in a mined cavity, similar to operations in progress at the Asse mine in West Germany; three plans for seabed disposal that exploit aspects of plate tectonics and rapid coverage characteristics of undersea regions with high sedimentation rates; and three plans for disposal on or below polar ice sheets.

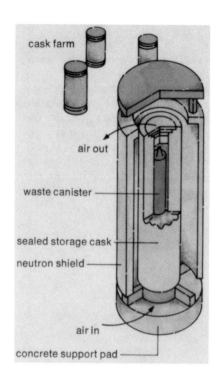

One method of temporary surface storage, the sealed storage cask concept, involves placement of solidified wastes in large steel canisters, which in turn would be encased in concrete cylinders and erected on barren land.

technical issues for the present. Such a scheme of retrievable storage is sensible if new technological developments related to ultimate disposal are anticipated.

The most recent proposal has been to pursue the study of geological disposal in salt beds of southeastern New Mexico, in shales, and in crystalline rocks such as granite in the hope that acceptable repositories may be developed in various parts of the country. This plan arose from a preliminary study of alternatives that included ten geological concepts, three concepts each for disposal in ice sheets and seabeds, three options for disposal in space, and three transmutation concepts. Given the reservation that analysis projected over a million or more years obviously cannot be verified experimentally, the study yielded several conclusions. (1) Full examination of any of the concepts may require 15–30 years. (2) The estimated cost of the most attractive methods will not hinder the development of commercial nuclear power. (3) Several geological concepts offer a low risk of serious environmental impact; *i.e.*, those offering retrievability, solidification before arrival at the disposal site, and storage without the involvement of melting rock by fission heat. (4) By comparison, all other proposed methods of storage are decidedly less secure.

Projection into space, at best feasible only for selected wastes, requires consideration of the possibility of an aborted mission, in which case the ability of the waste container to survive the incident intact must be guaranteed. Concepts for disposal in polar ice sheets, in which high-activity wastes would be allowed to melt into the ice of their own accord, possess some advantages for nuclides with relatively short half-lives, on the order of 700 years. The persistence of the transuranic nuclides and certain fission products, however, could allow them to outlive the estimated life expectancy of the ice sheet itself. In addition, the Antarctic currently is kept free of nuclear wastes by international treaty. With regard to disposal in the sea, only placement in prepared holes in the stable deep seabed is likely to be further pursued, and that only within a framework of international cooperation.

Transmutation concepts involve bombardment of wastes with neutrons to convert persistent nuclides to forms that are less toxic or of shorter half-life, or both. An objective would be to produce a waste that would not require absolute retention for more than about 1,000 years, thus diminishing one objection that million-year storage goes beyond the life span of civilizations. However, this method, as well as the extraterrestrial concepts, requires complex partitioning of the active wastes. It is quite possible that unavoidable releases at that step would exceed probable releases from sound geological disposal. (Even for untransmuted wastes, the necessity for secure million-year storage may be an overstatement; calculations have shown that the most active wastes would decay to the equivalent of a bed of uranium ore in several thousand years.)

The formulation of waste-disposal policies, practices, and goals in all nations that use nuclear reactors has been broadly similar. The principal practical differences lie in the aggressiveness with which improved methods have been installed. In France, for example, encasement of milder wastes in bitumen, far superior to incorporation in concrete, has been highly developed. As of late 1976, progress of a vitrification plant in southern France for the solidification of wastes was gratifying. Wastes and molten glass were to be mixed together, poured into stainless-steel containers, and temporarily stored in air-cooled underground concrete vaults. The United Kingdom, which has had excellent experience with liquid storage in stainless-steel tanks, is planning an industrial-scale vitrification plant for the mid-1980s. Canada has delayed its waste-management issues by not reprocessing fuel, choosing instead to retain fission wastes within spent fuel pieces. Japan, with limited land resources, looks favorably toward ocean dumping or shipment of its wastes to another country.

West Germany is the first nation to store low-level and intermediate-level solids in a salt mine. The Asse mine, near Braunschweig, is a former commercial salt and potash mine with 145 underground "rooms" or excavated caverns, each averaging 24,000 cubic meters. The rooms occupy galleries between 490 and 800 meters below ground. Dry waste is stored in drums at the lower levels with the intention to seal off a filled cavern with salt. Intermediate-level waste occupies a shielded cavern at the 490-meter level. Experiments conducted to study the heat and stress burdens of high-level solids were expected to lead to demonstration storage of quarter-million-curie glass blocks before 1980. German work on four methods involving solidification in glass ceramics was advancing in parallel.

It is truly difficult for the concerned citizen to reach a balanced judgment on the prospects for responsible waste management. There are advocates of nuclear power, and advocates of nuclear moratorium. One side considers that all the basic problems have been solved, and the other points to a nearly endless catalog of uncertainties, which range from arguments stressing the radiological hazard of the transuranium elements, risks of sabotage, diversion to illicit nuclear weapons, and the effects of geological changes to concerns about the lifetimes of civilizations and the risk of intrusion into forgotten disposal sites.

The most intense radioactivity of nuclear wastes is lost in a thousand years — a quite feasible span for a federal repository. Environmental damage from radioactive releases thus far has been quite small. If control methods continue to improve, it seems likely that results will continue to be good, although not necessarily perfect.

One course is clear. Waste-management research must have the attention and funding that it has received only belatedly in the last decade. Waste management can and must be upgraded and maintained at a high level throughout the nuclear age. It can, in fact, become the model for the way a responsible nation should handle hundreds of other carcinogenic and mutagenic substances upon whose beneficial qualities it also depends.

Courtesy, Gesellschaft für Strahlen-und Umweltforschung, Munich

View through protective lead-glass window into a storage chamber of the West German repository at the Asse mine reveals a pile of iron-banded drums of dry nuclear waste. Above the chamber and connected to it by vertical shafts is a control room from which drums of waste are lowered to their final location under surveillance of a closed-circuit TV system.

Expt 26.   Cyl. vol. dosage.
20.8.34.   Cyl.   Lgth 13.5 cm. diam 9. calc. as 14.5 × 10.
                    Outer Cyl.   54 mgs. ⎫   27 - 3 mgs.
                    Inner "   27 mgs. ⎬ 99 mgs.   6 - 2 s
750 mm.   Each end   9 -   ⎭   6 - 1 s
20°C.
            Construction as No: 24.
            Chamber V.1.   8 mins.
            Central section Dept.   2 cms. up.   4 cms. up.
                    0   26.5   27.0   28.0
                    1   28.0   26.0   28.0
                    2   29.0   26.0   25.0
                    3   26.0   24.5   24.0
                    4   25.4
                    4½   21.0   20.0   21.0

            6 cms. up.   End.
                0 ⎰ 26.5   24.0
                2 ⎨ 24.0   24.0
                4½ ⎱ 20.0   18.0

            Calcᵈ   5.28 r.
            F+bᵈ   4.73 r.

Expt 27.   Same mould with ends strengthened to
31.8.34.   Chr. 1.0.   10 mins.
755.   Central sectˡ   2 cms. up.
19°.   0 ⎰ 36.0   35.5
                2 ⎨ 35.1   34.0
                3 ⎱ 33.1   32.0
                4½   28.0   27.0

                                         End.
            6 cms up.   36.3
                0 ⎰ 38   40
                2 ⎨ 37   26
                4½ ⎱ 28

            Calcᵈ   7.2 r.
            Exptˡ   6.27 r.

PART VI
PICTORIAL MISCELLANY

at centre in 1½ hrs.
No gauzee.   86
gauzee   86
Lead shell.   64 in C. (45 mins)
No lead shell. 63 ?

3.3.36   Mrs Paterson's Big Expt.
         14 × 13·33 mgs
         4 × 6·66 mgs
         1 × 24·9
         1 × 23·0

10. E.Ts.
μ rings centred at 2·1 cm radius. measure 1·5 cms away
Ra.E. → 290 r/hr.

Ch. 1.
1st side at 1·75 cm. 3 mins.
99-15, 99-15.
at first 1·25 cm. 2½ mins.
99-12.
    Enters. at 1·5 cm. $\dfrac{29\cdot33 \times 7\cdot47}{6\cdot66}$ = 32·9 divs/min

Side 2.
                        $\dfrac{\frac{86}{3}\, 7\cdot47}{6\cdot66}$ = 32·1 div/min

55

Herbert M. Parker as a young man.

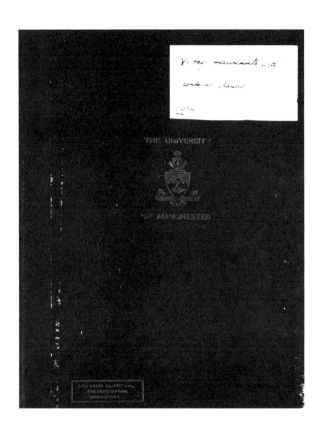

Cover of Parker's Manchester notebook 1934–1938.

Supervoltage X-RAY unit used by Parker at Swedish Hospital (Seattle) in the 1930s.

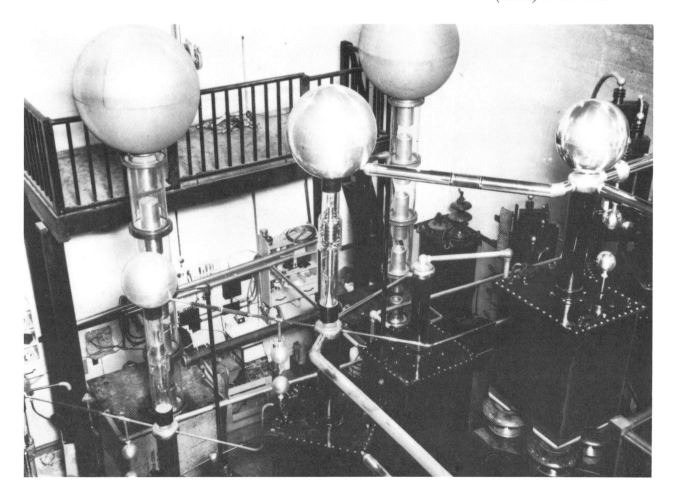

50

CVAC 172   Ground Scattering of Co^{60} γ
  C. H. Bernard
Single scattering & albedo data.

ORNL 1594
Track width of a heavy chgd ptle.
  Jacob Neufeld & W. S. Snyder

99% of energy used up within 5×10^{-7} cm of 'track'.

ORNL 1083
On the energy dissipation of moving ions in tissue
  W. S. Snyder, J. Neufeld

energy lost by (1) nuclear collisions (close to nucleus, disrupts molec)
              (2) electronic collisions — ioniz.

(2) accounts for nearly all energy at high ptle energies

Critical Energy at which nuclear coll loss = electronic collision

Z	Mev
1	0.0015
2	0.013
6	0.094
7	0.13
8	0.18

Pages 50 and 51 of Parker's personal notebook—1954.

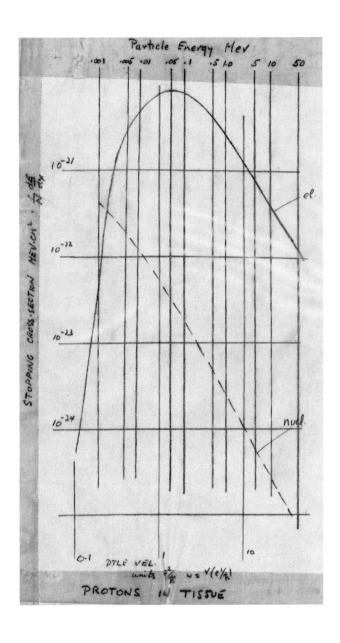

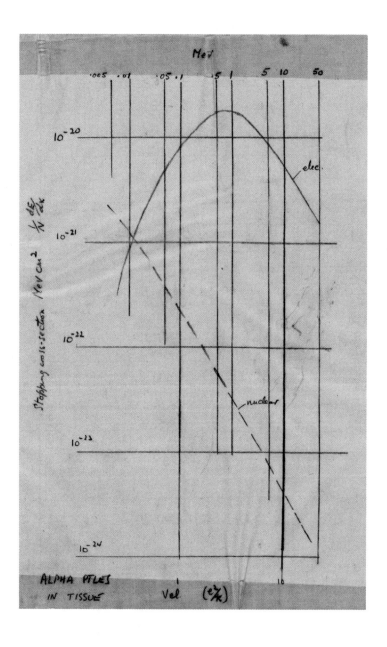

Pages 52 and 53 of Parker's personal notebook — 1954.

10/26 NP-3546 Permis dose & calc. risk
            Capt CF Behrens.

        Cosmic & natural activ   0.3 mr/24 hrs.
        1 back of luminous dial watch   100 mr/24 hrs
        Spectacle lens (U glass)   1-8 mr/hr to eyes
        Chest X ray conventional   50-250 mr/film
        Chest fluorogram   900 mr
        Dental. 500 mr/film
        Pelvis (obstetrical) film   200 mr/film

──

11/10   J G Hoffman.
        <u>Size & Growth of Tissue Cells</u>

        Human epidermis
        basal layer. 25,000 cells per mm² in a
                          monolayer.
nuclear
growth is
linear with time   cornified layer   large flat polygons
cytoplasmic growth   1250/mm²
exponential.
               no ironclad proof that basal layer
        is primary generating layer of all cells
        of human epidermis.

            abt 2×10¹¹ new red cells per day
                 5×10¹¹ total new cells per day.

         Duration of mitoses   Av turnover time (exfoliative growth)
Hum epidermis   15-30 min   Erythrocytes   120 days
Mouse   30-150"   Cornif. cells epiderm 10-20 "
Hum circ. granulocytes 37   Hair bulb   3
Chick embryo fibroblasts 30-80 Lymphocytes rat, dog 11·6 hrs
Pigeon crop gland   16   cat   8-48 hrs
Regen rat liver   49   Mononucl leucocytes rat 170 min
                               Polymorph   "   "   23 min

   Tumor doubling time   ~ 1-3 days.

OAK RIDGE NATIONAL LABORATORY
operated by
CARBIDE AND CARBON CHEMICALS COMPANY
A Division of Union Carbide and Carbon Corporation
Post Office Box P
Oak Ridge, Tennessee

July 30, 1953

To: _____

The attached material is contained in Chapter X of the book entitled HEALTH CONTROL AND NUCLEAR RESEARCH written by Karl Z. Morgan, of Oak Ridge National Laboratory, and Herbert M. Parker, of Hanford Works. Since it is expected that this book will be published sometime in the near future, I am asking that you do not reproduce any of the material from this chapter. Your compliance with this request will be greatly appreciated.

If suggestions or criticisms for improving the manuscript occur to you, it will be very helpful if you will let me have your comments.

Karl Z. Morgan, Director
Health Physics Division

KZM:mof:bmh

Letter for a book planned, but never published. The work was started in the 1940s, and several chapters were sent out for comment as early as January 1950. A 1955 prepublication version of chapters V, VI, IX, and X is known to exist.

Definition of "parker" in a standard reference work.

McGraw-Hill Dictionary of Scientific and Technical Terms

parker See rep.

rep [NUCLEO] A unit of ionizing radiation, equal to the amount that causes absorption of 93 ergs per gram of soft tissue. Derived from roentgen equivalent physical. Also known as parker; tissue roentgen.

Reprinted courtesy of McGraw-Hill

THE FIRST
INTERNATIONAL CONFERENCE ON
MEDICAL PHYSICS

I.C.M.P.

CONFERENCE DINNER

HOTEL MAJESTIC, HARROGATE

FRIDAY, SEPTEMBER 10th, 1965
7.30 for 8.00 p.m.

Signatures:
Herb Parker
Margaret Parker
Lowry Dobson
John Mallard
R. J. Gale
Carole Gale
John Laughlin
Barbara Laughlin
[illegible]
D. A. Innes

MENU

Cocktail d'Écrevisses

Consommé a la Reine

Saumon d'Écosse poche, sauce Hollandaise,
concombres au sel

Coeur de filet de boeuf roti garni
Sauce Raifort
Pommes Cocotte

Ananas en Surprise

Petits Fours

Café

Toastmaster: Mr. W. Gledhill

TOASTS

"The Queen"

Proposed by
The President,
Professor W. V. Mayneord

"Medical Physics"

Proposed by
Professor D. W. Smithers

Reply by
Mr. H. M. Parker

Closure

U.S. Transuranium Registry—U.S. Uranium Registry Advisory Committee, October 1984. Left to right: Langan Swint, Herbert Parker, Patricia Durbin, John Poston, Charles Mays, Robley Evans, George Voelz, and Newell Stannard.

Dedication of Life Sciences Laboratory at Hanford in 1971. Left to right: B.E. Vaughan, W.J. Bair, H.M. Parker, and H.A. Kornberg.